Capsaicinoids

Mallappa Kumara Swamy • Ajay Kumar
Editors

Capsaicinoids

From Natural Sources to Biosynthesis and their Clinical Applications

Springer

Editors
Mallappa Kumara Swamy
East West First Grade College (Bangalore University)
Bengaluru, India

Ajay Kumar
Central University of Kerala
Kasaragod, India

ISBN 978-981-99-7778-9   ISBN 978-981-99-7779-6 (eBook)
https://doi.org/10.1007/978-981-99-7779-6

This Springer imprint is published by the registered company Springer Nature Singapore Pte Ltd.
The registered company address is: 152 Beach Road, #21-01/04 Gateway East, Singapore 189721, Singapore

Paper in this product is recyclable.

DEDICATED TO OUR
PARENTS, FAMILY MEMBERS
AND TEACHERS
WHO TAUGHT US LIFE LESSONS

Preface

Plants being sessile produce numerous chemical compounds that are together recognized as specialized metabolites. These plant-derived compounds are not crucial for plant's growth and development; however they are chiefly synthesized to have ecological adaptation against both biotic and abiotic stress conditions. These specialized compounds are particularly important as defense molecules against attacks of biotrophic pathogens and herbivores. These compounds belong to different chemical classes, including phenolics, alkaloids, terpenes, steroids, and vanilloids. Remarkably, these compounds exhibit several pharmacological properties, and hence have become one of the sought after molecules of choice for the treatment of various ailments in humans.

Capsaicinoids are a group of important compounds that are particularly synthesized by various members of the genus *Capsicum* in their placenta. It is believed that the capsaicinoids have evolved in plants as a deterrent to grazers, herbivores, and pathogenic fungi. Interestingly, the capsaicinoids are not toxic to birds due to the absence of TRPV receptors in them. Evolutionary studies have suggested that the absence of TRPV1 receptor in birds must have favored the dispersal of their seeds. Capsaicinoids belong to vanilloid family of compounds such as vanillin (derived from vanilla), eugenol (from bay leaves and cloves), and zingerone (derived from ginger).

Capsaicin is the most abundant (69%) vanilloid compound among the different capsaicinoids in hot peppers. Other capsaicinoids are dihydrocapsaicin (22%), nordihydrocapsaicin (7%), homocapsaicin (1%), and homodihydrocapsaicin (1%). It is also known as *trans*-8-methyl-*N*-vanillyl-6-nonenamide; this chemical compound is crystalline, off-white solid, lipophilic, colorless, and odorless. It has a melting point of 62–65 °C and though not water soluble, it is soluble in ethanol, acetone, and fatty oils. The capsaicin has been proved as an important bioactive molecule with several properties against many ailments such as cancer, diabetes, obesity, and diseases of airway and urinary tracts.

Capsaicin interacts with TRPV1 receptors in humans. The Nobel Prize in Physiology or Medicine in 2021 awarded to David Julius for his groundbreaking discovery of how we sense cold, temperature, and mechanical stimuli. He used capsaicin for understanding how we sense various external stimuli. The burning sensation that we feel while eating capsicum is because of these chemical compounds. These compounds exert their functions by interacting with the TRPV receptors. There has been an increasing accumulation of literature surrounding capsaicin, especially after 2014 when two independent groups published the genome sequences of capsaicin-producing Capsicum plants that dissected the various genetic and micro regulators of capsaicin biosynthesis. After that, many studies have been going on in various parts of the world on the usage of capsaicin as a clinical molecule. Topics covering various aspects of capsaicinoids, from ethnobotany, conservation, biotechnology to the most important clinical significances, are of great importance. Scientists are also trying to find out the mechanisms of action of capsaicin and related compounds against various diseases. At present, improving capsaicinoids' bioavailability is another challenging issue. Since it is produced in the fruits and fruit maturation takes many months, genetic tools can be used to express some of the limiting genes in the leaves or other parts of capsicum. Capsicum plants also face production challenges due to many stresses and biotic pathogens. To meet the increasing demand for capsaicin, there is a need to develop high-yielding capsicum plants with disease and abiotic stress resistance.

This book entitled *Capsaicinoids: From Natural Sources to Biosynthesis and their Clinical Applications* comprises a wide-ranging topics on capsaicinoids. It includes 26 chapters contributed by researchers, technologists, and scholars from different parts of the world. Various topics on *Capsicum* species and capsaicinoids, such as ethnobotany, extraction and analysis, chemistry, capsaicin analogues, biosynthesis, industrial demand and applications, biotechnological strategies, including conservation, cell cultures, elicitation, breeding and omics, pharmacology, mechanisms of action studies, bioavailability, and nanotechnology, are comprehensively reviewed in this book. Largely, this book is an excellent volume for undergraduates, graduates, educationalists, and food and pharmaceutical industries involved in natural products and therapeutic research areas. We heartily thank all the contributors to this book volume for sharing their knowledge on capsaicinoids. We also thank the team at Springer for their cooperation at every stage of book publication.

Bengaluru, India Mallappa Kumara Swamy
Kasaragod, India Ajay Kumar

Contents

Editors and Contributors

About the Editors

Mallappa Kumara Swamy is working as a Professor in the Department of Biotechnology at East West First Grade College (Affiliated to Bangalore University), Bengaluru, India. He has completed his Postdoctoral Research at the Department of Crop Science, Faculty of Agriculture, Universiti Putra Malaysia (UPM), Serdang, Malaysia, from 2014 to 2018. Before that, he had worked as an Associate Professor and Head, Department of Biotechnology at Padmashree Institute of Management and Sciences, Bangalore University, Bengaluru, India. He obtained his PhD (Biotechnology) from Acharya Nagarjuna University, Guntur, India, in 2013. He has more than 20 years of teaching and research experience in the fields of plant biotechnology, microbiology, secondary metabolites production, phytochemistry, biomedicine, and bioactive studies. To his credit, he has published 65 research/review articles in peer-reviewed journals and 68 book chapters with reputed book publishers. So far, he has edited 17 books with Springer Nature Singapore Pte Ltd., Elsevier Inc., USA, CRC Press and Apple Academic Press (Taylor & Francis Group), USA, and Studium Press, India. He also serves as an editorial board member and reviewer for a few high-impact international journals. Presently, he is working in the area of natural products research, plant cell and tissue culture technology for bioactive compounds production and evaluation of their bioactivities. Also, his research is focused on nanobiotechnology for medical applications and plant omics studies.

Ajay Kumar is working as Assistant Professor of Plant Science at Central University of Kerala, India. He obtained his PhD from School of Life Sciences, Jawaharlal Nehru University, New Delhi, where he worked under the guidance of Dr. Nirala Ramchiary, a renowned crop scientist. He worked on various aspects of different species of Capsicum including capsaicinoids during his PhD degree and has published important papers on Capsicum and capsaicin. Dr. Ajay has more than 8 years of teaching and research experience and has published several research/review articles and book chapters in reputed journals and books, respectively. He has recently edited two books with Springer Nature. He has been a member of Academic Council and Court of the Central University of Kerala. He is currently a member of several academic bodies including Board of Studies, Department of Plant Science, and Department of Biochemistry and Molecular Biology, Central University of Kerala. His areas of research include phytochemical genomics, medicinal plants, capsaicinoids, Capsicum, food security, and abiotic stresses. He has successfully guided more than 30 postgraduate students for their dissertations.

Contributors

Abdul Jabbar Al-Rajab Etcetera Publications, Chesterville, ON, Canada

Abdel Rahman Al-Tawaha Department of Biological Sciences, Al-Hussein Bin Talal University, Ma'an, Jordan

S. Arpita Division of Vegetable Sciences, IARI, New Delhi, India

Prakhar Asthana Department of Biotechnology, Jaypee Institute of Information Technology, Noida, Uttar Pradesh, India

Nayana Veedu Aswathi Department of Plant Science, Tejaswini Hills, Central University of Kerala, Kasaragod, Kerala, India

Nikita Baliyan Division of Vegetable Science, ICAR-IARI, New Delhi, India

Satyaprakash Barik Department of Agriculture and Allied Sciences, CV Raman Global University, Bhubaneswar, India

Mandira Basumatary Department of Molecular Biology and Biotechnology, School of Sciences, Tezpur University, Tezpur, Assam, India

Ranjana Bhandari Pharmacological Research Laboratory, University Institute of Pharmaceutical Sciences, UGC-CAS, Panjab University, Chandigarh, India

Sayan Bhattacharya School of Ecology and Environment Studies, Nalanda University, Rajgir, India

Deepak Chandran Department of Veterinary Sciences and Animal Husbandry, School of Agricultural Sciences, Amrita Vishwa Vidyapeetham University, Coimbatore, Tamil Nadu, India

Angela Maria Chapa-Oliver Programa de Ingeniería en Biotecnología, Departamento de Ingeniería Agroindustrial, División de Ciencias de la Salud e Ingenierías, Campus Celaya-Salvatierra, Universidad de Guanajuato, Celaya, Mexico

Durgesh Nandini Chauhan Columbia College of Pharmacy, Raipur, Chhattisgarh, India

Nagendra Singh Chauhan Drugs Testing Laboratory, Avam Anusandhan Kendra, Raipur, Chhattisgarh, India

Sanjay J Daharwal University Institute of Pharmacy, Pandit Ravishankar Shukla University, Raipur, Chhattisgarh, India

Laxmi Devi Department of Biotechnology, Lovely Professional University Punjab, Jalandhar, India

Abhijit Dey Department of Life Sciences, Presidency University, Kolkata, West Bengal, India

Neelima Dhingra University Institute of Pharmaceutical Sciences, Panjab University, Chandigarh, India

Ashwini Kumar Dixit Department of Botany, Guru Ghasidas Vishwavidyalaya, Bilaspur, Chhattisgarh, India

Satish Dubey University Institute of Pharmacy, Pandit Ravishankar Shukla University, Raipur, Chhattisgarh, India

Debdarshan Dutta Department of Biotechnology, Jaypee Institute of Information Technology, Noida, Uttar Pradesh, India

Poornima D Vijendra Department of Food Technology and Biochemistry, Davangere University, Davangere, Karnataka, India

Shradha Devi Dwivedi University Institute of Pharmacy, Pandit Ravishankar Shukla University, Raipur, Chhattisgarh, India

Cemre Elmas Department of Nutrition and Dietetics, Faculty of Health Sciences, Cyprus International University, Nicosia, North Cyprus, Turkey

Robson Xavier Faria Evaluation and Promotion of the Ambiental Health Laboratory, Oswaldo Cruz Institute, Oswaldo Cruz Foundation, Rio de Janeiro, Brazil
Postgraduate Program in Sciences and Biotechnology, Institute of Biology, Federal Fluminense University, Niterói, Brazil

Pooja Gadkari University Institute of Pharmacy, Pandit Ravishankar Shukla University, Raipur, Chhattisgarh, India

Suvarna V. Gaikwad Department of Botany, Rayat Shikshan Sanstha's, Dada Patil Mahavidyalya, Karjat, Maharashtra, India

Ceren Gezer Department of Nutrition and Dietetics, Faculty of Health Sciences, Eastern Mediterranean University, Famagusta, North Cyprus, Turkey

Mimosa Ghorai Department of Life Sciences, Presidency University, Kolkata, West Bengal, India

Aparna Kozhampurathu Gokul Department of Plant Pathology, Kerala Agricultural University, Wayanad, Kerala, India

Maria Vanaina Souza Gonçalves Post-graduate Program in Sustainable Territorial Development, Universidade Federal do Paraná, Matinhos, Brazil

Ramón Gerardo Guevara-González C.A. Ingeniería de Biosistemas, Facultad de Ingeniería, Universidad Autónoma de Querétaro, Campus Amazcala, El Marqués, Querétaro, Mexico

Nami Jain Department of Biotechnology, Jaypee Institute of Information Technology, Noida, Uttar Pradesh, India

Devatman Jauhari Department of Biotechnology, Jaypee Institute of Information Technology, Noida, Uttar Pradesh, India

Mohammed Kashif School of Life Sciences, B.S. Abdur Rahman Crescent Institute of Science and Technology, Chennai, Tamil Nadu, India

Arshpreet Kaur Pharmacological Research Laboratory, University Institute of Pharmaceutical Sciences, UGC-CAS, Panjab University, Chandigarh, India

M. E. Kavya UASB, GKVK, Bengaluru, India

Anurag Kuhad Pharmacological Research Laboratory, University Institute of Pharmaceutical Sciences, UGC-CAS, Panjab University, Chandigarh, India

Ajay Kumar Central University of Kerala, Kasaragod, India

Sushil Kumar Department of Botany, Govt. Degree College Kishtwar, Kishtwar, Jammu and Kashmir, India

Roshani Kumari ICMR-National Institute for Research in Environmental Health, Bhopal, Madhya Pradesh, India

D. C. Lakshamanareddy Division of Basic Sciences, ICAR-Indian Institute of Horticultural Research, Bengaluru, India

Shisheng Li Department of Physiology and Membrane Biology, University of California at Davis, School of Medicine, Davis, CA, USA

K. Madhavi Reddy Division of Vegetable Crops, ICAR-Indian Institute of Horticultural Research, Bengaluru, India

Manisha Mangal Division of Vegetable Science, ICAR-IARI, New Delhi, India

Laura Mejía-Teniente C.A Biotecnología, Sustentabilidad e Ingeniería, Departamento de Ingeniería Agroindustrial, Universidad de Guanajuato, Campus Celaya-Salvatierra, Celaya, Guanajuato, Mexico

A. Mohan Rao UASB, GKVK, Bengaluru, India

Faisal Mushtaq Department of Botany, Govt. Degree College Kishtwar, Kishtwar, Jammu and Kashmir, India

Akila Muthuramalingam School of Life Sciences, B.S. Abdur Rahman Crescent Institute of Science and Technology, Chennai, Tamil Nadu, India

Mary Moses Mylabathula Department of Biotechnology, Krishna University, Machilipatnam, Andhra Pradesh, India

Urshila Naithani Department of Biotechnology, Jaypee Institute of Information Technology, Noida, Uttar Pradesh, India

Anitha Nandagopal Department of Pharmacology, Sultan-ul-Uloom College of Pharmacy, Hyderabad, Telangana, India
Department of Pharmacognosy, Sultan-ul-Uloom College of Pharmacy, Hyderabad, Telangana, India

Pavani Narigapalli Division of Vegetable Crops, ICAR-Indian Institute of Horticultural Research, Bengaluru, India

Lishma Nelliyaparambath Regional Agricultural Research Station, Kerala Agricultural University, Wayanad, Kerala, India

Sandeep R. Pai Department of Botany, Rayat Shikshan Sanstha's, Dada Patil Mahavidyalya, Karjat, Maharashtra, India

Devendra Kumar Pandey Department of Biotechnology, Lovely Professional University Punjab, Jalandhar, India

Ashok Kumar Pandurangan School of Life Sciences, B.S. Abdur Rahman Crescent Institute of Science and Technology, Chennai, Tamil Nadu, India

Karishma Pasupula Division of Vegetable Crops, ICAR-Indian Institute of Horticultural Research, Bengaluru, India

Umesh Kumar Patil Phytomedicine and Natural Product Research Laboratory, Department of Pharmaceutical Sciences, Dr. Harisingh Gour Vishwavidyalaya (A Central University), Sagar, Madhya Pradesh, India

Naresh Ponnam Division of Vegetable Crops, ICAR-Indian Institute of Horticultural Research, Bengaluru, India

K. N. Poornima Division of Basic Sciences, ICAR-IIHR, New Delhi, India
ICAR-IIHR, Hessarghatta, Bengaluru, India

Mansi Prashar University Institute of Pharmaceutical Sciences, Panjab University, Chandigarh, India

Nitin Puranik ICMR-National Institute for Research in Environmental Health, Bhopal, Madhya Pradesh, India

Rachana Department of Biotechnology, Jaypee Institute of Information Technology, Noida, Uttar Pradesh, India

Satyam Raj University Institute of Pharmaceutical Sciences, Panjab University, Chandigarh, India

Nemala Appala Raju Department of Pharmacology, Sultan-ul-Uloom College of Pharmacy, Hyderabad, Telangana, India

Department of Pharmacognosy, Sultan-ul-Uloom College of Pharmacy, Hyderabad, Telangana, India

Kratik Rana Department of Biotechnology, Jaypee Institute of Information Technology, Noida, Uttar Pradesh, India

Desam Nagarjuna Reddy School of Engineering and Applied Technology, BEST Innovation University, Gorantla, Andhra Pradesh, India

Saabira Banu Sahubar Sadik School of Life Sciences, B.S. Abdur Rahman Crescent Institute of Science and Technology, Chennai, Tamil Nadu, India

Juliana Pimenta Salles Laboratory of Studies in Experimental Pharmacology, Federal University of Rio de Janeiro, Rio de Janeiro, Brazil

M. Sarpras Mahatma Gandhi Government Arts College, Mahe, Pondicherry University, Mahe, Pondicherry, India

Anjani Saxena Institute of Biological Science, SAGE University, Indore, Madhya Pradesh, India

Kamal Shah Institute of Pharmaceutical Research, GLA University, Mathura, Uttar Pradesh, India

Mahebub Shaik Division of Vegetable Crops, ICAR-Indian Institute of Horticultural Research, Bengaluru, India

Aditi Sharma Department of Biotechnology, Lovely Professional University Punjab, Jalandhar, India

Radhika Sharma Department of Biotechnology, Jaypee Institute of Information Technology, Noida, Uttar Pradesh, India

Luiz Everson da Silva Post-graduate Program in Sustainable Territorial Development, Universidade Federal do Paraná, Matinhos, Brazil

Deependra Singh University Institute of Pharmacy, Pandit Ravishankar Shukla University, Raipur, Chhattisgarh, India

Manju Rawat Singh University Institute of Pharmacy, Pandit Ravishankar Shukla University, Raipur, Chhattisgarh, India

Swarn Singh Department of Botany, Padmshri Padma Sachdev Govt. PG College for Women Gandhi Nagar, Gandhi Nagar, India

Arpita Srivastava Division of Vegetable Science, ICAR-IARI, New Delhi, India

Mallappa Kumara Swamy East West First Grade College (Bangalore University), Bengaluru, India

Swetnisha Department of Microbiology, Patna Women's College (Autonomous), Patna, Bihar, India

Amit Talukda Department of Molecular Biology and Biotechnology, School of Sciences, Tezpur University, Tezpur, Assam, India

Anju Thattantavide Department of Plant Science, School of Biological Sciences, Central University of Kerala, Hosdurg, Kerala, India

Thuruthiyil Dennis Thomas Department of Plant Science, Tejaswini Hills, Central University of Kerala, Kasaragod, Kerala, India

Dipti Tripathi Department of Biotechnology, Jaypee Institute of Information Technology, Noida, Uttar Pradesh, India

Manasi Verma Pharmacological Research Laboratory, University Institute of Pharmaceutical Sciences, UGC-CAS, Panjab University, Chandigarh, India

Sandeep Kumar Verma Institute of Biological Science, SAGE University, Indore, Madhya Pradesh, India

Chandran Padikkal Krishna Vrundha Department of Plant Science, Tejaswini Hills, Central University of Kerala, Kasaragod, Kerala, India

Mohd Waseem School of Pharmacy, University of Maryland Eastern Shore, Princess Anne, MD, USA

Jie Zheng Department of Physiology and Membrane Biology, University of California at Davis, School of Medicine, Davis, CA, USA

Capsaicin-Producing Plants: Species Diversity, Distribution and Botanical Details

1

Sushil Kumar, M. Sarpras, Faisal Mushtaq, Swarn Singh, Anju Thattantavide, and Ajay Kumar

Abstract

The species of *Capsicum* are known to synthesise capsaicin. It has a molecular weight of 305.5 g/mol and the chemical formula $C_{18}H_{27}NO_3$. The IUPAC name of capsaicin is (*E*)-*N*-[(4-hydroxy-3-methoxyphenyl)methyl]-8-methylnon-6-enamide. It is responsible for the unique pungent taste of the *Capsicum*. It is crystalline, colourless, odourless and lipophilic. This molecule belongs to a special type of alkaloids called capsaicinoids. *Capsicum* spp. are known to produce several types of capsaicinoids such as capsaicin, homocapsaicin, dihydrocapsaicin, nordihydrocapsaicin, homodihydrocapsaicin and norcapsaicin. Capsaicin and dihydrocapsaicin constitute about 80–90% of capsaicinoids and hence are the major capsaicinoids of the *Capsicum*. The only genus that has been found to produce and accumulate capsaicinoids is *Capsicum*. Capsaicin concentration in plants varies and is generally affected by genetic, environmental and various other factors including biotic stresses. The capsaicin finds applications in diverse sectors including flavouring, food, cosmetics, defence sprays, medicine

Sushil Kumar and M. Sarpras contributed equally with all other contributors.

S. Kumar (✉) · F. Mushtaq
Department of Botany, Govt. Degree College Kishtwar, Kishtwar, Jammu and Kashmir, India

M. Sarpras (✉)
Department of Zoology, Mahatma Gandhi Govt. Arts College, Mahe, Pondicherry University, Mahe, Pondicherry, India

S. Singh
Department of Botany, Padmshri Padma Sachdev Govt. PG College for Women Gandhi Nagar, Gandhi Nagar, India

A. Thattantavide · A. Kumar (✉)
Department of Plant Science, School of Biological Sciences, Central University of Kerala, Kasaragod, Kerala, India

M. K. Swamy, A. Kumar (eds.), *Capsaicinoids*,
https://doi.org/10.1007/978-981-99-7779-6_1

and pharmacy. This book chapter provides a brief overview of the diversity of various capsaicin-producing plants along with their botanical, cytological and distribution details. This chapter is important for those who are interested in botany, taxonomy and diversity of the capsaicin-producing species.

Keywords

Capsicum · Capsaicin · Taxonomy · Capsaicinoids · Pungency · Hot peppers

1.1 Introduction

Capsicum spp. (Solanaceae) are known to synthesise and accumulate a class of alkaloids known as capsaicinoids (Hayman and Kam 2008; Reyes-Escogido et al. 2011). Capsaicinoids are first synthesised in the placenta of *Capsicum* fruits and from there they are transported to different plant organs (Estrada et al. 2002; Mazourek et al. 2009; González-Zamora et al. 2013; Sarpras et al. 2016; Fig. 1.1). Capsaicinoids give characteristic pungency to *Capsicum* and Scoville heat units (SHU) are used to determine the level of pungency (Zewdie and Bosland 2000; Hamed et al. 2019). Various techniques are recently developed to extract and quantitatively measure the amount of various capsaicin analogues in its various parts (Reyes-Escogido et al. 2011; Huang et al. 2013; Oğuzkan 2019). Capsaicin and dihydrocapsaicin are the major capsaicinoids found in the *Capsicum* fruits (80–90%) (Chapa-Oliver et al. 2016; Hayman and Kam 2008), followed by nordihydrocapsaicin (7%), homodihydrocapsaicin (1%) and homocapsaicin (1%) (Reyes-Escogido et al. 2011; Antonious 2018). The structure of various

Fig. 1.1 Structure of placenta which is the primary site of capsaicin biosynthesis and accumulation (black arrow) from Bhut jolokia (*C. chinense*)

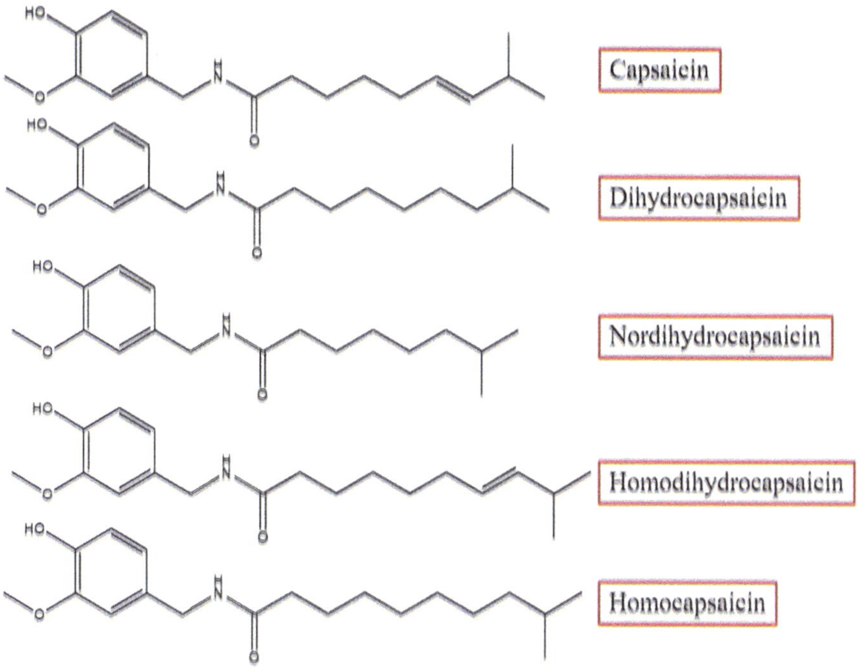

Fig. 1.2 Structure of important capsaicinoids (drawn using ChemDraw Software)

capsaicinoids are given in Fig. 1.2. Capsaicinoid content in *Capsicum* varies and it ranges from 0.1 mg/g to 60 mg/g (Parrish 1996; Al Othman et al. 2011). Various factors such as genotype of plant, plant/fruit developmental stage and growing conditions affect the concentration of capsaicin (Zewdie and Bosland 2000; Yaldiz et al. 2010). This means that there is a huge diversity of capsaicinoid content across the species. It is interesting to note that there are several species that are very less pungent or not even pungent. Either they do not contain capsaicinoids or, even if present, the quantity of the capsaicinoids is very low. Non-pungent sweet peppers contain another group of compounds known as capsinoids which do not cause burning sensation and therefore are not pungent. Vanillyl amine and branched fatty acid chains derived from the phenylpropanoid pathway and catabolism of amino acids, respectively, are the precursors for capsaicinoid biosynthesis (Reddy et al. 2014). The vanillylamine and a branched fatty acid undergo condensation leading to the biosynthesis of capsaicin and other capsaicinoid analogues are then derived from this through various enzymes (Aza-Gonzalez et al. 2011). With the help of the enzyme capsaicin synthase, vanillylamine and fatty acid moieties are combined to form capsaicin (Prasad et al. 2006). The synthesis of capsaicinoids begin after 20–30 days of fruit formation and are accumulated within the placental tissues (Castro-Concha 2016). This timing however varies across species and within species. Even stresses are known to cause changes in these synthesis patterns. Numerous genes have been shown to be involved in the synthesis of capsaicinoids. One such important gene is *Pun1* that plays a very important role in the biosynthesis as well as accumulation of capsaicinoids in *Capsicum* (Reddy et al. 2014). Capsaicinoid biosynthetic genes *Pun*1 and *AMT* can be genetically engineered to enhance production of capsaicinoids in *Capsicum* (Islam et al. 2023).

Capsicum plants generally produce capsaicin and other related compounds as a natural defence against infections like fungus and herbivores (Sharma et al. 2013). Other than its use in many cuisines, research has revealed that capsaicin has a variety of physiological impacts, including those on pain, respiratory, cardiovascular and urinary systems (O'Neill et al. 2012).

This book chapter describes various species of the genus *Capsicum* in which capsaicin has been reported. The distribution, ethnobotany and other important botanical details of capsaicin-producing *Capsicum* species are also briefly given. The production of capsaicin along with other related compounds is influenced by various factors, and the last section describes various factors affecting the production of capsaicinoids especially capsaicin.

1.2 Capsaicin-Producing Plants: Diversity, Cytology and Ethnobotanical Uses

1.2.1 Diversity and Distribution

Capsaicin is produced naturally by the species of the genus *Capsicum* L. (Barboza et al. 2022). Due to the extensive hybridisation that has taken place in *Capsicum* spp. during the course of its cultivation, several species have been described by various authors (Maga and Todd 1975). The genus *Capsicum* was thought to have 35 species (Yaldizet al. 2010; Carrizo Garcia et al. 2013); however, recently, Barboza et al. (2022) have mentioned about 43 species. WFO plant list also mentions 43 accepted species of the genus Capsicum (WFO 2023). *Capsicum* is regarded as the earliest spice used by humans, dating back to 7000 years, and it accounts for around 16% of the world spice industry (Krishna 2003; Igwe et al. 2019). This remarkable spice has its ancestral roots in South and Central America, especially Bolivia, while Brazil is considered as the centre of genetic diversity (Igwe et al. 2019). Following Christopher Columbus' first journey, chillies were brought into Europe and other parts of the world and after which they were distributed rapidly throughout Asia and Africa (Andrews 1984; Purseglove 1968; Mehta 2017). Several species of *Capsicum* are found in the wild but only five are cultivated throughout the world. The five cultivated species are *Capsicum annuum* var. *annuum, Capsicum frutescens, Capsicum chinense, Capsicum baccatum* var. *pendulum* and *Capsicum pubescence* (Ramchiary et al. 2014; Kumar et al. 2021, Carrizo García et al. 2016; Carrizo Garcia et al. 2022). Several different varieties have been developed as a result of hybridisation in *Capsicum* and as a result there are approximately 3000 different chilli pepper types (Antonious et al. 2006). Tropical and subtropical countries such as India, China, Mexico, Malaysia, Turkey and USA are some of the major chilli-growing areas of the world (Leff et al. 2004). The world's largest producer of chilli is India, and as per the data of 2020–2021, chilli production in India was about 19.88 lakh tonnes (www.indiastat.com). *Capsicum* generally grows well in the cool season but under controlled conditions can be grown throughout the year. A temperature of about 18–30 °C and humidity of 50–60% are optimum for its cultivation. Yield varies but generally it is 20–40 tonnes per hectare (Salehi et al. 2018).

1.2.2 Ethnobotany

Due to its nutritional and commercial benefits, *Capsicum* is one of the most extensively produced crops in the world (Batiha et al. 2020). Traditionally, they have been used as a source of spiciness and flavourant. However, several studies have demonstrated its ethnobotanical uses from various regions of the world. *Capsicum* spp. fruit acts as a circulatory stimulant and carminative and is also used in rheumatism and neuralgia (Sumner 2000; Nip 2004). *Capsicum* spp. have been in traditional medicines by Native from Mexico and Central America, since pre-Hispanic times (Salehi et al. 2018). As per the studies of botanical pharmacopoeia of native Mayan people of Mesoamerica, there are about 32 distinct health-related applications of *Capsicum* spp. such as treatment for rheumatism, arthritis, stomach problems and skin rashes, as well as for soothing snake as well as dog bites (Cichewicz and Thorpe 1996). In the classical text of Tibetan medicine, *Capsicum* fruits are employed as a remedy for oedema, haemorrhoids, protozoan parasites and leprosy as well as to raise the digestive warmth of the stomach (Basu and De 2003). *C. annuum* is taken to cure cold, cough and fever (Yanka et al. 2019). *C. chinense* is used for making pickles and also for the treatment of many ailments such as headache, asthma, fever, cold and cough, sore throat, arthritis and gastritis and also to relieve toothache and muscle pain in Assam (Baruah and Lal 2020; Yanka et al. 2019). *C. assamicum* has been traditionally used against asthma, various gastrointestinal disorders and toning of muscles after body exercise and also to remove pus from boils (Gogoi et al. 2018). The ethnobotanical survey of the *Capsicum* species of Maluku province in Indonesia documented that besides their use as condiment and spice, the leaves are consumed as a vegetable and fruits are added to palm wine. The fruits are used to cure wounds, swellings and boil (Yamamoto and Girsang 2021). Figure 1.3 outlines various ethnobotanical uses of *Capsicum* species. *Capsicum* leaves have been used for the treatment of boils and swellings, whereas fruits are applied to wounds (Yamamoto and Girsang 2021).

1.2.3 Chromosome Number

The chromosome number (2n) in most of the species of *Capsicum* is 24 but in *Capsicum annuum* L. a 26 number of diploid chromosomes have also been reported.

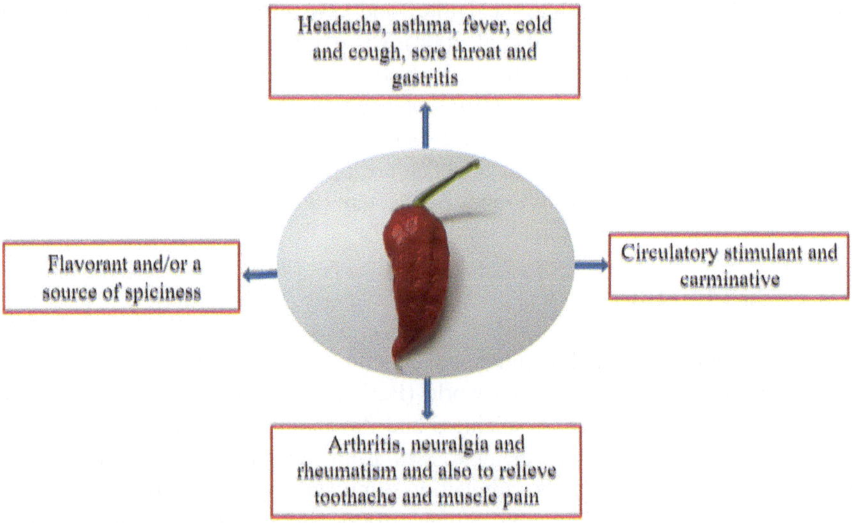

Fig. 1.3 Ethnobotany of *Capsicum* spp.

Table 1.1 Chromosome number of capsaicin-producing *Capsicum* spp.

Sl. No.	Name of species	Common name	Pungency level (in SHU)	Chromosome number
1	*C. annuum* L.	Red pepper	100,228.80 114,744.00	12,24,26 (CCDB)
2	*C. chinense* Jacq.	Bonnet pepper, Chinese Capsicum	146,823.20 (Nwokem et al. 2010)	24 (CCDB)
3	*C. frutescens* L.	Chilli Padi, bird's eye chilli	30,000–50,000 SHU. In some varieties it can reach up to 175,000 SHU (pepper scale)	24 (CCDB)
4	*C. baccatum* Buch.-Ham. ex Wall.	Aji, Bishop's-hat	30,000 to 50,000 SHU (pepper scale)	24 (CCDB)
5	*C. pubescence* Ruiz and Pav.	Rocoto pepper	May reach up to 50,000 to 250,000 SHU	24 (CCDB)

Tetraploidy ($2n = 4x = 48$) have also been reported in *C. annuum* var. *glabriusculum*. The estimated genome size in *Capsicum* is about 3.5 GB (Tripodi and Kumar 2019). The detail of chromosomal number of five common cultivated species is given below in Table 1.1.

1.2.4 Taxonomy of Capsaicin-Producing Plants

Capsicum L. belongs to the family Solanaceae (Dias et al. 2013; Kumar et al. 2021). Among different species reported in the genus *Capsicum*, six species are cultivated throughout the world: *C. assamicum, C. annuum, C. chinense, C. baccatum, C. frutescens* and *C. pubescens* (Kumar et al. 2021). *C. assamicum* is endemic to North-East India and it closely resembles *C. frutescens* and *C. chinense*. Branched habitat, yellow greenish corolla, anthers that are slightly pale blue in colour and orange-coloured fruits with dented skin as well as stem anatomy are some of the unique features that differentiate *C. assamicum* from *C. frutescens* and *C. chinense* (Purkayastha et al. 2012). But some authors, for example, Verma et al. (2013) and Roy (2016), pointed that *C. assamicum* is not a new species but is actually *C. chinense*. Based on geographical origin, morphological characteristics, reproductive behaviour, cytological and quantitative analysis as well as biochemical and information from molecular and biochemical markers, *Capsicum* species are divided into three major groups; these are *C. annuum, C. baccatum* and *C. pubescens*. These three groups are further organised into two main phylogenetic groups—the purple-flowered group and the white-flowered group including *C. pubescens,* and *C. annuum* and *C. baccatum,* respectively (Wahyuni et al. 2013). *Capsicum frutescens, C. annuum* and *C. chinense* are the members of the *C. annuum* complex. *C. baccatum* var. *pendulum, C. baccatum* var. *baccatum* and *C. praetermissum* are the members of the *C. baccatum* complex. *C. pubescens, C. eximium* and *C. cardenasii* are the members of the *C. pubescens* complex (Wahyuni et al. 2013).

The genus *Capsicum* L. has a very complex taxonomy and was originally characterised as having rotating corollas. But Heiser and Smith (1958) modified this description when they described *C. cardenasii* as having campanulate corollas. Later descriptions of other species such as *C. scolnikianum* (Hunziker 1961) and *C. friburgense* (Barboza and Bianchetti 2005) confirmed that campanulate corolla is found among other species of *Capsicum* L. (Eshbaugh 2012). The presence of capsaicin was another diagnostic feature of *Capsicum* species but Moscone et al. (2007) reported that *C. lanceolatum* and *C. rhomboideum* lack capsaicin, and hence, they do not possess pungency (Eshbaugh 2012). The current circumscription of the genus *Capsicum* into different species can vary greatly and is based on the features of flowers, fruits and leaves (Petters 2002). Generally, a combination of floral characteristics along with other characters as well as genetic characters are required to identify species in the genus *Capsicum* (Dias et al. 2013). Paredes-Andrade et al. (2020) have reported that flower characteristics such as their position, colour of the corolla and anthers were the most discriminant qualitative characteristics for the identification of *Capsicum* spp. and the length/width ratio of leaf; fruit features such as length, width and wall thickness; and height of the plant are the most discriminant quantitative characteristics for the identification of *Capsicum* species.

Eshbaugh (1993) indicated that for domestication, human populations selected several species of *Capsicum* unconsciously for a range of characteristics such as fruit shape, fruit size and colour, and due to the absence of any genetic analysis at that time, several names were given to these species. This list was subsequently sorted, and thus, reducing the long list of synonyms into four *Capsicum* species, these are *C. annuum* L., *C. frutescens* L., *C. chinense* Jacq. and *C. baccatum* L. Finally, *C. pubescens* R. was added to this list, thus leading to a total of five species (Esbaugh 2012). These five species also have a number of varieties that are grown in different regions of the world (Esbaugh 2012). As stated earlier it is very difficult to distinguish different species of *Capsicum*. Floral morphological features such as flower number per node, flower colour and calyx constriction are among the most used features for the identification of *Capsicum* spp. (Ince et al. 2010). Floral characteristics alone can be used to distinguish many *Capsicum* species such as *C. annuum, C. chinense* and *C. frutescens* (Paredes-Andrade et al. 2020). The flowers are solitary and creamish white in *C. annuum*, two to three whitish green flowers are present per leaf node with constriction at the base of the calyx in *C. chinense*, whereas in *C. frutescens* flowers are erect, white-green and without a calyx constriction (Paredes-Andrade et al. 2020). *C. baccatum* and *C. pubescens* are relatively easy to distinguish, *C. baccatum* is characterised by having flowers which are yellow or white-yellow in colour and petals have a greenish spot towards their base, whereas *C. pubescens* species are characterised by having flowers which are uniformly purple with a white base and seeds that are black and rough (Melgarejo et al. 2004; Chávez-Servia and Sevilla-Panizo 2006). The important features of the cultivated species of *Capsicum* L. is given in Table 1.2 (adapted from Moscone et al. 2007; Eshbaugh 2012 and Swamy 2023) and accepted names of *Capsicum* species are given in Table 1.3.

1.3 Biosynthesis of Capsaicin

The pungent taste of *Capsicum* is imparted due to the presence of capsaicinoids, which are produced by combining vanillylamine to a branched fatty acid (Curry et al. 1999; Aza-González et al. 2011). Although the pathway that produces capsaicinoids has been identified, all genes and enzymes involved have not been isolated and identified yet (Aza-González et al. 2011). Capsaicin biosynthetic pathway is relatively well characterised (Díaz et al. 2004) and its biosynthesis involves the product of two independent pathways—phenylpropanoid and fatty acid metabolism (Reyes-Escogido et al. 2011; Aza-González et al. 2011). In the phenylpropanoid pathway, phenylalanine is first converted into cinnamic acid with the help of the enzyme phenylalanine ammonia lyase (PAL), and subsequently, feruloyl-CoA is synthesised through various enzymatic steps (Zenk 1965; Naves et al. 2019). Then, it was proposed that a CoA-dependent beta-oxidative cleavage of feruloyl-CoA would produce vanilloyl-CoA, which would then be converted to vanillin. In any case, the vanillin will act as a precursor for the synthesis of vanillylamine and is predicted to be catalysed by the enzyme aminotransferase (Díaz et al. 2004). The biosynthesis of the branched fatty acid moiety, which is generated from valine, was confirmed by Leete and Louden (1968). Capsaicinoid synthase, an enzyme that catalyses the final generation of vanillylamine and branched-chain fatty acid, is the most elusive point

Table 1.2 Important morphological features of cultivated species of *Capsicum* (adapted from Moscone et al. 2007; Eshbaugh 2012 and Swamy 2023)

Capsicum spp.	Cultivars	Centre of origin	Growth form	Number of flowers per node	Shape and colour of corolla	Shape and colour of fruit	Seed colour
C. annuum L.	Jalapeno, Poblano, ancho, bell, big Jim, paprika, Sandia, serrano	Mexico/ Central America	Small shrubs Height—up to 2 m	1	Shape: stellate Colour: white or creamish; in exceptional case it can be violet	Shape: highly variable Colour: red	Yellowish
C. frutescens L. (very difficult to distinguish from *C. annuum* L. and *C. chinense* Jacq.)	Tabasco, African bird's eye, piri piri, Thai pepper, kambuzi, malagueta	Amazonia	Small shrub height—up to 2 m	1–3	Shape: stellate Colour: white or creamish	Shape: elongated Colour: red	Cream to yellowish
C. chinense Jacq.	Rocotillo, chilli blanko, habanero, 7-pot chilli Fatalii, Ají Chombo, Bhut jolokia cultivar 'Dorset', Naga pepper	Amazonia	Small stout shrub Height—up to 1.5 m	1–5	Shape: stellate Colour: white or creamish	Shape: spherical or conical Colour: various (such as red, orange, yellow, white or brown)	Cream to yellowish
C. baccatum L.	Sugar rush peach, Aji pineapple, Brazilian starfish, bishop crown, piquante pepper	Peru and Bolivia	Shrub Height—0.6–2 m	1–2	Shape: stellate Colour: white with green spots	Shape: ovoid or fusiform Colour: red	Yellowish
C. pubescens R.	Rocoto, manzano, locoto	Peru and Bolivia	Can grow horizontally up to a length of 18 m	1	Shape: rotate Colour: purple or white	Shape: elongated to spherical Colour: red orange and yellow	Dark brown or black seeds

Table 1.3 List of *Capsicum* species (accepted names from World Flora Online)

Sl. no.	Botanical name	Chromosome number (CCDB)
1	*C. annuum* L.	2n = 24
2	*C. baccatum* L.	$n = 12$
3	*C. caatingae* Barboza & Agra	$n = 12$
4	*C. caballeroi* M.Nee.	NA
5	*C. campylopodium* Sendtn.	2n = 26
6	*C. carassense* Barboza & Bianch.	NA
7	*C. ceratocalyx* M.Nee.	NA
8	*C. chacoense* Hunz.	2n = 24
9	*C. chinense* Jacq.	2n = 24
10	*C. ciliatum* (Kunth) Kuntze.	$n = 12$
11	*C. coccineum* (Rusby) Hunz.	$n = 12$
12	*C. cornutum* (Hiern) Hunz	2n = 26
13	*C. dimorphum* (Miers) Kuntze	$n = 12$
14	*C. eshbaughii* Barboza	NA
15	*C. eximium* Hunz.	2n = 24
16	*C. flexuosum* Sendtn.	2n = 24
17	*C. friburgense* Bianch. & Barboza	$n = 13$
18	*C. frutescens* L.	2n = 24
19	*C. galapagoense* Hunz.	$n = 12$
20	*C. geminifolium* (Dammer) Hunz.	$n = 12$
21	*C. hookerianum* (Miers) Kuntze	$n = 12$
22	*C. hunzikerianum* Barboza & Bianch.	NA
23	*C. lanceolatum* (Greenm.) C.V.Morton & Standl.	NA
24	*C. longidentatum* Agra & Barboza	$n = 12$
25	*C. lycianthoides* Bitter	NA
26	*C. minutiflorum* (Rusby) Hunz.	NA
27	*C. mirabile* Sendtn.	2n = 26
28	*C. parvifolium* Sendtn.	2n = 24
29	*C. pereirae* Barboza & Bianch.	NA
30	*C. pubescens* Ruiz & Pav.	2n = 24
31	*C. recurvatum* Witasek	$n = 13$
32	*C. rhomboideum* (Dunal) Kuntze	$n = 13$
33	*C. schottianum* Sendtn.	2n = 26
34	*C. scolnikianum* Hunz	$n = 12$
35	*C. tovarii* Eshbaugh, P.G.Sm. & Nickrent	NA
36	*C. villosum* Sendtn.	2n = 26

in this pathway (Díaz et al. 2004). Figure 1.4 presents an outline of the biosynthesis of capsaicin and the various enzymes involved.

The second pathway involved in capsaicin biosynthesis is the branched-chain fatty acid pathway which gives 8-methyl-6-noneonyl-CoA from pyruvic acid. Ketoacyl synthase (KAS) is an enzyme that facilitates the enzymatic synthesis of

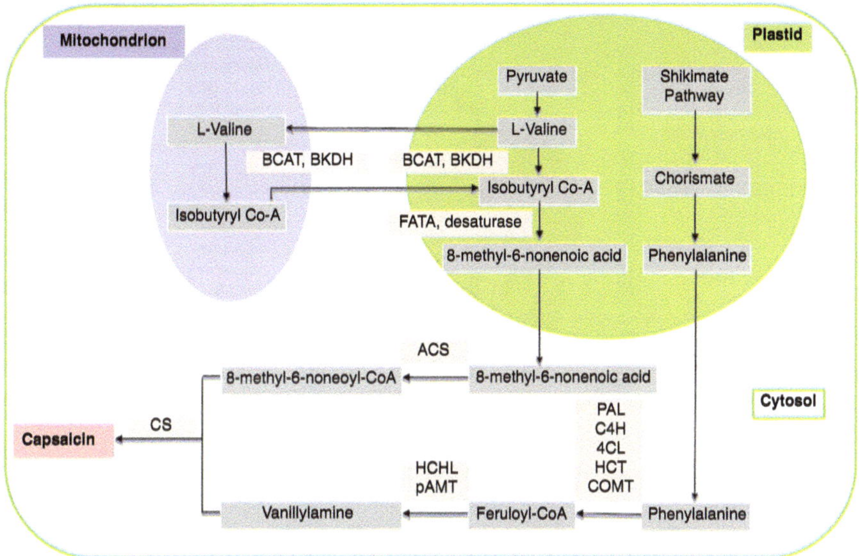

Fig. 1.4 Schematic representations of the main steps that take place during the production of capsaicinoids (adapted from Naves et al. 2019 and Zhang et al. 2022a, b)

8-methyl-nonanoic acid. Both vanillylamine and 8-methy-6-noneonyl-CoA are condensed to form capsaicin with the help of the enzyme capsaicin synthase (CS). Other enzymes involved in its biosynthesis include acyl-CoA synthetase (ACS), aminotransferase (AMT), branched-chain amino acid aminotransferase (BCAT), cinnamate 4-hydroxylase (C4H), 4-coumaroyl-CoA ligase (4CL), caffeoyl-CoA 3-O-methyltransferase (COMT), acyl-ACP thioesterase (FatA), hydroxycinnamoyl-CoA hydratase lyase (HCHL) and phenylalanine ammonia-lyase (PAL).

Various transcription factors have also been identified that regulate the biosynthesis of capsaicinoids; one such important transcription factor is MYB31 (Islam et al. 2023). In *C. chinense* MYB31 led to the upregulation of genes involved in capsaicinoid biosynthesis by recruiting another transcription factor WRKY9 (Zhang et al. 2022a, b). The other transcription factors that can upregulate the biosynthesis and accumulation of capsaicinoids include five bHLH (CabHLH07, CabHLH09, CabHLH026, CabHLH063 and CabHLH086), CcMYB48, CaMYB108 and CcERF2 (Liu et al. 2021; Sun et al. 2020; Wen et al. 2022).

1.4 Factors Affecting the Production of Capsaicinoids

The level of capsaicinoids in *Capsicum* spp. is highly dynamic and varies with fruit developmental stages, genotype and cultivation conditions (Zewdie and Bosland 2000). Figure 1.5 summarises various stresses and how they influence the

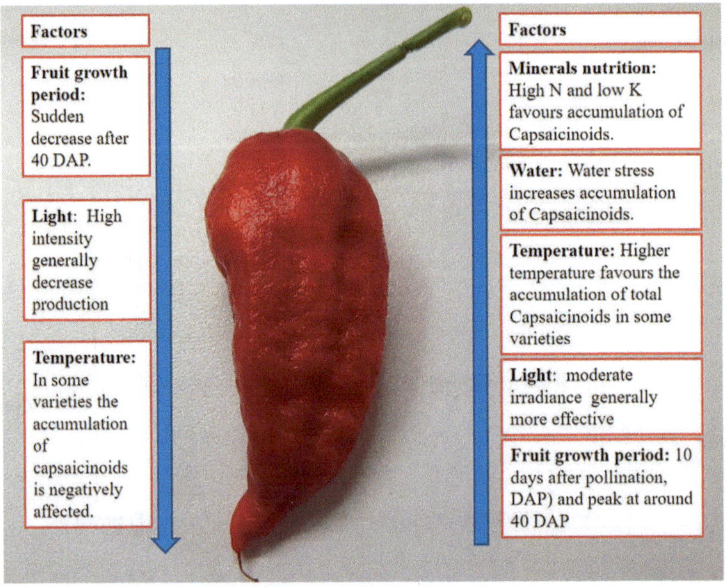

Fig. 1.5 Factors affecting capsaicin production

capsaicinoid production in *Capsicum*. The accumulation of capsaicinoids in the fruit starts from the early stages ~10 days after pollination (DAP) of fruit development and peaks at around 40 DAP followed by a sudden decrease (Barbero et al. 2014; Zhang et al. 2016; Naves et al. 2019). Various abiotic factors influence capsaicinoid biosynthesis and thus affect yield. Light is an important factor which determines capsaicinoid production and it generally stimulates capsaicinoid production in immature harvested fruits (Iwai et al. 1977). Further experiments have shown that moderate irradiance (150–350 mmol m^{-2} s^{-1}) was found to be more effective than high luminous intensities and high light intensity generally decreases production (Murakami et al. 2006; Uarrota et al. 2021). Temperature is another important factor and its effect on capsaicinoid production and accumulation varies from species to species and also from varieties to varieties (González et al. 2013). For some varieties, a higher temperature favours the production and accumulation of total capsaicinoids, whereas in others high temperature negatively affects the accumulation of capsaicinoids (Otha 1962; Naves et al. 2019). Water and mineral nutrition also affect the production and accumulation of capsaicinoids in *Capsicum* spp. Water stress decreases fruit size but increases accumulation of capsaicinoids in some varieties (Sung et al. 2005; Agyemang Duah et al. 2021). Mineral nutrition and most importantly nitrogen (N) and potassium (K) also affect capsaicinoid metabolism (Naves et al. 2019). A high nitrate nutrition generally enhances the level of capsaicin in *Capsicum* (Aldana-Iuit et al. 2015; Monforte-González et al. 2010). Pepper fruits that are produced under potassium deficiency showed increased accumulation of capsaicin contents (Freitas et al. 2021). The impact of various environmental factors on the accumulation of capsaicinoids is highly variable and it is very difficult to draw a definitive conclusion (Naves et al. 2019). This means that more studies are needed to understand the abiotic and biotic stress response mechanisms in *Capsicum* species particularly how the biosynthesis of capsaicinoids and their accumulation takes place in different species and cultivars. Research should be also integrated with genetic studies to understand the genetic dynamics and their relation with capsaicinoid accumulation. The stable expression of the genes is crucial for the optimum production of capsaicinoids. However, this is in reality too idealistic and may not be even possible, but efforts should be made to find out the regulators—both genetic and transcription factors—that can maintain a good production of the capsaicinoids under different conditions. Stresses may not influence the production in a similar fashion. Therefore, investigation of the genetic, enzymatic and metabolomic profiles under different stresses alone and in combination are important to devise strategies to improve capsaicinoid production in chilli peppers.

1.5 Capsaicin Isolation and Extraction

It is well-established that the most accepted way for the separation and purification of organic molecules is extraction methods. Among various extraction methods, solvent extractions are the most common and widely used method for capsaicin extraction. It is largely due to its fast, versatile and efficient performance and low cost. Since this method separates chemicals depending on its relative solubilities, researchers use liquid solvents, which are capable of dissolving required chemicals. This solid-liquid extraction process includes the mixing of two phases (dried *Capsicum* fruit and different solvents) as well as the separation and extraction of capsaicin molecules. Methanol, ethanol, acetonitrile and acetone are the commonly used solvents for capsaicin extraction. The other extraction methods and their properties are depicted in Table 1.4.

1.6 Techniques Used for Capsaicin Identification and Quantification

The preliminary methods for the estimation and quantification of capsaicin have been initiated by Todd (1977) using various chromatographic techniques. He determined the pungency and capsaicin contents from many chillies (red pepper) using gas-liquid chromatography. However, currently quantitative analysis of capsaicin from the different cultivated *Capsicum* species has been conducted by using many improvised techniques like nuclear magnetic resonance spectroscopy (NMR), high-performance liquid chromatography (HPLC), gas chromatography coupled with mass spectroscopy (GCMS) and liquid chromatography coupled with mass spectrometry (LCMS) (Kozukue et al. 2005; Ritota et al. 2010; Chinn et al. 2011; Chanthai et al. 2012; Giuffrida et al. 2013; Barbero et al. 2014; Sora et al. 2015 Sarpras et al. 2016). Various accessions (genotypes) of the *Capsicum* species were

Table 1.4 Different extraction methods of capsaicin and its properties

Methods	Properties	Capsaicin yield	Reference
Ultrasound-assisted extraction (UAE)	Reduce the solvent amount and capsaicin extraction time. Regulate the temperature	85.3 ± 1.4%	Lu et al. (2017)
Soxhlet extraction (SOX)	Lower capsaicin yield, longer extraction time, higher energy consumption	88.3 ± 1.03%	Bajer et al. (2015), Lu et al. (2017)
Liquids under pressure extraction (PLE)	High capsaicin yield, high capsaicin solubility, work in high temperatures and pressure, can be used for dry *Capsicum* samples	98.3 ± 1.5%	Barbero et al. (2016), Lu et al. (2017)
Microwave assisted extraction (MAE)	High capsaicin yield, reduce extraction time, moderately faster	86.4 ± 1.12%	Williams et al. (2007), Lu et al. (2017)
Solid phase microextraction (SPME)	No need for solvents, helps multiple sampling, extent capsaicin retention, reduces contamination, faster and automated	50–60%	Spencer and Almirall (2005)
Enzymatic treatment (ET)	Improve capsaicin yield and selectivity, *Capsicum* degrading enzymes easily and commercially available	>85%	Salgado-Roman et al. (2008)
Aqueous two-phase system (ATPS)	Can easily fractionate different capsaicinoids, more biocompatible, extract capsaicin from *Capsicum* biomass	85.6 ± 0.4%	Cienfuegos et al. (2017)
Supercritical fluid extraction (SFE)	High extract power with purity, low energy consumption and required mild temperature, non-expensive, nontoxic, CO_2 used as superficial solvent for capsaicin extraction	58.8–60.8%	Santos et al. (2015)

evaluated by numerous researchers to determine the degree of variation in capsaicin content (Islam et al. 2015; Sarpras et al. 2016).

1.6.1 Capsaicin Analysis Using HPLC

HPLC-based capsaicin quantification is the most accepted and widely used method among other analytical techniques. Since HPLC has established protocols available for sample preparations, both fresh and dried *Capsicum* samples can be used for analysis. The procedure described here has been used by various researchers for different *Capsicum* species, which includes both dry and fresh samples. For example, as an initial analysis, homogenise 0.5–1 g of *Capsicum* fruit in 20–40 mL of acetone at 50 °C for 30 min. Centrifuge the mixture followed by evaporating the sample to dry. Suspend dried powder in 2.5–5 mL of eluent and filter by using 0.2μL

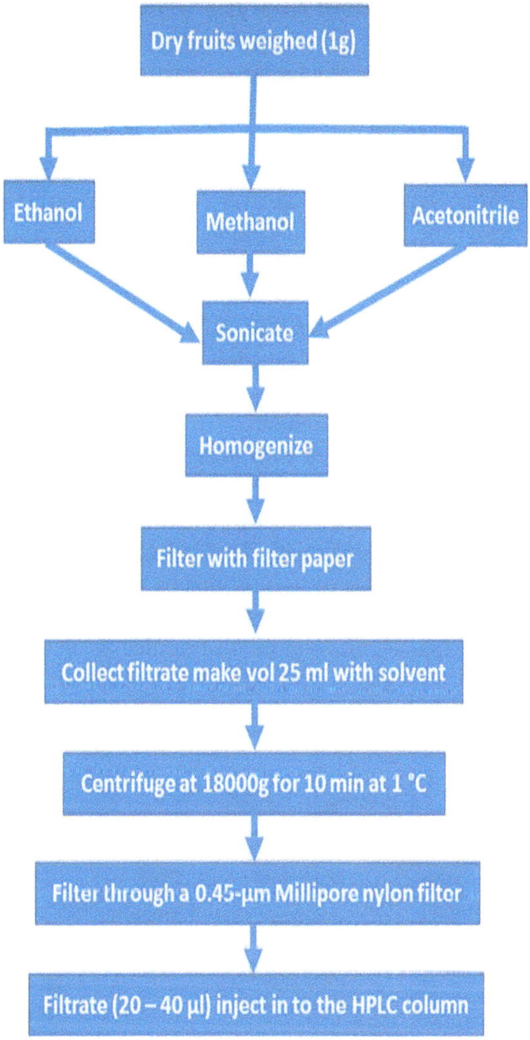

Fig. 1.6 Capsaicin extraction flowchart for HPLC analysis

nylon filter (Fig. 1.6). Two different-size columns were commonly used for the HPLC sample separation method (250 mm × 4.0 mm i.d. and 150 mm × 4.0 mm i.d.). The temperature of the column should be maintained at 30 °C and the flow rate kept at 1 mL/min. Importantly, the UV detection has to be set at 280 nm for proper detection of the peaks. Since acetonitrile quantifies different overlapped capsaicinoid peaks, it is considered as the primary mobile phase in many of the HPLC analyses. However, some laboratories have used two mobile phases, which include acetonitrile and 0.5% formic acid in water (45:55 v/v).

1.6.2 Capsaicin Analysis Using GC-MS

Many researchers, including our own previous studies, carry out capsaicin extraction in acetonitrile and quantified and separated by using GC-MS. The ripened and deseeded *Capsicum* fruits were homogenised in solvents, mostly methanol (1:10, w/v), by using a homogeniser or a mortar and pestle. Whatman paper no. 1 was used to filter the homogenate over Na_2SO_4. The filtrate is then dissolved in 10 mL of acetonitrile after being completely evaporated to dryness in a vacuum chamber (Chinn et al. 2011). Prior to injection into the GC-MS, the entire sample is centrifuged at 14,000 rpm for 10 min and then filtered through a 0.45-μm Millipore PTFE membrane filter (Fig. 1.7). For extraction and GC-MS analysis, samples are divided into independent triplicates. By using external standard calibration curves for dihydrocapsaicin and capsaicin, a quantitative analysis of the capsaicinoid complex can be carried out.

A well-established instrumental setup is available for GC-MS-based analytical estimation and quantification of capsaicinoids. GC-MS commonly equipped with the Rtx-5 MS capillary column (containing 0.25-mm internal diameter and 0.25-mm film thickness, along with 30 m in length) was commonly used for this purpose. The temperature of the oven must first be set at 100 °C for 2 min and then raised to 250 °C at a rate of 5 °C per minute and then to 280 °C at a rate of 10 °C per minute. In split mode, 1μL of the sample will suffice to inject into the column. Using external reference standards injected under identical circumstances, it is possible to determine the amounts of capsaicin and other capsaicinoids. Retention times and mass

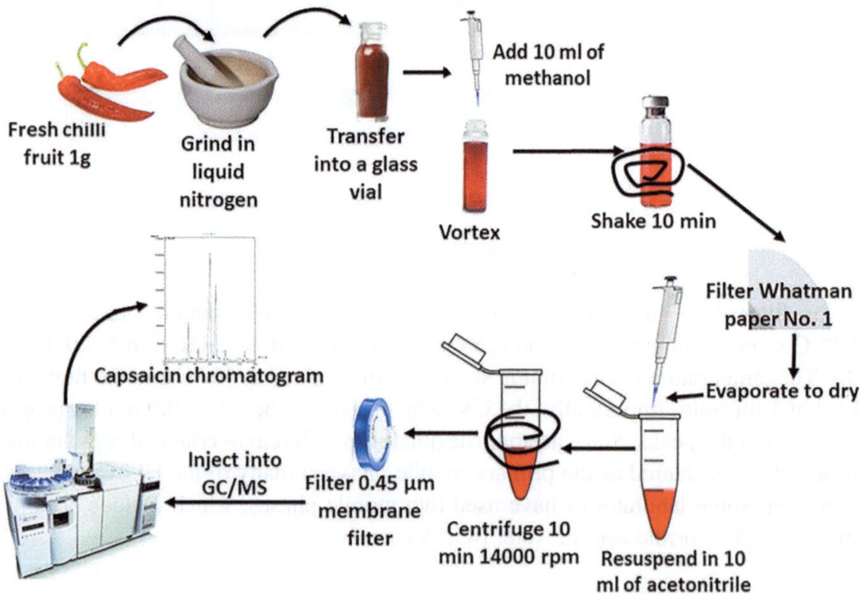

Fig. 1.7 Capsaicin extraction flowchart for GCMS analysis

measurements made under identical GC-MS conditions can be used for identification, while peak areas can be used to quantify the compound (Sarpras et al. 2016).

1.6.3 Capsaicin Analysis by Using LC-MS

One of the most popular analytical methods for identifying and quantifying semipolar plant chemical compounds is LC-MS. Its high-resolution and accurate mass spectra enable us to identify particular compounds and enhance the scope of subsequent analysis (De Vos et al. 2007). However, many researchers used LC-MS to investigate the detection and measurement of capsaicin and other individual capsaicinoids. The widely used column for capsaicinoid separation was the XDB-C18 column, (contains 4.6 × 150 mm width and length, along with 3.5-μm diameter). Aside from that, the two most popular LC-MS mobile phases are acetonitrile and distilled water with 0.5% formic acid (29:71 or 45:55, v/v) at 30 °C and flow rates of 1.0 or 0.75 mL/min. Selected-ion monitoring (SIM) was the operating mode chosen for the mass spectrometer with an electrospray as the ionisation source. The molecule protonated from capsaicin (m/z 306) was monitored. The other instrumental setup for LC-MS analysis is as follows: flow source voltage, maintained at 5.5 kV; capillary voltage, 27 V; current source, 80μA; multipole RF amplifier, 400 (Vp-p) maintained; and tube lens offset, 45 V (Kozukue et al. 2005).

1.6.4 Capsaicin Analysis by Using NMR

An important method for measuring the amount of capsaicinoids in different types of *Capsicum* is NMR-based analysis. It is fast, simple and one of the highly recommended tools for plant chemical analyses (Rasmussen et al. 2006). Since NMR functions according to the physical properties of the compound, it prevents the influence of external factors; hence, its results are considered to be highly reproducible (Kim et al. 2011). The sample preparation for NMR based capsaicinoid analysis is simple and can be performed quite rapidly. The prepared samples (25 mg) were inserted in a 50-μL spherical inserting unit of a 4-mm HRMAS rotor. After that, add 25 mL of 0.01 M phosphate buffer (pH 7.2). By spinning the samples at 7 kHz, operating at a 1-H frequency of 400.13 MHz and using a spectrometer with a 4-mm HRMAS dual-channel probe head, NMR spectra are obtained at 298 K. All ^{1}H NMR spectra can be manually or automatically baseline corrected, phased and aligned (Ritota et al. 2010).

1.7 Conclusions

Capsicum is the only genus in the plant kingdom that synthesises and accumulates capsaicinoids that are responsible for its pungency. Among all the capsaicinoids reported from *Capsicum* spp., capsaicin and dihydrocapsaicin are the most common, which accounts for about 80–90% of total capsaicinoids. Capsaicinoids, especially capsaicin, have a high nutritional and pharmaceutical value. The most important cultivated species of *Capsicum* are *C. annuum, C. frutescens, C. baccatum, C. chinense* and *C. pubescens*. The taxonomy of genus *Capsicum* is very complex and it is not easy to identify *Capsicum* species on the basis of morphology alone; thus, a combination of morphological as well as molecular approaches is needed for species identification. The level of capsaicin in *Capsicum* generally differs from one species to another as well as their genotype and environmental conditions. The entire genome sequence of pepper will provide an excellent opportunity to improve the nutritional and pharmaceutical value of the pepper and also to understand the biosynthetic pathway of capsaicinoids (Kim et al. 2014). By gaining information of all important genes that are important for capsaicinoids, biosynthesis and accumulation can play an important role in developing improved *Capsicum* cultivars. Thus, a combination of traditional and modern techniques such as omics approaches will enable us to better understand the taxonomy as well as to improve the commercially important traits of the fruit such as size, capsaicinoids content, yield and also resistance to abiotic stresses and pathogens of this important crop.

References

Agyemang Duah S, Silva e Souza C, Nagy Z, Pék Z, Neményi A, Daood HG, Vinogradov S, Helyes L (2021) Effect of water supply on physiological response and phytonutrient composition of chili peppers. Water 13(9):1284. https://doi.org/10.3390/w13091284

Al Othman ZA, Ahmed YB, Habila MA, Ghafar AA (2011) Determination of capsaicin and dihydrocapsaicin in Capsicum fruit samples using high performance liquid chromatography. Molecules 16(10):8919–8929. https://doi.org/10.3390/molecules16108919; PMID: 22024959; PMCID: PMC6264681

Aldana-Iuit JG, Sauri-Duch E, de Miranda-Ham ML, Castro-Concha LA, Cuevas-Glory LF, Vázquez-Flota FA (2015) Nitrate promotes capsaicin accumulation in *Capsicum chinense* immobilized placentas. Biomed Res Int 2015:794084. https://doi.org/10.1155/2015/794084

Andrews J (1984) Pepper -the domesticated *Capsicum*. University of Texas Press, Austin, TX

Antonious G, Jarret R (2006) Screening *Capsicum* accessions for capsaicinoids content. J Environ Sci Health B 41:717–729

Antonious GF (2018) Capsaicinoids and vitamins in hot pepper and their role in disease therapy. In: Capsaicin and its human therapeutic development. IntechOpen. https://doi.org/10.5772/intechopen.78243

Aza-González C, Núñez-Palenius HG, Ochoa-Alejo N (2011) Molecular biology of capsaicinoid biosynthesis in chili pepper (*Capsicum* spp.). Plant Cell Rep 30:695–706. https://doi.org/10.1007/s00299-010-0968-8

Bajer T, Bajerová P, Kremr D, Eisner A, Ventura K (2015) Central composite design of pressurised hot water extraction process for extracting capsaicinoids from chili peppers. J Food Compos Anal 40:32–38

Barbero GF, Carmelo MP, Barroso G (2016) Determination of capsaicinoids in peppers by microwave-assisted extraction–high-performance liquid chromatography with fluorescence detection. Anal Chim Acta 578:227–233

Barbero GF, Ruiz AG, Liazid A, Palma M, Vera JC, Barroso CG (2014) Evolution of total and individual capsaicinoids in peppers during ripening of the Cayenne pepper plant (*Capsicum annuum* L.). Food Chem 153:200–206. https://doi.org/10.1016/j.foodchem.2013.12.068

Barboza GE, Carrizo García C, de Bem BL, Romero MV, Scaldaferro M (2022) Monograph of wild and cultivated chili peppers (*Capsicum* L., Solanaceae). PhytoKeys 200:1–423

Barboza GE, De Bem Bianchetti L (2005) Three new species of *Capsicum* (Solanaceae) and a key to the wild species from Brazil. Syst Bot 30(4):863–871; http://www.jstor.org/stable/25064115

Baruah J, Lal M (2020) *Capsicum chinense* Jacq.: Ethnobotany, bioactivity and future prospects. In: Singh B (ed) Botanical leads for drug discovery. Springer, Singapore. https://doi.org/10.1007/978-981-15-5917-4_16

Basu SK, De AK (2003) Capsicum: historical and botanical perspectives. In: De AK (ed) Capsicum: the genus Capsicum. Taylor & Francis, New York, pp 1–15

Batiha GE, Alqahtani A, Ojo OA, Shaheen HM, Wasef L, Elzeiny M, Ismail M, Shalaby M, Murata T, Zaragoza-Bastida A, Rivero-Perez N, Magdy Beshbishy A, Kasozi KI, Jeandet P, Hetta HF (2020) Biological properties, bioactive constituents, and pharmacokinetics of some *Capsicum* spp. and capsaicinoids. Int J Mol Sci 21(15):5179. https://doi.org/10.3390/ijms21155179

Bayil Oğuzkan S (2019) Extraction of capsinoid and its analogs from pepper waste of different genotypes. Nat Prod Commun 14:1934578X1986567. https://doi.org/10.1177/1934578X19865673

Carrizo García C, Barboza GE, Palombo N, Weiss-Schneeweiss H (2022) Diversification of chiles (*Capsicum*, Solanaceae) through time and space: new insights from genome-wide RAD-seq data. Front Genet 13:1030536. https://doi.org/10.3389/fgene.2022.1030536

Carrizo García C, Barfuss MHJ, Sehr EM, Barboza GE, Samuel R, Moscone EA, Ehrendorfer F (2016) Phylogenetic relationships, diversification and expansion of chili peppers (*Capsicum*, Solanaceae). Ann Bot 118:35–51. https://doi.org/10.1093/aob/mcw079

Carrizo Garcia C, Sterpetti M, Volpi P, Ummarino M, Saccardo F (2013) Wild Capsicums: Identification and in situ analysis of Brazilian species. In: Lanteri S, Rotino GL (eds) Breakthroughs in the genetics and breeding of *Capsicum* and eggplant. Eucarpia, Turin, pp 205–213

Castro-Concha L, Baas-Espinola F, Ancona-Escalante W, Vázquez-Flota F, Miranda-Ham ML (2016) Phenylalanine biosynthesis and its relationship to accumulation of capsaicinoids during *Capsicum chinense* fruit development. Biol Plant 60:579–584. https://doi.org/10.1007/s10535-016-0608-4

Chanthai S, Juangsamoot J, Ruangviriyachai C, Techawongstien S (2012) Determination of capsaicin and dihydrocapsaicin in some chilli varieties using accelerated solvent extraction associated with solid-phase extraction methods and RP-HPLC-fluorescence. E-J Chem 9:1550–1561

Chapa-Oliver AM, Mejía-Teniente L (2016) Capsaicin: from plants to a cancer- suppressing agent. Molecules 21(8):931. https://doi.org/10.3390/molecules21080931

Chávez-Servia JL, Sevilla-Panizo R (eds) (2006) Seminario: Fundamentos Genéticos y Socioeconómicos para Analizar La Agrobiodiversidad en la Región de Ucayali, 16 de enero de 2003 Pucallpa, Perú. Bioversity International, Cali, p 93, ISBN 92-9043-738-3

Chinn MS, Sharma-Shivappa RR, Cotter JL (2011) Solvent extraction and quantification of capsaicinoids from *Capsicum chinense*. Food Bioprod Process 89:340–345

Cichewicz RH, Thorpe PA (1996) The antimicrobial properties of chile peppers (Capsicum species) and their uses in Mayan medicine. J Ethnopharmacol 52(2):61–70. https://doi.org/10.1016/0378-8741(96)01384-0

Cienfuegos NEC, Santos PL, García AR, Soares CMF, Lima AS, Souza RL (2017) Integrated process for purification of capsaicin using aqueous two-phase systems based on ethanol. Food Bioprod Process 106:1–10

Curry J, Aluru M, Mendoza M, Nevarez J, Melendrez M, O'Connell MA (1999) Transcripts for possible capsaicinoid biosynthetic genes are differentially accumulated in pungent and non-pungent *Capsicum* spp. Plant Sci 148:47–57 da Silva MPS, Mendonça Freitas MS, Pereira JAP, dos Santos PC, de Carvalho AJC, Vieira IJC, Rodrigues R (2021) Capsaicinoids and mineral composition of peppers produced under nutrient deficiencies. J Plant Nutr 44(6):845–853. https://doi.org/10.1080/01904167.2021.1871752

De Vos RCH, Moco S, Lommen A, Keurentjes JJB, Bino RJ, Hall RD (2007) Untargeted large-scale plant metabolomics using liquid chromatography coupled to mass spectrometry. Nat Protoc 2:778–791

Dias GB, Gomes VM, Moraes TM, Zottich UP, Rabelo GR, Carvalho AO, Moulin M, Gonçalves LS, Rodrigues R, Da Cunha M (2013) Characterization of *Capsicum* species using anatomical and molecular data. Genet Mol Res 12(4):6488–6501. https://doi.org/10.4238/2013

Díaz J, Pomar F, Bernal A, Merino F (2004) Peroxidases and the metabolism of capsaicin in Capsicum annuum L. Phyt Rev 3(1–2):141–157. https://doi.org/10.1023/B:PHYT.0000047801.41574.6e

Eshbaugh WH (2012) The taxonomy of the genus *Capsicum*. In: Russo VM (ed) Peppers: botany, production and uses. CAB International, pp 14–28

Eshbaugh WH (1993) Peppers: History and exploitation of a serendipitous new crop discovery. In: Janick J, Simon JE (eds) New crops. John Wiley & Sons, New York, NY, pp 132–139

Estrada B, Bernal MA, Díaz J, Pomar F, Merino F (2002) Capsaicinoids in vegetative organs of *Capsicum annuum* L. in relation to fruiting. J Agric Food Chem 50:1188–1191

Giuffrida D, Dugo P, Torre G, Bignardi C, Cavazza A, Corradini C, Dugo G (2013) Characterization of 12 *Capsicum* varieties by evaluation of their carotenoid profile and pungency determination. Food Chem 140:794–802

Gogoi J, Barman I, Murrey E (2018) Journey of *Capsicum* from wild chili to world hottest chili (Naga king chili). Int J Pharm Biol Sci 8(3):835–840

González-Zamora A, Sierra-Campos E, Luna-Ortega JG, Pérez-Morales R, Rodríguez Ortiz JC, García-Hernández JL (2013) Characterization of different *Capsicum* varieties by evaluation of their capsaicinoids content by high performance liquid chromatography, determination of pungency and effect of high temperature. Molecules 18(11):13471–13486. https://doi.org/10.3390/molecules181113471

Hamed M, Kalita D, Bartolo ME, Jayanty SS (2019) Capsaicinoids, polyphenols and antioxidant activities of *Capsicum annuum*: comparative study of the effect of ripening stage and cooking methods. Antioxidants 8(9):364. https://doi.org/10.3390/antiox8090364

Hayman M, Kam P (2008) Capsaicin: a review of its pharmacology and clinical applications. Curr Anaesth Crit Care 19:338–343

Heiser CB, Smith PG (1958) New species of *Capsicum* from South America. Brittonia 10(4):194. https://doi.org/10.2307/2804950

Huang XF, Xue JY, Jiang AQ, Zhu HL (2013) Capsaicin and its analogues: Structure-activity relationship study. Curr Med Chem 20(21):2661–2672. https://doi.org/10.2174/0929867311320210004

Hunziker AT (1961) Estudios sobre Solanaceae. III Kurtziana 1: 207–216

Igwe DO, Afiukwa CA, Acquaah G, Ude GN (2019) Genetic diversity and structure of *Capsicum annuum* as revealed by start codon targeted and directed amplified minisatellite DNA markers. Hereditas 156:32. https://doi.org/10.1186/s41065-019-0108-

Ince AG, Karaca M, Onus AN (2010) Genetic relationships within and between *Capsicum* species. Biochem Genet 48(1–2):83–95. https://doi.org/10.1007/s10528-009-9297-4

Islam K, Rawoof A, Kumar A, Momo J, Ahmed I, Dubey M, Ramchiary N (2023) Genetic regulation, environmental cues, and extraction methods for higher yield of secondary metabolites in *Capsicum*. J Agric Food Chem 71:9213. https://doi.org/10.1021/acs.jafc.3c01901; Epub ahead of print

Islam MA, Sharma SS, Sinha P, Negi MS, Neog B, Tripathi SB (2015) Variability in capsaicinoid content in different landraces of *Capsicum* cultivated in North-Eastern India. Sci Hortic 183:66–71

Iwai K, Suzuki T, Lee KR, Kobashi M, Oka S (1977) In vivo and in vitro formation of dihydrocapsaicin in sweet pepper fruits, *Capsicum annuum* L. var. *grossum*. Agric Biol Chem 41:1877–1882. https://doi.org/10.1271/bbb1961.41.1877

Kim HK, Choi YH, Verpoorte R (2011) NMR-based plant metabolomics: where do we stand, where do we go? Trends Biotechnol 29:267–275

Kim S, Park M, Yeom SI, Kim YM, Lee JM, Lee HA, Seo E, Choi J et al (2014) Genome sequence of the hot pepper provides insights into the evolution of pungency in *Capsicum* species. Nat Genet 46(3):270–278. https://doi.org/10.1038/ng.2877

Kozukue N, Han JS, Kozukue E, Lee SJ, Kim JA, Lee KR, Levin CE, Friedman M (2005) Analysis of eight capsaicinoids in peppers and pepper-containing foods by high-performance liquid chromatography and liquid chromatography– mass spectrometry. J Agric Food Chem 53: 9172–9181

De Krishna A (ed) (2003) *Capsicum*: the genus *Capsicum*, 1st edn. CRC Press. https://doi.org/10.1201/9780203381151

Kumar A, Kumar S, Anju T, Ramchiary N (2021) Genetic, epigenetic, and hormonal regulation of fruit development and ripening in *Capsicum* L. species. In: Roberts JA (ed) Annual plant reviews online, 1st edn. Wiley, pp 295–356

Leete E, Louden MCL (1968) Biosynthesis of capsaicin and dihydrocapsaicin in *Capsicum frutescens*. J Am Chem Soc 90:6837–6841

Leff B, Ramankutty N, Foley JA (2004) Geographic distribution of major crops across the world. Glob Biogeochem Cycles 18(1):2003GB002108. https://doi.org/10.1029/2003GB002108

Liu R, Song J, Liu S, Chen C, Zhang S, Wang J, Xiao Y, Cao B, Lei J, Zhu Z (2021) Genome-wide identification of the *Capsicum* BHLH transcription factor family: discovery of a candidate regulator involved in the regulation of species-specific bioactive metabolites. BMC Plant Biol 21(1):262

Lu M, Ho CT, Huang Q (2017) Extraction, bioavailability, and bioefficacy of capsaicinoids. J Food Drug Anal 25:27–36

Maga JA, Todd PH (1975) *Capsicum*. CRC Crit Rev Food Sci Nutr 6(2):177–199. https://doi.org/10.1080/10408397509527189

Mazourek M, Pujar A, Borovsky Y, Paran I, Mueller L, Jahn MM (2009) A dynamic interface for capsaicinoid systems biology. Plant Physiol 150(4):1806–1821. https://doi.org/10.1104/pp.109.136549

Mehta I (2017) Chillies—the prime spice—a history. IOSR J Humanit Soc Sci 22:32–36

Melgarejo L, Hernández M, Barrera J, Bardales X (2004) Caracterización y Usos Potenciales del Banco de Germoplasma de Ají Amazónico. Universidad Nacional de Colombia, Bogotá, CO, p 107, ISBN 958-97420-4-1

Monforte González M, Guzmán-Antonio A, Uuh-Chim F, Vázquez-Flota F (2010) Capsaicin accumulation is related to nitrate content in placentas of habanero peppers (*Capsicum chinense* Jacq.). J Sci Food Agric 90:764–768. https://doi.org/10.1002/jsfa.3880

Moscone EA, Scaldaferro MA, Grabiele M, Cecchini NM, Sánchez García Y, Jarret R, Daviña JR, Ducasse DA, Barboza GE, Ehrendorfer F (2007) The evolution of chili peppers (*Capsicum*—Solanaceae): a cytogenetic perspective. Acta Hortic 745:137–169. https://doi.org/10.17660/ActaHortic.2007.745.5

Murakami K, Ido M, Masuda M (2006) Fruit pungency of ‘Shishito’ pepper as affected by a dark interval in continuous fluorescent illumination with temperature alteration. Shokubutsu Kankyo Kogaku 18(4):284–289. https://doi.org/10.2525/shita.18.284

Naves ER, De Ávila SL, Sulpice R, Araújo WL, Nunes-Nesi A, Peres LEP, Zsögön A (2019) Capsaicinoids: Pungency beyond *Capsicum*. Trends Plant Sci 24(2):109–120. https://doi.org/10.1016/j.tplants.2018.11.001

Nip WK (2004) Handbook of food and beverage fermentation technology, vol 133. University of Hawaii at Manoa, Honolulu, HI, p 60

Nwokem CO, Agbaji EB, Kagbu JA, Ekanem EJ (2010) Determination of capsaicin content and pungency level of five different peppers grown in Nigeria. New York Science Journal 3(9):17–21

O'Neill J, Brock C, Olesen AE, Andresen T, Nilsson M, Dickenson AH (2012) Unravelling the mystery of capsaicin: A tool to understand and treat pain. Pharmacol Rev 64(4):939–971. https://doi.org/10.1124/pr.112.006163

Otha Y (1962) Physiological and genetical studies on the pungency of *Capsicum.* IV. Secretory organs, receptacles and distribution of capsaicin in the *Capsicum* fruit. Jpn J Breed 12:179–183

Paredes Andrade NJ, Monteros-Altamirano A, Tapia Bastidas CG, Sørensen M (2020) Morphological, sensorial and chemical characterization of chilli peppers (*Capsicum* spp.) from the CATIE genebank. Agronomy 10(11):1732. https://doi.org/10.3390/agronomy10111732

Parrish M (1996) Liquid chromatographic method for determining capsaicinoids in Capsicums and their extractives: Collaborative study. J AOAC Int 79:738–745. https://doi.org/10.1093/jaoac/79.3.738

Petters PJ (2002) Correlations between leaf structural traits and the densities of herbivorous insect guilds. Biol J Linn Soc Lond 77:43–65. https://doi.org/10.1046/j.1095-8312.2002.00091.x

Prasad BCN, Kumar V, Gururaj HB, Parimalan R, Giridhar P, Ravishankar GA (2006) Characterization of capsaicin synthase and identification of its gene (*csy1*) for pungency factor capsaicin in pepper (*Capsicum* sp.). Proc Natl Acad Sci 103(36):13315–13320. https://doi.org/10.1073/pnas.0605805103

Purkayastha J, Alam SI, Gogoi HK, Singh L (2012) *Capsicum assamicum* sp. nov. (Solanaceae), from Assam, Northeastern India. Ozean J Appl Sci 5:55–66

Purseglove JM (1968) Tropical Crops, vol 2. John Wiley and Sons, New York, NY

Ramchiary N, Kehie M, Brahma V, Kumaria S, Tandon P (2014) Application of genetics and genomics towards *Capsicum* translational research. Plant Biotechnol Rep 8:101–123. https://doi.org/10.1007/s11816-013-0306-z

Rasmussen B, Cloarec O, Tang H, Stærk D, Jaroszewski J (2006) Multivariate analysis of integrated and full-resolution 1 H-NMR spectral data from complex pharmaceutical preparations: St. John's Wort. Planta Med 72:556–563. https://doi.org/10.1055/s-2006-931567

Reddy UK, Almeida A, Abburi VL, Alaparthi SB, Unselt D, Hankins G, Park M, Choi D, Nimmakayala P (2014) Identification of gene-specific polymorphisms and association with capsaicin pathway metabolites in *Capsicum annuum* L. collections. PLoS One 9(1):e86393. https://doi.org/10.1371/journal.pone.0086393

Reyes-Escogido Mde L, Gonzalez-Mondragon EG, Vazquez-Tzompantzi E (2011) Chemical and pharmacological aspects of capsaicin. Molecules 16(2):1253–1270. https://doi.org/10.3390/molecules16021253

Ritota M, Marini F, Sequi P, Valentini M (2010) Metabolomic characterization of Italian sweet pepper (*Capsicum annuum* L.) by means of HRMAS-NMR spectroscopy and multivariate analysis. J Agric Food Chem 58:9675–9684

Roy A (2016) Bhut Jolokia (*Capsicum chinense* Jaqc): a review. Int J Pharm Sci Res 7(3):882–889

Salgado-Roman M, Botello-Álvarez E, Rico-Martínez R, Jiménez-Islas H, Cárdenas-Manríquez M, Navarrete-Bolaños JL (2008) Enzymatic treatment to improve extraction of capsaicinoids and carotenoids from chili (*Capsicum annuum*) fruits. J Agric Food Chem 56:10012–10018. https://doi.org/10.1021/jf801823m

Salehi B, Hernández-Álvarez AJ, Del Mar CM, Martorell M, Ramírez-Alarcón K, Melgar-Lalanne G, Matthews KR, Sharifi-Rad M, Setzer WN, Nadeem M, Yousaf Z, Sharifi-Rad J (2018) Potential phytopharmacy and food applications of Capsicum spp.: a comprehensive review. Nat Prod Commun 13(11):1934578X1801301. https://doi.org/10.1177/1934578X1801301133

Santos P, Aguiar AC, Barbero GF, Rezende CA, Martínez J (2015) Supercritical carbon dioxide extraction of capsaicinoids from Malagueta pepper (*Capsicum frutescens* L.) assisted by ultrasound. Ultrason Sonochem 22:78–88

Sarpras M, Gaur R, Sharma V, Chhapekar SS, Das J, Kumar A, Yadava SK, Nitin M, Brahma V, Abraham SK, Ramchiary N (2016) Comparative analysis of fruit metabolites and pungency candidate genes expression between *Bhut jolokia* and other *Capsicum* species. PLoS One 11(12):e0167791. https://doi.org/10.1371/journal.pone.0167791

Sharma SK, Vij AS, Sharma M (2013) Mechanisms and clinical uses of capsaicin. Eur J Pharmacol 720(1–3):55–62. https://doi.org/10.1016/j.ejphar.2013.10.053

Sora GTS, Haminiuk CWI, da Silva MV, Zielinski AAF, Gonçalves GA, Bracht A, Peralta RM (2015) A comparative study of the capsaicinoid and phenolic contents and in vitro antioxidant activities of the peppers of the genus *Capsicum*: An application of chemometrics. J Food Sci Technol 52:8086–8094

Spencer JO, Almirall JR (2005) Extraction of capsaicin in erosol defence sprays from fabrics. Talanta 67:377–382. https://doi.org/10.1016/j.talanta.2005.05.031

Sumner J (2000) The natural history of medicinal plants. Timber Press

Sun B, Zhou X, Chen C, Chen C, Chen K, Chen M, Liu S, Chen G, Cao B, Cao F, Lei J, Zhu Z (2020) Coexpression network analysis reveals an MYB transcriptional activator involved in capsaicinoid biosynthesis in hot peppers. Hortic Res 7:162. https://doi.org/10.1038/s41438-020-00381-2

Sung Y, Chang Y, Ting N (2005) Capsaicin biosynthesis in water-stressed hot pepper fruits. Bot Bull Acad Sin 46:35–42. https://doi.org/10.7016/BBAS.200501.0035

Swamy KRM (2023) Origin, distribution, taxonomy, botanical description, genetic diversity and breeding of Capsicum (Capsicum annuum L.). Int J Dev Res 13(03):61956–61977

Todd PH (1977) Determination of pungency due to *Capsicum* by gas-liquid chromatography. J Food Sci 42:660–665

Tripodi P, Kumar S (2019) The *Capsicum* crop: an introduction. In: Ramchiary N, Kole C (eds) The *Capsicum* genome. Compendium of plant genomes. Springer, Cham. https://doi.org/10.1007/978-3-319-97217-6_1

Uarrota VG, Maraschin M, de Bairros Âde FM, Pedreschi R (2021) Factors affecting the capsaicinoid profile of hot peppers and biological activity of their non-pungent analogs (Capsinoids) present in sweet peppers. Crit Rev Food Sci Nutr 61(4):649–665. https://doi.org/10.1080/10408398.2020.1743642

Verma PK, Rawat KK, Das N, Pradha BA (2013) A botanical enigma of India's hottest Chilli 'Bhoot Jolokia' (*Capsicum chinense* Jacq.). NY Sci J 6(11):49–51; http://www.sciencepub.net/newyork

Wahyuni Y, Ballester A-R, Sudarmonowati E, Bino RJ, Bovy AG (2013) Secondary metabolites of *Capsicum* species and their importance in the human diet. J Nat Prod 76(4):783–793. https://doi.org/10.1021/np300898z

Wen J, Lv J, Zhao K, Zhang X, Li Z, Zhang H, Huo J, Wan H, Wang Z, Zhu H, Deng M (2022) Ethylene-inducible AP2/ERF transcription factor involved in the capsaicinoid biosynthesis in *Capsicum*. Front Plant Sci 13:832669. https://doi.org/10.3389/fpls.2022.832669

WFO (2023) https://wfoplantlist.org/plant-list/. Accessed 9 May 2023

Williams OJ, Viyaya GS, Raghavan Orsat V, Dai J (2007) Microwave-assisted extraction of capsaicinoids from *Capsicum* fruit. J Food Biochem 28:113–122. https://doi.org/10.1111/j.1745-4514.2004.tb00059.x

Yaldiz G, Ozguven M, Sekeroglu N (2010) Variation in capsaicin contents of different *Capsicum* species and lines by varying drying parameters. Ind Crops Prod 32(3):434–438. https://doi.org/10.1016/j.indcrop.2010.06.013

Yamamoto S, Girsang W (2021) An ethnobotanical study on *Capsicum* peppers in Maluku Province. Indonesia Trop Agr Develop 65(1):1–9

Yanka H, Rinyo R, Das SK, Das TJ, Paul D, Gupta DD, Hui PK, Bansod S, Godugu C, Ananthan R, Jambhulkar S (2019) A brief cross-cultural ethnobotanical note on the Abotani tribes of Arunachal Pradesh, India. Pleione 13(2):269. https://doi.org/10.26679/Pleione.13.2.2019.269-283

Zenk MH (1965) Biosynthese von vanillin in Vanilla planifolia Andr Z Pflanzenphysiol 53:404–414

Zewdie Y, Bosland PW (2000) Evaluation of genotype, environment, and genotype- by- environment interaction for capsaicinoids in *Capsicum annuum* L. Euphytica 111:185–190. https://doi.org/10.1023/A:1003837314929

Zhang B, Hu F, Cai X, Cheng J, Zhang Y, Lin H, Hu K, Wu Z (2022a) Integrative analysis of the metabolome and transcriptome of a cultivated pepper and its wild progenitor Chiltepin (*Capsicum annuum* L. var. *glabriusculum*) revealed the loss of pungency during *Capsicum* domestication. Front Plant Sci 12:783496. https://doi.org/10.3389/fpls.2021.783496

Zhang W, Wu D, Zhang L, Zhao C, Shu H, Cheng S, Wang Z, Zhu J, Liu P (2022b) Identification and expression analysis of capsaicin biosynthesis pathway genes at genome level in *Capsicum chinense*. Biotechnol Biotechnol Equip 36(1):232–244

Zhang ZX, Zhao SN, Liu GF, Huang ZM, Cao ZM, Cheng SH, Lin SS (2016) Discovery of putative capsaicin biosynthetic genes by RNA-Seq and digital gene expression analysis of pepper. Sci Rep 6:34121. https://doi.org/10.1038/srep34121

Capsaicin: Its Sources, Isolation, Quantitative Analysis and Applications

2

Arpita Srivastava, Poornima K. N., Nikita Baliyan, and Manisha Mangal

Abstract

Capsaicin is a phenolic compound found in hot peppers and responsible for the characteristic pungent taste and burning sensation. The genus *Capsicum* includes hot peppers, which is the only genus producing the capsaicin compound. There are more than 32 species of *Capsicum* reported but five are domesticated, viz. *C. annuum*, *C. chinense*, *C. baccatum*, *C. frutescens* and *C. pubescens*. In general, *C. chinense* and *C. frutescens* have been reported to have much higher pungency when compared to other species of the genus *Capsicum*. Capsaicin is only present in the fruits of the *Capsicum* genus; however, it is not equally distributed in all parts of the fruit. It is mainly concentrated in the placental tissue of the fruit surrounding the seeds. Chemically, capsaicin is 8-methyl-*N*-vanillyl-6-nonenamide. It belongs to the group of capsaicinoids which also include dihydrocapsaicin, nordihydrocapsaicin, homocapsaicin and homodihydrocapsaicin. Eighty to ninety percent of the total pungency in chilli peppers is contributed by capsaicin and dihydrocapsaicin. Extraction of capsaicinoids from hot peppers starts with selection of appropriate solvents like methanol, ethanol, acetonitrile ether, benzene, dimethyl sulfoxide and acetone. The extraction techniques widely employed are enzymatic, microwave, ultrasound-assisted, Soxhlet, supercritical fluid and pressurized liquid extraction. Major analytical methods for quantification of capsaicinoids include high-performance liquid chromatography (HPLC), gas chromatography and UV spectrophotometry. Calorimetry has also been suggested for the quantification of extracted capsaicinoids after separating them from a chilli pepper by thin-layer

A. Srivastava (✉) · N. Baliyan · M. Mangal
Division of Vegetable Science, ICAR-IARI, New Delhi, India

P. K. N.
Division of Basic Sciences, ICAR-IIHR, New Delhi, India

M. K. Swamy, A. Kumar (eds.), *Capsaicinoids*,
https://doi.org/10.1007/978-981-99-7779-6_2

chromatography or paper chromatography. Capsaicin finds utility in the food, pharmaceutical and cosmetic industries. Capsaicin consumption is involved in body weight reduction, treating urinary disorders and antioxidant effects. Other pharmaceutical properties of capsaicin include pain alleviation, body temperature control and antimicrobial and anticancer activities.

Keywords

Capsicum · Spice crop · Vegetable · Pepper · Food industry · Flavour

2.1 Introduction

Pepper is one of the most significant vegetable and spice crops which are mainly cultivated in tropical and subtropical regions of the world. It belongs to the family Solanaceae and genus *Capsicum*. From the reported 35 species of the genus *Capsicum*, taxonomists recognize only five major cultivated species. *Capsicum annuum* is the most widely cultivated species in the world followed by other four species, viz. *C. frutescens*, *C. chinense*, *C. baccatum* and *C. pubescens* (Bosland et al. 1996). *Capsicum* fruits exhibit vast variability in size, shape, colour, flavour and pungency, which is the major reason for pepper being the most valued and commercially cultivated crop across the world according to regions' preference for colour, flavour and taste attributes (Baenasa et al. 2019).

Capsaicinoids are alkaloids responsible for pepper pungency and are exclusive to the genus *Capsicum*. The pungency of peppers is a valuable fruit attribute and one of the major reasons it is consumed. More recently, capsaicinoids have found utility in food, pharmaceutical and pesticide industries (Surh 2002). Twenty-three different groups of capsaicinoids have been reported so far, all of which share a common aromatic ring. The length of the chain and type of acyl moieties in terms of saturation conditions and the level of pungency (Krajewska and Powers 1988) are contributed by the particular capsaicinoid. Capsaicinoids mainly include groups like capsaicin, dihydrocapsaicin, homocapsaicin, homodihydrocapsaicin and nordihydrocapsaicin. The contribution of capsaicin and dihydrocapsaicin amongst the capsaicinoids is approximately 90% of the pungency in most *C. annuum* accessions (Suzuki et al. 1980). The variability of minor capsaicinoids, viz. nordihydrocapsaicin, homocapsaicin and homodihydrocapsaicin, in paprika peppers has been found to be more pronounced than for the major capsaicinoids when compared with chilli peppers. Quantities of these minor compounds constitute not more than 18% of the total capsaicinoids in peppers (Barbero et al. 2016). Production of capsaicinoids by pepper plants is for defence against fungi, bacteria and herbivores (Tewksbury et al. 2006).

Chemically, capsaicin has a molecular weight of 305.4 kDa and is a volatile, hydrophobic, odourless, colourless and crystalline to waxy alkaloid. Apart from its pungent nature, this compound is of interest to the academic community as well as researchers for its various pharmaceutical benefits like alleviation of chronic pain, antiobesity, anticancerous, important source of health-promoting bioactive compounds, etc. (Pino et al. 2007; Jaganathan et al. 2020). Capsaicin extracted from peppers is used as an ingredient in numerous products like self-defence sprays or insect repellents (Davis et al. 2007). Capsaicinoids also find application as animal feed additives in animal husbandry (Erdost et al. 2006). Food industry has registered significant upsurge in the demand of peppers as it is often used as a colouring (natural food dye) and flavouring agent in ketchup, soup, hams, sausages, snacks, candies, soda drinks and alcoholic liquors (Perkins et al. 2002).

2.2 *Capsicum* Genera: Source of Capsaicinoids

The *Capsicum* is a genus which is comprised of more than 35 species. Several new *Capsicum* species have been reported recently from Bolivia (Nee et al. 2006). There are strong opinions that exploration in southern Peru, Bolivia and southern Brazil will yield identification of new pepper species and in the distant future the *Capsicum* genus will include more than 40 species (Eshbaugh 2012). In the process of domestication, human populations unconsciously selected several *Capsicum* species of different fruit sizes, shape and colours which were later on grouped by taxonomists into five major species, namely, *C. annuum*, *C. frutescens*, *C. chinense*, *C. baccatum* and *C. pubescens* (Eshbaugh 1993).

Out of these three domesticated species, viz. *C. annuum*, *C. frutescens* and *C. chinense*, are related groups that mainly evolved in the lowland tropics of Latin America and the Caribbean. The introduction of *C. annuum* to Europe was done by explorers like Columbus, while Portuguese explorers introduced them to Eastern Europe, Africa and Asia (Eshbaugh et al. 1983; Andrews 1999). *C. annuum* species include genotypes of jalapeno, poblano, anaheim, ancho, Big Jim, bell peppers, serrano and cayenne types, while *C. chinense* include cultivars in the habanero, scotch bonnet, rocotillo and blanco type. *C. frutescens* encompasses tabasco, malagueta, Thai, bird eye, and piri piri pepper types. *C. baccatum* most commonly domesticated in Peru include aji, aji amarillo, cuerno de oro, cumbai and others, while *C. pubescens* include rocoto, locoto and Chile manzana types (Eshbaugh 2012).

Pungency levels (measurement of spiciness or heat) of *Capsicum* genus are usually measured using the Scoville scale and is recorded in Scoville heat units (SHUs). SHU calculation is elicited from the capsaicinoid concentration level, amongst which capsaicin is the predominant one. The measurement of capsaicinoid is detailed later in this chapter. SHU determination is based on the dilution factor of the capsaicin extract with water sugar solution to a point when connoisseurs cannot discern any pungency. The pepper genotype is then given an SHU value with respect to the dilution necessary for the burning sensation of capsaicin to no longer be observed. Ranking of some of the most common peppers on Scoville scale is as follows: habanero peppers, 100,000–350,000 SHU; cayenne pepper, 30,000–50,000 SHU; serrano pepper, 6000–23,000 SHU; chipotle pepper, 5000–10,000 SHU; jalapeño pepper, 2500–5000 SHU; and bell pepper, 0 SHU (Weiss 2002).

The capsaicin content usually ranges between 0.1 and 4.25 mg/g in different chilli genotypes (Al Othman et al. 2011). *C. frutescens, C. annuum* and *C. chinense* were found to carry 0.22–20 mg of capsaicinoids per gram of dry pepper tissue (Thomas et al. 1998; Duelund and Mouritsen 2016).

2.2.1 Variation in Capsaicinoid Content of Different *Capsicum* Species

Vast variability has been observed amongst the *Capsicum* species for capsaicinoid content. The intrinsic biological characteristic of each species is responsible for the capsaicinoid content variability amongst *Capsicum* species. Amongst the domesticated *Capsicum* species, *C. chinense* has been acclaimed as the most pungent species (Canto-Flick et al. 2008). When the capsaicinoid contents of *C. chinense* and the hybrid of *C. chinense* and *C. frutescens* were compared to *C. annuum* and *C. baccatum, C. chinense* and its hybrid with *C. frutescens* were found to have 3.5 to 5 times more capsaicinoid content (Zamljen et al. 2021a). In comparison to *C. baccatum, C. chinense* again contains more capsaicinoids with higher levels of pungency (Guillen et al. 2018). Pungency levels of six cultivars from *C. chinense*, *C. frutescens* and *C. annuum* were compared where the genotype 'Umorok' belonging to *C. chinense* had the highest pungency (Sanatombi and Sharma 2008). When comparative gene expression analysis of the capsaicinoid biosynthesis pathway candidate genes were carried out through qRT-PCR in *C. chinense*, *C. frutescens* and *C. annuum, C. chinense* divulged to have many fold higher expression of majority of the genes in comparison to *C. annuum* and *C. frutescens*, thereby confirming the highest pungent status of *C. chinense* (Sarpras et al. 2016).

The most pungent chillies reported till date in the world were all from the *C. chinense* group which further affirms its position as the most pungent group amongst the *Capsicum* genus. Red Savina belonging to *C. chinense* was recognized as the hottest chilli in 2006 with 577,000 Scoville heat units (SHU (Canto-Flick et al. 2008). However, in 2007, this position of the world's hottest known chilli pepper was given to 'Bhut Jolokia', which is a natural hybrid of *C. chinense* and *C. frutescens* with a pungency level of 879,953 to 1,001,304 SHU (Bosland and Baral 2007). The average pungency levels of habanero peppers (*C. chinense*) range from 100,000 to 350,000 SHU (Bosland and Baral 2007; Meghavansi et al. 2010). Popularly, 'Bhut Jolokia' is known by other names such as Naga Jolokia, Naga Morich, Bih Jolokia, Nagahari, Borbih Jolokia and Raja Mirchi (Purkayastha et al. 2012). Later, the title of hottest chilli was given to the genotype, namely, 'Trinidad Moruga Scorpion', originating from Trinidad and Tobago with heat levels of 2009, 231 SHU (Bosland et al. 2012). Another genotype, namely, 'Carolina Reaper', belonging to *C. chinense*, is currently the hottest pepper in the world with a pungency of 2.2 million SHU (Muñoz-Ramírez et al. 2018). It has been invented by US-based company Pucker Butt Pepper in South Carolina. Due to its highest pungency levels, it has been recommended as a genotype for higher capsaicinoid recovery and industrial use of capsaicinoids. Similarly, *C. chinense* and hybrids of *C. chinense* × *C. frutescens* cultivars can also be used for the above-cited purposes (Zamljen et al. 2021a).

Amongst *C. annuum* and *C. frutescens*, most genotypes of *C. frutescens* have been reported to be highly pungent (Greenleaf 1986; Bosland and Votava 2000). The capsaicin and dihydrocapsaicin profile of *C. baccatum* was compared with that of *C. chinense* and *C. frutescens* where the latter two had 5.9-fold much higher values than *C. baccatum* (Bogusz et al. 2018). Amongst the domesticated species of *Capsicum*, *C. pubescens* is the least explored one, especially with regard to its chemical architecture. Thirty-two different genotypes of *C. pubescens* were examined for capsaicinoid content where the total capsaicinoids ranged from 55 to 410 mg/100 g (proportional to 8400–60,000 SHU) (Meckelmann et al. 2015). Hence, *C. pubescens* is much mildly pungent when compared with *C. annuum, C. chinense* and *C. frutescens. C. annuum L. annuum* has higher capsaicin concentration than *C. annuum L. grossum* (Bhuvaneswari et al. 2013).

2.2.2 Variation in Capsaicinoid Content with Fruit Part

The placenta has been reported to have the highest content of capsaicinoids in various studies. Pericarp and seeds are other parts of chilli fruits manifesting the presence of capsaicinoids, but the capsaicinoid quantity is lower than that in the placenta (Nowaczyk et al. 2006a, b; Pandhair and Sharma 2008). A study by Guillen et al. (2018) reports the pericarp to be 63% of the fruit in *C. chinense* and 85% in *C. baccatum*. The placenta was ~10% of the fruit in both species which possessed the highest capsaicinoids (79% in *C. chinense* and 51% in *C. baccatum*). The seed constituted 23% of the fruit in *C. chinense* and 5% in *C. baccatum*. Majority of the chilli pungency factor is present in non-edible parts of the fruit like the placenta, seeds and interlocular septa. High contents of capsaicinoids, mostly capsaicin, accumulate in the placenta (Wahyuni et al. 2011). In a similar study, *C. chinense* and a hybrid of *C. chinense* and *C. frutescens* had the highest content of capsaicinoids in the placenta (2600 mg/100 g dry weight and 3004.3 mg/100 g dry weight) followed by the pericarp (1738 mg/100 g dry weight and 1809.2 mg/100 g dry weight) and seeds (748.7 mg/100 g dry weight and 1135.4 mg/100 g dry weight) (Zamljen et al. 2021a). Consumption of the placenta and pericarp has been suggested for greater health benefits as several metabolites including capsaicinoids are highest in amount in the placenta and pericarp.

The major enzymes of the capsaicin biosynthetic pathway include phenylalanine ammonia lyase (PAL), cinnamic acid 4 hydroxylase (Ca4H), *p*-coumaric acid-3-hydroxylase (Ca_3H) and caffeic acid-O-methyl transferase (CaOMT). The activity of these enzymes was studied in different fruit parts like the placenta, pericarp and seeds of *C. annuum*. The activity of PAL was found to be tenfold higher in the placenta, whereas the activities of the enzymes like Ca4H, p-Ca3H and CaOMT were about twofold more in the placenta when compared with other parts of the fruit (Pandhair and Sharma 2008). Further, the study also revealed low activity of these four enzymes of the capsaicin synthesis pathway during a period of 21 days from the fruit setting and exhibited maximal activity at the time of fruit elongation and thereafter.

Capsaicin content in fresh red pepper has been reported to be two to three times higher in mature green chillies (Hamed et al. 2019). There are also reports that pepper cultivars having a thick pericarp and a small placenta are comparatively less pungent when compared to cultivars which had a thin pericarp with a large portion of placenta. This was consistent with the observation of Balbaa et al. (1968), who reported that the fruits with thinner pericarps are generally more pungent than fruits with thicker pericarps. Capsaicinoids accumulate in oil cells present in the placenta where seeds are also attached. Chilli seeds have the lowest content of capsaicinoids. Capsaicin and other capsaicinoids have a role in causing seed dormancy resulting in the delay or stopping of seed germination completely (Barchenger and Bosland 2016). This may be the reason for seeds having the lowest content of capsaicinoids so that they can germinate quickly (Canto-Flick et al. 2008).

2.2.3 Capsaicinoid Biosynthesis

From the tenth to twentieth day of fruit setting biosynthesis of capsaicinoids occurs in the placenta. However, their presence is noticed in the pericarp and seeds of the fruits as after synthesis in the placenta they are eventually excreted (Cervantes-Hernández et al. 2019). There are reports quoting the synthesis of capsaicinoids in the pericarp and seeds also though in much smaller quantities (Ohta and Chuong 1975; Balbaa 1986; Estrada et al. 2002). This is evidenced by the presence of the enzymes for capsaicinoid biosynthesis in other parts of the fruit (Arora et al. 2011). Synthesis of capsaicinoids follows condensation of the vanillylamine molecule which is derived from phenylalanine to a fatty acid produced from either valine or leucine (Curry et al. 1999). Nearly ten different capsaicinoids have been reported till date (Mazourek et al. 2009). Out of these different capsaicinoids, 90% comprises of capsaicin and dihydrocapsaicin (Kozukue et al. 2005; Choi et al. 2006). Different parts of fruits report different capsaicinoid composition. The edible part pericarp has been reported to have 82% capsaicin, ~13% dihydrocapsaicin, ~2% nordihydrocapsaicin and homocapsaicin and 1% homodihydrocapsaicin. The placenta is composed of 77.5% capsaicin, ~12% dihydrocapsaicin, 5% nordihydrocapsaicin, 3.2% homocapsaicin and 2.5% homodihydrocapsaicin. Seeds contain ~84% capsaicin, ~12% dihydrocapsaicin, 2% nordihydrocapsaicin, 1.4% homocapsaicin and 0.8% homodihydrocapsaicin. Accumulation of capsaicinoids starts 20 days post fruit set in most of the species of the *Capsicum* genus (Iwai et al. 1979). Capsaicinoid synthesis happens in the epidermal cells of the placenta and finally accumulate within pockets located on the placental surface called 'blisters' (Suzuki et al. 1980; Stewart et al. 2007).

The capsaicinoid biosynthesis pathway was established by the end of the 1960s. The phenylpropanoid pathway of capsaicinoid synthesis starts with the production of phenylalanine, followed by other compounds like cinnamic, p-coumaric, caffeic and ferulic acids subsequently. The pathway finally results in formation of vanillin and vanillylamine which ultimately forms capsaicinoids (Bennett and Kirby 1968). The phenylpropanoid-mediated capsaicinoid biosynthesis pathway is catalysed by a set of enzymes, viz. PAL, C4H, C3H and COMT (Fujiwake et al. 1982; Sukrasno and Yeoman 1993). Later, researchers have proposed the involvement of few other enzymes such as 4-coumaroyl-CoA ligase (4CL), hydroxycinnamoyl transferase (HCT), caffeoyl-CoA O-methyltransferase (CCoAOMT) in place of COMT and hydroxycinnamoyl-CoA hydratase/lyase (HCHL) in capsaicinoid formation (Gasson et al. 1998; Hoffmann et al. 2003; Stewart et al. 2005; Merali et al. 2007; Mazourek et al. 2009).

An independent group of compounds called capsinoids has been identified which is quite similar to capsaicinoids. It has effects similar to capsaicinoids, but does not produce pungency (Hursel and Westerterp-Plantenga 2010). Instead of the amide group present in capsaicinoids, capsinoids have an ester group. The structural difference between capsaicinoids and capsinoids seems to be responsible for the lower pungency of capsinoids. Pungency of capsinoids is roughly 1000 times less when compared to capsaicinoids. Non-pungency of capsinoids is distinguished as an important characteristic as pungency of capsaicinoids make it hard for consumption by many people (Luo et al. 2011). Both capsaicinoids and capsinoids are products of capsaicin synthase in the phenylpropanoid-mediated capsaicinoid biosynthesis pathway. The vanillin compound produced in the capsaicinoid biosynthesis pathway is transformed to both vanillylamine and vanillyl alcohol which is successively converted to capsaicinoids and capsinoids, respectively (Sano et al. 2022). The synthesis of vanillylamine from vanillin is catalysed by the gene 'putative aminotransferase (p-AMT)' encoding for the aminotransferase enzyme. Mutations in the p-AMT gene lead to its loss of function as a consequence of which vanillyl alcohol was produced instead of vanillylamine from vanillin. Pepper genotypes carrying the mutant 'pamt' allele produce capsinoids over that of capsaicinoids and hence are non- or less pungent in nature (Vázquez-Espinosa et al. 2020).

The type of capsaicinoid present in the pepper fruits is genotype dependent. The 'Carolina Reaper' (*C. chinense*) pericarp has the highest content of capsaicin (2444 mg/100 g dry weight), dihydrocapsaicin (644 mg/100 g dry weight), nordihydrocapsaicin (126 mg/100 g dry weight) and homodihydrocapsaicin (~63 mg/100 g dry weight). In contrast, 'Naga Morich', again belonging to *C. chinense*, has the highest content of homocapsaicin (78 mg/100 g dry weight). The type of capsaicinoids also varies with the fruit part. 'Naga Morich' has the maximal content of homocapsaicin in the placenta (157 mg/100 g dry weight), while 'Carolina Reaper' has homodihydrocapsaicin (~93 mg/100 g dry weight) there. The highest level of capsaicin in the placenta was in 'chocolate habanero' (*C. chinense*) (2522 mg/100 g dry weight), while 'habanero orange' had higher levels of placental dihydrocapsaicin (580 mg/100 g dry weight) and nordihydrocapsaicin (178 mg/100 g dry weight) (González-Zamora et al. 2015). 'Cayenne' peppers belonging to *C. annuum* were reported to have higher levels of capsaicin, nordihydrocapsaicin, homocapsaicin and homodihydrocapsaicin. The pericarp of 'serrano' pepper has the highest content of dihydrocapsaicin. All five studied capsaicinoids was highest in the placenta of 'jalapeño' except for homodihydrocapsaicin which was found to be abundant in 'cayenne' seeds (~6 mg/100 g DW). The 'Aji Pineapple' cultivar of *C. baccatum* has been reported to have the highest content of capsaicin (~247 mg/100 g DW) and dihydrocapsaicin (~36 mg/100 g DW) in the pericarp, while the cultivar 'Lemon Drop' has higher amounts of nordihydrocapsaicin, homocapsaicin and homodihydrocapsaicin. In contrast to this, the placenta of 'Aji Pineapple' has the highest content of dihydrocapsaicin, nordihydrocapsaicin and homocapsaicin, while that of 'Lemon Drop' has capsaicin and homodihydrocapsaicin. All five capsaicinoids, viz. capsaicin, dihydrocapsaicin, nordihydrocapsaicin, homocapsaicin and homodihydrocapsaicin contents, were found to be high in seeds.

2.2.4 Factors Affecting Capsaicin Content

The pungency characteristic of peppers is highly variable and sensitive to the environment under which it is grown. Capsaicinoid quantity in pepper fruits is different which mainly depends on the species, genotype, developmental stage, growing conditions like geographical location, meteorological and environmental conditions and stage of harvest. A detailed explanation of variation in pungency amongst domesticated species of the *Capsicum* genus has been discussed in Sect. 2.4. Heat levels of chilli fruits vary amongst cultivars of the same species as well as within a single cultivar also (Zamljen et al. 2021b). Capsaicinoids showed considerable variation amongst landraces, viz. Bhut Jolokia, Naga Jolokia, Lota Bhut, Krishna Jolokia and Mem Jolokia belonging to *C. chinense* from North-east India (Islam et al. 2015). The pungency factor differs within a single cultivar and sometimes even in the same plot (Lindsey and Bosland 1996). The pungency of paprika peppers has been reported to be often mild when compared with chilli peppers (Joshi et al. 1993). Additionally, post-harvest practices also have a significant effect on capsaicinoid levels.

Several studies have endorsed that pungency is greatly influenced by the growing conditions of the plants which include climate and soil conditions. Amongst the three hottest chillies finding place in the Guiness Book of World Records, viz. Bhut Jolokia, Trinidad Moruga Scorpion and Carolina Reaper, when analysed for pungency levels after growing at Yucatan, an increase in their capsaicinoid content was observed at 48%, 15% and 27%, respectively. 'Carolina Reaper', which is the hottest chilli at present in the world, registered a pungency of 3,006,330 SHU when cultivated at Yucatan in comparison to the record levels mentioned as 200,000 SHU in the Guinness Book of World Records (Mun˜oz-Ramírez et al. 2018). The content of capsaicinoid and capsinoid compounds is influenced by different environmental factors like rainfall and humidity, light, temperature, growing conditions, mineral supply, etc. Lower humidity and rainfall resulted in 50% reduction in capsaicin and dihydrocapsaicin of King chilli (Tiwari et al. 2005). Accumulation of capsaicinoids in hot peppers (*Capsicum* spp.) is strongly affected by light intensity. Different genotypes respond differently to capsaicinoid accumulation upon shading. An increase of capsaicinoids was observed in Bhut Jolokia and

Akanee Pirote upon shading, while BGH1719 manifested a decrease in pungency levels upon shading (Jeeatid et al. 2017). An appropriate level of shading is recommended for the cultivation of hot peppers for a particular cultivar to increase capsaicinoid yield. High temperatures between 100 °C and 190 °C when applied to chilli for 15 min registered only <10% reduction in capsaicinoid content; however, at temperatures greater than 190 °C, significant reduction in capsaicin and dihydrocapsaicin has been observed (Wang et al. 2009). The rate of capsaicinoid reduction is faster at higher temperatures irrespective of initial concentration levels. For higher recovery of capsaicinoids under high temperature, selection of genotypes with high initial concentration is required as well as maintaining at neutral pH and also the high-temperature period should be short (Wenhui et al. 2014). The source of nutrients also affect the pungency levels. Nitrogen from organic sources like FYM has been reported to register higher levels of capsaicinoid in comparison to inorganic sources like urea (Pariari and Khan 2013). However, amongst the growing conditions, plant density does not influence the capsaicin, dihydrocapsaicin and total capsaicinoid content of chilli genotypes (Jeeatid et al. 2021). Disease in pepper plants report a reduction on capsaicinoid content. Anthracnose-infected fruits manifested reduction in capsaicin content in the range of 20–60% when compared to healthy fruits (Prathibha et al. 2013).

The stage of maturity of crop and the air temperature also affect the capsaicin content of pepper fruits. Harvests after the first one under high-temperature conditions produce higher capsaicin contents (El Saeid 1995; Dabrowska et al. 2001). In contrast to this capsaicin and dihydrocapsaicin contents of all the chilli varieties decreased significantly subsequently after each harvest (Kraikruan et al. 2008). At the early stage of development, chilli has a lesser number of fruits so the competition between fruits for nutrients to produce capsaicinoids is minimal (Sukrasno and Yeoman 1993). Storage condition and drying parameters also influence the capsaicin content of peppers. Drying peppers under solar tunnel dryer gave a higher capsaicin yield in comparison to normal sun drying (Yaldiz et al. 2010). Differences in post-harvest handling conditions can result in variation in the capsaicinoid content of peppers harvested from the same plant also (Iwai et al. 1979).

2.3 Extraction of Capsaicinoids

The extraction efficiency of capsaicinoids from peppers depends on the genotype, part of the fruit used for extraction and also the pre-extraction processes attempted. Extraction of capsaicinoids is usually attempted using organic solvents. Several capsaicin extraction methods have been devised which include solvent extraction (SE), microwave-assisted extraction (MAE), solid-phase micro-extraction (SPME), ultrasound-assisted extraction (UAE), SOX extraction, extraction with liquids under pressure (PLE), enzymatic treatment, aqueous two-phase system (ATPS) and supercritical fluid extraction.

2.3.1 Extraction Methods

Extraction is the method employed to separate and purify organic substances. The methods deployed must be quick, cheap, multifaceted and efficacious (selection of appropriate solvent being of critical importance). The conventional methods for extraction of capsaicinoids focus on using organic solvents by using different techniques such as solvent, Soxhlet (SOX) and enzymatic extraction (Table 2.1). Capsaicinoid recovery percent and quality after extraction depend on the extraction conditions. Conventional methods of extraction give a yield of 70–80% and the time of extraction decreases with increase in polarity of the solvent (Abdurahman and Olalere 2016). Enzymatic pre-treatment increased 14–22% capsaicin yield in the Soxhlet method of extraction (Baby and Ranganathan 2016) and 17% in the enzymatic method (de Farias et al. 2020).

With the implementation of more strict laws with respect to the health of industry employees by the government, new methods requiring minimal use of hazardous solvents became necessary. Reduction in solvent utilization would also decrease extraction cost as well as prevent pollution. Innovative technologies like pressurized liquid extraction (PLE), microwave-assisted extraction (MAE), ultrasound-assisted extraction (UAE), enzyme-assisted extraction and supercritical fluid extraction (SFE) have been developed which are much faster, cheaper and effective with good capsaicinoid yield. PLE technology requires small amounts of solvents based on the polarity of the target compound to be extracted. It basically involves use of high temperatures and pressure on the liquid solvent to enable its penetration within the solid matrix. The PLE method has been attempted to extract capsaicinoids by several workers (Barbero et al. 2006; Bajer et al. 2015; de Aguiar et al. 2020) (Table 2.1). MAE technology uses microwave energy to heat the solvents which increases its diffusivity within the matrix. Both PLE and MAE methods reduce the amount of solvent and time required in the extraction process (Nazari et al. 2007). Extraction of compounds on an industrial scale can be done using ultrasound-assisted extraction (UAE). Water can be used as a medium of reaction, thus avoiding the use of harmful solvents. The technology is based on the cavity principle which involves formation of small cavities in the matrix/solvent mixture to release energy (Esclapez et al. 2010). Chuichulcherm et al. (2013) observed the UAE method as the most optimum method for capsaicinoid extraction when compared to the MAE and Soxhlet method as it needed less energy consumption.

Enzyme-assisted extraction (EAE) is a convenient way to increase extraction yield by recovering the compounds from the cell wall also (Stoica et al. 2016; Sagar et al. 2018). Extraction of bioactive compounds using enzymatic treatment is an environmentally safe method as the solvent used is water which is more secure than organic solvents (Sagar et al. 2018). Cellulases, pectinases, beta-glucosidases or different combinations of enzymes are used to hydrolyze and break down cell wall components to release capsaicinoids attached to the cell wall (Walczak et al. 2018). Enzymatic pre-treatment before using any technology for extraction of compounds is another way to improve the extraction yield and quality of capsaicinoids. Pre-treatment with commercial enzymes like Viscozyme increased capsaicinoid

Table 2.1 Different methods used for the capsaicin extraction

Method of extraction	Extraction condition	Major solvents used	Yield	Advantages/disadvantages	Reference
Solvent extraction	(1) Ethanol, 3 h, 45 °C, 1:8 (solvent/fruit) (2) 80% ethanol, 80% methanol and 80% acetone (1:4; seed powder/solvent); 40 min, 40 °C	Ethanol, acetone and acetonitrile	87.4% ≈16 mg/dry g	*Disadvantages:* Longer extraction time, higher energy consumption and lower capsaicinoid yields	Boonkird et al. (2008), Liu et al. (2020b), Chinn et al. (2011)
Soxhlet (SOX extraction)	5 h, 78 °C, 1:8 (solvent/fruit)	Ethanol, methanol	88.31 ± 1.03%	*Disadvantages:* longer extraction time, higher energy consumption and lower capsaicinoid yields	Boonkird et al. (2008), Liu et al. (2013), Bajer et al. (2015), Santos et al. (2015)
Ultrasound-assisted extraction UAE	An ultrasonic wave passes through an organic solvent, producing energy to increase the mixing and penetration of the solvent into the sample matrix	Methanol	85.26 ± 1.35%	*Advantages:* reduction of solvent volume, temperature and extraction time	Lu et al. (2016), Barbero et al. (2008)
Extraction with liquids under pressure, PLE	Temp, 100 °C; pressure, 1500 psi	Methanol	98.31 ± 1.46%	*Advantages:* PLE is currently considered an advanced technique since it offers important benefits such as shorter extraction time, decreased solvent consumption, decreased sample handling and increased yield	Liu et al. (2013), Bajer et al. (2015), de Aguiar et al. (2020)
	Temperature 200 °C at a pressure of 20 MPa	Water	–		
	65 °C, 10 MPa, 60 min	Ethanol and water	77%		
Microwave-assisted extraction, MAE	Uses the energy produced by microwave radiation to heat the solvents and improve extraction kinetics	Ethanol	86.36 ± 1.12%	*Advantages:* doubled the yield of capsaicinoids and cut the extraction time in half as compared to conventional	Williams et al. (2004), Barbero et al. (2016)

(continued)

Table 2.1 (continued)

Method of extraction	Extraction condition	Major solvents used	Yield	Advantages/disadvantages	Reference
				reflux and stirring flask extraction methods	
SPE + US	Multiple sampling, sample retention, reduces the possibility of sample contamination due to the technique's simplicity, is also quicker than conventional techniques and can be easily automated	Does not use any solvent	58.8–60.8%	*Advantages:* simplicity of technique, quicker than conventional techniques and can be automated, no use of solvents, hence cheaper	Santos et al. (2015)
SPME		Methanol ethyl acetate chloroform methylene chloride	50–60% 44% 35% 30%	*Advantages:* simple, cheap having compatibility with analytical systems, automation and also solvent-free extraction	Spencer Jr and Almirall (2005)
Aqueous two-phase system (ATPS)	Uses ethylene oxide-propylene oxide (EOPO) copolymer, salt and ethanol ($EOPO/K_2HPO_4$)	Ethanol/salt	85.6 ± 0.4%	*Advantages:* provides a biocompatible environment for the extraction of sensitive and biologically active compounds	Cienfuegos et al. (2017)
TAPPIR	• New approach to overcome the limitations of ATPS • Eliminates the need for phase emulsification and separation by immobilizing one aqueous phase inside	Polyethylene glycol (PEG), 1-ethyl-3-methyl imidazolium acetate	85–92%	*Advantages:* improved version of the ATPS method By immobilizing one liquid phase inside, the need for phase emulsification and separation is eliminated. The impregnated solids are suspended in a second liquid	Lu and Cui (2019)

	porous solids prior to extraction			phase to achieve phase contact	
Enzymatic treatment	Commercially available enzymes were used to soften the tissues of *Capsicum* spp.	Enzymes (pectinase, cellulase), (carbohydrase, pectinesterase and arabanase) and tetrahydrofuran	>85%	*Advantages:* use of enzymes reduces the amount of solvent required for extraction and increases the yield of extractable compounds	Salgado-Roman et al. (2008)
Supercritical fluid extraction	The supercritical fluid used is CO_2 along with co-solvents such as ethanol or methanol used sometimes. Extraction conditions are above the critical temperature of 31 °C and critical pressure of 74 bar	Carbon dioxide (CO_2)	*0.68% to 17.1% (w/w)*	*Advantages:* Extraction is faster as the diffusivities are much faster in supercritical fluids than in liquids In addition, due to the lack of surface tension and negligible viscosities in case of liquids, the solvent can penetrate more into the matrix inaccessible to liquids	Santos et al. (2015)

extraction by 22% when compared with conventional solvent extraction (Baby and Ranganathan 2016). A major limitation of EAE is in terms of industrial scale-up processes and higher cost involved in using enzymes.

Supercritical fluid extraction is one of the latest technologies which is environment-friendly. It uses high pressure and temperatures to reach a threshold point for the solvent (carbon dioxide) to dissolve the compound to be extracted. Qualities of CO_2 like its non-toxicity, noncombustible nature and availability in highly pure form make it a solvent of choice for this method. However, this technique is more expensive compared to UAE and EAE. Optimization and standardization of conditions to enhance the yield of extraction with SFE are required. A specific combination of extraction conditions coupled with the use of assisted technologies can increase the extraction yield of capsaicinoids significantly (Santos et al. 2015; Stoica et al. 2016). These assisted technologies use ultrasound waves, microwaves, enzymatic pretreatment, etc. Simultaneous use of technologies can remarkably reduce the cost of compound extraction (de Aguiar et al. 2020).The extraction process is followed by purification of bioactive compounds. Purification methods like 'high-speed' counter-current chromatography (HSCCC) and reversed-phase preparative chromatography are currently used for purification but are very expensive for the industrial production of pure capsaicin (Goll et al. 2013). An aqueous two-phase system (ATPS) is also a well-defined process in recent decades to get pure compounds following extraction (Santos et al. 2015). The technique has the potential to separate several compounds from their complex form. Due to its higher water content, the technique allows compounds to preserve their chemical structure and key activities. Capsaicin of 92% purity was achieved after Tunable Aqueous Polymer-Phase Impregnated Resin (TAPPIR) extraction with recovery percent of 95.82% (Lu and Cui 2019). The combination of TAPPIR technology and reverse chromatography is a practical approach for extracting and purifying capsaicin from peppers.

2.3.2 Solvents Used for Extraction of Capsaicinoids

Several solvents have been experimented to extract capsaicinoids. Amongst acetone, chloroform, methanol, acidified methanol and acetonitrile tested for capsaicinoid extraction by Attuquayefio and Buckle (1987), acetone has been found to have the highest capsaicinoid recovery from dehydrated *Capsicum*. In subsequent studies, Collins et al. (1995) suggested the extraction with acetonitrile, while Barbero et al. (2006) found ethanol and methanol to be an efficient solvent in comparison to acetonitrile. Ethanol and acetonitrile were found to be more appropriate for capsaicin extraction from fresh samples, while acetone was better for dried pepper parts (Chinn et al. 2011). Nwokem et al. (2010) found methanol superior to acetone in extracting capsaicin. Methanol and ethanol was again found to be better as extraction solvents for capsaicin by Rafajlovska et al. (2011). Haejin et al. (2012) reported hexane as the best solvent for the extraction of capsaicin using the Soxhlet method as it gave the highest recovery percent. Ethanol is the single solvent which is nontoxic without any remaining toxic solvent (Xinrong et al. 2014). UAE technology was employed for capsaicinoid extraction using methanol as an extraction solvent (Barbero et al. 2008, 2014).

2.3.3 Quantification

In the past, pungency was determined using organoleptic tests like Scoville heat test in peppers (Scoville 1912). Based on Scoville heat units (SHUs), pungency has been categorized into five groups: non-pungent (0–700 SHU), mildly pungent (700–3000 SHU), moderately pungent (3000–25,000 SHU), highly pungent (25,000–70,000 SHU) and very highly pungent (>80,000 SHU) (Weiss 2002). In the last decade, emphasis has been given to the development of more reliable and accurate analytical methods which require less running time and averts the use of dangerous substances (Pena-Alvarez et al. 2009). With time the organoleptic method of pungency was replaced by several analytical methods like spectrophotometry (Anan et al. 1996), calorimetry (Gibss and O'Garro 2004), thin-layer chromatography (Sankarikutty et al. 1978), gas chromatography (Todd et al. 1977), liquid chromatography, GC/LC-MS (Iwai et al. 1979), NMR-flow probe analysis (Nyberg et al. 2001) and high-performance liquid chromatography (HPLC) (Weaver and Awde 1986). Different separation methods coupled with ultraviolet (UV) detection is employed for the quantification of capsaicinoids (Choi et al. 2006). Presently, high-performance liquid chromatography (HPLC) is the most widely used tool due to its reliability and rapidness for separation, identification and quantification of capsaicinoids from the extracts of chilli fruits. Acetonitrile or methanol and water acidified with formic, acetic or phosphoric acids have been used as the eluents in the HPLC–MS determination of capsaicinoids. Separation of capsaicinoids is performed on reversed-phase columns having a conventional 4.6 mm × 250 mm C-18 column (10-μm packing).

Detection of the presence of different compounds in a sample is done using different analytical techniques depending on the concentration of the target substance. For samples with a concentration greater than 700 ppm, UV absorption detection at a wavelength of 280 nm is preferred. On the other hand, for samples with a lower concentration of less than 700 ppm, fluorescence detection is the preferred method. This involves exciting the sample at a wavelength of 288 nm and monitoring the emission at 320 nm to detect the presence of the target substance (González-Zamora et al. 2015). Following the extraction process, preliminary purification is carried out before undergoing HPLC analysis for the detection of capsaicinoids. Various purification methods such as thin-layer chromatography (TLC) and column chromatography (CC) are employed to ensure the purity of the extract. However, in some cases, these methods take a considerable amount of time, and the separation of different isomers of HC is not feasible.

Recently, ultra-performance liquid chromatography (UPLC) coupled with MS has emerged as a more rapid and efficient method for the separation of capsaicinoids. This method ensures complete separation of all isomers of HC in a shorter timeframe, making it a preferred technique for the analysis of capsaicinoids (Hussein et al. 2015). An experiment was conducted to fractionate a pepper extract into NDC, C, DC, HC and HDC, using a 10-min runtime. Although this method was successful in separating the primary capsaicinoids, the separation of homo analogues was not satisfactory. Therefore, it has become necessary to develop an efficient protocol that can separate all analogues and isomers of capsaicin from spice red pepper products with a relatively short runtime.

2.4 Applications of Capsaicinoids

Pepper finds application in several industries including the pharmaceutical industry and cosmetic industries. The culinary and medicinal benefits of *Capsicum* have been brought out way back to 7000 BC (Meotti et al. 2014). Capsaicinoids and phenols are the main metabolites used in pharmaceutical industries. It is widely known for its anti-inflammatory, anticarcinogenic, antioxidant, antimicrobial and neurological properties. It contributes to weight loss treatments through adipose tissue metabolism, relieving pain and providing gastrointestinal and cardiovascular benefits, which are well studied (Anand and Bley 2011). Highly pungent pepper genotypes like 'Carolina Reaper', 'Borg 9 Pheno', 'Bhut Jolokia', etc. are reported to have high pungency due to higher capsaicinoid content and are suitable for the pharmaceutical industry. Peppers find application in the cosmetic industry due to the natural colour which it provides to the final product making them safe for consumer use (Baenasa et al. 2019). It impacts the thermoregulation of the body and is marked for its antioxidative and anticancerous properties. Capsaicin also finds utility as an ingredient in insect repellents. Details of its application are enlisted below.

2.4.1 Thermoregulation and Analgesic Properties

The threshold concentration of capsaicin in body induces warm sensation, while a higher concentration produces a burning pain. Capsaicin stimulates heat generation in the body thereby increasing the internal body temperature due to the increase in metabolic activity (Olszewska et al. 2011). Human populations of hot climate improve heat dissipation through skin vasodilation and perspiration resulting in a decrease in body temperature (Lee 1954). Capsaicin and capsiate stimulate thermogenesis thereby enhancing core body and skin temperature (Hachiya et al. 2007; Ludy and Mattes 2011) through increasing energy expenditure and substrate oxidation (Josse et al. 2010; Lee et al. 2010; Ludy and Mattes 2011).

Capsaicin-containing medicines commonly find utility in the management of chronic pain and is widely used for manufacturing analgesics (Seino and Miki 2003). Capsaicin has the ability to deplete nerves for the supply of P which is a neuropeptide supervising the transmission of pain signals to the brain. Depletion of subsistence P provides capsaicin the ability to treat pains and aches (Andrews 1999; Macfarlane et al. 2012). Capsaicin is an important ingredient of 'pepper sprays', a deterrent used for riot control and to ward off assailants (Adkins 2003). Capsaicin is also used to discourage encroachment as a repellent by bears, deer and other mammals (Yenigun and Thanassi 2019).

2.4.2 Anti-Inflammatory Actions

Several studies have shown that capsaicinoid compounds of chilli peppers exhibit anti-inflammatory activity (Hamed et al. 2019). Capsaicin effects anti-inflammatory activity through impediment of TNF, IL-1 and IL-6 (inflammatory cytokines) and NFκB (transcription factor) (Tanaka et al. 2014). Utilizing capsaicin as a therapeutic compound to develop medicines for treatment as well as the prevention of inflammatory disorders appears to be quite encouraging (Singh et al. 2022). However, validation of anti-inflammatory action of capsaicin and its therapeutic relevance still needs further investigation (Jolayemi and Ojewole 2013).

2.4.3 Weight Loss

Capsaicin is known to inhibit transcription of genes controlling buildup of adipose tissue (Joo et al. 2010) and activates transcription of genes producing proteins which generate heat. It is also induces fat reduction by decreasing adipose tissue content thereby assisting in body weight loss. It helps in heat generation by kindling the production of epinephrine and norepinephrine from the adrenal glands. The slimming effect of capsaicin is mainly due to reduction in appetite, increase in fat oxidation and activation of the sympathetic nervous system by catecholamines leading to higher energy consumption (Yoshioka et al. 1998; Belza et al. 2007).

2.4.4 Impact on the Cardiovascular System

Capsaicin shields the cardiovascular system by bringing down the blood pressure and alleviating damaged coronary blood vessels, while its antioxidative potential averts heart attack (Sun et al. 2016). In studies, it has been demonstrated that dietary capsaicin reduces the risk of buildup of plaque inside arteries, hypertension, cardiac enlargement and stroke (Juturu 2016). Regular consumption of chilli protects oxidation of lipoproteins in heart patients due to the antioxidant activity of capsaicin.

2.4.5 Antidiabetic Activities

Diabetic patients lose their natural ability to regulate blood sugar levels due to the inefficiency to respond to the insulin hormone. Capsaicin has been reported to have an antidiabetic response through suppressing the enzymes, viz. α-amylase and α-glucosidase, involved in polysaccharide hydrolyzation activities, thereby controlling body weight and reduction in levels of lipids in the blood (Thongnum and Chanthai 2018). It also maintains blood sugar levels through improved insulin sensitivity in liver cells, regulating glucose homeostasis and improved glucose metabolism. Glucose tolerance in obese patients is controlled by capsaicin through suppression of inflammatory responses in fat-storing cells. Regular use of capsaicin in the diet prevents corpulence arising from sugar intolerance (Zheng et al. 2017). Capsaicin-containing supplements have helped women with condition of gestational diabetes mellitus (GDM). GDM is a condition where a pregnant woman is unable to use insulin due to the placental secreted hormones resulting in the rise in blood sugar levels. This increase in blood sugar levels may have considerable impact on the lifelong health of the pregnant female.

2.4.6 Antibacterial and Antiviral Effects

The antimicrobial and antivirulence activity of capsaicin has recently enticed the researchers. The antibacterial effects of capsaicin have been reported against pathogens like *Helicobacter pylori* and *Pseudomonas aeruginosa* (Cowan 1999; Omolo et al. 2014), while the antivirulence activity was exhibited against *Vibrio cholerae*, *Staphylococcus aureus* and *Porphyromonas gingivalis* (Chatterjee et al. 2010; Kalia et al. 2012; Qiu et al. 2012; Zhou et al. 2014). This action of capsaicin is due to its ability to inhibit intracellular invasion and haemolytic activity. Capsaicin prevents the spread of infection to deep tissues and the formation of an intracellular reservoir of pathogens which are difficult to eradicate. Developing antibiotic resistance has raised an alarm against the overuse of antibiotics. Exploring novel antibiotic-independent ways to fight bacterial infection becomes indispensable. Investigating new natural compounds with antibacterial property has gained attention in pharmacological sciences. Capsaicin and its derivatives have proven to be an encouraging antimicrobial molecule which can complement or substitute antibiotics to fight bacterial diseases. Clinical trials validating the potential use of capsaicin as an antibiotic requires further investigations. This is particularly important in the pharmaceutical evaluation of animal infection models (Füchtbauer et al. 2021).

2.4.7 Effect of the Capsaicin on Milk Production and Feed Efficiency in Cattle

Studies evaluating feed additives containing capsaicinoids have been conducted to test the short-term effect (i.e. Latin square or change-over designs) on milk yield and nutrient digestibility (Tager and Krause 2011; Foskolos et al. 2020). Feed additive with different doses of encapsulated pepper (rich in capsaicin) was given to dairy to study its effect on the digestibility of nutrients, purine derivative excretion, milk composition and physiological parameters (respiration rate, rectal temperature and skin temperature in different regions) in dairy cows. Encapsulated pepper was found to improve nutrient digestibility, increase feed intake and milk yield and alters physiological parameters of dairy cows in a dose-dependent manner (Paulo et al. 2022).

2.4.8 Anticancer Effects

Capsaicin has displayed anticancerous activity in several human cancer studies (Díaz-Laviada and Rodríguez-Henche 2014; Chapa-Oliver and Mejía-Teniente 2016; Clark and Lee 2016). Capsaicin directly interacts with crucial molecules of the cytoplasmic, mitochondrial and metabolic pathways resulting in anticancer activity (Díaz-Laviada and Rodríguez-Henche 2014; Basith et al. 2016; Chapa-Oliver and Mejía-Teniente 2016; Srinivasan 2016). Capsaicin mediates suppression of cancer cells and also promotes their apoptosis by multiple mechanisms (Clark and Lee 2016; Friedman et al. 2017; Vendrely et al. 2017). The strong pungency and nociceptive effect of capsaicin enforced the patients to discontinue the use of capsaicin (Evangelista 2015), thereby emphasizing on the discovery and design of capsaicin-like compounds which can exhibit greater anticancer activity with mild side effects.

2.4.9 Role of Capsaicin in Parkinson's Disease

Neuronal cell death caused by oxidation and behavioural deterioration due to Parkinson's disease can be averted through the use of capsaicin. Capsaicin appears to be a potential compound for the management and treatment of neurological disorders resulting from Parkinson's disease. Several studies report capsaicin's ability to reduce degeneration of neurons and reduction in agileness caused by Parkinson's disease by decreasing microglial activation and reducing neuroinflammation (Park et al. 2012; Chung et al. 2017; Zhao et al. 2017; Kim et al. 2019; Liu et al. 2020a).

2.4.10 Treatment of Skin Diseases

Several studies identify traumatic changes in life provokes or aggravates various dermatological conditions starting from skin cancer to several types of inflammatory skin diseases such as acne, eczema, psoriasis, etc. Capsaicin is reported to treat several dermatological conditions like skin inflammation, psoriasis and atopic dermatitis and overcoming chronic pain amplification syndrome (Boyd et al. 2014).

2.4.11 Treating Headaches and Migraines

The capsaicin compound has been demonstrated to have a curative effect in idiopathic rhinitis, migraines and cluster headaches. The effect of persistent administration of intranasal capsaicin on chronic migraine was studied by Fusco et al. (2003). All patients treated with capsaicin reported amelioration in their migraine condition which was judged to be ranging between 50% and 80%. Statistical evaluation showed that capsaicin was significantly efficacious compared with placebo. Patients who consume hot peppers in their diets can benefit from shielding against cephalalgia condition, which is a condition of pain in the head.

2.4.12 Appetite Control Effects of Capsaicin and Capsiate

Researchers have reported capsaicin to play a role on appetite regulation (Reinbach et al. 2009, 2010; Ludy and Mattes 2011; Inoue et al. 2007). Capsaicin suppresses self-reported factors which stimulate appetite as well as the yearning to eat calorie dense foods (Ludy and Mattes 2011), spicy foods (Reinbach et al. 2010), sweetmeats (Ludy and Mattes 2011), salty foods (Ludy and Mattes 2011), craving to eat generally (Yoshioka et al. 1999), persistent thoughts about food (Yoshioka et al. 1999; Ludy and Mattes 2011) and hunger generally (Yoshioka et al. 1999; Westerterp-Plantenga et al. 2005).

2.5 Conclusions

Capsaicin is a phenolic compound found in hot peppers, which is responsible for the characteristic pungent taste and burning sensation. *Capsicum* is the only genus which produces this compound where *C. chinensis* is the species which is reported to have the higher capsaicin content when compared to other species. The amount of capsaicinoids in a chilli pepper is dependent on the species, genotype, environment where it is grown, fertilization levels, season of growing and the time of harvest. The amount of capsaicin in a particular genotype can vary, which is determined mainly by the amount of light intensity and temperature during crop growth period, the age of the fruit and the position of the fruit on the plant. With advancement in research, several novel technologies for the extraction and quantification of capsaicinoids have been developed. Capsaicin is now finding application in the pharmacological field extensively with the discovery of its potential to treat several medical conditions. Being natural in origin its further utility is also being explored to replace the harmful chemicals which are being used in conventional medicines.

2.6 Future Perspectives

With capsaicin becoming popular for its various pharmacological benefits, identifying genotypes with higher capsaicin content becomes important. As discussed in the chapter capsaicin content in genotypes is influenced by several factors which include genotype, environment (light, temperature, fertilization levels, etc.) and time and site of harvesting. Hence, efforts to standardize practices which will increase capsaicin content in a genotype are recommended. Further identification of places where cultivation of chilli improves the capsaicin content can also be looked into. Researches related to the development of newer technologies which increase the recovery of the capsaicin compound upon extraction are already underway. Techniques, involving the minimum usage of hazardous and cheap solvents, should be promoted.

References

Abdurahman HN, Olalere OA (2016) A comparative review of conventional and microwave assisted extraction in capsaicin isolation from chili pepper. Aust J Basic Appl Sci 10(10): 263–275

Adkins LD (2003) Oleoresin Capsicum: an analysis of the implementation of pepper spray into the law enforcement use of force continuum in a selected police department. Electronic Theses and Dissertations, Paper 779. https://dc.etsu.edu/etd/779

Al Othman ZA, Ahmed YB, Habila MA, Ghafar AA (2011) Determination of capsaicin and dihydrocapsaicin in Capsicum fruit samples using high performance liquid chromatography. Molecules 16:8919–8929. https://doi.org/10.3390/molecules16108919, PMID 22024959

Anan T, Ito H, Matsunaga H, Monma SY (1996) A simple method for determining the degree of pungency of peppers. Capsicum Eggplant Newsl 16:61–64

Anand P, Bley K (2011) Topical capsaicin for pain management: therapeutic potential and mechanisms of action of the new high-concentration capsaicin 8% patch. Br J Anaesth 107(4):490–502. https://doi.org/10.1093/bja/aer260; Epub 2011 Aug 17. PMID: 21852280; PMCID: PMC3169333

Andrews J (1999) Pepper trail: history and recipes from around the world. University of North Texas Press, Denton, TX, p 256

Arora R, Gill NS, Chauhan G, Rana AC (2011) An overview about versatile molecule capsaicin. Int J Pharm Sci Drug Res 3(4):280–286

Attuquaycfio VK, Buckle KA (1987) Rapid sample preparation method for HPLC analysis of capsaicinoids in Capsicum fruits and oleoresins. J Agric Food Chem 35:777–779

Baby KC, Ranganathan TV (2016) Effect of enzyme pretreatment on yield and quality of fresh green chilli (Capsicum annuum L) oleoresin and its major capsaicinoids. Biocatal Agric Biotechnol 7:95–101

Baenasa N, Belovic M, Ilic N, Moreno AD, García-Vigucra C (2019) Industrial use of pepper (*Capsicum annum* L.) derived products: technological benefits and biological advantages. Food Chem 274:872–885. https://doi.org/10.1016/j.foodchem.2018.09.047

Bajer T, Bajerová P, Kremr D, Eisner A, Ventura K (2015) Central composite design of pressurised hot water extraction process for extracting capsaicinoids from chili peppers. J Food Compos Anal 40:32–38. https://doi.org/10.1016/j.jfca.2014.12.008

Balbaa SI (1986) Chemistry of crude drugs laboratory manual. Faculty of Pharmacy, Cairo Univ, Egypt, p 195

Balbaa SI, Karawya MS, Girgis AN (1968) The capsaicin content of Capsicum fruits at different stages of maturity. Lloydia 31:272–274

Barbero GF, Palma M, Barroso CG (2006) Determination of capsaicinoids in pepper by microwave-assisted extraction—high-performance liquid chromatography with fluorescence detection. Anal Chim Acta 578:227

Barbero GF, Liazid A, Palma M, Barroso CG (2008) Ultrasound-assisted extraction of capsaicinoids from peppers. Talanta 75:1332–1337

Barbero GF, Ruiz AG, Liazid A, Palma M, Vera JC, Barroso CG (2014) Evolution of total and individual capsaicinoids in peppers during ripening of the Cayenne pepper plant *(Capsicum annuum* L.). Food Chem 153:200–206

Barbero GF, Liazid A, Azaroual L, Palma M, Barroso CG (2016) Capsaicinoid Contents in Peppers and Pepper-Related Spicy Foods. Int J Food Prop 19(3):485-493
Barchenger DW, Bosland PW (2016) Exogenous applications of capsaicin inhibits seed germination of *Capsicum annuum*. Sci Hortic 203:29–31
Basith S, Cui M, Hong S, Choi S (2016) Harnessing the therapeutic potential of capsaicin and its analogues in pain and other diseases. Molecules 21:1–28
Belza A, Frandsen E, Kondrup J (2007) Body fat loss achieved by stimulation of thermogenesis by a combination of bioactive food ingredients: a placebo-controlled, double-blind 8-week intervention in obese subjects. Int J Obes 31(1):121–130. https://doi.org/10.1038/sj.ijo.0803351; Epub 2006 Apr 25
Bennett DJ, Kirby GW (1968) Constitution and biosynthesis of capsaicin. J Chem Soc C Organ 2: 442–446
Bhuvaneswari G, Sivaranjani R, Reeth S, Ramakrishnan K (2013) Application of nitrogen and potassium efficiency on the growth and yield of chilli Capsicum annuum L. Int J Curr Microbiol App Sci 2(12):329–337
Bogusz SJ, Silvia Libardi SH, Dias FFG, Coutinho JP, Bochi VC, Rodrigues D, Melo AMT (2018) Brazilian *Capsicum* peppers: capsaicinoid content and antioxidant activity. J Sci Food Agric 98: 217–224
Boonkird S, Phisalaphong C, Phisalaphong M (2008) Ultrasound assisted extraction of capsaicinoids from *Capsicum frutescens* on a lab- and pilot-plant scale. Ultrason Sonochem 15:1075–1079
Bosland PW, Baral JB (2007) Bhut Jolokia'-the World's hottest known Chile pepper is a putative naturally occurring interspecific hybrid. HortScience 42(2):222–224
Bosland P, Votava E (2000) Peppers: vegetable and spice Capsicums. CABI Publishing, Cambridge
Bosland PW, Bailey AL, Iglesias-Olivas J (1996) Capsicum pepper varieties and classification. New Mexico Cooperative Extension Service, Las Cruces, NM
Bosland PW, Coon D, Reeves G (2012) Trinidad Moruga scorpion' pepper is the world's hottest measured Chile pepper at more than 2 million Scoville heat units. Hort Technol 22:534–538
Boyd K, Shea SM, Patterson JW (2014) The role of capsaicin in dermatology. Prog Drug Res 68: 293–306
Canto-Flick A, Balam-Uc E, Bello-Bello JJ, Lecona-Guzman C, Solís-Marroquin D, Aviles-Vinas-S, Gomez-Uc E, Lopez-Puc G, Santana-Buzzy N, Iglesias-Andreu LG (2008) Capsaicinoids content in habanero pepper (*Capsicum chinense* Jacq.): hottest known cultivars. HortScience 43:1344–1349
Cervantes-Hernández F, Alcalá-González P, Martínez O, Ordaz-Ortiz JJ (2019) Placenta, pericarp, and seeds of Tabasco chili pepper fruits show a contrasting diversity of bioactive metabolites. Metabolites 9(10):206
Chapa-Oliver AM, Mejía-Teniente L (2016) Capsaicin: from plants to a cancer-suppressing agent. Molecules 21:E931
Chatterjee S, Asakura M, Chowdhury N, Neogi SB, Sugimoto N, Haldar S et al (2010) Capsaicin, a potential inhibitor of cholera toxin production in *Vibrio cholerae*. FEMS Microbiol Lett 306: 54–60
Chinn MS, Sharma-Shivappa RR, Cotter JL (2011) Solvent extraction and quantification of capsaicinoids from *Capsicum chinense*. Food Bioprod Process 89:340–345
Choi SH, Suh BS, Kozukue E, Kozukue N, Levin CE, Friedman M (2006) Analysis of the contents of pungent compounds in fresh Korean red peppers and in pepper-containing foods. J Agric Food Chem 54(24):9024–9031
Chuichulcherm S, Prommakort S, Srinophakun P, Thanapimmetha A (2013) Optimization of capsaicin purification from *Capsicum frutescens* L. inn with column chromatography using Taguchi design. Ind Crop Prod 44:473–479
Chung YC, Baek JY, Kim SR, Ko HW, Bok E, Shin WH, Won SY, Jin BK (2017) Capsaicin prevents degeneration of dopamine neurons by inhibiting glial activation and oxidative stress in the MPTP model of Parkinson's disease. Exp Mol Med 49:e298

Cienfuegos NEC, Santos PL, García AR, Soares CMF, Lima AS, Souza RL (2017) Integrated process for purification of capsaicin using aqueous two-phase systems based on ethanol. Food Bioprod Process 106:1–10
Clark R, Lee SH (2016) Anticancer properties of capsaicin against human cancer. Anticancer Res 36:837–843
Collins MD, Wasmund LM, Bosland PW (1995) Improved method for quantifying capsaicinoids in Capsicum using high performance liquid chromatography. HortScience 30:37–139
Cowan MM (1999) Plant products as antimicrobial agents. Clin Microbiol Rev 12:564–582
Curry J, Aluru M, Mendoza M, Nevarez J, Melendrez M, O'Connell MA (1999) Transcripts for capsaicinoids biosynthetic genes are differentially accumulated in pungent and non-pungent *Capsicum* spp. Plant Sci 148:47–57
Dabrowska B, Suchorska-Tropiło K, Capecka E (2001) Presowing conditioning of hot pepper (*Capsicum annuum* L.) seeds and its results in a field growing. Part I. Effect on the vigour of seeds and seedlings. Ann Wars Agric Univ Hortic 22:3–7
Davis CB, Markey CE, Busch MA, Busch KW (2007) Determination of capsaicinoids in habanero peppers by chemometric analysis of UV spectral data. J Agric Food Chem 55:5925–5933
de Aguiar AC et al (2020) Economic evaluation of supercritical fluid and pressurized liquid extraction to obtain phytonutrients from Biquinho pepper: analysis of single and sequential-stage processes. J Supercrit Fluids 165:104935
de Farias VL, da Araújo ÍMS, da Rocha RFJ, dos Garruti DS, Pinto GAS (2020) Enzymatic maceration of Tabasco pepper: effect on the yield, chemical and sensory aspects of the sauce. Lwt 127:109311. https://doi.org/10.1016/j.lwt.2020.109311
Díaz-Laviada I, Rodríguez-Henche N (2014) The potential antitumor effects of capsaicin. Prog Drug Res 68:181–208
Duelund L, Mouritsen OG (2016) Contents of capsaicinoids in chilies grown in Denmark. Food Chem 221:913–918. https://doi.org/10.1016/j.foodchem.2016.11.074; PMID 27979294
El-Saeid HM (1995) Chemical composition of sweet and hot pepper fruits grown under plastic house conditions. Egypt J Hortic 22:11–18
Erdost H et al (2006) FSH and LH cells in the laying hens and cocks, fed with a diet containing red hot pepper. Int J Food Agric Environ 4:119–123
Esclapez MD, Sáez V, Milán-Yáñez D, Tudela I, Louisnard O, González-García J (2010) Sonoelectrochemical treatment of water polluted with trichloroacetic acid: from sonovoltammetry to pre-pilot plant scale. Ultrason Sonochem 17:1010–1020
Eshbaugh WH (1993) Peppers: history and exploitation of a serendipitous new crop discovery. In: Janick J, Simon JE (eds) New crops. Wiley, New York, NY, pp 132–139
Eshbaugh WH (2012) Peppers botany, production and uses. In: Russo VM (ed) The taxonomy of the genus Capsicum. CAB International, pp 14–28
Eshbaugh WH, Smith PG, Nickrent DL (1983) Capsicum tovarii (Solanaceae), a new species of pepper from Peru. Brittonia 35:55–60
Estrada B, Bernal MA, Díaz J, Pomar F, Merino F (2002) Capsaicinoids in vegetative, organs of *Capsicum annum* L. in relation to fruiting. J Agric Food Chem 50:1188–1191
Evangelista S (2015) Novel therapeutics in the field of capsaicin and pain. Expert Rev Clin Pharmacol 8:373–375
Foskolos A, Ferret A, Siurana L, Calsamiglia CS (2020) Effects of Capsicum and propyl-propane thiosulfonate on rumen fermentation, digestion, and milk production and composition in dairy cows. Animals 10:859
Friedman JR, Perry HE, Brown KC, Gao Y, Lin J, Stevenson CD, Hurley JD, Nolan NA, Akers AT, Chen YC et al (2017) Capsaicin synergizes with camptothecin to induce increased apoptosis in human small cell lung cancers via the calpain pathway. Biochem Pharmacol 129:54–66
Füchtbauer S, Mousavi S, Bereswill S, Heimesaat MM (2021) Antibacterial properties of capsaicin and its derivatives and their potential to fight antibiotic resistance—a literature survey. Eur J Microbiol Immunol (Bp) 11(1):10–17. https://doi.org/10.1556/1886.2021.00003; PMID: 33764892

Fujiwake H, Suzuki T, Iwai K (1982) Intracellular distributions of enzymes and intermediates involved in biosynthesis of capsaicin and its analogues in Capsicum fruits. Agric Biol Chem 46: 2685–2689

Fusco BM, Barzoi G, Agro F (2003) Repeated intranasal capsaicin applications to treat chronic migraine. Br J Anaesth 90(6):812. https://doi.org/10.1093/bja/aeg572

Gasson MJ, Kitamura Y, McLauchlan WR, Narbad A, Parr AJ, Parsons EHL, Payne J, Rhodes MJC, Walton NJ (1998) Metabolism of ferulic acid to vanillin. J Biol Chem 273:4163–4170

Gibss HAA, O'Garro LW (2004) Capsaicin content of West Indies hot pepper cultivars using colorimetric & chromatographic techniques. Hort Sci 39:132–135

Goll J, Frey A, Minceva M (2013) Study of the separation limits of continuous solid support free liquid–liquid chromatography: separation of capsaicin and dihydrocapsaicin by centrifugal partition chromatography. J Chromatogr 1284(2013):59–68

González-Zamora A, Sierra-Campos E, Pérez-Morales R, Vázquez-Vázquez C, Gallegos-Robles MA, López-Martínez JD, García-Hernández JL (2015) Measurement of capsaicinoids in Chiltepin hot pepper: a comparison study between spectrophotometric method and high performance liquid chromatography analysis. J Chem 1:10

Greenleaf WH (1986) Pepper breeding. In: Bassett MJ (ed) Breeding vegetable crops. AVI Publishing Co., Westport, CT, pp 69–127

Guillen NG, Tito R, Mendoza NG (2018) Capsaicinoides e pungênciaemfrutos de Capsicum chinense e Capsicum baccatum. Pesqui Agropecu Trop Goiânia 48(3):237–244

Hachiya S, Kawabata F, Ohnuki K, Inoue N, Yoneda H, Yazawa S, Fushiki T (2007) Effects of CH-19 sweet, a non-pungent cultivar of red pepper, on sympathetic nervous activity, body temperature, heart rate, and blood pressure in humans. Biosci Biotechnol Biochem 71:671–676

Haejin B, Jayaprakasha GK, Jifon J, Bhimanagouda PS (2012) Variation of antioxidant activity and the levels of bioactive compounds in lipophilic and hydrophilic extracts from hot pepper (*Capsicum* spp.) cultivars. Food Chem 134:1912–1918

Hamed M, Kalita D, Bartolo ME, Jayanty SS (2019) Capsaicinoids, polyphenols and antioxidant activities of *Capsicum annuum*: comparative study of the effect of ripening stage and cooking methods. Antioxidants (Basel) 8(9):364

Hoffmann L, Maury S, Martz F, Geoffroy P, Legrand M (2003) Purification, cloning, and properties of an acyl transferase controlling shikimate and quinate ester intermediates in phenylpropanoid metabolism. J Biol Chem 278:95–103

Hursel R, Westerterp-Plantenga MS (2010) Thermogenic ingredients and body weight regulation. Int J Obes 34:659–669. https://doi.org/10.1038/ijo.2009.299

Hussein GD, Halasz G, Palotás G, Palotás G, Bodai Z, Helyes L (2015) HPLC determination of capsaicinoids with cross-linked C18 column and buffer-free eluent. J Chromatogr Sci 53(1): 135–143

Inoue N, Matsunaga Y, Satoh H, Takahashi M (2007) Enhanced energy expenditure and fat oxidation in humans with high BMI scores by the ingestion of novel and non-pungent Capsaicin analogues (capsinoids). Biosci Biotechnol Biochem 71:380–389

Islam MD, Sharma SS, Sinha P, Negia MS, Neogb B, Tripathia SB (2015) Variability in capsaicinoid content in different landraces of Capsicum cultivated in North-Eastern India. Sci Hortic 183:66–71

Iwai K, Suzuki T, Fujiwake H (1979) Formation and accumulation of pungent principle of hot pepper fruits, capsaicin and its analogs, in *Capsicum annuum* var. *annuum* cv. Karayatsubusa at different growth-stages after flowering. Agric Biol Chem 43:2493–2498

Jaganathan R, Purushotham B, Radhakrishnan N, Swamy MK (2020) Capsaicin and its potential anticancer mechanisms of action. In: Plant-derived bioactives. Springer, Singapore, pp 301–321

Jeeatid N, Techawongstien S, Suriharn B et al (2017) Light intensity affects capsaicinoid accumulation in hot pepper (*Capsicum chinense* Jacq.) cultivars. Hortic Environ Biotechnol 58:103–110. https://doi.org/10.1007/s13580-017-0165-6

Jeeatid N, Techawongstien S, Chanthai S, Bosland PW, Techawongstien S (2021) Impact of plant density on capsaicinoid production of high pungent pepper (*Capsicum chinense* Jacq.) under Nethouse. Acta Hortic 1326:9–16. https://doi.org/10.17660/ActaHortic.2021.1326.2
Jolayemi AT, Ojewole JA (2013) Comparative anti-inflammatory properties of capsaicin and ethyl-A Acetate extract of *Capsicum frutescens* Linn. [Solanaceae] in rats. Afr Health Sci 13(2): 357–361
Joo JI, Kim DH, Choi JW, Yun JW (2010) Proteomic analysis for antiobesity potential of capsaicin on white adipose tissue in rats fed with a high fat diet. J Proteome Res 9:2977–2987
Joshi S, Thakur PC, Verma TS, Verma HC (1993) Selection of spice paprika breeding. Capsicum Eggplant Newsl 12:50–52
Josse AR, Sherriffs SS, Holwerda AM, Andrews R, Staples AW, Phillips SM (2010) Effects of capsinoid ingestion on energy expenditure and lipid oxidation at rest and during exercise. Nutr Metab 7:65
Juturu V (2016) Capsaicinoids modulating cardio metabolic syndrome risk factors: current perspectives. J Nutr Metab 2016:4986937
Kalia NP, Mahajan P, Mehra R, Nargotra A, Sharma JP, Koul S et al (2012) Capsaicin, a novel inhibitor of the NorA efflux pump, reduces the intracellular invasion of *Staphylococcus aureus*. J Antimicrob Chemother 67:2401–2408
Kim KI, Baek JY, Jeong JY, Nam J, Park ES, Bok E, Shin WH, Chung YC, Jin BK (2019) Delayed treatment of capsaicin produces partial motor recovery by enhancing dopamine function in MPP +-lesioned rats via ciliary neurotrophic factor. Exp Neurobiol 28:289–299
Kozukue N, Han JS, Kozukue E, Lee SJ, Kim JA, Lee KR et al (2005) Analysis of eight capsaicinoids in peppers and pepper-containing foods by high-performance liquid chromatography and liquid chromatography-mass spectrometry. J Agric Food Chem 53:9172
Kraikruan W, Sukprakarn S, Mongkolporn O, Wasee S (2008) Capsaicin and Dihydrocapsaicin contents of Thai chili cultivars. Kasetsart J (Nat Sci) 42:61–616
Krajewska AM, Powers JJ (1988) Sensory properties of naturally occurring Capsaicinoids. J Food Sci 53(3):902–905
Lee TS (1954) Physiological gustatory sweating in a warm climate. J Physiol 124:528–542. https://doi.org/10.1113/jphysiol.1954.sp005126; PMID 13175196
Lee P, Greenfield JR, Ho KK, Fulham MJ (2010) A critical appraisal of the prevalence and metabolic significance of brown adipose tissue in adult humans. Am J Physiol Endocrinol Metab 299:601–606
Lindsey K, Bosland PW (1996) A field study of environmental interaction on pungency. Capsicum Eggplant Newsl 14:36–38
Liu A, Han C, Zhou X, Zhu Z, Huang F, Shen Y (2013) Determination of three capsaicinoids in Capsicum annuum by pressurized liquid extraction combined with LC-MS/MS. J Sep Sci 36: 857e62
Liu J, Liu H, Zhao Z, Wang J, Guo D, Liu Y (2020a) Regulation of Actg1 and Gsta2 is possible mechanism by which capsaicin alleviates apoptosis in cell model of 6-OHDA-induced Parkinson's disease. Biosci Rep 40:BSR20191796
Liu T, Chen J, Li H, Li K (2020b) An integrated process for the separation and recovery of valuable metals from the spent LiNi0.5Co0.2Mn0.3O2 cathode materials. Sep Purif Technol 245:116869
Lu Y-m, Cui B (2019) Extraction and purification of capsaicin from Capsicum oleoresin using a combination of tunable aqueous polymer-phase impregnated resin (TAPPIR) extraction and chromatography technology. Molecules 24(21):3956
Lu M, Cao Y, Ho CT, Huang Q (2016) Development of organogel derived capsaicin nanoemulsion with improved bioaccessibility and reduced gastric mucosa irritation. J Agric Food Chem 64: 4735e41
Ludy MJ, Mattes RD (2011) The effects of hedonically acceptable red pepper doses on thermogenesis and appetite. Physiol Behav 102:251–258
Luo XJ, Peng J, Li YJ (2011) Recent advances in the study on Capsaicinoids and Capsinoids. Eur J Pharmacol 650(1):1–7

Macfarlane GJ, Ashraf EM, Bird H et al (2012) Complementary and alternative medicines report. Arthritis Researc, UZK

Mazourek M, PujarA BY, Paran I, Mueller L, Jahn MM (2009) A dynamic interface for Capsaicinoid systems biology. Plant Physiol 150:1806–1821

Meckelmann SW, Jansen C, Riegel DW, van Zonneveld M, Ríos L, Peña K, Mueller-Seitz E, Petz M (2015) Phytochemicals in native Peruvian Capsicum pubescens (Rocoto). Eur Food Res Technol 241:817

Meghavansi MK et al (2010) Naga chilli: a potential source of capsaicinoids with broad-spectrum ethnopharmacological applications. J Ethnopharmacol 132(1):1–14

Meotti FC, de Andrade EL, Calixto JB (2014) TRP modulation by natural compounds. Handb Exp Pharmacol 223:1177–1238. https://doi.org/10.1007/978-3-319-05161-119; PMID 24961985

Merali Z, Mayer MJ, Parker ML, Michael AJ, Smith AC, Waldron KW (2007) Metabolic diversion of the phenylpropanoid pathway causes cell wall and morphological changes in transgenic tobacco stems. Planta 225:1165–1178

Muñoz-Ramírez LS, Pen˜a-Yam LP, Aviles-Vinas SA, Canto-Flick A, Guzman Antonio AA, Santana-Buzzy N (2018) Behavior of the hottest chili peppers in the world cultivated in Yucatan. Mexico Hortsci 53(12):1772–1775

Nazari F, Ebrahimi SN, Talebi M, Rassouli A, Bijanzadeh HR (2007) Multivariate optimisation of microwave-assisted extraction of capsaicin from *Capsicum frutescens* L. and quantitative analysis by 1H-NMR. Phytochem Anal 18:333–340

Nee M, Bohs L, Knapp S (2006) New species of *Solanum* and *Capsicum* (Solanaceae) from Bolivia, with clarification of nomenclature in some Bolivian solanum. Brittonia 58:322–356

Nowaczyk P, Nowaczyk L, Banach M, Król I (2006a) Differences of capsaicinoids content in pericarp and paste of soft-flesh Capsicum spp. fruit Folia. Hortic 18(2):99–103

Nowaczyk P, Nowaczyk L, Banach M (2006b) The capsaicin and dihydrocapsaicin contents in soft-flesh fruit of *Capsicum frutescens* L. and *Capsicum annuum* L. hybrids. Herba Pol 52(1):38–42

Nyberg NT, Baumann H, Kenne L (2001) Application of solid-phase extraction coupled to an NMR flow-probe in the analysis of HPLC fractions. Magn Reson Chem 36:104–109

Nwokem CO, Agbaji EB, Kagbu JA, Ekanem EJ (2010) Determination of capsaicin Content and pungency level of five different peppers grown in Nigeria, New York. Sci J 3:9

Ohta Y, Chuong PV (1975) Hereditary changes in *Capsicum annum*. L.I. Induced by ordinary grafting. Euphytica 24:335–368

Olszewska D, Niklas-Nowak A, Nowaczyk P (2011) The assessment of doubled haploid lines obtained in pepper (*Capsicum annuum* L.) anther culture. Folia Hortic 23:93–99

Omolo MA, Wong ZZ, Mergen AK, Hastings JC, Le NC, Reiland HA et al (2014) Antimicrobial properties of chili peppers. J Infect Dis Ther 2:1–8. https://doi.org/10.4172/2332-0877.1000145

Pandhair V, Sharma S (2008) Accumulation of capsaicin in seed, pericarp and placenta of *Capsicum annuum* L. Fruit. J Plant Biochem Biotechnol 17:23

Pariari A, Khan SU (2013) Integrated nutrient management of chilli (Capsicum annuum L.) in Gangetic alluvial plains. J Crop Weed 9:128–130

Park ES, Kim SR, Jin BK (2012) Transient receptor potential vanilloid subtype 1 contributes to mesencephalic dopaminergic neuronal survival by inhibiting microglia-originated oxidative stress. Brain Res Bull 89:92–96

Paulo CV, Caio ST, Alanne TN, Rodrigo GC, Milena B, Guilherme G, TÃ¡ssia BP, Dias SMSS, Nathalia TS, Grigoletto FP, Renn Ã (2022) Feeding encapsulated pepper to dairy cows during the hot season improves performance without affecting core and skin temperature. J Dairy Sci 105(12):9542–9551

Pena-Alvarez A, Ramírez-Mayae E, Alvarado-Suárez LA (2009) Analysis of capsaicin and dihydrocapsaicin in peppers and pepper sauces by solid phase microextraction–gas chromatography–mass spectrometry. J Chromatogr A1216:2843–2847

Perkins B, Bushway R, Guthrie K, Fan T, Stewart B, Prince A, Williams M (2002) Determination of Capsaicinoids in salsa by liquid chromatography and enzyme immunoassay. J AOAC Int 85: 82–85

Pino J, Gonzalez M, Ceballos L, Centurionyah A, Trujilloaguirre J, Latourneriemoreno L, Sauriduch E (2007) Characterization of total Capsaicinoids, colour and volatile compounds of Habanero chilli pepper (Capsicum chinense Jack.) cultivars grown in Yucatan. Food Chem 104: 1682–1686

Prathibha VH, Mohan Rao A, Ramesh S, Nanda C (2013) Estimation of fruit quality parameters in anthracnose infected Chilli fruits. Int J Agric Food Sci Technol 4(2):57–60

Purkayastha J, Alam SI, Gogoi HK, Singh L, Veer V (2012) Molecular characterization of 'Bhut Jolokia' the hottest chili. J Biosci 37:757–768

Qiu J, Niu X, Wang J, Xing Y, Leng B, Dong J et al (2012) Capsaicin protects mice from community-associated methicillin-resistant *Staphylococcus aureus* pneumonia. PLoS One 7: e33032

Rafajlovska V, Slaveska-Raicki R, Klopcevska J, Srbinovska M (2011) Extraction of oleoresin from pungent red paprika under different conditions. Mass Transfer Chem Engineer Process 111:132

Reinbach HC, Smeets A, Martinussen T, Møller P, Westerterp-Plantenga MS (2009) Effects of capsaicin, green tea and CH-19 sweet pepper on appetite and energy intake in humans in negative and positive energy balance. Clin Nutr 28:260–265

Reinbach HC, Martinussen T, Moller P (2010) Effects of hot spices on energy intake, appetite, and sensory specific desires in humans. Food Qual Prefer 21:655–661

Sagar NA, Pareek S, Sharma S, Yahia EM, Lobo MG (2018) Fruit and vegetable waste: bioactive compounds, their extraction, and possible utilization. Compr Rev Food Sci Food Saf 17(3): 512–531. https://doi.org/10.1111/1541-4337.12330

Salgado-Roman M, Botello-Alvarez E, Rico-Martinez R, Jimenez-Islas H, Cardenas-Manriquez M, Navarrete Bolanos JL (2008) Enzymatic treatment to improve extraction of Capsaicinoids and carotenoids from chili (*Capsicum annuum*) fruits. J Agric Food Chem 56:10012e8

Sanatombi K, Sharma GJ (2008) Capsaicin content and pungency of different Capsicum spp. cultivars. Not Bot Horti Agrobot Cluj Napoca 36(2):88–90

Sankarikutty B, Sumathikutty MA, Nayaranan CSJ (1978) Standardization of extraction of pungency from whole chilli (Capsicum) for estimation of capsaicin. Food Sci Technol 15:126–127

Sano K, Uzawa Y, Kaneshima I, Nakasato S, Hashimoto M, Tanaka Y, Nakatani S, Kobata K (2022) Vanillin reduction in the biosynthetic pathway of capsiate, a non-pungent component of Capsicum fruits, is catalyzed by cinnamyl alcohol dehydrogenase. Sci Rep 12(1):12384

Santos P, Aguiar AC, Barbero JF, Rezende CA, Martínez J (2015) Super critical carbon dioxide extraction of capsaicinoids from malagueta pepper (*Capsicum frutescens* L.) assisted by ultrasound. Ultrason Sonochem 22:78–88

Sarpras M, Gaur R, Sharma V, Chhapekar SS, Das J, Kumar A, Yadava SK, Nitin M, Brahma V, Abraham SK, Ramchiary N (2016) Comparative analysis of fruit metabolites and pungency candidate genes expression between Bhut Jolokia and other Capsicum species. PLoS One 11(12):e0167791

Scoville WL (1912) Note on *Capsicum*. J Am Pharm Assoc 1:453–454. https://doi.org/10.1002/jps.3080010520

Seino S, Miki T (2003) Physiological and pathophysiological roles of ATP-sensitive K^+ channels. Prog Biophys Mol Biol 81:133–176

Singh S, Uddin M, Khan MMA, Singh S, Chishti AS, Bhat HU (2022) Therapeutic properties of capsaicin: a medicinally important bio-active constituent of Chilli pepper. Asian J Pharm Clin Res 15(7):47–48

Spencer O Jr, Almirall JR (2005) Extraction of Capsaicins in aerosol defense sprays from fabrics. Talanta 67:377–382. https://doi.org/10.1016/j.talanta.2005.05.031

Srinivasan K (2016) Biological activities of red pepper (*Capsicum annuum*) and its pungent principle capsaicin: a review. Crit Rev Food Sci Nutr 56:1488–1500

Stewart C, Mazourek M, Stellari GM, O'Connell M, Jahn M (2007) Genetic control of pungency in *C. chinense* via the Pun1 locus. J Exp Bot 58:979–991

Stewart C, Kang BC, Liu K, Mazourek M, Moore SL (2005) The Pun1 gene for pungency in pepper encodes a putative acyltransferase. Plant J 42:675–688
Stoica R, Moscovi M, Tomulescu C, Babeanu N (2016) Extraction and analytical methods of capsaicinoids-a review. Sci Bull 20:93–97
Sukrasno N, Yeoman MM (1993) Phenylpropanoid metabolism during growth and development of Capsicum frutescens fruits. Phytochemistry 32:839–844
Sun F, Xiong S, Zhu Z (2016) Dietary capsaicin protects Cardiometabolic organs from dysfunction. Nutrients 8(5):174
Surh YJ (2002) More than spice: capsaicin in hot chili peppers makes tumor cells commit suicide. JNCI 94:1263–1265
Suzuki T, Fujiwake H, Iwai K (1980) Intracellular localization of capsaicin and its analogues, capsaicinoid in Capsicum fruit 1. Microscopic investigation of the structure of the placenta of *Capsicum annuum* var. *annuum* cv. Karayatsubusa. Plant Cell Physiol 21:839–853
Tager LR, Krause KM (2011) Effects of essential oils on rumen fermentation, milk production, and feeding behavior in lactating dairy cows. J Dairy Sci 94:2455–2464
Tanaka T, Narazaki M, Kishimoto T (2014) IL-6 in inflammation, immunity, and disease. Cold Spring Harb Perspect Biol 6(10):a016295
Tewksbury JJ, Manchego C, Haak DC, Levey DJ (2006) Where did the chili get its spice? Biogeography of Capsaicinoids production in ancestral wild chili species. J Chem Ecol 32: 547–564
Thomas BV, Schreiber AA, Weisskopf CP (1998) Simple method for quantitation of Capsaicinoids in peppers using capillary gas chromatography. J Agric Food Chem 46:2655–2663. https://doi.org/10.1021/jf970695w
Thongnum K, Chanthai S (2018) Inhibitory reactivity of capsaicin with α-amylase and α-glucosidase related to antidiabetes using molecular docking and quantum calculation methods. Orient J Chem 34(5):2211
Tiwari A, Kaushik MP, Pandey KS, Dangi RS (2005) Adaptability and production of hottest chilli variety under Gwalior agroclimatic conditions. Curr Sci 88:1545–1546
Todd PH, Bensinger MG, Biftu T (1977) Determination of pungency due to Capsicum by gas–liquid chromatography. J Food Sci 42:660–665
Vázquez-Espinosa M, Fayos O, González-de-Peredo VA, Espada-Bellido E, Ferreiro-González M, Palma M, Garcés-Claver A, Barbero FG (2020) Content of Capsaicinoids and Capsiate in "Filius" pepper varieties as affected by ripening. Plants (Basel) 9(9):1222
Vendrely V, Peuchant E, Buscail E, Moranvillier I, Rousseau B, Bedel A, Brillac A, de Verneuil H, Moreau-Gaudry F, Dabernat S (2017) Resveratrol and capsaicin used together as food complements reduce tumor growth and rescue full efficiency of low dose gemcitabine in a pancreatic cancer model. Cancer Lett 390:91–102
Wahyuni Y, Ballester AR, Sudarmonowati E, Bino RJ, Bovy AG (2011) Metabolite biodiversity in pepper (*Capsicum*) fruits of thirty-two diverse accessions: variation in health-related compounds and implications for breeding. Phytochemistry 72:1358–1370. https://doi.org/10.1016/j.phytochem.2011.03.016
Walczak J, Buszewski B, Krakowska A, Rafinska K (2018) Enzyme-assisted optimized supercritical fluid extraction to improve Medicago sativa polyphenolics isolation. Ind Crop Prod 124: 931–940. https://doi.org/10.1016/j.indcrop.2018.08.004
Wang Y, Xia Y, Wang J, Luo F, Huang Y (2009) Capsaicinoids in Chili pepper (Capsicum annuum L.) powder as affected by heating and storage methods. Am Soc Agric Biol Eng 52(6): 2007–2010
Weaver KM, Awde DB (1986) Rapid high performance liquid chromatographic method for the determination of very low capsaicin levels. J Chromatogr 367:438–442
Weiss EA (2002) Spice crops. CABI Publishing International, New York, NY, p 411
Wenhui S, Sun WM, Zhen-Yu C, Hau YC (2014) Stability of Capsaicinoid content at raised temperatures. Nat Prod Commun 9(7):985–988

Westerterp-Plantenga MS, Smeets A, Lejeune MPG (2005) Sensory and gastrointestinal satiety effects of capsaicin on food intake. Int J Obes 29:682–688

Williams OJ, Viyaya-Raghavan GS, Orsat V, Dai JM (2004) Microwave-assisted extraction of Capsaicinoids from Capsicum fruit. J Food Biochem 28:113

Xinrong D, Xia L, Lei D, Fuzhi C, Zuojiao T, Zhonghua L (2014) Stage extraction of capsaicinoids and red pigments from fresh red pepper (Capsicum) fruits with ethanol as solvent. LWT 59:396–402

Yaldiz G, Ozguven M, Sekeroglu N (2010) Variation in capsaicin contents of different Capsicum species and lines by varying drying parameters. Ind Crop Prod 32(3):434–438

Yenigun OM, Thanassi M (2019) Capsaicin: an uncommon exposure and unusual treatment. Clin Pract Cases Emerg Med 3(3):219–221

Yoshioka M, St-Pierre S, Suzuki M, Tremblay A (1998) Effects of red pepper added to high-fat and high-carbohydrate meals on energy metabolism and substrate utilization in Japanese women. Br J Nutr 80:503–510

Yoshioka M, St-Pierre S, Drapeau V, Dionne I, Doucet E, Suzuki M, Tremblay A (1999) Effects of red pepper on appetite and energy intake. Br J Nutr 82:115–123

Zamljen T, Jakopič J, Hudina M et al (2021a) Influence of intra and inter species variation in chilies (*Capsicum* spp.) on metabolite composition of three fruit segments. Sci Rep 11:4932

Zamljen T, Medič A, Veberič R, Hudina M, Jakopič J, Slatnar A (2021b) Metabolic variation among fruits of different chili cultivars (Capsicum spp.) using HPLC/MS. Plants (Basel) 11(1):101

Zhao Z, Wang J, Wang L, Yao X, Liu Y, Li Y, Chen S, Yue T, Wang X, Yu W et al (2017) Capsaicin protects against oxidative insults and alleviates behavioral deficits in rats with 6-OHDA-induced Parkinson's disease via activation of TRPV1. Neurochem Res 42:3431–3438

Zheng J, Zheng S, Feng Q, Zhang Q, Xiao X (2017) Dietary capsaicin and its anti-obesity potency: from mechanism to clinical implications. Biosci Rep 37(3):BSR20170286

Zhou Y, Guan X, Zhu W, Liu Z, Wang X, Yu H et al (2014) Capsaicin inhibits *Porphyromonas gingivalis* growth, biofilm formation, gingivomucosal inflammatory cytokine secretion, and *in vitro* osteoclastogenesis. Eur J Clin Microbiol Infect Dis 33:211–219

[illegible]

Ethnobotany of Capsaicin-Producing Plants 3

Suvarna V. Gaikwad and Sandeep R. Pai

Abstract

The traditional system of medicine has made an important position in the first tier of healthcare in most parts of the world, which is continuously sourced by a rich heritage of floral diversity. Traditional usage of plants by local communities has served not only in the first tier of healthcare, but also pharma industries at large. Capsaicin, not a novel compound characterized especially by its burning sensation, is continuously under monitor due to its ever-increasing clinical vigour. This pungent alkaloid in *Capsicum* species is the major bioactive compound and has been broadly explored in the pharmacological point of view alongside its bio-potential. Until very recently, additional data are available on the exploitation of capsaicinoids in various biological activities, majorly anti-inflammatory, anti-lithogenic, cardioprotective and many others. There are five cultivated and ~30 wild species of *Capsicum*. This current chapter documents information on the significant ethnobotany of capsaicin-producing cultivated and wild plants. The taxonomy, distribution, usages, extractions and detection methods are discussed in this chapter. The chapter also discusses evidence of validation done through pharmacological claims. Such documentation shall promote further phytochemical and pharmacological investigations and lead to the development of new drugs.

Keywords

Ethnobotany · Capsaicin · Pharmacology · Validation · Bioactivities · Pungency

S. V. Gaikwad · S. R. Pai (✉)
Department of Botany, Rayat Shikshan Sanstha's, Dada Patil Mahavidyalya, Karjat, Maharashtra, India

M. K. Swamy, A. Kumar (eds.), *Capsaicinoids*,
https://doi.org/10.1007/978-981-99-7779-6_3

3.1 Introduction

Plants have been playing a significant role in the traditional medical system and in modern drug discovery (Sukumaran et al. 2020). Since the beginning of human evolution, the indigenous knowledge on medicinal plant utility to cure diseases and disorders has been a traditional and cultural practice (Mir et al. 2021). 'Traditional medicine' is the collection of theories, practices and skills that have been passed down through various societies and its utilization to maintain and improve not only the well-being but also to identify and treat varied instabilities (Rajalakshmi et al. 2019). The World Health Organization (WHO) says: in developing nations worldwide, about 80% of primary healthcare requirements are met by medicinal plants in the form of various formulations (Sukumaran et al. 2020; Rahman et al. 2022). Chander and co-workers (2014) reported that about 1.5 million people in India practice herbal medicine, and they employ ~25,000 effective plant-based preparations. About 6400 flowering plants in India have been reported for its medicinal potential, out of which only few (10%) have found place in pharmaceutical industries. Recently, medicinal plants have regaining importance in healthcare systems due to their low cost, availability and negligible side effects.

The scientific studies of plants in indigenous cultures for various daily routines is termed as ethnobotany; it may include medicine, building materials, tools for the home, musical instruments, food, ritualistic objects, firewood, pesticides, clothing, rituals and many more (Sargin 2015). Today, ethnobotanical research can serve as a preferred and trustworthy path to drug discovery, and therefore, scientists' focus has been diverted towards ethnomedicine and the use of herbal remedies due to the global revival of awareness in traditional health practises (Ahmed et al. 2014; Jamir et al. 2021). Capsaicin is one such important molecule produced by the members of the genus *Capsicum*, which has continuously been the topic of interest for researchers, because of its varied range of applications. Thus, we also see the use of different *Capsicum* species/varieties for various ethnomedicinal purposes across the nations and cultures. This current chapter documents information on the significant ethnobotany of capsaicin-producing cultivated and wild plants. The taxonomy, distribution, usages, extractions and detection methods are discussed in this chapter. The chapter also discusses evidence of validation done through pharmacological claims. Such documentation shall promote further phytochemical and pharmacological investigations and lead to the development of new drugs.

3.1.1 Capsaicin

Capsaicin is chemically known as *trans*-8-methyl-*N*-vanillyl-6-nonenamid and is known to give the hot characteristic flavour to chillies. Apart from the flavour it is also known to produce sensation of heat and pain in the person consuming it. The compound name capsaicin comes from the genus name *Capsicum*. All the species under this genus produce capsaicin except *C. annum* (Fattori et al. 2016). The amount of capsaicin may vary from species to species; even in an individual plant the distribution pattern of capsaicin is localized in higher concentration towards placental tissues in the fruits compared to any other part.

This pungent alkaloid has been widely explored for its pharmacological relevance. It is reported that capsaicin has been used to treat prolonged chronic body pain, shingles, arthritis, rhinitis and vasomotor and vasogenic facial pain. Apart from that it is also effective in the treatment of allergies, sore muscles, osteoporosis, kidney-related issues, nausea, vomiting and diabetes. In addition to that capsaicin also has therapeutic potential due to its analgesic, antitumorogenic, antiobesity, antioxidant, antiulcer, anticoagulant, anaesthetic, antiapoptotic, cardiovascular protective, gastroprotective, neuroprotective properties and others (Singh et al. 2022; Adetunji et al. 2022). Zhao and co-workers (2011) reported that simultaneous administration of capsaicin and isoflavone may additively promote the growth of hair in wild-type (WT) mice by boosting the release and production of CGRP (calcitonin gene-related peptide).

3.1.2 Capsaicin-Producing Plants

The genus *Capsicum* of the family Solanaceae is commonly known as 'pepper' which is widely used as a vegetable and spice crop. It includes approximately 35 species, but this is only an estimated number as new species are discovered regularly. Among them, only five, viz. *Capsicum annuum*, *C. baccatum*, *C. chinense*, *C. frutescens* and *C. pubescens*, are cultivated species and the remaining 30 are wild species (Tripodi and Kumar 2019; Guzman et al. 2019). Table 3.1 shows the cultivated and wild species of the genus *Capsicum*.

3.1.3 Distribution

Species of *Capsicum* are native to North and Central America (García et al. 2016). Kraft and co-workers (2014) in a separate study concentrate on the geographic origin of domesticated pepper using paleobilinguistics, archeobotanical and genetic data, concluding pepper originating from central-east Mexico. They are also cultivated in New Zealand, Australia, Nigeria, Malaysia, Sri Lanka and India. Some species of *Capsicum* are widely cultivated in all parts of India (Rudrapal and Sarwa 2020). Zhang and co-workers (2019) have worked on chemometrics with respect to the geographical origin using discrimination statistics in pepper. Similarly, global diversity of the genus based on single-nucleotide polymorphisms (SNPs) and passport information has been explored (Tripodi et al. 2021). Identification of landraces and germplasm to their adapted geographical conditions so as to mitigate drought-like effects has been studied in *Capsicum* (McCoy et al. 2022).

Table 3.1 List of known *Capsicum* species

Sl. no.	Species	Sl. no.	Species
1	*Capsicum annuum* L.[a]	19	*C. friburgense* Bianchetti & Barboza
2	*C. baccatum* var. *pendulum* (Willd.)[a]	20	*C. galapagoense* Hunz.
3	*C. chinense* Jacq.[a]	21	*C. geminifolium* (Dammer) Hunz.
4	*C. frutescens* L.[a]	22	*C. hookerianum* (Miers) Kuntze
5	*C. pubescens* Ruiz & Pav.[a]	23	*C. hunzikerianum* Bianchetti & Barboza
6	*C. buforum* Hunz.	24	*C. lanceolatum* (Greenm.) Morton & Standley
7	*C. caballeroi* M. Nee	25	*C. leptopodum* (Dunal) Kuntze
8	*C. campylopodium* Sendtn.	26	*C. minutiflorum* (Rusby) Hunz.
9	*C. cardenasii* Heiser & Smith	27	*C. parvifolium* Sendtn.
10	*C. ceratocalyx* M. Nee	28	*C. pereirae* Barboza & Bianchetti
11	*C. chacoense* Hunz.	29	*C. praetermissum* Heiser & Smith
12	*C. coccineum* (Rusby) Hunz.	30	*C. recurvatum* Witasek
13	*C. cornutum* (Hiern) Hunz.	31	*C. rhomboideum* (Dunal) Kuntze
14	*C. dimorphum* (Miers) Kuntze	32	*C. schottianum* Sendtn.
15	*C. dusenii* Bitter	33	*C. scolnikianum* Hunz.
16	*C. eshbaughii* Barboza	34	*C. tovarii* Eshbaugh, Smith & Nickrent
17	*C. eximium* Hunz.	35	*C. villosum* Sendtn.
18	*C. flexuosum* Sendtn.		

[a] Marked *Capsicum* species (sr. no. 1–5) are cultivated; all the rest (sr. no. 6–35) are wild.

3.1.4 Taxonomy of the Genus *Capsicum*

Capsicum belongs to the Solanaceae family. Plants are erect (*C. schottianum*, *C. geminifolium*) and compact (*C. chacoense*, *C. annuum*). They are subshrubs or shrubs, rarely trees (*C. rhomboideum*); species such as *C. chinense* are short-lived perennials or annual herbaceous like *C. annuum*; on the other hand, *C. coccineum* is supposed to be a sprawling vine. Woody stems and fissured bark and lenticels are observed in *C. rhomboideum* and *C. hookerianum*. Plants have sympodial growth which gives a 'zig-zag' appearance to the stem. *C. pereirae and C. hunzikerianum* have membranous ovate to elliptical leaves; small leaves are orbicular and sessile like in *C. dimorphum*. Entire leaf margins have asymmetric attenuate or truncate leaf base. A wide variety of leaf apices obtuse or acute to acuminate or long acuminate is observed in *C. hunzikerianum*. Petioles are longer in the cultivated species.

Flowers are axillary like solitary in *C. chacoense* or 1–2 flowered in *C. baccatum* or in 4–13 flowered in other species. In species like *C. ceratocalyx* inflorescence may have more than 18 flowers. These inflorescences in *C. ceratocalyx* and *C. coccineum* are unbranched, borne on short peduncle/epedunculate. Flowers are deciduous. All *Capsicum* flowers are pedicellate. The calyx is usually 5-merous and synsepalous. Some species bear key characters for identification like that in *C. chinense* where an annular constriction at the junction between the fruiting calyx and pedicel is present.

Capsicum species have 5-merous, sympetalous corollas. Most species have stellate corollas, whereas *C. lanceolatum* and *C. cardenasii* show campanulate, *C. rhomboideum* has broadly campanulate, *C. friburgense* has campanulate-urceolate, *C. flexuosum* and *C. baccatum* is rotate-stellate and *C. pubescens* bears rotate types of corollas. Dull white corollas are observed in *C. chinense* and *C. frutescens*; whereas it is white in *C. chacoense*, *C. galapagoense* and *C. annuum*, various other colours like yellow (*C. caballeroi*) and violet (*C. friburgense*) are also observed rarely in some species.

Androecium usually has five equal stamens. Longitudinally, dehiscent ellipsoid anthers are observed in cultivated as well as wild species of *Capsicum*. The colour of anthers may vary from yellow cream to blue, purple or even bluish-grey, for example, in *C. dimorphum*. Bicarpellate superior ovary having axile placentation with usually subglobose to ovoid, glabrous ovaries are observed. A simple straight to slightly curved, cylindrical glabrous style is observed in this genus. The style is commonly exerted beyond the anthers. In cultivated species, style length variations have been observed. However, the stigma is globose, rarely bilobed (e.g. *C. longifolium*, *C. lanceolatum*). Anatropous, numerous ovules are seen in *Capsicum*, which are tenuinucellate. Fruits of *Capsicum* are bicarpellate berry type; cultivated species may show variation up to 3–5 carpellate like in *C. annuum*, *C. chinense*, *C. pubescens* and *C. tovarii* (wild). Flat, angled subglobose seeds are numerous. They form a letter C or D shape and are rarely reniform (Barboza et al. 2022).

3.2 Ethnobotanical Uses of Different *Capsicum* Species Against Various Ailments

Capsicum species makes a special position in primary healthcare and the traditional system of medicine. It is used to treat sore throat, cough, toothache, muscle pain, stomach ailments, arthritis, rheumatism, skin rashes, dog or snakebite, wound healing, parasitic infections and many more (Table 3.2). The dried, ripe fruit of *Capsicum* species serves as a source of flavouring, cosmetic fragrances, external analgesics and repellent spray ingredients. Dried or raw fruits are edible in pulverized, dry powdered form or added in culinary preparations (Zimmer et al. 2011; Singletary 2011; Gogoi 2017; Mena et al. 2018). It is also beneficial in ulcers and diabetes (Amruthraj et al. 2013). Fruits of *Capsicum* are used to cure hallucinations, hangovers and stomach ache in the Amazon region. In Eastern Nicaragua, fruits and leaves are useful to treat skin rashes, sores, fever, postnatal pains and respiratory ailments. The tender leaves of the plant are applied on boils for removal of pus. Fruits are also useful in treating gastrointestinal disorders and muscle pain (Gogoi 2017; Valkova et al. 2021). Leaves are used to cure a strong headache, blood vomiting, sensation of cold body, epilepsy, breathing issues and also dark skin, whereas fruits are useful to prevent flu and other diseases (Valadeau et al. 2010). It reduces hypertension, asthma, muscle pain and stomach pain and prevents obesity (Elkhedir et al. 2022). An aboriginal community in Australia widely

Table 3.2 Ethnobotanical uses of *Capsicum*

Species name	Local name	Part used	Ethnomedicinal use	Country/state; tribe	Mode of preparation; administration/ consumption	References
Capsicum sp.	NM	Fruit	Sore throat, cough, toothache, stomach ailments, rheumatism, wound healing and parasitic infections, skin-conditioning agents, external analgesics, flavouring agents, cosmetic fragrances and repellent sprays	NM	Raw or powdered	Singletary (2011)
	NM	Leaves	Headache, madness, epilepsy, sensation of cold body, vomiting with blood, difficulty breathing, dark skin, strong headache	Yanesha	NM	Valadeau et al. (2010)
	NM	NM	Ulcers, diabetes and rheumatism		NM	Amruthraj et al. (2013)
	NM	NM	Arthritis, rheumatism, stomach ache, skin rashes, dog/snakebite and wounds	India, America and China	NM	Zimmer et al. (2011)
	NM	Fruit	Asthma, cough, sore throat, gastrointestinal disorders, to tone the muscle and alleviate pain after hard work	NM	NM	Gogoi (2017)
	NM		Toothache and muscle pain	NM	Hot infusion	
	NM	Leaves	Boils	NM	Topical paste	
	NM	Fruit	Stomach ache and to cure hangover or hallucination	Amazon	NM	

	NM	Fruit and leaves	Postpartum abdominal and back pain, fever, respiratory and pulmonary disorder, skin rashes and sores	Eastern Nicaragua	Decoction	
	NM	NM	To tone up body muscles after heavy workouts, toothache and muscle pain	Nagaland	Hot infusion	Mena et al. (2018)
	Chillies	Fruit	Colds, flu	Aboriginal community in northern New South Wales, Australia	NM	Packer et al. (2012)
	NM	Fruit	Fever, hypertension, asthma and rheumatism, prevent obesity, reduce muscle and stomach pain	NM	NM	Elkhedir et al. (2022)
	NM		Diarrhoea and dysentery	NM	NM	Kouassi et al. (2012)
	NM	Fruit	Gastrointestinal disorders, respiratory diseases, depression, pesticides in horticulture, as biochemical pest repellents and as pigments in the cosmetics industry	NM	NM	Valkova et al. (2021)
C. annuum L.	NM	Fruit	As aperitif, as cardiotonic	Central Italy	NM	Meghvansi et al. (2010)
	NM		Dog bite	Indian central Himalaya	NM	
	NM		Rubefacient for arthritis, stomachic, carminative,	Palestine	NM	

(continued)

Table 3.2 (continued)

Species name	Local name	Part used	Ethnomedicinal use	Country/state; tribe	Mode of preparation; administration/ consumption	References
			condiment, counter-irritant, rheumatism			
	NM		Stimulant	Enugu State of Nigeria	NM	
	NM		Antihypertensive	Southern Italy	NM	
	NM		Coccidiosis (in poultry)	Zimbabwe	NM	
	NM		Ascites	Lakhimpur, Assam (Northeast India)	NM	
	NM		Eye diseases of cattle	Kalahandi, Orissa (India)	NM	
	NM	Root	Snakebite	Central India	Root extract, orally	
	NM	Fruit			Fruit powder, topical	
	NM	Fruit	Dog bite	Uttarakhand Himalaya (India)	Fruit powder, topical	
	NM	Fruit	Skin disease of live stock		Fruit powder, orally	
	NM	Pulp	For dysentery, skin diseases, sensation of internal heat	Nawalparasi District, Nepal	Pulp orally	
	NM	Leaves	Headache, madness, epilepsy, sensation of cold body, vomiting with blood, difficult breath	Central region of Peru	Leaves in the form of a smoke bath/ steam bath	
	NM			North America		

		Blended with other herbs	In assisting memory, as well as for eczema, emphysema, asthma and other ailments of ageing		Blended with other herbs	
	NM	Fruits		Nagaland, India	Crushed fruits	
	NM	Fruits	Indigestion	Uttar Pradesh, India	Fresh fruits with other herbs, orally	
	NM	Fruits	Anti-fever	Northern Basilicata, Italy	Fruits with or without bulbs of *Leopoldia comosa*	
	NM	Fruits	Revulsive	Calabria region, Southern Italy	As poultice	
	NM	Fruits	Cuts and wounds	India	Dried fruits	
	NM	Stem	Dental and oral healthcare	Ekiti State, Nigeria	NM	
	NM	Fruit	Dog bite	Indian Central Himalaya	Fruit paste	
	NM	Fruits and leaves	Postpartum abdominal and back pain	Eastern Nicaragua	Decoction, topical	
	NM	Fruits	Ascariasis	Wonago Woreda, Ethiopia	Orally	
	NM	Fruits	Antihypertensive, antirheumatic, anti-fever, against the evil eye	Southern Italy	NM	
	NM	Not mentioned	Cough	Tlanchinol Hidalgo, Mexico	NM	
	NM	Fruits	Diarrhoea and stomach ache disorders in dogs	Eastern foothills of the Andes	NM	
	NM	Leaves	Painful menses		Orally	

(continued)

Table 3.2 (continued)

Species name	Local name	Part used	Ethnomedicinal use	Country/state; tribe	Mode of preparation; administration/consumption	References
				Dominican Republic		
	Pipariaddru	Fruit	To raise blood flow to superficial tissues, treat malaria	Calabria	NM	Tagarelli et al. (2010)
*C. annuu*m L. var. *acuminatum* Fing	Pipi	Fruit	Revulsive, treat malaria	Calabria	NM	Tagarelli et al. (2010)
C. annuum L.		Root	Colic or stomach discomfort due to overeating	Costa Rica	Decoction	
	Chile pepper	Fruit	Malaria fever	Nigeria	Infusion	Adeyemi et al. (2010)
	Pimiento (pimenton)	Fruit	Animal treatment of castrated cattle	Argentina	Milled, topical	Martinez and Lujan (2011)
		Leaves	Antiparasitic, gastric cancer, gastrointestinal disorders	Mexican	NM	Alonso-Castro et al. (2011)
	Maax iik	NM	Hard swelling	Yucatan Peninsula in Mexico	NM	Edgar et al. (2011)
	Gochu	Whole part	Veterinary ailments	Korea	Maceration, orally	Kim and Song (2011)
	Chillies	NM	Anticancer, stomachic, carminative, counter-irritant		Tea mixtures, orally	Kumar et al. (2011)
	Mirchi	Fruits	Blood purifier	Pakistan	Decoction, juice, orally	Ahmed et al. (2014)

			Carminative		Infusion, powder, orally	
			Cooling effect		Vegetable, paste, orally	
			Skin diseases		Topical	
			Snakebite		Topical	
	Pul biber	Fruits	Blood purification, arthritis, antirheumatism, heart stimulant	Turkey	Powder or tablet	Akbulut and Bayramoglu (2013)
	Marchakay	Fruits	Gastrointestinal disorders	North Pakistan	NM	Akhtar et al. (2013)
	NM	NM	Malaria	West Ethiopia	With *Allium sativum*, orally	Megersa et al. (2013)
	Subz mirch	Fruit	Anorexia, bronchitis, fever, indigestion, lameness, quidding, retention of urine, toxaemia, weakness (to treat equines—horses, donkeys and mules)	Punjab, Pakistan	Mixed with other herbs, topical	Goraya et al. (2013)
	Miroich	Leaves	Chest pain	Andaman and Nicobar Islands, India	NM	Chander et al. (2014)
			Circulatory stimulant		NM	Khan et al. (2014)
	Mirch, Marchi	Fruit	Appetizer, toothache	Jammu, Kashmir and Ladakh (India)	NM	Gairola et al. (2014)
	Peperoncino	Fruit	Gastritis	Campania, Southern Italy	As spice	Menale and Muoio (2014)
	Katumbal	Leaves	Insect bite; white spot, athlete's foot; skin rashes	Philippines	Topical	Ong and Kim (2014)

(continued)

Table 3.2 (continued)

Species name	Local name	Part used	Ethnomedicinal use	Country/state; tribe	Mode of preparation; administration/ consumption	References
			Appetite enhancer		Topical	
		Fruit	Abdominal pain		Topical	
	Pujay (Mende), red pepper (English)	Fruits	Pest control—storage weevils	Southern Sierra Leone	Dry fruits	Kanteh and Norman (2015)
	NM	NM	Childcare, digestive, ritual, women's health, spice	Gabon (Central Africa)		Towns et al. (2014)
	NM	Fruit	Headache	Macedonians and Albanians living in Golloborda, Eastern Albania	Lacto-fermented, topical	Pieroni et al. (2014)
	Pimento malagueta	NM	Boil, slow ageing	Brazil		Bieski et al. (2015)
	Qaara/Qariya	NM	Flu	Ethiopia	Fruit with other herbs orally	Kefalew et al. (2015)
		Fruit	Gastrointestinal disorders, amoebiasis		With *Zea mays*, orally	
	Biber, kırmızı biber, gırmızı biber, acıbibe	Fruit, seed	Analgesic, rheumatism	Turkey	Spice, topical	Sargin (2015)
	NM	NM	Enemas	Western African	Enema with other plants	Andel et al. (2015)
	Aji dulce	Fruit	Snakebites	Antioquia, Colombia	Orally	Vasquez et al. (2015)
	Seraimbsik	Fruit and seed	Malaria	Papua New Guinea	Orally	Koch et al. (2015)

		Fruit	Toothache, gastric activities, stimulant, carminative, neuralgia, rheumatism, uterine pain, painful muscle spasms, arthritis, lumbago, chilblains	Iraq	NM	Ali (2015)
		Aerial parts of plant	Fever and malaria	Guinea-Bissau, West Africa	NM	Catarino et al. (2016)
		Seed, fruit, leaves, flower	Diabetes	Central America	External and internal	Giovanninia et al. (2016)
	Marchakay	NM	Carminative, stomachic, diabetes, anticoagulant, hypertension	Pakistan		Shah et al. (2016)
		Fruit	Stomach problems and ulcer	Punjab Province, Pakistan	With other fruits, orally	Rehman et al. (2017)
	Shimla mirch	Fruit	Dyspepsia, asthma, diarrhoea, whooping cough, common cold	Punjab Province, Pakistan	Paste, orally	Rehman et al. (2017)
	Marchiya	Root	Snakebite	India	NM	Upasani et al. (2018)
	Bang-ga-la	Fruit	Antibacterial and leprosy, digestive and piles	Bhutan		Wangchuk et al. (2017)
	Karia/keto	Fruit	Malaria	Guba Lafto District, Northern Ethiopia	Orally	Chekole (2017)
	NM	Leaves	Heavy fever	Tamil Nadu—India	Paste with *Cleome gynandra* L., orally	Rajalakshmi et al. (2019)
	NM	Fruit	Haemorrhoids	Northern Angola		

(continued)

Table 3.2 (continued)

Species name	Local name	Part used	Ethnomedicinal use	Country/state; tribe	Mode of preparation; administration/ consumption	References
					Decoction, enema	Lautenschlager et al. (2018)
	Chilli	Fruit, seed	Poultice, topical or rubbing	Italy, Morocco	NM	Sadgrove (2018)
	Mopherefere (Sepedi), green pepper (English)	Roots	Mixed with food for flavour (all) and the roots are used to treat period pains	South Africa	NM	Mogale et al. (2019)
	Cayenne	NM	Stimulant, aphrodisiac	Bolivia and Southern Brazil	NM	Prescott and Khan (2020)
	Pimientos	Fruit	Made a poultice with spider web and powdered chili pepper	Castile-La Mancha, Spain	Apply	Rivera et al. (2019)
	NM	Fruit	Poultice, fresh	Eastern and South-Eastern Serbia	NM	Matejic et al. (2020)
	Pimentão doce	Fruit	NM	Brazil	NM	Teixeira et al. (2019)
	Felfel Hârr/soudania	Fruit	Diabetes	Morocco	Raw	Idm'hand et al. (2020)
	Divlji papric	Leaves	Gastritis	South Kosovo	Tea	Mustafa et al. (2020)
	Pachai milagai	Fruit	Tooth pain	Kani tribesof Pechiparai Hills, Western Ghats, India	Raw, topical	Sukumaran et al. (2020)
	Kanthari milagu	Fruit	Blood purification	Kani tribes of Pechiparai Hills,	Paste, orally	

				Western Ghats, India		
	Sili na bisaya	Fruit	Diabetes; hypertension	Manobo tribe of Agusan del Sur, Philippines	Orally	Dapar et al. (2020)
		Leaves	White spot, athlete's foot, appetite enhancer; boils, skin rashes and itchiness, psoriasis, dandruff, insect bites		Topical	
	NM	Leaves	NM	NM	NM	Afroz et al. (2020)
	Pul biber	Fruit	Cold, rheumatism, analgesic	Kahramanmaras, Ahir Mountain	Powder	Uzun and Koca (2020)
	NM	NM	Cold, sinus infection, sore throat and improve digestion, blood circulation, cancer, asthma, and cough, anorexia, haemorrhoids, liver congestion and varicose veins	NM	NM	Afroz et al. (2020)
	NM	Fruit	Skin, bladder cancer	Palestine	Decoction, orally	Aumeeruddy and Mahomoodally (2021)
		Leaves	Throat cancer	Ghana	Decoction	
	Tahmira	Fruit	Gastrointestinal	North of Morocco	Grind, topical	Chaachouay et al. (2022a, b)
			Dermal wound	Rif, Morocco	Topical	
	Lombok, paprika		Streamlining bowel movements, treating thrush, prevent anaemia, control blood sugar	Indonesia		Umartani and Nahdi (2021)
	Morok	Fruit	Gout	Manipur, Northeast India	Grind, topical	Laishram and Dey (2021)

(continued)

Table 3.2 (continued)

Species name	Local name	Part used	Ethnomedicinal use	Country/state; tribe	Mode of preparation; administration/ consumption	References
	NM	Fruit	Gastrointestinal disorders (appetizer)	Budgam, Jammu and Kashmir, India	Raw, oral	Mir et al. (2021)
	Hmarcha-te	Leaves, fruits	Fruits improve digestion; for fruits and leaves, juices applied to burn and snakebite. Fruits used as antihaemorrhoidal and antirheumatic	Mizoram, India	Juice, oral	Ralte et al. (2021)
	Rus. perets Kar	Fruit	Cuts, high temperature	Karelia	Tincture	Kolosova et al. (2022)
	Ball pepper	Fruits	Neurodegenerative diseases	Nigeria		Ayeni et al. (2022)
	Marchakay	Fruit	Digestive disorders, low blood pressure and heart diseases	Tribal communities of Buner Pakistan		Rahman et al. (2022)
	Pupaino	Fruit	Rheumatic pain	Southern Italy	Topical	Mautone et al. (2019)
	NM	Fruit	Cold and flu	NM	NM	Batsatsashvili et al. (2019)
	Sakaipilo	Fruit	Rheumatism—pain	Agnalazaha littoral forest (Southeastern Madagascar)	NM	Razafindraibe et al. (2013)
	Lombok	NM	Streamlining bowel movements, treating thrush	Boyolali	NM	Umartani and Nahdi (2021)
	Paprika	NM	Prevent anaemia, control blood sugar		NM	
	Hmarcha	Fruit			Paste	Laldingliani et al. (2022)

			Wasp sting, burn, toothache, boil, soothe and prevent scars from skin burns	Mizo tribes in Champhai district of Mizoram, India		
	NM	Fruit	Heartburn	Italy	NM	Monari et al. (2022)
	Ndongo (Sango)	Leaves	Hypertension	Central African Republic	Boiling, decoction	Odukoya et al. (2022)
	Tahmira	Fruit	Wounds, burns	Moulay Yacoub, northeast of Morocco	Powder	Khomsi et al. (2022)
			Soften hair, food		Maceration	
	Cabai merah	Fruit	Flu, hypotension, appetite	South Aceh District, Indonesia	Infusion, orally	Adnan et al. (2022)
	NM	Fruit, whole plant	Veterinary—mastitis	NM	Orally	Ajose et al. (2022)
	Marchangan	Fruit	Toothache	Jammu and Kashmir	Cooked	Balkrishna et al. (2022)
	Lazi	Fruits	Stimulate appetite and help digestion	China	Orally	Fan et al. (2022)
	Paprika	Fruits	Toothache	NM	NM	Gamage et al. (2022)
	NM	NM	Haemorrhoids, stimulant of appetite and nutrition	NM	NM	Gbogbo et al. (2022)
	Mhiripiri	Fruit	Veterinary—coccidiosis, internal parasites, diarrhoea, coughing, flu-like symptoms, Newcastle disease	NM	Orally	Gobvu et al. (2022)
	Cabik	NM	Sprue, tumour	South Bangka Regency, Indonesia	NM	Henri et al. (2022)
	Itshilisi	Fruits	Fever		NM	Maroyi (2022)

(continued)

Table 3.2 (continued)

Species name	Local name	Part used	Ethnomedicinal use	Country/state; tribe	Mode of preparation; administration/ consumption	References
				Eastern Cape Province in South Africa		
	Qariya	Fruit	Loss of appetite	Southern Ethiopia	Orally	Megersa and Woldetsadik (2022)
	Kariya	Fruit	Malaria	Jimma Town, Oromia, Ethiopia	Grinded and mixed with other plants, orally	Million et al. (2022)
	Biber	Fruit	Colds, rheumatism, blood purification, arthritis, heart stimulant	Turkey	Orally	Olcay et al. (2022)
	Jire khursani	Fruit	Chronic fever, typhoid fever and seasickness	Palpa District, Western Nepal	NM	Sigdel and Acharya (2022)
	Pipar, tsilli pipar, punane pipar, verikoder, moru pipar	Fruit	Cough, medicine, stomach pain	Estonia	Tincture	Soukand et al. (2022)
			Joint pain		Tincture, topical	
		Leaves	Treat burn wound	Ethiopia		Talwan et al. (2022)
	Miximixo	Fruits	Ascariasis	NM	Orally	Yirga et al. (2022)
	NM	NM	Pain reliever	NM	NM	Singh et al. (2022)
	NM	Fruit	Digestive disorders, low blood pressure and heart diseases	Tribal communities of Buner Pakistan	NM	Rahman et al. (2022)

Capsicum baccatum L. var. Pendulum (Willd.)	NM	Fruits	Postnatal care, inducing immunity	Kottoor Reserve Forest, Kerala, India	Dried fruits	Meghvansi et al. (2010)
	NM	Leaves	Against furunculous	Northeast of Brazil	NM	Chaachouay et al. (2022a, b)
	NM	Fruits	Against anorexia		NM	
	NM	Fruits	Treat discomforts and common ailments associated with pregnancy	Crocker Range Sabah, Malaysia		Meghvansi et al. (2010)
	NM	Fruit		Brazil		Zimmer et al. (2011)
	Pokok lada	Leaves	Ringworm	Ulu Kuang Village, Kuang, Gombak, Selangor	Parched and mashed, topical	Azliza et al. (2012)
	NM	NM	Antirheumatic, diaphoretic, digestive, irritant, rubefacient, sialagogue and tonic	NM	NM	Afroz et al. (2020)
	NM	Leaf extract	NM	NM	NM	
	Pimentinha	Fruit	As condiment	Brazil	**NM**	Teixeira et al. (2019)
	Aji quitucho	Fruit		Argentina		López et al. (2012)
	Aji, aji amarillo	Fruit	Antihaemorrhoidal, antirheumatic, digestive, irritant, rubefacient, sialagogue and tonic		Internal and external	Lim (2013)
Capsicum chinense Jacq.	**NM**	Fruits	Toothache, muscle pain	Northeast region of India	**NM**	Meghvansi et al. (2010)
	NM	Fruits		Amazon	NM	

(continued)

Table 3.2 (continued)

Species name	Local name	Part used	Ethnomedicinal use	Country/state; tribe	Mode of preparation; administration/ consumption	References
			Stomach ache remedy and a hallucinogenic 'hangover cure'			
	NM	Fruits and leaves	Postpartum abdominal and back pain, fever, aches and pains, respiratory and pulmonary disorders (cold, coughs, etc.), skin rashes and sores	Eastern Nicaragua	Decoction, topical	
	Pimenta-malaguetona	Leaves	Dermatologic problems—scabies, dermatophytes, pityriasis versicolor, itching and skin lesion	Unini River, AM, Brazil	NM	Santos et al. (2012)
	NM	Seed, fruit, leaves, flower	Diabetes	Nicaragua, Central America	NM	Giovanninia et al. (2016)
	Tasu	Fruit	Spider and caterpillar sting	Zeliangrong ethnic group of Manipur, northeast India	Raw, topical	Panmei et al. (2019)
	Naga king chilli or bhut jolokia	Fruit	Arthritis, asthma, cough and sore throat, headache, gastritis, toothache and muscle pain	Assam	NM	Baruah and Lal (2020)
			Gastrointestinal disorders	North east	NM	
	NM	Fruit	Toothache and muscle pain, digestive diseases and	NM	Hot infusion	

			chronic congestion, restore muscles			
	NM	Leaves	Remove pus from boils, summer heat remedy	NM	Grind, topical	
	NM		Asthma, gastrointestinal abnormalities, toothache and muscle pain, removal of pus from boils, arthritis	NM	NM	Afroz et al. (2020)
	Chiberachi	Leaves, fruit	High blood pressure, stomach ache and diarrhoea	NM	Decoction, oral	Jamir et al. (2021)
	Mpilipili	Fruit	Insect repellent	Kilombero Valley (Morogoro Region, Tanzania)	NM	Salinitro et al. (2017)
	NM	Pods	For tuning up body muscles, wrist pain, toothache, muscle pain, asthma, gastrointestinal disorders	India	Hot infusion	Rudrapal and Sarwa (2020)
	NM	Leaves	Boil, headache and night blindness		Paste	
Capsicum frutescens L.	NM	Chilli peppers	Mosquito repellent	Central-Southern Italy	Dried chilli peppers	Meghvansi et al. (2010)
	NM	Fruits	Stomach ache and diarrhoea	Taiwan and the Batanes Islands	NM	
	NM	Fruits	In the month immediately following childbirth to promote blood excretion	Taiwan and the Batanes Islands	NM	
	NM	Fruits	Applied to boils (animal healthcare)	Gohpur (Assam), Northeast India	Mixed with mustard oil, topical	
	NM	Fruits			NM	

(continued)

Table 3.2 (continued)

Species name	Local name	Part used	Ethnomedicinal use	Country/state; tribe	Mode of preparation; administration/ consumption	References
			Purgative, enhancer of other herbs, blood circulation	South-west Cameroon		
	NM	Fruits	Arthritis	Philippines	NM	
	NM	Fruits	Malaria	Southwest Nigeria	NM	
	NM	Fruits	Waist pain	Mayurbhanj, Orissa (India)	NM	
	NM	Leaves and fruits	Against furunculous	Northeast of Brazil	NM	
	NM	Leaves	Skin disorders (sores, boils)	Tonga, Samoa, Tahiti, and the Cook Islands	Crushed	
	NM	Leaves	Boils, abscesses and wounds	Fiji		
	NM	Fruits	To treat cough and cold	Agasthyamalai Biosphere Reserve, India	Fruits burnt and inhaled	
	NM	Leaves	Mumps in animals	Jajpur, Orissa (India)	With other plants orally	
	NM	Leaves and fruits	Fever and vaginal infections	Eastern Nicaragua	Decoction, topical	
	NM	Fruits and seeds	Dog bite	Edo State, Nigeria	Powdered, topical	
	NM	Fruits	Sexual impotence and erectile dysfunction	Western Uganda	Orally	
	NM	Leaves	Anhidrosis in horses	Trinidad and Tobago		

	NM	Fruit	Stomach ache	Dek Island, Ethiopia	Dry and powdered fruit	
	NM	Fruit	Rheumatism, artrosis	Eastern Cuba	Macerated with other plants	
	NM	Component of herbal formulation frucanga	Applied topically on boils	Cuba	Component of herbal formulation frucanga	
	NM	Fruits	Internal illness	Samoa		
	NM	Leaves, fruits and stem	Diuretic and in removing kidney stones	Mizoram, India	Decoction, orally	
	NM	Leaves	To remove water leech	Mizoram, India	Juice as ear drops	
	NM	Roots	Diarrhoea	Central Province, Cameroon	Decoction	
	NM	Fruits and leaves	Enteritis	Central Province, Cameroon	Decoction with other plants	
	NM	Fruits	Treating impotence and sterility	Eastern Cuba	Mixed with other plants	
	NM	Fruits	Peptic ulcers	Bangangte region, Western Cameroon	Mixed with other plants	
	NM	Fruits	Hypertension	Bafia, Cameroon	With other plants, orally	
	NM	Fruits	Rheumatism	Southern Spain		
	NM	Fruits and leaves	Malaria	Yanesha, Peru	Smoke of fruits and leaves	

(continued)

Table 3.2 (continued)

Species name	Local name	Part used	Ethnomedicinal use	Country/state; tribe	Mode of preparation; administration/consumption	References
	Piment rouge/red chilli	Fruit	Ear infection, stomach ache, diarrhoea, purgative, arthritis, waist pain, cough cold, malaria	Mauritius	As ear drops	Nunkoo and Mahomoodally (2012)
	Cipo—apui	Stalk	Musculoskeletal system—sprain and muscle strain	Unini River, AM, Brazil		Santos et al. (2012)
	Gaandhari menasu	Fruits	Common cold	Karnataka, India	Grind with other spices, orally	Lingaraju et al. (2013)
	NM	Fruits	Vermicide		Grind, topical	
	NM	NM	NM	West Ethiopia	Powdered with other herbs, orally	Megersa et al. (2013)
	Wael pima	Leaves	Muscle pain	Vanuatu (northern Australia)	Topical	Bradacs et al. (2011)
	NM	Fruit	Rheumatism			
	NM		Boils		Topical	
	Pilipililukwale	Root, bark	Wound	Tanzania	Crushed or pounded, topical	Amri and Kisangau (2012)
	NM	Fruit	Rheumatism	Granada province (Andalusia, southern Spain)	Maceration, topical	Benitez et al. (2010)
	Chili	Fruit and leaves	Earache, arthritis, rheumatism	Mauritius	Concoction, as ear drops	Sreekeesoon and

						Mahomoodally (2014)
	NM	Flower, all the plant	Cancer	NM	NM	
	NM	NM	External therapy in painful muscle spasms in areas of the shoulder, arm and spine; for treating arthritis, neuralgia, lumbago and chilblains, diabetes, blood pressure, bronchitis, burning feet; to increase circulation; relieve rheumatic pain; treat mouth sores and infected wounds; reduce blood clots; and aid digestion	Iraq	Topical	Ali (2015)
	NM	Flower	Ulcers, stomach pain with chest tightness, chest pain, stomach pain by ulcer	Marquesas Islands French Polynesia		Girardi et al. (2015)
	NM	NM	Stimulate gastric activities, increase circulation and enhance blood flow, asthma, pneumonia, toothache, pharyngitis and laryngitis	NM	NM	Kamal et al. (2015)
	NM	Fruit	Headache cure and for conjunctivitis, wound and sores as well as poultice for boils; relieve muscle, joint and toothache pain; to treat cough, asthma and sore throat; stimulant; and to treat	West Africa	NM	Titilayo and Morenikeji (2016)

(continued)

Table 3.2 (continued)

Species name	Local name	Part used	Ethnomedicinal use	Country/state; tribe	Mode of preparation; administration/ consumption	References
			stomach ache, sea-sickness and flatulence; improve natural warmth and blood circulation; it is taken in ordinary cold, fever and dysentery, malaria and gonorrhoea			
	NM		Breast cancer		Decoction	
	Cayenne pepper	NM	Postnatal care, as nutrition therapy, in pain management, to treat erectile dysfunction, as a circulatory tonic, as a remedy for rheumatic and arthritic pains, for the reduction of blood sugar levels, for the excretion of cholesterol	NM	NM	Otunola and Afolayan (2015)
	Pimenta	NM	Aphrodisiac	Brazil	NM	Santana et al. (2016)
	Ndongue	Seeds and roots	Tooth hygiene, antioxidative, anti-inflammatory	Gbaya ethnic group in Eastern Cameroon	Maceration, crushed	Fongnzossie et al. (2017)
	Pimenta-malagueta	Leaves and fruits	General infection, tetanus, cancer, skin infection, inflammation, vaginal	Brazil	Decoction, infusion, maceration, topical	Ribeiroa et al. (2017)

			discharge, wound healing, injury with swelling			
	NM	Fruit	Nutrition	Northern Angola	NM	Lautenschlager et al. (2018)
	Barbare	Fruit	Malaria	Ethiopia	Juice, orally	Suleman et al. (2018)
	Nunga (Bembe)—Makaya ma nungu (Lari)	Leaves	Haemorrhoids	Congo Basin	Infusion	Tchicaillat-Landou et al. (2018)
	Pherefere (Sepedi), chilis (children) (Sepedi), lebela bela (Sepedi), viriviri (Tsonga)	NM	Various medicinal uses	South Africa	Mixed with food for flavour	Mogale et al. (2019)
	Felfel har	Seeds	Diabetes	Algeria	Raw seeds	Hamza et al. (2019)
	Piripiri, malagueta, pimenteiro-de-caiena	Fruit	Digestive—haemorrhoids		Ingestion, orally	Vinagre et al. (2019)
	NM	Root, leaves	Malaria	Bukavu and Uvira areas in DR Congo		Mboni et al. (2020)
	NM	Fruit cover	Throat or breast or squamous cell carcinoma	Kenya	Infusion with other plants	Aumeeruddy and Mahomoodally (2021)
	NM	Flowers, whole plant	Cancer	Marquesas Islands	NM	
	NM	Leaves	Breast cancer	Ghana	Decoction	
	NM	Unripe fruit	Cancer	Pakistan	Paste, orally	
	NM	Not mentioned	Cancer	Brazil	NM	
	Sudaniya	Fruit	Endo- and ectoparasitic	North of Morocco	Crushing, inhalation	Chaachouay et al. (2022a, b)

(continued)

Table 3.2 (continued)

Species name	Local name	Part used	Ethnomedicinal use	Country/state; tribe	Mode of preparation; administration/ consumption	References
	Hmarchapui	Leaves, fruits	Improves digestion, burn and snakebite	Mizoram, India	Juice, topical	Ralte et al. (2021)
	NM	NM	Haemorrhoids, rheumatism	Congo	NM	Pathy et al. (2021)
	Sudaniya, Flfel Har	Fruit	Dermal wound	Rif, Morocco	Decoction	Chaachouay et al. (2022a, b)
	NM	NM	Ulcer protective activity	NM	NM	Baruah and Lal (2020)
	NM	Leaf and fruit extract	Antihaemorrhoidal, antirheumatic, digestive, antimicrobial	NM	NM	Afroz et al. (2020)
	Wild chilli pepper	Fruit—ripe and unripe	Memory enhancers and anti-ageing	Nigeria-West Africa	With other plants, orally	Ayeni et al. (2022)
	Pilipili (Kiswahili)	Fruit cover	Breast, throat, squamous cell carcinoma	Kenya	Powder with other plants, infusion	Omara et al. (2022)
	Edo, Nigeria	NM	Hypertension	NM	Decoction with other plants, orally	Alves-Silva et al. (2022)
	Ladoh	Leaves and fruits	Skin and common ailments in pregnancy	Malaysia	Crushed	Venmathi Maran et al. (2022)
	NM	Whole plant	Diabetes, gastrointestinal disorders, toothache, pain, muscle spasms, fever, infections	NM	NM	Ansari et al. (2022)
	Cabai rawit	Fruits	Flu, hypertension, appetite	South Aceh District, Indonesia	Infusion, orally	Adnan et al. (2022)

	Wild Chilli pepper	Fruits	Treat cancer, colds, coughs and headaches, bacterial diseases	Thenhipalam Village, Malappuram District, Kerala	NM	Anupriya and Danya (2022)
	Mixaamixuwa	Fruits	Respiratory tract infection, cold, loss of appetite	NM	Crushed with other plant, orally	Megersa and Woldetsadik (2022)
	Marchi	Fruits	Improve digestion	Bagh, Azad Jammu and Kashmir, Pakistan	NM	Maqsood et al. (2022)
	Comarim	Fruits	NM	Brazil	NM	Teixeira et al. 2019
	NM	Seed	Blackleg—in cattle	West Ethiopia	Powdered with other plants, orally	Megersa et al. (2013)
Capsicum pubescens Ruiz & Pav.	NM	Leaves	Headaches, tiredness, epilepsy or convulsions, paralysis	Yanesha (Amazonian Peruvian ethnic group)	NM	Valadeau et al. (2010)
Capsicum buforum Hunz.	NM	Fruits	As a preventive for flu and other diseases	NM	As smoke bath	
Capsicum campylopodium Sendtn.	Pimenta-do mato	NM	Boil, slow ageing	Brazil	NM	Bieski et al. (2015)
Capsicum chacoense Hunz.	Chemaxarai	Fruits	Snakebite (alexiteric)	Chaco (Northeast Argentina)	Milled, topical	Martinez and Barboza (2010)
	Aji del monte, puta pario	Fruits	NM	Argentina	NM	López et al. (2012)
	NM	Fruits	NM	Alto Paraguay Department	NM	Coronel et al. (2022)

NM not mentioned

uses the fruits of *Capsicum* to cure cold and flu (Packer et al. 2012). Kouassi et al. (2012) reported the use of *Capsicum* fruits in curing diarrhoea.

3.2.1 *Capsicum annuum* L.

The fruits of *Capsicum annuum* L. have been used as spices and condiments all over the world. In Central Italy and Philippines, the fruits are used as an appetizer and as cardiotonic, whereas in India and Albania, the leaves have high medicinal potency to cure a wide array of ailments from headache to dysentery and many more. The fruits are also taken to treat indigestion in India and Pakistan (Kim and Song 2011; Gairola et al. 2014; Goraya et al. 2013; Mir et al. 2021; Megersa and Woldetsadik 2022; Fan et al. 2022; Gbogbo et al. 2022; Adnan et al. 2022). The plant is also useful in the treatment of asthma, assisting memory, eczema, emphysema and dyspepsia in North America and Brazil (Meghvansi et al. 2010; Pieroni et al. 2014; Bieski et al. 2015; Rehman et al. 2017; Afroz et al. 2020). The fruits are given to relieve the symptoms of ascariasis in Ethiopia. The fruit powder is used to cure dog bite in Indian central Himalaya. In Palestine, Turkey and Iraq, the fruits are reported to cure arthritis (Meghvansi et al. 2010; Akbulut and Bayramoglu 2013; Ali 2015; Olcay et al. 2022; Yirga et al. 2022). Its fruits have been used as stomachic in Palestine and Pakistan (Kumar et al. 2011; Shah et al. 2016). The fruits and seeds have potential to cure rheumatism in Palestine, Turkey, Ahir Mountain and Southeastern Madagascar regions (Meghvansi et al. 2010; Akbulut and Bayramoglu 2013; Razafindraibe et al. 2013; Sargin 2015; Ali 2015; Uzun and Koca, 2020; Olcay et al. 2022). The leaves and roots are taken for painful menses in Dominican Republic. The fruits are not only used as effective cardiac stimulant in Nigeria, Turkey, Bolivia and Brazil but stimulant of appetite and nutrition also (Meghvansi et al. 2010; Akbulut and Bayramoglu 2013; Khan et al. 2014; Ali 2015; Mogale et al. 2019; Prescott and Khan 2020; Olcay et al. 2022; Gbogbo et al. 2022).

The fruits of *Capsicum* are potent antihypertensive agents and used in various parts such as Southern Italy, Pakistan and Philippines. In Central African Republic, the leaves are also useful in treating hypertension, whereas in Indonesia, the fruits are used to cure hypotension (Meghvansi et al. 2010; Shah et al. 2016; Dapar et al. 2020; Odukoya et al. 2022; Adnan et al. 2022). Recently, it is reported that the leaves and fruit juices of *C. annuum* are useful in healing burn, cuts and wounds and in softening hair in India, Morocco and Ethiopia (Ralte et al. 2021; Khomsi et al. 2022; Talwan et al. 2022; Chaachouay et al. 2022a, b; Laldingliani et al. 2022). The leaves are also useful in heartburn, lowering blood pressure and heart diseases (Monari et al. 2022; Rahman et al. 2022). The use of *C. annuum* leaves and fruits has also been documented to treat cold and flu worldwide (Kefalew et al. 2015; Rehman et al. 2017; Batsatsashvili et al. 2019; Uzun and Koca 2020; Afroz et al. 2020; Adnan et al. 2022; Olcay et al. 2022). The fruits are known to treat snakebites in India, Pakistan and Colombia (Ahmed et al. 2014; Vasquez et al. 2015; Upasani et al. 2018; Ralte et al. 2021). The fruits are widely used to heal skin diseases like wasp sting and soothe and prevent scars from skin burns in India and Pakistan, whereas the leaves are used to heal athlete's foot, psoriasis, itchiness, skin-related issues and dandruff (Meghvansi et al. 2010; Kim and Song 2011; Ahmed et al. 2014; Bieski et al. 2015; Dapar et al. 2020; Laldingliani et al. 2022). The fruits are used to raise blood flow to superficial tissues and as revulsive in Calabria, Nigeria and Italy (Tagarelli et al. 2010; Adeyemi et al. 2010; Megersa et al. 2013). The aerial parts of the plant are known to treat fever such as chronic fever, typhoid fever and seasickness in Italy, India, Pakistan and Africa (Goraya et al. 2013; Catarino et al. 2016; Rajalakshmi et al. 2019; Maroyi 2022; Sigdel and Acharya 2022). The fruits are reportedly used for treating different types of coughs such as whooping cough in Estonia, Mexico, Pakistan and Zimbabwe (Rehman et al. 2017; Afroz et al. 2020; Soukand et al. 2022; Gobvu et al. 2022).

The roots are known to be useful in treating stomach disorders in Costa Rica; on the other hand, the stems and fruits are useful in treating dental and oral healthcare such as toothache in Nigeria, Turkey and India (Meghvansi et al. 2010; Gairola et al. 2014; Ali 2015; Sukumaran et al. 2020; Laldingliani et al. 2022; Balkrishna et al. 2022; Gamage et al. 2022). The fruits and seeds of *C. annuum* species have been taken to cure malaria in Nigeria, New Guinea, Africa and Ethiopia (Tagarelli et al. 2010; Adeyemi et al. 2010; Megersa et al. 2013; Koch et al. 2015; Catarino et al. 2016; Chekole 2017; Million et al. 2022). The leaves and fruits serve to be effective in gastrointestinal troubles like gastric cancer, gastritis, ulcer and amoebiasis (Alonso-Castro et al. 2011; Akhtar et al. 2013; Menale and Muoio 2014; Kefalew et al. 2015; Ali 2015; Rehman et al. 2017; Mustafa et al. 2020; Chaachouay et al. 2022a, b). In Turkey, the fruits and seeds have been known to possess analgesic properties (Sargin 2015; Uzun and Koca 2020; Olcay et al. 2022). *Capsicum annuum* is also useful in enemas in Western Africa (Andel et al. 2015). Fruits are effective in treating skin and bladder cancer, whereas leaves in throat cancer (Aumeeruddy and Mahomoodally 2021). The fand seeds are used as poultice or cataplasm on dermal wounds (Sadgrove 2018; Rivera et al. 2019; Matejic et al. 2020; Afroz et al. 2020; Chaachouay et al. 2022a, b). It is known to be used in the cure of psilosis and tumour (Henri et al. 2022). The fruit is also used to cure digestive problems (Towns et al. 2014; Wangchuk et al. 2017; Rahman et al. 2022) and haemorrhoids (Lautenschlager et al. 2018; Afroz et al. 2020; Gbogbo et al. 2022). The fruits are reported to be used for relieving pain associated with neuralgia, joints, uterine pain, muscle spasms, lumbago and chilblains in Turkey (Ali 2015; Soukand et al. 2022; Singh et al. 2022). The leaves are used to relieve chest pain (Chander et al. 2014). The fruit is also effective in leprosy and piles treatment (Wangchuk et al. 2017). The plant is said to be an aphrodisiac and is widely used in Brazil for the said purpose (Prescott and Khan 2020). *Capsicum annuum* is also useful to relieve sinus infection, sore throat, blood circulation, anorexia, liver congestion and varicose veins and to improve digestion (Afroz et al. 2020). It is also used for the prevention of anaemia, streamlining bowel movements and treating thrush (Umartani and Nahdi 2021). The leaves, flowers, fruits and seeds are considered to be useful in the treatment of diabetes in Central America, Pakistan, Morocco and Philippines (Giovanninia et al. 2016; Shah et al. 2016; Idm'hand et al. 2020; Dapar et al. 2020; Umartani and Nahdi 2021). The leaves are also applied to cure hard swellings

(Caamal-Fuentes et al. 2011). The fruits have been used as a blood purifier and to give a cooling effect. (Ahmed et al. 2014; Akbulut and Bayramoglu 2013; Sukumaran et al. 2020). They are also an effective pest control agent in Sierra Leone (Kanteh and Norman 2015). In Central Africa, the plant has been traditionally used for childcare, rituals and women's health (Towns et al. 2014).

Capsicum is also utilized in veterinary medicine for treating diarrhoea, skin and eye diseases, stomach ache in dogs, treatment of castrated cattle and many other veterinary ailments. It is also useful in the treatment of anorexia, indigestion, lameness, bronchitis, retention of urine and weakness in equines—horses, donkeys and mules (Meghvansi et al. 2010; Martinez and Lujan 2011; Kim and Song 2011; Goraya et al. 2013; Ajose et al. 2022; Gobvu et al. 2022).

3.2.2 *Capsicum baccatum* var. *pendulum* (Willd.)

The fruits of *C. baccatum* are used as a condiment in Brazil (Teixeira et al. 2019). In India and Malaysia, the fruits are reported to be used for treating discomforts and common ailments associated with pregnancy; they are also useful for inducing immunity and postnatal care (Meghvansi et al. 2010). In Brazil, the fruits and leaves of the species are used against furunculous and anorexia (Chaachouay et al. 2022a, b). The leaves are being used in treating ringworm (Azliza et al. 2012). The fruits of the species are useful for the production of pepper sprays, toxic and repellent effects against spider mites (Lim 2013).

3.2.3 *Capsicum chinense* Jacq.

The fruits and leaves of *C. chinense* Jacq. are believed to be effective in postpartum abdominal pain, back pain and fever. They are also useful to cure aches and pains, skin rashes, sores and respiratory and pulmonary disorders such as sore throat, cold, coughs, asthma, etc. (Meghvansi et al. 2010; Baruah and Lal 2020; Afroz et al. 2020; Rudrapal and Sarwa 2020). In Brazil, the leaves can be used to cure dermatologic problems (Santos et al. 2012). The leaves, flower, fruit and seeds are said to be effective in the management of diabetes in America (Giovanninia et al. 2016). The fruits have been used against spider and caterpillar sting (Panmei et al. 2019). The fruits are used in the treatment of toothache and muscle pain (Meghvansi et al. 2010; Baruah and Lal 2020; Afroz et al. 2020). The fruits are also used in treating gastrointestinal disorders, stomach ache and diarrhoea (Baruah and Lal 2020; Afroz et al. 2020; Rudrapal and Sarwa 2020). The fruits are good for headache (Baruah and Lal 2020; Rudrapal and Sarwa 2020). The pods of *C. chinense* are reported to be effective for tuning up body muscles, treating wrist pain (Baruah and Lal 2020; Rudrapal and Sarwa 2020). The leaves are used as an effective summer heat remedy and for the removal of puss from boils (Afroz et al. 2020; Rudrapal and Sarwa 2020). The fruits are given to cure arthritis (Afroz et al. 2020; Baruah and Lal

2020). The leaves and fruits are also useful in high blood pressure (Jamir et al. 2021). In Tanzania, the fruits are used as insect repellent (Salinitro et al. 2017).

3.2.4 *Capsicum pubescens* Ruiz & Pav.

The leaves of *C. pubescens* are reported to be used against headaches, tiredness, epilepsy or convulsions and paralysis by the ethnic people of Amazon (Valadeau et al. 2010).

3.2.5 *Capsicum frutescens* L.

Dried chilli peppers are an effective mosquito repellent in Italy. The Taiwanese use the fruits to relieve stomach ache and to promote blood excretion immediately following childbirth. In veterinary medicine, the fruits are applied to boils, whereas the leaves are useful in mumps and anhidrosis in horses. The leaves and fruits are considered to be effective against furunculous in Brazil. The roots and fruits are known to be effective in diarrhoea. They are also valued as purgative and enhancer of other herbs (Meghvansi et al. 2010; Afroz et al. 2020). The species is used for improving blood circulation in Cameroon and West Africa (Kamal et al. 2015; Titilayo and Morenikeji 2016; Otunola and Afolayan 2015; Ali 2015). The fruits are used to relieve pain associated with the waist in India and Mauritius (Nunkoo and Mahomoodally 2012). The fruits are also used to treat arthritis (Nunkoo and Mahomoodally 2012; Sreekeesoon and Mahomoodally 2014; Ali 2015). The roots, leaves and fruits serve to cure malaria in Nigeria, Peru, Mauritius, West Africa, Ethiopia and DR Congo (Meghvansi et al. 2010; Nunkoo and Mahomoodally 2012; Titilayo and Morenikeji 2016; Suleman et al. 2018; Mboni et al. 2020). In Fiji, the leaves are applied to skin disorders like boils, abscesses and wounds. In India, the fruits are treated as a cough remedy (Nunkoo and Mahomoodally 2012; Titilayo and Morenikeji 2016; Anupriya and Danya 2022; Chaachouay et al. 2022a, b). The fruits are helpful to treat cold in India and Africa (Lingaraju et al. 2013; Titilayo and Morenikeji 2016; Anupriya and Danya 2022; Megersa and Woldetsadik 2022). The leaves and fruits of the species are used in treating vaginal infections (Titilayo and Morenikeji 2016; Ansari et al. 2022). In Nigeria, the fruits and seeds are used against dog bite. In Western Uganda, the fruits are consumed for sexual impotence and erectile dysfunction. The plant is also effective in treating rheumatism and artrosis (Benitez et al. 2010; Sreekeesoon and Mahomoodally 2014; Pathy et al. 2021; Otunola and Afolayan 2015; Afroz et al. 2020).

The leaves, stem and fruits possess a diuretic property and are believed to be effective in removing kidney stones in India. In Brazil, the stalks and leaves are considered to be effective in sprain and muscle strain, whereas the fruits are used to treat earache and infection (Bradacs et al. 2011; Nunkoo and Mahomoodally 2012; Santos et al. 2012; Sreekeesoon and Mahomoodally 2014; Ansari et al. 2022). The fruits show vermicidal property. The leaves are applied for the treatment of boils

(Bradacs et al. 2011; Titilayo and Morenikeji 2016). The fruits are also effective in treating internal illness. The use of flowers, fruits and leaves has also been documented to cure enteritis and ulcers (Girardi et al. 2015; Kamal et al. 2015; Baruah and Lal 2020). The flowers and fruits are considered to be effective in stomach pain with chest tightness, chest pain and hypertension (Girardi et al. 2015; Alves-Silva et al. 2022; Adnan et al. 2022). The plant is known to stimulate gastric activities and treat gastrointestinal disorders (Ansari et al. 2022). It is also used to cure respiratory tract infections, asthma, pneumonia, pharyngitis and laryngitis (Kamal et al. 2015; Megersa and Woldetsadik 2022).

In West Africa and Kenya, the leaves, flowers and fruits are reported to be used in breast, throat and squamous cell carcinoma (Aumeeruddy and Mahomoodally 2021; Omara et al. 2022). They are also used for headache, conjunctivitis, sores and poultice. It relieves joint and toothache pain (Ansari et al. 2022; Anupriya and Danya 2022). They treat sore throat seasickness and flatulence. They are also used to treat gonorrhoea (Titilayo and Morenikeji 2016). The plant is reportedly used for postnatal care and as nutrition therapy (Lautenschlager et al. 2018). The seeds and fruits are also effective in the management of pain, diabetes and erectile dysfunction and for the excretion of cholesterol (Otunola and Afolayan 2015; Ali 2015; Hamza et al. 2019; Ansari et al. 2022). In Brazil, the leaves and fruits are used to treat general infection, tetanus, cancer, skin infection, inflammation, vaginal discharge, wound and injury with swelling (Ribeiroa et al. 2017; Ansari et al. 2022; Anupriya and Danya 2022). The leaves and fruits improve digestion and also treat haemorrhoids (Ali 2015; Tchicaillat-Landou et al. 2018; Vinagre et al. 2019; Afroz et al. 2020; Pathy et al. 2021; Ralte et al. 2021; Maqsood et al. 2022).

The leaves and fruit rind are effective in the treatment of teeth-related diseases, i.e. odontalgy. The leaves are reported to be useful in contusion and injury (Benitez et al. 2010). The plant is believed to be effective in relaxing muscle spasms. The species is used in various forms to treat neuralgia, lumbago, bronchitis and burning feet and also to reduce blood clots (Ali 2015). The plant also possesses an aphrodisiac property (Santana et al. 2016). In India, the leaves and fruits are applied to cure burns and snakebite (Ralte et al. 2021). Ripe and unripe fruits are effective memory enhancers and possess anti-ageing properties (Ayeni et al. 2022). In Malaysia, the leaves and fruits are used to cure skin and common ailments during pregnancy (Venmathi Maran et al. 2022). The fruits can be used to cure flu and stimulate appetite (Adnan et al. 2022; Megersa and Woldetsadik 2022).

3.2.6 *Capsicum buforum* Hunz.

The fruits of *C. buforum* are considered to be effective in the prevention of flu and other diseases (Jarret et al. 2019).

3.2.7 *Capsicum campylopodium* Sendtn.

In Brazil, the plant is beneficial for the treatment of boils and for slowing the ageing process (Bieski et al. 2015).

3.2.8 *Capsicum chacoense* Hunz.

In Central Chaco (Argentina), the fruits are applied to cure snakebite (Martinez and Barboza 2010).

3.3 Mode of Preparation for Ethnomedicinal Purpose

The fruit of the *Capsicum* spp. has been widely used in many ethnobotanical uses. The fruits are generally consumed raw. They are also available in ground form, or in food that has been added *Capsicum oleoresin* as a supplement (Keith Singletary 2011). Muscle pain and toothaches are treated with hot infusions of the fruit (Balkrishna et al. 2022). It is also common practice to apply a fine paste made from the plant's tender leaves to boils in order to facilitate the easy drainage of pus. The decoction form of the fruits and leaves in Eastern Nicaragua is used to treat fever, back pain, respiratory disorders and many others (Gogoi 2017). After strenuous exercise, *Capsicum* spp. were used to tone up body muscles in Nagaland. For muscle pain and toothaches, hot infusions are used (Mena et al. 2018). The Gaddi community of Jammu and Kashmir, India, ground roots in locally produced alcohol, which are applied topically to the skin as an antimicrobial agent (Dutt et al. 2015).

3.3.1 *Capsicum annuum* L.

In Indian Central Himalaya, the fruit powder is used externally over wounds caused by dog bites. In Central India, extract from the root is taken orally, whereas the fruit powder is applied to wounds for the treatment of snakebite (Vasquez et al. 2015). *Syzygium aromaticum* and *Piper nigrum* flowers are combined with the fruit powder of *C. annuum* and given orally as a liquid to cure skin disease of livestock. The pulp is taken in the treatment of various skin diseases, dysentery, etc. in Nepal, India. The people of Peru use leaves in the form of smoke bath to cure epilepsy, cold body sensation, etc. In North America, *C. annuum* is mixed with other herbs like *Ginkgo biloba*, *Clematis* sp., *Centella asiatica* and others against emphysema, asthma eczema, etc. Crushed fruits are used as antipruritic and counter-irritant in Nagaland, India. The fresh fruits of *Capsicum* are used orally for indigestion alongside ginger and onion. In some parts of Italy, the fruits are used against fever along with other herbs. The fruits are not only used as poultice, but applied on cuts and wounds. The fruit paste is applied against dog bite in Indian Central Himalaya.

In Eastern Nicaragua, a decoction made from fruits and leaves is taken to relieve postpartum abdominal and back pain. The fruits and leaves are also externally applied for treating it. Fresh or dry fruits are used to chew and swallow for the treatment of ascariasis in Wonago Woreda, Ethiopia (Yirga et al. 2022). In Dominican Republic, tea made from leaves is consumed during painful menses, whereas in Costa Rica, root decoction is used to cure colic discomfort caused by overeating (Meghvansi et al. 2010). Calabrian people used fruits to rise blood flow to superficial tissues, as revulsive and to treat malaria (Tagarelli et al. 2010). Likewise, fruits and seeds are cooked and taken orally against malaria in New Guinea (Koch et al. 2015). An infusion which is made up of fruits of *C. annuum* and *Alstonia boonei* is given to cure malaria fever in Nigeria (Adeyemi et al. 2010). Fresh fruits are sliced and taken orally for treating malaria in Guba Lafto District, Northern Ethiopia (Chekole 2017). In Argentina, milled fruits are externally applied for the treatment of castrated cattles (Martinez and Lujan 2011). Maceration made up from all parts of *C. annuum* is given orally for veterinary ailments in the southern mountainous region of Korea (Kim and Song 2011). In Punjab Province, Pakistan, a decoction made up of fruits is taken orally as a blood purifier. An infusion made up of powder of *C. annuum* is given orally as carminative. Paste made up from the vegetable gives a cooling effect. Additionally, the fruits are applied topically against snakebite (Ahmed et al. 2014).

In Turkey, the fruits are taken in the form of powder or tablet as a blood purifier and heart stimulant and also to cure arthritis and rheumatism (Akbulut and Bayramoglu 2013). People of West Ethiopia use bulbs of onion with *Capsicum* against malaria (Megersa et al. 2013). A mixture of ginger rhizome (50 g), *pepper* fruit (125 g), onion bulb (250 g) and common salt (125 g) is ground and administered for the treatment of equine diseases of horses, donkeys and mules (Goraya et al. 2013). In the Campania region of Southern Italy, the fruits are used as spice for food and to cure gastritis (Menale and Muoio 2014). The leaves are crushed and rubbed on the skin or mixed with vinegar and applied externally over skin diseases, insect bite and athlete's foot. The leaves are crushed and rubbed on the stomach as an appetite enhancer. Additionally, the fruits are ground and rubbed on the stomach against abdominal pain in the Philippines (Ong and Kim 2014). In Southern Sierra Leone, the dry fruits of *Capsicum* are used to control weevils in stored seeds and grains (Kanteh and Norman 2015).

Macedonian and Albanian people living in Golloborda, Eastern Albania, externally applied lacto-fermented fruits on the forehead to relieve headache (Pieroni et al. 2014). In Ada'a District of Ethiopia, the fruits of *Capsicum annuum* along with the leaves of *Ocimum basilicum* and the roots of *Aloe macrocarpa* are cooked together and taken as a remedy for flu. Additionally, the dried and powdered fruits of *C. annuum* are taken with bread baked of *Zea mays* to treat amoebiasis (Kefalew et al. 2015). They are added to the meal three times per day. They are applied externally twice a day for treating rheumatism. They are also useful as an analgesic (Sargin 2015). In West Africa, *C. annuum* is mixed with *Zingiber officinale* or more additional species. This mixture is boiled with tap water, filtered and then syphoned into enema devices for insertion (Andel et al. 2015). The seeds, fruits, leaves and flowers are consumed as a remedy for diabetes (Giovanninia et al. 2016; Idm'hand et al. 2020). To treat dyspepsia and cough, cut the fruits with soybean oil and salt, with *Allium sativum* bulb paste, *Zingiber officinale* root paste and *Allium sativum* bulb paste and take orally. The fruits of *C. annuum* and *Lycopersicon esculentum* with the leaves of *Coriandrum sativum* are cooked and taken two times per day for stomach problems and ulcer in the Punjab Province region of Pakistan (Rehman et al. 2017). The leaf paste of *C. annuum* (10%) blended with 20 g of *Cleome gynandra* and salt is taken two times a day for 3 days for the treatment of heavy fever. This remedy is commonly used in the Tamil Nadu region of India (Rajalakshmi et al. 2019).

A decoction made up of fruits has been used for the treatment of haemorrhoids. Fruits are also used in the form of nose drops for treating madness. The fruits are also used as a nutrition supplement in Northern Angola (Lautenschlager et al. 2018). The fruits of *C. annuum* are powdered and combined with spiderweb. This is applied for treating muscle pain associated with myositis in Castile-La Mancha region of Spain (Rivera et al. 2019). A tea which is made up of the leaves of *C. annuum* is taken against gastritis in South Kosovo (Mustafa et al. 2020). Kani tribes of Western Ghats, India, applied raw fruits topically for treating tooth pain. Furthermore, they consumed fruit paste orally for blood purification (Sukumaran et al. 2020). The Manobo tribe of Philippines either consumed the fresh fruit or used it as a seasoning in recipes for treating diabetes and hypertension—seven fruits once or twice a day or as required. Moreover, three to five leaves are crushed and applied externally once or twice a day for the treatment of white spot, athlete's foot, boils, skin rashes, itchiness, psoriasis, dandruff and insect bites. They are also used as an appetite enhancer (Dapar et al. 2020). The fruits are powdered and used as a remedy for cold and rheumatism. They are also used as an analgesic in the Kahramanmaras region of Ahir Mountain (Uzun and Koca 2020). Using four fruits (100 g) of *C. annuum* boiled with 200 mL water for 20 min, five drops of this decoction is given orally every 8 h for patients suffering from skin and bladder cancer; likewise leaf decoction is useful for throat cancer. This is widely used by people of Palestine (Aumeeruddy and Mahomoodally 2021).

In the north of Morocco, the fruits are ground and used as dermal for animal therapy (sheep) to treat gastrointestinal disorders. The fruits have also been used as cataplasm for dermal wounds of animals (Chaachouay et al. 2022a, b). The fruit is applied on the infected part against gout in Manipur, Northeast India (Laishram and Dey 2021). Raw fruits are taken orally in Jammu and Kashmir, India, as an appetite stimulant (Mir et al. 2021). In Mizoram, India, the fruit is used as a spice and condiment. The leaves fermented with pork are taken as vegetables. The fruit and leaf juices are used topically to treat snakebites and burns. The fruits are also used as an antihaemorrhoidal and antirheumatic (Ralte et al. 2021). The fruits of the species are used for the treatment of cuts (Kolosova et al. 2022). In Southern Italy, a macerate made up from fruits of *C. annuum* and olive oil is effective in antirheumatic massages (Mautone et al. 2019). This macerate is also effective for softening hair in Moulay Yacoub, northeast of Morocco. Additionally, the fruit powder is used against wounds and burns (Khomsi et al. 2022). The Mizo tribes in India apply the fruit paste to treat wasp sting, burn, toothache and boil and soothe and prevent scars from skin burns (Laldingliani et al. 2022). The leaf decoction is used against hypertension (Odukoya et al. 2022).

An infusion made up of fruits is taken against flu and hypotension and as an appetite stimulant in South Aceh District, Indonesia (Adnan et al. 2022). Fifty-gram fruits or the whole plant is sèrved for the control of veterinary mastitis (Ajose et al. 2022). In China, the fruits can be used in different cold dishes and seasoning. This is an effective remedy to stimulate appetite and help digestion (Fan et al. 2022). It is common practice in Zimbabwe to crush fruits in water and give to animals to treat veterinary coccidiosis and internal parasites. Moreover, crushed fruits are added to a sugar solution and given to birds to treat diarrhoea, cough, flu-like symptoms and Newcastle disease (Gobvu et al. 2022). The fruits are regularly consumed as an appetizer together with other foods in Southern Ethiopia (Megersa and Woldetsadik 2022). Ground fruits are combined with *Allium cepa* and *A. sativum*. This mixture is then consumed as it is or with food to treat malaria (Million et al. 2022). In Turkey, the fruits are taken to cure colds and rheumatism. They are also effective as a blood purifier and heart stimulant (Olcay et al. 2022). Fruits in the form of tincture are effective in cough and stomach pain. It is also applied topically to heal joint pain in Estonia (Soukand et al. 2022).

3.3.2 *Capsicum baccatum* var. *pendulum* (Willd.) Eshbaugh

It is common practice to give dried fruits of *C. baccatum* for postnatal care and for inducing immunity in Kerala, India (Meghvansi et al. 2010). In Selangor, the leaves are parched and mashed with slaked lime. This is applied topically to treat ringworm (Azliza et al. 2012). The fruit is given internally to treat cold-stage fevers, old age or convalescence debility, varicose veins, asthma and digestive issues. Additionally, it is applied topically to cure sprains, neuralgia, unbroken chilblains and pleurisy (Lim 2013).

3.3.3 *Capsicum chinense* Jacq.

Fruit and leaf decoctions are widely used for the treatment of respiratory disorders (cold, coughs, etc.) and fever. This decoction is also applied topically to cure aches, pains, postpartum abdominal, back pain, skin rashes and sores (Meghvansi et al. 2010). Raw fruits are applied topically to treat spider and caterpillar sting in Manipur, India (Panmei et al. 2019). The fruit infusion is used to treat chronic congestion, digestive disorders, toothaches and muscle pain. It helps muscles recover after strenuous exercise. Additionally, leaves are ground mixed with coconut oil and is used to drain pus from boils. It is also used for combating summer heat (Baruah and Lal 2020). Decoction of leaves and fruits is taken orally in case of high blood pressure, stomach ache and diarrhoea (Jamir et al. 2021). It has also been reported that *C. chinense* pod hot infusion can relieve toothache and muscle discomfort. The leaf paste is applied to treat boils. Small doses of *C. chinense* pods are taken on a regular basis for several gastrointestinal disorders (Rudrapal and Sarwa 2020).

3.3.4 *Capsicum frutescens* L.

The 'Garo' ethnic group of Bangladesh used the dried fruits of the *C. frutescens* to make the fermenting medium for their traditional alcoholic beverage. Dried chilli peppers are used as mosquito repellent in Central-Southern Italy. Fruits and mustard oil are applied to boils of animals. In Tonga, Samoa, Tahiti and the Cook Islands, leaves of the species are crushed with coconut oil used in skin disorders. It is common practice in Agasthyamalai Biosphere Reserve, India, to burn the fruits and inhale for the treatment of cough and cold. *Crinum asiaticum* bulbs with leaves of *Capsicum* are given orally to cure mumps in animals. In Eastern Nicaragua, decoction of leaves and fruits is used for fever, whereas leaves and fruits are applied topically for vaginal infections. Ripe fruits and seeds are ground into a powder, combined with palm oil, and applied to wounds and cuts caused by dog bite in Edo State, Nigeria. In Dek Island, Ethiopia, dry and powdered fruits are taken to relieve stomach ache. A mixture of fruits of *C. frutescens* macerated with the seeds of *Illicium* sp. and the bark of *Cinnamomum verum* is consumed in case of rheumatism and artrosis. The leaf, stem and fruit decoction acts as a diuretic and is helpful in removing kidney stones. In Mizoram (India), the leaf juice of the species is used as ear drops to remove water leech. The fruits of *C. frutescens* blended with *Myristica fragrans* are taken with eggs, milk, and wine to treat impotence and sterility. In Bangangte region, Western Cameroon, the fruits are combined with leaves of *Aspilia africana*, *Emilia coccinea* and *Centella asiatica*. This mixture is known to be useful to cure peptic ulcers. In Bafia region of Cameroon, the fruits consumed with the leaves of *Vigna unguiculata* for the treatment of hypertension. The smoke of fruits and leaves are reported to be effective in treating malaria in Yanesha, Peru (Meghvansi et al. 2010).

In Mauritius, a few drops of the oil-fried chilies added to treat ear infection (Nunkoo and Mahomoodally 2012). The seeds of *C. frutescens*, along with other plant species are given to affected cattle against blackleg (Megersa et al. 2013). Tribals in Kodagu district, Karnataka, India, grind fruits of *C. frutescens* with garlic, ginger, black pepper and salt. They consume it to treat common cold. The fruits of *Capsicum* are ground with salt and charcoal to remove maggots from cattle wounds (Lingaraju et al. 2013). In Vanuatu region of Northern Australia, the leaves rubbed on the body to relieve muscle pain. Additionally, the fruits are rubbed on the body to treat rheumatism and boil (Bradacs et al. 2011). In Tanzania, the root and bark of the plant are crushed and applied topically on wounds (Amri and Kisangau 2012). In Granada province region of Southern Spain, macerated fruits are used topically for treating rheumatism (Benitez et al. 2010). The dried chilies' seeds are taken out, and then the oil is used to make a mixture. One to two drops of this mixture are used for earache in Mauritius (Sreekeesoon and Mahomoodally 2014). The soup made from the fruit of *C. frutescens* is used to relieve childbirth-related uterine pain. It is also effective in muscle spasms (Ali 2015). In West Africa, a decoction of the fruit in water is reported to be used for the treatment of breast cancer (Titilayo and Morenikeji 2016).

The Gbaya ethnic group in Eastern Cameroon commonly used macerated seeds and roots for tooth hygiene (Fongnzossie et al. 2017). Infusions, decoctions and macerated leaves and fruits are used to cure general infection, tetanus, cancer, skin infection, inflammation and vaginal discharge in Brazil. The leaves and fruits are also applied as poultice for wound healing and treating injury with swelling (Ribeiroa et al. 2017). In Ethiopia, the fresh or dried fruit juice is consumed for treating malaria (Suleman et al. 2018). Infusion of leaves and fresh or dried fruits is reported to be used for the treatment of haemorrhoids (Tchicaillat-Landou et al. 2018; Vinagre et al. 2019). The raw seeds are consumed in case of diabetes in Algeria (Hamza et al. 2019). The fruit cover of *C. frutescens* is shed dried, powdered and filtered. A capsule is made from this capsule and 1 capsule is taken per day for 10 days. Generally, it is blended with the stem bark of *Spathodea campanulata* ssp. *nilotica*, *Juniperus procera*; the leaves of *Microglossa pyrifolia* and the root and leaves of *Carica papaya* are infused and are believed to be effective for the treatment of throat, breast and squamous cell carcinoma in Kenya (Aumeeruddy and Mahomoodally 2021; Omara et al. 2022).

Decoction of the fruit is effective for healing dermal wound. In the north region of Morocco, crushed fruits are inhaled for animal therapy. The remedy is useful in endo- and ectoparasitic issues of sheep (Chaachouay et al. 2022a, b). The fruit and leaf juices are applied over burns and snakebite in Mizoram, India (Ralte et al. 2021). *Capsicum frutescens* is cooked with *Cleome gynandra* leaves and taken as pot soup. Furthermore, the leaves and stem bark of *Dalbergia lacteal* is ground with *C. frutescens* and taken orally to enhance memory and as an anti-ageing component (Ayeni et al. 2022). The fruit cover is powdered and made into capsules. This infusion is used to treat breast, throat and squamous cell carcinoma. A decoction is made with the bulb of *Allium cepa,* and *C. frutescens*, snail water and salt are filtered and taken orally two times a day as a remedy for hypertension (Alves-Silva et al. 2022). In Malaysia, crushed leaves and fruits are used for skin and common ailments in pregnancy (Venmathi Maran et al. 2022). An infusion made up from fruit is taken orally for flu, hypertension and loss of appetite in South Aceh District, Indonesia (Adnan et al. 2022). The fruit of *C. frutescens* is crushed with the bulb of *A. sativum* and cooked and 2–3 spoons of this concoction is taken every day for the treatment of respiratory tract infection, cold and loss of appetite (Megersa and Woldetsadik 2022).

3.3.5 *Capsicum buforum* Hunz.

The fruits are placed over burning charcoal for a smoke bath. This smoke bath is used as a preventive for flu and other diseases (Jarret et al. 2019).

3.3.6 *Capsicum chacoense* Hunz.

In Chaco region of Northeast Argentina, the fruits of *C. chacoense* are milled and applied as poultice for the treatment of snakebite (Martinez and Barboza 2010).

3.4 Pharmacological Validation for Ethnobiomedicinal Uses of *Capsicum*

Different *Capsicum* species are popularly used in the traditional system of medicine in different parts of the world; however, there are few studies in the literature that provide pharmacological support for these claims (Table 3.3). The plant is valued for its anticancerous, antibacterial and anticoagulant properties (Kumar et al. 2011; Kouassi et al. 2012; Shah et al. 2016; Wangchuk et al. 2017). The fruits of *Capsicum* species are also used as an antipruritic in India and a carminative in Palestine, Pakistan and Turkey (Meghvansi et al. 2010; Kumar et al. 2011; Akhtar et al. 2013; Ahmed et al. 2014; Ali 2015; Shah et al. 2016). These fruits have been widely used as a counter-irritant in Palestine and India (Meghvansi et al. 2010; Kumar et al. 2011). The leaves also possess antiparasitic properties (Alonso-Castro et al. 2011). The leaves and fruits of *Capsicum* species are used as antirheumatic, antihaemorrhoidal, diaphoretic, digestive, irritant, rubefacient, sialagogue, antinociceptic/analgesic, antioxidant, antimicrobial and tonic (Lim 2013; Afroz et al. 2020). The seeds and roots of *C. frutescens* possess antioxidative and anti-inflammatory properties (Fongnzossie et al. 2017). *Capsicum frutescens* also show carminative, antispasmodic, diaphoretic, sialagogue, antibiotic and antiseptic effects (Sreekeesoon and Mahomoodally 2014; Afroz et al. 2020). The fruits of *C. annuum* possess antioxidant, antiseptic, antidyspeptic and anti-inflammatory properties (Tagarelli et al. 2010; Ralte et al. 2021). Majority of *Capsicum* fruits have significant antioxidant levels that are principally responsible for their antibacterial, antifungal, antiviral, anti-inflammatory and immunomodulatory activities (Pappalardo et al. 2021). *Capsicum annuum* also possesses antimicrobial, antioxidant, anticancer, anti-inflammatory and antiviral properties (Khan et al. 2014). The fruits and leaves of *C. annuum, C. baccatum* and *C. frutescens* have antimicrobial, antioxidant and anti-inflammatory, antihyperlepidaemic, antiviral, anticancer and analgesic potential (Wangchuk et al. 2017; Teixeira et al. 2019; Afroz et al. 2020; Laldingliani et al. 2022; Talwan et al. 2022). Similarly, the fruits of *C. baccatum* have antioxidant properties, whereas the fruits of *C. frutescens* have anti-inflammatory, antimicrobial and antioxidant potential (Teixeira et al. 2019).

3.4.1 Anti-inflammatory Activity

Zimmer and co-workers (2011) observed anti-inflammatory and antioxidant activities of *C. baccatum*. The carrageenan-induced pleurisy model in mice for

Table 3.3 Pharmacological studies in *Capsicum* species

Species name	Pharmacological activities	References
Capsicum sp.	Anticancerous, antibacterial, anticoagulant	Kumar et al. (2011), Kouassi et al. (2012), Shah et al. (2016), Wangchuk et al. (2017)
	Antipruritic, carminative	Meghvansi et al. (2010), Kumar et al. (2011), Akhtar et al. (2013), Ali (2015), Shah et al. (2016)
	Fruits—counter-irritant	Meghvansi et al. (2010), Kumar et al. (2011)
	Leaves—antiparasitic	Alonso-Castro et al. (2011)
	Leaves and fruits—antirheumatic, antihaemorrhoidal, diaphoretic, digestive, irritant, rubefacient, sialagogue, antinociceptic or analgesic, antioxidant, antimicrobial and tonic	Lim (2013), Afroz et al. (2020)
	Fruits—antioxidant, antibacterial, antifungal, antiviral, anti-inflammatory, and immunomodulatory	Pappalardo et al. (2021)
C. annuum L.	Fruits—antioxidant, antiseptic, antidyspeptic and anti-inflammatory	Tagarelli et al. (2010), Ralte et al. (2021)
	Antidiabetic and antihypertension (in vitro enzyme assays for alpha-glucosidase, alpha-amylase and angiotensin I-converting enzyme [ACE] inhibitory activities)	Idm'hand et al. (2020)
	Antimicrobial, antioxidant, anticancer, anti-inflammatory and antiviral	Khan et al. (2014)
	Fruits and leaves—antimicrobial, antioxidant, and anti-inflammatory, antihyperlipidaemic, antiviral, anticancer and analgesic	Wangchuk et al. (2017), Teixeira et al. (2019), Laldingliani et al. (2022), Talwan et al. (2022)
	Antimicrobial	Afroz et al. (2020)
	Analgesic and anti-inflammatory —Swiss albino mice of (20–25 g) and Wistar albino rats (150–180 g) Analgesic—acetic acid-induced writhing, tail immersion and hot plate tests, carrageenan-induced paw oedema in rats	Mazhar et al. (2021)
	Antioxidant (DPPH and ABTS), anticancer (aqueous extract against prostate [PC-3] and breast [MCF-7] carcinoma cell lines in vitro)	Shaimaa et al. (2016)

(continued)

Table 3.3 (continued)

Species name	Pharmacological activities	References
	Anticancer, ethanolic extract, leaves, cell line NA (neuro-2A) mouse neuroblastoma	Alonso-Castro et al. (2011)
	Antiobesity (aqueous extract, mice preadipocytes)	Narang et al. (2017)
	Antibacterial (methanol and aqueous extract)	Kouassi et al. (2012)
	Rheumatoid arthritis (alcoholic extract, male rats, ibuprofen as a reference)	Boiko et al. (2019)
C. baccatum L. var. *pendulum* (Willd.)	Fruits—antioxidant	Teixeira et al. (2019)
	Antimicrobial	Afroz et al. (2020)
	Antioxidant (DPPH) and anti-inflammatory (using the carrageenan-induced pleurisy model in mice)	Zimmer et al. (2011)
	Anti-inflammatory (2 models, AA-induced ear oedema in mice and carrageenan-induced paw oedema in rats, indomethacin as reference [2 mg/ear])	López et al. (2012)
C. frutescens L.	Seeds and roots—antioxidative, anti-inflammatory	Fongnzossie et al. (2017)
	Fruits—antimicrobial, antioxidant, and anti-inflammatory	Teixeira et al. (2019)
	Carminative, antispasmodic, diaphoretic, sialagogue, antibiotic, antiseptic	Sreekeesoon and Mahomoodally (2014), Afroz et al. (2020)
	Antimicrobial	Afroz et al. (2020)
	Antioxidant (DPPH and ABTS), anticancer (aqueous extract-against prostate [PC-3] and breast [MCF-7] carcinoma cell lines in vitro)	Shaimaa et al. (2016)
	Antidiabetic (male albino rats of Wistar strain [300–350 g] as animal model, glibenclamide as reference), analgesic, antimicrobial, anti-inflammatory, immunological	Otunola and Afolayan (2015)
	Antidiabetic	Ansari et al. (2022)
	Anthelmintic (methanolic extract, levamisole [1 mg/mL] as standard reference, distilled water as control), antioxidative, hypocholesterolaemic, hypolipidaemic, immune-modulatory and anti-mutagenic	Kamal et al. (2015)

(continued)

Table 3.3 (continued)

Species name	Pharmacological activities	References
	Diuretic (white rats)	Maramag (2013)
	Anti-neurodegenerative (in vitro neuronal) activities in rat brain tissue with a lethal dose of 190 mg/kg, using rotenone-intoxicated Wistar male albino rat brain as a model with a lethal dose of 500–5000 mg/kg	Ayeni et al. (2022)
	Anti-Alzheimer (ethanolic extract, in vitro, donepezil as control)	Perumal and Chong (2022)
	Antiulcer (ethanolic extract, albino mice)	Baruah and Lal (2020)
	Antibacterial (methanol and aqueous extract)	Kouassi et al. (2012)
	Antifungal (aqueous)	Soumya and Nair (2012)
C. chacoense Hunz.	Anti-inflammatory (two models, AA-induced ear oedema in mice and carrageenan-induced paw oedema in rats, indomethacin as reference [2 mg/ear])	López et al. (2012)
	Antioxidant (ABTS, TPC), anthelmintic (methanolic extract, Swiss albino mice)	Coronel et al. (2022)
C. chinense	Anticancer—in vitro, against human hepatocellular carcinoma cells, acetonitrile extract	Amruthraj et al. (2014)

anti-inflammatory activity and the DPPH scavenging abilities were reported in ethanolic and butanol extracts of *Capsicum baccatum*.

Polar to non-polar extracts of *C. chacoense* and *C. baccatum* using AA-induced ear oedema in mice and carrageenan-induced paw oedema in rats for inflammation were studied (López et al. (2012). *Capsicum chacoense* extracts showed significant inhibition (46% and 38%), whereas *C. baccatum* caused an inhibition of 52% and 35% in both the assays tested. However, the anti-inflammatory effect using carrageenan-induced paw oedema did not produce results when given orally (López et al. (2012).

3.4.2 Analgesic Activities

The analgesic activity of four different varieties of bell pepper using Swiss albino mice and Wistar albino rats were investigated by Mazhar et al. (2021). The 95%

ethanol extracts and effects were compared with aspirin and other anti-inflammatory drugs. Acetic acid-induced writhing, hot plate tests and tail immersion methods were used to examine the analgesic effect. The results concluded that the fruit extracts of all the four varieties of bell pepper had promising analgesic properties (Mazhar et al. 2021).

3.4.3 Anticancer Potential

Shaimaa et al. (2016) assessed the antioxidant and anticancer properties of aqueous extracts of *Capsicum frutescens* and *C. annuum*. The total antioxidant capacity using DPPH and ABTS scavenging models was measured, whereas the anticancer activity was examined against PC-3 (prostate) and MCF-7 (breast) carcinoma cell lines in vitro. The extracts showed high antioxidant potential ranging from 77.73% to 93.11% scavenging for ABTS and 96.95% to 98.64% scavenging for DPPH. Besides this, chilli pepper had a stronger anticancer activity against the breast cell line, whereas the later species showed higher anticancer activity against the prostate cell line. Furthermore, it is reported that the ethanolic extract of leaves of *C. annuum* is used for cancer treatment. The cell line used for this study is NA (neuro-2A) mouse neuroblastoma (Alonso-Castro et al. 2011).

Amruthraj et al. (2014) from Chennai investigated the anticancer potency of *C. chinense* using human hepatocellular carcinoma cells. The cytotoxicity was confirmed using cell viability MTT assay and enzyme markers such as lactate dehydrogenase, nitric oxide and lipid peroxidation. They reported that the treatment with acetonitrile extracts revealed decrease in cell viability. It also significantly reduced the release of studied enzymes. These findings clearly indicated the potential of *C. chinense* as an anticancer agent (Amruthraj et al. 2014).

3.4.4 Antidiabetic Activity

In Kwon et al. (2007), phenolic phytochemicals from nine pepper varieties were evaluated for in vitro enzyme assays like alpha-amylase, alpha-glucosidase and angiotensin I-converting enzyme (ACE). Fruit extracts prepared in water showed high a-glucosidase inhibitory activity. Out of the tested long hot and green peppers showed poor alpha-amylase activity (Idm'hand et al. 2020).

Another study was conducted on the hypoglycaemic activity of a water extract of ginger, garlic and pepper fruits (*C. frutescens*) using Male albino rats of Wistar strain animal models. The results demonstrated in alloxan-induced diabetes; the water extracts had strong ability against hypoglycaemic conditions. Also, these extracts possess potential as organ-protective and hypolipidaemic. Additionally, orally administrated aqueous extracts not only altered organ weights, decreased hyperglycaemia and attenuated blood and cellular toxicity measures, but also was effective in restoring pancreatic and kidney damage. They further reported that the fruits of *C. frutescens* are used as an analgesic with reports of anti-inflammatory, antimicrobial and immunological effects (Otunola and Afolayan 2015). It has been studied that *C. frutescens* (whole plant) increases insulin secretion, lowers blood glucose, improves tolerance against glucose and inhibits pro-inflammatory cytokines (Ansari et al. 2022).

3.4.5 Anthelmintic Activity

Kamal and co-workers (2015) evaluated the anthelmintic activity of *C. frutescens* fruit-methanolic extract on aquarium worm (*Tubifex tubifex*). Levamisole was used as standard reference and distilled water as control. The results showed considerable anthelmintic action at higher concentration (10 mg/mL). Other activities such as hypocholesterolaemic, hypolipidaemic, antimutagenic, immune-modulatory and antioxidative potential were also studied (Kamal et al. 2015).

In another study, Coronel et al. (2022) observed the nutritional composition, anthelmintic activity, antioxidant activity and acute toxicity of *C. chacoense* fruit-methanolic extracts. The findings showed that methanolic extracts of *C. chacoense* exhibited greater in vitro activity and anthelmintic potential than the albendazole standard. Additionally, these extracts also showed no signs of acute toxicity under test conditions and they were nontoxic up to a dose of 2000 mg/kg (Coronel et al. 2022).

3.4.6 Diuretic Potential

A separate study was conducted to determine the diuretic potential of *C. frutescens* decoction on albino rats using the rat metabolic cage method. To assess their potency as a diuretic, measurements of urine volume excreted after 5 h were taken. It was found that albino rats produced more urine when exposed to a concentration of 10 mL/kg. In the result, decoction was comparable to furosemide which is a commercial diuretic in inducing urine. The findings clearly indicated the diuretic potential of *C. frutescens* (Maramag 2013).

3.4.7 Anti-neurodegenerative Potential

The ripe and unripe fruits of *C. frutescens* are studied for in vitro neuronal activities in rat brain tissue with a lethal dose of 190 mg/kg. Furthermore, the fruits of *C. annuum* are also studied by using rotenone-intoxicated Wistar male albino rat brain as a model with a lethal dose of 500–5000 mg/kg. Both plants confirmed potent anti-neurodegenerative medicinal potential in West Africa (Ayeni et al. 2022).

The acetylcholinesterase (AChE) and beta-secretase enzyme inhibition abilities of ethanolic extract of *Capsicum frutescens* (EECF) investigated in vitro methods by QuantiChrom™ and BACE-1 FRET assay kits. In the result, there was a considerable rise in the percentage of inhibition against acetylcholinesterase and BACE-1 when evaluated with an increasing concentration of EECF. Donepezil showed greater IC50 values against both of the tested major enzymes than EECF, but it also showed essential inhibitory activities in both experiments. Additionally, EECF showed greater effectiveness in preventing the progression of Alzheimer disease via inhibiting acetylcholinesterase. The study revealed greater anti-Alzheimer effects of EECF were seen at concentrations of 1 and 2 mg/mL through inhibition of AChE and BACE-1 (Perumal and Chong 2022).

3.4.8 Antiulcer Activity

Baruah and Lal (2020) reported the antiulcer potential of *C. frutescens* ethanolic extracts. They used albino mice for this experiment. According to the study, the plant's ethanolic extract decreased stomach ulcers by lowering pepsin levels and raising mucus secretion (Baruah and Lal 2020).

3.4.9 Antiobesity Activity

The aqueous extracts of seven types of *C. annuum* studied by Korean researchers Baek et al. for antiobesity activities showed that Oyee gochu, green, red and yellow paprika tend to rise lipoprotein lipase levels, whereas Kwari putgochu, Putgochu and Cheongyang gochu reduced lipoprotein lipase levels (Narang et al. 2017).

3.4.10 Antimicrobial Activity

A study was conducted by Kouassi et al. (2012) to determine the efficacy of *C. annuum* L. and *C. frutescens* fruit extracts for their antibacterial activities using methanolic and aqueous extracts. Comparing the two, the methanolic extract of *C. annuum* exhibited higher antibacterial effect. The minimal inhibitory concentration (MIC) results confirmed the antibacterial potential of both extracts.

A study was conducted by Soumya and Nair (2012) to assess the antifungal activity of the leaf and fruit of *Capsicum frutescens* aqueous extracts against fungus strains with respect to groundnut storage. Extracts were tested to determine their MIC and minimum fungicidal concentrations (MFCs). The results revealed that in the well diffusion method, the leaf extract demonstrated a strong activity against *Aspergillus flavus* at MIC (88.06%), whereas the fruit extract demonstrated a strong activity against *A. niger* (88.33%). A greater incidence of fungal inhibition was seen in groundnut seeds treated with the fruit extract from *C. frutescens* (10 mg/mL). According to the findings, *Arachis hypogea* seeds treated with fruit extracts of *C. frutescens* have potential in preventing fungal infection.

3.4.11 Arthritis Treatment

Boiko et al. (2019) from Ukraine investigated the efficacy of alcoholic extract of *C. annuum* in the treatment of arthritis. Changes in the morphological markers of inflammation, levels of alpha-1-acid glycoprotein in serum, differential white blood cell counts and acetylcholinesterase activity were observed. The alcoholic *C. annuum* extracts used to lessen the intensity of non-specific inflammation resulted in reduction in oedema by 50–60%. Additionally, the development of autoimmune inflammation was prevented by the administration of alcoholic extracts of *C. annuum*, confirming its efficacy in the treatment of rheumatoid arthritis.

3.5 Conclusion and Future Prospects

Thus, the chapter herein presents various ethnobotanical and pharmacological claims from different researchers all over the world. In the current scenario of scientific study, it is very important to have a comprehensive and collective evaluation of such important genus. The study clearly indicated the ethnomedicinal role of this important genus *Capsicum* in the treatment of a wide array of ailments. The scientific literature justifies its pharmacological actions. This chapter is suggestive of shortcomings, that is, few works suggest the presence or absence of capsaicin in *C. annuum*. The ambiguity should be definitely due to pungency levels in the different varieties tested by the workers using different extraction and detection methods. It is important to note that the species studied together along with single extraction and detection methods is most essential to obtain an exact view. In conclusion, the authors want to draw attention towards other *Capsicum* species (Table 3.4) which are yet unexplored to the potential of ethnobotanical or pharmacological purview. The chapter reveals that there are around 31 such species in which pharmacological studies were not claimed and 25 species in which ethnobotanical claims were sparse. There is no doubt on the ethnomedicinal utility of *Capsicum* worldwide as a first line of treatment in the primary healthcare system. This traditional knowledge combined with modern scientific explorations and validation will not only provide a better understanding of the genus but also will broaden avenues to pharma industries. Thus, a cumulative big data needs to be generated on this important and widely used genus.

Table 3.4 List of *Capsicum* species with less or no references available

Sl. no.	Name of species
1.	*C. pubescens* Ruiz & Pav.
2.	*C. buforum* Hunz.
3.	*C. caballeroi* M. Nee
4.	*C. campylopodium* Sendtn.
5.	*C. cardenasii* Heiser & Smith
6.	*C. ceratocalyx* M. Nee
7.	*C. coccineum* (Rusby) Hunz.
8.	*C. cornutum* (Hiern) Hunz.
9.	*C. dimorphum* (Miers) Kuntze
10.	*C. dusenii* Bitter
11.	*C. eshbaughii* Barboza
12.	*C. eximium* Hunz.
13.	*C. flexuosum* Sendtn.
14.	*C. friburgense* Bianchetti & Barboza
15.	*C. galapagoense* Hunz.
16.	*C. geminifolium* (Dammer) Hunz.
17.	*C. hookerianum* (Miers) Kuntze
18.	*C. hunzikerianum* Bianchetti & Barboza
19.	*C. lanceolatum* (Greenm.) Morton & Standl.
20.	*C. leptopodum* (Dunal) Kuntze
21.	*C. minutiflorum* (Rusby) Hunz.
22.	*C. mirabile* Mart.
23.	*C. parvifolium* Sendtn.
24.	*C. pereirae* Bianchetti & Barboza
25.	*C. praetermissum* Heiser & Smith
26.	*C. recurvatum* Witasek
27.	*C. rhomboideum* (Dunal) Kuntze
28.	*C. schottianum* Sendtn.
29.	*C. scolnikianum* Hunz.
30.	*C. tovarii* Eshbaugh et al.
31.	*C. villosum* Sendtn.

Acknowledgments The authors are indebted to the head of the department and institution for the support during the work.

References

Adetunji TL, Olawale F, Olisah C, Adetunji AE, Aremu AO (2022) Capsaicin: a two-decade systematic review of global research output and recent advances against human cancer. Front Oncol 12:908487

Adeyemi AA, Gbolade AA, Moody JO, Ogbole OO, Fasanya MT (2010) Traditional anti-fever phytotherapies in Sagamu and Remo North Districts in Ogun state, Nigeria. J Herbs Spices Med Plants 16(3–4):203–218

Adnan A, Navia ZI, Silvia M, Antika M, Suwardi AB, Baihaqi B, Yakob M (2022) Diversity of herbs and spices plants and their importance in traditional medicine in the South Aceh District, Indonesia. Biodiversitas 23(7):3836–3843

Afroz M, Akter S, Ahmed A, Rouf R, Shilpi JA, Tiralongo E, Sarker SD, Goransson U, Uddin SJ (2020) Ethnobotany and anti-microbial peptides from plants of the solanaceae family: an update and future prospects. Front Pharmacol 11:565. https://doi.org/10.3389/fphar.2020.00565

Ahmed N, Mahmood A, Mahmood A, Tahir SS, Bano A, Malik RN, Hassan S, Ishtiaq M (2014) Relative importance of indigenous medicinal plants from Layyah district, Punjab Province, Pakistan. J Ethnopharmacol 155(1):509–523. https://doi.org/10.1016/j.jep.2014.05.052

Ajose DJ, Oluwarinde BO, Abolarinwa TO, Montso KP, Fri J, Aremu AO, Ateba CN (2022) Combating bovine mastitis in the dairy sector in an era of anti-microbial resistance: ethnoveterinary medicinal option as a viable alternative approach. Front Vet Sci 9:800322. https://doi.org/10.3389/fvets.2022.800322

Akbulut S, Bayramoglu MM (2013) The trade and use of some medical and aromatic herbs in Turkey. Stud Ethno-Med 7(2):67–77. https://doi.org/10.1080/09735070.2013.11886446

Akhtar N, Rashid A, Murad W (2013) Diversity and use of ethno-medicinal plants in the region of Swat, North Pakistan. J Ethnobiol Ethnomed 9:25

Ali EA (2015) The pharmacological importance of *Capsicum* species (*Capsicum annuum* and *Capsicum frutescens*) grown in Iraq. J Pharm Biol 5(3):124–142

Alonso-Castro AJ, Villarreal ML, Salazar-Olivo LA, Gomez-Sanchez M, Dominguez F, Garcia-Carranca A (2011) Mexican medicinal plants used for cancer treatment: pharmacological, phytochemical and ethnobotanical studies. J Ethnopharmacol 133(3):945–972. https://doi.org/10.1016/j.jep.2010.11.055

Alves-Silva JM, Zuzarte M, Girao H, Salgueiro L (2022) Natural products in cardiovascular diseases: the potential of plants from the Allioideae subfamily (Ex-Alliaceae Family) and their sulphur-containing compounds. Plants 11(15):1920

Amri E, Kisangau DP (2012) Ethnomedicinal study of plants used in villages around Kimboza forest reserve in Morogoro, Tanzania. J Ethnobiol Ethnomed 8:1. https://doi.org/10.1186/1746-4269-8-1

Amruthraj NJ, Raj JP, Lebel LA (2013) Polar aprotic extraction of capsaicinoids from *Capsicum chinense* Bhut jolokia fruit for anti-microbial activity. Int J Biol Pharm Res 4(12):959–964

Amruthraj NJ, Preetam Raj JP, Saravanan S, Lebel LA (2014) In vitro studies on anti-cancer activity of capsaicinoids from *Capsicum Chinense* against human hepatocellular carcinoma cells. Int J Pharm Pharm Sci 6(4):254–558

Andel TV, Onselen SV, Myren B, Towns A, Quiroz D (2015) "The medicine from behind": the frequent use of enemas in western African traditional medicine. J Ethnopharmacol 174:637–643. https://doi.org/10.1016/j.jep.2015.06.040

Ansari P, Akther S, Hannan JMA, Seidel V, Nujat NJ, Abdel-Wahab YH (2022) Pharmacologically active phytomolecules isolated from traditional anti-diabetic plants and their therapeutic role for the management of diabetes mellitus. Molecules 27(13):4278. https://doi.org/10.3390/molecules27134278

Anupriya KT, Danya U (2022) A survey of native medicinal plants in the Kandaripadam, Thenhipalam Village, Malappuram district, Kerala. Ymer 21(5):1265–1276

Aumeeruddy MZ, Mahomoodally MF (2021) Global documentation of traditionally used medicinal plants in cancer management: a systematic review. South Afr J Bot 138:424–494. https://doi.org/10.1016/j.sajb.2021.01.006

Ayeni EA, Gong Y, Yuan H, Hu Y, Bai X, Liao X (2022) Medicinal plants for anti-neurodegenerative diseases in West Africa. J Ethnopharmacol 285:114468. https://doi.org/10.1016/j.jep.2021.114468

Azliza MA, Ong HC, Vikineswary S, Noorlidah A, Haron NW (2012) Ethno-medicinal resources used by the Temuan in Ulu Kuang village. Stud Ethno-Med 6(1):17–22. https://doi.org/10.1080/09735070.2012.11886415

Balkrishna A, Nag S, Dabas A, Arya V (2022) Analgesic potential of medicinal plants used by the tribes of Western Himalayan region of India: a systematic review. Med Plants-Int J Phytomed Relat Ind 14(3):382–404. https://doi.org/10.5958/0975-6892.2022.00043.0

Barboza GE, García CC, Bianchetti LB, Romero MV, Marisel SM (2022) Monograph of wild and cultivated chili peppers (*Capsicum* L., Solanaceae). PhytoKeys 200:1–423. https://doi.org/10.3897/phytokeys.200.71667

Baruah J, Lal M (2020) *Capsicum chinense* Jacq.: ethnobotany, bioactivity and future prospects. In: Botanical leads for drug discovery. Springer, Berlin, pp 349–362. https://doi.org/10.1007/978-981-15-5917-4_16

Batsatsashvili K, Bussmann RW, Kikvidze Z, Paniagua-Zambrana NY, Khutsishvili M, Maisaia I, Sikharulidze S, Tchelidze D (2019) *Capsicum annuum* L. Solanaceae. In: Ethnobotany of the mountain regions of far Eastern Europe. Springer, Berlin. https://doi.org/10.1007/978-3-319-77088-8_35-2

Benitez G, Gonzalez-Tejero MR, Molero-Mesa J (2010) Pharmaceutical ethnobotany in the western part of Granada province (southern Spain): ethnopharmacological synthesis. J Ethnopharmacol 129:87–105. https://doi.org/10.1016/j.jep.2010.02.016

Bieski IG, Leonti M, Arnason JT, Ferrier J, Rapinski M, Violante IMP, Martins DT (2015) Ethnobotanical study of medicinal plants by population of valley of Juruena region, legal Amazon, Mato Grosso, Brazil. J Ethnopharmacol 173:383–423. https://doi.org/10.1016/j.jep.2015.07.025

Boiko YA, Shandra AA, Boiko IA, Kravchenko IA (2019) Experimental study of the effectiveness the *Capsicum annuum* L. extracts for treatment of the rheumatoid arthritis. J Phytopharmacol 8(2):46–51

Bradacs G, Heilmann J, Weckerle CS (2011) Medicinal plant use in Vanuatu: a comparative ethnobotanical study of three islands. J Ethnopharmacol 137:434–448. https://doi.org/10.1016/j.jep.2011.05.050

Caamal-Fuentes E, Torres-Tapia LW, Simá-Polanco P, Peraza-Sánchez SR, Moo-Puc R (2011) Screening of plants used in Mayan traditional medicine to treat cancer-like symptoms. J Ethnopharmacol 135(3):719–724. https://doi.org/10.1016/j.jep.2011.04.004

Catarino L, Havik PJ, Romeirasa MM (2016) Medicinal plants of Guinea-Bissau: therapeutic applications, ethnic diversity and knowledge transfer. J Ethnopharmacol 183:71–94. https://doi.org/10.1016/j.jep.2016.02.032

Chaachouay N, Azeroual A, Bencharki B, Douira A, Zidane L (2022a) Ethnoveterinary medicines plants for animal therapy in the Rif, north of Morocco. South Afr J Bot 147:176–191. https://doi.org/10.1016/j.sajb.2021.12.037

Chaachouay N, Benkhnigue O, Zidane L (2022b) Ethnobotanical and ethnomedicinal study of medicinal and aromatic plants used against dermatological diseases by the people of Rif, Morocco. J Herb Med 32:100542. https://doi.org/10.1016/j.hermed.2022.100542

Chander MP, Kartick C, Gangadhar J, Vijayachari P (2014) Ethno medicine and healthcare practices among Nicobarese of Car Nicobar—an indigenous tribe of Andaman and Nicobar Islands. J Ethnopharmacol 158(Part A):18–24. https://doi.org/10.1016/j.jep.2014.09.046

Chekole G (2017) Ethnobotanical study of medicinal plants used against human ailments in Gubalafto District, Northern Ethiopia. J Ethnobiol Ethnomed 13:55. https://doi.org/10.1186/s13002-017-0182-7

Coronel E, Mereles L, Caballero S, Alvarenga N (2022) Crushed *Capsicum chacoense* Hunz Fruits: a food native resource of Paraguay with anti-oxidant and anthelmintic activity. Int J Food Sci 2022:1512505. https://doi.org/10.1155/2022/1512505

Dapar MLG, Alejandro GJD, Meve U (2020) Quantitative ethnopharmacological documentation and molecular confirmation of medicinal plants used by the Manobo tribe of Agusan del Sur, Philippines. J Ethnobiol Ethnomed 16:14. https://doi.org/10.1186/s13002-020-00363-7

Dutt HC, Bhagat N, Pandita S (2015) Oral traditional knowledge on medicinal plants in jeopardy among Gaddi shepherds in hills of northwestern Himalaya, J&K, India. J Ethnopharmacol 168:337–348. https://doi.org/10.1016/j.jep.2015.03.076

Edgar CF, Luis WT, Paulino SP, Sergio RP, Rosa MP (2011) Screening of plants used in Mayan traditional medicine to treat cancer-like symptoms. J Ethnopharmacol 135:719–724. https://doi.org/10.1016/j.jep.2011.04.004

Elkhedir AE, Iqbal A, Zogona D, Mohammed HH, Murtaza A, Xu X (2022) Apigenin glycosides from green pepper enhance longevity and stress resistance in *Caenorhabditis elegans*. Nutr Res 102:23–34. https://doi.org/10.1016/j.nutres.2022.02.003

Fan Y, Cheng Z, Zhang Q, Xiong Y, Li B, Lu X, He L, Jiang X, Tan Q, Long C (2022) Meizi-consuming culture that fostered the sustainable use of plum resources in Dali of China: an ethnobotanical study. Biology (Basel) 11:832. https://doi.org/10.3390/biology11060832

Fattori V, Hohmann MS, Rossaneis AC, Pinho-Ribeiro FA, Verri WA (2016) Capsaicin: current understanding of its mechanisms and therapy of pain and other pre-clinical and clinical uses. Molecules 21(7):844. https://doi.org/10.3390/molecules21070844

Fongnzossie EF, Tize Z, Fogang NPJ, Nyangono BCF, Bouelet NIS, Dibong SD, Nkongmeneck BA (2017) Ethnobotany and pharmacognostic perspective of plant species used as traditional cosmetics and cosmeceuticals among the Gbaya ethnic group in Eastern Cameroon. South Afr J Bot 112:29–39. https://doi.org/10.1016/j.sajb.2017.05.013

Gairola S, Sharma J, Bedi YS (2014) A cross-cultural analysis of Jammu, Kashmir and Ladakh (India) medicinal plant use. J Ethnopharmacol 155(2):925–986. https://doi.org/10.1016/j.jep.2014.06.029

Gamage DG, Dharmadasa RM, Abeysinghe CD, Saman RG, Prathapasinghe GA, Someya T (2022) Global perspective of plant-based cosmetic industry and possible contribution of Sri Lanka to the development of herbal cosmetics. Evid Based Complement Alternat Med 2022:9940548. 26 pages. https://doi.org/10.1155/2022/9940548

García CC, Barfuss MHJ, Sehr EM, Barboza GE, Samuel R, Moscone EA, Ehrendorfer F (2016) Phylogenetic relationships, diversification and expansion of chili peppers (Capsicum, Solanaceae). Ann Bot 118(1):35–51. https://doi.org/10.1093/aob/mcw079

Gbogbo M, Agbo EA, Loukou AL, Brou K, Yapo PA (2022) Evaluation of appetite-stimulating properties of *Solanum torvum* berries in albino Wistar rats. Int J Biosci 20(2):39–46. http://www.innspub.net

Giovanninia P, Howesc MR, Edwards SE (2016) Medicinal plants used in the traditional management of diabetes and its sequelae in Central America: a review. J Ethnopharmacol 184:58–71. https://doi.org/10.1016/j.jep.2016.02.034

Girardi C, Butaud JF, Ollier C, Ingert N, Weniger B, Raharivelomanana P, Moretti C (2015) Herbal medicine in the Marquesas Islands. J Ethnopharmacol 161:200–213. https://doi.org/10.1016/j.jep.2014.09.045

Gobvu V, Pote W, Poshiwa X, Benhura MA (2022) Medicinal plants used for the treatment of poultry diseases in Zimbabwe: a systematic review. Res Square. https://doi.org/10.21203/rs.3.rs-1646051/v1

Gogoi B (2017) *Capsicum chinense* Jacq. (Bhut Jolokia)—rich source of capsaicin with wide application and economic potential. Ann Plant Sci 6:1664–1667. https://doi.org/10.21746/aps.2017.08.003

Goraya K, Iqbal Z, Sajid M, Muhammad G, Ain QU, Saleem M (2013) Diversity of flora used for the cure of equine diseases in selected peri-urban areas of Punjab, Pakistan. J Ethnobiol Ethnomed 9(1):70. https://doi.org/10.1186/1746-4269-9-70

Guzma'n FA, Moore S, Vicente MC, Jahn MM (2019) Microsatellites to enhance characterization, conservation and breeding value of *Capsicum* germplasm. In: Genetic resources and crop evolution, vol 67. Springer, Berlin, p 569. https://doi.org/10.1007/s10722-019-00801-w

Hamza N, Berke B, Umar A, Cheze C, Gind H, Moore N (2019) A review of Algerian medicinal plants used in the treatment of diabetes. J Ethnopharmacol 238:111841. https://doi.org/10.1016/j.jep.2019.111841

Henri H, Sari DP, Hakim L (2022) Medicinal plants for traditional treatment used by the Malays in South Bangka Regency, Indonesia. Biosaintifika 14(1):125–134

Idm'hand E, Msanda F, Cherifi K (2020) Ethnopharmacological review of medicinal plants used to manage diabetes in Morocco. Clin Phytosci 6:18. https://doi.org/10.1186/s40816-020-00166-z

Jamir K, Seshagirirao K, Meitei MD (2021) Indigenous oral knowledge of wild medicinal plants from the Peren district of Nagaland, India in the Indo Burma hot-spot. Acta Ecol Sinica 42(3): 206–223. https://doi.org/10.1016/j.chnaes.2021.04.001

Jarret RL, Barboza GE, Batista FRC, Berke T, Chou YY, Kemp AH, Alejo NO, Tripodi P, Veres A, Garcia CC, Csillery G, Huang YK, Kiss E, Kovacs Z, Kondrak M, Arce-Rodriguez ML, Scaldaferro MA, Szoke A (2019) *Capsicum*—an abbreviated compendium. J Am Soc Hort Sci 144(1):3–22. https://doi.org/10.21273/JASHS04446-18

Kamal AM, Chowdhury KA, Chy MM, Shill LK, Chowdhury SR, Chy MA, Habib MZ (2015) Evaluation of anthelmintic activity of seeds of *Sesamum indicum* L. and fruits of *Capsicum frutescens* L. J Pharmacogn Phytochem 3:256–259

Kanteh SM, Norman JE (2015) Diversity of plants with pesticidal and medicinal properties in Southern Sierra Leone. Biol Agri Hortic 31(1):18–27. https://doi.org/10.1080/01448765.2014.945621

Kefalew A, Asfaw Z, Kelbessa E (2015) Ethnobotany of medicinal plants in Ada'a District, East Shewa zone of Oromia regional state, Ethiopia. J Ethnobiol Ethnomed 11:25. https://doi.org/10.1186/s13002-015-0014-6

Khan FA, Mahmood T, Ali M, Saeed A, Maalik A (2014) Pharmacological importance of an ethnobotanical plant: *Capsicum annuum* L. Nat Prod Res 28(16):1267–1274

Khomsi EM, Dandani Y, Chaachouay N, Hmouni D (2022) Ethnobotanical study of plants used for medicinal, cosmetic, and food purposes in the region of Moulay Yacoub, Northeast of Morocco. J Pharm Pharmacogn Res 10(1):13–29

Kim H, Song MJ (2011) Analysis and recordings of orally transmitted knowledge about medicinal plants in the southern mountainous region of Korea. J Ethnopharmacol 134:676–696. https://doi.org/10.1016/j.jep.2011.01.024

Koch M, Kehop DA, Kinminja B (2015) An ethnobotanical survey of medicinal plants used in the East Sepik province of Papua New Guinea. J Ethnobiol Ethnomed 11:79. https://doi.org/10.1186/s13002-015-0065-8

Kolosova V, Pashkova T, Muslimov M, Soukand R (2022) Historical review of ethnopharmacology in Karelia (1850s–2020s): herbs and healers. J Ethnopharmacol 282:114565. https://doi.org/10.1016/j.jep.2021.114565

Kouassi CK, Koffi-Nevry R, Guillaume LY, Yesse ZN, Koussemon M, Kablan T, Athanase KK (2012) Profiles of bioactive compounds of some pepper fruit (*Capsicum* L.) varieties grown in Cote d'Ivoire. Innov Rom Food Biotechnol 11:23–31

Kraft KH, Brown CH, Nabhan GP, Luedeling E, Ruiz JDL, d'Eeckenbrugge GC, Hijmans RJ, Gepts P (2014) Multiple lines of evidence for the origin of domesticated chili pepper, Capsicum annuum, in Mexico. Proc Natl Acad Sci 11(17):6165–6170. https://doi.org/10.1073/pnas.1308933111

Kumar S, Jawaid T, Dubey SD (2011) Therapeutic plants of Ayurveda; a review on anti-cancer. Pharmacogn J 3(23):1. https://doi.org/10.5530/pj.2011.23.1

Kwon YI, Apostolidis E, Shetty K (2007) Evaluation of pepper (*Capsicum annuum*) for management of diabetes and hypertension. J Food Biochem 31(3):370–385. https://doi.org/10.1111/j.1745-4514.2007.00120

Laishram J, Dey M (2021) Evaluation of plants used in traditional medicine in Bishnupur district of Manipur, Northeast India. J Himal Ecol Sustain Dev 16. ISSN 0973-7502

Laldingliani TBC, Thangjam NM, Zomuanawma R (2022) Ethnomedicinal study of medicinal plants used by Mizo tribes in Champhai district of Mizoram, India. J Ethnobiol Ethnomed 18:22. https://doi.org/10.1186/s13002-022-00520-0

Lautenschlager T, Monizi M, Pedro M (2018) First large-scale ethnobotanical survey in the province of Uige, northern Angola. J Ethnobiol Ethnomed 14:51. https://doi.org/10.1186/s13002-018-0238-3

Lim TK (2013) *Capsicum baccatum* var. *pendulum*. In: Edible medicinal and non-medicinal plants. Plants: vol 6, fruits. Springer, Cham. https://doi.org/10.1007/978-94-007-5628-1_29

Lingaraju DP, Sudarshana MS, Rajashekar N (2013) Ethnopharmacological survey of traditional medicinal plants in tribal areas of Kodagu district, Karnataka, India. J Pharm Res 6(2):284–297. https://doi.org/10.1016/j.jopr.2013.02.012

López P, Susana G, Cristina A, Rosario A, Graciela F (2012) Chemical study and anti-inflammatory activity of *Capsicum chacoense* and *C. baccatum*. Rev Bras Farmacogn Braz J Pharmacogn 22(2):455–458

Maqsood T, Munawar T, Bibi Y, Askary EA, Gharib A, Elmissbah T, Elesawy B, Qayyum A (2022) Study of plant resources with ethnomedicinal relevance from district Bagh, Azad Jammu and Kashmir, Pakistan. Open Chem 20(1):146–162. https://doi.org/10.1515/chem-2022-0129

Maramag RP (2013) Diuretic potential of *Capsicum frutescens* Linn. *Corchorus oliturius* Linn. and *Abelmoschus esculentus* Linn. Asian J Nat Appl Sci 2(1):60–69

Maroyi A (2022) Traditional uses of wild and tended plants in maintaining ecosystem services in agricultural landscapes of the eastern Cape Province in South Africa. J Ethnobiol Ethnomed 18(1):1–20. https://doi.org/10.1186/s13002-022-00512-0

Martinez GJ, Barboza GE (2010) Natural pharmacopoeia used in traditional Toba medicine for the treatment of parasitosis and skin disorders (Central Chaco, Argentina). J Ethnopharmacol 132(1):86–100. https://doi.org/10.1016/j.jep.2010.07.049

Martinez GJ, Lujan MC (2011) Medicinal plants used for traditional veterinary in the Sierras de Cordoba (Argentina): an ethnobotanical comparison with human medicinal uses. J Ethnobiol Ethnomed 7:23. https://doi.org/10.1186/1746-4269-7-23

Matejic JS, Stefanovic N, Ivkovic M, Zivanovic N, Marin PD, Dzamic AM (2020) Traditional uses of autochthonous medicinal and ritual plants and other remedies for health in Eastern and South-Eastern Serbia. J Ethnopharmacol 261:113186. https://doi.org/10.1016/j.jep.2020.113186

Mautone M, De Martino L, De Feo V (2019) Ethnobotanical research in cava de' Tirreni area, Southern Italy. J Ethnobiol Ethnomed 15:50. https://doi.org/10.1186/s13002-019-0330-3

Mazhar N, Baig SG, Ahmed S, Hasan MM, Palla A, Ishrat G (2021) Analgesic and anti-inflammatory potential of four varieties of bell pepper (*Capsicum annuum* L.) in rodents. Pak J Pharm Sci 34(4):1369–1376

Mboni HM, Keymeulen F, Ngezahayo J, Amuri SB, Mutombo EK, Byanga JK, Duez P, Stevigny C, Simbi JL (2020) Anti-malarial herbal remedies of Bukavu and Uvira areas in DR Congo: an ethnobotanical survey. J Ethnopharmacol 249:112422. https://doi.org/10.1016/j.jep.2019.112422

McCoy JE, McHale LK, Kantar M, Barbolla LJ, Mercer KL (2022) Environment of origin and domestication affect morphological, physiological, and agronomic response to water deficit in Chile pepper (Capsicum sp.). PLoS One 17(6):e0260684. https://doi.org/10.1371/journal.pone.0260684

Megersa M, Woldetsadik S (2022) Ethnobotanical study of medicinal plants used by local communities of Damot Woyde district, Wolaita zone, Southern Ethiopia. Nusantara Biosci 14(1):10–24. https://doi.org/10.13057/nusbiosci/n140102

Megersa M, Asfaw Z, Kelbessa E, Beyene A, Woldeab B (2013) An ethnobotanical study of medicinal plants in Wayu Tuka District, East Welega Zone of Oromia Regional State, West Ethiopia. J Ethnobiol Ethnomed 9(1):68. https://doi.org/10.1186/1746-4269-9-68

Meghvansi MK, Siddiqui S, Khan MH, Gupta VK, Vairale MG, Gogoi HK, Singh L (2010) Naga chilli: a potential source of capsaicinoids with broad-spectrum ethnopharmacological applications. J Ethnopharmacol 132(1):1–14. https://doi.org/10.1016/j.jep.2010.08.034

Mena E, Warade SD, Ansari MT, Ramjan MD (2018) Evaluation of capsaicin, ascorbic acid, α-carotene and ß-carotene in Bhut Jolokia (*Capsicum chinense* Jacq.) genotypes from North East India. Pharma Innov J 7(7):93–97

Menale B, Muoio R (2014) Use of medicinal plants in the south-eastern area of the Partenio Regional Park (Campania, Southern Italy). J Ethnopharmacol 153(1):297–307. https://doi.org/10.1016/j.jep.2014.02.039

Million E, Mulugeta T, Umeta B (2022) Traditional medicine practice and its role in the management of malaria in Jimma town, Oromia, Ethiopia. Infect Drug Resist 15:2187–2198. https://doi.org/10.2147/IDR.S339782

Mir TA, Jan M, Khare RK (2021) Ethnomedicinal application of plants in Doodhganga forest range of district Budgam, Jammu and Kashmir, India. Eur J Integr Med 46:101366. https://doi.org/10.1016/j.eujim.2021.101366

Mogale MP, Raimondo DC, Wyk V (2019) The ethnobotany of Central Sekhukhuneland, South Africa. South Afr J Bot 122:90–119. https://doi.org/10.1016/j.sajb.2019.01.001

Monari S, Ferri M, Salinitro M, Tassoni A (2022) Ethnobotanical review and dataset compiling on wild and cultivated plants traditionally used as medicinal remedies in Italy. Plants 11(15):2041. https://doi.org/10.3390/plants11152041

Mustafa B, Hajdari A, Pulaj B, Quave CL, Pieroni A (2020) Medical and food ethnobotany among Albanians and Serbs living in the Shterpce/Strpce area, South Kosovo. J Herb Med 22:100344. https://doi.org/10.1016/j.hermed.2020.100344

Narang N, Jiraungkoorskul W, Jamrus P (2017) Current understanding of anti-obesity property of capsaicin. Pharmacogn Rev 11:23–26. https://doi.org/10.4103/phrev.phrev_48_16

Nunkoo DH, Mahomoodally MF (2012) Ethnopharmacological survey of native remedies commonly used against infectious diseases in the tropical Island of Mauritius. J Ethnopharmacol 143(2):548–564. https://doi.org/10.1016/j.jep.2012.07.013

Odukoya JO, Mmutlane EM, Ndinteh DT (2022) Ethnopharmacological study of medicinal plants used for the treatment of cardiovascular diseases and their associated risk factors in sub-Saharan Africa. Plants 11(10):1387. https://doi.org/10.3390/plants11101387

Olcay B, Gul R, Kultur Ş (2022) An ethnobotanical study of Elmasuyu village, Elazıg (eastern Anatolia) in Turkey. Phytol Balc 28(1):103–118. https://doi.org/10.7546/PhB.28.2022.9

Omara T, Odero MP, Obakiro SB (2022) Medicinal plants used for treating cancer in Kenya: an ethnopharmacological overview. Bull Natl Res Cent 46:148. https://doi.org/10.1186/s42269-022-00840-x

Ong HG, Kim YD (2014) Quantitative ethnobotanical study of the medicinal plants used by the Ati Negrito indigenous group in Guimaras Island, Philippines. J Ethnopharmacol 157:228–242. https://doi.org/10.1016/j.jep.2014.09.015

Otunola GA, Afolayan AJ (2015) Anti-diabetic effect of combined spices of *Allium sativum*, *Zingiber officinale* and *Capsicum frutescens* in alloxan-induced diabetic rats. Front Life Sci 8(4):314–323. https://doi.org/10.1080/21553769.2015.1053628

Packer J, Brouwer N, Harrington D, Gaikwad J, Heron R, Ranganathan S, Subramanyam V, Joanne J (2012) An ethnobotanical study of medicinal plants used by the Yaegl Aboriginal community in northern New South Wales, Australia. J Ethnopharmacol 139:244–255. https://doi.org/10.1016/j.jep.2011.11.008

Panmei R, Gajurel PR, Singh B (2019) Ethnobotany of medicinal plants used by the Zeliangrong ethnic group of Manipur, Northeast India. J Ethnopharmacol 235:164–182. https://doi.org/10.1016/j.jep.2019.02.009

Pappalardo I, Santarsiero A, De LM, Acquavia MA, Todisco S, Caddeo C, Bianco G, Infantino V, Martelli G, Vassallo A (2021) Exploiting the anti-inflammatory potential of white *Capsicum* extract by the nanoformulation in phospholipid vesicles. Antioxidants 10:1683. https://doi.org/10.3390/anti-ox10111683

Pathy KK, Flavien NB, Honore BK (2021) Ethnobotanical characterization of medicinal plants used in Kisantu and Mbanza-Ngungu territories, Kongo-Central Province in DR Congo. J Ethnobiol Ethnomed 17:5. https://doi.org/10.1186/s13002-020-00428-7

Perumal P, Chong DYH (2022) Pharmacological evaluation of green chilli in Alzheimer's disease. Pharm Res 14(1):61–70. https://doi.org/10.5530/pres.14.1.10

Pieroni A, Cianfaglione K, Nedelcheva A (2014) Resilience at the border: traditional botanical knowledge among Macedonians and Albanians living in Gollobordo, Eastern Albania. J Ethnobiol Ethnomed 10:31. https://doi.org/10.1186/1746-4269-10-31

Prescott H, Khan I (2020) Medicinal plants or herbal supplements as female aphrodisiacs: does any evidence exist to support their inclusion or potential in the treatment of FSD? J Ethnopharmacol 251:112464. https://doi.org/10.1016/j.jep.2019.112464

Rahman SU, Ullah Z, Ali A, Aziz MA, Alam N, Sher H, Ali I (2022) Traditional knowledge of medicinal flora among tribal communities of Buner Pakistan. Phytomed Plus 2(3):100277. https://doi.org/10.1016/j.phyplu.2022.100277

Rajalakshmi S, Vijayakumar S, Arulmozhi P (2019) Ethnobotanical survey of medicinal plants in Thanjavur and its surrounding (Tamil Nadu - India). Acta Ecol Sin 39(5):380–397. https://doi.org/10.1016/j.chnaes.2018.09.010

Ralte L, Bhardwaj U, Singh YT (2021) Traditionally used edible Solanaceae plants of Mizoram, India have high anti-oxidant and anti-microbial potential for effective phytopharmaceutical and nutraceutical formulations. Heliyon 7(9):e07907. https://doi.org/10.1016/j.heliyon.2021.e07907

Razafindraibe M, Kuhlman AR, Rabarison H (2013) Medicinal plants used by women from Agnalazaha littoral forest (southeastern Madagascar). J Ethnobiol Ethnomed 9:73. https://doi.org/10.1186/1746-4269-9-73

Rehman MN, Ahmad M, Sultana S, Zafar M, Edwards S (2017) Relative popularity level of medicinal plants in Talagang, Punjab Province, Pakistan. Rev Bras 27:751–775. https://doi.org/10.1016/j.bjp.2017.09.004

Ribeiroa RV, Bieskia IGC, Baloguna SO, Martins O (2017) Ethnobotanical study of medicinal plants used by Ribeirinhos in the North Araguaia microregion, Mato Grosso, Brazil. J Ethnopharmacol 205:69–102. https://doi.org/10.1016/j.jep.2017.04.023

Rivera D, Verde A, Fajardo J, Obon C, Consuegra V, Garcia-Botia J, Rios S, Alcaraz F, Valdes A, Moral AD, Laguna E (2019) Ethnopharmacology in the Upper Guadiana river area (Castile-La Mancha, Spain). J Ethnopharmacol 241:111968. https://doi.org/10.1016/j.jep.2019.111968

Rudrapal M, Sarwa KK (2020) *Capsicum*: chemistry and medicinal properties of indigenous Indian varieties. IntechOpen, London. SN-978-1-83880-941-6. https://doi.org/10.5772/intechopen.92241

Sadgrove NJ (2018) The new paradigm for androgenetic alopecia and plant-based folk remedies: 5α-reductase inhibition, reversal of secondary microinflammation and improving insulin resistance. J Ethnopharmacol 227:206–236. https://doi.org/10.1016/j.jep.2018.09.009

Salinitro M, Vicentini R, Bonomi C (2017) Traditional knowledge on wild and cultivated plants in the Kilombero Valley (Morogoro Region, Tanzania). J Ethnobiol Ethnomed 13:17. https://doi.org/10.1186/s13002-017-0146-y

Santana BF, Voeks RA, Funch LS (2016) Ethnomedicinal survey of a Maroon community in Brazil's Atlantic tropical forest. J Ethnopharmacol 181:37–49. https://doi.org/10.1016/j.jep.2016.01.014

Santos JL, Pagani E, Ramos J, Rodrigues E (2012) Observations on the therapeutic practices of riverine communities of the Unini River, AM, Brazil. J Ethnopharmacol 142:503–515. https://doi.org/10.1016/j.jep.2012.05.027

Sargin SA (2015) Ethnobotanical survey of medicinal plants in Bozyazı district of Mersin, Turkey. J Ethnopharmacol 173:105–126. https://doi.org/10.1016/j.jep.2015.07.009

Shah SA, Shah NA, Ullah S, Alam MM, Badshah H, Ullah S, Mumtaz AS (2016) Documenting the indigenous knowledge on medicinal flora from communities residing near Swat River (Suvastu) and in high mountainous areas in swat Pakistan. J Ethnopharmacol 182:67–79. https://doi.org/10.1016/j.jep.2016.02.008

Shaimaa GA, Mahmoud MS, Mohamed MR, Emam AA (2016) Phytochemical screening, antioxidant activities and in vitro anti-cancer potential of Egyptian *Capsicum* spp. Biochem Pharmacol (Los Angel) 5:205. https://doi.org/10.4172/2167-0501.1000205

Sigdel S, Acharya S (2022) Ethnomedicinal study of home garden species in Palpa District-Western Nepal. Asian J Pharmacogn 1(4):14–26

Singh S, Uddin M, Khan M, Singh S, Chishti AS, Bhat UH (2022) Therapeutic properties of Capsaicin: a medicinally important bio-active constituent of Chilli pepper. Asian J Pharm Clin Res 15(7):47–58

Singletary K (2011) Red pepper overview of potential health benefits, culinary science. Nutr Today 46(1):33

Soukand R, Kalle R, Pieroni A (2022) Homogenisation of biocultural diversity: plant ethnomedicine and its diachronic change in Setomaa and Võromaa, Estonia in the last century. Biology (Basel) 11(2):192. https://doi.org/10.3390/biology11020192

Soumya SL, Nair BR (2012) Anti-fungal efficacy of *Capsicum frutescens* L. extracts against some prevalent fungal strains associated with groundnut storage. J Agri Technol 8(2):739–750

Sreekeesoon DP, Mahomoodally MF (2014) Ethnopharmacological analysis of medicinal plants and animals used in the treatment and management of pain in Mauritius. J Ethnopharmacol 157: 181–200. https://doi.org/10.1016/j.jep.2014.09.030

Sukumaran S, Sujin RM, Geetha VS, Jeeva S (2020) Ethnobotanical study of medicinal plants used by the Kani tribes of Pechiparai Hills, Western Ghats, India. Acta Ecol Sinica 41(5):365–376

Suleman S, Tufa TB, Kebebe D, Bele S, Mekonnen Y, Gashe F, Musa S, Wynendaele E, Duchateau L, Spiegeleer BD (2018) Treatment of malaria and related symptoms using traditional herbal medicine in Ethiopia. J Ethnopharmacol 213:262–279. https://doi.org/10.1016/j.jep.2017.10.034

Tagarelli G, Tagarelli A, Piro A (2010) Folk medicine used to heal malaria in Calabria (southern Italy). J Ethnobiol Ethnomed 6:27. https://doi.org/10.1186/1746-4269-6-27

Talwan MP, Kumar S, Gautam MD (2022) Method of treatment of burns caused by acid and boiled warnings: a natural opportunity. Ymer 21(5):517–533

Tchicaillat-Landou M, Petit J, Gaiani C, Miabangana ES, Kimbonguila A, Nzikou JM, Scher J, Matos L (2018) Ethnobotanical study of medicinal plants used by traditional healers for the treatment of oxidative stress-related diseases in The Congo Basin. J Herb Med 13:76–90. https://doi.org/10.1016/j.hermed.2018.05.002

Teixeira N, Jean CSM, Luiz FB, Juliana PS, Pãmella F, Maria GLB (2019) Edible fruits from Brazilian biodiversity: a review on their sensorial characteristics versus bioactivity as tool to select research. Food Res Int 119:325–348. https://doi.org/10.1016/j.foodres.2019.01.058

Titilayo GI, Morenikeji ES (2016) A review of twenty ethnobotanicals used in the management of breast cancer in Abeokuta, Ogun State, Nigeria. Afr J Pharm Pharmacol 10(27):546–564. https://doi.org/10.5897/AJPP2015.4327

Towns AM, Quiroz D, Guinee L, Boer HD, Andel TV (2014) Volume, value and floristic diversity of Gabon's medicinal plant markets. J Ethnopharmacol 155(2):1184–1193. https://doi.org/10.1016/j.jep.2014.06.052

Tripodi P, Kumar S (2019) The *Capsicum* crop: an introduction. In: Ramchiary N, Kole C (eds) The *Capsicum* genome, compendium of plant genomes. Springer, Berlin. https://doi.org/10.1007/978-3-319-97217-6_1

Tripodi P, Rabanus-Wallace MT, Barchi L, Kale S, Esposito S, Acquadro A, Schafleitner R, van Zonneveld M, Prohens J, Diez MJ, Börner A, Salinier J, Caromel B, Bovy A, Boyaci F, Pasev G, Brandt R, Himmelbach A, Portis E, Finkers R, Lanteri S, Paran I, Lefebvre V, Giuliano G, Stein N (2021) Global range expansion history of pepper (*Capsicum* spp.) revealed by over 10,000 genebank accessions. Proc Natl Acad Sci U S A 118(34):e2104315118. https://doi.org/10.1073/pnas.2104315118

Umartani LA, Nahdi MS (2021) Ethnobotanical study of edible plant communities on the slopes of mount Merapi and Merbabu, Indonesia. Biol Med Nat Prod Chem 10(1):33–39. https://doi.org/10.14421/biomedich.2021.101.33-39

Upasani MS, Upasani SV, Beldar VG, Beldar CG, Gujarathi PP (2018) Ethnomedicinal plants used for snakebites in India: an overview. Integr Med Res 7(1):9–26. https://doi.org/10.1016/j.imr.2017.10.003

Uzun SP, Koca C (2020) Ethnobotanical survey of medicinal plants traded in herbal markets of Kahramanmaras. Plant Divers 42:443–454. https://doi.org/10.1016/j.pld.2020.12.003

Valadeau C, Castillo JA, Sauvain M, Lores AF, Bourdy G (2010) The rainbow hurts my skin: medicinal concepts and plants uses among the Yanesha (Amuesha), an Amazonian Peruvian ethnic group. J Ethnopharmacol 127(1):175–192. https://doi.org/10.1016/j.jep.2009.09.024

Valkova V, Duranova H, Ivanisova E, Galovicova L, Godocikova L, Borotova P, Kunova S, Miklasova K, Lopasovsky L, Mnahoncakova E, Kacaniova M (2021) Anti-oxidant and anti-microbial activities of fruit extracts from different fresh chili peppers. Acta Sci Pol Technol Aliment 20(4):465–472. https://doi.org/10.17306/J.AFS.2021.0997

Vasquez J, Alarcon JC, Jimenez SL, Jaramillo GI, Gomez-Betancur IC, PaolaRey-Suarez J, Jaramillo KM, Munoz DC, Marin DM, Romero JO (2015) Main plants used in traditional medicine for the treatment of snake bites n the regions of the department of Antioquia, Colombia. J Ethnopharmacol 170:158–166. https://doi.org/10.1016/j.jep.2015.04.059

Venmathi Maran BA, Iqbal M, Gangadaran P, Ahn BC, Rao PV, Shah MD (2022) Hepatoprotective potential of Malaysian medicinal plants: a review on phytochemicals, oxidative stress and anti-oxidant mechanisms. Molecules 27(5):1533. https://doi.org/10.3390/molecules27051533

Vinagre C, Vinagre S, Carrilho E (2019) The use of medicinal plants by the population from the protected landscape of "Serra de Montejunto", Portugal. J Ethnobiol Ethnomed 15:30. https://doi.org/10.1186/s13002-019-0309-0

Wangchuk P, Yeshi K, Jamphel K (2017) Pharmacological, ethnopharmacological and botanical evaluation of subtropical medicinal plants of Lower Kheng region in Bhutan. Integr Med Res 6(4):372–387. https://doi.org/10.1016/j.imr.2017.08.002

Yirga ST, Mersa A, Sisiay B, Ashebir R, Akliku B (2022) Ethnomedicinal uses of Ethiopian traditional medicinal plants used to manage some of human helminthic and parasitic disease: a review. J Traditi Med Clin Naturopathy 11(1):306

Zhang J, Yang R, Chen R, Li YC, Peng Y, Wen X (2019) Geographical origin discrimination of pepper (*Capsicum annuum* L.) based on multi-elemental concentrations combined with chemometrics. Food Sci Biotechnol 28(6):1627–1635. https://doi.org/10.1007/s10068-019-00619-3

Zhao J, Harada N, Kurihara H, Nakagata N, Okajima K (2011) Dietary isoflavone increases insulin-like growth factor-I production, thereby promoting hair growth in mice. J Nutr Biochem 22: 227–233. https://doi.org/10.1016/j.jnutbio.2010.01.008

Zimmer AR, Leonardi B, Mirona D, Schapoval E, Oliveira JR, Gosmann G (2011) Anti-oxidant and anti-inflammatory properties of *Capsicum baccatum*: from traditional use to scientific approach. J Ethnopharmacol 139:228–233. https://doi.org/10.1016/j.jep.2011.11.005

Extraction of Capsaicin and Related Compounds by Using Conventional and Contemporary Technologies

4

Aditi Sharma, Laxmi Devi, Mallappa Kumara Swamy, and Devendra Kumar Pandey

Abstract

Bioactive phytocompounds are produced as secondary metabolites, which are not needed for the plant functioning every day, but are essential for competition, attraction, defence and signalling. Capsaicin, the main secondary metabolite in *Capsicum* species is responsible for the pungent flavour of the fruit and posses several health benefits. Other compounds such as carotenoids and phenolics found in *Capsicum* also exhibit significant bioactivities. During the last few decades, several high-yielding extraction techniques for capsaicin from pepper plants have been utilised. This chapter analyses the conventional and contemporary techniques used for the extraction of capsaicin and related compounds from real sources. Main approaches highlighted in this chapter are solvent extraction, Soxhlet extraction, ultrasound-assisted extraction, microwave-assisted extraction, supercritical fluid extraction, aqueous two-phase system and enzymatic treatments.

A. Sharma · L. Devi · D. K. Pandey (✉)
Department of Biotechnology, Lovely Professional University Punjab, Jalandhar, India

M. K. Swamy
East West First Grade College (Bangalore University), Bengaluru, India

M. K. Swamy, A. Kumar (eds.), *Capsaicinoids*,
https://doi.org/10.1007/978-981-99-7779-6_4

Keywords

Food · Chili · Extraction · Bioactivity · Medicine · Enzyme · Solvent · Technology

4.1 Introduction

Secondary metabolites are produced from plants and are chemically very diverse. They serve many purposes throughout the plant's life cycle including mediating the interactions of the plants with their abiotic and biotic surroundings, for example, plant-plant, plant-microbe, and/or plant-animal interactions. Phytochemicals can be used for therapeutic applications and could be used as precursors in the synthesis of valuable medications. Moreover, plant secondary metabolites are the vital source for preparing several medications, food additives, scents and pigments. (Swamy and Sinniah 2016; Arumugam et al. 2016; Mohanty et al. 2017; Kirubakari et al. 2019). The retrieval, purification and identification of plant secondary metabolites are very difficult and challenging, due to an enormous range of chemical components found in plants and often low quantity of the target secondary metabolites (Zhang et al. 2011; Swamy et al. 2017).

The species of *Capsicum* are widely grown for their fruits and are expected to possess little protein, carbohydrate and fat content. The pungency of fruits is the reason for their domestication, and a set of phenolic alkaloids, capsaicinoids, are the compounds that contribute to spiciness. Among different capsaicinoids, 8-methyl-N-vanillyl-6-nonenamide (capsaicin) is the most pungent contributing compound, accounting up to 90%. It is followed by N-(4-hydroxy-3-methoxybenzyl)-8-methylnonanamide (dihydrocapsaicin) (Barbero et al. 2008a). A constituent called capsiate, possessing a chemical structure, is extremely comparable to capsaicin; however, only a minor pungency was initially identified in the *C. annuum* 'CH-19 Sweet' variety (Yazawa et al. 1989). Pungency causes an irritation in the sensory nerve endings and results in a sense of pain and hotness, allowing it to be used in culinary applications as an efficient way to remove insipidity and attribute new flavours to meals. *Capsicum* fruits have been used by a wide range of industries, including food and pharmaceutical industries. They are used as spices, food colourants, pain relievers, cardiovascular protectors, anti-inflammatories, antioxidants and anticancer agents (Antonio et al. 2018). Capsaicinoids, predominantly capsaicin, improve diabetes, gastric syndrome, obesity, heart diseases, cancers, asthma, immunity related diseases and main viral diseases such as the latest COVID-19 (Cortés-Ferré et al. 2021). Phytochemical studies on *Capsicum* have discovered the occurrence of numerous other related compounds like flavonoids, oleoresin, phenolic compounds and carotenoids that exhibit significant bioactivities (Sricharoen et al. 2017).

The occurrence of capsaicin quantity in particular cultivars might vary on the basis of factors, including the age of the fruits, location of fruits on the plant, the temperature and the light intensity at which it is produced. The Scoville test, created by Wilbur Scoville in 1912, is the leading test designed to measure pungency.

Scoville heat units (SHUs) are used to categorise pungency into five stages: (1) non-pungent (0–700 SHU), (2) mildly pungent (700–3000 SHU), (3) moderately pungent (3000–25,000 SHU), (4) very pungent (25000–70,000 SHU) and (5) extremely highly pungent (>80,000 SHU). Nevertheless, the present days' chromatographic procedures, which are thought to be more trustworthy and accurate, have essentially taken the role of the Scoville organoleptic test (Othman et al. 2011).

Various extraction methods are used for obtaining bioactive phytocompounds from a particular plant species. The sample preparation is one of the crucial aspects for regulating the kind and quantity of phytocompounds to be extracted. Extraction processes may be characterised into two types, i.e. the conventional and the advanced/contemporary techniques (Sagar et al. 2018; Cortés-Ferré et al. 2021). The extraction of capsaicinoids from *Capsicum* species is typically from their fruits using different organic solvents. The other techniques include extraction through maceration, hydro-distillation and Soxhlet apparatus use. Remarkably, pharmacological activities of the extracted compound are affected subject to the extraction settings, the extraction yield and the quality. The widely employed method for recovering capsaicinoids is the maceration extraction, which is dependent on the use of solvents (hexane, methanol, acetonitrile and ethanol) that are highly noxious and yield residues, which may lead to compromised quality (Martins et al. 2017; Lu et al. 2017). The other conventional approach includes hydro-distillation, the simplest techniques for obtaining capsaicinoids; however, it yields a low quantity of capsaicinoids. Additionally, the extraction using Soxhlet apparatus is broadly used for attaining better extract yields (Jiang et al. 2013). However, in spite of providing increased extraction yields and capsaicinoid content, these approaches require more quantity of solvents and long-time extractions, increasing the total budget. Furthermore, these approaches restrict their applications because of the environmental and health issues (Cortés-Ferré et al. 2021). To overcome these problems, modern methods have been developed that involve the minimal usage of organic solvents and reduced extraction time, thus making them more attractive in recent times, because of them being cost-effective and environmental-friendly. Due to the increased demand from industries, more established processes are adopted to increase their extraction yield and characterisation. Therefore, the extraction of capsaicinoids and related compounds is highly important because of its industrial and pharmacological benefits.

This chapter analyses the conventional and contemporary techniques used for capsaicin and related compound extraction from real sources; solvent extraction, Soxhlet extraction, ultrasound-assisted extraction, microwave-assisted extraction, supercritical fluid extraction, aqueous two-phase system and enzymatic treatment approaches.

4.2 Extraction Methods

Extraction is the route of moving a component or constituents from a liquid or solid phase into a liquid phase that is immiscible or partially miscible with the first phase. It is a vital process, a challenging step that significantly impacts on the obtaining of biologically active components that frequently occur in the lower concentration in foodstuffs and natural resources. Analysis of a compound from a sample generally involves extraction, isolation and recovery of the compound. There are numerous methods, including conventional and contemporary technologies for extracting capsaicin, such as solvent extraction (SE), microwave-assisted extraction (MAE), ultrasound-assisted extraction (UAE), SOX extraction, extraction with liquids under pressure (PLE), enzyme-assisted extraction (EAE), surface-mediated extraction, aqueous two-phase system (ATPS) and supercritical fluid extraction (SFE).

During the last several decades, many methods for extracting capsaicinoids from pepper plants have been developed. The main stage in designing an extraction procedure is choosing an acceptable organic solvent that can provide a high yield of the target chemical. Hexane, acetonitrile, methanol, ethanol and water are the most often utilised extraction solvents for capsaicinoids (Barbero et al. 2008b; Lu et al. 2017). Along with solvent selection, several other factors also influence extraction efficiency. These parameters include the quantity of samples, volume of solvents, temperature, extraction time and reproducibility of the procedures (Lu et al. 2017). Markedly, the yield of capsaicinoids vary on the basis of different pepper species used for extraction and how the extraction is processed. Currently, the maceration extraction approach is the common and simple procedure used for extracting capsaicin. Modern extraction techniques include electro-technologies, ultrasound, microwave and high-pressure extraction as well as supercritical and subcritical fluid extraction (Castro-Muñoz et al. 2022).

4.2.1 Conventional Techniques

Among the conventional techniques, the solvent extraction/maceration technique is the most extensively utilised one. The following steps are involved in the extraction of phytocompounds using the solvent extraction method: (1) the penetration of solvents into solid matrices, (2) dissolving of the solutes in the solvents, (3) diffusing the solute out of the solid matrices, and (4) collection of the removed solutes (Zhang et al. 2018). The conventional extraction practices like maceration, percolation and reflux extraction involve the use of organic solvents, need large quantities of solvents and usually need a lengthy time for completion. The selection of solvents is very crucial for any solvent extraction approaches. During the selection, it is required to scrutinise the solubility, price and safety. As per the principle of similarity and intermiscibility, the polarity values of a solvent closer to its solute accomplish a better extract yield. Ethanol and methanol are multipurpose solvents and utilised very commonly in conventional extractions for obtaining phytocompounds. A

Table 4.1 Different extraction methods used in *Capsicum* species

Species	Solvent	Method	Compound	References
C. frutescens *C. chinense* *C. baccatum*	–	Supercritical carbon dioxide (SC-CO2)	Capsaicin, dihydrocapsaicin	de Aguiar et al. (2013)
C. chinense (biquinho pepper)	Hexane, acetone	Soxhlet extraction (SE)	Capsiate	de Aguiar et al. (2014)
C. frutescens L. (malagueta pepper)	–	Supercritical fluid extraction (SFE) assisted by ultrasound	Capsaicinoids	Santos et al. (2015)
Capsicum baccatum L. var. pendulum (dedo de moca pepper)	–	Supercritical carbon dioxide (SC-CO2)	Capsaicin	Dias et al. (2016)
Capsicum annuum L. (hot pepper)	Water-methanol	Ultrasound-assisted extraction (UAE)	Capsaicin	Sricharoen et al. (2017)
Capsicum annuum L. (hot pepper)	Methanol	Solvent-assisted extraction (SAE)	Capsaicin, flavonoids	Marincaş et al. (2018)
Capsicum frutescens L. (malagueta pepper)	Acetonitrile, acetic acid, water	Supercritical fluid extraction (SFE)	Capsaicinoids	de Aguiar et al. (2018)
Green, yellow and red bell pepper	SC-CO_2 with 5% methanol (modifier)	Supercritical fluid extraction (SFE)	Capsaicin	Hamada et al. (2019)
C. chinense (biquinho pepper)		Supercritical carbon dioxide (SC-CO2)	Capsinoids	de Aguiar et al. (2019)
C. annuum	Ethyl acetate	Surface-mediated extraction	Capsaicin	Waqas et al. (2022)

summary of the various extraction methods used for capsaicin and other related constituents is presented in Tables 4.1 and 4.2.

4.2.1.1 Maceration

Maceration is the process of preparing the extract by using solvent extraction in combination with temperature treatments. This simple approach has several drawbacks, including a lengthy extraction duration and poor extraction efficacy. It could be employed for extracting phytocompounds that are heat-labile. Fernández-Ronco et al. (2012) achieved the extraction of *C. annuum* oleoresin by maceration using hexane as a solvent. Carbonic maceration (CM) pre-treatment effects were achieved on the drying kinetics of *C. annuum* and the dehydrating outcomes

Table 4.2 Extraction techniques for bioactive-related compound in *Capsicum* species

Species	Solvent	Extraction method	Compound	References
C. frutescens (red pepper)	–	Supercritical carbon dioxide (SC-CO2)	Oleoresin	Silva and Martínez (2014)
C. baccatum L. var. pendulum (dedo de moca pepper)	Carbon dioxide	Supercritical fluid extraction (SFE)	Oleoresin	Dias et al. (2016)
C. baccatum L. (dedo de moca pepper)	Hexane, ethyl acetate, methanol, ethanol	Soxhlet extraction	Phenolic compounds, capsaicinoids	Dias et al. (2017)
C. baccatum L. (dedo de moca pepper)	Methanol, ethanol	Ultrasound-assisted extraction (UAE)	Phenolic compounds, capsaicinoids	Dias et al. (2017)
C. chinense (biquinho pepper)	–	Supercritical carbon dioxide (SC-CO2)	Capsiate-rich oleoresin	de Aguiar et al. (2019)
C. annuum L. (red peppers)	–	Supercritical carbon dioxide (SC-CO_2)	Oleoresin, carotenoids	Jimenez et al. (2021)
C. annuum L. (red peppers)	Ethanol (100%) and ethanol-water (75%)	Pressurised liquid extraction (PLE)	Phenolic compounds	Jimenez et al. (2021)
C. chinense (habanero pepper)	Methanol (MeOH)	Ultrasound-assisted extraction (UAE)	Phenolic compounds	Herrera-Pool et al. (2021)
C. annuum L.	Acetone, ethanol & water	Accelerated solvent extraction (ASE)	Phenolic compounds	Ahmad et al. (2021)
C. annuum L.	–	Supercritical carbon dioxide (SC-CO_2)	Carotenoids	Uquiche et al. (2022)

revealed that the usual dehydrating rate for CM trials was as high as 150–185% (Liu et al. 2014).

4.2.1.2 Percolation

Percolation is a constant procedure, wherein the saturated solvents are constantly being refilled by new solvents. This approach is relatively more effective compared to maceration. Tandon et al. (1964) revealed that one of the finest solvents for oleoresin extraction by the percolation technique is ether. Using the pericarp for extraction avoids the challenge of separating the fatty oil from the oleoresin, since it contains far less fatty oil than the seeds do. Extraction of resins from *C. annuum* var. *longum* by the percolation was studied by employing 95% alcohol. They observed the presence of resins in the extract as confirmed from the preliminary test for resins (Frial-McBride 2016). Sharma et al. (2021) evaluated the study comprising the percolation of the crushed, dried, ripened fruits of C. *annuum* to extract the oleoresin. Their investigation established that *Capsicum* oleoresin contains a large number of fatty acids and their amides, esters, terpenes and phytosterols. The authors also demonstrated noteworthy antimicrobial activities of *Capsicum* oleoresin against disease-causing microbes.

4.2.1.3 Reflux Extraction

Reflux extraction is a solid-liquid extraction technique, involving continuous solvent evaporation and condensation for a defined period without losing solvents. This method is commonly utilised in the herbal industries, because of its effectiveness and affordability and it is easy to handle (Chua et al. 2016). Compared to maceration and percolation methods, the reflux extraction method is very efficient and utilises a lesser amount of solvents and extraction duration. Dong et al. (2014) developed a method to extract phytocompounds as well as red pigments from red ripe fruits of *C. annuum* by using heat refluxing extraction. Initially, capsaicinoids were obtained from red peppers using ethanol (40–50%) without a dehydrating procedure. Later, red pigments without spiciness were attained from the remains once the capsaicinoids were removed using ethanol (95%) as solvent, dehydrating at 90 °C in a water bath, and the extraction time was 2 h. Later, the solution after extraction was accumulated and subjected to further extraction with similar conditions till the capsaicinoids were drained out. All solutions were analysed for capsaicinoid content using high pressure liquid chromatography. Sharma et al. (2021) developed the reflux method of extracting oleoresin *Capsicum* and optimised the conditions to attain a higher percentage of yield. The conditions required for improved extraction of oleoresin were observed to be the use of acetone as a solvent, temperature of 40 °C and 5 h of extraction period. These conditions maintained during a reflux process yields the highest percentage of oleoresin (3.7% w/w).

4.2.1.4 Soxhlet Extraction

The Soxhlet extraction technique employs the reflux and siphoning principles for a continuous extraction of herbs using new solvents, and it combines the benefits of both percolation and reflux extraction. In comparison to maceration or percolation, the Soxhlet extraction is an automated uninterrupted extraction technique with higher extraction efficacy, and it requires short time and less amount of solvents. However, thermal degradation is more likely observed as the Soxhlet extractions involve high temperature and a lengthy extraction period (Zhang et al. 2018).

Extracts containing bioactive components were produced from dried out and pulverised *C. baccatum* (dedo de moça pepper) using the Soxhlet extraction method (Dias et al. 2017). Four organic solvents, namely, hexane, ethanol, methanol and ethyl acetate, were used, and refluxing was set for 6 h. The obtained extract yield was weighed and analysed for phenolic compounds and antioxidants. They obtained a substantial quantity of extract yield and total phenolics with good antioxidant capabilities. De Aguiar et al. (2014) investigated the best conditions for the total yield of capsaicinoids using a Soxhlet apparatus using different extraction solvents (hexane, acetone, ethanol and methanol). For extraction around 5 g of freeze-dried pepper was used, i.e. packed within filter papers and implanted in the extractor. About 150 mL of solvent was poured, the Soxhlet apparatus was heated up till boiling, and refluxing was retained for 6 h. Later, the solvent was vaporised under vacuum, and the extract was recovered. From their results, it was found that methanol was the best solvent for extraction, which yielded 0.42 g extract/g freeze-dried.

4.2.2 Merits and Demerits of Conventional Extraction Techniques

Effective, cost-effective and favourable extraction methods are crucial for recovering bioactive compounds without negotiating on the quality of extracts. There exists distinct differences between maceration, Soxhlet, percolation and reflux extraction methods; however, each method has its own unique advantages, and their usage is mainly dependent on the plant sample, solvents, time of extraction, etc. The merits associated with the above-mentioned traditional methods include simplicity of set-up, less chemical modification of extract, simple recycling of solvents, less contamination from organic solvents and complete extraction of compound with high molecular weight. Some of the demerits include more solvent requirement, lesser extract yield, high extraction cost, contamination of solvents and a considerably long duration to complete extraction procedure (Shams et al. 2015; Brovko et al. 2015; Mohammad Azmin et al. 2016; Bitwell et al. 2023).

4.2.3 Contemporary Technologies

Traditional or conventional techniques do not efficiently and rationally extract the target constituent; therefore, new enhanced technologies must be found (Brovko et al. 2015; Bitwell et al. 2023). Some of the contemporary technologies used for extracting capsaicinoids include ultrasound-assisted extraction (UAE), microwave-assisted extraction (MAE), pressurised liquid extraction (PLE), supercritical fluid extraction (SFE), aqueous two-phase extraction (ATPS) and enzyme-assisted extraction (EAE).

4.2.3.1 Ultrasound-Assisted Extraction (UAE)

The UAE, often called as the sonication method, works on the principle of cavitation phenomenon, i.e. passing of ultrasonic waves that pass over the solvent system generates energy that enhances the blending and permeation of solvents into the sample matrices (Karnka et al. 2002). The major advantages of UAE include requiring less energy and solvents and reduced extraction time and temperature. Moreover, this approach is well suited for extracting unstable and thermolabile chemicals and often used in the extraction of several different phytocompounds (Boonkird et al. 2008; Zhang et al. 2018). The effect of operational factors like the type of solvent, solvent to powder ratio and temperature on the yield of capsaicinoids from dried *C. frutescens* was studied by Boonkird et al. (2008) using the UAE method. According to them, the appropriate parameters for extracting capsaicinoid is subjecting samples (1 g dry powder and 95% 5 mL ethanol) to ultrasonic waves (35-kHz frequency) at 45 °C. These parameters could be able to obtain 85% of the capsaicinoids from the raw material within 3 h' time. Overall, their study showed that the UAE can significantly reduce the extraction duration at a low working temperature as compared to the hot maceration procedure. Dias et al. (2017) investigated the application of the UAE method to extract capsaicinoids from *C. baccatum*. They showed that the use of parameters, i.e. ultrasonic intensities (150, 300 and 450 W/cm^2), alcohols (methanol and ethanol) as solvents and temperatures (40, 50 and 60 °C), significantly increased good yield (up to 26%), which was more than the yield obtained by the Soxhlet method. It is noteworthy to mention here that higher intensities of ultrasound waves improved total phenolic compound yield and antioxidative properties; however, it reduced the recovery of capsaicinoids. Herrera-Pool et al. (2021) studied the influence of the polarity of different solvents in the UAE method to obtain total phenolic compounds from *C. chinense* leaves. They reported the fast recovery of phenolic compounds and noticed that solvent polarity can strongly influence the retrieval of phenolic compounds. The maximum recovery of total phenolic compounds (24 mg GAE/g dry wt) was achieved with the use of 50% methanol as solvents.

4.2.3.2 Microwave-Assisted Extraction (MAE)

The MAE technique is established via combining the conventional solvent extraction and microwave-assisted extraction methods, and this approach uses the energy produced due to microwave irradiation to heat the solvent and enhance the kinetics of extraction (Lu et al. 2017). The heating process used by MAE dramatically shortens the extraction time (often from a few seconds to 30 min) as compared to other methods (Tatke and Jaiswal 2011). This has the benefit of preventing oxidation and heat deterioration. Moreover, MAE is a non-toxic approach to obtain better extraction yield with improved efficiency. It uses fewer solvents that are simple for recycling and may be applied to larger quantities. It is one of the easiest methods and most effective among various existing techniques. The associated maintenance expenses for this approach remains considerable (Lu et al. 2017). The MAE method is being used by several researchers previously for extracting capsaicinoids from pepper species. Williams et al. (2004) investigated the use of microwave radiations for supporting capsaicinoid extraction from *Capsicum* fruits. The process was accomplished by irradiating 2 g of dried samples in a closed vessel. Later, gas chromatography was performed to identify capsaicinoid derivatives. The optimal settings for extraction were found to be the use of 30% acetone for 7 min, regardless if pulverised or complete tissue. This process could be able to recover around 95% of the total capsaicinoid portion within 15 min as compared to 60 min observed in the reflux and 1 h in the maceration methods. They suggested this technique for a rapid extraction of capsaicinoids from bulky materials as it offers great savings in terms of cost and time.

Barbero et al. (2006) developed an MAE extraction method to extract capsaicinoids from peppers. According to them, the ideal conditions required for

MAE are as follows: 25 mL of ethanol (100%) as solvent, freshly triturated peppers (0.5 g), temperature of 125 °C and 5 min of extraction time. The recovered capsaicinoids were stable under the amended extraction settings, and degradation was not noticed up to 150 °C. According to Chuichulcherm et al. (2013), the best method to extract capsaicinoids from *C. frutescens* is the UAE method, which yielded 5.28 capsaicinoids per gram of dried out chili powder. However, the use of the MAE approach recovered only about 4.01 capsaicinoids per gram of dried out chili powder. Hussain et al. (2022) developed the MAE process to extract capsaicin from chili. They used two deep eutectic solvents, i.e. choline chloride-urea (CCU) and choline chloride-glycerol (CCG). They observed maximum extract yield in both solvents at an irradiation power of 600 W and the solvent to solid ratio was found to be 30 mL/g for CCU 18.88 mg/g and for CCG 14.49 mg/g.

4.2.3.3 Supercritical Fluid Extraction (SFE)

The SFE is a selective technique and employs pressurised fluids/liquids as solvents. A supercritical fluid is formed when a fluid is subjected to pressures and temperatures that exceed its critical point, causing the liquefied and gaseous phases to become indistinguishable from one another. Supercritical fluids have both liquid and gas physicochemical properties (Reverchon and De Marco 2006). They have greater transport qualities than standard organic solvents and can easily permeate through samples, enhancing the efficiency of the extraction process and the target molecules' overall yield. A mild critical temperature (31.3 °C) and pressure (72.9 atm) is seen in supercritical carbon dioxide. It is one of the widely used supercritical fluids. To recover and stabilise vitamins from *C. frutescens* by-products, the SFE method was used by Romo-Hualde et al. (2012). The highest yield was found at 60 °C, 24 MPa, and the extract recovered was a red-coloured oil. Likewise, de Aguiar et al. (2018) economically evaluated oleoresin and total capsaicinoid extraction from malagueta pepper employing the SFE approach. The cost estimated for extracting oleoresin was reported to be 125.41 US$ per kg of extract and suggested the use of SFE for a large-scale production. Soldan et al. (2021) evaluated the SFE to obtain oleoresin from jalapeno pepper industrial waste. For the SFE process, carbon dioxide at dissimilar conditions, temperatures (40 and 60 °C), pressures (200 and 250 bar) and the absence and presence of ethanol were adopted. The total biomass yields were found to vary from 9.38% to 10.08%. The extracts were discovered with several bioactives, including phenolics, flavonoids, carotenoids and fatty acids; however, they failed to exhibit any noteworthy antioxidative properties.

4.2.3.4 Pressurised Liquid Extraction (PLE)

In the PLE approach, solvents at higher temperature, i.e. lower than their critical point and beyond their boiling point, and enough pressure is applied to keep them liquid. This allows increased penetration of solvents into the sample matrices (Lu et al. 2017). It is one of the advanced extraction technique with several advantages over traditional extraction processes. It is also referred to as pressurised fluid extraction, high-pressure high-temperature solvent extraction, high-pressure solvent extraction and subcritical solvent extraction (Alvarez-Rivera et al. 2020). A pressurised liquid extraction method was reported by Liu et al. (2013) to extract capsaicinoids from *C. annuum* and identify them by LC-MS/MS analysis. They used methanol as solvent for extraction, 100 °C was maintained during the process, and 1500 psi pressure was applied. Overall, PLE contributed for higher yields, and the extract was shown to contain three capsaicinoids. De Aguiar et al. (2019) proposed a sequential high-pressure extraction for recovering bioactive compounds from biquinho peppers. The non-polar part was attained by means of SFE using supercritical carbon dioxide, and the resulting extract was subjected to PLE using a mixture of 75% ethanol and water at varying temperatures, between 45 °C and 65 °C for recovering bioactives of higher polarity. The SFE extracted 4.75% of the total capsaicinoid yield with 8.67 mg/g oleoresin concentration. Another study by de Aguiar et al. (2020) evaluated the financial requirement by different extraction methods (single-stage SFE and PLE and sequential SFE + PLE) to obtain phenolic compounds and capsiate-rich oleoresin from biquinho peppers. According to them, single-stage extraction methods showed the lowest cost of manufacturing for capsiate and phenolics.

4.2.3.5 Enzyme-Assisted Extraction (EAE)

The EAE is a unique and convenient technique for recovering cell wall-associated compounds and enhance their extraction yield. When enzyme-assisted extraction is coupled with solvent extraction techniques including UAE and SFE, the extraction yields can be further increased, thus making it a more efficient and promising approach to recover high percentages of phytoconstituents, mainly from fruits and seeds (Sagar et al. 2018). The factors that can influence the extraction by this technique include the type of catalysts, molecular sizes of plant substances and the hydrolysis period. Importantly, the EAE is an eco-friendly approach for extracting oils and biologically active constituents, because it uses water as the solvent as a substitute to organic solvents (Sagar et al. 2018; Cortés-Ferré et al. 2021). Several nutraceutical compounds are associated with the plant cell wall, for example, phenolic compounds, carotenoids, capsaicinoids and many other compounds. In the EAE method, different enzymes like cellulases, beta-glucosidases, pectinases or the blend of enzymes are used for hydrolysing and degrading cell wall components and recovering valuable bioactives of interest (Cortés-Ferré et al. 2021). The addition of a commercial enzyme, Viscozyme L (beta-glucanase), during the process of EAE significantly enhanced the extraction yield of capsaicinoids (up to 22%) from green chili peppers (Baby and Ranganathan 2016). Similarly, Nath et al. (2016) used three carbohydrases, i.e. pectinase, Viscozyme L and cellulase, to test their liquefaction properties and capabilities for recovering higher carotenoids in the aqueous extract of red *Capsicum*. Overall, the enzymatic liquefaction was shown to enhance the total soluble solids and extract yield up to threefold when compared to control. *Capsicum* pomace powder contained greater carotenoid content of 279.83 mg/100 g. As reviewed by Lu et al. (2017), the use of pectinases and carbohydrases can increase the extract yield up to 88% from dried chili pepper fruits. However, it is required to maintain an exact temperature and agitation speed during the process of extraction. With the use of cellulases of *Aspergillus niger* ATCC 10864, Gamarra Mendoza et al. (2020) increased five times higher extraction yield of carotenoids and oleoresin as compared to the conventional method. Recently, a study by Cortes-Ferre et al. (2022) extracted the anti-inflammatory bioactives, such as phenolic compounds, capsaicin and dihydrocapsaicin, from chili pepper seeds using cellulase-assisted extraction, involving water as a solvent.

4.2.3.6 Aqueous Two-Phase Extraction (ATPS)

The ATPS is a liquid-liquid fractionation technique, which is acknowledged for its adaptability and reliability. It is a simple process used for recovering and purifying a variety of biomolecules (Nadar et al. 2017). The aqueous two-phase systems were developed by using acetonitrile and cholinium-based ionic liquids for recovering capsaicin from chilli pepper (Santos et al. 2016). By using ATPS composed of an ethylene oxide-propylene oxide (EOPO) copolymer, ethanol and salt, researchers extracted and purified the capsaicin from *Capsicum* and obtained yield was observed to be 95.5% (Fan et al. 2017). Likewise, the ATPS set up based on ethanol and sodium salts was used for obtaining capsaicin from *Capsicum chinense* var. Cumari-do-Pará with an extraction efficiency of nearly 85.6% (Cienfuegos et al. 2017).

4.3 Conclusions

The extraction of plant secondary metabolites is a vital process that significantly affects the recovery of biologically active components as they occur relatively in lower concentration. The extraction procedures significantly alter purification and quality of bioactives. As discussed above, the methods employed for recovering capsaicinoids from *Capsicum* species include both conventional and contemporary technologies, such as SE, MAE, UAE, SOX PLE, EAE, ATPS and SFE. Traditional extraction techniques like maceration and percolation have been replaced by more advanced techniques in order to increase the selectivity and efficiency of the extraction of bioactive compounds and to keep up with the rising demand on the market. A variety of extraction methods, including microwaves, ultrasonic waves, supercritical fluids, enzymes, pressured liquids, electric fields, etc., are used in advanced procedures. Compared to conventional extraction methods, these cutting-edge extraction techniques enable final extracts that are selectively rich in desired chemicals without the development of artifacts. They are also often direct, quick, environment-friendly and totally automated approaches.

References

Ahmad R, Ahmad N, Alkhars S, Alkhars A, Alyousif M, Bukhamseen A, Abuthayn S, Aqeel M, Aljamea A (2021) Green accelerated solvent extraction (ASE) with solvent and temperature effect and green UHPLC-DAD analysis of phenolics in pepper fruit (Capsicum annum L.). J Food Compos Anal 97:103766

Alvarez-Rivera G, Bueno M, Ballesteros-Vivas D, Mendiola JA, Ibanez E (2020) Pressurized liquid extraction. In: Liquid-phase extraction. Elsevier, Amsterdam, pp 375–398

Antonio AS, Wiedemann LSM, Veiga Junior VF (2018) The genus Capsicum: a phytochemical review of bioactive secondary metabolites. RSC Adv 8(45):25767–25784

Arumugam G, Swamy MK, Sinniah UR (2016) Plectranthus amboinicus (Lour.) Spreng: botanical, phytochemical, pharmacological and nutritional significance. Molecules 21(4):369

Baby KC, Ranganathan TV (2016) Effect of enzyme pretreatment on yield and quality of fresh green chilli (*Capsicum annuum L*) oleoresin and its major capsaicinoids. Biocatal Agric Biotechnol 7:95–101. https://doi.org/10.1016/j.bcab.2016.05.010

Barbero GF, Liazid A, Palma M, Barroso CG (2008a) Fast determination of capsaicinoids from peppers by high-performance liquid chromatography using a reversed phase monolithic column. Food Chem 107(3):1276–1282

Barbero GF, Liazid A, Palma M, Barroso CG (2008b) Ultrasound-assisted extraction of capsaicinoids from peppers. Talanta 75(5):1332–1337

Barbero GF, Palma M, Barroso CG (2006) Determination of capsaicinoids in peppers by microwave-assisted extraction–high-performance liquid chromatography with fluorescence detection. Anal Chim Acta 578(2):227–233

Bitwell C, Sen IS, Luke C, Kakoma MK (2023) A review of modern and conventional extraction techniques and their applications for extracting phytochemicals from plants. Sci Afr 19:e01585

Boonkird S, Phisalaphong C, Phisalaphong M (2008) Ultrasound-assisted extraction of capsaicinoids from Capsicum frutescens on a lab-and pilot-plant scale. Ultrason Sonochem 15(6):1075–1079

Brovko OS, Palamarchuk IA, Boytsova TA, Ivakhnov AD, Bogolitsyn KG, Valchuk NA (2015) Comparative analysis of traditional and modern extraction methods of usnic acid from raw lichen material. Fundament Res 11(4):659–663

Castro-Muñoz R, Gontarek-Castro E, Jafari SM (2022) Up-to-date strategies and future trends towards the extraction and purification of capsaicin: a comprehensive review. Trends Food Sci Technol 123:161–171. https://doi.org/10.1016/j.tifs.2022.03.014

Chua LS, Abd Latiff N, Mohamad M (2016) Reflux extraction and cleanup process by column chromatography for high yield of andrographolide enriched extract. J Appl Res Med Aromat Plants 3(2):64–70

Chuichulcherm S, Prommakort S, Srinophakun P, Thanapimmetha A (2013) Optimization of capsaicin purification from Capsicum frutescens Linn. With column chromatography using Taguchi design. Ind Crop Prod 44:473–479

Cienfuegos NEC, Santos PL, García AR, Soares CMF, Lima AS, Souza RL (2017) Integrated process for purification of capsaicin using aqueous two-phase systems based on ethanol. Food Bioprod Process 106:1–10

Cortés-Ferré HE, Guajardo-Flores D, Romero-De La Vega G, Gutierrez-Uribe JA (2021) Recovery of capsaicinoids and other phytochemicals involved with TRPV-1 receptor to re-valorize chili pepper waste and produce nutraceuticals. Front Sustain Food Syst 4:588534

Cortes-Ferre HE, Antunes-Ricardo M, Gutiérrez-Uribe JA (2022) Enzyme-assisted extraction of anti-inflammatory compounds from habanero chili pepper (Capsicum chinense) seeds. Front Nutr 9:2029. https://doi.org/10.3389/fnut.2022.942805

de Aguiar AC, da Fonseca Machado AP, Angolini CFF, de Morais DR, Baseggio AM, Eberlin MN, Junior MRM, Martinez J (2019) Sequential high-pressure extraction to obtain capsinoids and phenolic compounds from biquinho pepper (Capsicum chinense). J Supercrit Fluids 150:112–121

de Aguiar AC, dos Santos P, Coutinho JP, Barbero GF, Godoy HT, Martínez J (2014) Supercritical fluid extraction and low pressure extraction of Biquinho pepper (Capsicum chinense). LWT-Food Sci Technol 59(2):1239–1246

de Aguiar AC, Osorio-Tobón JF, Silva LPS, Barbero GF, Martinez J (2018) Economic analysis of oleoresin production from malagueta peppers (Capsicum frutescens) by supercritical fluid extraction. J Supercrit Fluids 133:86–93
de Aguiar AC, Osorio-Tobon JF, Vigano J, Martinez J (2020) Economic evaluation of supercritical fluid and pressurized liquid extraction to obtain phytonutrients from biquinho pepper: analysis of single and sequential-stage processes. J Supercrit Fluids 165:104935
De Aguiar AC, Sales LP, Coutinho JP, Barbero GF, Godoy HT, Martínez J (2013) Supercritical carbon dioxide extraction of capsicum peppers: global yield and capsaicinoid content. J Supercrit Fluids 81:210–216
Dias ALB, Arroio Sergio CS, Santos P, Barbero GF, Rezende CA, Martínez J (2016) Effect of ultrasound on the supercritical CO2 extraction of bioactive compounds from dedo de moça pepper (Capsicum baccatum L. var. pendulum). Ultrason Sonochem 31:284–294
Dias ALB, Sergio CSA, Santos P, Barbero GF, Rezende CA, Martínez J (2017) Ultrasound-assisted extraction of bioactive compounds from dedo de moça pepper (Capsicum baccatum L.): effects on the vegetable matrix and mathematical modeling. J Food Eng 198:36–44
Dong X, Li X, Ding L, Cui F, Tang Z, Liu Z (2014) Stage extraction of capsaicinoids and red pigments from fresh red pepper (Capsicum) fruits with ethanol as solvent. LWT-Food Sci Technol 59(1):396–402
Fan Y, Lu YM, Yu B, Tan CP, Cui B (2017) Extraction and purification of capsaicin from Capsicum oleoresin using an aqueous two-phase system combined with chromatography. J Chromatogr B 1063:11–17
Fernández-Ronco MP, Gracia I, de Lucas A, Rodríguez JF (2012) Extraction of Capsicum annuum oleoresin by maceration and ultrasound-assisted extraction: influence of parameters and process modeling. J Food Process Eng 36(3):343–352
Frial-McBride MG (2016) Extraction of resins from Capsicum annuum var. longum (Siling haba) for the study of their potential anti-microbial activities. J Chem Pharm Res 8(3):117–127
Gamarra Mendoza N, Velasquez Rodriguez S, Roque Lima B (2020) Improvement of the extraction of carotenoids and capsaicinoids of chili pepper native (*Capsicum baccatum*), assisted with cellulolytic enzymes. Rev Peru Biol 27:55–060. https://doi.org/10.15381/rpb.v27i1.17588
Hamada N, Hashi Y, Yamaki S, Guo Y, Zhang L, Li H, Lin JM (2019) Construction of on-line supercritical fluid extraction with reverse phase liquid chromatography–tandem mass spectrometry for the determination of capsaicin. Chin Chem Lett 30(1):99–102
Herrera-Pool E, Ramos-Díaz AL, Lizardi-Jiménez MA, Pech-Cohuo S, Ayora-Talavera T, Cuevas-Bernardino JC, García-Cruz U, Pacheco N (2021) Effect of solvent polarity on the ultrasound assisted extraction and antioxidant activity of phenolic compounds from habanero pepper leaves (Capsicum chinense) and its identification by UPLC-PDA-ESI-MS/MS. Ultrason Sonochem 76: 105658
Hussain M, Qamar MT, Ahmed D (2022) Microwave-and ultrasound-assisted extraction of capsaicin from Capsicum annuum using deep eutectic solvents. Int J Veg Sci 28(4):312–319
Jiang ZT, Feng X, Li R, Wang Y (2013) Composition comparison of essential oils extracted by classical hydro distillation and microwave-assisted hydrodistillation from Pimenta dioica. J Essent Oil Bear PL 16:45–50. https://doi.org/10.1080/0972060X.2013.764178
Jimenez D, Vardanega R, Salinas F, Espinosa-Álvarez C, Bugueno-Munoz W, Palma J, Meireles MAA, Ruíz-Domínguez MC, Cerezal-Mezquita P (2021) Effect of drying methods on biorefinery process to obtain capsanthin and phenolic compounds from Capsicum annuum L. J Supercrit Fluids 174:105241
Karnka R, Rayanakorn M, Watanesk S, Vaneesorn Y (2002) Optimization of high-performance liquid chromatographic parameters for the determination of capsaicinoid compounds using the simplex method. Anal Sci 18(6):661–665
Kirubakari B, Shanmugapriya ST, Vijayarathna S, Chen Y, Kanwar JR, Herng C, Shin LN, Swamy MK, Subramaniam S, Sasidharan S (2019) Antibacterial and antifungal agents of higher plants. In: Akhtar M, Swamy M, Sinniah U (eds) Natural bio-active compounds. Springer, Singapore. https://doi.org/10.1007/978-981-13-7154-7_16

Liu A, Han C, Zhou X, Zhu Z, Huang F, Shen Y (2013) Determination of three capsaicinoids in Capsicum annuum by pressurized liquid extraction combined with LC-MS/MS. J Sep Sci 36(5): 857–862

Liu L, Wang Y, Zhao D, An K, Ding S, Wang Z (2014) Effect of carbonic maceration pre-treatment on drying kinetics of chilli (Capsicum annuum L.) flesh and quality of dried product. Food Bioprocess Technol 7:2516–2527

Lu M, Ho CT, Huang Q (2017) Extraction, bioavailability, and bioefficacy of capsaicinoids. J Food Drug Anal 25:27–36. https://doi.org/10.1016/j.jfda.2016.10.023

Marincaş O, Feher I, Magdas DA, Puşcaş R (2018) Optimized and validated method for simultaneous extraction, identification and quantification of flavonoids and capsaicin, along with isotopic composition, in hot peppers from different regions. Food Chem 267:255–262

Martins FS, Borges LL, Ribeiro CSC, Reifschneider FJB, Conceição CC (2017) Novel approaches to extraction methods in recovery of capsaicin from habanero pepper (CNPH 15.192). Pharmacogn Mag 13:S375–S379. https://doi.org/10.4103/0973-1296.210127

Mohammad Azmin SN, Abdul Manan Z, Wan Alwi SR, Chua LS, Mustaffa AA, Yunus NA (2016) Herbal processing and extraction technologies. Sep Purif Rev 45(4):305–320

Mohanty SK, Swamy MK, Sinniah UR, Anuradha M (2017) Leptadenia reticulata (Retz.) Wight & Arn.(Jivanti): botanical, agronomical, phytochemical, pharmacological, and biotechnological aspects. Molecules 22(6):1019

Nadar SS, Pawar RG, Rathod VK (2017) Recent advances in enzyme extraction strategies: a comprehensive review. Int J Biol Macromol 101:931–957

Nath P, Kaur C, Rudra SG, Varghese E (2016) Enzyme-assisted extraction of carotenoid-rich extract from red Capsicum (*Capsicum annuum*). Agric Res 5:193–204. https://doi.org/10.1007/s40003-015-0201-7

Othman ZAA, Ahmed YBH, Habila MA, Ghafar AA (2011) Determination of capsaicin and dihydrocapsaicin in capsicum fruit samples using high performance liquid chromatography. Molecules 16(10):8919–8929

Reverchon E, De Marco I (2006) Supercritical fluid extraction and fractionation of natural matter. J Supercrit Fluids 38(2):146–166

Romo-Hualde A, Yetano-Cunchillos AI, González-Ferrero C, Sáiz-Abajo MJ, González-Navarro CJ (2012) Supercritical fluid extraction and microencapsulation of bioactive compounds from red pepper (Capsicum annum L.) by-products. Food Chem 133(3):1045–1049

Sagar NA, Pareek S, Sharma S, Yahia EM, Lobo MG (2018) Fruit and vegetable waste: bioactive compounds, their extraction, and possible utilization. Compr Rev Food Sci Food Saf 17:512–531. https://doi.org/10.1111/1541-4337.12330

Santos P, Aguiar AC, Barbero GF, Rezende CA, Martínez J (2015) Supercritical carbon dioxide extraction of capsaicinoids from malagueta pepper (Capsicum frutescens L.) assisted by ultrasound. Ultrason Sonochem 22:78–88

Santos PL, Santos LNS, Ventura SPM, de Souza RL, Coutinho JAP, Soares CMF, Lima AS (2016) Recovery of capsaicin from Capsicum frutescens by applying aqueous two-phase systems based on acetonitrile and cholinium-based ionic liquids. Chem Eng Res Des 112:103–112

Shams KA, Abdel-Azim NS, Saleh IA, Hegazy MF, El-Missiry MM, Hammouda FM, Bohouth E, Tahrir E (2015) Green technology: economically and environmentally innovative methods for extraction of medicinal & aromatic plants (MAP) in Egypt. J Chem Pharm Res 7(5):1050–1074

Sharma PK, Fuloria S, Alam S, Sri MV, Singh A, Sharma VK, Kumar N, Subramaniyan V, Fuloria NK (2021) Chemical composition and antimicrobial activity of oleoresin of Capsicum annuum fruits. Mindanao J Sci Technol 19(1):1

Sharma S, Kushwaha RS, Naman S, Patil UK, Baldi A (2021) Optimized extraction of oleoresin Capsicum and analytical method validation for capsaicin using HPLC. Indian J Nat Prod 35(1): 38–45. https://doi.org/10.5530/ijnp.2021.1.8

Silva LPS, Martínez J (2014) Mathematical modeling of mass transfer in supercritical fluid extraction of oleoresin from red pepper. J Food Eng 133:30–39

Soldan ACF, Arvelos S, Watanabe EO, Hori CE (2021) Supercritical fluid extraction of oleoresin from Capsicum annuum industrial waste. J Clean Prod 297:126593

Sricharoen P, Lamaiphan N, Patthawaro P, Limchoowong N, Techawongstien S, Chanthai S (2017) Phytochemicals in Capsicum oleoresin from different varieties of hot chilli peppers with their antidiabetic and antioxidant activities due to some phenolic compounds. Ultrason Sonochem 38: 629–639

Swamy MK, Arumugam G, Kaur R, Ghasemzadeh A, Yusoff MM (2017) Sinniah UR (2017) GC-MS based metabolite profiling, antioxidant and antimicrobial properties of different solvent extracts of Malaysian Plectranthus amboinicus leaves. Evid Based Complement Alternat Med 2017:1

Swamy MK, Sinniah UR (2016) Patchouli (Pogostemon cablin Benth.): botany, agrotechnology and biotechnological aspects. Ind Crop Prod 87:161–176

Tandon GL, Dravid SV, Siddappa GS (1964) Oleoresin of Capsicum (red chilies)—some technological and chemical aspects. J Food Sci 29(1):1–5

Tatke P, Jaiswal Y (2011) An overview of microwave assisted extraction and its applications in herbal drug research. Res J Med Plant 5(1):21–31

Uquiche E, Millao S, del Valle JM (2022) Extrusion affects supercritical CO2 extraction of red pepper (Capsicum annuum L.) oleoresin. J Food Eng 316:110829

Waqas M, Ahmed D, Qamar MT (2022) Surfactant-mediated extraction of capsaicin from Capsicum annuum L. fruit in various solvents. Heliyon 8:e10273

Williams OJ, Raghavan GV, Orsat V, Dai J (2004) Microwave-assisted extraction of capsaicinoids from capsicum fruit. J Food Biochem 28(2):113–122

Yazawa S, Suetom N, Okamoto K, Namiki T (1989) Content of capsaicinoids and capsaicinoid-like substances in fruit of pepper (Capsicum annuum L.) hybrids made with′ CH-19 sweet′ as a parent. J Jpn Soc Hortic Sci 58(3):601–607

Zhang QW, Lin LG, Ye WC (2018) Techniques for extraction and isolation of natural products: a comprehensive review. Chin Med 13:1–26

Zhang HF, Yang XH, Wang Y (2011) Microwave assisted extraction of secondary metabolites from plants: current status and future directions. Trends Food Sci Technol 22(12):672–688

5 Analysis of Capsaicin and Related Compounds by Modern Chromatographic Methods

Aditi Sharma, Laxmi Devi, Mallappa Kumara Swamy, Sayan Bhattacharya, and Devendra Kumar Pandey

Abstract

Capsaicin is a major bioactive of chilli pepper, belonging to the genus *Capsicum* (family Solanaceae). It is the most pungent compound of hot peppers with marked pharmacological properties. Chilli pepper also possesses other chemical compounds, such as dihydrocapsaicin, nordihydrocapsaicin, carotenoids, oleoresin, flavonoids, etc., and the selectivity of these bioactive compounds has been done to meet up the increasing market demand. An accurate analysis of a sample's chemical composition can provide priceless data that will help solve chemical issues and guarantee the quality of a chemical formulation or product. There is an increased demand for novel analytical techniques over the last 10 years that are more reliable, accurate and quick to implement. Chromatographic techniques are regularly utilized during analytical processes for separating, determining and identifying wide-ranging analytes existing in samples having multifaceted and sometimes varying matrices. For analysing capsaicin and related compounds, various modern techniques, including high-performance liquid chromatography, ultra-performance liquid chromatography, high-performance thin-layer

A. Sharma · L. Devi · D. K. Pandey (✉)
Department of Biotechnology, Lovely Professional University Punjab, Jalandhar, India

M. K. Swamy
East West First Grade College (Bangalore University), Bengaluru, India

S. Bhattacharya
School of Ecology and Environment Studies, Nalanda University, Rajgir, India

M. K. Swamy, A. Kumar (eds.), *Capsaicinoids*,
https://doi.org/10.1007/978-981-99-7779-6_5

chromatography, gas chromatography–mass spectrometry, etc., have been extensively used. This chapter therefore discusses on different chromatographic approaches used for separating and/or identifying the capsaicin contents in a mixture.

Keywords

Capsicum · Analysis · Techniques · Separation · Chromatography · Molecules

5.1 Introduction

The fruit of the *Capsicum* species, generally named as pepper, is extensively used in many different cookeries throughout the globe for its distinct palates, which may be hot, spicy, sweetened or even bitter. Several members of the universal populace, particularly those who lived near the Mediterranean, Southern America and Asia, developed a need for these distinctive flavours and began consuming them on a regular basis in the form of raw, cooked or processed foods (Pugliese et al. 2013). The genus *Capsicum* is recognized by its foremost domesticated species, *C. chinense*, *C. annuum*, *C. frutescens*, *C. pubescens* and *C. baccatum*. A group of alkaloid compounds occurring in *Capsicum* species are known as capsaicinoids, accounting for the pungent and spicy flavour, and have potentially valuable pharmacological properties. Capsaicin ($C_{18}H_{27}NO_3$) is the chief bioactive compound of chilli peppers. The other major valuable compounds include nordihydrocapsaicin, dihydrocapsaicin, homodihydrocapsaicin and homocapsaicin. All these compounds put together are known as capsaicinoids (Sricharoen et al. 2017; Jang et al. 2021). Most reports state that capsaicinoids are kept in vesicles on the surface of the placenta's epidermis, although seeds sometimes absorb capsaicinoids attributable to their intimacy to the placenta and the substance's low concentration in fruit flesh (Moreno et al. 2012). The pungent property of capsaicinoids is a consequence of the ability of mammals to perceive capsaicin and other related compounds, named collectively vanilloids. In fact vanilloid receptor subtype 1 (VR1), a capsaicin receptor, has been cloned and identified. Due to this pungency perception in mammals, capsaicin has many pharmacological properties (Díaz et al. 2004). The primary hot substance in chilli peppers is capsaicin, whose pungency is often gauged in Scoville heat unit (SHU). The Scoville scale, established by the American pharmacist, Wilbur Scoville in the early 1900s, is a purely subjective measurement based on the sensitivity of those who have consumed spicy peppers before to capsaicin (Othman et al. 2011).

Lately, CH-19 (a non-pungent cultivar of red pepper), has been shown to have another collection of capsaicin analogues known as capsinoids. It consists of capsiate, dihydrocapsiate, nordihydrocapsiate, etc., having a similar structure to capsaicinoids, but with minor pungency. Due to their wide range of therapeutic properties, such as analgesic, anti-inflammation and antioxidant effects, capsaicinoids and capsinoids have both received a lot of interest amongst researchers, drug manufacturers and food technologists (luo et al. 2011).

Chilli peppers, having several minerals, amino acids and vitamins, are indispensable for human healthiness, growth and development. The nutritive value of *Capsicum* is high and comprises high intensities of antioxidants like flavonoids, vitamin A and vitamin C (Palma et al. 2015). Peppers are also rich in carotenoids like lutein, capsanthin, capsorubin, β-carotene and violaxanthin. The chromophore that gives each colour its characteristics, known as a carotenoid, is a tetraterpene composed of a core polyenic chain with 9 conjugated double bonds and several end groups (Mohd Hassan et al. 2019).

Presently, estimation of capsaicinoid contents in *Capsicum* species is of great attention for food technologists, nutritionists and plant breeders. Besides, more interest is paved towards assessing breeding lines for genetic selection drives and monitoring capsaicin levels in the course of processing or storage. This interest has emphasized the requirement for quick and effective quantification approaches (González-Zamora et al. 2015). Capsaicin and other related compounds can be analysed by different chromatographic techniques. These include spectrophotometry; thin-layer chromatography (TLC); high-performance liquid chromatography (HPLC) and HPLC methods with UV, electrochemical, fluorescence and mass spectroscopy (MS) and gas chromatography with flame ionization detection; ultra-performance liquid chromatography (UPLC); ultra-high-performance liquid chromatography (UHPLC); and high-performance thin-layer chromatography (HPTLC) (Chuichulcherm et al. 2013; Cai 2014; Nag et al. 2017; Rochat 2018; Oliveira Junior et al. 2019; Emanuel et al. 2019; Kim and Jin 2020; Nadi et al. 2020; Sousa et al. 2021 Biradar et al. 2022). This chapter will therefore discuss on different chromatographic approaches used for separating and/or identifying the capsaicin contents in a mixture.

5.2 Chromatography Techniques

The term 'chromatography' refers to a variety of methods based on the partitioning and/or adsorption of the molecules to be separated between mobile phases and stationary phases. The effect of varying adsorption energies and molecule distributions between the two phases is separation. Mobile phases are often liquids or gases, while stationary phases are typically solids (inorganic or organic). The characteristics of the mobile phases and stationary phases, the shape of the stationary phase and the forces that drive separation are all used to categorize chromatographic procedures. Chromatographic methods are extensively utilized in the food industries to characterize and quantify diverse food constituents (Sarker et al. 2006; Kumari et al. 2022; Mir-Cerdà et al. 2023). Chromatography has also developed methods for analysing nanoparticles. Based on variations in their sizes, this method may be employed to distinguish certain nanomaterials and other intrusive particles (Brewer and Striegel 2011). Chromatographic techniques have been more widely accepted and used in food science and technology because of their great separation capacity for the quantitative study of a wide range of molecules, including natural food components, legal or illegal additions, contaminants, etc. Chromatography techniques can be classified based on the physical state of phases, the bed and the separation approaches (Kumari et al. 2022).

While there are several analytical methods that come under the field of organic chemistry, gas chromatography (GC) and liquid chromatography (LC) are two of the most often utilized environmental materials. Both methods use a chromatography column for separating mixture components, but as their names imply, one does it while the components are in the gaseous form and the other while they are in the liquid state.

Setting up an analytical technique for environmental samples is a difficult procedure, since there are so many factors to be considered. The sample's suitability for analysis, how it is brought into the system, how chromatographic separation is accomplished and how the detector is set up to track results are all factors that must be taken into account while developing a technique.

5.2.1 Sample Preparation

A sample has to be ready for analysis, so that any separation may be performed on it. An effective sample preparation aims to transform the sample, whether it be a solid, liquid or gas, via a procedure that will leave it in a suitable container and condition so that it may be put into the analytical system for analysis. This method may vary greatly in complexity and can sometimes be as easy as transferring a part of the material into a vial that can be examined. The kind of sample matrix to be evaluated, the kind of sample introduction procedure to be utilized and the nature of the chemicals of interest will all influence how a sample is prepared for analysis.

5.2.2 Analytical Methods

The previous 10 years has witnessed an increase in demand for novel analysing techniques, which are very precise and dependable, need less time to operate, are less expensive and utilize and produce less hazardous compounds (Stoica et al. 2016; Kumar et al. 2022). Traditional methods like use of panel tasters are employed to determine the pungency level or capsaicin concentration. One of such best examples is the Scoville organoleptic testing method. A summary of recent advances in analytical techniques is shown in Table 5.1.

Qualitative analysis is usually chromatographic data that is shown as a chromatogram, which is a graph of detector response (y-axis) versus retention time (x-axis). This gives a spectrum of peaks for a sample that shows which analytes are in the sample and when they elute from the column (Valcárcel et al. 2007). In quantitative analysis, for example, chromatographic analysis, quantitative measurements are made by measuring the peak height or peak area of a sample whose concentration is unknown.

Table 5.1 Summary of modern analytical techniques for the analysis of capsaicin and related compounds

Species	Mobile phase	Techniques	Compound/s	References
C. chinense jacq.	Acetonitrile in water (50:50)	HPLC	Capsaicin, dihydrocapsaicin	Kehie et al. (2013)
C. chinense	(acetonitrile/water) (60:40) isocratic	UHPLC–DAD–MS/MS Thermo LCQ Fleet system.	Capsaicinoids	Sganzerla et al. (2014)
C. annuum L.	Benzene/ethyl acetate/methanol (75:20:5)	HPTLC	Capsaicin	Gantait et al. (2012)
C. chinense. C. annuum C. frutescens	Acetonitrile/water (50:50) isocratic	Thermo HPLC system, photodiode array (PDA) detector, column BetaSil C18	Capsaicin, dihydrocapsaicin and nordihydrocapsaicin	Othman et al. (2011), Sganzerla et al. (2014)
Capsicum spp.	Acetonitrile (A)/ water 0.1% formic acid (B). Gradient started with 40% A changing to 50% of A in 8 min	UPLC–MS Waters Acquity column BEH C18	Capsaicin, dihydrocapsaicin, nordihydrocapsaicin, homocapsaicin and homodihydrocapsaicin	Alothman et al. (2012), Sganzerla et al. (2014)
C. chinense	Methanol/water (75:25)	HPLC	Capsaicinoids (capsaicin, dihydrocapsaicin) Capsinoids (capsiate, dihydrocapsiate)	Tanaka et al. (2015)
C. chinense (Bhut Jolokia)	Methanol/acetic acid/hexane (2.85: 0.15:0.15)	HPTLC	Capsaicin	Kundu et al. (2015)
C. chinense C. annuum C. frutescens	Ethyl acetate	HPLC–ESI–MS/MS	Capsiate, dihydrocapsiate	Fayos et al. (2019)
C. annuum	Acetonitrile/water (70:30)	HPLC	Capsiate	Zunun-Pérez et al. (2017)
C. annuum (red bell pepper)	Chloroform/ethyl acetate/formic acid (5:4:1) Toluene/ethyl acetate/formic acid (5:4:1) Benzene/ isopropanol (9:1)	HPTLC	Phenols Flavonoids Carotenoids	Prabakaran et al. (2017)
C. annuum L.	Chloroform/ methanol/acetic acid (9.5:0.5:0.1)	HPTLC	Capsaicin, ascorbic acid	Pradhan et al. (2018)

(continued)

Table 5.1 (continued)

Species	Mobile phase	Techniques	Compound/s	References
C. chinense	Water (0.1% acetic acid)/acetonitrile (0.1% acetic acid)	HPLC with a C18 reverse-phase fused-core column	Capsaicin, dihydrocapsaicin	Stipcovich et al. (2018)
C. chinense	Hexane, ethyl acetate, acetone, methanol	UPLC–PDA–ESI–MS/MS	Phenolic compounds (N-caffeoyl putrescine, apigenin, luteolin and diosmetin derivatives)	Herrera-Pool et al. (2021)
C. annuum	Water/acetonitrile (1:1)	HPLC	Capsaicin, dihydrocapsaicin, nordihydrocapsaicin, homocapsaicin	Genovese et al. (2021)
C. annuum	Acetone/ethanol (1:1)	UHPLC–DAD	Phenolic compound	Ahmad et al. (2021)
C. chinense (Naga jolokia V.)	Water and acetonitrile both acidified by adding 0.1% acetic acid)	UHPLC–DAD	Capsaicinoids and capsinoids	Vázquez-Espinosa et al. (2021)

5.3 Column Chromatography (CC)

One of the easy and commonly used approach of separation and purification is column chromatography (CC). It can be utilized for separating and purifying both liquid and solid materials (Jiang et al. 2021). In CC, immobile solid phases work along with liquid mobile phases for adsorbing and separating the components moving through it. Compounds are adsorbed on the basis of their chemical composition and eluted according to their varying adsorption of materials by the adsorbent. For CC analysis, various immobile phases, such as silica, calcium phosphate, alumina, starch, calcium carbonate and magnesia, are used. Likewise, different solvent compositions are utilized considering chemical types to be isolated and separated (Regenstein and Regenstein 1984).

Chuichulcherm et al. (2013) identified the optimum capsaicin separation and purification from *C. frutescence* Linn. using CC having the Taguchi design (Chuichulcherm et al. 2013). Likewise, CC was used for purifying the capsicum extract by making use of a two-solvent system up until the decontaminated capsaicin was identified and collected (Al-Samydai et al. 2019). Kim and Jin (2020) used C-18 column chromatography for the purification isorhamnetin-7-*O*-glucopyranoside (flavonoid) from *C. chinense* with an isocratic elution, i.e. a solvent system of 50% methanol–water.

5.4 Thin-Layer Chromatography (TLC)

TLC is like column chromatography, wherein the solvents flow down through the adsorbent in the column. Unlike CC, it is a rapid, sensitive and cheap method requiring a small sample (in micrograms) for an effective analysis. TLC is often utilized to (1) figure out how many parts are in mixture/s, (2) confirm the distinctiveness and pureness of compounds, (3) track reactions' progress, (4) resolve the arrangement of solvents for a preparative separation and (5) analyse the chemical fractions from CC (Cai 2014).

(Al-Samydai et al. 2019) determined the isolated and purified capsaicin using a combination of CC and TLC with the two-solvent system. In the first column, petroleum ether/*n*-hexane (1:1 v/v) was utilized for de-fatting, and later methanol was gradually poured (1% increment up until 8%). Nadi et al. (2020) evaluated the capsaicin in *C. annuum* Linn. ethanolic extract, and their results showed the level of capsaicin in powder as 0.36%. In the ethanol extract, it was observed to be 1.84%. In their study, the extract was located on a silica gel plate (60 F254) and incubated in the filled space having a solution (chloroform, toluene, acetone) for 2s, 4s and 30s, correspondingly. The plate comprising a dry solution was measured at 228-nm wavelength.

5.5 Gas Chromatography

Gas chromatography (GC) is a sophisticated analytic method preferred by researchers. GC has been used to identify per- and polyfluoroalkyl substances (PFASs); however, it is less often employed than liquid chromatography. GC coupled to a flame ionization detector (GC–FID) and an electron capture detector (GC–ECD) was used in an early investigation on the analytical measurement of PFASs. For identifying and quantifying samples in various environmental matrices, gas chromatography (GC) coupled with a mass spectrometer (MS) is utilized preferentially (Al Amin et al. 2020).

Maokam et al. (2014) determined the major and minor constituents of capsaicinoids in seven leading *Capsicum* varieties of fruits cultivated in the north-eastern part of Thailand using GC–MS. Oliveira Junior et al. (2019) evaluated the natural capsaicin in pepper spray by using GC–MS/FID as an alternate to using oleoresin *Capsicum*, and the same effect was maintained, but there was better chemical control.

5.6 Liquid Chromatography

Liquid chromatography (LC) is one of the best and most common analytical techniques to find and separate proteins, peptides and other biomolecules in which the mobile phase is liquid. LC has shown itself to be very selective and flexible. This distinctive combination makes it possible to get a good mass yield and a high level of biological activity, as well as a very high purity (Afeyan et al. 1992). The application of combined liquid chromatography with mass spectroscopy (LC–MS) has been seen as a great success. Although LC–MS equipment is widely utilized in many other domains of application, including biochemical and biotechnological ones, it plays a significant role in environmental analysis (Niessen 1999). The revolution is sparked by the introduction of new liquid-phase atmospheric pressure ionization (API) methods, including electrospray (ES) ionization. The application of both qualitative and quantitative LC–MS approaches to highly polar compounds has been substantially eased by LC–ES–MS (Gelpí 1995; Rochat 2018).

5.6.1 High-Performance Liquid Chromatography (HPLC)

The modern liquid chromatography includes high-performance liquid chromatography (HPLC), which guarantee high sensitivity, and also has a gas analogue. Like liquid chromatography (LC), liquid–solid chromatography (LSC) and liquid–liquid chromatography (LLC), HPLC operates on the same fundamental principles.

Nag et al. (2017) authenticated the capsaicin in Indian *Capsicum* species through the RP-HPLC method, which was performed in reverse-phase C_{18} column with water and acetonitrile (35: 65 v/v) as mobile phase at a 1 mL/min flow rate. Oliveira Junior et al. (2019) used HPLC to determine the natural capsaicin in pepper spray, and the result showed the ratio of capsaicin and dihydrocapsaicin as 2:1, which agreed with the typical ratio. Caleb et al. (2021) quantified three capsaicinoids (capsaicin, dihydrocapsaicin, nordihydrocapsaicin) with a high-performance liquid chromatography–diode array detector (HPLC–DAD) from pepper (*C. annuum* L.) using an optimum isocratic mobile-phase composition, i.e. methanol (55% v/v) and acetic acid (45% v/v) at a 1.2 ml/min flow rate and 22 °C.

5.6.2 Supercritical Fluid Chromatography

In supercritical fluid chromatography (SFC), the supercritical fluid is mobile phase, which is between a gas and a liquid. So, SFC has qualities that are a mix of LC and GC. Carbon dioxide is an example of a supercritical fluid; at ≥31.1 °C and high pressure (7.4 MPa), it exhibits supercritical behaviour (Sousa et al. 2021). It is regularly employed for separations that involve thermally labile or non-volatile substances that are difficult to separate by LC or GC. The viscosities and solute diffusivities of supercritical mobile phases (usually CO_2) fall somewhere in the middle of those of gases and liquids. By applying great pressure to the gas, supercritical CO_2 is produced (White 2017).

Giuffrida et al. (2017) determined the apocarotenoids in red habanero pepper (*C. chinense* Jacq. Cv) by using novel supercritical fluid chromatography–triple-quadrupole/mass spectrometry (SFC–QqQ/MS) methodology, and in their result a total of 25 different apocarotenoids (11 apocarotenoid fatty acids esters and 14 free apocarotenoids) were identified.

5.6.3 Ultra-High-Pressure Liquid Chromatography (UHPLC)

In several application sectors, ultra-high-performance liquid chromatography (UHPLC) has already developed into a mature and admired separation technique. Practitioners are drawn to the vast spectrum of UHPLC equipment with columns packed with superficially porous particles (less than 2 μm), because of their high efficiency, high resolution, speed of analysis, robustness, dependability and commercial availability. For the analysis of complex materials, combining UHPLC with mass spectrometry (MS) offers added advantages in terms of sensitivity, selectivity and high throughput (Nováková et al. 2017).

Schmidt et al. (2017) determined the capsaicin and dihydrocapsaicin by the UHPLC method from Bhut jolokia (*C. chinense*) within 1.7 min using the mobile phase of formic acid and acetonitrile. Biradar et al. (2022) evaluated the capsaicinoids and capsinoids from peppers (*Capsicum* spp.) by the UHPLC method using a phenyl–hexyl stationary phase with acidified water and methanol as a mobile phase at column temperature (55 ° C) with a 0.5 mL/min flow rate in 5 min.

5.6.4 Ultra-Performance Liquid Chromatography (UPLC)

Ultra-performance liquid chromatography (UPLC) is a very attractive method mainly because of these reasons: sensitivity, speed and resolution. It is one of the contemporary techniques that gives a new direction to liquid chromatography. UPLC works on the same basic ideas as HPLC. The main difference is that the particles in the column material are less than 2 m in size (Trenerry and Rochfort 2010).

Emanuel et al. (2019) identified the phenolic compounds in root and leaf samples of peppers (*C. chinense*) by UPLC analysis using photodiode array detector (PDA) and mass spectroscopy in the mobile phase of water and methanol. Cortés-Ferré et al. (2022) determined the phenolic profile from chilli pepper seeds (*C. chinense*) by UPLC–MS-QqQ using mobile-phase formic acid and acetonitrile at a 0.3 mL/min flow rate.

5.6.5 High-Performance Thin Layer Chromatography

High-performance thin-layer chromatography (HPTLC) is often used in the pharmaceutical industry for procedure development, identifying adulterants and detecting herbal raw material, determining pesticides and mycotoxin contents and quality check of herbal and health food items. Due to its high throughput (many samples may be studied at once), an economy in the use of mobile phase, and affordability, HPTLC has reportedly been shown to provide more advantages over the standard HPLC technology (Mishra et al. 2018). It works on the same principle as adsorption chromatography, i.e. separations based on adsorption principle.

Hema et al. (2021) quantified capsaicin in transferosome formulation by the HPTLC method consisting of N-hexane: tert-iso-butyl-methyl ether (5:15 v/v). Foudah et al. (2022) determined the ascorbic acid in the solvent extract of pepper (*C. annuum*) by the green HPTLC method using a mobile phase of water and ethanol (70:30, v/v) and the method was linear in the 25–1200 ng/band range.

5.7 Detectors

After the chemicals in a sample have been separated using gas or liquid chromatography, a detector of some form must be used to identify the components of the mixture that has been separated. The kind of detector utilized will be greatly influenced by the type of analysis that is needed. A broad range of detectors is available that may be paired with either an LC or a GC system to enable a chemical analysis to be accomplished. Certain detector types, such as fluorescence detectors, are only appropriate for LC systems, whereas others, like flame ionization detectors (FID), may only be utilized for GC. There are detectors that can operate with either method, and mass spectrometry (MS) is a popular illustration of one of them. Since gas chromatography–mass spectrometry (GC–MS) and liquid chromatography–mass spectrometry (LC–MS) are two extremely potent and popular methods for environmental organic analysis, a great deal of research and development has been done on these methods to improve their selectivity and lower thresholds for substance detection.

5.8 Conclusion

The choice of chromatographic technology for refolding relies on the physicochemical characteristics of samples, such as solubility, hydrophobicity, isoelectric point, molecular weight and stability conditions. When it comes to analysing environmental samples that include complex combinations of substances, both liquid and gas chromatography are very helpful procedures. The best procedures may be selected according to the sample type and the chemistry of the chemicals of interest, and when they are used in conjunction with an effective sample extraction procedure and a functional detector, accurate analysis can be accomplished. To meet up the market demand, purification and characterization of capsaicin and related compounds (dihydrocapsaicin, nordihydrocapsaicin, carotenoids) is achieved by different modern analytical techniques like TLC, HPLC, GC–MS, HPTLC, UHPLC and UPLC. The above discussion suggests that HPLC, UPLC and HPTLC are the best techniques for characterizing capsaicin and its related compounds. A substantial improvement has been achieved in the recent past regarding selection of columns, solvents choices and process modelling. However, ample prospects remain further for a systematic evaluation of technology choices and expenses and selection of solvents in identifying capsaicinoids. More emphasis should be given on designing rapid and cost-effective chromatographic approaches to detect capsaicin, which is of the chief source in demand by the food and pharma industries.

References

Afeyan NB, Gordon NF, Fulton SP, Regnier FE (1992) Perfusion chromatography: a novel tool for protein purification and analysis. In: Techniques in protein chemistry III. Academic, Cambridge, MA, pp 135–149. https://doi.org/10.1016/B978-0-12-058756-8.50021-3

Ahmad R, Ahmad N, Alkhars S, Alkhars A, Alyousif M, Bukhamseen A, Abuthayn S, Aqeel M, Aljamea A (2021) Green accelerated solvent extraction (ASE) with solvent and temperature effect and green UHPLC-DAD analysis of phenolics in pepper fruit (Capsicum annum L.). J Food Compos Anal 97:103766

Al-Samydai A, Aburjai T, Alshaer W, Azzam H, Al-Mamoori F (2019) Qualitative and quantitative analysis of capsaicin from Capsicum annum grown in Jordan. Int J Res Pharm Sci 10(4): 3768–3774

Al Amin M, Sobhani Z, Liu Y, Dharmaraja R, Chadalavada S, Naidu R, Fang C (2020) Recent advances in the analysis of per-and polyfluoroalkyl substances (PFAS)—a review. Environ Technol Inno 19:100879

Al Othman ZA, Ahmed YBH, Habila MA, Ghafar AA (2011) Determination of capsaicin and dihydrocapsaicin in capsicum fruit samples using high performance liquid chromatography. Molecules 16(10):8919–8929

Alothman ZA, Wabaidur SM, Khan MR, Ghafar AA, Habila MA, Ahmed YBH (2012) Determination of capsinoids in Capsicum species using ultra performance liquid chromatography–mass spectrometry. J Sep Sci 35(21):2892–2896

Biradar K, Singh J, Pillai SS, Crosby KM, Patil BS (2022) Separation of nordihydrocapsiate from capsiate and major capsaicinoid analogues using ultra high performance liquid chromatography. Food Chem 382:132585

Brewer AK, Striegel AM (2011) Characterizing the size, shape, and compactness of a polydisperse prolate ellipsoidal particle via quadruple-detector hydrodynamic chromatography. Analyst 136(3):515–519

Cai L (2014) Thin layer chromatography. Curr Protoc Essent Lab Tech 8(1):6–3

Caleb J, Alshana U, Hanoğlu A, Çaliş İ (2021) Dispersive liquid-liquid microextraction for the isolation and HPLC-DADdetermination of three major capsaicinoids in Capsicum annuum L. Turk J Chem 45(2):420–429

Chuichulcherm S, Prommakort S, Srinophakun P, Thanapimmetha A (2013) Optimization of capsaicin purification from *Capsicum frutescens* Linn. With column chromatography using Taguchi design. Ind Crop Prod 44:473–479

Cortés-Ferré HE, Martínez-Avila M, Antunes-Ricardo M, Guerrero-Analco JA, Monribot-Villanueva JL, Gutiérrez-Uribe JA (2022) In vitro evaluation of anti-inflammatory activity of "habanero" chili pepper (Capsicum chinense) seeds extracts pretreated with cellulase. Plant Foods Hum Nutr 8(1):109–116

Díaz J, Pomar F, Bernal A, Merino F (2004) Peroxidases and the metabolism of capsaicin in Capsicum annuum L. Phytochem Rev 3(1):141–157

Emanuel HP, Jesús PV, Ana RD, Teresa AT, Neith P (2019) Extraction and identification of phenolic compounds in roots and leaves of Capsicum chinense by UPLC–PDA/MS. J Bioeng 3(2):17–27

Fayos O, Savirón M, Orduna J, Barbero GF, Mallor C, Garcés-Claver A (2019) Quantitation of capsiate and dihydrocapsiate and tentative identification of minor capsinoids in pepper fruits (Capsicum spp.) by HPLC-ESI-MS/MS(QTOF). Food Chem 270:264–272

Foudah AI, Alam P, Shakeel F, Alam A, Salkini MA, Alshehri S, Ghoneim MM, Ross SA (2022) A simple, cost-effective, and green HPTLC method for the estimation of ascorbic acid in solvent and ultrasound-assisted extracts of *Phyllanthus emblica*, *Capsicum annuum*, and *Psidium guajava*. Agronomy 12(5):1016

Gantait A, Maji A, Barman T, Banerji P, Venkatesh P, Mukherjee PK (2012) Estimation of capsaicin through scanning densitometry and evaluation of different varieties of capsicum in India. Nat Prod Res 26(3):216–222

Gelpí E (1995) Biomedical and biochemical applications of liquid chromatography-mass spectrometry. J Chromatogr A 703(1–2):59–80

Genovese S, Epifano F, Marchetti L, Bastianini M, Cardellini F, Spogli R, Fiorito S (2021) Pre-concentration of capsaicinoids from different cultivars of Capsicum annuum after extraction in heterogenous mixtures. J Food Compos Anal 102:104052

Giuffrida D, Zoccali M, Giofre SV, Dugo P, Mondello L (2017) Apocarotenoids determination in Capsicum chinense Jacq. cv. Habanero, by supercritical fluid chromatography-triple-quadrupole/mass spectrometry. Food Chem 231:316–323

González-Zamora A, Sierra-Campos E, Pérez-Morales R, Vázquez-Vázquez C, Gallegos-Robles MA, López-Martínez JD (2015) García-Hernández JL (2015) measurement of capsaicinoids in chiltepin hot pepper: a comparison study between spectrophotometric method and high performance liquid chromatography analysis. J Chem 2015:1

Hema NS, Shivamurthy MV, Karunakar P (2021) Novel simultaneous identification of capsaicin and it's quantification in transferosome formulation by HP-TLC technique. Curr Pharm Anal 17(1):172–183

Herrera-Pool E, Ramos-Díaz AL, Lizardi-Jiménez MA, Pech-Cohuo S, Ayora-Talavera T, Cuevas-Bernardino JC, García-Cruz U, Pacheco N (2021) Effect of solvent polarity on the ultrasound assisted extraction and antioxidant activity of phenolic compounds from habanero pepper leaves (Capsicum chinense) and its identification by UPLC-PDA-ESI-MS/MS. Ultrason Sonochem 76: 105658

Jang S, Park M, Lee DG, Lim JH, Jung JW, Kang BC (2021) Breeding Capsicum chinense lines with high levels of capsaicinoids and capsinoids in the fruit. Agriculture 11(9):819

Jiang T, Ghosh R, Charcosset C (2021) Extraction, purification and applications of curcumin from plant materials-a comprehensive review. Trends Food Sci Technol 112:419–430

Kehie M, Kumaria S, Tandon P (2013) In vitro plantlet regeneration from cotyledon segments of Capsicum chinense Jacq. cv. Naga king chili, and determination of capsaicin content in fruits of in vitro propagated plants by high performance liquid chromatography. Sci Hortic 164:1–8

Kim JH, Jin CH (2020) Inhibitory activity of flavonoids, chrysoeriol and luteolin-7-o-glucopyranoside, on soluble epoxide hydrolase from Capsicum chinense. Biomol Ther 10(2): 180

Kumar S , Ahanger MA , Alshaya H , Jan BL , Yerramilli V, (2022) Salicylic acid mitigates salt induced toxicity through the modifications of biochemical attributes and some key antioxidants in capsicum annuum. Saudi J Bio Sci 29(3):1337–1347

Kumari VC, Patil SM, Ramu R, Shirahatti PS, Kumar N, Sowmya BP, Egbuna C, Uche CZ, Patrick-Iwuanyanwu KC (2022) Chromatographic techniques: types, principles, and applications. In: Analytical techniques in biosciences. Academic, Cambridge, MA, pp 73–101

Kundu SU, Das AN, Ghosh BI (2015) Modulation of pungency and major bioactive compounds in pepper due to agro-climatic discrepancy: a case study with Capsicum chinense 'Bhut Jolokia'fruit. Int J Pharm Pharm Sci 7:294

Luo XJ, Peng J, Li YJ (2011) Recent advances in the study on capsaicinoids and capsinoids. Eur J Pharmacol 650(1):1–7

Maokam C, Techawongstien S, Chanthai S (2014) Determination of major and minor capsaicinoids in different varieties of the Capsicum fruits using GC-MS and their inhibition effect of the chilli extract on [alpha]-amylase activity. Int Food Res J 21(6):2237

Mir-Cerdà A, Nunez O, Granados M, Sentellas S, Saurina J (2023) An overview of the extraction and characterization of bioactive phenolic compounds from agri-food waste within the framework of circular bioeconomy. TrAC Trends Anal Chem 2023:116994

Mishra J, Sharma RK, Misra K (2018) Characterization techniques for herbal products. In: Management of high altitude pathophysiology. Academic, Cambridge, MA, pp 171–202

Mohd Hassan N, Yusof NA, Yahaya AF, Mohd Rozali NN, Othman R (2019) Carotenoids of capsicum fruits: pigment profile and health-promoting functional attributes. Antioxidants 8(10): 469

Moreno E, Fita A, González-Mas MC, Rodríguez-Burruezo A (2012) HS-SPME study of the volatile fraction of Capsicum accessions and hybrids in different parts of the fruit. Sci Hortic 135:87–97

Nadi MS, Fikri F, Purnama MTE (2020) Determination of capsaicin levels in Capsicum annum Linn Ethanolic extract using thin layer chromatography analysis. Drugs 70(14):1831–1842

Nag M, Chanda J, Biswas R, Al-Dhabi NA, Duraipandiyan V, Banerji P, Mukherjee PK (2017) Validation of capsaicin in Indian Capsicum species through RP-HPLC. Indian J Pharm Educ Res 51(2):337–342

Niessen WMA (1999) State-of-the-art in liquid chromatography–mass spectrometry. J Chromatogr A 856(1–2):179–197

Nováková L, Svoboda P, Pavlík J (2017) Ultra-high performance liquid chromatography. In: Liquid chromatography. Elsevier, Amsterdam, pp 719–769

Oliveira Junior N, Padulla V, Flores I, Annunciação D, Neiva M, Brant V (2019) Evaluation of natural capsaicin (N. Cap) in pepper spray by GC-MS/FID, NMR and HPLC as an alternative to the use of oleoresin Capsicum (OC). Hum Factors Mech Eng Def Saf 3:1–10

Palma JM, Sevilla F, Jiménez A, del Río LA, Corpas FJ, Alvarez de Morales P, Camejo DM (2015) Physiology of pepper fruit and the metabolism of antioxidants: chloroplasts, mitochondria, and per-oxisomes. Ann Bot 116:627–636

Prabakaran S, Ramu L, Veerappan S, Pemiah B, Kannappan N (2017) Effect of different solvents on volatile and non-volatile constituents of red bell pepper (Capsicum annuum L.) and their in vitro antioxidant activity. J Food Meas Charact 11(4):1531–1541

Pradhan K, Nandi A, Das A, Sahu N, Senapati N, Mishra SP, Patnaik A, Pandey G (2018) Quantification of capsaicin and ascorbic acid content in twenty four Indian genotypes of chilli (Capsicum annuum L.) by HPTLC and volumetric method. Int J Pure Appl Biosci 6:1322–1327

Pugliese A, Loizzo MR, Tundis R, O'Callaghan Y, Galvin K, Menichini F, O'Brien N (2013) The effect of domestic processing on the content and bioaccessibility of carotenoids from chili peppers (Capsicum species). Food Chem 141(3):2606–2613

Regenstein JM, Regenstein CE (1984) Column chromatography. In: Food protein chemistry. Academic, Cambridge, MA, pp 144–167

Rochat B (2018) Quantitative and qualitative LC-high-resolution MS: the technological and biological reasons for a shift of paradigm. In: Ince M, Ince OK (eds) Recent advances in analytical chemistry. IntechOpen, London. https://doi.org/10.5772/intechopen.81285

Sarker SD, Latif Z, Gray AI (2006) An introduction to natural products isolation. In: Natural products isolation. Springer, Cham, pp 1–25

Schmidt A, Fiechter G, Fritz EM, Mayer HK (2017) Quantitation of capsaicinoids in different chilies from Austria by a novel UHPLC method. J Food Compos Anal 60:32–37

Sganzerla M, Coutinho JP, de Melo AMT, Godoy HT (2014) Fast method for capsaicinoids analysis from Capsicum chinense fruits. Food Res Int 64:718–725

Sousa BC, Wakelam MJ, Lopez-Clavijo AF (2021) Methods of lipid analysis. In: Biochemistry of lipids, lipoproteins and membranes. Elsevier, Amsterdam, pp 53–83

Sricharoen P, Lamaiphan N, Patthawaro P, Limchoowong N, Techawongstien S, Chanthai S (2017) Phytochemicals in Capsicum oleoresin from different varieties of hot chilli peppers with their antidiabetic and antioxidant activities due to some phenolic compounds. Ultrason Sonochem 38: 629–639

Stipcovich T, Barbero GF, Ferreiro-González M, Palma M, Barroso CG (2018) Fast analysis of capsaicinoids in Naga Jolokia extracts (Capsicum chinense) by high-performance liquid chromatography using fused core columns. Food Chem 239:217–224

Stoica RM, Moscovici M, Tomulescu C, Băbeanu N (2016) Extraction and analytical methods of capsaicinoids-a review. Sci Bull Ser F Biotechnol 20:93–98

Tanaka Y, Sonoyama T, Muraga Y, Koeda S, Goto T, Yoshida Y, Yasuba K (2015) Multiple loss-of-function putative aminotransferase alleles contribute to low pungency and capsinoid biosynthesis in Capsicum chinense. Mol Breed 35(6):1–13

Trenerry VC, Rochfort SJ (2010) Natural products research and Metabolomics. Elsevier, Amsterdam, pp 595–628

Valcárcel M, Cárdenas S, Simonet BM, Carrillo-Carrión C (2007) Principles of qualitative analysis in the chromatographic context. J Chromatogr A 1158(1–2):234–240

Vázquez-Espinosa M, González-de-Peredo AV, Espada-Bellido E, Ferreiro-González M, Barbero GF, Palma M (2021) Simultaneous determination by UHPLC-PDA of major capsaicinoids and capsinoids contents in peppers. Food Chem 356:129688

White RL (2017) Chromatography-IR, methods and instrumentation. In: Lindon JC, Tranter GE, Koppenaal DW (eds) Encyclopedia of spectroscopy and spectrometry, vol 1, 3rd edn. Academic, Cambridge, MA, pp 251–255. https://doi.org/10.1016/B978-0-12-409547-2.10997-7. ISBN:978–0–12-803225-1

Zunun-Pérez AY, Guevara-Figueroa T, Jimenez-Garcia SN, Feregrino-Pérez AA, Gautier F, Guevara-González RG (2017) Effect of foliar application of salicylic acid, hydrogen peroxide and a xyloglucan oligosaccharide on capsiate content and gene expression associated with capsinoids synthesis in *Capsicum annuum* L. J Biosci 42:245–250

Biosynthesis of Capsaicinoids in Plants

6

K. N. Poornima, M. E. Kavya, S. Arpita, and A. Mohan Rao

Abstract

Plant secondary metabolites have played a major role in providing special traits to different plants. Chilli peppers have their pungency due to the secondary metabolites called capsaicinoids, with capsaicin as the principal component. There are two pathways involved in its biosynthesis, one mediated by phenylpropanoid and the other by fatty acid metabolism. Capsaicin concentration in plants is influenced by genetic, environment and crop management factors. Its synthesis can however be enhanced by using elicitors. Capsaicin has been used as a food additive and in pharmaceutical applications as cancer preventive agent. Utilization of cell cultures for the production of biochemicals has been generally low due to low levels of product formation either in the plant or via tissue culture techniques. The media composition can be manipulated to increase culture yield that is on par or higher than mother plant. A comparative study of regulation of accumulation of capsaicin in suspension cultures, immobilized cell cultures and whole fruit gives us an insight about its biosynthetic requirements and efficiency. Efficient capsaicin production has been achieved by application of salicylic acid and calcium channel modulators in suspension cultures. Some of the techniques like virus-induced gene silencing have identified genes to be involved in capsaicinoid synthesis. The study of expression profiles of some genes has shown a relative correlation with pungency. The genome sequence information of hot pepper has been a very potent resource for clear understanding of

K. N. Poornima (✉)
ICAR-IIHR, Hessarghatta, Bengaluru, India

M. E. Kavya · A. Mohan Rao
UASB, GKVK, Bengaluru, India

S. Arpita
Division of Vegetable Sciences, IARI, New Delhi, India

M. K. Swamy, A. Kumar (eds.), *Capsaicinoids*,
https://doi.org/10.1007/978-981-99-7779-6_6

capsaicinoid pathway, and it also helps study evolution of secondary metabolites in plants. This chapter is focused to giving all information related to capsaicinoids, biosynthetic pathways, factors affecting capsaicinoid content and their enhanced production by the use of elicitors.

Keywords

Secondary metabolites · Capsaicinoids · Biosynthetic pathways · Elicitors

6.1 Introduction

Organic molecules like secondary metabolites with their unique carbon skeleton structure offer plants the capacity to compete and thrive in their own environment (Teoh 2015). With a low molecular weight, these secondary metabolites ensure continued existence of the plants by aiding them to interact with their surroundings. In the plant kingdom so far, over 50,000 secondary metabolites have been discovered (Teoh 2015) and are found to assist plants in several modes like for defending against predators and pathogens, for imparting allelopathic effect and nutrient cycling, for dispersal of seeds, for attracting pollinators or for a particular taste or flavour (Pagare et al. 2015). Being the by-products of primary metabolism in plants, secondary metabolites are broadly categorized into three chemically distinct classes, viz., terpenes, phenolics and nitrogen and sulphur–containing compounds (Pagare et al. 2015). Hence, they can be exclusively harnessed by humans for fabrication of drugs, flavours, fragrances, insecticides and dyes.

An ongoing process encompassing academic and pharmaceutical institutions looks for novel secondary products in plants with the intention of finding new products or, even better, new ways for the treatment of sickness. Capsaicinoids are one such secondary metabolites, present as an active component of chili peppers of the genus *Capsicum* and found to have enormous uses in food, in terms of flavour, colouring and preservation, as well as in medicines (Zhang et al. 2021). Until now, a number of studies had been carried out to appraise the chemical structure of capsaicin, pathways involved in the synthesis of capsaicinoids and factors affecting the capsaicin synthesis in chilli.

This chapter is focused to giving all information related to capsaicinoids, their biosynthetic pathways, factors affecting capsaicinoid content and their enhanced production by the use of elicitors.

6.2 Secondary Metabolites in Chilli Peppers

Capsicum sp., enjoying the position in both vegetable and spice groupings (Cho et al. 2020; Kumar et al. 2021), is one of the most consumer-admired and cultivated plants worldwide. *Capsicum* genus comprises of around 30 different species (Wahyuni et al. 2013; Antonio et al. 2018) out of which only 5 species, i.e. *C. annuum* L., *C. frutescens*, *C. chinense*, *C. baccatum* L. and *C. pubescens*, have been cultivated after domestication (Wahyuni et al. 2013). Probably no other cultivated species is used in as many ways as the red pepper like vegetable, spice, condiment, ornament and medicine (Wahyuni et al. 2013; Kumar et al. 2021). Because pepper can be used to fulfil many purposes as food, medicine, dye, etc., it attracted researchers to analyse its chemical composition in detail (Antonio et al. 2018). Secondary metabolites have to be credited for the varied characteristics of chilli pepper, which makes it accessible to consumers in different versions.

A chilli fruit harbours about 200 different volatile compounds and secondary metabolites like capsaicinoids, carotenoids, flavonoids, tocopherols, phytosterols and ascorbic acid (Antonio et al. 2018; Kumar et al. 2021). Upon ripening, the concentration of capsaicinoids rises and gets concentrated in the placenta epidermal cells of the fruit (Naves et al. 2019; Antonio et al. 2018), which decides the pungency it offers to the end users. The irritation offered by capsaicin is only for mammals, but not for birds, thereby adding the advantage of seed dispersal to plants from shifting predation to mutualism (Naves et al. 2019). In spite of being attributed as responsible for the hot taste of chilli, capsaicinoids can also influence non-pungent flavour in sweet pepper, and that analogue of capsaicinoids is called as capsinoid (Antonio et al. 2018; Kumar et al. 2021). Carotenoids, one of the most abundant secondary metabolites present in chilli along with flavonoids, are responsible for vivid colours of the fruit (Antonio et al. 2018). Red colour of chilli owes to the presence of two major carotenoids called capsanthin and capsorubin, while the presence of pigments violaxanthin, β-carotene, lutein, antheraxanthin and zeaxanthin results in yellow-coloured fruits. Quercetin and luteolin contributing to about 41% of the total flavonoids in chilli plant are responsible for UV protection, growth regulation and attraction to pollinators in addition to their chromogenic property. Compared to different sources of vitamin C among fruits and vegetables, chilli takes an upper hand by harbouring a generous amount of Vitamin C (Kumar and Tata 2009), conferring the water-soluble antioxidant property (Howard et al. 1994). Vitamin E active compounds like phytosterols and tocopherols are another class of antioxidants that enhances the nutritional quality of chilli seed oil (Kumar et al. 2021).

Chilli pepper can be called as a functional food due to its therapeutic values such as antioxidant, anti-inflammatory, antitumoral and weight loss properties. All these qualities of chilli can be attributed to capsaicinoids (Naves et al. 2019). Among all other secondary metabolites found in chilli, capsaicin can be extensively exploited for food and therapeutic industry, and it has been the topic of focus for many researchers recently (Antonio et al. 2018).

6.3 Capsaicinoid Structure

In 1816, Christian F. Bucholz extracted a compound from chilli fruits in partially purified crystalline form for the first time (Basith et al. 2016, Kaiser et al. 2017). A year later, was the first to use the term "capsaicin" after noticing that the pungent constituents could form salts with alkalis. After 60 years, Thresh isolated the same compound in pure crystalline form and named it as capsaicin (Basith et al. 2016; Ilie et al. 2019). Nelson, in 1919, partially discovered the chemical structure of capsaicin, which was then artificially synthesized by Spath and Darling in 1930, and later Japanese chemists referred to such compounds as capsaicinoids.

Capsaicinoids, also called as N-vanillyl-acylamides (Kobata et al. 2014), are one among the members of the secondary plant nontoxic alkaloid metabolites family that occur in chilli pepper (Kaiser et al. 2017), where more than 20 different capsaicinoids can be discovered (Yang et al. 2016). In chilli fruits, all these identified capsaicinoids are the vanillamides of branched fatty acids with 9–11 carbon atoms (Sato et al. 1999). Capsaicin (69%) is the most abundant capsaicinoid followed by dihydrocapsaicin (22%), which together contributes for about 80–90% (Kaiser et al. 2017) of the pungency. Other active ingredients of capsaicinoids are nordihydrocapsaicin (7%), homocapsaicin (1%) and homodihydrocapsaicin (1%) (Bennet and Kirby 1968; Kaiser et al. 2017; Popelka et al. 2017). The common structural feature that can be noticed among all capsaicinoids is the presence of an aromatic ring possessing a methoxy and hydroxyl substituents (Rios and Olivo 2014), whereas the differentiating structural feature is the length of aliphatic side chain (C8–C13) along with the way it terminates (linear, iso- or anteiso-series) (Popelka et al. 2017) and degree of saturation in the alkyl side chain region (Basith et al. 2016).

Capsaicin, also called as trans-8-methyl-N-vanillyl-6-nonenamide, is a pale white, crystalline, lipophilic, highly volatile, colourless and odourless solid protoalkaloid (Chairat 2013; Ilie et al. 2019; Rezazadeh et al. 2021) with a molecular weight and melting temperature of 305.4 kDa and 62–65 °C (Basith et al. 2016), respectively. When capsaicin is compared with the compounds such as vanillin, eugenol and zingerone, there is a resemblance in their structure as they belong to the same family vanilloids having an aromatic ring and a long hydrophobic chain with a polar amide group (Ilie et al. 2019). To be specific, the structure of capsaicin consists of an A-region possessing vanillyl (methylcatechol) head group connected to the C-region of hydrophobic aliphatic tail by the central amide bond called B-region (Basith et al. 2016). The pungency offered by capsaicin is owed to the presence of vanillyl moiety in A-region (Rios and Olivo 2014; Basith et al. 2016). An alkaloid called capsinoid, responsible for non-pungency in sweet pepper, is almost similar to capsaicin in its structure; the only difference is that the B-region consists of an ester bond instead of amide (Rios and Olivo 2014; Basith et al. 2016).

Capsaicin and dihydrocapsaicin is comprised of nine-carbon acyl chain (Reyes-Escogido et al. 2011; Rios and Olivo 2014), which is the optimum length required to instigate intense sensation of pungency in humans. Capsaicinoids exhibit a property that approximately half of the pungency will be lost upon addition or deletion of a single carbon atom to the chain until it reaches 17 carbon atoms, and when there are more than 17 carbons, this property cannot be observed (Rios and Olivo 2014). The molecular structure of capsaicinoids is such that it is resistant to ionizing radiation and it readily decomposes due to the cleavage of carbon-nitrogen bond, when exposed to over 80 °C temperature (Uarrota et al. 2021).

When underlying pathways for capsaicinoid synthesis were investigated using radiotracer at the end of 1960s, it was discovered that the two important components of capsaicinoids, vanillamine moiety and branched chain fatty acids, were derived from phenylalanine and valine, respectively. (Aza-González et al. 2011).

6.4 Biosynthetic Pathways

We are already aware of the fact that phenylalanine and valine are the precursors for the synthesis of two main components of capsaicinoids, vanillylamine moiety and branched chain fatty acid, respectively (Aza-González et al. 2011; Ferri et al. 2018; Yap et al. 2021). Hence, biosynthesis of capsaicin occurs in the chilli fruit placental tissue (Díaz et al. 2004; Aza-González et al. 2011) following two separate pathways using capsaicin synthase, one is the condensation of vanillylamine, which is a phenylpropanoid pathway intermediate, and the other is condensation of branched chain fatty acid moieties (8-methyl-6-nonenoic acid) (Prasad et al. 2005; Fernandes et al. 2016; Kaiser et al. 2017). Mainly three nitrogenous molecules are involved in capsaicinoid biosynthesis. These are phenylalanine, providing the phenolic moiety; either one of these two branched amino acids, leucine or valine, which contributes with the lateral aliphatic chain; and GABA (γ-aminobutyric acid), a four-carbon non-protein amino acid, which is the amino donor necessary for forming vanillylamine from vanillin (Gururaj et al. 2012; Fig. 6.1). Therefore, as you would expect, capsaicinoid content in *Capsicum* sp. organs respond to the availability of nitrogen externally (Aldana-Iuit et al. 2015).

In the first pathway where the vanillylamine moiety is synthesized via phenylpropanoid shikimate/arogenate, initially L-phenylalanine is sequentially converted into cinnamic, *p*-coumaric, caffeic and ferulic acids and then to vanillin, which is finally transformed to vanillylamine (Kaiser et al. 2017). Different enzymes that are repeatedly reported to be involved in vanillin synthesis are phenylalanine ammonia-lyase (PAL), cinnamate 4-hydroxylase (C4H), coumarate 3 hydroxylase (C3H), caffeic acid *O*-methyltransferase (COMT) and hydroxycinnamoyl-CoA hydratase/lyase (HCHL) (Walton et al. 2003; Kaiser et al. 2017). The synthesized vanillin is then converted into vanillylamine using an aminotransferase (Aza-González et al. 2011; Kaiser et al. 2017).

Curry et al. (1999), through Northern blot analysis, revealed that the transcript level of *Pal, C4h* and *Comt* coding genes was in correspondence with pungency level as they are involved in capsaicinoid biosynthesis. Two more genes showing the same pattern of transcription as that of *Pal, C4h* and *Comt* genes, one which has sequence homologous to 3-keto-acyl-ACP synthase (*Kas* gene) and the other with high homology to a putative aminotransferase (*pAmt* gene), were identified in chilli pepper. These *Kas* and *pAmt* genes, homologous to newly identified genes, were formerly reported as the putative genes involved in the biosynthesis of branched-chain fatty acid and vanillylamine, respectively, and were observed to express specifically in the placental tissue of chilli fruit.

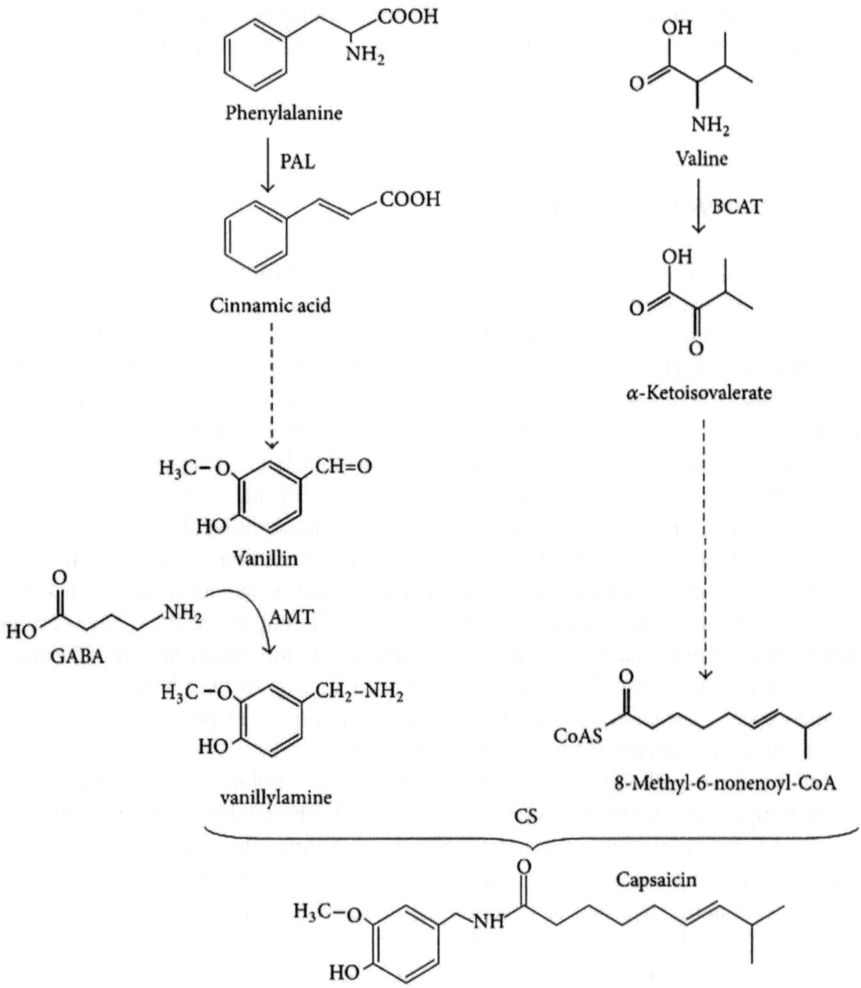

Fig. 6.1 Involvement of nitrogen or nitrogenous metabolites in capsaicin biosynthesis. *AMT* vanillin aminotransferase, *BCAT* branched amino acid aminotransferase, *PAL* phenylalanine ammonia-lyase (Adapted from Aldana-Iuit et al. 2015)

One more pathway comprises of branched chain fatty acid synthesis from valine. Leete and Louden (1968) reported that valine is converted into fatty acid moiety via formation of isobutyryl-coenzyme A with the aid of fatty acid synthase (FAS) enzyme. Acyl carrier protein (Acl), A thioesterase (Fat) and a beta-ketoacyl-ACP synthase (Kas) are the three enzymes that might be involved in the fatty acid synthesis pathway. At last capsaicinoid synthase (CS) assists in the condensation of vanillylamine to branched chain fatty acid moiety, resulting in the final product, capsaicinoid. Capsaicinoid synthase enzyme, with a molecular weight of 38 kDa (Uarrota et al. 2021), is believed to be encoded by a gene called *At3* (Stewart et al.

2005). In chilli pepper, capsaicinoids are synthesized and start accumulating 10–20 days post-anthesis (DPA) (Iwai et al. 1979; Aza-González et al. 2011), reaching maximum accumulation at 20–40 DPA, and then it decreases (Arce-Rodríguez and Ochoa-Alejo 2017; Arce-Rodríguez and Ochoa-Alejo 2019). Hence, the pattern of capsaicinoid biosynthesis and accumulation might be regulated at genetic, transcriptional, translational and functional protein levels as it is dependent on many enzymes that are produced in the placental tissues of fruits at specific developmental stages.

6.5 Factors Affecting Capsaicinoid Content

The synthesis of secondary metabolites and their accumulation in plants is affected by a number of environmental factors (Verma and Shukla 2015; Arce-Rodríguez and Ochoa-Alejo 2019). An optimum light intensity of 1400 μmol/m^2 s is required for chilli pepper, as it is one of the important factors affecting the capsaicinoid biosynthesis (Uarrota et al. 2021). Higher intensity than the optimum reduces the synthesis of capsaicinoids (Gurung et al. 2011; Uarrota et al. 2021) while a lower intensity may cause either an increase or a decrease of capsaicinoid accumulation (Jeeatid et al. 2017; Uarrota et al. 2021). As chilli pepper belongs to the family where plants prefer to grow and flower in the night and shade, shading also affects capsaicinoid by enhancing its synthesis (Jeeatid et al. 2017).

According to Arce-Rodríguez and Ochoa-Alejo (2019), when exposure to light, moisture, CO_2 and altitude during chilli cultivation is increased, enhancement in the synthesis of capsaicinoids is observed. The optimum temperature required for capsicum species is 25–30 °C, which when fluctuated will influence the amount of capsaicinoid biosynthesis and accumulation (Rahman and Inden 2012). High temperatures may affect the activity of the enzyme capsaicin synthase, which could contribute to the accumulation of increased quantities of capsaicin (Uarrota et al. 2021). Similarly, variation in water stress conditions will also enhance the extent of capsaicinoids synthesized (Jeeatid et al. 2018). Upon nitrogen fertilization, pungency level of chilli pepper is enhanced, whereas potassium application will not influence the pungency (Arce-Rodríguez and Ochoa-Alejo 2019).

The pungency in chilli pepper, due to the accumulation of capsaicinoids, is an inherited characteristic, which depends on pre-harvest factors like species, cultivar, genotype, climatic conditions, growing locations, agronomic factors, harvest factors and many other internal factors (Uarrota et al. 2021; Aza-González et al. 2011). Pungency is quantified as a phenotypic trait by assessing capsaicin and dihydrocapsaicin content, and this pungency trait exhibited heterosis, suggesting the existence of epistasis, overdominance or dominance complementation (Uarrota et al. 2021).

Fruit traits, like age of the fruit and the position at which the fruit is present on the plant, also affect the pungency level of the fruit (Popelka et al. 2017). The mean pungency of fruits taken from the second node was higher than that of fruits taken from the other node positions. As the node position grew, pungency of the fruit reduced (Zewdie and Bosland 2000; Uarrota et al. 2021). With chilli pepper fruit development or maturation, capsaicinoids were seen to enhance (Estrada et al. 2000). Early stages of plant development result in fewer fruits per plant and possibly less fruit competition for a substrate resulting in higher concentration of capsaicinoids, whereas later at later stages, capsaicinoid accumulation is decreased (Zewdie and Bosland 2000).

Capsaicinoids are synthesized and accumulated specifically in the placental tissue of capsicum species at a particular stage of fruit development, where their biosynthesis can be regulated at the level of transcription itself by transcriptional factors like *bZip* genes (Blum et al. 2003). Acetyl transferase encoding gene, *Pun1*, might also regulate the capsaicinoid synthesis pathway (Stewart et al. 2005). Another class of transcription factor called myeloblastosis (MYB) transcription factor is also an important regulator of capsaicin production in chilli pepper (Wang et al. 2020). Once synthesized, the stability of capsaicinoids is influenced by various post-harvest stages and food-processing operations (Tiwari and Cummins 2013; Uarrota et al. 2021).

The commercial significance of capsaicinoids has enabled extensive research activities about its in vitro production. The effectiveness of chilli pepper cells, tissues and organs to synthesize capsaicinoids in vitro has been harnessed, and researchers have identified many factors that influence the amount of synthesis and accumulation in vitro. Kehie et al. (2015) reported that in vitro production of capsaicinoids can be enhanced by manipulation of culture strategies, viz. cell immobilization, feeding with precursors, selection of cell, treatments with elicitors, osmotic stress, influence of calcium channel modulators, nutrient stress, pH stress, etc., enhancing the accumulation of capsaicinoids in in vitro cultures of chili pepper.

6.6 Elicitors Used in Enhancing Capsaicin Concentration

Since plants are exposed to various biotic and abiotic stresses, their defence mechanism activates and results in the production of hydrogen peroxide (H_2O_2) and reactive oxygen species (ROS). This oxidative burst finally leads to the biosynthesis of several secondary metabolites as a part of defence reaction against stress condition. However, the signalling cascade so formed can also be activated by the use of elicitors to augment the synthesis of secondary metabolites (Chapa-Oliver et al. 2016). Elicitors can be aptly defined as stress factors that directly or indirectly trigger the inducible defence changes in a plant system, which results in an activation of array of protection mechanisms, including induction or expansion of biosynthesis of fine chemicals, which do have a major role in the adaptation of plants to the stressful environment (Goel et al. 2011; Naik and Al-Khayri 2016).

Based on their nature, elicitors can be classified into two types, biotic and abiotic. Polysaccharides and microorganisms come under biotic elicitors, which are the substances of biological origin, while abiotic elicitors are further classified into physical, chemical and hormonal factors, which are of non-biological origin (Naik and Al-Khayri 2016).

Plant cell, tissue and organ cultures are the alternative methods for the large-scale production of plant-specific valuable metabolites like capsaicinoids. Elicitors proliferate the productivity of worthy metabolites in cell, tissue or organ cultures by modifying cell metabolism, in such a way that the overall cultivation period is reduced (Naik and Al-Khayri 2016).

The activities of enzymes involved in capsaicinoid biosynthesis like phenylalanine ammonia lyase and coumarate O-methyltransferase were positively impacted by salicylic acid or methyl jasmonate elicitor, which inhibits the action of ethylene biosynthesis (Gutiérrez-Carbajal et al. 2010), in a way similar to the action of silver nitrate as an elicitor (Gammoudi et al. 2019). Although salicylic acid, methyl jasmonate and jasmonic acid (JA) elevated the production of capsaicinoids in the cell cultures, the increments were lesser when compared to the enhancement by chitosan (Kabita et al. 2020). The treatment of abiotic elicitors like $NaCl_2$ and polyethylene glycol (PEG) at optimum concentration to cell or callus cultures of chilli influences capsaicinoid production to a certain extent (Gammoudi et al. 2019). Among the different culture strategies employed to enhance capsaicin biosynthesis in cell cultures (Kehie et al. 2015), capsaicin production was elicited when curdlan and xanthan were used as elicitors (Johnson et al. 1991). Supplementing MS medium with 1.5% *A. niger* and 100 M of methyl jasmonate increased the rate of capsaicin accumulation significantly (Taha and El-Ghit 2018).

C. chinense has been the best for making chilli sauces and for capsaicin extraction. About 59 genes were identified by Zhang et al. 2022 to be associated with shikimic acid, phenylpropane, unsaturated fatty acids, phenylalanine, valine, leucine, isoleucine and other metabolic pathways on examining the pepper genome. Of the 59 capsaicin synthesis genes, 58 genes were located unevenly on 11 chromosomes, and several genes with the same number of exons and high homology were found to be clustered in a duplication region of the chromosome. On analysing the promoter regions of these genes, they identified MYB binding sites, indicating further exploration into MYB target genes. Further, these capsaicin synthesizing genes were examined for their expression levels in different transcriptome studies. It was deciphered that expression levels varied with different developmental stages in different tissues, which may be playing important role in growth and development of pepper plants. Comparative genomics studies shall help in understanding capsaicin synthesis mechanisms more accurately and also help pepper breeding.

6.7 Conclusions

Capsaicinoids including capsaicin have been of great interest to researchers due to their bioactive properties, making them a pharmaceutically important compound. Although almost all *Capsicum* species harbour capsaicin, only few have been used as a source of their extraction and utilization. The general understanding of the biosynthetic pathways and the elicitors involved in capsaicinoid production paves the way to scale up their utilization. The understanding about the genes involved in their biosynthesis and also gene expression studies would further escalate this process. New strategies to increase this secondary metabolite can also be deciphered for good crop health and also capsaicin yields and purification for use in therapies for diseases like cancer.

References

Aldana-Iuit JG, Sauri-Duch E, Miranda-Ham MD, Castro-Concha LA, Cuevas-Glory LF, Vázquez-Flota FA (2015) Nitrate promotes capsaicin accumulation in Capsicum chinense immobilized placentas. Biomed Res Int 2015:1. https://doi.org/10.1155/2015/794084

Antonio AS, Wiedemann LSM, Junior VV (2018) The genus Capsicum: a phytochemical review of bioactive secondary metabolites. RSC Adv 8(45):25767–25784

Arce-Rodríguez ML, Ochoa-Alejo N (2017) An R2R3-MYB transcription factor regulates capsaicinoid biosynthesis. Plant Physiol 174(3):1359–1370

Arce-Rodríguez ML, Ochoa-Alejo N (2019) Biochemistry and molecular biology of capsaicinoid biosynthesis: recent advances and perspectives. Plant Cell Rep 38:1017–1030

Aza-González C, Núñez-Palenius HG, Ochoa-Alejo N (2011) Molecular biology of capsaicinoid biosynthesis in chili pepper (Capsicum spp.). Plant Cell Rep 30:695–706

Basith S, Cui M, Hong S, Choi S (2016) Harnessing the therapeutic potential of capsaicin and its analogues in pain and other diseases. Molecules 21(8):966

Bennet DJ, Kirby GW (1968) Constitution and biosynthesis of capsaicin. J Chem Soc 4:442–446

Blum E, Mazourek M, O'Connell M, Curry J, Thorup T, Liu K, Jahn M, Paran I (2003) Molecular mapping of capsaicinoid biosynthesis genes and quantitative trait loci analysis for capsaicinoid content in Capsicum. Theor Appl Genet 108:79–86

Chairat W (2013) Development of novel microemulsions for capsaicin transdermal delivery (doctoral dissertation)

Chapa-Oliver, Angela M, Mejía-Teniente L (2016) Capsaicin: from plants to a cancer-suppressing agent. Molecules 21(8):931

Cho SY, Kim HW, Lee MK, Kim HJ, Kim JB, Choe JS et al (2020) Antioxidant and anti-inflammatory activities in relation to the flavonoids composition of pepper (Capsicum annuum L.). Antioxidants 9(10):986

Curry J, Aluru M, Mendoza M, Nevarez J, Melendrez M, O'Connell MA (1999) Transcripts for possible capsaicinoid biosynthetic genes are differentially accumulated in pungent and non-pungent Capsicum spp. Plant Sci 148(1):47–57

Díaz J, Pomar F, Bernal A, Merino F (2004) Peroxidases and the metabolism of capsaicin in Capsicum annuum L. Phytochem Rev 3(1–2):141–157

Estrada B, Bernal M, Díaz J, Pomar F, Merino F (2000) Fruit development in Capsicum annuum: changes in capsaicin, lignin, free phenolics, and peroxidase patterns. J Agric Food Chem 48(12): 6234–6239. https://doi.org/10.1021/jf000190x

Fernandes ES, Cerqueira ARA, Soares AG, Costa SK (2016) Capsaicin and its role in chronic diseases. In: Drug discovery from mother nature. Springer, Cham, pp 91–125

Ferri M, Gruarin N, Barbieri F, Tassoni A (2018) Capsicum spp. in vitro liquid cell suspensions: a useful system for the production of capsaicinoids and polyphenols. Plant Biosyst 152(3): 436–444

Gammoudi N, Zerria K, Nagaz K, Ferchichi A (2019) Enhancement of capsaicinoids in vitro production by abiotic elicitors in placenta-derived callus of Capsicum annuum L. Tunisian var. 'Baklouti Medenine'. Biologia 74:725–732

Goel MK, Mehrotra S, Kukreja AK (2011) Elicitor-induced cellular and molecular events are responsible for productivity enhancement in hairy root cultures: an insight study. Appl Biochem Biotechnol 165(5–6):1342–1355. https://doi.org/10.1007/s12010-011-9351-7

Gurung TS, Techawongstien B, Suriharn S, Techawongstien (2011) Impact of environments on the accumulation of Capsaicinoids in Capsicum spp. HortScience 46(12):1576–1581. https://doi.org/10.21273/HORTSCI.46.12.1576

Gururaj HB, Padma MN, Giridhar P, Ravishankar GA (2012) Functional validation of Capsicum frutescens aminotransferase gene involved in vanillylamine biosynthesis using Agrobacterium mediated genetic transformation studies in Nicotiana tabacum and Capsicum frutescens calli cultures. Plant Sci 195:96–105

Gutiérrez-Carbajal MG, Monforte-González M, Miranda-Ham MDL, Godoy-Hernández G, Vázquez-Flota F (2010) Induction of capsaicinoid synthesis in Capsicum chinense cell cultures by salicylic acid or methyl jasmonate. Biol Plant 54(3):430–434

Howard L, Smith RT, Wagner AB, Villalon B, Burns EE (1994) Provitamin A and ascorbic acid content of fresh pepper cultivars (Capsicum annuum) and processed jalapenos. J Food Sci 59(2): 362–365

Ilie MA, Caruntu C, Tampa M, Georgescu SR, Matei C, Negrei C et al (2019) Capsaicin: physicochemical properties, cutaneous reactions and potential applications in painful and inflammatory conditions. Exp Ther Med 18(2):916–925

Iwai K, Suzuki T, Fujiwake H (1979) Formation and accumulation of pungent principle of hot pepper fruits, capsaicin and its analogs, in Capsicum annuum var. annuum cv. Karayatsubusa at different growth-stages after flowering. Agric Biol Chem 43:2493–2498

Jeeatid N, Suriharn B, Techawonstien S, Chanthai S, Bosland PW, Techawongstien S (2018) Evaluation of the effect of genotype by- environment interaction on capsaicinoid production in hot pepper hybrids (*Capsicum chinense* Jacq.) under controlled environment. Sci Hortic 235: 334–339

Jeeatid N, Techawongstien S, Suriharn B, Bosland P, Techawongstien S (2017) Light intensity affects Capsaicinoid accumulation in hot pepper (Capsicum chinense Jacq.) cultivars. Hortic Environ Biotechnol 58(2):103–110. https://doi.org/10.1007/s13580-017-0165-6

Johnson TS, Ravishankar GA, Venkataraman LV (1991) Elicitation of capsaicin production in freely suspended cells and immobilized cell cultures of Capsicum frutescens Mill. Food Biotechnol 5(2):197–205

Kabita KC, Sanatombi K, Sharma SK (2020) Efficient enhancement of capsaicinoids biosynthesis in cell suspension cultures of Capsicum chinense Jacq. cv.'Umorok'by elicitors and differential gene expression analysis of elicited cultures. Plant Cell Tiss Organ Cult 141:145–154

Kaiser M, Higuera I, Goycoolea FM (2017) Capsaicinoids: occurrence, chemistry, biosynthesis, and biological effects. Fruit and vegetable phytochemicals: chemistry and human Health, 2nd edn. Wiley, Hoboken, NJ, pp 499–514

Kehie M, Kumaria S, Tandon P, Ramchiary N (2015) Biotechnological advances on in vitro capsaicinoids biosynthesis in Capsicum: a review. Phytochem Rev 14:189–201

Kobata K, Takemura I, Tago G, Moriya T, Kubota K, Nakatani S, Watanabe T (2014) Formation of long-chain N-vanillyl-acylamides from plant oils. Biosci Biotechnol Biochem 78(7):1242–1245

Kumar A, Kumar S, Anju T, Ramchiary N (2021) Genetic, epigenetic, and hormonal regulation of fruit development and ripening in Capsicum L species. Annu Plant Rev Online 4:295–356

Kumar OA, Tata SS (2009) Ascorbic acid contents in chili peppers (Capsicum L.). Not Sci Biol 1(1):50–52

Leete E, Louden MCL (1968) Biosynthesis of capsaicin and dihydrocapsaicin in *Capsicum frutescens*. J Am Chem Soc 90:6837–6841

Naik PM, Al-Khayri JM (2016) Impact of abiotic elicitors on in vitro production of plant secondary metabolites: a review. J Adv Res Biotechnol 1(2):1–7

Naves ER, de Ávila SL, Sulpice R, Araújo WL, Nunes-Nesi A, Peres LE, Zsögön A (2019) Capsaicinoids: pungency beyond Capsicum. Trends Plant Sci 24(2):109–120

Pagare S, Bhatia M, Tripathi N, Pagare S, Bansal YK (2015) Secondary metabolites of plants and their role: overview. Curr Trends Biotechnol Pharm 9(3):293–304

Popelka P, Jevinova P, Smejkal K, Roba P (2017) Determination of capsaicin content and pungency level of different fresh and dried chilli peppers. Folia Vet 61(2):11–16

Prasad BCN, Shrivastava R, Ravishankar GA (2005) Capsaicin: a promising multifaceted drug from Capsicum spp. Evid Based Integr Med 2:147–166

Rahman MJ, Inden H (2012) Effect of nutrient solution and temperature on capsaicin content and yield contributing characteristics in six sweet pepper (*Capsicum annuum* L.) cultivars. J Food Agric Environ 10(1):524–529

Reyes-Escogido MDL, Gonzalez-Mondragon EG, Vazquez-Tzompantzi E (2011) Chemical and pharmacological aspects of capsaicin. Molecules 16(2):1253–1270

Rezazadeh A, Hamishehkar H, Ehsani A, Ghasempour Z, Moghaddas KE (2021) Applications of capsaicin in food industry: functionality, utilization and stabilization. Crit Rev Food Sci Nutr 63: 1–17

Rios MY, Olivo HF (2014) Natural and synthetic alkamides: applications in pain therapy. Stud Nat Prod Chem 43:79–121

Sato K, Sasaki SS, Goda Y, Yamada T, Nunomura O, Ishikawa K, Maitani T (1999) Direct connection of supercritical fluid extraction and supercritical fluid chromatography as a rapid quantitative method for capsaicinoids in placentas of Capsicum. J Agric Food Chem 47(11): 4665–4668

Stewart C, Kang BC, Liu K, Mazourek M, Moore SL, Yoo EY, Kim BD, Paran I, Jahn MM (2005) The Pun1 gene for pungency in pepper encodes a putative acyltransferase. Plant J 42:675–688

Taha HS, El-Ghit HMA (2018) Implement of biotic and abiotic stress for enhancement and production of capsaicin in suspension cultures of Capsicum annum spp. Pak J Biol Sci 21(6): 292–299

Teoh ES (2015) Secondary metabolites of plants. In: Medicinal Orchids of Asia, vol 5. Springer, Cham, pp 59–73. https://doi.org/10.1007/978-3-319-24274-3_5

Tiwari U, Cummins E (2013) Factors influencing levels of phytochemicals in selected fruit and vegetables during pre- and post-harvest food processing operations. Food Res Int 50(2): 497–506. https://doi.org/10.1016/j.foodres.2011.09.007

Uarrota VG, Maraschin M, de Bairros ADFM, Pedreschi R (2021) Factors affecting the capsaicinoid profile of hot peppers and biological activity of their non-pungent analogs (Capsinoids) present in sweet peppers. Crit Rev Food Sci Nutr 61(4):649–665

Verma N, Shukla S (2015) Impact of various factors responsible for fluctuation in plant secondary metabolites. J Appl Res Med Aromat Plants 2:105–113

Wahyuni Y, Ballester AR, Sudarmonowati E, Bino RJ, Bovy AG (2013) Secondary metabolites of Capsicum species and their importance in the human diet. J Nat Prod 76(4):783–793

Walton NJ, Mayer MJ, Narbad A (2003) Vanillin. Phytochemistry 63:505–515

Wang J, Liu Y, Tang B, Dai X, Xie L, Liu F, Zou X (2020) Genome-wide identification and capsaicinoid biosynthesis-related expression analysis of the R2R3-MYB gene family in Capsicum annuum L. Front Genet 11:598183

Yang Q, Zhu J, Ma F, Li P, Zhang L, Zhang W et al (2016) Quantitative determination of major capsaicinoids in serum by ELISA and time-resolved fluorescent immunoassay based on monoclonal antibodies. Biosens Bioelectron 81:229–235

Yap ESP, Uthairatanakij A, Laohakunjit N, Jitareerat P, Vaswani A, Magana AA et al (2021) Plant growth and metabolic changes in 'Super Hot'chili fruit (Capsicum annuum) exposed to supplemental LED lights. Plant Sci 305:110826

Zewdie Y, Bosland P (2000) Evaluation of genotype, environment, and genotype-by-environment interaction for Capsaicinoids in *Capsicum annuum* L. Euphytica 111(3):185–190s

Zhang D, Sun X, Battino M, Wei X, Shi J, Zhao L et al (2021) A comparative overview on chili pepper (Capsicum genus) and Sichuan pepper (zanthoxylum genus): from pungent spices to pharma-foods. Trends Food Sci Technol 117:148–162

Zhang W, Wu D, Zhang L, Zhao C, Shu H, Cheng S et al (2022) Identification and expression analysis of capsaicin biosynthesis pathway genes at genome level in Capsicum chinense. Biotechnol Biotechnol Equip 36(1):232–244

Conventional Propagation and Micropropagation of Capsaicin-Containing Plants

7

Nayana Veedu Aswathi, Chandran Padikkal Krishna Vrundha, and Thuruthiyil Dennis Thomas

Abstract

Capsaicin is a secondary metabolite with various medicinal properties, which is high in demand in the pharmaceutical industry. This compound is mainly derived from the genus *Capsicum* L., which belongs to the family Solanaceae. The genus *Capsicum* comprises of economically important vegetable crops around the world and are cultivated for their ornamental values too. Natural propagation of some species of this genus is restricted due to the short duration of viability and low seed germination rate. Moreover, some seeds are highly prone to fungal and viral pathogen attack. The natural propagation of chilli by vegetative means is limited, so to conserve genetic purity, it is very important to use techniques like micropropagation. The plants are recalcitrant toward in vitro propagation and genetic transformation efforts, even though there are various attempts to improve the propagation. Micropropagation of *Capsicum* is an alternative mode of rapid propagation of elite clones to meet the high demand from agriculture industry. This chapter consolidates the information available on both conventional and in vitro propagation of capsaicin containing plants.

Keywords

Capsicum · Vegetative propagation · Micropropagation · Recalcitrant · Plant tissue culture

N. V. Aswathi · C. P. K. Vrundha · T. D. Thomas (✉)
Department of Plant Science, Central University of Kerala, Tejaswini hills, Periye, Kasaragod, Kerala, India

M. K. Swamy, A. Kumar (eds.), *Capsaicinoids*,
https://doi.org/10.1007/978-981-99-7779-6_7

7.1 Introduction

Capsicum L. is a valuablė genus of Solanaceae family, which is cultivated for thousands of years in America, and cultivation has spread worldwide (Kothari et al. 2010). They are known as one of the leading spice crops around the world, and India has a vast pepper diversity. The genus comprises of about 5 cultivated and 25 wild species (Sanatombi and Sharma 2007). *C. annuum* L., *C. frutescens* L., *C. chinense* Jacq., *C. baccatum* L., and *C. pubescens* R. & P. are the five cultivated species (Sanatombi and Sharma 2007). The species of *Capsicum* possess various medicinal properties like analgesic, rubefacient, antimicrobial, antifungal, anti-cancer, and carminative activities. The genus is high in anti-inflammatory and antioxidant potential and has anti-mutagenic and hypoglycaemic activities also (Ahmad et al. 2006). Some species of the genus are used in the treatment of rheumatism, lumbago, and neuralgia and is routinely used in Ayurveda and Homeopathy systems (Gayathri et al. 2015).

The pungency of fruits is the main characteristic of pepper plants that makes them valuable commercially. The group of compounds named capsaicinoids is responsible for the pungency of *Capsicum* (Bosland and Votava 2012), which finds extensive use in pharmaceutics, cosmetics, and food industries (Antonious 2018). Although it is a self-pollinating crop, the genus *Capsicum* should be deemed facultative cross-pollinating species as the out-crossing rates range from 7 to 91% (Odland and Porter 1941; Tanksley 1984; Bosland and Votava 2012). The natural vegetative propagation is limited in chilli. Loss of genetic purity and elite varieties or cultivars results in the loss of superior characteristics during cross-pollination. Propagation of pepper plants using seeds is restricted because of high vulnerability to fungal and viral pathogens and short window of viability with low germination rate (Ahmad et al. 2006; Sanatombi and Sharma 2007). Apart from that, capsaicin content has been found decreasing in species like *C. chinense* due to cross-pollination and genetic mutation (Bora et al. 2019). The high price of seeds along with all these factors suggests that the development of efficient plant regeneration methods for *Capsicum* cultivars will help in producing uniform young plants and cut the relying on nursery plant production thereby reducing seed prices (Swamy et al. 2014). In some countries like Mexico, the production of this vegetable does not satisfy the internal and external demand, primarily due to the poor quality of the seeds as well as the low level of technology used for cultivation (Jaiswal et al. 2021).

Although morphogenesis is accomplished easily in Solanaceae family members, strong recalcitrance is exhibited by pepper, and reports suggest that the low regeneration rate of many cultivars is due to the genotype dependence (Kothari et al. 2010). The genus *Capsicum* is recalcitrant toward micropropagation, which results in inefficient application of genetic modification through recombinant DNA technologies and creates difficulties in developing pest and disease-resistant varieties (Ashrafuzzaman et al. 2009). Also, the formation of rosette shoots and sensitivity to ethylene gas make in vitro propagation and genetic transformation studies difficult in *Capsicum* spp. (Monteiro do Rêgo et al. 2016). In this chapter, we discuss the natural propagation methods of the genus, existing difficulties in the conventional propagation methods of chilli peppers, as well as efforts made to overcome these problems to improve the conventional propagation of the genus. The incorporation of new techniques like micropropagation and genetic manipulation has highly improved the breeding and production of chilli peppers. Here we discuss the advancements in micropropagation of different species of the genus and the hurdles needed to be overcome in achieving them.

7.2 Conventional Propagation of the Genus *Capsicum*

The propagation of vegetable crops is of different types including sexual (seed propagation), asexual (vegetative propagation), and in vitro propagation (tissue culture). Propagation through seed is of two types. Seeds can be sown directly to the plot or raise the seedlings in the nursery and transplant the healthiest seedling to the field (Thamburaj and Singh 2000). In chilli pepper, nursery seedlings are produced from the seeds and transplanted to the soil at six to eight leaves age or 50–60 days after sawing (Pinto et al. 2016).

The rate of germination in chilli seeds is affected by physiological or endogenous dormancy, which leads to a delay in seed germination and reduced seedling vigour during the growing season (Alcalá-Rico et al. 2019). The non-uniform and delayed germination of chilli affects crop establishment, especially under salinity and temperature stress conditions which leads to crop loss (Demir and Okcu 2004). Different studies reported an improvement in the germination rate of chilli seeds with the treatment of gibberellic acid (GA_3) at lower concentrations (Debbarma et al. 2018; Muhammad et al. 2017). Priming with 50 ppm GA_3 and 1.1 MPa polyethylene glycol (PEG) for 24 h and 12 h of hydro-priming was observed optimum for the germination percentage and seedling growth in chilli by Debbarma et al. (2018). A high germination rate with a better seedling vigour index was achieved by the pre-treatment of seeds with 72.25 μM GA_3. This method can be applied for the commercial cultivation of chilli peppers (Junaidy and Shahruddin 2022).

Air layering (marcotting) is a mode of vegetative propagation by rooting the stem while still attached to the mother plant. This allows the rapid multiplication of true-to-type plants with desirable qualities and is widely applied in forestry and agriculture (Thirunanvoukkarasu et al. 2004). Successful air layering in *C. annuum* and *C. frutescens* clones has been achieved with the use of sphagnum moss as a rooting medium soaked in 5.71 μM indole-3-acetic acid (IAA). Here over 50% of plants survived having desirable characteristics of the mother plant (Mohamed et al. 2014).

Grafting is a technique of combining the shoot of one plant (scion) and the root of another plant (rootstock) of different species or varieties to produce a plant that has the desirable quality of both plants combined (Goldschmidt 2014). Grafting is a conventional breeding method that does not require genetic manipulation and develops vegetable crops of high-yielding various biotic and abiotic stress-resistant plants, which reduce dependence on agrochemicals and overall maintenance (Keatinge et al. 2014; Cardarelli et al. 2020; Birkas et al. 2021). *Capsicum* plants are susceptible to various biotic and abiotic stresses that restrict survival of plants and potential yield. Several successful grafting were conducted in different *Capsicum* species against various soil-borne pathogens and abiotic stress (Pico et al. 2017). Mild salinity-resistant (Penella et al. 2015), higher-temperature-resistant (López-Marín et al. 2013), *Phytophthora nicotianae* (Saadoun and Allagui 2013), *Phytophthora capsica*, *Ralstonia solanacearum* (Jang et al. 2012), *Meloidogyne incognita*-resistant (Ros-Ibanez et al. 2014) and higher fruit yield (Attia et al. 2003; Donas-Ucles et al. 2014) grafts were developed in *C. annuum*. Tolerance to mild salinity in *C. chinense* and *C. baccatum* (Penella et al. 2015), better yield in hot wet and dry conditions in *C. frutescens* (Palada and Wu 2008), and root-knot nematode resistance in *C. chinense* (Oka et al. 2004) were also achieved using the grafting technique. Foliar application of abscisic acid (ABA) increased survival of old-age grafted *C. annuum* plants by the increase in stomatal resistance and water potential and decreased defoliation (Johkan et al. 2009).

Vegetative propagation using stem cutting is an easy, cost-effective, and time-saving procedure that ensures elite parental qualities without genetic changes and an abundance of material available for propagation (Alvarez et al. 2016). Formation of the adventitious root is an important part of the procedure, which is influenced by the age of the stem, season, nutrient solution, and agronomic management. The condition of the formed roots, irrigation, and fertilization have impacts on seedling quality and its field establishment (Hassanein 2013). The production of quality *C. annuum* seedlings from stem cutting can be attained by the use of Steiner nutrient solution at electrical conductivities of 1.25 dS m^{-1} (Carballo-Méndez et al. 2022). In *C. annuum*, cutting off the mainstem above the highest node, harvesting primary and secondary lateral shoots, and increasing fertilization increased the frequency of stem cuttings or scion obtaining for propagation from 1.5 times from week to 4.5 times from weeks. The rooted scion can be transplanted to the field (Johkan et al. 2014). The application of honey and indole-3-butyric acid (IBA) improved rooting and rate of survival of *C. annuum* stem cuttings. The highest result was obtained by the treatment with 100 ppm IBA (Dogra and Sharma 2022).

7.3 Micropropagation of the Genus *Capsicum*

Tissue culture is the technique of propagation of plants in vitro from explants like tissues and organs by exploring the totipotency of plant cells (Haberlandt 1902). Despite the fact that the genus *Capsicum* is highly recalcitrant toward micropropagation, several efforts have been done to improve micropropagation protocols for the genus (Table 7.1).

Abolition of bacterial diseases and viruses, selection and cloning of elite genotypes, the rescue of embryos formed from intra and interspecific crosses, conservation of genetic variability, and breeding for abiotic stress tolerance have been achieved through the application of tissue culture techniques (Monteiro do Rêgo et al. 2016). The choice of explant, type of plant growth regulators (PGRs) used, composition of medium, and carbon source are very important for the success of in vitro propagation protocols along with physical factors including light,

Table 7.1 Various reports of micropropagation of *Capsicum* by using different explants

Explant[a]	Mode of response[a]	Medium and hormone	References
C. annuum			
C	DO	MS + BA 8.88 µM+ IAA 2.85–5.71 µM	Gunay and Rao (1978)
ZE		MS+ BA 22.2 µM	Agrawal and Chandra (1983)
P	DO	MS+ BA (2.22–44.4 µM)	Phillips and Hubstenberger (1985)
C, H, L, R, S, SA, ZE	DO, IO	MS+ BA 22.2 µM + IAA (2.85–5.71 µM)	Batista et al. (2013)
C, H	DO	MS+ BA (2.22; 4.40; 8.80 µM) + IAA (0.57; 5.71 µM)	Agrawal et al. (1989)
P	DO	MS+ BA 22.2 µM + NAA 0.54 µM	Arroyo and Revilla (1991)
MS	DO	MS without regulators	Ebida and Hu (1993)
IZE	DO	MS without regulators	Ezura et al. (1993)
IZE	DSE	MS+ 2–10% sucrose +10% coconut water +2,4-D (4.53–22.6 µM)	Gatz (2014)
MZE	DSE	MS+ 6–10% sucrose +2,4-D 9.0 µM	Harini and Sita (1993)
H	DSE	WPM+ 80 g/L sucrose, 1.0 mg/L $AgNO_3$ + TDZ 5.1 µM or 2,4-D 9.1 µM	Aboshama (2011)
C	ISE	MS+ 3% sucrose +2.4-D 452 µM	Binzel et al. (1996)
SA	DO	MS+ BA 8.8 µM+ IAA 2.85 µM + $AgNO_3$ 5.8 µM	Büyükalaca and Mavituna (1996)
H	DO	MS+ BA 8.8 µM	Hyde and Phillips (1996)
C	DO	MS+ BA 66.6 and 88.8 µM	Madhuri and Rajam (1993)
ZE	DO	MS+ BA 13.35 µM + IAA (3.4–5.9 µM) + EBr (0.1 µM)	Ramírez-Malagón and Ochoa-Alejo (1996)
P, EE	DO	MS+ BA (22.2–31.0 µM) + PAA 14.0 µM	Dabauza and Pena (2001)
C, L	DO	MS+ BA 22.2 µM+ NAA 5.37 µM	Husain et al. (1999)
C	DO	MS+ TDZ (4.5–9.0 µM)	Arous et al. (2001)
C, H, L	DO	MS+ TDZ (4.5–13.5 µM)	Dabauza and Pena (2001)
		MS+ BA 22.2 µM + PAA 14.7 µM	Venkataiah et al. (2003)
		MS+ BA 8.8 µM + IAA 11.4 µM	Joshi and Kothari (2007)
		MS+ BA 22.2 µM + kinetin 4.6 µM	Sanatombi and Sharma (2008)

(continued)

Table 7.1 (continued)

Explant[a]	Mode of response[a]	Medium and hormone	References
			Swamy et al. (2014)
C	DO	MS+ BA 26.4 μM + IBA 4.92 μM	Zemene and Worku (2018)
A	SE	MS+ 6.0% sucrose +2,4-D 10–8 M	Morrison et al. (1986)
A	DO	MS+ kinetin 0.46 μM	Olszewska et al. (2014)
A	DO	MS+ 2,4-D 0.45 μM + kinetin 0.46 μM	Grozeva et al. (2021)
A	DO	MS+ zeatin 2.5 μM + IAA 5 μM	Supena and Custers (2011)
A	DO	Modified W14+ 9% maltose, 1000 mg/L glutamine, kinetin 2.33 μM+ 2,4-D 2.27 μM	Lantos et al. (2009)
A	DO	Nitsch + zeatin 2.5 μM and IAA 5.0 μM	Supena et al. (2006)
A	DO	MS+ NAA 1.07 μM + kinetin 2.33 μM	Cheng et al. (2013)
ST	DO	MS + 1 μM TIBA + BA 44.4 μM or kinetin 69.7 μM	Christopher and Rajam (1994)
C, H, L	DO	MS+ IAA 5.7 μM + BA 13.3 μM or 22 μM or 44 μM	Christopher and Rajam (1996)
C, H	DO	MS + BA 22.2 μM + IAA 2.88 μM+ $CuSO_4$ 0.47 mg/L	Martínez-López et al. (2021)
SM	DO	MS+ TDZ 4.54 μM	Venkataiah et al. (2006)
N, ST	NA	MS+ BA 78.14 μM + IAA 2.88 μM MS+ zeatin 36.4 μM MS+ BA 13.32 μM + kinetin 9.3 μM	Hailu et al. (2015)
H	NA	Demre pepper (MS+ BA 17.6 μM + IAA 2.85 μM) Charleston pepper (MS+ BA 13.32 μM + IAA 2.85 μM) Capia pepper, bell pepper (MS + BA 22.2 μM + IAA 2.85 μM)	İzgü et al. (2020)
H, ST	DO	MS+ 500 ÂμM salicylic acid+ TDZ 10.2 μM	Bhutia et al. (2018)
I	IO	MS + 2,4-D 10 μM + BA 2.0 μM for callus MS+ BA 5.0 μM or TDZ 2.5 μM for shoot induction	Khan et al. (2011)
H	DO	MS+ zeatin 34.12 μM + GA_3 5.78 μM	Hegde et al. (2017)
C. frutescens			
C	DO	MS+ BA 8.88 μM + IAA (2.85–5.71 μM)	Gunay and Rao (1978)
SA	DO	MS+ BA 22.2 μM + kinetin 4.6 μM	Sanatombi and Sharma (2007)
C, H, L	DO	MS+ BA 8.8 μM + IAA 11.4 μM	Sanatombi and Sharma (2008)
H	DO	MS+ BA 22.2 μM + NAA 0.54 μM	Chee et al. (2018)
SM	DO	MS+ TDZ 4.54 μM	Venkataiah et al. (2006)

(continued)

Table 7.1 (continued)

Explant[a]	Mode of response[a]	Medium and hormone	References
H	IO	MS+ 2,4-D 6.83 μM for callus MS+ BA 13.32 μM for shoot induction	Hasnat et al. (2007)
CN, ST	DO	MS+ kinetin 4.65 μM + IBA 4.92 μM	Gururaj et al. (2004)
SL	DO	MS+ 2- (N-morpholine) ethane sulphonic acid buffer+ IAA 2.28 μM + 10 μM $AgNO_3$+ BA 26.63 μM	Kumar et al. (2007)
C. chinense			
C, H, L	DO	MS+ BA 8.8 μM + IAA 11.4 μM	Sanatombi and Sharma (2008)
		MS+ 18.16 μM TDZ	Kehie et al. (2013)
		MS + 35 μM BA +35 μM kinetin	Gogoi et al. (2014)
			Rego et al. (2013)
SL	DO	MS+ BA 22.2 μM + IAA 2.85 μM	Raj et al. (2015)
SS	IO	MS+ BA 13.32 μM + NAA 5.37 μM	Raj et al. (2015)
ST	DO	MS+ zeatin 91.2 μM or BA 31.1 μM + kinetin 4.7 μM	Sanatombi and Sharma (2008)
N, SS	DO	MS+ kinetin 31.78 μM	Santana-Buzzy et al. (2005)
ZE	ISE	MS+ NAA 8.9 μM, IAA 11.4 μM + BA 8.9 μM	Solís-Ramos et al. (2010)
C. baccatum			
C	DO	MS+ BA 13.3 μM + IAA 5.71 μM+ BA 44.4 μM, 22.2 μM	Christopher and Rajam (1994)
C, H	DO	MS + BA 22.2 μM + IAA 2.855 μM + 0.47 mg/L $CuSO_4$	Martínez-López et al. (2021)
C, L	SE	MS+ 2,4-D 9.1 μM, kinetin 2.33 μM and MS + BA 4.44 μM	Venkataiah et al. (2016)
RS	IO	MS+ BA 22.2 μM + IAA 5.7 μM + GA_3 6.1 μM	Valera-Montero and Phillips (2005)
SM	DO	MS+ TDZ 9.18 μM	Venkataiah et al. (2006)
C, H, L	DO	MS+ IAA 5.7 μM + BA 13.3 μM or 22.0 μM or 44.0 μM	Christopher and Rajam (1996)
CP	SE	MS+ BA 13.32 μM + IAA 2.85 μM	Shu et al. (2022)
C. praetermissum			
ST	DO	MS + BA 22.2–133.2 μM or kinetin 46.5–185.9 μM or MS+ TIBA 1 μM + BA 22.2 μM or kinetin 46.5 μM	Christopher and Rajam (1994)
SM	DO	MS+ TDZ 4.54 μM	Venkataiah et al. (2006)
C, H, L	DO	MS+ IAA 5.7 μM + BA 13.3 μM or 22.0 μM or 44.0 μM	Christopher and Rajam (1996)

(continued)

Table 7.1 (continued)

Explant[a]	Mode of response[a]	Medium and hormone	References

[a]*A* anther, *C* cotyledon, *CN* cotyledonary node, *CP* cotyledons with petiole, *DO* direct organogenesis, *DSE* direct somatic embryogenesis, *EE* embryonic explant, *H* hypocotyl, *I* internode, *IO* indirect organogenesis, *ISE* indirect somatic embryogenesis, *IZE* immature zygotic embryo, *L* leaf, *MS* mature seed, *MZE* mature zygotic embryo, *N* node, *NA* not available, *P* plantlets, *R* root, *RS* radicle-side of half-seed, *S* stem, *SA* stem apex, *SE* somatic embryogenesis, *SL* seedling, *SM* shoot meristem, *SS* stem segment, *ST* shoot tip, *ZE* zygotic embryo

humidity, and temperature (Shahzad et al. 2017; García-Gonzáles et al. 2010). In the case of *Capsicum* genus, the use of responsive genotype also is fundamental for the success of micropropagation (Ochoa-Alejo and Ireta-Moreno 1990; Kothari et al. 2010). The first report on *Capsicum* micropropagation was by Smith and Heiser Jr (1957), via the in vitro germination of well-developed zygotic embryos of the *C. pendulum* × *C. annuum* hybrid to obtain F1 plants. Although regeneration via somatic embryogenesis or protoplasts and production of haploids by anther culture has been explored, primarily regeneration of peppers has been achieved through organogenesis (Kothari et al. 2010). Organogenesis is the de novo formation of shoot or root from the explant and is the most frequent pathway for plant regeneration in vitro (Duclercq et al. 2011). This could be achieved through differentiation of non-meristematic tissues (adventitious shoot formation) or pre-existing meristematic tissues (axillary shoot formation). Organogenesis has been achieved in *C. frutescens* from cotyledon (Gunay and Rao 1978; Sanatombi and Sharma 2008), leaf and hypocotyl (Sanatombi and Sharma 2008) and leaf, cotyledon, and hypocotyl of *C. chinense* (Sanatombi and Sharma 2008). In *C. baccatum*, direct organogenesis of cotyledon is successfully achieved by Christopher and Rajam (1994). In *C. annuum*, several protocols have been developed for organogenesis from different explants like cotyledon (Gunay and Rao 1978; Madhuri and Rajam 1993; Arous et al. 2001), stem apex (Büyükalaca and Mavituna 1996), zygotic embryo (Ramírez-Malagón and Ochoa-Alejo 1996), plantlets, and embryonic explant (Dabauza and Pena 2001). In addition to organogenesis, somatic embryogenesis has been effectively utilized in the micropropagation of *Capsicum* genus. It is the process of development of somatic cells to somatic embryos, which behave like normal zygotic embryos and develop into individual plantlets (George et al. 2008). In *Capsicum* genus, direct somatic embryogenesis of *C. annuum* cv. California Wonder from immature zygotic embryo explants is the first report on successful somatic embryogenesis (Harini and Sita 1993). Later, several researchers successfully employed the technique in the genus (Binzel et al. 1996; Büyükalaca and Mavituna 1996; Gatz 2014). A method toto continuously obtain somatic embryos from bioreactor by means of embryonic suspension culture, with a higher multiplication rate, developed by (Binzel et al. 1996). Büyükalaca and Mavituna (1995) achieved synthetic seed production by encapsulating somatic embryos. The formation of somatic embryos is considered expedient as it reduces multiplication time; eliminates stages like shoot induction, elongation, and rooting; and also provides high genetic uniformity (Stasolla and Yeung 2003).

Anther culture and haploid production have great plant breeding potential (Sarvesh et al. 1993) and are accomplished in *Capsicum* species by various researchers where haploid embryos are induced from anther or microspore culture. Double haploids result in fully homozygous plants in a very short period, which has a predominant role in creating genetic diversity and acts as a base for introducing higher-quality traits (Irikova et al. 2011). Anther culture is extensively used in tissue culture to produce double haploid lines, which was first developed in *Datura innoxia* (Guha and Maheshwari 1964). Haploid plants spontaneously occur naturally in *Capsicum* species (Pochard and Dumas de Vaulx 1979). Wang et al. (1973) were the first to regenerate *C. annuum* from anthers in China and George and Narayanaswamy (1973) in India. Later, several scientists achieved successful regeneration from anther culture in the genus *Capsicum* (Dumas de Vaulx et al. 1981; Dolcet-Sanjuan et al. 1997; Sibi et al. 1979; Koleva-Gudeva 2003a, b; Koleva-Gudeva et al. 2007; Rodeva 2001). Recently, Supena (2021) developed a protocol with double-layered media to induce double haploid plants in *C. annuum*. Direct embryogenesis was obtained by Phuong (2021) in anther culture of *C. annuum*. The anther culture and production of haploid plants through androgenesis is highly influenced by the species, genotype, cultivar, and hybrid as well as the growth conditions of donor plants, appropriate microspore stage in the selected flower, medium composition, and physical conditions like light, humidity, and temperature (Irikova et al. 2011; Monteiro do Rêgo et al. 2016). Even if embryos are induced from anthers of *Capsicum*, most of them fail to develop into fully developed plants, and most of the acclimatized plants lack fertility (Irikova et al. 2011). Micropropagation of various species of *Capsicum* is described in the following headings.

7.3.1 *Capsicum annuum*

C. annuum is an herbaceous annual or perennial plant with colourful, fleshy fruits in variable shapes and sizes. The fruit attracts users for its colour, texture, and relatively less pungent to moderately pungent; thus, the bell peppers and chilli peppers are mostly eaten fresh or are cooked and used to enhance other dishes (Ribes-Moya et al. 2018). Chilli pepper is highly nutritious with low calories and has a vast number of bioactive phytochemicals like phenols, flavonoids, and carotenoids (Anaya-Esparza et al. 2021).

C. annuum is one of the most studied *Capsicum* species for in vitro propagation. The first report on the plant was in 1978 where direct organogenesis from cotyledon using Murashige and Skoog basal medium (MS, Murashige and Skoog 1962) augmented with 8.88 μM 6-benzylaminopurine (BA) and 2.85–5.71 μM IAA (Gunay and Rao 1978). Direct shoot organogenesis from hypocotyl explants of different *C. annuum* lines were described by İzgü et al. (2020). The maximum regeneration response was achieved in MS medium with 19.08 μM BA+ 2.85 μM

IAA in demre pepper, 13.32 μM BA+ 2.85 IAA in charleston pepper, 22.2 μM BA +2.85 μM IAA capia pepper, and 22.2 μM BA+ 2.85 μM IAA in bell pepper. Direct shoot organogenesis from shoot tips and nodes of *C. annuum* cv. Mareko Fana was optimized. The highest shoot induction from nodal explant in MS+ 19.98 μM BA +2.85 μM IAA and shoot tip explants in 13.65 μM zeatin were obtained. For shoot multiplication of both the explants, MS + 13.32 μM BA +9.3 μM kinetin was found optimum (Hailu et al. 2015).

Salicylic acid treatment for the direct shoot induction from hypocotyl, cotyledon, and shoot tip explants of *C. annuum* cv. Dalle Khursani was conducted. MS medium with 500 ÂμM salicylic acid and 10.2 μM TDZ was found optimum for all the explants (Bhutia et al. 2018). Cotyledon on MS medium augmented with 26.64 μM BA, and 5.71 μM IAA produced the highest organogenesis compared to other explants (root, shoot tip, and hypocotyl; Zemene and Worku 2018).

Direct shoot induction from different explants like zygotic embryo (Agrawal and Chandra 1983; Dabauza and Pena 2001; Batista et al. 2013; Gatz 2014), plantlets (Phillips and Hubstenberger 1985; Dabauza and Pena 2001), cotyledon (Arous et al. 2001; Venkataiah et al. 2003; Joshi and Kothari 2007; Batista et al. 2013), hypocotyl (Sanatombi and Sharma 2008; Batista et al. 2013; Hegde et al. 2017), leaf (Batista et al. 2013; Dabauza and Pena 2001), shoot tip (Büyükalaca and Mavituna 1996; Venkataiah et al. 2006; Batista et al. 2013), and mature seeds (Ebida and Hu 1993) was reported. Indirect shoot organogenesis from internode explant in *C. annuum* was reported by Khan et al. (2011).

Harini and Sita (1993) were the first to report somatic embryogenesis in *C. annuum* through direct embryogenesis from immature zygotic embryos. Indirect somatic embryogenesis and regeneration of *C. annuum* somatic embryo in liquid media are first reported by (Büyükalaca and Mavituna 1996). For somatic embryogenesis, MS medium with 3–10% sucrose and 4–18 μM 2,4-dichlorophenoxyacetic acid (2,4-D) was found suitable, whereas cytokinin showed no significant effect. MS medium supplemented with GA_3 or TDZ alone or in combination induced germination of somatic embryos. The somatic embryo maturation was promoted using ABA (Binzel et al. 1996; Büyükalaca and Mavituna 1996). The synthetic seed production in *C. annuum* via encapsulation of somatic embryo was first reported by Büyükalaca et al. (1995). Woody plant medium (WPM) with 80 g/L sucrose, 1.0 mg/L of $AgNO_3$, and 5.1 μM TDZ or 2,4-D induced highest number of somatic embryos directly from hypocotyl explants (Aboshama 2011).

Different protocols have been developed for the haploid plant production in *C. annuum* including Wang et al. (1973); Dumas de Vaulxet al. (1981); Morrison et al. 1986; Supena et al. (2006); Irikova et al. (2011); Supena and Custers (2011); Cheng et al. (2013); Olszewska et al. (2014), and Barroso et al. (2015). Grozeva et al. (2021) analysed the effectiveness of anther culture and microspore embryogenesis in *C. annuum* for breeding purposes. Double-layered media reported better than microspore culture on liquid medium and anther culture on solid medium for the generation of doubled haploid plants (Supena 2021). Co-culture with ovary tissues of *Capsicum* or wheat showed an improvement in microspore culture in *Capsicum* (Lantos et al. 2009).

7.3.2 *Capsicum chinense*

C. chinense, commonly known as habanero peppers are native to amazone and are adapted to diverse climates all over the world. They are mostly consumed for its hotness, colour, unique flavour, and medicinal properties (Lannes et al. 2007). *C. chinense* have some of the hottest chilli varieties as this species contains more capsaicin content compared to others. The pungency of capsicum is due to capsaicinoid compounds, and *C. chinense* is a valuable source of economically and pharmaceutically important capsaicin (Sganzerla et al. 2014).

C. chinense is highly recalcitrant toward in vitro morphogenesis. MS medium supplemented with 91.2 μM zeatin or 31.1 μM BA and 4.7 μM kinetin produced optimum regeneration from shoot tip explant (Sanatombi and Sharma 2008). An indirect somatic embryogenesis protocol from zygotic embryos has been standardized. Semi-solid MS medium supplemented with 8.9 μM 1-naphthaleneacetic acid (NAA), 11.4 μM IAA, and 8.9 μM BA produced embryogenic calli. The calli converted into somatic embryos, which later developed into plantlets (Solís-Ramos et al. 2010). MS basal supplemented with 22.2 μM BA and 2.85 μM IAA is more effective for multiple shoot regeneration and for callus induction; MS medium in combination with 13.32 μM BA and 5.37 μM NAA was achieved in *C. chinense* using stem segments (Raj et al. 2015).

7.3.3 *Capsicum frutescens*

C. frutescens has wide distribution in Africa, Asia, and South America and is very pungent with a distinct flavour that makes it an integral part of local food (Fig. 7.1a–c). Traditionally, the fruit is used to treat sore throat, diabetes, back pain, inflammation, and nausea (Gurnani et al. 2016). Seed dispersal is mostly by the birds, which eat its attractive, red, and small fruits (Yamamoto and Nawata 2005). Though genetic transformation studies of bird eye chilli (*C. frutescens*) for the improved

Fig. 7.1 Habit of *Capsicum frutescens* with fruits. (**a**) Habit of *C. frutescens* plant, (**b**, **c**) Different stages of fruit development

production of capsaicin contents are frequently done, micropropagation is less frequently reported in *C. frutescens.*

A protocol standardized for direct organogenesis from shoot tip and cotyledonary node was achieved in MS medium augmented with 4.65 μM kinetin and 4.92 μM IBA by Gururaj et al. (2004). But here, only a single shoot was formed from nodal explants with three to five nodes. Direct organogenesis was also attained from shoot meristem (Venkataiah et al. 2006), hypocotyl (Chee et al. 2018; Sanatombi and Sharma 2008), cotyledon (Gunay and Rao 1978; Sanatombi and Sharma 2008), stem apex (Sanatombi and Sharma 2007), and leaf (Sanatombi and Sharma 2008). Indirect organogenesis was achieved through callus induction and subsequent shoot regeneration from hypocotyl explants by Hasnat et al. (2007).

In vitro flower bud formation was observed on MS supplemented with 5.3 μM NAA in vented autoclavable plant tissue culture containers. When 5.0 μg/mL silver thiosulfate, 5.3 μM NAA, and 3% sucrose were added onto the surface of earlier-mentioned medium, in vitro fruit formation and ripening were observed (Haque et al. 2016). High-frequency flowering and fruiting, which could be a pioneering step toward in vitro hybrid production, were successfully induced in ten cultivars of three species of *Capsicum*, namely, *C. frutescens*, *C. annuum*, and *C. chinense* by Haque et al. (2016). The authors suggest a strong synergism between silver ions and cytokinin in the improvement of in vitro flowering. Here, BA (3.33 μM) in combination with 30.0 μM silver thiosulphate was most optimum for flowering (Haque et al. 2016). Bodhipadma and Leung (2003) achieved in vitro flowering and fruiting. Exogenous administration of 30 μM $AgNO_3$ or cobalt chloride ($CoCl_2$) significantly increased shoot organogenesis from the in vitro nodal explants and subsequent in vitro flowering (Sharma et al. 2008).

The application of triacontanol (TRIA), a naturally occurring plant growth regulator in vitro derived shoot tips of *C. frutescens*, resulted in improved levels of shoot multiplication and rooting, and the highest axillary shoot induction was obtained on the application of 2.0 μg/L TRIA (Obul Reddy et al. 2002). Kumar et al. (2007) reported that the incorporation of exogenous polyamines resulted in increased shoot bud induction from seedling explants of *C. frutescens* and completely inhibited organogenesis when the polyamine inhibitors are added to the media.

7.3.4 *Capsicum praetermissum*

C. praetermissum is commonly known as comari, cumari, pimenthinha, and pimenta, which is a wild species of chilli with small round fruits (Heiser and Smith 1958). The species is usually considered as a cultivar of the species *C. baccatum* but later identified as different species (Heiser and Smith 1958).

Christopher and Rajam (1994) were successful in multiple shoot organogenesis from shoot tip explants on MS medium with BA (22.2–133.2 μM) or kinetin (46.5–185.9 μM) in *C. praetermissum.* A higher number of shoots without chromosomal aberrations was obtained when 2,3,5-triiodobenzoic acid (TIBA), an antiauxin, was added to the medium in comparison to BA or kinetin alone. Here, MS supplemented with 1.0 μM TIBA and 22.2 μM BA or 46.5 μM kinetin was the most optimum combination. Venkataiah et al. (2006) reported that TDZ at a concentration of 4.54 μM was the most appropriate for direct organogenesis from shoot meristem explants and is superior over adenine (Ad), BA, kinetin, and zeatin. Successful direct organogenesis from cotyledon, leaf, and hypocotyl explants were obtained by Christopher and Rajam (1996).

7.3.5 *Capsicum baccatum*

C. baccatum or Ajíes/Peruvian hot pepper is divided into two major groups, wild *C. baccatum* L. var. Baccatum and domesticated *C. baccatum* var. Pendulum (Willd.) Eshbaugh (Albrecht et al. 2012). They are mostly found in Latin America and is an important part of Andean cuisine. Compared with *C. chinense*, *C. baccatum* is mildly pungent and has a more vegetable-like aroma as in bell peppers (Kollmannsberger et al. 2011).

Mostly regeneration is obtained through direct organogenesis in *Capsicum* spp., and this may be because the time window for the regeneration is very narrow and after the regeneration potential rapidly decreases. Therefore, there are only a few reports claiming indirect regeneration in the whole genus. Successful organogenic callus induction was reported by Valera-Montero and Phillips (2005) in MS medium with BA (22.2 μM), IAA (5.7 μM) and GA_3 (6.1 μM) from radicle-side half-seed explants of *C. baccatum*, where the shoot buds on the calli eventually elongated and these shoots, when transferred to PGRs free MS basal medium, developed into whole plants. The calli developed retained the shoot regeneration efficiency for more than 3 years, overcoming the narrow time window.

Direct organogenesis is the most utilized attempt in chilli peppers. Venkataiah et al. (2016) established an optimum protocol for the direct somatic embryogenesis from *C. baccatum* cotyledon and leaf explants. Here, the highest response was obtained in MS medium containing 9.1 μM 2,4-D and 2.32 μM kinetin. The embryos were successfully converted into whole plantlets after transferring to MS medium fortified with 4.44 μM BA. Leaf-derived somatic embryos are more responsive in induction, maturation, and conversion into complete plantlets than those from cotyledon explants. Shoot regeneration from shoot meristem explants on MS medium augmented with 9.18 μM TDZ was observed by Venkataiah et al. (2006). They also observed that *C. baccatum* produced the lowest number of shoots when compared to *C. annuum*, *C. frutescens*, and *C. praetermissum*. Similarly, Manzur et al. (2013) observed lowest in vitro germination response exhibited by *C. baccatum* when compared to *C. annuum*, *C. frutescens*, *C. pubescens*, and *C. chinense*, in the in vitro immature embryo culture.

7.4 Difficulties Associated with the Micropropagation of Chilli Pepper

The genus *Capsicum* exhibits natural recalcitrance to morphogenesis. The failure of organs, tissues, or cells to respond to an in vitro culture is recalcitrance. Plant response to in vitro culture is highly influenced by the physiology of the donor plant from which the explant is taken, in vitro stress physiology, and manipulation (Benson 2000). Even if there are some studies reporting the relative success of micropropagation, the low morphogenetic potential delimits the genetic engineering of these species (Ochoa-Alejo and Ramirez-Malagon 2001). A confined time window in the regenerative process may be one of the possible reasons behind the morphogenetic recalcitrance of the *Capsicum* genus (Valera-Montero and Phillips 2005).

Another common problem associated with the micropropagation of *Capsicum* genus is the formation of rosette shoots, which are clustered shoots that fail to elongate (Monteiro do Rêgo et al. 2016). Adventitious shoot regeneration from different explants like cotyledons, shoot tips, and hypocotyl has been achieved in the genus earlier, but the results were often not satisfying due to the development of stunted indistinct buds with a cauliflower appearance and had distorted leaves (Liu et al. 1990; Ochoa-Alejo and Ireta-Moreno 1990; Arroyo and Revilla 1991). Mezghani et al. (2007) observed hyper-hydricity, hypertrophy, and the formation of rosettes of leaves in *Capsicum*. Kumar et al. (2012) reported that cotyledons and leaf discs resulted in the formation of indistinct shoot buds and rosette leaves. Similar observations by Verma et al. (2013) in bell pepper report the formation of rosette shoots with distinct leaves was obtained, which failed to elongate. Scientists have made quite a few attempts to overcome rosette formation and shoot-elongation impediments. Arroyo and Revilla (1991) developed a method by rooting rosettes and transplanting to field, which resulted in elongation. The introduction of various additives like silver nitrate ($AgNO_3$; Hyde and Phillips 1996); α-lactone; 24-epibrassinolide (EBr; Franck-Duchenne et al. 1998); phenylacetic acid (PAA; Husain et al. 1999); and copper (Cu; Joshi and Kothari 2007) into the medium overcomes the bud and shoot-elongation problem. There are also reports on the successful elongation of rosettes in shoot elongation media with the addition of PGRs like GA_3 and BA (Mezghani et al. 2007). But the problem persists and is yet to be solved in many species of peppers.

In the genus *Capsicum*, a major factor influencing micropropagation is genotype dependence. The difficulty to establish broad-spectrum protocols that are efficient and reliable for the members of this genus is associated with a wide array of in vitro culture responses exhibited by the great diversity of the existing genotypes (Kothari et al. 2010). A clear specificity exists between the genotypes and regenerative capacity of different cultivars. There is a different need for each cultivar of the genus. So, designing cultivar-specific media formulations is needed (Dabauza and Pena 2001). This signifies an important limiting factor in the development of micropropagation protocols for the genus (Fari 1988; Steinitz et al. 1999; Kothari et al. 2010). Recent studies have backed the evidence of genotype-dependent organogenesis of *Capsicum* genus. *C. annuum* cv. Byadagi Dabbi showed superior in vitro response in comparison with cv. Arka Lohit (Mathew 2002). A variable degree of regeneration response depending upon the genotype was observed in *C. annuum* cultivars (Kumar et al. 2012). Orlinska and Nowaczyk (2015) reported considerably higher organogenesis response in interspecific hybrid (*C. frutescens* L. × *C. annuum*), than the comparable responses of three *C. annuum* L. genotypes. Venkataiah et al. (2003) recorded the dependence of organogenetic response on genotype specificity in ten *Capsicum* cultivars in a study on thidiazuron (TDZ)-induced organogenesis. Similarly, many observations were made on genotype-dependent variations in the genus *Capsicum* (Christopher and Rajam 1996; Hasnat et al. 2007; Manzur et al. 2013; Shu et al. 2022).

Sensitivity to ethylene gas is a major obstacle faced by researchers in the micropropagation of the *Capsicum* genus. Ethylene is a gaseous plant hormone responsible for various physiological and morphological responses in plants like abscission, fruit ripening induction, leaf and flower senescence elevation, and cell expression inhibition (An et al. 2010). The *Capsicum* genus is highly sensitive to ethylene during in vitro culture (Monteiro do Rêgo et al. 2016). Chlorosis, abscission of leaf primordia, and loss of vigour are common symptoms associated in response to ethylene sensitivity (Santana-Buzzy et al. 2005, 2006; Núñez-Pastrana et al. 2011). Maintaining the plants in higher gas exchange containers is one way to overcome this problem. Another strategy is to use inhibitors to prevent its biosynthesis like aminoethoxyvinylglycine (AVG), silver ions, carbon dioxide, and polyamines (Monteiro do Rêgo et al. 2016).

7.5 Conclusions and Future Perspectives

Chilli is an important spice crop used in food and pharmaceutics worldwide. It is particularly important for the production of capsaicinoids. Therefore, the propagation of chilies is a crucial step for further downstream harvesting and extraction steps. Conventional vegetative propagation is limited in *Capsicum* genus, and seed propagation is insufficient due to various reasons. Capsaicin content has been declining due to cross-pollination and mutation in certain species of *Capsicum*. This demands for the enhancement of seed germination and alternative methods for the vegetative propagation of the genus. Noticeable progress has been achieved in the conventional and in vitro propagation of the generally recalcitrant *Capsicum* genus. This will increase the availability of elite planting materials. As the in vitro organogenesis is an obligatory tool for genetic engineering studies, these developments in the micropropagation will accelerate the genetic improvement of the crop. We believe that these advancements will aid in the development of pest, pathogen, and abiotic stress-resistant plants with higher yield and will make an impact on chilli production and its market.

References

Aboshama HM (2011) Direct somatic embryogenesis of pepper (*Capsicum annuum* L.). World J Agric Sci 7:755–762

Agrawal S, Chandra N (1983) Differentiation of multiple shoot buds and plantlets in cultured embryos of *Capsicum annuum* L. var. Mathania. Curr Sci 52:645–646

Agrawal S, Chandra N, Kothari SL (1989) Plant regeneration in tissue cultures of pepper (*Capsicum annuum* L. cv. Mathania). Plant Cell Tissue Organ Cult 16:47–55

Ahmad N, Siddique I, Anis M (2006) Improved plant regeneration in *Capsicum annuum* L. from nodal segments. Biol Plant 50:701–704

Albrecht E, Zhang D, Saftner RA, Stommel JR (2012) Genetic diversity and population structure of *Capsicum baccatum* genetic resources. Genet Resour Crop Evol 59:517–538

Alcalá-Rico JS, López-Benítez A, Vázquez-Badillo ME, Sánchez-Aspeytia D, Rodríguez Herrera SA, Pérez-Rodríguez MÁ, Ramírez-Godina F (2019) Seed physiological potential of *Capsicum annuum* var. Glabriusculum genotypes and their answers to pre-germination treatments. Agronomy 9:325. https://doi.org/10.3390/agronomy9060325

Alvarez C, Valledor L, Sáez P, Sánchez-Olate M, Ríos D (2016) Proteomic analysis through adventitious rooting of *Pinus radiata* stem cuttings with different rooting capabilities. Am J Plant Sci 7:1888–1904

An F, Zhao Q, Ji Y, Li W, Jiang Z, Yu X, Zhang C, Han Y, He W, Liu Y, Zhang S (2010) Ethylene-induced stabilization of ETHYLENE INSENSITIVE3 and EIN3-LIKE1 is mediated by proteasomal degradation of EIN3 binding F-box 1 and 2 that requires EIN2 in Arabidopsis. Plant Cell 22:2384–2401

Anaya-Esparza LM, Mora ZV, Vázquez-Paulino O, Ascencio F, Villarruel-López A (2021) Bell peppers (*Capsicum annum* L.) losses and wastes: source for food and pharmaceutical applications. Molecules 26:5341. https://doi.org/10.3390/molecules26175341

Antonious GF (2018) Capsaicinoids and vitamins in hot pepper and their role in disease therapy. In: Mozsik G (ed) Capsaicin and its human therapeutic development. Intech Open, London. https://doi.org/10.5772/intechopen.78243

Arous S, Boussaid M, Marrakchi M (2001) Plant regeneration from zygotic embryo hypocotyls of Tunisian chili (*Capsicum annuum* L.). J Appl Hortic 3:17–22

Arroyo R, Revilla M (1991) In vitro plant regeneration from cotyledon and hypocotyl segments in two bell pepper cultivars. Plant Cell Rep 10:414–416

Ashrafuzzaman M, Hossain MM, Ismail MR, Haque MS, Shahidullah SM, Uz-zaman S (2009) Regeneration potential of seedling explants of chilli (*Capsicum annuum*). Afr J Biotech 8:591–596

Attia MF, Arafa AM, Moustafa MA, Mohamed MA (2003) Pepper grafting, a method of controlling soil-borne diseases and enhancement of fruit yield: 1. Improvement of pepper resistance to *Fusarium* wilt. Egypt J Phytopathol 31:151–165

Barroso PA, Rego MM, Rego ER, Soares WS (2015) Embryogenesis in anthers from ornamental pepper plants (*Capsicum annum* L.). Gen Mol Res 14:13349–13363

Batista DS, Dias LL, Macedo AF, do Rêgo MM, do Rêgo ER, Floh EI, Finger FL, Otoni WC (2013) Suppression of ethylene levels promotes morphogenesis in pepper (*Capsicum annuum* L.). In Vitro Cell Dev Biol-Plant 49:759–764

Benson EE (2000) Special symposium: in vitro plant recalcitrance: do free radicals have a role in plant tissue culture recalcitrance? In Vitro Cell Dev Biol-Plant 36:163–170

Bhutia KL, Meetei NT, Khanna VK (2018) In vitro direct regeneration of Dalle Khursani (*Capsicum annum*) from salicylic acid treated explants. J Pharmacogn Phytochem 7:1008–1012

Binzel ML, Sankhla N, Joshi S, Sankhla D (1996) Induction of direct somatic embryogenesis and plant regeneration in pepper (*Capsicum annuum* L.). Plant Cell Rep 15:536–540

Birkas Z, Balazs GZ, Kokai Z (2021) Effect of grafting and growing media on the chosen fruit quality compounds and sensory parameters of sweet pepper (*Capsicum annuum* L.). Acta Aliment 50:1–12

Bodhipadma K, Leung DW (2003) In vitro fruiting and seed set of *Capsicum annuum* L. cv. Sweet Banana. In Vitro Cell Dev Biol-Plant 39:536–539

Bora G, Gogoi HK, Handique PJ (2019) Influence of silver nitrate and glutamine on in vitro organogenesis of Lota Bhot (*Capsicum chinense* Jacq.), an indigenous pungent pepper variety of Assam. J Appl Biol Biotech 7:2–8

Bosland PW, Votava EJ (2012) Peppers: vegetable and spice capsicums. CABI publishing, New York, p 91

Büyükalaca S, Mavituna F (1995) Artificial seeds of pepper somatic embryos. Acta Hortic 412: 106–110

Büyükalaca S, Mavituna F (1996) Somatic embryogenesis and plant regeneration of pepper in liquid media. Plant Cell Tissue Organ Cult 46:227–235

Büyükalaca S, Mavituna F, Gomez-Guillamon ML (1995) Artificial seeds of pepper somatic embryos. Acta Hortic 412:106–110

Carballo-Méndez FJ, Urrestarazu M, Carlos Rodríguez-Ortiz J, Morales I (2022) Electrical conductivity of the nutrient solution on the vegetative propagation of bell pepper and tomato. Cienc Rural 53:1–7

Cardarelli M, Rouphael Y, Kyriacou MC, Colla G, Pane C (2020) Augmenting the sustainability of vegetable cropping systems by configuring rootstock-dependent rhizomicrobiomes that support plant protection. Agronomy 10:1185. https://doi.org/10.3390/agronomy10081185

Chee MJ, Lycett GW, Chin CF (2018) Development of a direct transformation method by GFP screening and in vitro whole plant regeneration of *Capsicum frutescens* L. Electron J Biotechnol 34:51–58

Cheng Y, Ma R, Jiao Y, Qiao N, Li T (2013) Impact of genotype, plant growth regulators and activated charcoal on embryogenesis induction in microspore culture of pepper (*Capsicum annuum* L.). S Afr J Bot 88:306–309

Christopher T, Rajam MV (1994) In vitro clonal propagation of *Capsicum* spp. Plant Cell Tissue Organ Cult 38:25–29

Christopher T, Rajam MV (1996) Effect of genotype, explant and medium on in vitro regeneration of red pepper. Plant Cell Tissue Organ Cult 46:245–250

Dabauza M, Pena L (2001) High efficiency organogenesis in sweet pepper (*Capsicum annuum* L.) tissues from different seedling explants. Plant Growth Regul 33:221–229

Debbarma A, Devi J, Barua M, Sarma D (2018) Germination performance of chilli (*Capsicum annuum* L.) and coriander (*Coriandrum sativum* L.) as affected by seed priming treatments. J Pharmacogn Phytochem 7:2648–2652

Demir I, Okcu G (2004) Accelerated hydration treatment for improving germination and seedling growth in aubergine (*Solanum melongena*) and pepper (*Capsicum annuum*). Ann Appl Biol 144:121–123

Dogra V, Sharma P (2022) Effect of indole butyric acid (IBA) and honey on root parameters of different sized stem cuttings in bell pepper. Himachal J Agric Res 47:256–259

Dolcet-Sanjuan R, Claveria E, Huerta A (1997) Androgenesis in *Capsicum annuum* L.-effects of carbohydrate and carbon dioxide enrichment. J Am Soc Hortic Sci 122:468–475

Donas-Ucles F, Jimenez-Luna M, Gongora-Corral JA (2014) Influence of three rootstocks on yield and commercial quality of Italian sweet pepper. Ciencia Agrotec 38:538–545

Duclercq J, Sangwan-Norreel B, Catterou M, Sangwan RS (2011) De novo shoot organogenesis: from art to science. Trends Plant Sci 16:597–606

Dumas de Vaulx R, Chambonnet D, Pochard E (1981) In vitro anther culture in red pepper (*Capsicum annuum* L.): improvement of the rate of plant production in different genotypes by treatments at 35 C. Agronomie 1:859–864

Ebida AI, Hu CY (1993) In vitro morphogenetic responses and plant regeneration from pepper (*Capsicum annuum* L. cv. Early California wonder) seedling explants. Plant Cell Rep 13:107–110

Ezura H, Nishimiya S, Kasumi M (1993) Efficient regeneration of plants independent of exogeneous growth regulators in bell pepper (*Capsicum annumm* L.). Plant Cell Rep 12:676–680

Fari M (1988) Pepper (*Capsicum annuum* L.). In: Bajaj YPS (ed) Biotechnology in agriculture and forestry, vol 2. Springer, Berlin, pp 345–362

Franck-Duchenne M, Wang Y, Ben Tahar S, Beachy RN (1998) In vitro stem elongation of sweet pepper in media containing 24-epi-brassinolide. Plant Cell Tissue Organ Cult 53:79–84

García-Gonzáles R, Quiroz K, Carrasco B, Caligari P (2010) Plant tissue culture: current status, opportunities and challenges. Int J Agric Nat Res 37:5–30

Gatz A (2014) Morpho-histological aspects of adventitious shoot formation without plant growth regulators in seed explants of *Capsicum annuum* L., and impact of preculture on regeneration. Acta Sci Pol Hortorum Cultus 13:135–150

Gayathri N, Gopalakrishnan M, Sekar T (2015) In vitro micropropagation of *Capsicum chinense* Jacq. (Naga king chili). Asian J Plant Sci Res 5:13–18

George EF, Hall MA, De Klerk GJ (2008) Plant propagation by tissue culture. In: The background, vol 1. Springer, Cham, pp 205–226

George L, Narayanaswamy S (1973) Haploid *Capsicum* through experimental androgenesis. Protoplasma 78:467–470

Gogoi S, Acharjee S, Devi J (2014) In vitro plantlet regeneration of *Capsicum chinense* Jacq. cv.'Bhut jalakia': hottest chili of northeastern India. In Vitro Cell Dev Biol-Plant 50:235–241

Goldschmidt EE (2014) Plant grafting: new mechanisms, evolutionary implications. Front Plant Sci 5:727. https://doi.org/10.3389/fpls.2014.00727

Grozeva S, Todorova V, Nankar AN (2021) Creation of pepper doubled haploids and morphological characterization of androgenic plants. Euphytica 217:1–16

Guha S, Maheshwari SC (1964) In vitro production of embryos from anthers of *Datura*. Nature 204: 497

Gunay AL, Rao PS (1978) In vitro plant regeneration from hypocotyl and cotyledon explants of red pepper (*Capsicum*). Plant Sci Lett 11:365–372

Gurnani N, Gupta M, Mehta D, Mehta BK (2016) Chemical composition, total phenolic and flavonoid contents, and in vitro antimicrobial and antioxidant activities of crude extracts from red chilli seeds (*Capsicum frutescens* L.). J Taibah Univ Sci 10:462–470

Gururaj HB, Giridhar P, Sharma A, Prasad BC, Ravishankar GA (2004) In vitro clonal propagation of bird eye chilli (*Capsicum frutescens* Mill.). Indian J Exp Biol 42:1136–1140

Haberlandt G (1902) Culturversuche mit isolierten Pflanzenzellen. Sitz-Ber Math-Nat Kl Kais Akad Wiss Wien 111:69–92

Hailu T, Abera B, Daksa J (2015) *In vitro* direct organogenesis protocol for mass propagation of an elite Ethiopian hot pepper (*Capsicum annuum* L.) cultivar: Mareko Fana. Am J Plant Sci 6: 1435–1443

Haque SM, Paul S, Ghosh B (2016) High-frequency in vitro flowering, hand-pollination and fruit setting in ten different cultivars of *Capsicum* spp. (*C. annuum, C. chinense,* and *C. frutescens*): an initial step towards in vitro hybrid production. Plant Cell Tissue Organ Cult 127:161–173

Harini I, Sita GL (1993) Direct somatic embryogenesis and plant regeneration from immature embryos of chilli (*Capsicum annuum* L.). Plant Sci 89:107–112

Hasnat R, Abbasi NA, Ahmad T, Hafiz IA (2007) Induction and regeneration of hypocotyls derived calli in hot chilli (*Capsicum frutescens* L.) varieties. Pak J Bot 39:1787–1795

Hassanein AM (2013) Factors influencing plant propagation efficiency via stem cuttings. J Hortic Sci Ornamen Plants 5:171–176

Hegde V, Partap PS, Yadav RC (2017) Plant regeneration from hypocotyl explants in Capsicum (*Capsicum annuum* L.). Int J Curr Microbiol App Sci 6:545–557

Heiser CB, Smith PG (1958) New species of Capsicum from South America. Brittonia 10:194–201

Husain S, Jain A, Kothari SL (1999) Phenylacetic acid improves bud elongation and in vitro plant regeneration efficiency in *Capsicum annuum* L. Plant Cell Rep 19:64–68

Hyde CL, Phillips GC (1996) Silver nitrate promotes shoot development and plant regeneration of Chile pepper (*Capsicum annuum* L.) via organogenesis. In Vitro Cell Dev Biol-Plant 32:72–80
Irikova T, Grozeva S, Rodeva V (2011) Anther culture in pepper (*Capsicum annuum* L.) in vitro. Acta Physiol Plant 33:1559–1570
İzgü T, İlbi H, Mendi YY (2020) Optimization of plant regeneration in different pepper (*Capsicum annuum* L.) lines. Turk J Agri Food Sci Tech 8:471–477
Jaiswal V, Gahlaut V, Kumar N, Ramchiary N (2021) Genetics, genomics and breeding of chili pepper *Capsicum frutescens* L. and other *Capsicum* species. In: Al-Khayri JM, Jain SM, Johnson DV (eds) Advances in plant breeding strategies: vegetable crops. Springer, Cham, pp 59–86
Jang Y, Yang E, Cho M, Um Y, Ko K, Chun C (2012) Effect of grafting on growth and incidence of Phytophthora blight and bacterial wilt of pepper (*Capsicum annuum* L.). Hortic Environ Biotech 53:9–19
Johkan M, Mitukuri K, Yamasaki S, Mori G, Oda M (2009) Causes of defoliation and low survival rate of grafted sweet pepper plants. Sci Hortic 119:103–107
Johkan M, Okuta M, Takizawa K, Nakagawa T, Maruo T, Tsukagoshi S, Hohjo M, Shinohara Y (2014) A procedure to increase the number of cuttings available for vegetative propagation of sweet pepper plants. Hortic Res (Japan) 13:119–124
Joshi A, Kothari SL (2007) High copper levels in the medium improves shoot bud differentiation and elongation from the cultured cotyledons of *Capsicum annuum* L. Plant Cell Tissue Organ Cult 88:127–133
Junaidy RB, Shahruddin S (2022) Germinability and seedling growth performance of chilli (*Capsicum annuum*) seeds in response to different gibberellic acid concentrations pre-treatment. Agro Tech-Food Sci Tech Environ 1:10–16
Keatinge JDH, Lin L-J, Ebert AW, Chen WY, Jd'A H, Luther GC, Wang J-F, Ravishankar M (2014) Overcoming biotic and abiotic stresses in the Solanaceae through grafting: current status and future perspectives. Biol Agric Hortic 30:272–287
Kehie M, Kumaria S, Tandon P (2013) In vitro plantlet regeneration from cotyledon segments of *Capsicum chinense* Jacq. cv. Naga king chili, and determination of capsaicin content in fruits of in vitro propagated plants by high performance liquid chromatography. Sci Hortic 164:1–8
Khan H, Siddique I, Anis M, Khan PR (2011) In vitro organogenesis from internode derived callus cultures of *Capsicum annuum* L. J Plant Biochem Biotechnol 20:84–89
Koleva-Gudeva L (2003a) The effect of incubation treatment on the pepper (*Capsicum annuum* L.) androgenesis. J Agric Plant Sci 3:87–94
Koleva-Gudeva L (2003b) Anther culture in pepper (*Capsicum annuum* L.). J Agric Plant Sci 3:95–102
Koleva-Gudeva LR, Spasenoski M, Trajkova F (2007) Somatic embryogenesis in pepper anther culture: the effect of incubation treatments and different media. Sci Hortic 111:114–119
Kollmannsberger H, Rodríguez-Burruezo A, Nitz S, Nuez F (2011) Volatile and capsaicinoid composition of ají (*Capsicum baccatum*) and rocoto (*Capsicum pubescens*), two Andean species of Chile peppers. J Sci Food Agric 91:1598–1611
Kothari SL, Joshi A, Kachhwaha S, Ochoa-Alejo N (2010) Chilli peppers-a review on tissue culture and transgenesis. Biotechnol Adv 28:35–48
Kumar RV, Sharma VK, Chattopadhyay B, Chakraborty S (2012) An improved plant regeneration and *Agrobacterium*-mediated transformation of red pepper (*Capsicum annuum* L.). Physiol Mol Biol Plants 18:357–364
Kumar V, Sharma A, Narasimha Prasad BC, Gururaj HB, Giridhar P, Ravishankar GA (2007) Direct shoot bud induction and plant regeneration in *Capsicum frutescens* Mill.: influence of polyamines and polarity. Acta Physiol Plant 29:11–18
Lannes SD, Finger FL, Schuelter AR, Casali VW (2007) Growth and quality of Brazilian accessions of *Capsicum chinense* fruits. Sci Hort 112:266–270

Lantos C, Juhász AG, Somogy G, Ötvös K, Vági P, Mihály R, Kristóf Z, Somogyi N, Pauk J (2009) Improvement of isolated microspore culture of pepper (*Capsicum annuum* L.) via co-culture with ovary tissues of pepper or wheat. Plant Cell Tissue Organ Cult 97:285–293

Liu W, Parrott WA, Hildebrand DF, Collins GB, Williams EG (1990) Agrobacterium induced gall formation in bell pepper (*Capsicum annuum* L.) and formation of shoot-like structures expressing introduced genes. Plant Cell Rep 9:360–364

López-Marín J, González A, Pérez-Alfocea F, Egea-Gilabert C, Fernández JA (2013) Grafting is an efficient alternative to shading screens to alleviate thermal stress in greenhouse-grown sweet pepper. Sci Hortic 149:39–46

Madhuri V, Rajam MV (1993) Apical shoot meristem culture in red pepper (*Capsicum annuum* L.). J Plant Biochem Biotechnol 2:67–68

Manzur JP, Penella C, Rodríguez-Burruezo A (2013) Effect of the genotype, developmental stage and medium composition on the in vitro culture efficiency of immature zygotic embryos from genus *Capsicum*. Sci Hortic 161:181–187

Martínez-López M, García-Pérez A, Gimeno-Páez E, Prohens J, Vilanova S, García-Fortea E (2021) Screening of suitable plant regeneration protocols for several *Capsicum* spp. Through direct organogenesis. Horticulturae 7:261–278

Mathew D (2002) In vitro shoot and root morphogenesis from cotyledon and hypocotyl explants of hot pepper cultivars Byadagi Dabbi and Arka Lohit. Capsicum Eggplant Newsl 21:69–72

Mezghani N, Jemmali A, Elloumi N, Gargouri-Bouzid R, Kintzios S (2007) Morpho-histological study on shoot bud regeneration in cotyledon cultures of pepper (*Capsicum annuum*). Biologia 62:704–710

Mohamed M, Bridgemohan P, Singh K (2014) A method of rapid propagation of hot pepper using marcotting technique to maintain clonal characteristics. Afr J Food Sci 5:96–99

Monteiro do Rêgo M, Ramalho do Rêgo E, Barroso PA (2016) Tissue culture of *Capsicum* spp. In: do Rêgo ER, do Rego MM, Finger FL (eds) Production and breeding of chilli peppers (*Capsicum* spp.). Springer, Cham, pp 97–127

Morrison RA, Koning RE, Evans DA (1986) Anther culture of an interspecific hybrid of Capsicum. J Plant Physiol 126:1–9

Muhammad NA, Abdullah SNAI, Hashim SN (2017) A preliminary study on pre- treatment solutions towards chili seeds germination. J Academia 5:61–68

Murashige T, Skoog F (1962) A revised medium for rapid growth and bio assays with tobacco tissue cultures. Physiol Plant 15:473–497

Núñez-Pastrana R, Arcos-Ortega GF, Souza-Perera RA, Sánchez-Borges CA, Nakazawa-Ueji YE, García-Villalobos FJ, Guzmán-Antonio AA, Zúñiga-Aguilar JJ (2011) Ethylene, but not salicylic acid or methyl jasmonate, induces a resistance response against *Phytophthora capsica* in habanero pepper. Eur J Plant Pathol 131:669–683

Obul Reddy B, Giridhar P, Ravishankar GA (2002) The effect of triacontanol on micropropagation of *Capsicum frutescens* and *Decalepis hamiltonii* W & A. Plant Cell Tissue Organ Cult 71:253–258

Ochoa-Alejo N, Ireta-Moreno L (1990) Cultivar differences in shoot-forming capacity of hypocotyl tissues of chilli pepper (*Capsicum annuum* L.) cultured in vitro. Sci Hortic 42:21–28

Ochoa-Alejo N, Ramirez-Malagon R (2001) In vitro chili pepper biotechnology. In Vitro Cell Dev Biol-Plant 37:701–729

Odland ML, Porter AM (1941) A study of natural crossing in pepper (*Capsicum frutescens* L.). J Am Soc Hortic Sci 38:585–588

Oka Y, Offenbach R, Pivonia S (2004) Pepper rootstock graft compatibility and response to *Meloidogyne javanica* and *M. incognita*. J Nematol 36:137–141

Olszewska D, Kisiala A, Niklas-Nowak A, Nowaczyk P (2014) Study of in vitro anther culture in selected genotypes of genus *Capsicum*. Turk J Biol 38:118–124

Orlinska M, Nowaczyk P (2015) In vitro plant regeneration of 4 *Capsicum* spp. genotypes using different explant types. Turk J Biol 39:60–68

Palada MC, Wu DL (2008) Evaluation of chili rootstocks for grafted sweet pepper production during the hot-wet and hot-dry seasons in Taiwan. Acta Hortic 767:167–174

Penella C, Nebauer SG, Quinones A, San Bautista A, López-Galarza S, Calatayud A (2015) Some rootstocks improve pepper tolerance to mild salinity through ionic regulation. Plant Sci 230:12–22

Phillips GC, Hubstenberger JF (1985) Organogenesis in pepper tissue cultures. Plant Cell Tissue Organ Cult 4:261–269

Phuong NT (2021) Direct embryogenesis from anther culture of hot chilli *Capsicum annuum* L. Ho Chi Minh City Open University J Sci Eng Technol 11:3–10

Pico MB, Thompson AJ, Gisbert C et al (2017) Genetic resources for rootstock breeding. In: Colla G, Perez-Alfocea F, Schwarz D (eds) Vegetable grafting: principles and practices. CAB, Oxfordshire, pp 22–69

Pinto CM, Santos IC, Araujo FF, Silva TP (2016) Pepper importance and growth (*Capsicum* spp). In: do Rêgo ER, do Rego MM, Finger FL (eds) Production and breeding of chilli peppers (*Capsicum* spp.). Springer, Cham, pp 1–25

Pochard E, Dumas de Vaulx R (1979) Haploid parthenogenesis in Capsicum annuum L. In: Hawkes JG, Lester RN, Skelding AD (eds) The biology and taxonomy of the Solanaceae. Academic, London, pp 455–472

Raj RP, Glint VD, Babu KN (2015) In vitro plant regeneration in *Capsicum chinense* Jacq. (Naga chili). J Appl Biol Biotech 3:30–33

Ramírez-Malagón R, Ochoa-Alejo N (1996) An improved and reliable chili pepper (*Capsicum annuum* L.) plant regeneration method. Plant Cell Rep 16:226–231

Rego MM, Rego ER, Soares WS, Nascimento KS, Barroso PA, Ferreira KT (2013) Effects of TDZ on multiple shoots in chili pepper. In: Lanteri S, Rotino GL (eds) Breakthroughs in the genetics and breeding of *Capsicum* and eggplant. Università degli Studi di Torino, Torino, pp 541–544

Ribes Moya AM, Raigón MD, Moreno-Peris E, Fita A, Rodríguez-Burruezo A (2018) Response to organic cultivation of heirloom Capsicum peppers: variation in the level of bioactive compounds and effect of ripening. PLoS One 13:e0207888. https://doi.org/10.1371/journal.pone.0207888

Rodeva V (2001) In vitro regeneration in anther culture of pepper (*Capsicum annuum* L.). Sci Works Agricult Univ Plovdiv 3:211–214

Ros-Ibanez C, Robertson L, del Carmen M-LM, Cano-García A, Lacasa-Plasencia A (2014) Development of virulence to *Meloidogyne incognita* on resistant pepper rootstocks. Span J Agri Res 12:225–232

Saadoun M, Allagui MB (2013) Management of chili pepper root rot and wilt (caused by *Phytophthora nicotianae*) by grafting onto resistant rootstock. Phytopathol Mediterr 52:141–147

Sanatombi K, Sharma GJ (2007) Micropropagation of *Capsicum frutescens* L. using axillary shoot explants. Sci Hortic 113:96–99

Sanatombi K, Sharma GJ (2008) In vitro propagation of *Capsicum chinense* Jacq. Biol Plant 52: 517–520

Santana-Buzzy N, Canto-Flick A, Barahona-Pérez F, del Carmen M-PM, Zapata-Castillo PY, Solís-Ruiz A, Zaldívar-Collí A, Gutiérrez-Alonso O, de Lourdes Miranda Ham M (2005) Regeneration of habanero pepper (*Capsicum chinense* Jacq.) via organogenesis. Hort Sci 40:1829–1831

Santana-Buzzy N, Canto-Flick A, Iglesias-Andreu LG, Montalvo-Peniche MD, López-Puc G, Barahona-Pérez F (2006) Improvement of in vitro culturing of habanero pepper by inhibition of ethylene effects. Hort Sci 41:405–409

Sarvesh A, Reddy TP, Kishor PB (1993) Embryogenesis and organogenesis in cultured anthers of an oil yielding crop Niger (*Guizotia abyssinica* Cass). Plant Cell Tissue Organ Cult 35:75–80

Sganzerla M, Coutinho JP, de Melo AM, Godoy HT (2014) Fast method for capsaicinoids analysis from *Capsicum chinense* fruits. Food Res Int 64:718–725

Shahzad A, Sharma S, Parveen S, Saeed T, Shaheen A, Akhtar R, Yadav V, Upadhyay A, Ahmad Z (2017) Historical perspective and basic principles of plant tissue culture. In: Abdin M, Kiran U,

Kamaluddin Ali A (eds) Plant biotechnology: principles and applications. Springer, Singapore, pp 1–36

Sharma A, Kumar V, Giridhar P, Ravishankar GA (2008) Induction of in vitro flowering in *Capsicum frutescens* under the influence of silver nitrate and cobalt chloride and pollen transformation. Electron J Biotechnol 11:84–89

Shu H, Zhang Y, He C, Altaf MA, Hao Y, Liao D, Li L, Li C, Fu H, Cheng S, Zhu G (2022) Establishment of in vitro regeneration system and molecular analysis of early development of somatic callus in *Capsicum chinense* and *Capsicum baccatum*. Front Plant Sci 13:1025497. https://doi.org/10.3389/fpls.2022.1025497

Sibi M, Dumas de Vaulx R, Chambonnet D (1979) Obtention of haploid plants through in vitro androgenesis in sweet pepper (*Capsicum annuum* L.). Annales de l'Amelioration des Plantes 29: 583–606

Smith PG, Heiser CB Jr (1957) Taxonomy of *Capsicum chinense* Jacq. And the geographic distribution of the cultivated *Capsicum* species. Bull Torrey Bot Club 1:413–420

Solís-Ramos LY, Nahuath-Dzib S, Andrade-Torres A, Barredo-Pool F, González-Estrada T, de la Serna EC (2010) Indirect somatic embryogenesis and morpho-histological analysis in *Capsicum chinense*. Biologia 65:504–511

Stasolla C, Yeung EC (2003) Recent advances in conifer somatic embryogenesis: improving somatic embryo quality. Plant Cell Tissue Organ Cult 74:15–35

Steinitz B, Wolf D, Matzevitch-Josef T, Zelcer A (1999) Regeneration in vitro and genetic transformation of pepper (*Capsicum* spp.): the current state of art. Capsicum Eggplant Newsl 18:9–15

Supena ED (2021) An efficient anther culture on double-layered media to produce doubled haploid plants of pepper (*Capsicum annuum*). In: Segui-Simarro JM (ed) Doubled haploid technology. Humana, New York, pp 267–278

Supena EDJ, Custers JBM (2011) Refinement of shed-microspore culture protocol to increase normal embryos production in hot pepper (*Capsicum annuum* L.). Sci Hortic 130:769–774

Supena EDJ, Suharsono S, Jacobsen E, Custers JBM (2006) Successful development of a shed microspore culture protocol for doubled haploid production in Indonesian hot pepper (*Capsicum annuum* L.). Plant Cell Rep 25:1–10

Swamy S, Krupakar A, Chandran DS, Koshy EP (2014) Direct regeneration protocols of five *Capsicum annuum* L. varieties. Afr J Biotech 13:307–312

Tanksley SD (1984) High rates of cross-pollination in Chile pepper. Hort Sci 19:580–582

Thamburaj S, Singh N (2000) Textbook of vegetables, Tubercrops and spices. Indian Council of Agricultural Research, New Delhi, p 469

Thirunanvoukkarasu M, Brahman M, Dhal NK (2004) Vegetative propagation of *Hymenea courbaril* by air layering. J For Trop Sci 16:268–270

Valera-Montero LL, Phillips GC (2005) Long-lasting *Capsicum baccatum* 'organogenic callus' formation. In Vitro Cell Dev Biol-Plant 41:470–476

Venkataiah P, Bhanuprakash P, Kalyan SS, Subhash K (2016) Somatic embryogenesis and plant regeneration of *Capsicum baccatum* L. J Genet Eng Biotechnol 14:55–60

Venkataiah P, Bhanuprakash P, Subhash K (2006) In vitro shoot multiplication and plant regeneration in four *Capsicum* species using thidiazuron. Sci Hortic 107:117–122

Venkataiah P, Christopher T, Subhash K (2003) Thidiazuron induced high frequency adventitious shoot formation and plant regeneration in *Capsicum annuum* L. J Plant Biotech 5:245–250

Verma S, Dhiman K, Srivastava DK (2013) Efficient in vitro regeneration from cotyledon explants in bell pepper (*Capsicum annuum* L. cv. California wonder). Int J Adv Biotech Res 4:391–396

Wang YY, Sun CS, Wang CC, Chien NJ (1973) The induction of pollen plantlets of triticale and *Capsicum annuum* anther culture. Sci Sinica 16:147–151

Yamamoto S, Nawata E (2005) *Capsicum frutescens* L. in Southeast and East Asia, and its dispersal routes into Japan. Econ Bot 59:18–28

Zemene A, Worku A (2018) Protocol optimization for micro-propagation of green pepper (*Capsicum annum* L.) cultivated in Ethiopia. J Med Plants Stud 6:229–234

8 Biotechnological Approaches for Enhanced Production of Capsaicin

Swet Nisha

Abstract

Capsaicin compound that is in high demand in the market, owing to its wide pharmacological applications. Apart from drug industry, it is also utilized in the food industry, armed forces and personal defence. Industries producing high-purity research and pharmacological-grade capsaicin still rely on extraction and purification from chili plants. Various biotechnological approaches such as in vitro callus cultivation, elicitor treatments, abiotic and biotic stress, addition of precursors, immobilized cell systems and application of plant growth regulators have been found to be successful in enhancing capsaicin production. Various studies have shown that such approaches can be utilized for industrial-scale production. These are advantageous over enzymatic and chemical synthesis as they are free of hazardous substrates or by-products, give high yield and ensure round-the-year availability of capsaicin. The chapter focuses on chemistry, characterization, pharmacokinetics and action mechanism of capsaicin, various biotechnological approaches for enhanced capsaicin production along with their advantages over traditional methods and future prospects of biotechnology-based production.

Keywords

Capsaicin · Cell immobilization · Suspension culture · Elicitation · Abiotic and biotic stress

S. Nisha (✉)
Department of Microbiology, Patna Women's College, Patna, Bihar, India
e-mail: swetnisha.mbio@patnawomenscollege.in

M. K. Swamy, A. Kumar (eds.), *Capsaicinoids*,
https://doi.org/10.1007/978-981-99-7779-6_8

8.1 Introduction

The chili pepper fruit is used worldwide for culinary purposes, owing to its characteristic spicy flavour (Lee et al. 2011). The compounds responsible for pungency of *Capsicum* belong to a group of alkaloids known as capsaicinoids, of which capsaicin is present in the highest amount in capsicum followed by dihydrocapsaicin, nordihydrocapsaicin, homodihydrocapsaicin and homocapsaicin (De Lourdes et al. 2011). Capsaicin biosynthesis is the result of the merging of two pathways, namely, phenylpropanoid pathway and branched-chain fatty acid pathway. The final product of both the pathways, i.e. vanillylamine and 8-methyl-6-nonenoic-S CoA, respectively, undergoes a condensation reaction to synthesize capsaicin, which is brought about by capsaicin synthase enzyme (CS) and requires Mg^{2+}, ATP and coenzyme A (CoA) for its activity (Ochoa-Alejo and Gomez-Peralta 1993; De Lourdes et al. 2011). Different forms of capsaicinoids differ from each other by number of lateral chain carbon (R) or degree of unsaturation. The lateral chain length varies from 9 to 11 carbons with double bonds present at different positions (De Lourdes et al. 2011). Capsaicin and dihydrocapsaicin are the most prominent capsaicinoids that structurally differ from each other only in acyl group saturation (Bosland 1996). Distribution of capsaicin in chilli pepper is not uniform but is localized in the placental region. Approximately 89% of the total capsaicin produced is present in placental partition, whereas only 5–6% is found in pericarp (Andrews 1984; Walpole et al. 1996; Taira et al. 2012).

Capsaicin (trans-8-methyl-*N*-vanillyl-6-nonenamide) is a crystalline, lipophilic compound with a molecular weight of 305.80 g/mol. The molecular structure of capsaicin was first elucidated by Nelson and Dawson in (1923). *Cis*/*trans* isomerism is displayed by capsaicin as the presence of double bonds prevents internal rotation. However, in *cis* form, –CH $(CH_3)_2$ and the region to the other side of the double bond would come closer, thereby causing steric hindrance; this additional strain makes *trans* isomer less stable than the *cis* isomer (De Lourdes et al. 2011) (Fig. 8.1).

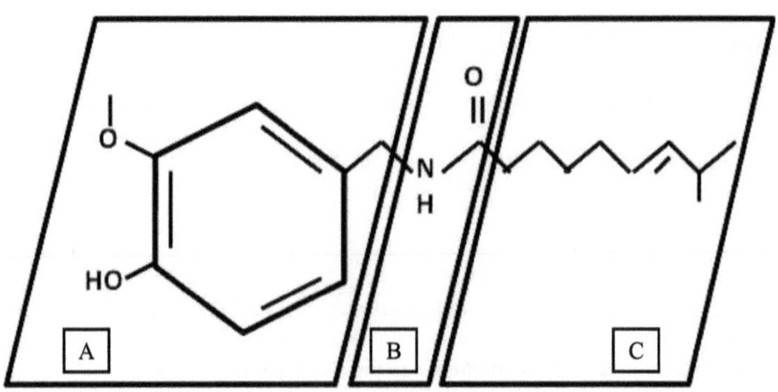

Fig. 8.1 Regions of capsaicin molecule: (**a**) aromatic ring; (**b**) amide bond; (**c**) hydrophobic side chain

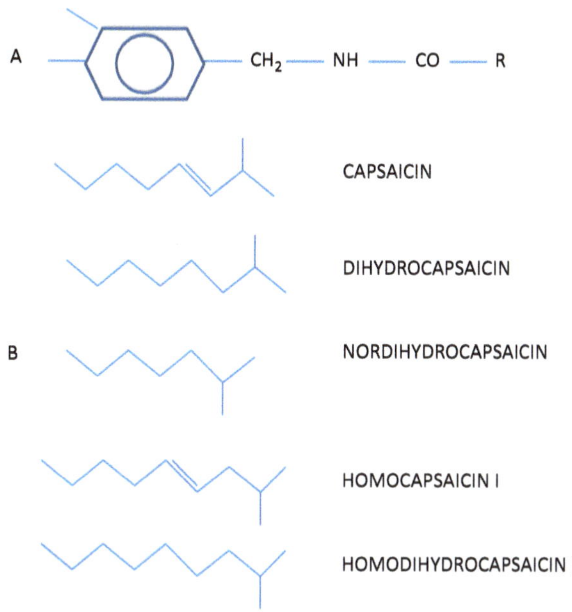

Fig. 8.2 (**a**) Chemical structure of capsaicinoid, (**b**) side chain (R) of various analogues

By dividing capsaicin molecule in three distinct regions, the structure activity relationship (SAR) of capsaicin agonist has been elucidated. The potent agonist activity has been attributed to substituents of position 3 and 4 of aromatic ring; the 4-OH group acts as H-donor, which is vital for its agonist activity. A hydrophobic group in C-region is required for high potency (Katritzky et al. 2003). The lateral chain length has also been found to be linked with capsaicin bioactivity (Barbero et al. 2010) (Figs. 8.2 and 8.3).

Capsaicin is a versatile molecule that has many pharmacological aspects, making it a compound of high demand in pharmaceutical industry. Some of the important clinical applications of capsaicin include its role as analgesic and anti-cancerous compound, gastrointestinal effects etc.

Capsaicin has been shown to be beneficial in relieving pain in rheumatoid arthritis, neuralgia and diabetic neuropathy. Upon topical application, it increases concentration of Ca^{2+} in the cell and inhibits pain signalling pathway (Reynolds 1999; Braun and Walker 1996; Butera 2007). Capsaicin has demonstrated anti-cancer properties as well. Application of capsaicin in in vitro cultured cells demonstrated that it was able to prevent cell growth and increased the number of cells that are arrested in G1 phase of cell cycle. The cells also demonstrated induction of autophagy and apoptosis in cancer cells after exposure to capsaicin (Oh et al. 2008). Migration of cancer cell was inhibited in breast cancer cell lines grown in vitro upon application of capsaicin; it also inhibited the migration of breast cancer cells in immunodeficient mice (Thoennissen et al. 2010). TGF-α and EGFR expression is upregulated by STAT protein Stat3 in the event of squamous cell

Fig. 8.3 Advantages of suspension culture for secondary metabolite production

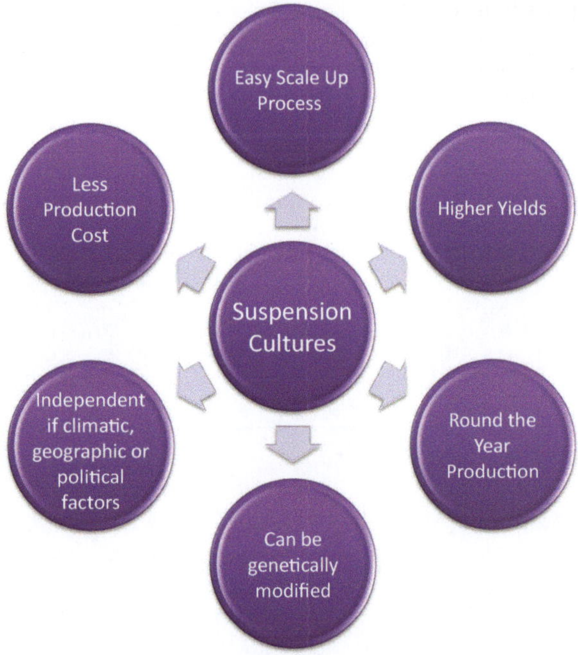

carcinoma of the head and neck, which is inhibited by application of capsaicin, thus preventing cancer progression (Song and Grandis 2000).

Consumption of chilli has been associated with increase in calorie burning and lipid oxidation (Yoshioka et al. 1999) and is also responsible for reducing food cravings (Belza et al. 2007). A study on glucose intolerance in obese subjects showed that after consumption of capsaicin, there is enhancement in fatty acid oxidation in adipose tissue and liver (Kang et al. 2009). A regular consumption of chilli has been seen to improve meal-induced hyperinsulinemia (Ahuja and Ball 2006). Capsaicin also demonstrates antioxidant activity by decreasing the production of superoxide in rat brain (Dairam et al. 2008).

Capsaicinoids have demonstrated cardiovascular properties as well and have proven beneficial in treatment of hypertension, myocardial infraction, coronary heart disease and arthrosclerosis (Harada and Okajima 2009; Peng and Li 2010). A regular consumption of chilli has been found to be associated with decrease in oxidation of serum lipoprotein (Ahuja and Ball 2006). Gastrointestinal effect of capsaicin is dosage dependent; in high concentration, they damage the sensory nerves associated with the gastrointestinal tract. However, in low concentrations, they increase the secretion of mucous and promote antimicrobial activity (Nishihara et al. 2002).

Due to these pharmacological applications, capsaicin has been the centre of research and is in high demand in the R&D sector as well as in pharmaceutical industries.

8.2 Chemical and Enzymatic Synthesis

Owing to its numerous medicinal properties and potential application in pharmaceutical industry, there has been increased interest in capsaicin. Numerous researches are being conducted to enhance capsaicin production by modifying cultivation methods, chemical and enzymatic synthesis and biotechnological approaches such as plant tissue culture (Sung et al. 2005; Kaga et al. 1989; Castillo et al. 2007; Ochoa-Alejo and Ramírez-Malagón 2001). These studies have been discussed in detail in subsequent sections.

Various attempts have been made to chemically synthesize capsaicin on industrial scale. Capsaicin and its analogues can be synthesized by chlorinated fatty acids and amines by heating at temperature 140–170 °C under moderate pressure (Kaga et al. 1996). An analogue of capsaicin, 1-hydroxy-2-pyridone, has been synthesized using bioisosterism; this analogue has the same biological activities and effects of capsaicin (Choi and Yoon 1999). A study conducted by Appendino et al. (2002) developed an expeditious method of synthesizing N-acyl vanillamines. Acylation of vanillylamines was achieved by using phosphoric anhydrides, and activation was done by using diethyl phosphorocyanidate and propane phosphonic acid anhydride under mild conditions. The resultant mixed phosphoric anhydrides has the same activity as capsaicinoids and simplified the whole reaction process to one-pot operation. Selective esterification of phenolic alcohols has also been achieved by using cerium chlorate (III) yielding 70% vanillyl nonanoate (Appendino et al. 2002). A non-pungent capsaicin analogue, dihydrocapsaicin-β-D-glucopyranoside, has also been synthesized using chemical enzymatic process (Torregiani et al. 2005). Although chemical synthesis has simplified the capsaicinoid synthesizing process, it is not extensively used due to involvement of toxic reagents, thus making enzymatic synthesis more appealing.

However, enzyme-catalysed synthesis is advantageous over chemical synthesis because of the use of non-toxic reagents and substrate specificity of the former.

A number of trials conducted by Kobata et al. (1999) for synthesizing capsaicin analogues with lipase catalysed reactions showed that condensation of vanillylamine with fatty acid derivatives yielded 40–59% capsaicin and 2–44% other analogues. In another study, various natural oils were used as a substrate for production of non-pungent capsaicin analogues (Kobata et al. 1998). Different-pungent and low-pungency analogues of capsaicin can be synthesized though lipase catalysed condensation of fatty acyl chains of varied length and vanillyamine in *n-hexane* and 2-methyl-2-butanol (Castillo et al. 2007). Lipase catalysed transacylation with methyl tetradecanoate in supercritical CO_2 resulted in 54% yield of capsaicin analogue vanillyl tetradecanamide (Kobata et al. 2003).

8.3 Extraction of Capsaicin from Natural Sources

Extraction of capsaicin oleoresins from *Capsicum* is the most widely used method of capsaicin production on industrial scale. Extraction can be performed in various organic polar and non-polar solvents. Chemical extraction works on the principle that the sample produces fractions with varying properties under the influence of an extraction agent. The separation process depends upon heterogeneity of sample matrix; for heterogeneous mixtures, mechanical separation techniques would suffice, whereas for homogeneous mixtures, physiochemical properties need to be properly evaluated. Various physical factors like density, viscosity and surface tension also need to be considered. Capsaicin yield largely depends upon the extraction method employed. Soxhlet extraction yields 9 times more capsaicin as compared to extraction through vacuum filtration (Amruthraj et al. 2014). The solvent system used for extraction also determines the amount of capsaicin. Various polar solvents, alone or in different ratio, have been used for extraction giving varied yield (Dang et al. 2018; Lu and Cui 2019; Gudeva et al. 2013).

8.4 Capsaicin Production Through Biotechnological Approaches

In vitro cultivation of plant cells, tissues and organs has been used for large-scale production of desired compound through genetic manipulation. Advancements in molecular tools and techniques have enabled researchers to explore newer avenues for production of plant-based products. The rising demand of natural and herbal products with minimum side effects has given rise to the necessity to develop other techniques for production of various phytochemicals, cutting out any dependence upon plants. Alternatively, plant tissue culture has become the method of choice for production of such phytochemicals (Ravishankar and Venkataraman 1990; Alfermann and Petersen 1995; Dicosmo and Misawa 1995; Dornenburg and Knorr 1997; Ravishankar et al. 1999). The totipotent nature of plant cells makes them the best manufacturing unit for commercial production of phytochemicals.

Establishment of cell suspension culture is the most promising method for production of secondary metabolites (Walpole et al. 1996). This technique has numerous benefits over traditional methods for secondary metabolite production. With the use of appropriate growth condition and optimized culture conditions, it is possible to obtain 20–30 times higher yield.

Though plant tissue culture technique is a promising method for production of capsaicin, there are a number of factors that have to be optimized in order to achieve high accumulation of capsaicin. Numerous studies have been conducted to enhance capsaicin accumulation by modulating culture conditions, precursor feeding, biotransformation, elicitation, permeabilization and many more. All of these techniques have been discussed in subsequent sections.

8.4.1 Modulation of Culture Conditions

As secondary metabolites are produced in response to an environmental stimulus, modification in physical and chemical environment of culture system is the principal methodology adopted to enhance its accumulation. The immobilized cells of *Capsicum frutescens* Mill cv. Annum, when grown in the absence of orthophosphate, showed enhanced production of capsaicin by 15 days of culture; when grown in media devoid of both nitrate and phosphate, an increased accumulation of capsaicin was observed by day 7 of culture (Lindsey 1985). The level of capsaicin in chili peppers is correlated to the expression level of capsaicin synthase gene. Addition of capsaicin ionophore A23187 to cell suspension cultures of *Capsicum frutescens* resulted in enhanced capsaicin accumulation. However, addition of calcium modulators verapamil and chlorpromazine reduced capsaicin production showing that calcium ion is involved in capsaicin biosynthetic pathway (Sudha and Ravishankar 2002). Capsaicin accumulation is enhanced in immobilized placental tissues cultured in vitro with media containing increased concentration of nitrate. Capsaicin concentration can be increased up to the level produced by intact placenta (Aldana-Iuit et al. 2015).

8.4.2 Selection of High-Yielding Cell Lines

In order to obtain maximum production of capsaicin, high capsaicin yielding cell lines need to be selected. Callus induced from different parent plant are screened, and a heterogeneous population of high-yielding cell clones is selected (Ogino et al. 1978). The presence of differential biochemical activity within the cell clones is estimated to screen those with high yield. Biochemical techniques, such as HPLC and RIA, have widely been used for clone selection (Zvara et al. 2006; Matsumoto et al. 1980). Many studies have also employed mutation techniques for this purpose. A cytotoxic inhibitor or an environmental stress is employed that would allow only the resistant cells to grow. 5-Methyltryptophan, glyphosate, biotin etc. can be used as selective agents (Wataneba et al. 1982; Widholm 1974; Amrhein et al. 1985).

8.4.3 Precursor Feeding

Many studies have shown that by adding the precursors or intermediates of the biosynthetic pathway, it can result in enhanced accumulation of desired compound (Benjamin et al. 1987; Jayapaul et al. 2005). Addition of ferulic acid, one of the intermediates in capsaicin biosynthesis, to the immobilized cells of capsicum resulted in twofold increase in capsaicin yield (Ravishankar et al. 1999). In another study conducted by Pandhair and Gosal (2009), vanillin (one of the compounds involved in condensation reaction used for capsaicin synthesis in chilli) was added to cell suspension cultures, which resulted in up to sevenfold increase in capsaicin production.

8.4.4 Elicitation for Increasing Capsaicin Yield

Elicitor is a compound which, when added to the culture media, results in enhanced production of a specific compound, and the mechanism is referred to as elicitation. Elicitors can either be biotic (derived from living cells) or abiotic (chemical compounds, ions, rays). There is a correlation between secondary metabolite biosynthetic pathway and biochemical reactions prompted by elicitors. Different elicitors have different underlying pathways and present variations pertaining to their origin, specificity, concentration, physiochemical environment etc.

With the addition of salicylic acid (SA) and methyl jasmonate (MeJA) to cell suspension cultures of *Capsicum frutescens*, accumulation of capsaicin was found to increase; however, no increment in capsaicin accumulation was observed when the two elicitors were used in combination (Sudha and Ravishankar 2003). In another study, administration of curdlan and tyrosine enhanced capsaicin yield by 9 times as compared to un-elicitated cultures (Ravishankar et al. 1999). Harishchandra et al. (2012) observed that cells treated with laminarin produced 2 times higher the amount of capsaicin than untreated cells. Islek et al. (2014) used cellulase for elicitation of callus derived from *Capsicum annum* L. and observed enhanced capsaicin production both in suspension cultures and immobilized cells.

8.4.5 Use of Immobilized Cell System

One of the latest advances in secondary metabolite production is the use of SPIC (Surface Immobilized plant cell) (Zhu 2007). Some of the prospective benefits presented by this technique are presented in Fig. 8.4. Immobilization methods include entrapment of cells within a gel (agar, agarose, alginate or carrageenan) or membranes. Immobilization of plant cell was first reported by Brodelius et al. (1979) when they used calcium alginate for entrapment. Hollow-fibre membranes of cellulose acetate silicone polycarbonate have been demonstrated to entrap cells while maintaining their viability (Yeoman 1987). These immobilized cells are maintained in liquid media, and their permeable nature allows the passage of nutrients to the cells (Fig. 8.5). Higher yields of capsaicin and dihydrocapsaicin were obtained when precursors of capsaicin biosynthesis were added to immobilized placental tissues of *Capsicum frutescens* (Lindsey et al. 1983).

8.5 Challenges Associated with Production of Capsaicin Through Plant Tissue Culture

Though plant tissue culture is a highly promising mechanism of capsaicin production, there are numerous bottlenecks faced while transferring this technology from laboratory to industry. There is a lack of interest among the industrialists to take initiative to produce secondary metabolites through plant tissue culture. There is also inadequacy of skilled manpower and infrastructure for commercialization of this

Fig. 8.4 Advantages of immobilized cell systems for secondary metabolite production

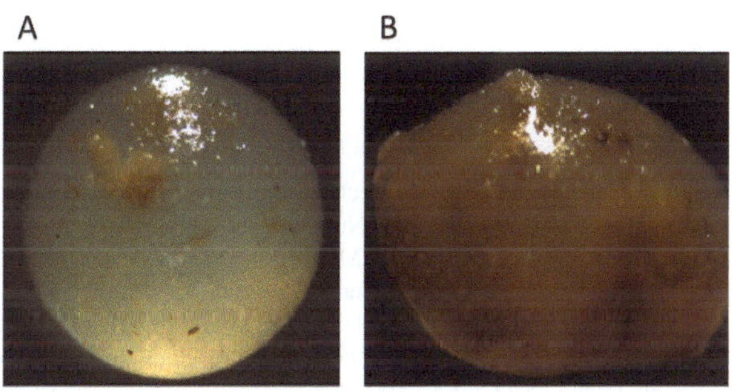

Fig. 8.5 Callus immobilized in calcium alginate. (**a**) On the day of immobilization; (**b**) seventh day from the day of immobilization

technology. Biotechnology is still in its nascent stage when it comes to the use of cell culture systems for plant-based compound production. There is a need to develop scientific temperament in this direction. Moreover, many research conducted in this field is done for high-impact research publication and is not designed to develop a technology that can be translated into a viable product. Scientists and researchers along with industrialists must work in this direction to fully exploit this technology that would also contribute to sustainable development as well.

8.6 Future Prospects

Plant tissue culture is not just a science but is also an art. Production strategy must be optimized for each plant product. Capsaicin, which is a wonder compound as it has multifaceted application, has high demand in international market. There is a need to develop an optimized culture system where capsaicin can be produced in high quantities at a reduced cost. There exists a huge gap between laboratory-based research and industrial production on a commercial scale. This gap needs to be filled, and there is an ample scope of further research.

8.7 Conclusion

Plant tissue culture presents a very promising scope for producing high amounts of capsaicin, which can be used for commercial production. It presents the prospect of capsaicin production without any dependency on agricultural produce. It is also free from the drawbacks of artificial synthesis and produces capsaicin without any release or involvement of toxic compounds. The knowledge of biosynthetic and molecular pathway for capsaicin biosynthesis in plants enables us to modify the cultured cells on a molecular level in order to obtain high amounts of capsaicin. In short, plant tissue culture allows capsaicin production without any dependency on agriculture, climate or geographical conditions.

References

Ahuja KD, Ball MJ (2006) Effects of daily ingestion of chilli on serum lipoprotein oxidation in adult men and women. Br J Nutr 96:239–242

Aldana-Iuit JG, Sauri-Duch E, Miranda-Ham, Castro-Concha LA, Cuevas-Glory LF, Vázquez-Flota FA (2015) Nitrate promotes capsaicin accumulation in *Capsicum chinense* immobilized placentas. Biomed Res Int 2015:1. https://doi.org/10.1155/2015/794084

Alfermann AW, Petersen M (1995) Natural products formation by plant cell biotechnology. Plant Cell Tissue Org Cult 43:199–205

Amrhein N, Johanning D, Smart CC (1985) A glyphosate-tolerant plant tissue culture. In: Neumann KH, Barz W, Reinhard E (eds) Primary and secondary metabolism of plant cell cultures. Springer, Berlin, pp 355–361

Amruthraj NJ, Preetam Raj JP, Antoine LL (2014) Comparative studies on the extraction of capsaicinoids from *Capsicum chinese* and their analysis by phosphomolybdic acid and reduction and HPLC. Int J Pharm Sci Rev Res 44:247–252

Andrews J (1984) Peppers: the domesticated capsicums. University of Texas Press, Austin, Texas

Appendino G, Minassi A, Morello AS, De Petrocellis L, Di Marzo V (2002) *N*-acylvanillamides: development of an expeditious synthesis and discovery of new acyl templates for powerful activation of the vanilloid receptor. J Med Chem 45:3739–3745

Barbero GF, Molinillo JMG, Varela RM, Palma M, Macias FA, Barroso CG (2010) Application of Hansch's model to capsaicinoids and capsinoids: a study using the quantitative structure-activity relationship. A novel method for the synthesis of capsinoids. J Agric Food Chem 58:3342–3349

Belza A, Frandsen E, Kondrup J (2007) Body fat loss achieved by stimulation of thermogenesis by a combination of bioactive food ingredients: a placebo controlled, double-blind 8-week intervention in obese subjects. Int J Obesity 31:121–130
Benjamin BD, Roja P, Heble MR, Chadha M (1987) Multiple shoot cultures of *Atropa belladonna* effect of physicochemical factors on growth and alkaloid formation. J Plant Nutr 129:129–135
Bosland PW (1996) Innovative uses of an ancient crop. ASHS Press, Arlington, VA
Braun DM, Walker JC (1996) Plant transmembrane receptors: new pieces in the signaling puzzle. Trends Biochem Sci 21:70–73
Brodelius P, Deus B, Mosbach K, Zenk MH (1979) Immobilized plant cells for the production and transformation of natural products. FEBS Lett 103:93–97
Butera JA (2007) Current and emerging targets to treat neuropathic pain. J Med Chem 50:2543–2546
Castillo E, Lopez-Gonzalez I, De Regil-Hernandez R, Reyes-Duarte D, Sánchez-Herrera D, López-Munguía A, Darszon A (2007) Enzymatic synthesis of capsaicin analogs and their effect on the T-type Ca2+ channels. Biochem Biophys Res Comm 356:424–430
Castillo E, Torres-Gavilán A, Severiano P, Arturo N, López-Munguía A (2007) Lipase-catalyzed synthesis of pungent capsaicin analogues. Food Chem 100:1202–1208
Choi HY, Yoon SH (1999) Bioisoster of capsaicin: synthesis of 1-hydroxy-2-pyridone analogue. Bull Kor Chem Soc 20:857–859
Dairam, A., Ronen, F., Santy, D., Janice, L. Antioxidant and iron-binding properties of curcumin, capsaicin, and sallylcysteinereduce oxidative stress in rat brain homogenate. Journal of Agricultural and Food Chemistry 2008;56:3350–3356.
Dang Y, Zhang H, Xiu Z (2018) Three-liquid-phase extraction and separation of Capsanthin and capsaicin from *Capsicum annum* L. Czech J Food Sci 32:109–114
De Lourdes Reyes-Escogido, Gonzalez-Mondragon EG, Vazquez-Tzompantzi E (2011) Chemical and pharmacological aspects of capsaicin. Molecules 16:1253–1270
DiCosmo F, Misawa M (1995) Plant cell and tissue culture: alternatives for metabolite production. Biotechnol Adv 13:425–435
Dornenburg H, Knorr D (1997) Challenges and opportunities for metabolite production from plant cell and tissue cultures. Food Technol 51:47–54
Gudeva LK, Mitrev L, Maksimova V, Spasov D (2013) Content of capsaicin extracted from hot pepper (Capsicum annuum ssp. microcarpum L.) and its use as an ecopesticide. Hem Ind 67: 671–675. https://doi.org/10.2298/HEMIND120921110K
Harada N, Okajima K (2009) Effects of capsaicin and isoflavone on blood pressure and serum levels of insulin-like growth factor-I in normotensive and hypertensive volunteers with alopecia. Biosci Biotechnol Biochem 73:1456–1459
Harishchandra BG, Parvatam G, Ravishankar GA (2012) Laminarin as a potential non-conventional elicitor for enhancement of capsaicinoid metabolites. Asian J Plant Sci Res 2:490–495
Islek C, Ustun AS, Koc E (2014) The effects of cellulase on capsaicin production in freely suspended cells and immobilized cell cultures of *Capsicum annuum* L. Pak J Bot 46(5): 1883–1887
Jayapaul K, Kavi Kishor PB, Reddy KJ (2005) Production of pyrroquinzoline alkaloid from leaf and petiole derived callus cultures of *Adhatoda zeylanica. In Vitro Cell Dev Biol Plant* 41:682–685
Kaga H, Goto K, Takahashi T, Hino M, Tokuhashi T, Orito K (1996) A general and stereoselective synthesis of the capsaicinoids via the orthoester Claisen rearrangement. Tetrahedron 52:8451–8470
Kaga H, Miura M, Orito K (1989) A facile procedure for synthesis of capsaicin. J Org Chem 54: 3477–3478
Kang, J. H., Tsuyoshi, G., Han, I. S., Kawada, T., Kim, Y. M., Yu, R. Dietary capsaicin reduces obesity-induced insulinresistance and hepatic steatosis in obese mice fed a high-fat diet. Obesity 2009;18:780–787.

Katritzky AR, Xu YJ, Vakulenko AV, Wilcox AL, Bley KR (2003) Model compounds of caged capsaicin: design, synthesis, and photoreactivity. J Org Chem 68:9100–9104

Kobata K, Kobayashi M, Kinpara S, e.t al. (2003) Supercritical CO_2 as a reaction medium for synthesis of capsaicin analogues by lipase-catalyzed transacylation of capsaicin. Biotechnol Lett 25:1575–1578. https://doi.org/10.1023/A:1025428410952

Kobata K, Kobayashi M, Tamura Y, Miyoshi S, Ogawa S, Watanabe T (1999) Lipase-catalyzed synthesis of capsaicin analogs by transacylation of capsaicin with natural oils or fatty acid derivatives in n-hexane. Biotechnol Lett 21:547–550

Kobata K, Toyoshima M, Kawamura M, Watanabe T (1998) Lipase-catalyzed synthesis of capsaicin analogs using natural oils as an acyl donor. Biotechnol Lett 20:781–783

Lee J, Li Y, Li C, Li D (2011) Natural products and body weight control. North Am J Med Sci 3:13–19

Lindsey K (1985) Manipulation, by nutrient limitation, of the biosynthetic activity of immobilized cells of *Capsicum frutescens* Mill. cv. annuum. Planta 165:126–133

Lindsey K, Yeoman MM, Black GM, Mavituna F (1983) A novel method for the immobilization and culture of plant cells. FEBS Lett 155:143–149

Lu Y, Cui B (2019) Extraction and purification of capsaicin from Capsicum oleoresin using a combination of tunable aqueous polymer-phase impregnated resin (TAPPIR) extraction and chromatography technology. Molecules 24(21):3956. https://doi.org/10.3390/molecules24213956. PMID: 31683728; PMCID: PMC6866130

Matsumoto T, Ikeda T, Kanno N, Kisaki T, Noguchi M (1980) Selection of high ubiquinone 10-producing strain of tobacco cultures by cell cloning technique. Agric Biol Chem 44:967–969

Nelson, E.K.; Dawson, L.E. The constitution of capsaicin, the pungent principle of Capsicum. III.J. Am. Chem. Soc. 1923, 45, 2179-2181.

Nishihara K, Nozawa Y, Nakano M, Ajioka H, Matsuura N. Sensitizing effects of lafutidine on CGRP-containing afferentnerves in the rat stomach. Br. J. Pharmacol. 2002;135:1487–1494.

Ochoa-Alejo N, Gomez-Peralta JE (1993) Activity of enzymes involved in capsaicin biosynthesis in callus tissue and fruits of chili pepper (*Capsicum annuum* L.). J Plant Physiol 141:147–152

Ochoa-Alejo N, Ramírez-Malagón R (2001) *In vitro* chili pepper biotechnology. In Vitro Cell Dev Biol Plant 37:701–729

Ogino T, Hiraoka N, Tabata M (1978) Selection of high nicotine producing cell lines of tobacco callus by single cell cloning. Phytochemicals. 22:2447–2450

Oh SH, Kim YS, Lim SC, Hou YF, Chang IY, You HJ (2008) Dihydrocapsaicin (DHC), a saturated structural analogue of capsaicin, induces autophagy in human cancer cells in a catalase-regulated manner. Autophagy 4:1009–1019

Pandhair V, Gosal SS (2009) Capsaicin production in cell suspension cultures derived from placenta of *Capsicum annum* L. fruit. Indian J Agri Biochemist 22(2):78–82

Peng J, Li YJ (2010) The vanilloid receptor TRPV1: role in cardiovascular and gastrointestinal protection. Eur J Pharmacol 627:1–7

Ravishankar GA, Bhyalakshmi N, Ramachandra RS (1999) Production of food additives. In: Ramawat KG, Merillon JM (eds) Biotechnology: secondary metabolites. New Delhi, Oxford IBH, pp 89–110

Ravishankar GA, Venkataraman LV (1990) Food applications of plant cell cultures. Curr Sci 59: 914–920

Reynolds JEF (1999) Martindale: the extra pharmacopoeia, 32nd edn. Royal Pharmaceutical Society, London

Song JI, Grandis JR (2000) STAT signaling in head and neck cancer. Oncogene 19:2489–2495

Sudha G, Ravishankar GA (2002) Influence of calcium channel modulators in capsaicin production by cell suspension cultures of *Capsicum frutescens* Mill. Curr Sci 83(4):480–484

Sudha G, Ravishankar GA (2003) Influence of methyl jasmonate and salicylic acid in the enhancement of capsaicin production in cell suspension cultures of *Capsicum frutescens* Mill. Curr Sci 85:1212–1217

Sung Y, Chang YY, Ting NL (2005) Capsaicin biosynthesis in water-stressed hot pepper fruits. Bot Bull Acad Sin 46:35–42

Taira S, Shimma S, Osaka I, Kaneko D, Ichiyanagi Y, Ikeda R, Konishi-Kawamura Y, Zhu S, Tsuneyama K, Komatsu K (2012) Mass spectrometry imaging of the capsaicin localization in the capsicum fruits. Int J Biotechnol Wellness Ind 1:61–65

Thoennissen NH, O'Kelly J, Lu D, Iwanski GB, La DT, Abbassi S, Leiter A, Karlan B, Mehta R, Koeffler HP (2010) Capsaicin causes cell cycle arrest and apoptosis in ER- positive and-negative breast cancer cells by modulating the EGFR/HER-2 pathway. Oncogene 29: 285–296

Torregiani E, Seu G, Minassi A, Appendino G (2005) Cerium (III) chloride promoted chemoselective esterification of phenolic alcohols. Tetrahedron Lett 46:2193–2196

Walpole CS, Bevan S, Bloomfield G, Breckenridge R, James IF, Ritchie T, Szallasi A, Winter J, Wrigglesworth R (1996) Similarities and differences in the structure-activity relationships of capsaicin and resiniferatoxin analogues. J Med Chem 39:2939–2952

Wataneba K, Yano SI, Yamada Y (1982) Selection of cultured plant cell lines producing high levels of biotin. Phytochemicals 21:513–516

Widholm JM (1974) Evidence for compartmentation of tryptophan in cultured plant tissues. Free tryptophan levels and inhibition of anthranilate synthetase. Physiol Plant 30:323–326

Yeoman MM (1987) Cell structure and somatic cell genetics of plants. In: Constabel F, Vasil IK (eds) Cell culture in phytochemistry: techniques, characteristics, properties and commercial potential of immobilized plant cells, vol 4. Academic, San Diego, pp 197–215

Yoshioka M, St-Pierre S, Drapeau V, Dionne I, Doucet E, Suzuki M, Tremblay A (1999) Effects of red pepper on appetite and energy intake. Br J Nutr 82:115–123

Zhu Y (2007) Immobilized cell fermentation for production of chemicals and fuels. In: Shang-Tian Y (ed) Bioprocessing for value added products from renewable resources. Elsevier, Amsterdam, pp 373–396

Zvara A, Bencsik P, Fodor G, Csont T Jr, Hackler L, Dux M, Furst S, Jancso G, Puskas LG, Ferdinandy P (2006) Capsaicin-sensitive sensory neurons regulate myocardial function and gene expression pattern of rat hearts: a DNA microarray study. FASEB J 20:160–162

Transcriptional Regulation of Capsaicin Biosynthesis

9

Swet Nisha

Abstract

The total amount of RNA transcribed in a tissue at specific point of time under constant environmental condition is referred to as transcriptome. Transcriptomic analysis is vital in understanding the developmental process in plants. The transcription factors are crucial in coordinating and regulating various processes in plants including the level of pungency. Neither the capsaicin biosynthetic pathway is well characterized nor is the regulation mechanism fully understood. The capsaicinoid biosynthetic pathway is presumed to involve more than 50 genes. Transcriptomic studies are helpful in understanding their expression pattern that would moderate the amount of capsaicin produced. An analysis of expression profile of various transcription factors (TFs) in different *Capsicum* varieties provides insights into regulatory mechanism of capsaicin biosynthetic pathway. Numerous studies have been conducted on capsaicinoid biosynthetic genes in different *Capsicum* varieties. The present chapter aims at developing an understanding of how various molecular mechanisms are involved in regulating capsaicin biosynthesis. It has also been explained why certain species are more pungent than others.

Keywords

Transcriptome · Capsaicin · Transcription factors · Expression analysis · Hot peppers · Transcriptional regulation

S. Nisha (✉)
Department of Microbiology, Patna Women's College (Autonomous), Patna, India
e-mail: swetnisha.mbio@patnawomenscollege.in

M. K. Swamy, A. Kumar (eds.), *Capsaicinoids*,
https://doi.org/10.1007/978-981-99-7779-6_9

9.1 Introduction

Capsaicinoids belong to the alkaloid group that are synthesized in the fruits of *Capsicum* species and are responsible for their pungency. The biosynthesis of capsaicinoid is a highly regulated and coordinated pathway that occurs during fruit ripening. The ripening of fruit is characterized by various biochemical and physiological changes that are genetically coordinated and occurs during various stages of fruit development (Giovannoni 2011, 2004; Dubey et al. 2009).. The capsaicinoid biosynthetic pathway involves the unification of two independent biochemical pathways, i.e. phenylpropanoid pathway and branched-chain fatty acid pathway (Aza-González et al. 2011). With the evolution of modern sequencing techniques, it has been elucidated that genetic regulation of capsaicinoid biosynthesis is affected by interaction between the genotype and environmental factors (Qin et al. 2014). It has been identified that the expression of capsaicin-synthesizing genes are highly localized in chilli peppers, especially in the placenta and pericarp; the biosynthesizing genes' expression is highly dependent upon developmental phase (Liu et al. 2013). With the help of genomic analysis of various cultivars of *Capsicum*, structural genes involved in the biosynthesis of capsaicinoids have been identified. Some of the genes identified include phenylalanine ammonia lyase (*PAL*) and cinnamic acid 4-hydroxylase (*Ca4H*). Other genes that are crucial in the biosynthesis of capsaicin are 4-coumarate-CoA ligase (*4CL*), hydroxycinnamoyl transferase (*HCT*) and caffeic acid *O*-methyltransferase (*Comt*). Hydroxycinnamoyl-CoA hydratase/ligase and a putative acyltransferase (*AT3*) are also involved in the biosynthesis of capsaicin (Liu et al. 2013; Curry et al. 1999; Kim et al. 2001; Stewart Jr et al. 2005; Mazourek et al. 2009). Another study conducted by Zhang et al. (2022) revealed that downregulation of various transcription factors (TFs) (PAL, BCAT, ACLd, Kas, FatA and HCHL) in domesticated variety as compared to the wild-type variety has higher pungency level as compared to the former.

Many studies aimed at elucidating capsaicinoid biosynthetic pathway and identifying the genes responsible in the process have been conducted; however, the transcriptional regulation of the pathway is still an underexplored domain. Study conducted by Keyhaninejad et al. (2014) identified two ethylene response factor TFs, whose expression pattern was found to be correlated with that of the pungency level of all the chilli pepper cultivars studied. The ERF family transcription factors *Erf* and *Jerf* were found to be expressed during early developmental stages of chilli pepper and were involved in pungency regulation.

The MYB family of TFs is expressed in numerous plant species where they have been involved in the production of different secondary metabolites (Dubos et al. 2010; Hatlestad et al. 2015; Yin et al. 2017; Gigolashvili et al. 2008). CaMYB48 TF belonging to the MYB family has been found to act as an activator in capsaicinoid biosynthesis. The C-terminal domain of MYB containing 168–191 acidic amino acids is associated with the transcription activation (Soler et al. 2015). Deletion of this motif has been associated with the loss of activation (Sun et al. 2020). Another TF MYB31 has tissue specific expression in *Capsicum* sp., which is specifically localized in placental region. An expression of placenta-specific activator WRKY9 is responsible for increasing *MYB31* expression via binding to its promoter leading to upregulation of capsaicinoid biosynthetic genes (Zhu et al. 2019). There are many TFs and activators that are choreographed together to regulate the production of capsaicinoid. The present review aims at integrating all the knowledge available regarding the transcription factors involved in biosynthetic pathway of capsaicinoid.

9.2 Capsaicin Biosynthesis in Chilli Peppers

Capsaicin biosynthesis is a highly coordinated process that involves merging of two independent pathways, namely, phenylpropanoid pathway and branched-chain fatty acid pathway (Fig. 9.1). The phenylpropanoid pathway initiates with phenylalanine, whereas the amino acids valine or leucine act as precursor for branched-chain fatty acid pathway. Capsaicin is the result of condensation reaction (catalysed by enzyme *capsaicin synthase*) between the final product of both the pathways, i.e. vanillylamine and 8-methyl-6-nonenoyl-CoA, respectively (Prasad et al. 2005). These two key pathways are involved in biosynthesis of all the different types of capsaicinoids (Aza-González et al. 2011). Capsaicinoid biosynthesis takes place in placental tissues with its accumulation starting around 20 days post-anthesis (DPA),

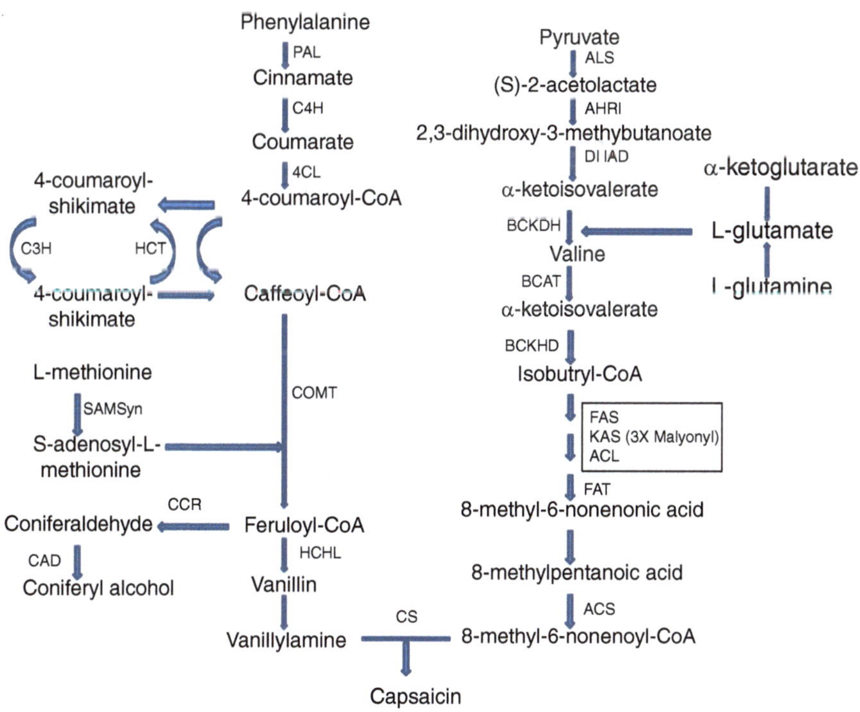

Fig. 9.1 Capsaicin biosynthetic pathway (Adopted from Zhang et al. 2016)

reaching its maximum concentration at around 40 days, after which it starts decreasing (Arce-Rodríguez and Ochoa-Alejo 2017).

9.3 Genomic Characterization of Capsaicin Synthase

The genomic studies conducted on loss of function mutation in the two key genes involved in capsaicin biosynthesis reveal the importance of capsaicin synthase and putative aminotransferase in biosynthetic pathway (Arce-Rodríguez and Ochoa-Alejo 2017). A dominant gene (*Pun 1*) responsible for pungency in chilli peppers was the first one to be identified in *Capsicum annum* (Deshpande 1935). The *Pun1* locus encodes AT3, an acyltransferase catalysing the final step in capsaicin biosynthetic pathway. Later, studies conducted by Stellari et al. (2010) and Han et al. (2019) confirmed another pungency related loci *pun2* and *Pun3.* A comparative study of AT3 gene sequence isolated from various pungent and non-pungent varieties of chilli has revealed a 2.5 kb deletion including the putative promoter and a part of the first exon in non-pungent varieties. This recessive allele was named *pun1.* A second recessive allele $pun1^2$ was discovered in non-pungent *C. chinense* harbouring a 4 bp deletion in the first exon generating a pruned AT3 protein (Stewart et al. 2007). Another study involving AT3 allele present in non-pungent *C. frutescens* showed that the allele harbours deletion in its second exon and the stop codon. No transcription was observed in this allele (Stellari et al. 2010).

A non-sense mutation caused by incorporation of a thymine nucleotide followed by GA in *pAmt* gene led to the formation of a new stop codon. Though transcription is not affected, translation of *pAmt* is affected, thereby rendering the enzyme ineffective in producing vanillylamine (Lang et al. 2009). A base substitution mutation in *pAmt* allele in another chili cultivar substituting arginine for cysteine leads to the production of ineffective enzyme. A 5 bp insertion mutation in eighth exon of *pAmt* allele present in Belize Sweet cultivar of chili makes the allele non-functional. In the case of Zavory Hot and Aji Dulce cultivars, mutation is caused by transposable elements in introns of *pAmt*, making it nonfunctional (Tanaka et al. 2010a, b). A study on nonfunctional alleles of *pAmt* conducted by Koeda et al. (2014) discovered insertion mutation in exons of two cultivars of *C. chinense*. Both mutations affect the PLP domain of pAMT enzyme required for its aminotransferase activity. Park et al. (2015) identified a 12 bp insertion in the seventh exon of *pAmt* allele of *C. frutescens* that prevents translation of the gene. This mutation makes the S3212 cultivar of *C. frutescens* a low-pungency cultivar. They also identified a hAT family transposon insertion in *pAmt* alleles in two different cultivars of *C. chinense.* This mutation downregulates the expression of *pAmt* gene (Park et al. 2015).

9.4 Transcription Factors Involved in Capsaicin Biosynthesis

Transcription regulatory proteins, known as transcription factors, are the regulatory machinery involved in all metabolic pathways. Capsaicinoid biosynthesis and accumulation are largely correlated to fruit developmental stage and hence are regulated at transcriptional level. Capsaicinoid accumulation is also determined by various biotic (wounding, pathogenic or insect attack) and abiotic factors (temperature, light, water, level of CO_2, nutrient availability). It has been seen that the abiotic factors play a key role in upregulation of capsacinoid biosynthetic genes (Arce-Rodriguez and Ochoa-Alejo 2019; Burgos-Valencia et al. 2020). Naves et al. (2019) reported that capsaicin level changes in response to varied temperature; capsaicinoid accumulation increased in response to water level as well. Nitrogen and potassium concentration are also determining factors in capsaicinoid accumulation (Naves et al. 2019). Furthermore, the genotype of various Capsicum species is responsible for pungency level.

Numerous studies aimed at identifying TFs involved in regulation of the expression of genes that encode proteins involved in capsaicin biosynthesis (Table 9.1). Transcriptional regulation of capsaicinoid biosynthesis is involved in regulation of metabolic pathway and precursor synthesis, thereby determining capsaicinoid accumulation.

9.4.1 CaMYB Family

Zhang et al. (2016) study the expression profile of genes of *C. frutescens* cv. Guijianwang and found nearly 20 enzyme encoding genes that could have participated in capsaicinoid biosynthesis (Zhang et al. 2016). The MYB TF is highly involved in numerous secondary metabolite production pathways (Du et al. 2009). They are composed of two domains—an N-terminal conserved DNA-binding

Table 9.1 Transcription factors involved in capsaicin biosynthesis

Transcription Factor	Consensus Motif	Gene(s)	Reference
bHLH (basic helix-loop-helix	CANNTG	CabHLH009 CabHLH032 CabHLH048 CabHLH095 CabHLH100	Liu et al. (2021)
ERF	TAAGAGCCGCC	CaERF102, CaERF53, CaERF111, CaERF92	Ohme-Takagi and Shinshi (1995), Song et al. (2020)
MYB	TAACTAAC	CaMYB31, 48, 103, 108, 115 and CaDIV14	Jin and Martin (1999), Arce-Rodríguez and Ochoa-Alejo (2017), Sun et al. (2019), Sun et al. (2020), Arce-Rodriguez et al. (2021)

domain having 52 amino acids (R2R3-type domain in plants) and a C-terminal variable region conferring the regulatory activity (Ambawat et al. 2013; Arce-Rodríguez and Ochoa-Alejo 2017). Characterization of *R2R3-MYB* family gene *CaMYB31* isolated from *C. annum* illustrated the presence of positive correlation between the expression level of *CaMYB31* and capsaicin biosynthetic enzymes *Ca4H*, *Comt*, *pAmt*, *AT3* and *Kas* (Arce-Rodríguez and Ochoa-Alejo 2017). Furthermore, the expression of these genes as well as its pungency level decreased after silencing of *CaMYB31*. The study on the relation between the above-mentioned genes and pungency level in *C. annuum* cv Tampiqueño 74, *C. annuum* cv California Wonder and *Capsicum chinense* cv Habanero showed the pungency level to be collaborating with the expression levels of *Kas*, *pAmt* and *AT3*. However, *Comt* and *Ca4H* genes were also found to be expressed in non-pungent cultivar *C. annuum* cv California Wonder and also in the vegetative tissues of root, stem and leaves where capsaicinoid is not produced, indicating that only *Kas*, *pAmt*, and *AT3* genes have roles in the production of capsaicinoids (Arce-Rodríguez and Ochoa-Alejo 2017). Various biotic and abiotic factors including wounding, presence of phytohormones, temperature and light also affect the pungency level in chilli pepper (Lindsey and Bosland 1995; Suresh and Ravishankar 2005; Kim et al. 2009; Gutiérrez-Carbajal et al. 2010). Wounding of plant is associated with production of jasmonic acids (JA) and decrease in amount of salicylic acid (SA) (Lee et al. 2004) Application of JA has shown to be correlated with downregulation of *CaMYB31*, *Kas* and *pAmt*, whereas application of SA has resulted in increase in the expression level of all the three genes. Expression of *CaMYB31*, *Kas* and *pAmt* also increased in response to lower temperatures (4 °C) as compared to higher temperature (37 °C). These studies show the effect of important environmental factors on regulation of genes involved in capsaicinoid biosynthesis (Arce-Rodríguez and Ochoa-Alejo 2017). Therefore, the CaMYB31 TF plays a central role in capsaicinoid biosynthesis. Moreover, genomic analysis of *C. annum* has revealed the presence of a homologous *CaMYB31* gene in both pungent and non-pungent varieties. This recessive gene locus was termed *Pun3* (Han et al. 2019). Although present in both pungent (cv. Tean) and non-pungent (cv. YMC334) cultivars, *Pun3* was only found to be expressed in the pungent variety. Sequence analysis of *Pun3* in *C. annum* cv. Tean has revealed the presence of a mutation generating a premature stop codon; as a result, the capsaicinoid biosynthetic genes are downregulated as compared to the pungent cultivars (Han et al. 2019).

A comparative genomic analysis of different *Capsicum* species was performed, which revealed the presence of *Cap1* locus that encodes for MYB31 TF. A differential expression of MYB31 TF was seen in placental region where the highest concentration of capsaicinoids is found. The expression of this TF is correlated with the pungency level as it has been seen to be involved in expression of genes relating to capsaicinoid biosynthesis. The upregulation of *MYB31* in response to environmental factors leads to its binding to WRKY9 transcriptional activator that further enhances the expression of capsaicinoid biosynthetic genes. The WKRY9 binds W-box in *MYB31* promoter in *C. chinense*, thereby activating its transcription.

The upregulation of *MYB31* leads to higher expression of capsaicinoid biosynthetic genes (Zhu et al. 2019; Kondo et al. 2021).

Arce-Rodriguez et al. (2021) described CaMYB108 TF in *C. annum* (Arce-Rodriguez et al. 2021). Upon virus-induced silencing of this TF, it was observed that expression levels of genes involved in capsaicinoids biosynthesis were drastically reduced and correlated with the capsaicinoid content (Sun et al. 2019). An analysis of expression pattern of CaR2R3-MYB genes and capsaicinoid biosynthetic genes has shown the involvement of six CaR2R3-MYB genes in regulation of capsaicinoid biosynthesis (Wang et al. 2020). However, of all the TFs described (namely, CaMYB31, 47, 48, 64, 73, 74, 87, 92 and CaMYB108), only CaMYB31 showed positive correlation with expression level of genes of capsaicinoid biosynthesis (Arce-Rodriguez et al. 2021). Moreover, a comparative investigation of enzyme-producing genes and TF genes revealed three TFs, namely, CaMYB103, CaMYB115 and CaDIV14, whose transcription profile was positively correlated with that of AT3 (Arce-Rodriguez et al. 2021).

Nearly 103 *CaMYB*-related genes have been identified and described and have been shown to contain 0–22 introns. All the 103 genes have been divided into nine subfamilies. Structural analyses of genes have shown the presence of ten conserved motifs (Liu et al. 2022). *CaMYB31* has a reading frame of 747 bp encoding R2R3-MYB proteins of 249 amino acids with a 5' UTR of 148 bp and a 3' UTR of 189 bp. On the basis of cDNA sequence analysis, exons and introns of this TF have been identified as shown in Fig. 9.2 (Arce-Rodriguez et al. 2021).

A differential expression analysis of *CaMYB31* and *Kas*, *pAmt*, *Comt* and *Ca4H* performed in different tissues of chilli pepper fruit and plant of different cultivars of *C. annum* revealed that *Kas* and *pAmt* genes were expressed at high levels in placental tissues. However, the transcription pattern showed that more expression level was higher in immature plants. Positive correlation between transcription level and pungency level was also found. On the contrary, low level of *Comt* and *Ca4H* was expressed in seed, pericarp and vegetative tissues (Arce-Rodriguez et al. 2021).

Virus-mediated gene silencing of *CaMYB31* demonstrated reduction in transcription level of various capsaicinoid biosynthetic genes (Arce-Rodriguez et al. 2021). The reduction in the expression of these genes varied from 30.3% (*Acl*) *to* 88.6% (*BCAT*). The expression of other genes was 81.8% for *Ca4H*. For *4CL*, it was 80.6%, and for *C3H*, it was 76.4%. Further, the reduction for *HCT*, *Comt*, *pAmt*, *BCKDH* and *Kas*, respectively, was 49%, 42.7%, 71%, 48.5% and 70%. Test plants in which *CaMYB31* TF has been silenced, the concentrations of capsaicin and dihydrocapsaicin were reduced by 74.2% and 73% respectively. These results show the importance of *CaMYB31* in the regulation of the biosynthesis of capsaicin and dihydrocapsaicin. Various environmental factors also affect the expression pattern of *CaMYB31* TF. Wounded plants and those exposed to high temperature and dark conditions showed significant decrease in expression of *CaMYB31*, *Kas* and *pAmt* (Arce-Rodriguez et al. 2021).

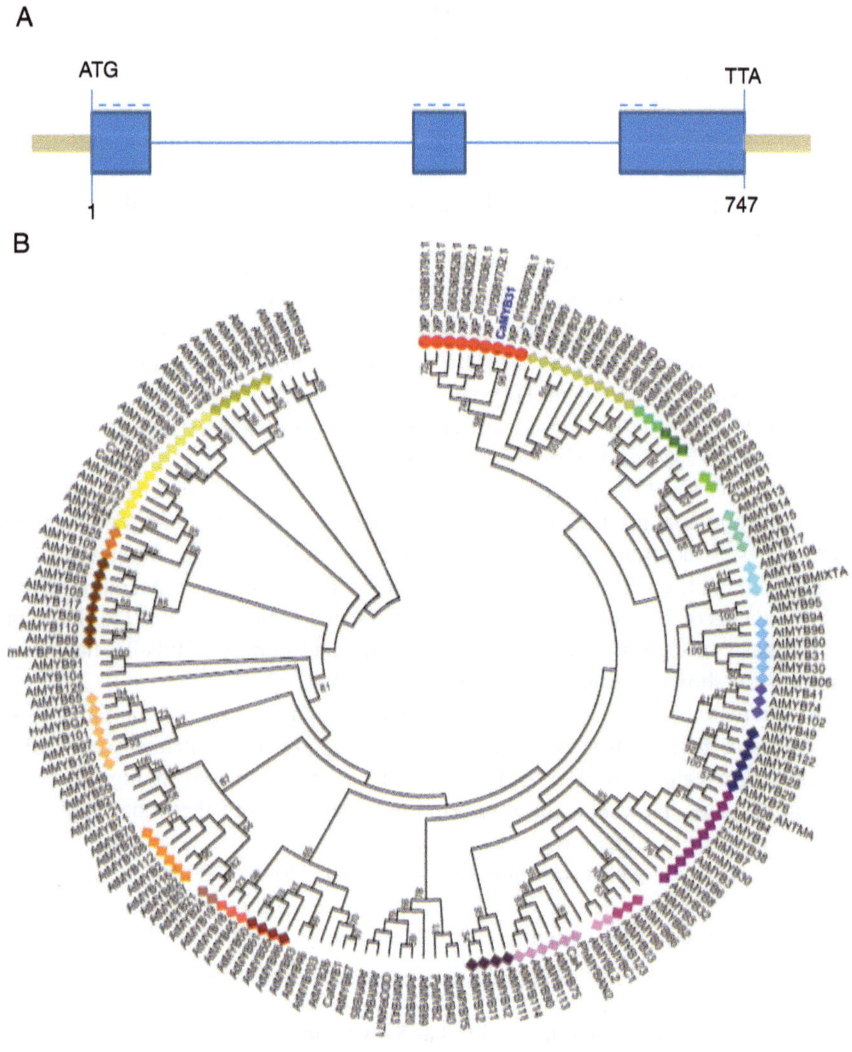

Fig. 9.2 *CaMYB31* gene and its phylogenetic relationship with other MYB TFs; (**a**) represents introns and exons in *CaMYB31*, black boxes represent exons, solid lines represent intron, grey boxes represent 5′ and 3' UTR, dotted lines represent R2R3-MYB domain; (**b**) phylogenetic tree depicting relationship of *CaMYB31* and other MYBTFs (Adapted from Arce-Rodriguez et al. 2021)

9.4.2 Ethylene Response Factor TF Family

Ethylene Response Factor (ERF) TF family are also associated with transcription of capsaicinoid biosynthetic genes. ERFs are a part of apetala2/ERF (AP2/ERF) superfamily of TFs, which consist of a single DNA binding domain AP2/ERF. The AP2/ERF domain consists of 60–70 amino acids. The genes belonging to this do not always depend upon ethylene action but also regulate numerous growth and developmental processes in response to biotic and abiotic factors (Ohme-Takagi and Shinshi 1995; Nakano et al. 2006; Huang et al. 2016). Analysis of transcription profile of Erf and Jerf in nine different cultivars of chilli with varying degree of pungency revealed their positive correlation with pungency. Highest expression of both genes was found between 16 and 20 days after anthesis prior to capsaicinoid accumulation in the placenta of the fruits (Keyhaninejad et al. 2014). Further analysis demonstrated that expression of genes encoding ERF TFs, namely, CaERF102, CaERF53, CaERF111 and CaERF92, coincided with capsaicinoid accumulation and thus can be said to be involved in regulation of capsaicinoid biosynthetic pathway (Song et al. 2020). Wen et al. (2022) studied the expression profile of various capsaicinoid biosynthetic genes (*PAL*, *Ca4H*, *CoMT*, *pAMT*, *CS*, *BCAT*, *KAS*, *ACL*, *FAT* and *ACS*) along with that of AP2/ERF TF ERF2. They found that the expression levels of ERF2 increased with the concentration of capsaicin and dihydrocapsaicin. The expression level of capasaicinoid biosynthetic genes also followed the same pattern indicating the importance of *ERF2* in capsaicinoid biosynthesis.

An ERF2 silencing experiment was performed by VIGS vector pTRV-*CcERF2* in *Capsicum chinense*. The chilli fruit infected with this vector construct resulted in the downregulation of *CcERF*. This reduction in expression only slightly affected the expression of genes involved in fatty acid pathway (*KAS*, *ACL*, *FAT* and *ACS*), whereas the expression of those involved in phenylpropanoid pathway (*PAL*, *Ca4H*, *CoMT*, *pAMT*) and capsaicin synthase gene was significantly reduced (Wen et al. 2022; del Rosario Abraham-Juárez et al. 2008). The results suggest that *CcERF2* regulates only certain genes in capsaicinoid biosynthesis.

9.4.3 Basic/Helix-Loop-Helix (bHLH) Family

This family of TFs consists of proteins with a conserved domain, which is approximately of 60 amino acids. There are two functionally distinctive regions: an N-terminal basic region (15 amino acids, most of which are basic) with DNA binding activity and a C-terminal HLH region (made up of hydrophobic residues that act as dimerization domain) (Toledo-Ortiz et al. 2003). The expression of bHLH genes CabHLH007, CabHLH009, CabHLH026, CabHLH063 and CabHLH086 has shown positive regulation with capsaicin accumulation. It has been seen that the expression levels of these TFs peaked from 20 DPA to 30 DPA, which coincides with increase in capsaicinoid accumulation in placenta (capsaicinoid accumulation increases from 16 DPA and peaks at 25 DPA) (Liu et al. 2021).

9.5 Future Prospects

For several decades, studies on biosynthesis and regulation of capsaicin have been the topic of extensive research; however, there are numerous gaps in our understanding of its regulation on molecular level. More studies need to be conducted to understand the regulation of biosynthetic enzymes. The role of environmental factors and physical stimulus in regulating expression of transcription factors involved in capsaicinoid biosynthesis is not fully understood. In-depth understanding of how capsaicinoid biosynthesis is regulated is required to enhance biosynthesis. Omics is an important tool that can be used to gain insight into regulation mechanism. Transcriptomics can be used to identify transcription factor genes involved in capsaicinoid biosynthesis pathway and to determine correlation between transcription pattern of structural and regulatory genes. More studies need to be conducted to improve our understanding of relation between various biotic and abiotic factors and expression of regulatory factors involved in capsaicinoid biosynthesis.

9.6 Conclusions

Capsaicinoid is an important group of alkaloids with numerous pharmaceutical and industrial applications. Transcription factors are the central regulatory mechanism in controlling development and gene expression in plants. Several transcription factors, namely, bHLH, MYB and ERF, have been identified in regulation of capsaicinoid biosynthesis. Insights into the transcriptional regulation of biosynthetic pathway would allow enhancing capsaicinoid accumulation. Comprehensive genomics, transcriptomics and proteomics study provide valuable information about transcription pattern of various structural and regulatory genes. Correlation analysis between the TFs and biosynthetic genes and their expression pattern in different fruit parts can be useful in identifying candidates for functional gene evaluation. There is scope of conducting more studies through omics, silencing through VIGS and gene editing through CRISPR-Cas9 for studying interaction between different TFs and structural genes. This would also help in identifying new TFs involved in capsaicinoid biosynthesis.

References

Ambawat S, Sharma P, Yadav NR, Yadav RC (2013) MYB transcription factor genes as regulators for plant responses: an overview. Physiol Mol Biol Plants 19:307–321

Arce-Rodriguez ML, Martinez O, Ochoa-Alejo N (2021) Genome-wide identification and analysis of the MYB transcription factor gene family in chili pepper (*Capsicum spp.*). Int J Mol Sci 22: 2229

Arce-Rodríguez ML, Ochoa-Alejo N (2017) An R2R3-MYB transcription factor regulates capsaicinoid biosynthesis. Plant Physiol 174:1359–1370

Arce-Rodriguez ML, Ochoa-Alejo N (2019) Biochemistry and molecular biology of capsaicinoid biosynthesis: recent advances and perspectives. Plant Cell Rep 38:1017–1030

Aza-González C, Núñez-Palenius HG, Ochoa-Alejo N (2011) Molecular biology of capsaicinoid biosynthesis in chili pepper (*Capsicum* spp.). Plant Cell Rep 30:695–706

Burgos-Valencia E, Echevarría-Machado I, Narváez-Zapata JA, Martínez-Estévez M (2020) Gene expression related to the capsaicinoids biosynthesis in the Capsicum genus: molecular and transcriptomic studies. Braz J Bot 43:201–212

Curry J, Aluru M, Mendoza M, Nevarez J, Melendrez M, O'Connell MA (1999) Transcripts for capsaicinoids biosynthetic genes are differentially accumulated in pungent and non-pungent *Capsicum* spp. Plant Sci 148:47–57

del Rosario Abraham-Juárez M, del Carmen Rocha-Granados M, López MG, Rivera-Bustamante RF, Ochoa-Alejo N (2008) Virus-induced silencing of Comt, pAmt and Kas genes results in a reduction of capsaicinoid accumulation in chili pepper fruits. Planta 227(3):681–695

Deshpande RB (1935) Studies in Indian chillies: inheritance of pungency *in Capsicum annuum* L. Indian J Agric Sci 5:513–516

Du H, Zhang L, Liu L, Tang XF, Yang WJ, Wu YM, Huang YB, Tang YX (2009) Biochemical and molecular characterization of plant MYB transcription factor family. Biochemist 74:1–11

Dubey M, Jaiswal V, Rawoof A, Kumar A, Nitin M, Chhapekar SS, Kumar N, Ahmad I, Islam K, Brahma V, Ramchiary N (2009) Identification of genes involved in fruit development/ripening in Capsicum and development of functional markers. Genomics 111(6):1913–1922. https://doi.org/10.1016/j.ygeno.2019.01.002

Dubos C, Stracke R, Grotewold E, Weisshaar B, Martin C, Lepiniec L (2010) MYB transcription factors in Arabidopsis. Trends Plant Sci 15(10):573–581

Gigolashvili T, Engqvist M, Yatusevich R, Müller C, Flügge UI (2008) HAG2/MYB76 and HAG3/MYB29 exert a specific and coordinated control on the regulation of aliphatic glucosinolate biosynthesis in *Arabidopsis thaliana*. New Phytol 177(3):627–642

Giovannoni J (2004) Genetic regulation of fruit development and ripening. Plant Cell 16:S170–S180

Giovannoni J (2011) Molecular biology of fruit maturation and ripening. Annu Rev Plant Physiol Plant Mol Biol 52:725–749

Gutiérrez-Carbajal MG, Monforte-González M, Miranda-Ham MDL, Godoy-Hernández G, Vázquez-Flota F (2010) Induction of capsaicinoid synthesis in *Capsicum chinense* cell cultures by salicylic acid or methyl jasmonate. Biol Plant 54(3):430–434

Han K, Jang S, Lee JH, Lee DG, Kwon JK, Kang BC (2019) A MYB transcription factor is a candidate to control pungency in *Capsicum annuum*. Theor Appl Genet 132:1235–1246

Hatlestad G, Akhavan N, Sunnadeniya R et al (2015) The beet *Y* locus encodes an anthocyanin MYB-like protein that activates the betalain red pigment pathway. Nat Genet 47:92–96. https://doi.org/10.1038/ng.3163

Huang PY, Catinot J, Zimmerli L (2016) Ethylene response factors in *Arabidopsis* immunity. J Exp Bot 67:1231–1241

Jin H, Martin C (1999) Multifunctionality and diversity within the plant MYB-gene family. Plant Mol Biol 41:577–585

Keyhaninejad N, Curry J, Romero J, O'connell MA (2014) Fruit specific variability in capsaicinoid accumulation and transcription of structural and regulatory genes in Capsicum fruit. Plant Sci 21:59–68. https://doi.org/10.1016/j.plantsci.2013.10.013

Kim M, Kim S, Kim S, Ki BD (2001) Isolation of cDNA clones differentially accumulated in the placenta of pungent pepper by suppression subtractive hybridization. Mol Cells 11(2):213–219

Kim JS, Park M, Lee DJ, Kim BD (2009) Characterization of putative capsaicin synthase promoter activity. Mol Cells 28(4):331–339

Koeda S, Sato K, Tomi K, Tanaka Y, Takisawa R, Hosokawa M, Doi M, Nakazaki T, Kitajima A (2014) Analysis of non-pungency, aroma, and origin of a *Capsicum chinense* cultivar from a Caribbean Island. J Jpn Soc Hort Sci 3:244–251

Kondo F, Hatakeyama K, Sakai A, Minami M, Nemoto K, Matsushima K (2021) The pungent-variable sweet chili pepper 'Shishito' (*Capsicum annuum*) provides insights regarding the relationship between pungency, the number of seeds, and gene expression involving capsaicinoid biosynthesis. Mol Genet Genomics 296:591–603

Lang Y, Kisaka H, Sugiyama R, Nomura K, Morita A, Watanabe T, Tanaka Y, Yanzawa S, Miwa T (2009) Functional loss of pAMT results in biosynthesis of capsinoids, capsaicinoid analogs, in *Capsicum annuum* cv. CH-19 sweet. Plant J 59:953–961

Lee A, Cho K, Jang S, Rakwal R, Iwahashi H, Agrawa GK, Shim J, Han O (2004) Inverse correlation between jasmonic acid and salicylic acid during early wound response in rice. Biochem Biophys Res Commun 318(3):734–738
Lindsey K, Bosland PW (1995) A field study of environmental interaction on pungency. Capsicum Eggplant Newsl 14:36–38
Liu S, Li W, Wu Y, Chen C, Lei J (2013) De novo transcriptome assembly in chili pepper (Capsicum frutescens) to identify genes involved in the biosynthesis of capsaicinoids. PLoS One 8:e48156. https://doi.org/10.1371/journal.pone.0048156
Liu R, Son J, Liu S, Chen C, Zhang S, Wang J, Xiao Y, Cao B, Lei J, Zhu Z (2021) Genome-wide identification of the Capsicum bHLH transcription factor family: discovery of a candidate regulator involved in the regulation of species-specific bioactive metabolites. BMC Plant Biol 21:262
Liu Y, Zhang Z, Fang K, Shan Q, He L, Dai X, Zou X, Liu F (2022) Genome-wide analysis of the MYB-related transcription factor family in pepper and functional studies of *CaMYB37* involvement in capsaicin biosynthesis. Int J Mol Sci 23:11667. https://doi.org/10.3390/ijms231911667
Mazourek M, Pujar A, Borovsky Y, Paran I, Mueller L, Jahn MM (2009) A dynamic interface for capsaicinoid systems biology. Plant Physiol 150(4):1806–1821. https://doi.org/10.1104/pp.109.136549
Nakano T, Suzuki K, Fujimura T, Shinshi H (2006) Genome-wide analysis of the ERF gene family in *Arabidopsis* and rice. Plant Physiol 140:411–432
Naves ER, de Avila SL, Sulpice R, Araujo WL, Nunes-Nesi A, Peres LEP, Zsogon A (2019) Capsaicinoids: pungency beyond Capsicum. Trends Plant Sci 24:109–120
Ohme-Takagi M, Shinshi H (1995) Ethylene-inducible DNA binding proteins that interact with an ethylene-responsive element. Plant Cell 7:173–182
Park YJ, Nishikawa T, Minami M, Nemoto K, Iwasaki T, Matsushima K (2015) A low- pungency S3212 genotype of *Capsicum frutescens* caused by a mutation in the putative aminotransferase (*pAMT*) gene. Mol Genet Genomics 290:2217–2224
Prasad BCN, Shrivastava R, Ravishankar GA (2005) Capsaicin. Evid Based Integrative Med 2: 147–166. https://doi.org/10.2165/01197065-200502030-00006
Qin C, Yu C, Shen Y, Fang X, Chen L, Min J et al (2014) Whole-genome sequencing of cultivated and wild peppers provides insights into Capsicum domestication and specialization. Proc Natl Acad Sci U S A 111:5135–5140. https://doi.org/10.1073/pnas.1400975111
Soler M, Camargo ELO, Carocha V, Cassan-Wang H, San Clemente H, Savelli B, Hefer CA, Paiva JAP, Myburg AA, Grima-Pettenati J (2015) The *Eucalyptus grandis* R2R3-MYB transcription factor family: evidence for woody growth-related evolution and function. New Phytol 206(4): 1364–1377
Song J, Chen C, Zhang S, Wang J, Huang Z, Chen M, Cao B, Zhu Z, Lei J (2020) Systematic analysis of the Capsicum ERF transcription factor family: identification of regulatory factors involved in the regulation of species-specific metabolites. BMC Genomics 21:573
Stellari GM, Mazourek M, Jahn MM (2010) Contrasting modes for loss of pungency between cultivated and wild species of Capsicum. Heredity 104:460–471
Stewart C Jr, Kang BC, Liu K, Mazourek M, Moore SL, Yoo EY, Kim BD, Paran I, Jahn MM (2005) The Pun1 gene for pungency in pepper encodes a putative acyltransferase. Plant J 42(5): 675–688. https://doi.org/10.1111/j.1365-313X.2005.02410.x
Stewart C, Mazourek M, Stellari GM, O'Connell M, Jahn M (2007) Genetic control of pungency in *C. chinense* via the *Pun1* locus. J Exp Bot 58:979–991
Sun B, Zhou X, Chen C, Chen C, Chen K, Chen M, Liu S, Chen G, Cao B, Cao F, Lei J, Zhu Z (2020) Coexpression network analysis reveals an MYB transcriptional activator involved in capsaicinoid biosynthesis in hot peppers. Hortic Res 7:162. https://doi.org/10.1038/s41438-020-00381-2
Sun B, Zhu Z, Chen C, Chen G, Cao B, Chen C, Lei J (2019) Jasmonate-inducible R2R3-MYB transcription factor regulates capsaicinoid biosynthesis and stamen development in Capsicum. J Agric Food Chem 67:10891–10903

Suresh B, Ravishankar GA (2005) Methyl jasmonate modulated biotransformation of phenylpropanoids to vanillin related metabolites using Capsicum frutescens root cultures. Plant Physiol Biochem 43(2):125–131

Tanaka Y, Hosokawa M, Miwa T, Watanabe T, Yazawa S (2010a) Newly mutated putative-aminotransferase in nonpungent pepper (*Capsicum annuum*) results in biosynthesis of capsinoids, capsaicinoid analogues. J Agric Food Chem 58:1761–1767

Tanaka Y, Hosokawa M, Miwa T, Watanabe T, Yazawa S (2010b) Novel loss-of-function putative aminotransferase alleles cause biosynthesis of capsinoids, nonpungent capsaicinoid analogues, in mildly pungent chili peppers (*Capsicum chinense*). J Agric Food Chem 58:11762–11767

Toledo-Ortiz G, Huq E, Quail PH (2003) The Arabidopsis basic/helix-loop-helix transcription factor family. Plant Cell 15:1749–1770

Wang J, Liu Y, Tang B, Dai X, Xie L, Liu F, Zou X (2020) Genome-wide identification and capsaicinoid biosynthesis-related expression analysis of the R2R3-MYB gene family in *Capsicum annuum* L. Front Genet 11:598183

Wen J, Lv J, Zhao K, Zhang X, Li Z, Zhang H, Huo J, Wan H, Wang Z, Zhu H, Deng M (2022) Ethylene-inducible AP2/ERF transcription factor involved in the Capsaicinoid biosynthesis in Capsicum. Front Plant Sci 13:832669. https://doi.org/10.3389/fpls.2022.832669

Yin L, Chen H, Cao B, Lei J, Chen G (2017) Molecular characterization of MYB28 involved in aliphatic glucosinolate biosynthesis in Chinese kale (Brassica oleracea var. alboglabra bailey). Front Plant Sci 8:1083

Zhang ZX, Zhao SN, Liu GF, Huang ZM, Cao ZM, Cheng SH, Lin SS (2016) Discovery of putative capsaicin biosynthetic genes by RNA-Seq and digital gene expression analysis of pepper. Sci Rep 6:34121. https://doi.org/10.1038/srep34121

Zhang B, Hu F, Cai X, Cheng J, Zhang Y, Lin H, Hu K, Wu Z (2022) Integrative Analysis of the Metabolome and Transcriptome of a Cultivated Pepper and Its Wild Progenitor Chiltepin (Capsicum annuum L. var. glabriusculum) Revealed the Loss of Pungency During Capsicum Domestication. Front. Plant Sci. https://doi.org/10.3389/fpls.2021.783496

Zhu Z, Sun B, Cai W, Zhou X, Mao Y, Chen C, Wen J, Cao B, Chen C, Chen G et al (2019) Natural variations in the MYB transcription factor MYB31 determine the evolution of extremely pungent peppers. New Phytol 223:922–938

Challenges in Capsaicin Production Due to Abiotic Factors 10

Angela María Chapa-Oliver, Ramón Gerardo Guevara-González, and Laura Mejía-Teniente

Abstract

Capsaicin is a secondary metabolite that is biosynthesized in the placenta of the chili fruit, and has a high added value in both the food and pharmaceutical industries. Recently, capsaicin has taken importance due to its effect against various types of cancer and its extensive use as phytobiotic for animal nutrition. Therefore, attention has been focused on the different production forms of this bioactive compound, either by chili crops managing in the field or through the use of cell cultures. In both cases, the production of capsaicin can be induced by abiotic stress factors, either by physical-environmental factors or by the application of eliciting molecules of chemical or hormonal origin. However, one of the great challenges in the production of capsaicin is the modulation of stress, which directly impacts the performance of this bioactive compound. Therefore, in this chapter, we address the abiotic factors that have a positive impact on the capsaicin biosynthesis, as well as the stress modulation strategies, which have a molecular and biochemical impact on capsaicin production.

Keywords

Capsaicin · Elicitor · Abiotic stress · Stress Modulation · Environmental factors · Capsaicin biosynthesis

A. M. Chapa-Oliver · L. Mejía-Teniente (✉)
Health Sciences and Engineering Division, Department of Agroindustrial Engineering, Biotechnology Engineering Program, University of Guanajuato-Campus Celaya-Salvatierra, Celaya, Guanajuato, Mexico
e-mail: laura.mejia@ugto.mx

R. G. Guevara-González
Center of Applied Research in Biosystems, School of Engineering, Autonomous University of Queretaro-Campus Amazcala, El Marqués, Querétaro, Mexico

M. K. Swamy, A. Kumar (eds.), *Capsaicinoids*,
https://doi.org/10.1007/978-981-99-7779-6_10

10.1 Introduction

One of the most ancient and widely utilized vegetables, condiments, and spices in the world is the chilli pepper, which comes under the family Solanaceae and the genus *Capsicum*. It is one of the earliest domesticated plants in the Western Hemisphere and has about 25 species (Kim et al. 2014). Since ancient times, the plant had an extensive history of use as a vegetable, natural dye, and folk remedy (Hernández-Pérez et al. 2020). This crop has a high nutritional value because it possesses of a wide variety of health-protective chemical compounds such as flavonoids, carotenoids, minerals, vitamins C and E, and capsaicinoids, which include capsaicin. The latter are the ones that give chilli peppers their characteristically pungent oral sensation (Materska and Perucka 2005; Hernández-Pérez et al. 2020). It is produced in the placental tissue and interlocular septum of maturing *Capsicum* through the convergence of the metabolic pathways of the branched fatty acids and the phenylpropanoids. These compounds have wide applications in the field of food, medicine, cosmetic, and pharmacy (Chapa-Oliver and Mejía-Teniente 2016). Additionally, it has been shown that capsaicin alters the expression of several genes responsible for metastasis, angiogenesis, growth arrest, and survival of cancer cells (Bley et al. 2012).

Around 90% of capsaicinoids present in *Capsicum* fruit pericarp, placenta, and seed tissues consist of dihydrocapsaicin and capsaicin (Prasad et al. 2005; Gurung et al. 2011), while less than 20% of the total amount of capsaicinoids include homodihydrocapsaicin, nordihydrocapsaicin, and homocapsaicin (Hernández-Pérez et al. 2020). Capsaicin (trans-8-methyl-N-vanillyl-6-nonenamide) is a homovanillic acid derivative that forms part of the wealth of bioactive compounds synthesized in the different species of *Capsicum* (Ramakrishna and Ravishankar 2011). The difference in the production of capsaicinoids can be related to the environmental factors and inherent genetic traits of each genotype (Hamed et al. 2019). The quantity of capsaicinoids, particularly capsaicin, may also vary depending on the stage of fruit maturation. Under controlled stress, the concentration of this metabolite can be increased, similar to other secondary metabolites found in plants. According to reports, pungency rises in response to environmental stress such as excessive temperatures, drought, flooding, and uneven soil fertility (Gurung et al. 2011).

An abrupt change in the physiological, biochemical, and molecular state of the plant caused by environmental factors that have a negative impact on the plant's growth and development can be defined as plant stress. Plant homeostasis is disturbed by stress, which has a negative impact on vital physiological, morphological, and metabolic procedures. These processes limit energy production and disturb cellular integrity (Rivero et al. 2021; Kaur et al. 2023). Under stress, plants undergo metabolic and physiological changes that affect the shoot and root biomass, a decrease in photosynthesis and nutrient uptake, a suppression of flowering and seed germination, and an incline in the synthesis of secondary metabolites synthesis that enable plants to adapt to the stress (Kaur et al. 2023). Nevertheless, the generation of reactive oxygen species (ROS) is within the primary responses that plants experience due to whatever type of stress or stressor to which they are exposed. Consequently, ROS and mainly hydrogen peroxide guide a series of intracellular events that end in the expression of genes related to the biotic or abiotic stress response, as well as the generation and accumulation of secondary metabolites (Mejía-Teniente et al. 2013; Chapa-Oliver and Mejía-Teniente 2016). Thus, secondary metabolite biosynthesis plays protective roles with respect to abiotic and biotic stress, and plants can be stressed to increase their ability to produce secondary metabolites. Therefore, it may be possible to use biotic and abiotic factors that increased secondary metabolite production as a tool to overproduce advantageous phytochemicals for a variety of applications (Gupta et al. 2019; Vázquez-Hernández et al. 2019; Cisneros-Zevallos and Jacobo-Velázquez 2020). Therefore, the contribution of this chapter is to show the effect of the abiotic stress on the biosynthesis of capsaicin as well as the stress modulation generated by elicitors (eustressors) in *Capsicum* spp. under different cultivation methods.

10.2 Abiotic Factors Influencing Capsaicin Production

The phenylpropanoid and the branched fatty acid pathway are the secondary metabolic pathways that produce capsaicinoids. Both abiotic components and genetic regulation act as key factors in the biosynthesis of capsaicinoids (Tripodi et al. 2019). Therefore, environmental factors like salinity, high or low temperature, drought, and nutrient stress may have an impact on the formation of capsaicinoids in pepper plants, although these may vary interspecifically (Ruiz-Lau et al. 2011). For plants to grow and develop, abiotic factors are essential, and each species of plant requires a particular environment in order to flourish (Gupta et al. 2019). Numerous environmental factors, including extremes in temperature, alkalinity, drought, salinity, pathogen infection, and UV stress, have the potential to harm plants (Ramakrishna and Ravishankar 2011). Pepper plants are well-known for being sensitive to moisture stress during the reproductive phase of the flower and fruit set, and plants that are under water stress frequently lose their blossoms and immature pods (Ruiz-Lau et al. 2011). However, plants under water stress typically yield more pungent pods. These findings demonstrate the dual influences of genetics and interactions between plants and their environment on pepper pungency levels (Fig. 10.1) (Ruiz-Lau et al. 2011). The main abiotic stresses that generate an effect on the synthesis of capsaicinoids are discussed below and are summarized in Fig. 10.1.

10.2.1 Salt Stress

High soil salinity creates a stressful environment for most plant species. One of the major stresses, especially in the arid and semi-arid regions, is salt stress in soil or water, which can significantly retard the growth and development of plants (Ramakrishna and Ravishankar 2011). Salt-induced cellular dehydration resulted

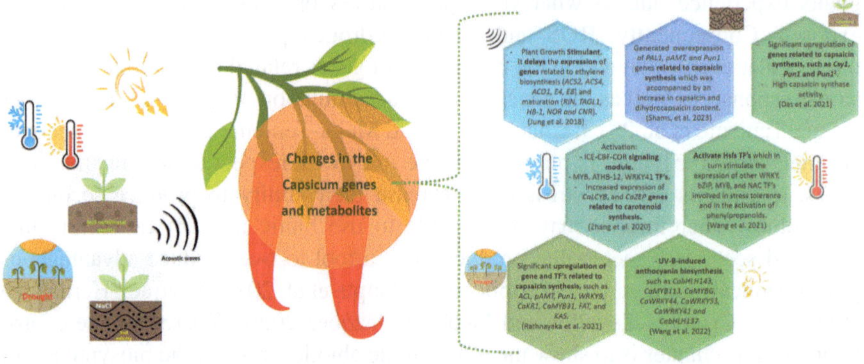

Fig. 10.1 The main abiotic stresses that affect the synthesis of capsaicinoids and other metabolites

in osmotic stress and water removal for the cytoplasm, which decreased the cytosolic and vacuolar volumes. It additionally causes ionic stress and osmotic stress in plants, which can lead to a variation in the number of secondary metabolites in plants (Ramakrishna and Ravishankar 2011). Salination of soil and water is an ongoing issue that is made worse by the unwise use of agrochemicals and fertilizers, excessive groundwater pumping for the irrigation of farmland, and general crop intensification (Arrowsmith et al. 2011). When plants are grown in environments with high sodium chloride (NaCl) concentrations, they frequently experience morphological, biochemical, physiological, and metabolic changes that have a substantial negative effect on growth, productivity, and the quality of the plant. However, it has been demonstrated that salinity, a chemical stressor, affects both the physical quality and chemical makeup of a variety of vegetable commodities (Rouphael et al. 2018).

Purnama et al. (2022) investigated the capsaicin content and fruit development of hot peppers grown on five various salinity media. They were 15.20 dS/m for A, 5.70 dS/m for B, 1.10 dS/m for C, 2.85 dS/m for D, and 3.25 dS/m for E. Gas chromatography-mass spectrometry was used to measure the capsaicin content at 14 and 35 days after flowering. They discovered that in planting mediums A and D, plants cannot survive for longer than 2 weeks. Instead, plants can thrive, produce flowers, and gradually bear fruit in planting mediums B, C, and E (Purnama et al. 2022). They showed that brownish-green fruits have higher capsaicin content than green fruits in all survival-growing mediums. Capsaicin levels in green and brownish-green fruits were lower in medium B than in medium C and medium E in terms of salinity. Therefore, salinity negatively affected capsaicin content, and this depends on the fruit ripening stage. The above may be due to the salinity effect on the pericarpium and the placenta thickness, which was reduced in the media with higher saline content. In comparison to the other two growing mediums, C and E, medium B's hot pepper placenta also had less development (Purnama et al. 2022). Therefore, the placenta and pericarpium are the most important site of metabolite synthesis in *Capsicum* (Materska 2014). The placenta is the primary site of synthesis of capsaicinoid, primarily capsaicin, while both the seed and pericarp synthesis crucial enzymes for the production of capsaicinoid. Therefore, if the development of the placenta is affected, the production of capsaicin will be affected (Puruma et al. 2020). The production of fruits and other biomass would become more challenging as salt stress increased, resulting in higher capsaicin levels. Therefore, plants would spend more energy in the production of capsaicin in fruits at higher salinity, which increases the possibility of seed dispersal through the animals who consume the seeds (Arrowsmith et al. 2011).

In general, biotic or abiotic stresses stimulate the synthesis and accumulation of polyphenols. Navarro et al. (2006) found that red peppers had an increased total phenolic content at moderately salinity level. According to Marn et al. (2009), pepper plants with low irrigation frequency and an increase in nutrient solution EC up to 30 mM NaCl had higher vitamin C content in their red fruit than their green fruit, which showed no discernible effect. *Capsicum annum* cv. Jalapeño pepper plants were treated with 0, 0.5, 1, and 1.5% NaCl solutions by Arrowsmith et al. (2011) to investigate the connection between abiotic stress and capsaicin content of fruit. According to their findings, plants at the two highest salinities contained more capsaicin than those in the control and lowest salinity groups. Additionally, as salinity increased, the total plant biomass did not decrease; although modest, fruit production was consistent across treatments, and plants exposed to higher salinity developed more fruits than those of low salinity (Arrowsmith et al. 2011).

To evaluate the impact of salinity on capsaicin content and growth, Zahra et al. (2015) used pepper callus. By adding sodium chloride to MS medium to produce media with electric conductivities of 6, 9, and 12 dSm-1, they were able to create a salt stress. They discovered that capsaicin levels increased with salinity, peaking at 9 dSm-1 with a concentration of 36.49 g/g. But at 12 dSm-1, capsaicin decreased to 32.74 g/g, and calli lost its capacity for regeneration. Furthermore, they demonstrated that placenta-derived calli produced the highest amount of capsaicin, 53.11 g/g, at 9 dSm-1, while the root calli produced the lowest amount, 6.09 g/g, at 3 dSm-1 (Zahra et al. 2015). According to study reports, cultivation techniques such as the application of salinity could be beneficial tools for a wide range of quality management strategy of vegetable crops that manipulates the contents of phytochemicals and antioxidants (Rouphael et al. 2018).

10.2.2 Drought Stress

Crop production has been threatened by declining water resource availability brought on by regional and global climate change (Mahmood et al. 2021). Drought stress develops when the amount of water in the soil reaches critical levels, and atmospheric factors contribute to ongoing water loss. All plants can withstand drought stress, but the degree of their tolerance varies between species (Ramakrishna and Ravishankar 2011). One of the most important abiotic stresses that affect plant growth and development is drought stress. Malondialdehyde, the end product of lipid peroxidation in cellular membranes and a marker of oxidative membrane damage, can be produced as a result of the accumulation of reactive oxygen species (ROS), which can cause cellular disruptions and damage to redox homeostasis in plant cells (Airaki et al. 2012; Kopta et al. 2020). Antioxidant enzymes increased in the leaves and fruits of the chilli pepper under drought (Kopta et al. 2020).

There are a minimum of two main processes by which drought may affect the metabolism of phytochemicals. First, in most cases, drought causes a decrease in leaf stomatal conductance, which lowers net photosynthesis. Fruits are the primary source of the precursors needed to synthesize phenolic compounds, carotenoids, and ascorbate, so if net photosynthesis declines, less primary metabolites may be transported to the fruits (González-Chavira et al. 2018; Seleiman et al. 2021). Second, oxidative stress and oxidative signalling may be activated by drought. It is well-known that oxidative stress can both directly and indirectly affect these compounds' biosynthetic pathways. In addition, a lack of water may affect the metabolism of phytochemicals that are beneficial to health by accelerating fruit development (González-Chavira et al. 2018).

The ascorbic acid and pungency of mature fruits were elevated under drought stress, though the pungency relies on the genetic potential of various genotypes and the extent of the stress period (Zahra et al. 2015; Mahmood et al. 2021). The antioxidant enzymes and other enzymes in pepper fruits and leaves were more active due to the drought. Phenylalanine ammonia-lyase activity and capsaicinoid content were substantially increased among the various hot pepper cultivars. PAL is an essential enzyme involved in the biosynthesis of capsaicin under drought stress. According to research, capsaicin is either upregulated or downregulated in some cases when plants are stressed by drought; this depends on the genetics of the cultivars and the stress levels (Mahmood et al. 2021). Numerous studies have been conducted in plants to determine how drought affects the amount of secondary metabolites. The studies recorded elevated accumulation of secondary metabolite under water stress, including phenolics, terpenoids, and compounds containing N (glucosinolates, alkaloids, and cyanogenic glucosides). However, in water-stressed situations, growth of the plant is decreased in contrast to the amount of secondary metabolites (Gupta et al. 2019).

In comparison to control plants, more capsaicinoids were documented from the water-stressed Padrón pepper (*C. annuum* L.) fruits, especially under low-water treatments (Estrada et al. 1999). Their results point towards the possibility of competition between the production of capsaicinoids and other phenylpropanoid metabolites. The high-pungency Habanero cultivar resulted in an increase in capsaicinoid content when subjected to extreme water stress. Habanero pepper (*Capsicum chinense* Jacq) production of capsaicin and dihydrocapsaicin was examined by Ruiz-Lau et al. (2011). They proved that a lack of water in the soil has a significant impact on plant growth and the accumulation of total dry matter. Stress condition substantially decreased plant height, but had no discernible impact on the production of flowers or fruits in Habanero pepper plants. They measured the concentrations of capsaicin and dihydrocapsaicin in the seed, pericarp, and placenta of the fruit to examine the impact of water stress on capsaicinoid concentration.

These assessments were carried out 25 and 45 days after the anthesis of the fruits. They discovered that stress caused by a water shortage significantly raised the levels of capsaicin and dihydrocapsaicin in the pericarp and seed. In contrast to controls, treatments significantly increased the concentration of capsaicin and dihydrocapsaicin in the placenta (Ruiz-Lau et al. 2011). The authors also discovered that capsaicin synthase activity was decreased in highly potent chilli peppers when subjected to water stress, and this effect depended on the severity of the stress and maturity of the fruit. The Scoville Heat Units (SHU) of the pungency of leaves and fruits and enzymatic antioxidants of four chilli peppers under drought stress were examined by Kopta et al. (2020). The most productive drought application period for capsaicinoid synthesis was 21 days than 34 days according to their study. The cultivar referred to as 'Jolokia', which is considered to be very pungent, responded most strongly to a lack of water by producing more capsaicin. It was the plant that responded to stress by becoming more pungency sensitive to drought than any other (Kopta et al. 2020). The endogenous antioxidant enzyme activity of leaves and fruits increased during drought, with 'Jolokia' having the highest antioxidant activity in relation to capsaicinoids among the other cultivars tested (Kopta et al. 2020). These findings demonstrate that the medium drought stress exertion conditions to pot-grown chilli peppers triggered physiological and biochemical adaptation mechanisms in cultivar-dependent manner, and variations in the biosynthesis pathways due to the drought were different in the leaves and fruits (Kopta et al. 2020).

Although low-pungency cultivars appear to be more sensitive, drought is known to be a stressor that increases capsaicinoid accumulation in pepper fruits. According to Jeeatid et al. (2018), *Capsicum chinense* Jacq. plants with varying levels of pungency produce capsaicinoids under various water regimes. After anthesis, they subjected the four hot pepper cultivars to various water regimes: daily irrigation (control), every 2 days, every 3 days, and every 4 days. The fruit yield of cultivars was reduced by the water stress, and it varied greatly based on the water stress level and cultivar type. Additionally, they found that under water stress, cultivars with bigger fruit and more pungency produced more capsaicinoids (Jeeatid et al. 2018). A crucial enzyme for the production of capsaicin, phenylalanine ammonia-lyase (PAL), was also found to be more active in *C. chinense* fruit under drought stress (Phimchan et al. 2014). Regarding the developmental stage, *Capsicum* spp. are more vulnerable to the effects of drought during the vegetative stage than during the reproductive stages (Jeeatid et al. 2018).

Phimchan et al. (2012) found that under drought stress, the capsaicin quantity in the low-medium pungent cultivars was higher than in control plants. Nine different hot pepper cultivars were used, each with different growth habits, leaf and fruit traits, and levels of pungency. According to their initial level of pungency, cultivars were divided into low, medium, and high categories. Drought gradually stressed cultivars during the flowering stage. At 10, 20, and 30 days after flowering (DAF), plants in the drought stress group received less water supply of 25%, 50%, and 75% by volume. According to their claims, the low pungency group's yield performances decreased under drought stress, while there was no change for the group with higher pungency. The drought-stressed plants of all low-pungent cultivars had significantly lower dry fruit yields than the control plants. Except for the cultivars in the high pungency group, all cultivars studied experienced an increase in capsaicinoid levels under drought stress. They found that all of the cultivars they studied were affected by drought stress because the stress began during the flowering stage and lasted until 40 to 55 days after flowering. This is due to the fact that, depending on the cultivar, capsaicinoids start to accumulate early in fruit development and peak at 30 to 50 days after flowering (Phimchan et al. 2012). These findings clarified how a genotype-environment interaction affected capsaicinoid content.

In a different study, Gurung et al. (2011) discovered that cultivars with high pungency were highly stable, while cultivars with lower pungency were extremely sensitive to environmental changes. The high-pungency cultivars they used produced smaller fruits than the low- and medium-pungency cultivars, which may mean that the capsaicinoids in these smaller fruits are less sensitive to drought. Therefore, plant development and yield and content of capsaicinoid are less influenced by the environment (Phimchan et al. 2012; Gurung et al. 2011). They found that all high-pungency cultivars and some medium-pungency cultivars have smaller leaf sizes and less physiological, growth, yield, and capsaicinoid responses to drought stress than the other cultivars (Phimchan et al. 2012). There is a lot of interest in making drought events as tools to enhance crop quality while minimizing yield losses (González-Chavira et al. 2018).

10.2.3 Cold and Heat Stress

Temperature increases have an impact on plant growth and secondary metabolite production because they change the metabolic pathways that control physiology, signalling, and defence mechanisms (Gupta et al. 2019). Low temperature, which affects photosynthesis and the uptake of water and nutrients, is a significant environmental factor that affects plant growth. The production and quality of subtropical plants are impacted by cold sensitivity as well (Airaki et al. 2012). One of the most detrimental abiotic stresses affecting temperate plants is low temperature. In the autumn, these species alter their metabolism to produce more of various kinds of cryo-protective compounds that enhance their resistance to cold temperatures (Ramakrishna and Ravishankar 2011).

The impact of low temperatures on the diversity, distribution, productivity, and developmental processes of plants is one of the most severe stresses on plant systems (Gupta et al. 2019). Plants adjust their coping mechanisms to deal with cold stress by redistributing resources and growth retardation. They can alter based on the type of plant and/or the length and severity of the stress exposure. Numerous genes have been shown to be regulated by low temperatures, and biochemical changes have been shown to have an impact on the levels of numerous proteins, lipids, and metabolites (Airaki et al. 2012). Since they are native to tropical areas, pepper plants need high temperatures for proper growth. Therefore, the ideal temperature for growth ranged from 25 °C to 30 °C, with the result that temperature changes have an impact on various physiological processes and morphological development. Temperatures below 15 °C cause a reduction in pepper growth as well as a halt to bloom and fruit production (Airaki et al. 2012).

Heat stress is among the most harmful abiotic stressors that have an impact on pepper development, colouration, nutritional value, and yield of fruit. Due to global warming, the incidence and frequency of heat stress have significantly increased in modern times. Temperatures above 32 °C for an extended period of time can lead to serious development issues such as dropping of fruits and buds, as well as blemished fruits due to the formation of plate spots and poor quality (Lin et al. 2021). However, heat regularly affects the synthesis of secondary metabolites, including phenolic substances, terpenes, and alkaloids (Yang et al. 2018; Wang et al. 2021). There are reports that capsaicinoids accumulate more quickly at high temperatures. Compared to plants grown in fields at 21 to 24 °C, plants grown at 30 °C day and night temperatures contained significantly more capsaicinoids (Otha 1960).

The pepper fruits were subjected to heat stress, and Lin et al. (2021) recorded the multi-omics results of stress response such as transcriptomics, metabolomics, and proteomics. In order to mimic the high-temperature conditions of summer, pepper plants that were bearing fruits were placed in a light incubator and heated to 36 °C. Fruit pericarps at different time points were sampled. In response to heat treatments, they discovered 39 differentially accumulated metabolites (DAMs), 1999 differentially abundant proteins (DAPs), and 14,513 differentially expressed genes (DEGs) (Lin et al. 2021). They also determined the capsaicin content in fruit pericarps and found that this was substantially impaired by heat stress and correlated with the transcriptome results (Lin et al. 2021). According to their findings, heat stress decreased the amount of capsaicin, and this was possibly brought on by fatty acid metabolism being inhibited and by affecting a number of transcription factors that might control the expression of vital genes involved in the pathway that produces capsaicin (Lin et al. 2021).

10.2.4 Influence of Light (UV Radiation)

The physical influence of light on plant growth and the synthesis of primary and secondary metabolites is well-known. There are evidences for the link between rising light levels and secondary metabolite concentrations (Zhang et al. 2020a, 2020; Castillejo et al. 2022). The ultraviolet light region is the region of the electromagnetic spectrum between 10 and 400 nm. Numerous fruits, vegetables, and ornamental crops can perform better and have higher nutritional values when exposed to natural levels of ultraviolet radiation of the sunlight (Neugart and Schreiner 2018). The growth, photosynthesis, and secondary metabolites of important horticultural and agricultural crops were impacted by UVB (280–315 nm) and UVA (315–400 nm) radiation. Additionally, UV light can be used to stimulate the expression of antioxidant enzymes, which increases the biosynthesis of carotenoids, vitamins and minerals, flavonoids, phenolic acids, and related phenolic compounds, which serve as antioxidants and UV protection components (Scotta et al. 2017; Dou et al. 2019; Castillejo et al. 2022). UVB, UVA, and visible radiation affect the biosynthesis of flavonoids (Neugart and Schreiner 2018).

The latest studies point towards the possibility of utilisation of UVC light to stimulate the production of secondary metabolites as an abiotic stress factor. UVC (ultraviolet-C) light is a novel technology used for surface and water disinfection (Alothman et al. 2009). Low doses of UVB and C radiation have been shown to have hormetic behaviour, which is characterized by a favourable biological response (Cisneros-Zevallos and Jacobo-Velázquez 2020). Regarding this, UVA, UVB, and UVC light have been applied to a variety of fruits and vegetables, including blueberry, strawberry, apple, grape, papaya, broccoli, carrot, tomato, and spinach, and the results have shown an increase in bioactive compounds and antioxidant enzymes (Ávila-Sosa et al. 2016).

UV light treatment increased the amount of carotenoids, flavonoids, and antioxidant activity in red pepper (*C. annuum* L.) during storage. During a period of refrigerated shelf-life, Castillejo et al. (2022) assessed the impact of 6 kJ/m^2 UVB or UVC and 6 kJ/m^2 UVB + UVC on the primary bioactive compounds of red bell peppers. After 8 and 14 days at 7 °C, they examined the change in carotenoids, phenolics, and flavonoids and discovered that exposure to UV radiation caused carotenoid and flavonoid accumulation (Castillejo et al. 2022). Similar research by Rodoni et al. (2015) revealed that fresh-cut red peppers treated with UVC light had a small rise in the quantity of phenolic compounds. They examined the impact of UVC treatments on the ability of freshly cut red peppers to maintain quality. Red pepper sticks (5x1 cm) were exposed to UVC radiation at doses of 1.5, 3, 5, 6, 10, and 20 kJ/m^2 on either the inner or outer fruit surface. They found that 10 kJ/m^2 UVC treatments on both surfaces were the most effective and had no effect on the sugars, colour, acidity, or antioxidant capacity. Furthermore, UVC treatment significantly decreased pectin solubilization, softening, weight loss, and decay. Their findings demonstrated that UVC can increase phenolic surface accumulation, which could be used in conjunction with low-temperature storage to increase the shelf life of freshly cut red peppers (Rodoni et al. 2015).

The effect of blue and UVC light as abiotic stress on the antioxidant capacity and bioactive compounds of habanero pepper (*C. chinense*) stored under tight packaging and low temperature (4–5 °C) was evaluated in a study conducted by Pérez-Ambrocio et al. (2018). In their experiment, the total capsaicin and antioxidant capacity of peppers were assessed for 30 days. They discovered that the synthesis of all bioactive compounds examined was stimulated by both lights, which also increased the antioxidant capacity. Total capsaicin concentration was 17.7 mg/kg at first, which was correlated with higher antioxidant activity of 21.3 mg Trolox equivalent/kg. In comparison to untreated habanero pepper, all treated habanero pepper displayed a significantly higher level of total capsaicin and antioxidant capacity after UVC treatment. AN increased value of 39.4 mg/kg of capsaicin and 102.3 mg Trolox equivalent/kg antioxidant capacity was recorded (Pérez-Ambrocio et al. 2018).

Other reports revealed comparable outcomes. Six genotypes of habanero (*C. chinense*) grown in three different Texas locations were examined by Butcher et al. (2012) for fruit size and capsaicinoid and flavonoid concentrations: Weslaco, Uvalde, and College Station. Average solar radiation in these places was monitored from May to August, with July being the month with the maximum solar radiation in all places tested (Weslaco: 40 MJ/m^2; Uvalde: 25 MJ/m^2; College Station: 20 MJ/m^2). They claimed that compared to peppers grown in Uvalde or College Station, those grown in Weslaco were larger and contained more capsaicin and dihydrocapsaicin. However, flavonoid levels varied and were low across all genotypes and environments (Butcher et al. 2012).

10.2.5 Acoustic Stress

The advantages of biophysical treatments, such as acoustic ones, for agriculture and biotechnology, have increased their popularity. However, its uses in relation to advantages based on frequency, intensity, and duration of application are still being investigated (Caicedo-Lopez et al. 2020). Acoustic emissions, or sound vibrations, are given the most focus as exogenous treatments. It has been demonstrated that these physical factors, which are vibrational waves that speakers emit, significantly boost secondary metabolism while assisting plants in withstanding drought in *Arabidopsis* and rice plants (Jeong et al. 2014; López-Ribera and Vicient 2017). Plants can recognize the sounds made by herbivores, which can trigger a defence response (Mishra and Bae 2019). According to natural frequencies, vibration patterns function as information brokers that transmit information between plants and between plants and their environment, as well as acts as a participant in the process of adaptability behaviour signalling (Fernandez-Jaramillo et al. 2018). However, additional study is necessary to support this. As stated in these chapters, drought causes pepper plants to accumulate more capsaicinoids. The performance of plants on a genetic, biochemical, and physiological level may be enhanced by biophysical factors such as acoustic emissions (Fernandez-Jaramillo et al. 2018). According to studies that suggested sound frequencies could cause determinate plant changes, the vibration generated by plants under hydric stress can be artificially exerted on healthy and properly watered plants and be perceived by them as a real "harm alarm" of hydric stress. These results develop changes in the plant similar to changes during the water deficiency period (Caicedo-Lopez et al. 2020).

Acoustic emissions patterns (AEP), which are most similar to the causes of naturally hydric stressed plants, elicited the capsaicin contents, which were 8.33 times higher than the negative control without AEP. Pepper plants can be easily elicited by AEPs. In this respect, Caicedo-López et al. (2021) obtained vibrational signals generated by hydric-stressed *C. annuum* L. plants and used them as AEP. A laser beam was used to record these vibrational signals. The other experimental conditions included constant pressure level of 80 dB in an anechoic chamber with an enveloped field, and AEP treatment for a short period was applied. They assessed the levels of capsaicin and dihydrocapsaicin in fruits as well as the expression of two gene markers for the biosynthesis of capsaicinoids, *keto acyl synthase* (*kas*) and *capsaicin synthase* (*at3*) genes (Caicedo-López et al. 2021). They claimed that treatment of pepper plants with AEPs in the reproductive developmental stage resulted in an accumulation of capsaicin, illustrating the potential for selectiveness connection on the secondary metabolism towards some AEPs. In contrast to the negative controls of white noise and silence, their findings indicated that plants exposed to AEPs displayed noticeable variations in capsaicinoids levels (Caicedo-López et al. 2021). Moreover, a 2.3-fold upregulation was observed in *at3* and *kas* genes with contrasting differences in the capsaicinoid content after AEP treatments. Their results show a positive correlation between the molecular response of these genes and capsaicin content (Caicedo-López et al. 2021). The information provided suggests that AEP vibrations oscillate at frequencies that may increase ability of plants to produce secondary metabolites. Therefore, understanding the patterns that plants use to detect and emit mechanical signals might assist in the management of abiotic stress by activating intricate defence mechanisms that affect biochemical and physiological pathways. In order to improve commercial crops' secondary metabolism for antioxidant defence and subsequent drought tolerance, AEPs have emphasized the potential value of sensible applications on these crops (Kollasch et al. 2020). To increase secondary metabolite yields or even give plants a protective response to biotic or abiotic stress, more study is required on biological factors influencing frequency selection for sound treatments.

10.2.6 Soil Quality

The effect of soil characteristics on the genetic control of pungency and capsaicin formation in chilies is poorly understood. According to Medina-Lara et al. (2008), potassium fertilizer application had no impact on growth or productivity, while nitrogen fertilizer substantially boosted plant development and fruit yield while maintaining high levels of capsaicin. Yadav et al. (2021) figured that soil had the greatest impact on plant secondary metabolism of all environmental factors. Researchers have found that plants produce secondary metabolites in response to stress related to nutrient deficiency, water shortage, heavy metal pollution, etc. *C. frutescens* cv. Nodaria and *C. annuum* cv. Balijuri, two endemic chilli cultivars, were grown in both alluvial and lateritic soil by Das et al. (2021) to assess the amount of capsaicin and the activity of the enzyme in the capsicum fruits. It was observed that chilies grown in alluvial soil had significantly higher levels of capsaicin, pungency, and capsaicin synthase activity than those grown in lateritic soil (Das et al. 2021). Additionally, a strong positive correlation between soil organic carbon and capsaicin content was found by correlation analysis (Das et al. 2021). Their findings demonstrated that alluvial soil's bulk density and water holding capacity were superior to lateritic soil in terms of crop production. Although both soils were acidic, the extent of the acidity was greater in the alluvial than the lateritic. The pH and physical characteristics of the soil were not significantly changed by chilli cultivation, but both soils showed a sharp decrease in bulk density and an increase in water holding capacity when compared to their initial values (Das et al. 2021). In comparison to lateritic soil, alluvial soil was richer in nutrients (N, P, and K). Additionally, the former one had a richer microbial population. These findings showed that soil moisture content, soil organic carbon, and nutrient availability (N, P, and K) have a significant impact on crop productivity. Soil organic carbon, humic acid carbon, fulvic acid carbon, total N, microbial biomass carbon, and pH were the most important factors for enhancing capsaicin production among the various soil quality attributes (Das et al. 2021).

10.3 Capsaicin Synthesis by Application of Elicitors

As described, capsaicin is a secondary metabolite that forms part of the wealth of bioactive compounds synthesized in the different species of *Capsicum*. The accumulation of associated natural substances influences the quantity of various bioactive compounds (Ramakrishna and Ravishankar 2011). Nevertheless, the accumulation of these secondary metabolites can be achieved through the application of elicitors, which are molecules of biotic or abiotic origin, of low molecular weight that trigger the hypersensitivity response of the plant, generating physiological, genetic, and metabolic changes as part of the plant defence system (Mejía-Teniente et al. 2010, 2019; Narayani and Srivastava 2017; Ramirez-Estrada et al. 2015). Figure 10.2 summarizes the elicitor effects on the plant or cell culture of *Capsicum*.

10.3.1 Chemical Elicitors

Mainly, in different investigations, the role of chemical elicitors in plant defence and secondarily in the production of bioactive compounds has been evaluated. However, chemical elicitors are also used to induce metabolites other than capsaicin. For example, in non-hot peppers such as sweet bell peppers, they have been used to

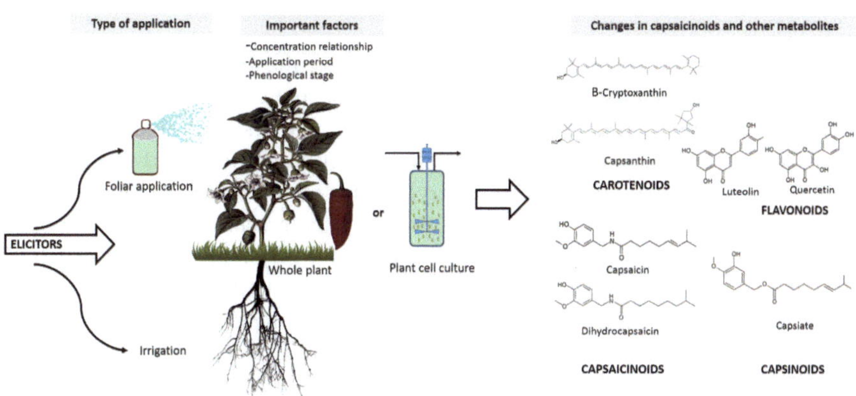

Fig. 10.2 Summary of the elicitor effects on the plant or its cell cultures

induce carotenoid synthesis and increase antioxidant capacity (Garcia-Mier et al. 2015). Garcia-Mier et al. (2015) observed that the foliar application of elicitors mixture containing jasmonic acid (JA, 1 μM), hydrogen peroxide (H_2O_2, 27 mM), and chitosan (CH, 0.75%) during crop development had a positive effect on the accumulation of flavonoids, condensed tannins, and carotenoids in mature fruits as compared to control plants. They also indicated that of the carotenoids evaluated, lycopene is the most representative with a production of 13.447 ± 1.629 μg/g, and in mature fruits, the antioxidant activity had no significant difference with control plants. Similarly, Jimenez-Garcia et al. (2018) showed that application of 36 mM H_2O_2 and mix of salicylic acid (SA, 0.1 mM) and 36 mM H_2O_2 positively changed the sensory quality traits of two commercial sweet bell pepper varieties. Another way to introduce chemical elicitors is through fertilization. In this sense, Medina-Lara et al. (2008) showed that behaviour of capsaicin content in response to N and K fertilization was not dependent or linear to the concentration of these elements. That is, concentrations less than or equal to 1 mM of N generated little content of this metabolite, but very high concentrations also generated a decrease in capsaicin content. They showed that capsaicin content increased only in 7.5 mM and 15 mM N treatments, ≈ 9 mg g^{-1} DW and ≈ 12 mg g^{-1} DW, respectively. However, Stan et al. (2021) compare chemical and organic fertilization and find that the type of fertilizer used as well as the chilli pepper variety had an impact on the capsaicinoids content. The findings of Stan et al. (2021) demonstrated a high accumulation of capsaicin (0.83 mg/g) and dihydrocapsaicin (0.53 mg/g) in fruits of the Chorbadzhiyski variety treated with chemicals (235 kg/ha NPK), in contrast to the amounts found for the untreated De Cayenne variety (0.52 mg/g capsaicin and 0.33 mg/g dihydrocapsaicin). Regarding the organic fertilizer, they reported a production of 0.23 mg/g of dihydrocapsaicin and 0.40 mg/g of capsaicin. From the above studies, we can infer that chemical elicitors can be applied in crop to increase secondary metabolites without compromising the fitness of the crops.

10.3.2 Hormonal Elicitors

The assessment on effect of hormonal elicitors on the production of secondary metabolites is primarily performed by suspension cultures. Similar to chemical elicitors, protective effect against different types of biotic and abiotic stress has been evaluated on plants. In this sense, Shams et al. (2018) showed that the foliar application of 50 and 100 μM of Benzyl amino purine (BAP, cytokinin) every 10 days after the fifth day of the establishment of the greenhouse culture favoured the increase of the capsaicin content. Phenological, physiological, and metabolic developmental processes, as well as the regulation of the plant cell cycle, are all influenced by cytokinins, which are chemical messengers (hormones) that are specific to plants (Schmülling 2004). Shams et al. (2018) reported the production of 27 and 30 mgkg $^{-1}$ of capsaicin in 50 and 100 μM BAP treatments, significantly higher than non-hormone treatment. However, as the hormone concentration increases, fruit production decreases (Shams et al. 2018). According to research findings, BAP only affects plant growth and the production of secondary metabolites in plants (Shams et al. 2018). For instance, it has been reported that the application of BAP increased the camptothecin levels in *Ophiorrhiza rugosa* var. *decumbens* shoots and roots, an alkaloid with significant antitumor activity to ovarian, lung, breast, stomach, and pancreatic cancers (Hatefi and Amsden 2002; Vineesh et al. 2007). Other example was reported by Ding et al. (2004) where BAP application increased the shikonin formation – a naphthoquinone that exhibits powerful anticancer activities for various cancer cells – in *Onosma paniculatum* cultured cell.

In suspension culture, Gutierrez-Carbajal et al. (2010) created a cell line from the Habanero pepper (*C. chinense* Jacq.) and monitored capsaicinoid biosynthesis following the addition of methyl jasmonate and salicylic acid. Elicitation of the suspension culture with 5, 50, and 500 μM of salicylic acid (SA) or methyl jasmonate (MeJa) every 12 h until completing a period of 72 h resulted in increased capsaicinoid content within the first 24 h (0.250 mmol g^{-1} dm), reaching maximum to 36 h (0.375 mmol g^{-1} dm), and it slightly dropped over the course of the experiment. Also noted was the fact that these cell cultures' capsaicin content was 200 times lesser than that of fruit placentas (Monforte-González et al. 2007). Similar to this, Kehie et al. (2016) found that suspension cultures of the Naga King Chilli (*C. chinense* Jacq.) exposed to various concentrations of SA (0.1–2 mM) produced the highest capsaicin content (567.4 ± 8.1 lgg^{-1}f wt). Plants use SA, a potent signalling molecule, to coordinate specific responses to biotic and abiotic stresses (Krantev et al. 2006; Mejía-Teniente et al. 2019). In essence, they show that, in contrast to fruits with high levels of capsaicin, cell cultures have an inherent limitation related to their low ability to produce capsaicin under in vitro conditions.

10.3.3 Nanoparticles as Elicitors

Nanoparticles (NPs) are part of the wide range of nanomaterial structures that have relevancy by their different applications in agriculture as nanofertilizer, nanoprotector against diseases caused by pathogens, nanocarriers of bio-elements to improve nutrition, as well as eliciting activity. They are defined as particles within the size range of 1 and 100 nm (Anjum et al. 2019; Khan et al. 2019). The shape, size, construction material, penetration level, and application form of nanoparticles determine the toxic and beneficial activity to the plant (Rico et al. 2011). Commonly used nanoparticles in agricultural applications are of zinc oxide (ZnO), silver (Ag), titanium dioxide (TiO_2), silicon dioxide (SiO_2), and copper oxide (CuO) (Boxi et al. 2016). According to several studies, nanoparticles are capable of disrupting the electron transport chains in chloroplasts and mitochondria. This could cause an oxidative burst and the release of reactive oxygen species (ROS) (Marslin et al. 2017; Rastogi et al. 2017). The expression of genes encoding crucial enzymes of secondary metabolism, such as PAL, is required to direct signal transduction via MAP kinases and produce a response. These genes include those related to oxidative stress, such as superoxide dismutase (SOD), catalase (CAT), and ascorbate peroxidase (Marslin et al. 2017). In *Capsicum*, García-López et al. (2019) showed that foliar application of zinc sulphate ($ZnSO_4$) and zinc oxide (ZnO) nanoparticles had a positive effect on phenological development of *Capsicum chinense* in 1000–2000 mg L^{-1} concentrations. In relation to capsaicin and dihydrocapsaicin content, García-López et al. (2019) showed that 2000 mg L^{-1} of ZnO NPs generated a maximum of 952.15 ± 8.17 mg kg^{-1} of the capsaicinoids, which is 16% more than the control (625.44 mg kg^{-1}), while $ZnSO_4$ treatments were statistically the same with control. The same behaviour was observed with the Scoville Heat Units (SHU). Likewise, Bhat and Bhat (2016) reported that use of 50 ppm silver nanoparticles (Ag-NPs) capsaicin increased levels in *C. annuum* suspension cultures. Bhat and Bhat (2016) indicate that elicitation was applied in callus cultures only one time and that capsaicin production was evaluated in three times, differing after elicitation. They show that capsaicin production is linear in suspension cultures, of 16.043 mg g^{-1} in day 3 until 32.695 mg g^{-1} in day 9, and discontinuous when compared with capsaicin from fruits of *C. frutescens* and *C. annuum* without eliciting, having a maximum of 6.148 mg g^{-1} and 5.962 mg g^{-1}, respectively. Likewise, they showed that control of the suspension culture without elicitation had a capsaicin lower production (0.787 mg g^{-1}). However, it also mentions that biosynthetic production, via suspension culture, is not yet standardized to meet the demand for this metabolite at an industrial level. On the other hand, although González-García et al. (2021) do not show the direct production of capsaicin, they do show an enzymatic modification and positive production of other important secondary metabolites in bell pepper (*C. annuum* L.) when Se (10 and 50 mg L^{-1}), Si (200 and 1000 mg L^{-1}) and Cu (100 and 500 mg L^{-1}) nanoparticles were applied directly on the substrate five times in a month. González-García et al. (2021) show that, compared to control plants, plants treated with nanoparticles significantly increased the enzymatic activity of APX, GPX, CAT, and PAL, mainly involved in ROS detoxification, which was beneficial in fruit quality through the elevated biosynthesis of phytochemicals such as GSH, phenols, carotene, flavonoids, and yellow carotenoids. They also noticed that the application of NP considerably raised the amount of lycopene, GPX activity, and chlorophyll in the leaves.

10.4 Stress Modulation

Different biological, physical, or chemical stress factors called stressors activate signalling pathways that increase the amount of bioactive compounds and other desirable qualities in plant products. As a result, its use in modern horticulture could be a helpful tool for meeting the demands of the agro-food industry and consumers for high-quality and useful vegetable products (Rouphael et al. 2018). High-pungency pepper cultivar growers are advised to maintain an appropriate pre-harvest temperatures and/or application of certain compounds, such as nano-Se or GA3, at the time before harvesting of the paper in order to produce high-quality fruits (Lin et al. 2021). In peppers, both genotype and environmental factors affect the expression of certain phytochemicals. To produce the highest quality product, choosing the right environmental conditions to grow a particular genotype of pepper is crucial. The amount of various phytochemicals present in fruit tissue as well as the size of marketable fruit can change depending on the environment (Butcher et al. 2012). Salinity is regarded as a chemical stressor that has been demonstrated to affect both the physical quality and chemical composition of a variety of vegetable commodities, according to reports in the literature (Rouphael et al. 2018). In order to manipulate the antioxidant and phytochemical contents of vegetable crops and to increase the positive health effects of vegetable products, cultivation techniques like salinity eliciting may be useful (Rouphael et al. 2018).

By choosing hot pepper varieties with high capsaicinoid content and enhancing cultural production techniques, more capsaicinoid can be produced in hot peppers. Environmental stressors like water stress cause *C. annuum* to accumulate more capsaicin (Jeeatid et al. 2018). Due to changes in water status, irrigation frequency every 7 or 9 days increased the high-pungency Habanero cultivar's accumulation of capsaicinoid compared to daily irrigation frequency (Ruiz-Lau et al. 2011). The level of drought stress, cultivar, and species affect the amount of capsaicin that is produced when there is water stress (Phimchan et al. 2012). Water stress that was lower had no negative effects on crop productivity, and higher capsaicinoid content was required for high capsaicinoid production. Water stress that was lower or had no negative effects on crop productivity and higher capsaicinoid content were required for high capsaicinoid production (Jeeatid et al. 2018). Therefore, growers can produce chilli with the pungency necessary to meet the standards of the food and pharmaceutical industries by choosing an appropriate cultivar and adjusting the stress level.

10.4.1 Molecular Impact

The alkaloid capsaicin, which is only present in the hot fruit of the genus *Capsicum*, is what gives chilli fruits their pungent flavour. By condensation of vanillylamine (phenylpropanoid pathway) and the acid 8 methyl-6-nonen (fatty acid pathway), capsaicin synthase activity produces this compound (Kim et al. 2019; Lozada et al. 2021). Chilli fruit pungency is controlled by the interaction of *Csy1*, *Pun1*, and its recessive allele $Pun1^2$ (Lozada et al. 2021). The gene *Csy1* is essential, because it controls capsaicin synthase activity (Kim et al. 2019). Research shows that the capsaicin synthase enzyme is particularly localized in the fruit's placenta (Narasimha et al. 2006). The work performed by Ruiz-Lau et al. (2011) proved that the synthase activity in the placenta of Habanero peppers subjected to water stress was quantified. Their findings demonstrated that fruit age affected the impact of water stress on capsaicin synthase activity. The placenta of 25 days after anthesis fruits had significantly higher enzyme activity, whereas the placenta of 45 days after anthesis fruits had lower enzyme activity. They discovered that the degradation of capsaicinoids involved peroxidases (Ruiz-Lau et al. 2011; Phimchan et al. 2014). Additionally, Sung et al. (2005) discovered that fruits of *C. annuum* L. var. *annuum* exposed to a water deficit have lower peroxidase activity 50 days after anthesis. Thus, when plants experience water stress, the oxidation of capsaicin may be decreased.

Additionally, according to Ruiz-Lau et al. (2011), when plants experience a water deficit, the activity of peroxidases found in the fruits of habanero peppers decreases. Therefore, the accumulation of capsaicin in stressed fruit can only be explained by a decrease in its degradation.

According to reports, fatty acid esters and the amount of capsaicin in pepper fruits exhibit a strong correlation (Wahyuni et al. 2013). According to Lin et al. (2021), heat treatment reduced the amount of capsaicin and a number of metabolites of fatty acids and conjugates. There was a positive correlation between the amount of capsaicin and a number of fatty acids, including cis-9-palmitoleic acid, arachidic acid, and 16-hydroxypalmitic acid (Lin et al. 2021). CaMYB31, CaMYB108, and CaMYB48 were the only transcription factors identified as master regulators that regulate directly the expression of structural genes in the pathway that produces capsaicin (Lin et al. 2021). Three of the transcription factors they discovered were predicted to be able to directly bind the promoters of several genes involved in capsaicin biosynthesis. They discovered ten transcription factors that strongly correlated with capsaicin content both at the transcript and protein levels. Only one of them, CA02g14840, a member of the MYB transcription factor family, demonstrated a strong positive correlation with the amount of capsaicin both at the transcriptional level and in the proteins; as a result, it is most likely a positive regulator of capsaicin synthesis (Sun et al. 2020; Lin et al. 2021).

It's interesting to note that the *Pun1* gene controls the pungency of chilli, whereas *Pun1*^{2}, a recessive allele of *Pun1*, controls the non-pungency (Das et al. 2021; Stewart Jr et al. 2007). Capsaicin builds up in the vesicles of the pungent cultivars as a result of a *Pun1*-induced substance (Lozada et al. 2021). As previously mentioned, *Csy1*, *Pun1*, and its recessive allele, *Pun1*^{2}, work together to control the pungency of chilli fruits. In order to understand how genotype and soil quality interact with the expression of the genes that cause pungency in chilies, Das et al. (2021) conducted a research on the expression of the genes responsible for pungency (*Csy1*, *Pun1*, and *Pun1*^{2}) and the protein capsaicin synthase. Under various soil types, they carried out quantitative real-time PCR on two varieties, *C. frutescens* cv. *Nodaria* and *C. annuum* cv. *Balijuri*. They found that *Pun1* expression was diminished in the chosen chilli cultivars grown in alluvial soil, but the fruits' pungency was noticeably higher than in lateritic soil (Das et al. 2021). This might be caused by the lateritic soil's low organic carbon content. In general, growing chilies in carbon-rich soil significantly improves their genetic performance. Expression of *Pun1*^{2} was greater than *Pun1* in both cultivars. They also observed that soil carbon pool had an important effect on the expression of genes that control the pungency of capsaicin and that *Csy1* was upregulated more in carbon-, phosphorus-, nitrogen-, and potassium-enriched alluvial soil than in carbon-fatigued lateritic soil. Their findings demonstrate that high pungency in chilies grown in alluvial soil is caused by suppression of the recessive allele *Pun1*^{2} (Das et al. 2021).

The expression of numerous genes and transcription factors involved in capsaicin biosynthesis is altered by biotic stress such as drought and salinity (Fig. 10.3). Shams et al. (2023) used Habanero and Maras genotypes to analyse the effect of salinity on the capsaicinoid biosynthesis gene expression. In the fruits of the Maras and

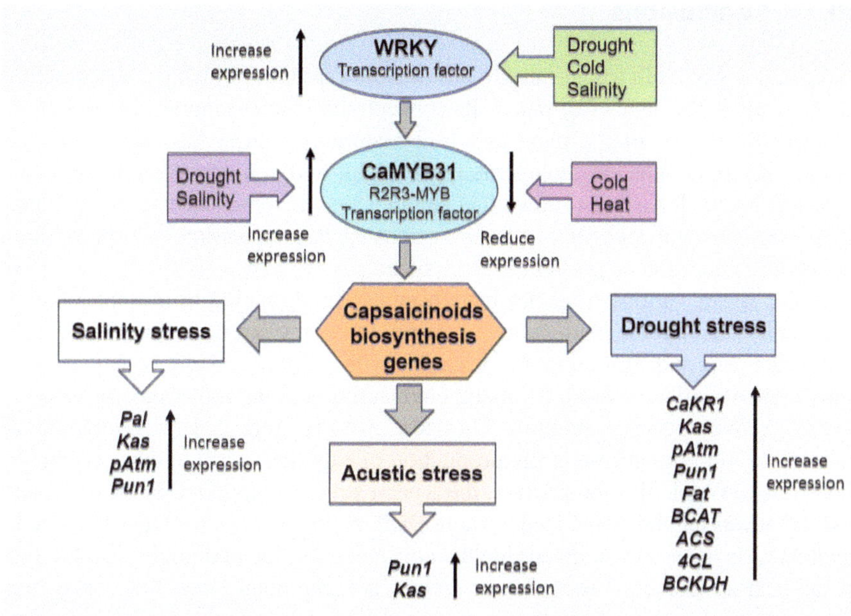

Fig. 10.3 Capsaicinoid biosynthesis gene expression under some abiotic stress (own elaboration). Arrows indicate increased and decreased expression of these genes under the stress condition (Aza-González et al. 2011; Arce-Rodríguez and Ochoa-Alejo 2017; Zhu et al. 2019; Rathnayaka et al. 2021)

Habanero genotypes, they reported that the salinity treatment of 5 dS m-1 increased the capsaicin content by 35.11% and 37.00%, as well as the dihydrocapsaicin content by 30.82% and 72.89%, respectively. Intriguingly, they discovered that both genotypes of the *PAL1*, *pAMT*, and *PUN1* genes were overexpressed when exposed to salinity stress (Shams et al. 2023). In contrast, Rathnayaka et al. (2021) perform an experiment to ascertain the impact of drought stress on the pungency and the levels of expression of genes that produce capsaicin in chilli peppers. They cultivated Japanese chilli pepper cultivars called 'Shishito' and 'Sapporo' on plants under drought stress and carried out a gene expression analysis using a RT-qPCR for 18 capsaicinoid biosynthesis genes and found that at 20 DAF, both cultivars displayed higher levels of gene expression for *pAMT*, *ACL*, *Pun1*, *CaKR1*, *WRKY9*, *CaMYB31*, *FAT*, and *KAS I*, while only one cultivar simultaneously displayed higher levels for *KAS III*, *BCKDH*, *ACS*, *BCAT*, and *4CL* in the drought stress condition. Additionally, neither cultivar's *C3H*, *PAL*, *HCT*, *COMT*, or *C4H* genes displayed any appreciable differences in gene expression during drought stress. These results correlated with the capsaicinoid concentration in plants under drought stress, which was higher in plants subjected to stress than in control plants in both cultivars (Rathnayaka et al. 2021).

10.5 Conclusions

The pungency of chilli fruits is due to the alkaloid capsaicin, which is only present in the hot fruit of the genus *Capsicum*. Its concentration varies among cultivars and is affected by the age of the plant and fruit maturation stage. Also, environmental factors like light intensity, temperature, drought, and soil composition influence capsaicin levels. It has been shown that abiotic stress increases the capsaicin content of the fruit in chilli pepper plants by increasing the expression of capsaicinoids biosynthesis genes, thus promoting their synthesis.

In this sense, various strategies have been explored ranging from improving the soil quality, the organic matter content, using different types of fertilization, elicitation strategies, and searching for varieties with a high content of capsaicin to improvement varieties through genetic engineering and the novel incorporation of nanobiotechnology in agriculture. Therefore, we can infer that stress modulation plays a very important role in the production of capsaicinoids and other secondary metabolites. Within this modulation, the type of stress is important, biotic or abiotic, and if that was natural or induced through elicitors. In this case, response induced by elicitors, it is necessary to determine the concentration, the periodicity and the type of application, foliar or root. Those factors are important since they have been observed to influence the synthesis of bioactive compounds due to the generation of ROS and modification of gene expression and key enzyme activity involved in their biosynthesis pathways. Therefore, a theoretical foundation for the pre-harvest management of collecting superior quality pepper fruits at a higher capsaicin level can be provided by understanding both the molecular mechanism of pepper fruit response to various types of biotic and abiotic stress and the modulable factors.

Acknowledgments We thank Dr. Adan Topiltzin Vargas for the design and elaboration of Fig. 10.3 in this chapter.

Attribution Figure 10.2 was made with free download vectors from Creazilla, Vecteezy, and Flaticon.

References

Airaki M, Leterrer M, Mateos RM, Valderrama R, Chaki M, Barroso JB, Del Río LA, Palma JM, Corpas FJ (2012) Metabolism of reactive oxygen species and reactive nitrogen species in pepper (*Capsicum annuum* L.) plants under low temperature stress. Plant Cell Environ 35:281–295. https://doi.org/10.1111/j.1365-3040.2011.02310.x

Alothman M, Bhat R, Karim AA (2009) Effects of radiation processing on phytochemicals and antioxidants in plant produce. Trends Food Sci Tech 20:201–212. https://doi.org/10.1016/j.tifs.2009.02.003

Anjum S, Anjum I, Hano C, Kousar S (2019) Advances in nanomaterials as novel elicitors of pharmacologically active plant specialized metabolites: current status and future outlooks. RSC Adv 9:40404–40423. https://doi.org/10.1039/C9RA08457F

Arce-Rodríguez ML, Ochoa-Alejo N (2017) An R2R3-MYB transcription factor regulates Capsaicinoid biosynthesis. Plant Physiol 174(3):1359–1370. https://doi.org/10.1104/pp.17.00506

Arrowsmith A, Todd P, Egan TP, Forrest J, Meekins JF, Powers D, Metcalfe M (2011) Effects of salt stress on capsaicin content, growth, and fluorescence in a Jalapeño cultivar of Capsicum annuum (Solanaceae). Bios 83(1):1–7. https://doi.org/10.1893/0005-3155-83.1.1

Ávila-Sosa R, Navarro-Cruz AR, Vera-López O, Hernández-Carranza P, OchoaVelasco CE (2016) Ultraviolet light stimulation of bioactive compounds with antioxidant capacity of fruits and vegetables. In: Siddiqui MW, Bansal V, Prasad K (eds) Plant secondary metabolites, vol 2. Apple Academic Press, Palm Bay, FL. https://doi.org/10.1201/9781315366319

Aza-González C, Núñez-Palenius HG, Ochoa-Alejo N (2011) Molecular biology of capsaicinoid biosynthesis in chili pepper (Capsicum spp.). Plant Cell Rep 30(5):695–706. https://doi.org/10.1007/s00299-010-0968-8

Bhat P, Bhat A (2016) Silver nanoparticles for enhancement of accumulation of capsaicin in suspension culture of Capsicum sp. J Exp Sci 7:1–6. https://doi.org/10.19071/jes.2016.v7.3001

Bley KB, Boorman G, Mohammad B, McKenzie D, Babbar S (2012) Comprehensive review of the carcinogenic and anticarcinogenic potential of capsaicin. Toxicol Pathol 40(6):847–873. https://doi.org/10.1177/0192623312444471

Boxi SS, Mukherjee K, Paria S (2016) Ag doped hollow TiO2 nanoparticles as an effective green fungicide against Fusarium solani and Venturia inaequalis phytopathogens. Nanotechnology 27:085103. https://doi.org/10.1088/0957-4484/27/8/085103

Butcher JD, Crosby KM, Sun Too K, Patil BS (2012) Environmental and genotypic variation of capsaicinoid and flavonoid concentrations in habanero (Capsicum chinense) peppers. HortScience 47:574–579. https://doi.org/10.21273/HORTSCI.47.5.574

Caicedo-Lopez LH, Contreras-Medina LM, Guevara-Gonzalez RG, Perez-Matzumoto AE, Ruiz-Rueda A (2020) Effects of hydric stress on vibrational frequency patterns of Capsicum annuum plants. Plant Signal Behav 15:7. https://doi.org/10.1080/15592324.2020.1770489

Caicedo-López LH, Guevara- Gonzalez RG, Andrade JE, Esquivel-Delgado A, Perez Matzumoto E, Torres-Pacheco I, Contreras-Medina LM (2021) Effect of hydric stress related acoustic emission on transcriptional and biochemical changes associated with a water deficit in Capsicum annuum L. Plant Physiol Biochem 165:251–264. https://doi.org/10.1080/15592324.2020.1770489

Castillejo N, Martínez-Zamora L, Artés-Hernández F (2022) Postharvest UV radiation enhanced biosynthesis of flavonoids and carotenes in bell peppers. Postharvest Biol Technol 184:111774. https://doi.org/10.1016/j.postharvbio.2021.111774

Chapa-Oliver A, Mejía-Teniente L (2016) Capsaicin: from plants to a cancer-suppressing agent. Molecules 21:931 945. https://doi.org/10.3390/molecules21080931

Cisneros-Zevallos L (2003) The use of controlled postharvest abiotic stresses as a tool for enhancing the nutraceutical content and adding-value of fresh fruits and vegetables. J Food Sci 68(5):1560–1565. https://doi.org/10.1111/j.1365-2621.2003.tb12291.x

Cisneros-Zevallos L, Jacobo-Velázquez DA (2020) Controlled abiotic stresses revisited: from homeostasis through Hormesis to extreme stresses and the impact on nutraceuticals and quality during pre- and postharvest applications in horticultural crops. J Agric Food Chem 68:11877–11879. https://doi.org/10.1021/acs.jafc.0c06029

Das S, Sarkar S, Das M, Banik P, Bhattacharya SS (2021) Influence of soil quality factors on capsaicin biosynthesis, pungency, yield, and produce quality of chili: an insight on Csy1, Pun1, and Pun1 [2] signaling responses. Plant Physiol Biochem 166:427–436. https://doi.org/10.1016/j.plaphy.2021.06.012

Ding J, Shi S, Jiang BH, Yang YH, Huang J, Shen HG, Xia K, Zhang J, Jiang X (2004) Effects of methyl jasmonate with indole-3-acetic acid and 6-enzylaminopurine on the secondary metabolism of cultured Onosma paniculatum cells. In Vitro Cell Dev Biol Plant 40(6):581–585. https://doi.org/10.1079/IVP2004578

Dou H, Niu G, Gu M (2019) Pre-harvest UV-B radiation and photosynthetic photon flux density interactively affect plant photosynthesis, growth, and secondary metabolites accumulation in basil (Ocimum basilicum) plants. Agronomy 9:434. https://doi.org/10.3390/agronomy9080434

Estrada AB, Pomar F, Dıáz J, Merino F, Bernal A (1999) Pungency level in fruits of Padrón pepper with different water supply. Sci Hort 81:385–396. https://doi.org/10.1016/S0304-4238(99)00029-1

Fernandez-Jaramillo AA, Duarte-Galvan C, Garcia-Mier L, Jimenez-Garcia SN, Contreras-Medina LM (2018) Effects of acoustic waves on plants: an agricultural, ecological, molecular and biochemical perspective. Sci Hortic 235:340–348. https://doi.org/10.1016/j.scienta.2018.02.060

García-López JI, Niño-Medina G, Olivares-Sáenz E, Lira-Saldivar RH, Barriga-Castro ED, Vázquez-Alvarado R, Rodríguez-Salinas PA, Zavala-García F (2019) Foliar application of zinc oxide nanoparticles and zinc sulfate boosts the content of bioactive compounds in habanero peppers. Plan Theory 8(8):254. https://doi.org/10.3390/plants8080254

Garcia-Mier L, Jimenez-Garcia SN, Guevara-González RG, Feregrino-Perez AA, Contreras-Medina LM, Torres-Pacheco I (2015) Elicitor mixtures significantly increase bioactive compounds, antioxidant activity, and quality parameters in sweet bell pepper. J Chem 2015:1. https://doi.org/10.1155/2015/269296

González-Chavira MM, Herrera-Hernández MG, Guzmán-Maldonado H, Pons-Hernández JL (2018) Controlled water deficit as abiotic stress factor for enhancing the phytochemical content and adding-value of crops. Sci Hortic 234:354–360. https://doi.org/10.1016/j.scienta.2018.02.049

González-García Y, Cárdenas-Álvarez C, Cadenas-Pliego G, Benavides-Mendoza A, Cabrera-de-la-Fuente M, Sandoval-Rangel A, Valdés-Reyna J, Juárez-Maldonado A (2021) Effect of three nanoparticles (Se, Si and Cu) on the bioactive compounds of bell pepper fruits under saline stress. Plan Theory 10:217. https://doi.org/10.3390/plants10020217

Gupta A, Singh PP, Singh P, Singh K, Singh AV, Singh SK, Kumar A (2019) Medicinal plants under climate change: impacts on pharmaceutical properties of plants. In: Choudhary KK, Kumar A, Singh AK (eds) Climate change and agricultural ecosystems. Woodhead Publishing, New Delhi, pp 181–209. https://doi.org/10.1016/B978-0-12-816483-9.00008-6

Gurung T, Techawongstien S, Suriharn B, Techawongstien S (2011) Impact of environments on the accumulation of Capsaicinoids in Capsicum spp. HortScience 46(12):1576–1581. https://doi.org/10.21273/HORTSCI.46.12.1576

Gutierrez-Carbajal MG, Monforte-Gonzalez M, Miranda-Ham ML, Godoy-Hernandez G, Vazquez-Flota F (2010) Induction of capsaicinoid synthesis in Capsicum chinense cell cultures by salicylic acid or methyl jasmonate. Biol Plant 54:430–434. https://doi.org/10.1007/s10535-010-0078-z

Hatefi A, Amsden B (2002) Camptothecin delivery methods. Pharm Res 19:1389–1399. https://doi.org/10.1023/A:1020427227285

Hamed M, Kalita D, Bartolo ME, Jayanty SS (2019) Capsaicinoids, polyphenols and antioxidant activities of Capsicum annuum: comparative study of the effect of ripening stage and cooking methods. Antioxidants 8(9):364. https://doi.org/10.3390/antiox8090364

Hernández-Pérez T, Gómez-García MDR, Valverde ME, Paredes-López O (2020) Capsicum annuum (hot pepper): An ancient Latin-American crop with outstanding bioactive compounds and nutraceutical potential. A review. Compr Rev Food Sci Food 19(6):2972–2993. https://doi.org/10.1111/1541-4337.12634

Jeeatid N, Techawongstien S, Suriharn B, Chanthai S, Bosland PW (2018) Influence of water stresses on capsaicinoid production in hot pepper (Capsicum chinense Jacq.) cultivars with different pungency levels. Food Chem 245:792–797. https://doi.org/10.1016/j.foodchem.2017.11.110

Jeong MJ, Il CJ, Park SH, Kim KH, Lee SK, Kwon TR, Park SC, Siddiqui ZS (2014) Sound frequencies induce drought tolerance in rice plant. Pakistan J Bot 46:2015–2020

Jimenez-Garcia SN, Vázquez-Cruz MA, Miranda-Lopez R, Garcia-Mier L, Guevara-González RG, Feregrino-Perez AA (2018) Effect of elicitors as stimulating substances on sensory quality traits in color sweet bell pepper (Capsicum annuum L. cv. Fascinato and Orangela) grown under greenhouse conditions. Pol J Food Nutr Sci 68:359–365. https://doi.org/10.2478/pjfns-2018-0003

Kaur S, Tiwari V, Kumari A, Chaudhary E, Sharma A, Ali U, Garg M (2023) Protective and defensive role of anthocyanins under plant abiotic and biotic stresses: an emerging application in sustainable agriculture. J Biotechnol 361:12–29. https://doi.org/10.1016/j.jbiotec.2022.11.009

Kehie M, Kumaria S, Tandon P (2016) Biotechnological enhancement of capsaicin biosynthesis in cell suspension cultures of Naga King Chili (Capsicum chinense Jacq.). Bioprocess Biosyst Eng 39:205–210. https://doi.org/10.1007/s00449-015-1504-6

Khan I, Saeed K, Khan I (2019) Nanoparticles: properties, applications and toxicities. Arab J Chem 12:908–931. https://doi.org/10.1016/j.arabjc.2017.05.011

Kim JH, Jo YD, Jin CH (2019) Isolation of soluble epoxide hydrolase inhibitor of capsaicin analogs from Capsicum chinense Jacq. cv. Habanero. Int J Biol Macromol 135:1202–1207. https://doi.org/10.1016/j.ijbiomac.2019.06.028

Kim S, Park M, Yeon SI, Kim YM, Lee JM, Seo E, Choi J, Cheong K, Kim KT, Jung K et al (2014) Genome sequence of the hot pepper provides insights into the evolution of pungency in Capsicum species. Nat Genet 46:270–279. https://doi.org/10.1038/ng.2877

Kollasch AM, Abdul-Kafi AR, Body MJA, Pinto CF, Appel HM, Cocroft RB (2020) Leaf vibrations produced by chewing provide a consistent acoustic target for plant recognition of herbivores. Oecologia 194:1–13. https://doi.org/10.1007/s00442-020-04672-2

Kopta T, Agnieszka SA, Pokluda R, Ferby V, Caruso G (2020) Screening of Chilli pepper genotypes as a source of Capsaicinoids and antioxidants under conditions of simulated drought stress. Plan Theory 9:364. https://doi.org/10.3390/plants9030364

Kopta T, Sekara A, Pokluda R, Ferby V, Caruso G (2020) Screening of chilli pepper genotypes as a source of capsaicinoids and antioxidants under conditions of simulated drought stress. Plants 9 (3):364. https://doi.org/10.3390/plants9030364

Krantev A, Yordanova R, Popova L (2006) Salicylic acid decreases Cd toxicity maize plants. Gen Appl Plant Physiol 1:45–52. https://doi.org/10.1007/s00449-015-1504-6

Lin TH, Lin SW, Wang YW, van Zonneveld M, Barchenger DW (2021) Growing environment and heat treatment effects on intra- and interspecific pollination in Chile pepper (Capsicum spp.). Agronomy 11:1275. https://doi.org/10.3390/agronomy11071275

López-Ribera I, Vicient CM (2017) Drought tolerance induced by sound in Arabidopsis plants. Plant Signal Behavior 12:e138938. https://doi.org/10.1080/15592324.2017.1368938

Lozada DN, Coon DL, Guzmán I, Bosland PW (2021) Heat profiles of 'superhot' and new Mexican type Chile peppers (Capsicum spp.). Sci Hortic 283:110088. https://doi.org/10.1016/j.scienta.2021.110088

Mahmood T, Rana RM, Ahmar S, Saeed S, Gulzar A, Khan MA, Wattoo FM, Wang X, Branca F, Mora-Poblete F et al (2021) Effect of drought stress on capsaicin and antioxidant contents in pepper genotypes at reproductive stage. Plan Theory 10:1286. https://doi.org/10.3390/plants10071286

Materska M (2014) Bioactive phenolics of fresh and freeze-dried sweet and semi-spicy pepper fruits (Capsicum annuum L.). J Funct Foods:7269–7277. https://doi.org/10.1016/j.jff.2014.02.002

Marslin G, Sheeba CJ, Franklin G (2017) Nanoparticles Alter secondary metabolism in plants via ROS burst. Front Plant Sci 8:832. https://doi.org/10.3389/fpls.2017.00832

Materska M, Perucka I (2005) Antioxidant activity of the main phenolic compounds isolated from hot pepper fruit (Capsicum annuum L.). J Agric Food Chem 53(5):1750–1756. https://doi.org/10.1021/jf035331k

Medina-Lara F, Echevarría-Machado I, Pacheco-Arjona R, Ruiz-Lau N, Guzmán-Antonio A, Martinez-Estevez M (2008) Influence of nitrogen and potassium fertilization on fruiting and capsaicin content in habanero pepper (Capsicum chinense Jacq.). HortSci 43:1549–1554. https://doi.org/10.21273/HORTSCI.43.5.1549

Mejía-Teniente L, Durán-Flores FDD, Chapa-Oliver AM, Torres-Pacheco I, Cruz-Hernández A, González-Chavira MM, Ocampo-Velázquez RV, Guevara-González RG (2013) Oxidative and molecular responses in Capsicum annuum L. after hydrogen peroxide, salicylic acid and chitosan foliar applications. Int J Mol Sci 14:10178–10196. https://doi.org/10.3390/ijms140510178

Mejía-Teniente L, Durán-Flores BA, Torres-Pacheco I, González-Chavira MM, Rivera-Bustamante RF, Feregrino-Perez AA, Pérez-Ramírez I, Rocha-Guzmán NE, Reynoso-Camacho R, Guevara-González RG (2019) Hydrogen peroxide protects pepper (Capsicum annuum L.) against pepper golden mosaic geminivirus (PepGMV) infections. Physiol Mol Plant Pathol 106:23–29. https://doi.org/10.1016/j.pmpp.2018.11.008

Mejía-Teniente L, Torres-Pacheco I, González-Chavira MM, Ocampo-Velazquez RV, Herrera-Ruiz G, Chapa-Oliver AM, Guevara-González RG (2010) Use of elicitors as an approach for sustainable agriculture. Afr J Biotechnol 9:9155–9162

Mishra RC, Bae H (2019) Plant cognition: ability to perceive 'touch' and 'sound. In: Sopory S (ed) Sensory biology of plants. Springer, Singapore, pp 137–162. https://doi.org/10.1007/978-981-13-8922-1_6

Monforte-González M, Medina-Lara F, Gutiérrez-Carbajal G, Vázquez-Flota F (2007) Capsaicinoid quantitation by in situ densitometry of thin layer chromatography plates. J Liq Chromatogr Relat Technol 30:1697–1704. https://doi.org/10.1080/10826070701225041

Narasimha PBC, Kumar V, Gururaj HB, Parimalan R, Giridhar P, Ravishankar GA (2006) Characterization of capsaicin synthase and identification of its gene (csy1) for pungency factor capsaicin in pepper (Capsicum sp.). Proc Natl Acad Sci U S A 103:13315–13320. https://doi.org/10.1073/pnas.0605805103

Narayani M, Srivastava S (2017) Elicitation: a stimulation of stress in in vitro plant cell/tissue cultures for enhancement of secondary metabolite production. Phytochem Rev 16:1227–1252. https://doi.org/10.1007/s11101-017-9534-0

Navarro JM, Flores P, Garrido C, Martinez V (2006) Changes in the contents of antioxidant compounds in pepper fruits at ripening stages, as affected by salinity. Food Chem 96:66–73. https://doi.org/10.1016/j.foodchem.2005.01.057

Neugart S, Schreiner M (2018) UVB and UVA as eustressors in horticultural and agricultural crops. Sci Hortic 234:370–381. https://doi.org/10.1016/j.scienta.2018.02.021

Otha Y (1960) Physiological and genetical studies on the pungency of Capsicum IV. Pungency under various growing conditions. Rep Kihara Inst Biol Res 11:73–74. https://doi.org/10.1270/jsbbs1951.12.179

Pérez-Ambrocio A, Guerrero-Beltránb JA, Aparicio-Fernándezc X, Ávila-Sosad R, Hernández-Carranzad P, Cid-Pérezd S, Ochoa-Velascod CE (2018) Effect of blue and ultraviolet-C light irradiation on bioactive compounds and antioxidant capacity of habanero pepper (Capsicum chinense) during refrigeration storage. Postharvest Biol Technol 135:19–26. https://doi.org/10.1016/j.postharvbio.2017.08.023

Phimchan P, Chanthai S, Bosland PW, Techawongstien S (2014) Enzymatic changes in phenylalanine ammonia-Lyase, Cinnamic-4- hydroxylase, capsaicin synthase, and peroxidase activities in Capsicum under drought stress. J Agric Food Chem 62(29):7057–7062. https://doi.org/10.1021/jf4051717

Phimchan P, Techawongstien S, Chanthai S, Bosland PW (2012) Impact of drought stress on the accumulation of capsaicinoids in Capsicum cultivars with different initial capsaicinoid levels. HortScience 47:1204–1209. https://doi.org/10.21273/HORTSCI.47.9.1204

Prasad NBC, Shrivastava R, Ravishankar GA (2005) Capsaicin as multifaceted drug from Capsicum spp. Evid Based Integrative Med 2:147–166. https://doi.org/10.2165/01197065-200502030-00006

Purnama PC, Sumardi I, Nugroho LH (2022) Fruit development and capsaicin content of hot pepper (Capsicum annuum) plant cultivated in different soil salinity stress. Nus Biosci 14(2):166–171. https://doi.org/10.13057/nusbiosci/n140206

Ramakrishna A, Ravishankar GA (2011) Influence of abiotic stress signals on secondary metabolites in plants. Plant Signal Behav 6:1720–1731. https://doi.org/10.4161/psb.6.11.17613
Ramirez-Estrada K, Osuna L, Moyano E, Bonfill M, Tapia N, Cusido RM, Palazon J (2015) Changes in gene transcription and taxane production in elicited cell cultures of Taxus media and Taxus globosa. Phytochemistry 117:174–184. https://doi.org/10.1016/j.phytochem.2015.06.013
Rastogi A, Zivcak M, Sytar O, Kalaji HM, He X, Mbarki S, Brestic M (2017) Impact of metal and metal oxide nanoparticles on plant: a critical review. Front Chem 5:78. https://doi.org/10.3389/fchem.2017.00078
Rathnayaka RMSMB, Kondo F, Prabandaka SS, Nemoto K, Matsushima K (2021) Drought stress induced an increase in the pungency and expression of Capsaicinoid biosynthesis genes in chili pepper (Capsicum annuum L.). Hortic J 90(4):410–419. https://doi.org/10.2503/hortj.UTD-282
Rico CM, Majumdar S, Duarte-Gardea M, Peralta-Videa JR, Gardea-Torresdey JL (2011) Interaction of nanoparticles with edible plants and their possible implications in the food chain. J Agric Food Chem 59:3485–3498. https://doi.org/10.1021/jf104517j
Rivero RM, Mittler R, Blumwald E, Zandalinas SI (2021) Developing climate-resilient crops: improving plant tolerance to stress combination. Plant J 109:373–389. https://doi.org/10.1111/tpj.15483
Rodoni LM, Zaro MJ, Hasperu'e JH, Concellón A, Vicente AR (2015) UV-C treatments extend the shelf life of fresh-cut peppers by delaying pectin solubilization and inducing local accumulation of phenolics. LWT 63:408–414. https://doi.org/10.1016/j.lwt.2015.03.042
Rouphael Y, Petropoulos SA, Cardarelli M, Colla G (2018) Salinity as eustressor for enhancing quality of vegetables. Sci Hortic 234:361–369. https://doi.org/10.1016/j.scienta.2018.02.048
Ruiz-Lau N, Medina-Lara F, Minero-García Y et al (2011) Water deficit affects the accumulation of Capsaicinoids in fruits of Capsicum chine nse Jacq. HortScience 46(3):487–492. https://doi.org/10.21273/HORTSCI.46.3.487
Schmülling T (2004) Cytokinin. In: Lennarz W, Lane MD (eds) Encyclopedia of biological chemistry. Academic, Cambridge, MA
Scotta G, Rupara M, Fletchera AGD, Dickinsona M, Shamab G (2017) A comparison of low intensity UV-C and high intensity pulsed polychromatic sources as elicitors of hormesis in tomato fruit. Postharvest Biol Technol 125:52–58. https://doi.org/10.1016/j.postharvbio.2016.10.012
Seleiman MF, Al-Suhaibani N, Ali N, Akmal M, Alotaibi M, Refay Y, Dindaroglu T, Abdul-Wajid HH, Battaglia ML (2021) Drought stress impacts on plants and different approaches to alleviate its adverse effects. Plants 10(2):259. https://doi.org/10.3390/plants10020259
Shams M, Yildirim E, Ekinci M, Agar G, Turan M, Kul R (2018) Exogenous cytokinin application increased the capsaicin and ascorbic acid content in pepper fruit. Sci Papers 62:507–511
Shams M, Yuksel EA, Agar G, Ekinci M, Kul R, Turan M, Yildirim E (2023) Biosynthesis of capsaicinoids in pungent peppers under salinity stress. Physiol Plant 175(2):e13889. https://doi.org/10.1111/ppl.13889
Stan T, Munteanu N, Teliban G-C, Cojocaru A, Stoleru V (2021) Fertilization management improves the yield and Capsaicinoid content of chili peppers. Agriculture 11:181. https://doi.org/10.3390/agriculture11020181
Stewart C Jr, Mazourek M, Stellari GM, O'Connell M, Jahn M (2007) Genetic control of pungency in C. chinense via the Pun1 locus. J Exp Bot 58:79–991. https://doi.org/10.1093/jxb/erl243
Sun BM, Zhou X, Chen CM, Chen CJ, Chen KH, Chen MX, Zhu ZS (2020) Coexpression network analysis reveals an MYB transcriptional activator involved in capsaicinoid biosynthesis in hot peppers. Hortic Res 7(1):162. https://doi.org/10.1038/s41438-020-00381-2
Sung Y, Chang YY, Ting NL (2005) Capsaicin biosynthesis in water-stressed hot pepper fruits. Bot Bull Acad Sinica 46:35–42. https://doi.org/10.7016/BBAS.200501.0035
Tripodi P, Ficcadenti N, Rotino GL, Festa G, Bertone A, Pepe A, Lo Scalzo R (2019) Genotypic and environmental effects on the agronomic, health-related compounds and antioxidant

properties of chilli peppers for diverse market destinations. J Sci Food Agric 99:4550–4560. https://doi.org/10.1002/jsfa.9692
Vázquez-Hernández MC, Parola-Contreras I, Montoya-Gómez LM, Torres-Pacheco I, Schwarz D, Guevara-González RG (2019) Eustressors: chemical and physical stress factors used to enhance vegetables production. Sci Hortic 250:223–229. https://doi.org/10.1016/j.scienta.2019.02.053
Vineesh V, Jelly CL, Fijesh PV, Jaimsha VK, Padikkala J (2007) Effect of N 6-benzyl amino purine and naphthalene acetic acid on camptothecin production through in vitro propagation of Ophiorrhiza rugosa Wall. Var. decumbens (Gardn. Ex Thw.) Deb & Mondal. Indian J Nat Prod Radiance 6:405–409
Wahyuni Y, Ballester AR, Tikunov Y, de Vos RCH, Pelgrom KTB, Maharijaya A, Bovy AG (2013) Metabolomics and molecular marker analysis to explore pepper (Capsicum sp.) biodiversity. Metabolomics 9:130–144. https://doi.org/10.1007/s11306-012-0432-6
Wang F, Yin Y, Yu C, Li N, Shen S, Liu Y, Gao S, Jiao C, Yao M (2021) Transcriptomics analysis of heat stress-induced genes in pepper (Capsicum annuum L.) seedlings. Horticulturae 7:339. https://doi.org/10.3390/horticulturae7100339
Yadav B, Jogawat A, Rahman MS, Narayan OP (2021) Secondary metabolites in the drought stress tolerance of crop plants: a review. Gene Rep 23:101040. https://doi.org/10.1016/j.genrep.2021.101040
Yang L, Wen KS, Ruan X, Zhao YX, Wei F, Wang Q (2018) Response of plant secondary metabolites to environmental factors. Molecules 23:762. https://doi.org/10.3390/molecules23040762
Zahra N, Hattab A, Saadon A, Ajeel, Ekhlas A, Kaaby E (2015) Effect of salinity stress on Capsicum annuum callus growth, regeneration and callus content of capsaicin, phenylalanine, proline and ascorbic acid. J Life Sci 9:304–310. https://doi.org/10.17265/1934-7391/2015.07.002
Zhang X, Bian Z, Yuan X, Chen X, Lu C (2020a) A review on the effects of lightemitting diode (LED) light on the nutrients of sprouts and microgreens. Trends Food Sci Technol 99:203–216. https://doi.org/10.1016/j.tifs.2020.02.031
Zhang RX, Zhu WC, Cheng GX, Yu YN, Li QH, Ul Haq S, Said F, Gong ZH (2020b) A novel gene, CaATHB-12, negatively regulates fruit carotenoid content under cold stress in Capsicum annuum. Food Nutr Res 64:3729. https://doi.org/10.29219/fnr.v64.3729
Zhu Z, Sun B, Cai W, Zhou X, Mao Y, Chen C et al (2019) Natural variations in the MYB transcription factor MYB31 determine the evolution of extremely pungent peppers. New Phytol 223(2):922–938. https://doi.org/10.1111/nph.15853

Breeding for Resistance to Biotic and Abiotic Stresses and Quality Traits in Chilli

11

Madhavi Reddy K, Naresh Ponnam, Satyaprakash Barik, Mahebub Shaik, Karishma Pasupula, Pavani Narigapalli, and Lakshmana Reddy D C

Abstract

The *Capsicum* genus originated in Central America and includes approximately 35 species, which includes five domesticated species (*C. annuum*, *C. pubescens*, *C. frutescens*, *C. chinense*, and *C. baccatum*). Fruits owing to their bioactive compounds have various applications in spice, food, phytogenic feed, cosmeceutical, and pharmaceutical industries. Due to the rise in demand and climate change, the chili crop is affected due to the emergence of various pests and diseases, and there is a need to breed suitable varieties/hybrids with required fruit quality traits along with resistance to biotic and abiotic stresses. There exists vast germplasm collected and conserved globally. Genetic resources of *Capsicum* for various traits have been identified and have been explored in developing commercial cultivars suitable for different segments of the market. Further wild species that were not explored must be deployed for the identification of novel alleles and their use in crop improvement programs. Conventional breeding through stringent phenotypic selection has successfully led to the development of resistant varieties/hybrids. However, with the advancement of genomic technologies, fast and precise marker discovery is enabling marker-assisted selection.

Madhavi Reddy K (✉) · N. Ponnam · M. Shaik · K. Pasupula · P. Narigapalli
Division of Vegetable Crops, ICAR-Indian Institute of Horticultural Research, Bengaluru, India

S. Barik
Department of Agriculture and Allied Sciences, CV Raman Global University, Bhubaneswar, India

L. R. D C
Division of Basic Sciences, ICAR-Indian Institute of Horticultural Research, Bengaluru, India

M. K. Swamy, A. Kumar (eds.), *Capsaicinoids*,
https://doi.org/10.1007/978-981-99-7779-6_11

Keywords

Capsicum · Abiotic stress · Biotic stress · Disease resistance · Chili breeding · Crop improvement

11.1 Introduction

The term "chilli" is frequently and interchangeably used with "peppers (hot/ sweet/ bell)', "chile", "aji", and "paprika" to refer to various species of the "chilli pepper plants", which all belong to *Capsicum* genus. Chilli is believed to be originated in 7000 BC in Mexico and Central America. There are about 35 species reported in *Capsicum* (Carrizo García et al. 2016), out of which 5 are domesticated (*C. annuum*, *C. frutescens*, *C. chinense*, *C. pubescens*, and *C. baccatum*). The wild progenitor of chilli is *C. annuum* var. *glabriusculum* (Aguilar-Meléndez et al. 2009). India dominates the global dry chilli market with an annual production of 1.2 million tonnes (FAOSTAT 2020). Chilli is mainly grown for its fruits, which are the source of capsaicinoids (pungency) and carotenoids (colour). The fruits have high industrial applications in food, spice, phytogenic feed, pharmaceutical and cosmeceutical applications. Capsaicin and other capsaicinoids (dihydrocapsaicin, nor-dihydrocapsaicin) have vast pharmaceutical applications to treat neurological complications and musculoskeletal oxidative and inflammatory diseases (Review: Barik et al. 2022).

The biochemical composition of fruit colour and pungency are carotenoids and capsaicinoids, respectively. Pepper's carotenoids are employed as natural colouring agents, while capsaicinoids, which are only found in *Capsicum* species, have numerous uses in food, spice, pharmaceutical, and medical industries as well as in warfare. Major chilli alkaloids, i.e. capsaicinoids, capsinoids, and carotenoids, have several uses in pharmaceutical industries (Baenas et al. 2019). Capsaicin as Qutenza® is a popular remedy against neuropathic pain (Aitken et al. 2017). A new therapeutic approach of capsaicin to combat obesity *via* increased fat oxidation has been reported (Zsiboras 2018). There has been an increase in research into bioactive substances and plant extracts as natural components for usage in cosmetics and pharmaceuticals as a result of consumers and companies adopting an eco-friendly lifestyle. Natural ingredients could be used in place of artificial preservatives to ensure the safety of pharmaceutical and cosmetic products while avoiding contact allergies. Paprika oleoresin has vast utility as colorant in cosmetical products such as bath oils, shampoo, and other beauty products such as eye makeup and lipsticks (Baenas et al. 2019). Chilli have also been used in phytogenic feed industries. Capsaicin exhibits a positive effect on feed intake, average egg weight, total egg yield, and the albumin index of broilers as well as Japanese quails in the poultry industry (Tripathi et al. 2013). This chapter covers the comprehensive data on application of different breeding strategies for improving the quality traits of chilli species, including resistance against biotic and abiotic stresses.

11.2 Breeding for Resistance Against Viral Diseases

11.2.1 Chilli Leaf Curl Disease

In the present climate scenario, whitefly-transmitted begomoviruses is one of the major devastating diseases (Yadav et al. 2022). Distinct begomoviruses are associated with ChLCD in different locations. In India, major begomoviruses causing ChLCD are *Chilli leaf curl virus*, *Tomato leaf curl New Delhi virus*, *Chilli leaf curl Ahmedabad virus*, and *Tomato leaf curl Joydebpur virus*. Being single-stranded DNA viruses, there exists a high possibility of genetic recombination. Chlorosis, curling and a reduction in the size of the leaf and fruit are indications of the highly devastating plant disease known as ChLCD, which ultimately renders the fruit unmarketable. The fact that farmers in India have stopped growing chili due to a nearly 100% output loss illustrates the disease's potential for damage (Kumar et al. 2015). Identification of resistant sources will help in breeding resistant varieties/ hybrids. Through artificial challenge inoculation, resistant sources such as S-343 for ChLCD with predominant *Tomato leaf* curl *Joydebpur virus* (ToLCJoV) with monogenic dominant nature of gene action were identified (Thakur et al. 2019). Further, two SSR markers on chromosome 6 were reported associated with resistance for marker-assisted selection (Thakur et al. 2020). Recently, three resistant sources, IHR4615, IHR4630, and IHR4517, of resistance to Chilli leaf curl virus-Raichur isolate were identified, and genetic studies showed monogenic dominant nature of resistance (Yadav et al. 2022). Some sources with recessive gene control resistance also have been reported (Rai et al. 2014). DLSel.10 was reported as a good source of resistance with monogenic recessive control of resistance (Maurya et al. 2019).

11.2.2 Aphid-Transmitted Viruses

More than 45 viruses infect peppers (Green and Kim 1991); out of these, aphid-transmitted cucumber mosaic virus, pepper veinal mottle virus, and chilli veinal mottle virus (ChVMV) are causing severe yield losses. Asian Vegetable Network (AVNET) initially screened a large number of germplasm, and resistant sources for both these viruses were identified. Results were inconsistent across the regions (Duriat et al. 1997). This inconsistency might be due to the strain diversity of viruses (Tsai et al. 2008). Several researchers identified different sources for resistance for local isolates of CMV and ChVMV (Lee et al. 2013). Genetic inheritance studies revealed the polygenic nature of CMV resistance in the Indian line Perennial (Chaim et al. 2001), monogenic recessive nature (Singh and Thakur 1977), and partially dominant nature of resistance (Lapidot et al. 1997). These varied results are due to different accessions and isolates used in their studies. Major and several minor QTLs were reported in cv. Vania (Caranta et al. 2002), and in BJ0747–1–3-1, two major QTLs were reported (Yao et al. 2013). In cv. Bukang, monogenic dominant gene action was reported (Kang et al. 2010) against Korean isolates. For potyvirus resistance, several recessive *pvr* loci and some dominant loci have been identified (Lee et al. 2013; Venkatesh et al. 2018). Recently, Venkatesh et al. (2018) reported that dominant *Pvr4* and *Pvr7* are the same genes and are mapped on chr 10. For ChVMV resistance, two major dominant QTLs *Cvr2-1* and *Cvr2-2* resistant loci on chr 6 and 10 were identified, explaining 51.52% of phenotypic variation were identified (Lee et al. 2017). Recently, Naresh et al. (2022) mapped ChVMV resistance loci in *C. annuum* cv. IHR 2451 on chromosome 9 through genotyping by sequencing approach.

11.2.3 Tospoviruses

These viruses are transmitted by *Thrips* spp. (Vijayalakshmi et al. 1995). Groundnut bud necrosis virus (Reddy et al. 2008) is considered one of the most economically destructive plant viruses. Pavithra et al. (2020) reported eight accessions to be highly tolerant to GBNV, but these lines should be further validated. The current GBNV management strategy through pesticides has limited success and is challenging, and there is a clear and urgent need to breed resistant cultivars.

11.3 Breeding for Resistance Against Bacterial Pathogens

11.3.1 Bacterial Wilt

Plants die suddenly and irreversibly from bacterial wilt caused by *Ralstonia solanacearum*. IIHR-B-HP 130 is reported to be resistant to bacterial wilt (Nischay et al. 2021), and its genetic studies showed resistance is controlled by digenic and complementary gene action with other minor genes (Naveena et al. 2020). A major QTL (*Bw1*) was reported in LS2451 (Mimura et al. 2009). Sequence analysis of CAMS451 marker associated with QTL *Bw1* showed that it corresponds to ORF1 58 of RRP41 genes (Mathew 2020). Further, a major QTL, *qRRs-10.1*, accounted for 19.01% phenotypic variation in BVRC 25 (susceptible) × BVRC-1 (Bacterial wilt resistant) F_2 population, and a cluster of five predicted resistant genes and three defence-related genes, which are closely located near the markers ID10-196208712 or ID10–194305124, are reported to be candidate genes for bacterial wilt resistance in chilli (Du et al. 2019). Through GBS approach in resistant line KC352 x 14F6002–14 (Susceptible) F_2 population, on chromosome 1, a major QTL(pBWR-1) explaining 20.13 to 25.15% of phenotypic variation was reported (Chae et al. 2022).

11.3.2 Bacterial Leaf Spot

It is caused by *Xanthomonas* species (*X. euvesicatoria* (*Xe*), *X. vesicatoria*, and *X. gardneri* (*Xg*)) (Osdaghi et al. 2021). This drastically affects the crop by causing severe defoliation. Major dominant non-allelic genes have been reported, i.e. *Bs1* (PI 163192, *C. annuum*), *Bs3* (PI 271322, *C. annuum*), *Bs2* (PI 260435, *C. chacoense*), *Bs4* (PI 235047, *C. pubescens*), and *Bs7* (UNEF1556, *C. pubescens*) (Cook and Guevara 1984; Wai et al. 2015). *bs5* and *bs6* genes, which are monogenic recessive in nature, have been reported in PI 271322 and PI 163192, respectively, for *X. euvesicatoria* resistance (Jones et al. 2004). *bs5* and *bs6* have been mapped on chromosomes 3 and 6, respectively (Sharma et al. 2022). Further, PI 163192 (*C. annuum*) has been identified as the resistant source for *X. gardneri*, and the major recessive resistant locus designated as bs8 conferring resistance was mapped on chromosome 11 (Anuj Sharma et al. 2022).

11.4 Breeding for Resistance Against Fungal Diseases

11.4.1 Phytophthora Root Rot

The damaging pathogen *Phytophthora capsici*, a fungal oomycete found in soil, causes root rot. Due to its constant resistance to a variety of diverse and aggressive strains of *Phytophthora capsici*, accession CM334 was the preferred option among breeders (Sy et al. 2008). A significant QTL *Pc 5.1* in INRA pepper maps (H3 X Vania, Perennial X Yolo Wonder and Yolo wonder X CM334) was mapped on chromosome 5 (Mallard et al. 2013). Additionally, they identified three major meta-QTLs on chromosome 5, namely, *Meta Pc5.2*, *Meta Pc5.3*, and *Meta Pc5.1*. The *Pc 5.1* QTL described by Mallard et al. (2013), the Phyt-1 QTL discovered by Sugita et al. (2006), three QTLs discovered by Truong et al. (2012), and the QTL reported by Minamiyama et al. (2006) are some of the important QTLs demonstrating broad-spectrum resistance that makes up Meta *Pc5.1*. *Pc5.2* reported by Mallard et al. (2013), Phyto-U reported by Ogundiwin et al. (2005), and QTL between CDI78 and CDI25 were all included in *Meta Pc5.2*. The *Meta Pc 5.3* included *Pc5.3* (Mallard et al. 2013), three QTLs detected by Truong et al. (2012). Recently, SSR markers HpmsE034 and HpmsE062 associated with resistance against the Indian isolate were identified (Kumar et al. 2022a, b). QTL mapping using 1973 polymorphic SNP markers identified major genomic regions on chromosomes 5, 8, and 9, explaining phenotypic variation of 19.7 to 30.4% (Lozada et al. 2021). These QTLs will help in marker-assisted breeding.

11.4.2 Anthracnose Fruit Rot

It is caused by *Colletotrichum* spp. (*C. gloeosporioides*and *C. truncatum*). It causes both pre- and post-harvest fruit rot and aggravates if the fruit ripening stage coincides with rains. The cultivated *C. annuum* types are most susceptible to this disease. Among *Capsicum* spp. resistant sources have been identified mainly in PBC932 (*C. chinense*) and PI594137, PBC81, and PBC80 (*C. baccatum*) (Gniffke et al. 2013). *C. baccatum* is cross incompatible with *C. annuum*, which is the main hindrance to practical breeding. Genetic inheritance studies showed that resistance is reported to be monogenic recessive nature (Mahasuk et al. 2009), duplicate gene ation (Lin et al. 2007) and polygenic QTLs (Lee et al. 2010). Yoon and Park (2005) reported triple cross bridge cross method (*C. annuum* x *C. chinense*) × *C. baccatum* PBC80 for introgression into *C.annuum* by bridge cross method breeding. However, there are not much reports of the same method in the development of resistant varieties/hybrids (no information is available in public domains). Embryo rescue was used by Suwor et al. (2017), coupled with use of *SSR-HpmsE032* and *SCAR-Indel* marker-associated resistance from PBC 80 to *C. annuum* background for mainly Taiwan isolates of *Colletotrichum* spp. QTLs have been identified using two populations including Bangchang x PBC932 and PBC80 x CA1316; identified flanking SNP-based markers were reported (Mahasuk et al. 2016). The molecular mechanism study revealed that the temporal expression of 17 defence-related genes is correlated with the multiple defence signalling pathways along with the increased level of jasmonic acid and ethylene-responsive gene accumulation in the resistant cultivar (Bhut jolokia) compared to susceptible cultivar (Teja jhal) for chilli anthracnose fruit rot caused by *C. truncatum* (Mishra et al. 2017). The transcriptome analysis of the resistant genotype IIVRC-452 and susceptible cv. Pusa Jwala revealed five key candidate genes, i.e. *CaLAG-1*, *CaLOX*, *CaCYP76A2*, *CaPG*, and *CaSAP-13*, associated with resistance (Kumar et al. 2022a, b).

11.4.3 Powdery Mildew

Powdery mildew is caused by an obligate ascomycete's plant fungus, *Leveillula taurica*. This fungus causes premature defoliation and drastically reduces yields. *C. chinense*, *C. frutescens,* and *C. baccatum* species are most often resistant, while the majority of *C. annuum* are susceptible to powdery mildew (De Souza and Cafe Filho 2003). 'HV-12', 'H3', and '4638', 'IHR 703', KC 604,605,608 (*C. baccatum*); KC 616 (*C. chinense*); and KC 638,640,641,642,643,644 (*C. pubescens)* are reported resistant sources to *L. taurica* (De Souza and Cafe Filho 2003). Inheritance studies found that powdery mildew resistance is a major dominant to polygenic nature (Blat et al. 2005). The *C. frutescens* line 'IHR 703' is thought to have three pairs of incompletely dominant genes for resistance (Anand et al. 1987). Recently *PMR1*, major dominant gene on chromosome 4, harbouring two potential NBS-LRR-type disease resistance genes, were identified in VK515R-resistant line (Jo et al. 2017).

11.5 Breeding for Resistance Against Root Knot Nematodes

Numerous studies on genetic resistance have demonstrated that resistance predominates in nature. There are currently known nine separate resistance genes (*Mechl*, *Mech2*, *N*, *Me1*, *Me2*, *Me3*, *Me4*, *Me5*, and *Me7*) that control resistance (Wang and Bosland 2006). A variety of *Meloidogyne* species, including

M. arenaria, *M. javanica*, and *M. incognita*, are susceptible to *Me1*, *Me3*, *and Me7* genes (Djian-Caporalino et al. 2007). The *Mech1*, *Mech2*, *Me1*, *Me3*, *Me4*, and *Me7* genes are concentrated on chromosome P9, according to comparative mapping, and markers (SCAR, SSCP, and CAPs) associated with the Me genes were created for marker-assisted selection (Djian-Caporalino et al. 2007). The N gene was linked to a SCAR marker by Wang et al. (2009), and subsequent research revealed that this N gene was also allelic to the Me7 gene and co-localized in the *Me* gene cluster on chromosome 9 reported by Djian-Caporalino et al. (2007), suggesting the potential of N gene-associated SCAR marker for marker-assisted selection.

11.6 Breeding for Resistance Against Insect Pests

11.6.1 *Thrips* spp.

Thrips are polyphagous; damage directly by feeding on leaves, fruits, or flowers; and cause indirect damage by transferring viruses. Leaves and fruits develop light-brown to silvery spots and deformation upon feeding by thrips. There are several species of thrips that are known to infect chilli. Among them are *Scirtothrips dorsalis*, *F. occidentalis*, and recently *T. parvispinus*. Genetic inheritance of *Thrips parvispinus* resistance is controlled by a dominant gene (Maharijaya et al. 2011). For *Scirtothrips dorsalis*, chilli germplasm BC-7-2-1 and BC 25 were identified as resistant (Praveen et al. 2021). Two thrips species, *F. occidentalis* and *T. tabaci*, were examined in resistant and susceptible *Capsicum annuum* backgrounds. Plants that carry the QTL region containing CGN16975 allele in the homozygous state on chromosome 6 was identified as a valid source for thrips resistance against both species (van Haperen et al. 2020). For *Thrips palmi*, in a resistant line *C. annuum* AC1979, four QTLs have been identified in an F_2 population of *C. annuum* AC 1979 (highly resistant) x *C. annuum* Berceo (highly susceptible) M238, M106 on chromosome 3 and M171, M1445 on chromosome 12 were colocalized using SNP markers (Saeko and Pongjaroenkit 2020). Maharijaya et al. (2011) reported resistance against *T. parvispinus* and *Frankliniella occidentalis* in *Capsicum annuum* lines AC 1979, Bisbas, Keystone Resistant Giant, CM 331 and *Capsicum baccatum* lines No. 1553, Aji Blanco, Christal.

11.6.2 Mites

In addition to the downward curling and crinkling of leaves that give them an inverted boat-shaped look, *Polyphagotarsonemus latus* also causes slowed development, scarring of fruits, and reduced fruit size. Young foliage and growing tips are sucked for their sap by both nymphs and adults. The peak activity of pests is observed in dry humid weather. So far, no reports have been found regarding the genetic inheritance of mites' resistance in chilli. The genotypes DCC-109, DCC-185, DCC- 3, and DCC-89 were identified as moderately resistant (Latha and Hunumanthraya 2018).

11.6.3 Whitefly

Whiteflies (*Bemesia tabaci*) feeds by sucking on the leaves, resulting in nutrient loss for the plant and physiological issues. The insect's honeydew excretions on the leaf lamina promote the establishment of black sooty mould, which hinders the ability of plants to use sunlight for photosynthesis and, in turn, reduces production. The genetic inheritance of whitefly resistance is not yet studied. IIHR 4338 and IIHR 4300 have been identified as resistant sources (Yadav et al. 2022). Some reported resistant lines against whitefly are *C. annuum* accessions Bisbas, CM331, and California Wonder 300 (Firdaus et al. 2011) and Amaxito, Tabaquero, and Simojovel (Latournerie-Moreno et al. 2015). CA 9, CA 28, and ACC 05 showed antixenotic defence mechanism against *B. tabaci* (Jeevanandham et al. 2018), while GCh-3 and GCh-1 (Rajput et al. 2017) and IAC-1579, IAC-1545, and IAC-1544 showed antibiosis mechanism for whitefly (Pantoja et al. 2018).

11.7 Breeding for Abiotic Stresses Tolerance

11.7.1 Breeding for Heat Tolerance

The current climate change scenario demands the exploitation of tolerance to high temperatures in cultivated crops for sustainable crop production. High temperature is a major abiotic stress-limiting factor affecting crop yields in chilli (Lin et al. 2022). A temperature range of 18–30 °C is ideal for the optimum growth and development of chilli, and drastic reduction in vigour and yield of the crop occurs when the mean temperature exceeds 30^{0}c (Berke et al. 2005). Manifestation of heat stress during the critical stages especially during the reproductive stage of chilli leads to a severe negative impact on the crop (Aloni et al. 1992). Reduction in fruit set occurs due to the high sensitivity of pollen to elevated temperature (Hedhly et al. 2009; Giorno et al. 2013). General field observations indicate that both low and high humidity have an impact on fruit set; at lower humidity with increasing temperatures (dry heat), pollen desiccates fast, and under high humidity conditions and high temperatures (humid heat), pollen forms clumps and does not stick properly to stigma.

To mitigate heat stress, ecogeographic characterization of pepper has been undertaken by WorldVeg in collaboration with CIAT, which revealed the existence of adaptability mechanisms against climatic extremes in the wild relatives of pepper (Khoury et al. 2020). However, little progress has been obtained in profiling these available resources for tolerance to heat due to the lack of efficient and effective screening and selection strategies.

For the efficient evaluation of pollen activity under heat stress conditions, the World Vegetable Center–Taiwan recently developed a standardized protocol utilizing an impedance flow cytometer (IFC) (Ampha Z32, Amphasys, Lucerne, Switzerland) (Lin et al. 2020). By using this methodology, significant progress has been obtained in phenotyping for pollen traits in *Capsicum*, including using IFC to evaluate pollen activity in 500 breeding lines and gene bank accessions, leading to the identification of 20 high-performing heat tolerance lines (Lin et al. 2020). Intraspecific and interspecific hybridizations and self-pollination of wild and domesticated *Capsicum* species were also carried out to validate the role of high temperature on fruit set and to validate the IFC findings (Lin et al. 2020). The physiological and biochemical investigations revealed that there is a constant photosynthetic rate via increased stomatal conductivity and transpiration rate at high temperatures resulting in improved leaf cooling. The constant photosynthetic rate and high proline content in high temperatures enable seedlings of heat-tolerant hot pepper plants to recover more quickly from heat damage, resulting in a stable fruit set and high yield stage (Rajametov et al. 2020).

Pollen is highly heat-sensitive, and fruit set is reduced or entirely impeded in the absence of viable pollen. High temperatures have been shown to have a negative effect on flower tissues, especially pollen, which has led to a decrease in the success of pollination. The wild progenitor *C. annuum* var. *glabriusculum*, PBC 1970, and PBC 1969 have been identified as potential sources for heat tolerance (Lin et al. 2021). Seven stages of flower development were screened, and stages 3, 4, and 5 were identified to be the best stages to combine high pollen concentration and activity. The flowers of the three stages were subjected to different temperature regimes and found that accessions PBC 831 and PBC 507 were reported as the novel heat tolerant lines, based on using IFC revealing efficiency and accuracy of the method in assessing pollen traits in pepper (Lin et al. 2022).Based on the cumulative temperature response index, Treasures Red, Explosive Ember, Thai Ho, Medusa, and Chilly Chili were identified as heat-tolerant lines (Gajanayake et al. 2011). Based on the heat susceptibility index, hot pepper Pepsi-17-2 was found to be heat tolerant and a potential donor (Kaur et al. 2016). In the recent Temperature Induction Response and field screening of germplasm, Punjab Guchhedar and Ajeet 1 were reported as heat-tolerant donors in chilli (Mishra et al. 2020).

Evaluation of genetic studies was carried out, and good-to-excellent heat-tolerant lines have been identified, i.e. AVPP9905, AVPP0105, AVPP9812, AVPP0514, AVPP0201, AVPP0307 AVPP0805, AVPP0803, AVPP0702, AVPP0506, AVPP0513, AVPP9805, AVPP0705, and AVPP0907 as potential heat-tolerant donors in chilli (Usman et al. 2015). Recently, Usman et al. (2018) publicized the incorporation of the heat-tolerant genes (*Hsps*) from AVPP0702 into the cultivar Kulai through marker-assisted backcrossing. Through marker-assisted breeding, 12 new heat-tolerant chilli lines were developed. Improved heat-tolerant lines in Kulai background were able to retain most of their morphological and agronomical traits, as well as their ability to withstand high temperatures compared to the Kulai variety (Usman et al. 2020).

11.7.2 Moisture Stress Tolerance

Moisture stress significantly affects plant growth and yield of chilli. *Capsicum chinense* cv. IHR 4502 has been reported to be deep rooted and drought tolerant.

In a genetic analysis of an interspecific population of *Capsicum annuum* (IHR3453) × IHR 4502 (*C. chinense*), it was found that root traits are mainly controlled by dominance component, indicating bulk pedigree and single seed descent method as most efficient strategies for improving drought tolerance (Naresh et al. 2017). Through colchicine treatment in GVC-111 (*C. annuum* L.), tetraploids with stronger and robust system were developed, which can be explored for breeding to aid in drought tolerance (Kulkarni and Borse 2010).

11.8 Breeding for Quality Traits

Chilli fruits are the richest source of several biochemical compounds. Wahyuni et al. (2011) identified major capsaicinoids, carotenoids, as well as flavonoid glycosides upon high-performance liquid chromatography (HPLC)-mediated biochemical analysis of 32 *Capsicum* sp. accessions. The capsaicinoid content in fruit pericarp varied between 0.07 and 80 mg/100 g fruit weight, while vitamin C and vitamin E contents were estimated to be 200 mg/100 g and 16 mg/100 g fruit weight, respectively. The metabolite analysis revealed chili fruits are rich sources of vitamins A, C, and E. Sarpras et al. (2016) evaluated 136 different genotypes procured from the northeastern part of India for 65 fruit metabolites and found significant variations for fruit metabolites among the genotypes. Metabolite diversity was analysed through GC-MS and found that the metabolites comprised of carboxylic acids, fatty acids and esters, hydrocarbons, aldehydes, terpenoids, alcohol, and vitamin E. Specific genotypes rich in these metabolites were identified. Seventeen *C. chinense* genotypes were identified as highly pungent accessions with more than 0.9 million SHU surpassing the previous reports of pungent peppers. This study indicated the existence of wide variability in metabolites, which can be explored further in breeding programmes for nutritionally rich chillies.

During ripening, chilli contains carotenoid pigments, which give chilli fruits their array of vibrant colours, including green, yellow, orange, and red. Crytocapsin, capsorubin, and capsanthin, as well as ß-carotene, ß-cryptoxanthin, violaxanthin, Lutein, antheraxanthin, and zeaxanthin, are only found in pepper fruits out of the 30 pigments that have been detected in chilli (Matus et al. 1991). Giuffrida et al. (2013) observed a significant variation in the composition of 52 carotenoids identified in 12 different *Capsicum* cultivars. The red colour in pepper is due to the carotenoids capsanthin (60% of the total carotenoids) and capsorubin, while the yellow-orange colour is due to ß-carotene and violaxanthin. Harkay-Vinkler (1974) and Kanner et al. (1979) found that capsanthin and capsorubin content enhance proportionally with the advancement of the fruit development stage, with capsanthin being the more stable of the two. The most abundant carotenoids (precursors of vitamin A) in ripe pepper fruits are capsorubin, capsanthin, and β-carotene. Guzman et al. (2010) studied the carotenoid content of orange-coloured *Capsicum* accessions and defined them along with the allelic variability of the biosynthetic enzymes. They cloned and sequenced four genes, viz. *Lcyb*, *Psy*, *Ccs*, and *CrtZ*-2, responsible for the biosynthesis of carotenoids from seven orange fruit cultivars.

Six candidate genes contribute to the carotenoid biosynthesis in chilli and among these two genes, *Psy* and *Ccs* contribute to its ripe fruit colour (Jeong et al. 2019). Apart from carotenoids, chilli fruits are a rich bundle of capsaicinoids and vitamins (C and E)(Gómez-García and Ochoa-Alejo 2013). Studies also show that the majority of pungent peppers synthesize a trace amount of "capsinoids", which is a non-pungent analogue of capsaicin (Uarrota et al. 2021).

Genetic approaches have been used in several studies to understand the relationship between the presence of structural genes that cause carotenoid biosynthesis in ripe fruit colours in chilli (Lefebvre et al. 1998). They reported that the deletion or absence of the *Ccs* gene results in peppers that are orange or yellow. Bouvier et al. (1994), Popovsky and Paran (2000), and Lang et al. (2004) also found similar results in yellow-coloured golden summer and Jaune de pignerolle. In contrast to these reports, Ha et al. (2007) reported two structural mutations in *Ccs* gene resulting in the yellowing of fruits. The *Ccs* gene coding area was found in the orange cultivars examined, according to Guzman et al. (2010), and a unique variant, Ccs-3, was discovered in a ß-carotene-rich cultivar Fogo. This variant had a cytosine deletion at 1283 base pairs, which led to a frameshift in the coding sequence and the creation of 423 amino acid sequences. Candidate gene sequencing of a yellow pepper line, LCA 1068, indicated that the emergence of yellow colour during ripening is not required for *Ccs* gene deletion. The *Ccs* coding did not contain any introns, but it did reveal variations among pepper accessions that changed the amino acid. Bydagi dabbi, an Indian landrace with the highest ASTA colour value, displayed unexpected allelic variations.

In order to find allelic differences, Jeong et al. (2019) used single-molecule real-time sequencing of six candidate genes in 94 accessions. The PSY1, PSY2, Lcyb, CrtZ-2, ZEP, and CCS genes' altered coding regions resulted in a total of 12, 3, 7, 9, 16, and 10 unique alleles. These allelic differences could be used to create potential molecular markers to evaluate the colouring of pepper fruits. As the genes targeted here are smaller than 10 kb, they stated that SMRT sequencing is an effective NGS technology that gives greater read lengths and that sequences can be discovered in a single run (Rhoads and Au 2015).

Capsaicin and dihydrocapsaicin are major alkaloids among capsaicinoids, used in the pharmaceutical, medicine, food, and other industries worldwide (Baenas et al. 2019). The earliest report of the high pungency of Bhut Jolokia (8,55,000 SHU) was reported by Mathur et al. (2000). Later, Bosland and Baral (2007) reported a maximum of 9,27,199 SHU in Bhut Jolokia. Islam et al. (2015) found high variability of capsaicinoids in 92 accessions of *Capsicum* collected from northeastern states of India, which ranged from 1,91,135 to 11,52,832 SHU. Currently, Carolina Reaper is the world's most highly pungent pepper, which is a cross between really nastily hot x Naga Viper (Buchanan 2020). Two *Capsicum acyltransferase* genes (*Catf-1 and Catf-2)* are putative genes reported to be associated with pungency (Lang et al. 2006). Ninety-four *Capsicum annuum* accessions underwent genome-wide association analysis for capsaicinoids, and 30 and 56 SNPs were linked to the contents of capsaicin and dihydrocapsaicin, respectively (Nimmakayala et al. 2016). Further, five candidate genes, i.e. *CA10g13930*, *CA00g8691*, *CA04g13110*,

CA06g15270, and *CA02g0461*, were reported to be associated with high capsaicin content (Zhang et al. 2016). Five major QTLs on chromosomes 1, 3, 2, 4, and 10 were identified to be associated with capsaicinoid content over multi-locations in two RIL populations (Han et al. 2018).

Capsinoids, non-pungent capsaicinoid analogs, were first extracted from CH-19 Sweet line. Through conventional breeding, Dieta series lines having high level of capsinoids were developed (Seki et al. 2020). Very recently, QTLs associated with capsinoid content (capsiate, nordihydrocapsiate, and dihydrocapsiate) were reported on chromosomes 6 and 10 in SNU11-001 x Bhut Jolokia F2 population; further, *C. chinense* lines having high levels of capsinoids and capsaicinoids were developed through marker-assisted breeding (Jang et al. 2021). These developed lines can be explored for industrial uses.

11.9 Conclusions

Chilli is a unique crop with fruits producing capsaicinoids, capsinoids, and carotenoids, which have wider application in several industries especially in spice, cosmeceutical, and pharmaceutical industries. Due to emerging demand, there is a need to develop high-quality varieties/hybrids resistant/tolerant to biotic and abiotic stresses in the present climate change scenario. The available genetic resources must be fully explored in breeding programmes. Application of genomic approaches will enable marker-assisted selection to accelerate crop improvement programmes. Further, there is a need to explore the possible utilization of wild species of *Capsicum* in identification of novel alleles for resistance and incorporate into cultivated species to breed resistance against emerging serious pests, diseases, and abiotic stresses.

References

Aguilar-Meléndez A et al (2009) Genetic diversity and structure in semiwild and domesticated chillies (*Capsicum annuum*; Solanaceae) from Mexico. Am J Bot 96(6):1190–1202

Aitken E, McColl G, Kingsmore D (2017) The role of Qutenza® (topical capsaicin 8%) in treating neuropathic pain from critical ischemia in patients with end-stage renal disease: an observational cohort study. Pain Med 18(2):330–340

Aloni B, Karni L, Daie J (1992) Effect of heat stress on the growth, root sugars, acid invertase and protein profile of pepper seedlings following transplanting. J Hortic Sci 67:717–725

Anand N, Deshpande AA, Sridhar TS (1987) Resistance to powdery mildew in an accession of Capsicum frutescens and its inheritance pattern. Capsicum Eggplant Newsletter 6:77–78

Baenas N, Belović M, Ilic N, Moreno DA, García-Viguera C (2019) Industrial use of pepper (*Capsicum annuum* L.) derived products: technological benefits and biological advantages. Food Chem 274:872–885

Barik S et al (2022) Breeding peppers for industrial uses: Progress and prospects. Ind Crop Prod 178:114626

Berke T et al (2005) Suggested cultural practices for chilli pepper. International CoOperator's Guide

Blat SF, da Costa CP, Vencovsky R, Sala FC (2005) Inheritance of reaction to Leveillula taurica (Lev.) Arn. in Capsicum baccatum. Crop Breed Appl Biotechnol 5:467–472

Bosland PW, Baral B (2007) 'Bhut Jolokia'-the world's hottest known Chile pepper is a putative naturally occurring interspecific hybrid. HortSci 42:222–224
Bouvier F et al (1994) Xanthophyll biosynthesis in chromoplasts: isolation and molecular cloning of an enzyme catalyzing the conversion of 5,6-epoxycarotenoid into ketocarotenoid. Plant J 6: 45–54. https://doi.org/10.1046/j.1365-313X.1994.6010045.x
Buchanan M (2020) Some like it hot. Nat Phys 16(1). https://doi.org/10.1038/s41567-019-0766-3
Caranta C et al (2002) QTLs involved in the restriction of Cucumber mosaic virus (CMV) long-distance movement in pepper. Theor Appl Genet 104:586–591
Carrizo García C et al (2016) Phylogenetic relationships, diversification and expansion of chili peppers (capsicum, Solanaceae). Ann Bot 118(1):35–51
Chae SY et al (2022) QTL mapping of resistance to bacterial wilt in pepper plants (Capsicum annuum) using genotyping-by-sequencing (GBS). Horticulturae 8(2):115
Chaim BA, Grube RC et al (2001) Identification of quantitative trait loci associated with resistance to cucumber mosaic virus in Capsicum annuum. Theor Appl Genet 102:1213–1220
Cook AA, Guevara YG (1984) Hypersensitivity in Capsicum chacoense to race 1 of the bacterial spot pathogen of pepper (No. RESEARCH)
De Souza, VL, Cafe Filho AC (2003) Effect of chemical control on the progress of sweet pepper powdery mildew under greenhouse conditions. Summa Phytopathol *(Brazil)*
Djian-Caporalino C, Fazari A, Arguel MJ, Vernie T, Vande Casteele C, Faure I, Brunoud G, Pijarowski L, Palloix A (2007) Root-knot nematode (Meloidogyne spp.) Me resistance genes in pepper (Capsicum annuum L.) are clustered on the P9 chromosome. Theor Appl Genet 114 3:473–486
Du H, Wen C, Zhang X, Xu X, Yang J, Chen B, S. (2019) Geng identification of a major QTL (*qRRs-10.1*) that confers resistance to *Ralstonia solanacearum* in pepper (*Capsicum annuum*) using SLAF-BSA and QTL mapping. Int J Mol Sci 20(23):5887
Duriat AS, Gunaeni N, Sulastrini I (1997) Further studies on screening pepper varieties for resistance to CMV strains. In: AVRDC Collaborative vegetable research in Southeast Asia proceedings of the AVNET-II final workshop, Publication No. 119–124 (pp 451). Thailand: Bangkok
Faostat (2020) FAOSTAT statistical database Publisher. FAO (Food and Agriculture Organization of the United Nations), Rome, Italy
Firdaus S, Van Heusden A, Harpenas A, Supena ED, Visser RG, Vosman B (2011) Identification of silverleaf whitefly resistance in pepper. Plant Breed 130:708–714. https://doi.org/10.1111/j.1439-0523.2011.01894.x
Gajanayake B, Trader BW, Reddy KR, Harkess RL (2011) Screening ornamental pepper cultivars for temperature tolerance using pollen and physiological parameters. HortSci 46(6):878–884
Giorno F, Wolters-Arts M, Mariana C, Rieu I (2013) Ensuring reproduction at high temperatures: the heat stress response during anther and pollen development. Plan Theory 2:489–506
Giuffrida D, Dugo P, Torre G, Bignardi C, Cavazza A, Corradini C, Dugo G (2013) Characterization of 12 Capsicum varieties by evaluation of their carotenoid profile and pungency determination. Food Chem 140(4):794–802
Gniffke PA et al (2013) Pepper research and breeding at AVRDC—the World Vegetable Center. In: Proceedings of XV EUCARPIA meeting on genetics and breeding of capsicum and eggplant, Turin, Italy, 2–4 September: 305–311
Gómez-García MDR, Ochoa-Alejo N (2013) Biochemistry and molecular biology of carotenoid biosynthesis in chili peppers (Capsicum spp.). Int J Mol Sci 14(9):19025–19053
Green SK, Kim JS (1991) Characteristics and control of viruses infecting peppers: a literature review, vol 18. Asian Vegetable Research and Development Center, Shanhua, Taiwan
Guzman I, Hamby S, Romero J, Bosland PW, O'Connell MA (2010) Variability of carotenoid biosynthesis in orange colored Capsicum spp. Plant Sci 179(1–2):49–59. https://doi.org/10.1016/j.plantsci.2010.04.014
Ha SH, Kim JB, Park JS, Lee SW, Cho KJ (2007) A comparison of the carotenoid accumulation in Capsicum varieties that show different ripening colours: deletion of the capsanthin-capsorubin

synthase gene is not a prerequisite for the formation of a yellow pepper. J Exp Bot 58(12):3135–3144. https://doi.org/10.1093/jxb/erm132

Han K et al (2018) QTL mapping and GWAS reveal candidate genes controlling capsaicinoid content in Capsicum. Plant Biotechnol J 16(9):1546–1558

Harkay-Vinkler M (1974) Storage experiments with raw material of seasoning paprika with particular reference to the red color pigment components. Acta Alim Acad Sci Hung 3:239–249

Hedhly A, Hormaza JI, Herrero M (2009) Global warming and sexual plant reproduction. Trends Plant Sci 14(1):30–36

Islam MA et al (2015) Variability in capsaicinoid content in different landraces of capsicum cultivated in North-Eastern India. Sci Hortic 183:66–71

Jang S, Park M, Lee DG, Lim JH, Jung JW, Kang BC (2021) Breeding capsicum chinense lines with high levels of capsaicinoids and capsinoids in the fruit. Agriculture (Switzerland) 11(9). https://doi.org/10.3390/agriculture11090819

Jeevanandham N et al (2018) Levels of plant resistance in chillies Capsicum spp against whitefly, Bemisia tabaci. Int J Curr Microbial 7(1):1419–1441

Jeong H et al (2019) Single-molecule real-time sequencing reveals diverse allelic variations in carotenoid biosynthetic genes in pepper (Capsicum spp.). Plant Biotechnol J 17(6):1081–1093

Jo J et al (2017) Molecular mapping of PMR1, a novel locus conferring resistance to powdery mildew in pepper (Capsicum annuum). Front Plant Sci 8:2090

Jones JB et al (2004) Reclassification of the xanthomonads associated with bacterial spot disease of tomato and pepper. Syst Appl Microbiol 27(6):755–762

Kang WH et al (2010) Molecular mapping and characterization of a single dominant gene controlling CMV resistance in peppers (Capsicum annuum L.). Theor Appl Genet 120:1587–1596

Kanner J et al (1979) Content and stability of. alpha.-tocopherol in fresh and dehydrated pepper fruits (Capsicum annuum L.). J Agric Food Chem 27(6):1316–1318

Kaur N, Dhaliwal MS, Jindal S, Singh P (2016) Evaluation of hot pepper (Capsicum annuum L.) genotypes for heat tolerance during reproductive phase. Int J Bio-resour 7(1):126–129

Khoury CK et al (2020) Modeled distributions and conservation status of the wild relatives of Chile peppers (Capsicum L.). Divers Distrib 26(2):209–225

Kulkarni M, Borse T (2010) Induced polyploidy with gigas expression for root traits in Capsicum annuum (L.). Plant Breed 129:461–464. https://doi.org/10.1111/j.1439-0523.2009.01696.x

Kumar M et al (2022a) Identification of molecular marker linked to resistance gene loci against Indian isolate of Phytophthora capsici L. causing root rot in chilli (Capsicum annuum L.). Australas Plant Pathol 51(2):211–220

Kumar R et al (2022b) De novo assembly, differential gene expression and pathway analyses for anthracnose resistance in chilli (Capsicum annuum L.). J Plant Biochem Biotechnol 31(1): 124–138

Kumar RV, Singh AK, Singh AK, Yadav T, Basu S, Kushwaha N, Chattopadhyay B, Chakraborty S (2015) Complexity of begomovirus and betasatellite populations associated with chilli leaf curl disease in India. J Gen Virol 96:3143–3158

Lang Y, Yanagawa S, Sasanuma T, Sasakuma T (2004) Orange fruit colour in Capsicum due to deletion of capsanthin–capsorubin synthesis gene. Breed Sci 54:33–39. https://doi.org/10.1270/jsbbs.54.33

Lang Y, Yanagawa S, Sasanuma T, Sasakuma T (2006) A gene encoding a putative acyl-transferase involved in pungency of Capsicum. Breed Sci 56(1):55–62. https://doi.org/10.1270/jsbbs.56.55

Lapidot M et al (1997) Tolerance to Cucumber mosaic virus in pepper: development of advanced breeding lines and evaluation of virus level. Plant Dis 81:185–188

Latha S, Hunumanthraya L (2018) Screening of chilli genotypes against chilli thrips (Scirtothrips dorsalis Hood) and yellow mite [Polyphagotarsonemus latus (banks)]. J Entomol Zool Stud 6: 2739–2744

Latournerie-Moreno L, Ic-Caamal A, Ruiz-Sánchez E et al (2015) Survival of Bemisia tabaci and activity of plant defense-related enzymes in genotypes of Capsicum annuum L. Chil J Agric 75(1):71–77

Lee HR et al (2013) Development of a novel codominant molecular marker for Chili veinal mottle virus resistance in *Capsicum annuum* L. Euphytica 193:197–205

Lee J, Hong JH, Do JW, Yoon JB (2010) Identification of QTLs for resistance to anthracnose to two Colletotrichum species in pepper. J Crop Sci Biotechnol 13:227–233

Lee JH et al (2017) Identification and molecular genetic mapping of Chili veinal mottle virus (ChiVMV) resistance genes in pepper (*Capsicum annuum*). Mol Breed 37:121

Lefebvre V, Kuntz M, Camara B, Palloix A (1998) The capsanthin-capsorubin synthase gene: a candidate gene for the y locus controlling the red fruit colour in pepper. Plant Mol Biol 36(5): 785–789. https://doi.org/10.1023/A:1005966313415

Lin SW, Gniffke PA, Wang TC (2007) Inheritance of resistance to pepper anthracnose by Colletotrichum acutatum. Acta Hort 760:329–334

Lin SW, Lin YW, Wang, Barchenger DW (2020) High throughput pollen phenotyping using impedance flowcytometry for heat tolerance selection in pepper (Capsicum spp.). Hort Sci 55(9):S156. (abstr.)

Lin SW et al (2022) Impedance flow cytometry for selection of pollen traits under high temperature stress in pepper. HortSci 57(2):181–190

Lin TH et al (2021) Growing environment and heat treatment effects on intra-and interspecific pollination in Chile pepper (Capsicum spp.). Agronomy 11(7):1275

Lozada DN, Nunez G, Lujan P, Dura S, Coon D, Barchenger DW, Sanogo S, Bosland PW (2021) Genomic regions and candidate genes linked with Phytophthora capsici root rot resistance in chile pepper (Capsicum annuum L.). BMC Plant Biol 21:601

Maharijaya A, Vosman B, Steenhuis-Broers G, Harpenas A, Purwito A, Visser RGF, Voorrips RE (2011) Screening of pepper accessions for resistance against two thrips species (Frankliniella occidentalis and Thrips parvispinus). Euphytica 177:401–410

Mahasuk P, Struss D, Mongkolporn O (2016) QTLs for resistance to anthracnose identified in two Capsicum sources. Mol Breed 36:10. https://doi.org/10.1007/s11032-016-0435-5

Mahasuk P, Taylor PWJ, Mongkolporn O (2009) Identification of two new genes conferring resistance to Colletotrichum acutatum in Capsicum baccatum. Phytopathology 99:1100–1104

Mallard S et al (2013) A key QTL cluster is conserved among accessions and exhibits broad-spectrum resistance to *Phytophthora capsici*: a valuable locus for pepper breeding. Mol Breed 32:349–364

Mathew D (2020) Analysis of QTL Bw1 and marker CAMS451 associated with the bacterial wilt resistance in hot pepper (*Capsicum annuum* L.). Plant Gene 24.100260

Mathur R, Dangi RS, Dass SC, Malhotra RC (2000) The hottest chili variety in India. Curr Sci 79(3):287–288

Matus Z, Deli J, Szabolcs J (1991) Carotenoid composition of yellow pepper during ripening: isolation of. beta.-cryptoxanthin 5, 6-epoxide. J Agric Food Chem 11:1907–1914

Maurya PK et al (2019) Genetic analysis for resistance to leaf curl disease in chilli peppers (*Capsicum annuum* L.) under specific situations. Indian J Genet Plant Breed 79(04):741–748

Mimura Y, Yoshikawa M, Hirai M (2009) Property of resistance to bacterial wilt in *Capsicum* line LS2341. Hortic Res 1:231

Minamiyama Y, Tsuro M, Kubo T, Hirai M (2006) An SSR-based linkage map of *Capsicum annuum*. Mol Breed 18:157–169

Mishra R et al (2017) Differential expression of defense-related genes in chilli pepper infected with anthracnose pathogen Colletotrichum truncatum. Physiol Mol 97:1–10

Mishra S, Laxman RH, Reddy KM, Venugopalan R (2020) TIR approach and stress tolerance indices to identify donor for high-temperature stress tolerance in pepper (Capsicum annuum L.). Plant Genet Res 18(1):19–27

Naresh P, Bhatt RM, Venkatachalapathi V, Gangadharrao P, Madhavi Reddy K (2017) Inheritance of root traits in an interspecific cross of *Capsicum annuum* × *C. chinense* in the presence of low moisture. Int J Veg Sci 23:575–583

Naresh P et al (2022) Molecular mapping of Chilli veinal mottle virus (ChiVMV) resistance in hot pepper (*Capsicum annuum* L.). Plant Gene 33:100396

Naveena GS et al (2020) Genetic analysis of bacterial wilt resistance in hot pepper (Capsicum annuum L.). J Veg Sci 47(1):153–156

Nimmakayala P et al (2016) Genome-wide diversity and association mapping for capsaicinoids and fruit weight in *Capsicum annuum* L. Sci Rep 6(1):1–14. https://doi.org/10.1038/srep38081

Nischay PK et al (2021) Identification of potential chilli (*Capsicum annuum* L.) accession as a rootstock for managing bacterial wilt disease in bell pepper. J Veg Sci 48(2):246–249

Ogundiwin EA et al (2005) Construction of 2 intraspecific linkage maps and identification of resistance QTL for Phytophthora capsici root-rot and foliar-blight diseases of pepper (*Capsicum annuum* L.). Genome 48:698–711

Osdaghi E et al (2021) A centenary for bacterial spot of tomato and pepper. Mol Plant Pathol 22(12):1500

Pantoja KF et al (2018) Identification of capsicum accessions tolerant to tomato severe rugose virus and resistant to Bemisia tabaci Middle East-Asia Minor 1 (MEAM1). Trop Plant Pathol 43(2): 138–145

Pavithra BS et al (2020) Identification of resistant sources in chilli (*Capsicum* sp.) genotypes to Groundnut bud necrosis virus (GBNV). Australas Plant Pathol 49(1):15–23

Popovsky S, Paran I (2000) Molecular genetics of the y locus in pepper: its relation to capsanthin–capsorubin synthase and to fruit color. Theor Appl Genet 101:86–89. https://doi.org/10.1007/s001220051453

Praveen SL et al (2021) Field screening of chilli germplasm collections for resistance to chilli thrips, Scirtothrips dorsalis Hood. Pest Manage Hortic Ecosyst 27(1):70–76

Rai VP et al (2014) Monogenic recessive resistance to pepper leaf curl virus in Capsicum. Sci Hort 172:34–38

Rajametov S, Yang EY, Cho MC, Chae SY, Jeong HB (2020) Morpho-physiological response to heat stress during seedling stage and interrelation with agronomical traits in pepper (Capsicum Annuum L.) cultivars differing in heat tolerance. Res Square. https://doi.org/10.21203/rs.3.rs-120894/v1

Rajput LB, Gilal AA, Bukero A, Wahocho NA (2017) Impact of different oils as biopesticide against sucking insect pests in cotton. J Basic Appl 13:34–40

Reddy BA et al (2008) Detection of a tospovirus-infecting tomato (*Solanum lycopersicon* L.). Indian J Virol 19:1–5

Rhoads A, Au KF (2015) PacBio sequencing and its applications. Genom Proteom Bioinform 13: 278–289. https://doi.org/10.1016/j.gpb.2015.08.002

Saeko A, Pongjaroenkit S (2020) *DEVELOPMENT of SNP MARKERS SPECIFIC TO THRIPS (THRIPS PALMI) RESISTANCE IN PEPPER* (Doctoral dissertation, Maejo University)

Sarpras MS et al (2016) Comparative analysis of fruit metabolites and pungency candidate genes expression between bhut jolokia and other *Capsicum* species. PLoS One 11:1–19

Seki T, Ota M, Hirano H, Nakagawa K (2020) Characterization of newly developed pepper cultivars (Capsicum chinense) 'Dieta0011- 0301', 'Dieta0011-0602', 'Dieta0041-0401', and 'Dieta0041-0601'containing high capsinoid concentrations and a strong fruity aroma. Biosci Biotechnol Biochem 84:1870–1885

Sharma A et al (2022) Identification and mapping of bs8, a novel locus conferring resistance to bacterial spot caused by Xanthomonas gardneri. Phytopathology 112(8):1640–1650

Singh J, Thakur MR (1977) Genetics of resistance to Tobacco mosaic virus, Cucumber mosaic virus and leaf curl virus in hot pepper (Capsicum annuum L.). In: Capsicum 77: 3rd Eucarpia Meet, 119–123

Sugita T et al (2006) QTL analysis for resistance to *Phytophthora* blight (*Phytophthora capsici* Leon.) using an intraspecific doubled-haploid population of *Capsicum annuum*. Breed Sci 56(2):137–145

Suwor P, Sanitchon J, Thummabenjapone P, Kumar S, Techawongstien S (2017) Inheritance analysis of anthracnose resistance and marker-assisted selection in introgression populations of chili (Capsicum annuum L.). Sci Hort 220:20–26

Sy O, Steiner R, Bosland PW (2008) Recombinant inbred line differential identifies race-specific resistance to phytophthora root rot in *Capsicum annuum*. Phytopathology 98:867–870

Thakur H, Jindal SK, Sharma A, Dhaliwal MS (2019) A monogenic dominant resistance for leaf curl virus disease in chilli pepper (Capsicum annuum L.). Crop Prot 116:115–120

Thakur H, Jindal SK, Sharma A, Dhaliwal MS (2020) Molecular mapping of dominant gene responsible for leaf curl virus resistance in chilli pepper (*Capsicum annuum* L.). 3Biotech 10(4):182

Tripathi D et al (2013) Effect of ajwain, hot red pepper and black pepper on the performance of Japanese quail. Indian J Anim Nutr 30(4):431–433

Truong HTH et al (2012) Identification of isolate-specific resistance QTLs to *Phytophthora* root rot using an intraspecific recombinant inbred line population of pepper (*Capsicum annuum*). Plant Pathol J 61(1):48–56

Tsai WS, Huang YC, Zhang DY, Reddy K, Hidayat SH, Srithongchai W, Green SK, Jan FJ (2008) Molecular characterization of the CP gene and 3′ UTR of Chilli veinal mottle virus from South and Southeast Asia. Plant Pathol 57:408–416

Uarrota VG, Maraschin M, de Bairros ÂFM, Pedreschi R (2021) Factors affecting the capsaicinoid profile of hot peppers and biological activity of their non-pungent analogs (Capsinoids) present in sweet peppers. Crit Rev Food Sci Nutr 61(4):649–665. https://doi.org/10.1080/10408398.2020.1743642

Usman MG et al (2015) Expression of target gene Hsp70 and membrane stability determine heat tolerance in chili pepper. J Am Soc Hortic 140(2):144–150

Usman MG, Rafii MY, Martini MY, Yusuff OA, Ismail MR, Miah G (2018) Introgression of heat shock protein (Hsp70 and sHsp) genes into the Malaysian elite chilli variety Kulai (Capsicum annuum L.) through the application of marker-assisted backcrossing (MAB). Cell Stress Chaperones 23(2):223–234

Usman MG et al (2020) Molecular confirmation of candidate Hsp70 gene associated with heat tolerance in BC3F2 advanced backcross lines and their phenotypic resemblance with recurrent chilli Kulai. Acta Agric Scand B—Soil Plant Sci 70(3):252–264

van Haperen P, Voorrips RE, Lucatti AF, Schellart W, van Loon JJA, Vosman B (2020) The effect of a thrips resistance QTL in different Capsicum backgrounds. Euphytica 216:187

Venkatesh J et al (2018) Fine mapping of the dominant potyvirus resistance gene *Pvr7* reveals a relationship with *Pvr4* in *Capsicum annuum*. Phytopathology 108:142–148

Vijayalakshmi K, Wightman JA, Reddy DVR, Ranga Rao GV, Buiel AAM, Reddy DDR (1995) Transmission of Peanut bud necrosis virus by Thrips palmi in India. In: Parker BL, Skinner M, Lewis T (eds) Thrips biology and management. Plenum Press, New York, pp 179–184

Wahyuni Y, Ballester AR, Sudarmonowati E, Bino RJ, Bovy AG (2011) Metabolite biodiversity in pepper (Capsicum) fruits of thirty-two diverse accessions: variation in health-related compounds and implications for breeding. Phytochemistry 72(11-12):1358–1370

Wai KP, Pa MI, Siddique H-SM, Yoo HJ, Byeon S-E, Jegal Y, Mekuriaw AA, Kim B-S (2015) Pathotypes of bacterial spot pathogen infecting Capsicum peppers in Korea. Plant Pathol J 31 (4):428

Wang D, Bosland PW (2006) The genes of capsicum. HortSci 41:1169–1187

Wang LH et al (2009) A SCAR marker linked to the *N* gene for resistance to root knot nematodes (*Meloidogyne* spp.) in pepper (*Capsicum annuum* L.). Sci Hortic 122:318–322

Yadav RK et al (2022) Evaluation of chilli genotypes and understanding biochemical basis of whitefly (Bemisia tabaci Genn.) resistance. S Afr J Bot 151:433–444

Yao M, Li N, Wang F, Zhibiao YE (2013) Genetic analysis and identification of QTLs for resistance to Cucumber mosaic virus in chili pepper (Capsicum annuum L.). Euphytica 193:135–145

Yoon JK, Park HG (2005) Trispecies bridge crosses, (Capsicum annuum x C. chinense) x C. baccatum, as an alternative for introgression of anthracnose resistance from C. baccatum into C. annuum. Hort Environ Biotechnol 46:5–9

Zhang ZX, Zhao SN, Liu GF (2016) Discovery of putative capsaicin biosynthetic genes by RNA-Seq and digital gene expression analysis of pepper. Sci Rep 6:34121. https://doi.org/10.1038/srep34121

Zsiboras C (2018) Capsaicin and capsiate could be appropriate agents for treatment of obesity: a meta-analysis of human studies. Crit Rev Food Sci Nutr 58(9):1419–1427

Industrial Demand and Applications of Capsaicin

12

Desam Nagarjuna Reddy, Mary Moses Mylabathula, and Abdul Jabbar Al-Rajab

Abstract

Capsaicin occurs naturally in hot pepper (*Capsicum annuum*) fruits normally known as chillies. It has been utilised since antiquity as a food vegetable, natural colourant, flavouring component, and conventional medicine. The genus *Capsicum* (*Cap*) is an extensively suitable plant for cultivation as it is widely consumed in the world. Because of their broad diversity, peppers are an important vegetable and spice crop, and the products of which contain various bioactive chemical constituents, including capsinoids, capsaicinoids, carotenoids, and oleoresin. Growing attention has been paid to the physiologically active chemical components of pepper due to the ecologically friendly plant products. Red pepper has important commercial uses in the cosmetic, pharmaceutical, and agro-food industries. Chilli contains remarkable levels of bioactive chemical constituents, including capsaicinoids, carotenoids, vitamins, and minerals; its pigments include lutein, chlorophyll, anthocyanin, etc. Capsaicin accounts for the strong or pervasive or irritating flavour of these families. These species have been demonstrated to perform a constructive responsibility in well-being. Furthermore, capsaicin is utilised in sauces to provide sharpness, flavour, and colour; other chemical constituents of peppers can be utilised for the conservation and expansion of commercial products' life expectancy and used for their antioxidant, anticancer, anti-obesity, urinary disorders, and antimicrobial activities. The utilisation of

D. N. Reddy (✉)
School of Engineering and Applied Technology, BEST Innovation University, Gorantla, Andhra Pradesh, India

M. M. Mylabathula
Department of Biotechnology, Krishna University, Machilipatnam, AP, India

A. J. Al-Rajab
Etcectera Publications, Chesterville, ON, Canada

M. K. Swamy, A. Kumar (eds.), *Capsaicinoids*,
https://doi.org/10.1007/978-981-99-7779-6_12

advanced compounds derived from chilli in medicinal procedures for the therapy of erythrogenic as well as torment-associated situation is also likely, because peppers consist of vitamin A and vitamin C, carotenoids, capsaicinoids, phenolic compounds, and minerals, for example, calcium (Ca) and iron (Fe), possessing a well-being potentiality. Furthermore, investigations on significant extricate procedures, bioactivity, reliability as well as security are important to determine modern pepper chemical constituents within food stuffs, non-medicinal or cosmetic, and pharmaceutical applications. This chapter covers the importance of industrial demand and uses of capsaicin in the agro-food, pharma, as well as cosmetic industries.

Keywords

Capsicum annuum · Microorganisms · Capsaicin · Encapsulation · Capsinoids · Technological properties · Phenolic compounds

12.1 Introduction

Capsicum is an established vegetable spice and herb. It is the most commercially significant species in the Solanaceae family, comprising almost 30 genera (Dong et al. 2014). Its fruits, prominently known as chillies, chilli pepper, cayenne pepper, Christmas pepper, African chillies, and tabasco pepper, are all different varieties of chilli plants. Chilies are presumed to have originated in the Americas (specifically South America), where it was utilised for clinical as well as dietary purposes (Barceloux 2009). Mosqueda-Solís et al. (2021) found that chillies are versatile crops that are consumed daily in different forms, particularly in Indonesia, Mexico, China, and Turkey. Lu et al. (2020) reported that red peppers, which are featured for their strong and pungent flavour, are a well-known spice used all over the world. The chili pepper's spicy chemical constituents include capsaicinoids, capsaicin (for pungency), and dihydrocapsaicin. Chapa-Oliver and Mejía-Teniente (2016) observed that chili peppers consist of 0.1–1.0% of capsaicin and dihydrocapsaicin and proportions of 1:1–2:1 for 80–90% capsaicinoids, exhibiting about 2 times the fieriness to savour and irritation as the minor capsaicinoids. Shah et al. (2020) observed that capsaicin is 7–10 times more abundant in dried red chillies when compared with fresh red chillies. Furthermore, capsaicin as well as dihydrocapsaicin peppers consist an assortment of natural abundant capsaicinoids, which could generate a burning sensation when eaten, such as homodihydrocapsaicin and nordihydrocapsaicin (Srinivasan 2016). Capsaicin is a constituent of the vanilloid group of chemical constituents, which incites their biological activity. Further, as a vanilloid, capsaicin is fundamentally comprised of a hydrophobic carbon tail, amide functional group, and a benzene ring. It is a lipophilic, odourless, crystalline, and colourless compound partially soluble in water (at 25 °C 10.3 mg/L); it is more dissolvable in acetone, carbon tetrachloride (CCl_4), ethyl alcohol (C_2H_5OH) as well as methanol (CH_3OH) and also stable in alkali and acids at 25 °C (Shah et al. 2020; Wang et al. 2014).

Capsaicin is generally utilised worldwide as a food additive as well as colouring agent (Liu et al. 2015). Because of its antibacterial and antifungal activity as well as antioxidation properties (Si et al. 2021), it is utilised for food preservation (Zhao et al. 2020). Capsaicin can be used in different foods as paste, hot sauce, green chilli sauce, pickles, kimchi, salted seafood, and instant noodles (Cho and Kwon 2020; Lu et al. 2020; Sharif et al. 2018; Yang and Du 2018). Basith et al. (2016) have reported that capsaicin exhibits biological activities, including antimicrobial activity, anticancer activity, anti-inflammatory activity, cardiovascular protection, skin disease treatment, and digestive performance medication. Capsaicin is lipophilic in nature as well as less soluble in water; hence it could be utilised in humidity-restricted foods (Reyes-Escogido et al. 2011). Likewise, capsaicin's enhances its biological activity (Feng et al. 2018). Inborn acerbity, stability as well as the losses of flavour with exposure to intense light are the other technical issues with the utilisation of capsaicin (Costanzo et al. 2014; Reyes-Escogido et al. 2011; Sanatombi and Rajkumari 2020). Furthermore, a high oral capsaicin intake could cause discomfort, incontinence, burning sensation, diarrhoea, vomiting, and gastrointestinal squeezing (Hayman and Kam 2008). In addition, Bode and Dong (2011) report that excessive capsaicin intake may be related to cancer-causing consequences. According to Hayman and Kam (2008), capsaicin contact with skin and eyes could trigger wound pain and burning. Hence, the amount of capsaicin utilised plays a major role that could influence either the beneficial impacts or toxicity.

Chillies are commonly bell-shaped and display several colours, including yellow, green, orange as well as red. Chillies have been used by humans as both a spice and an herb. Indigenous Mexicans used them in their daily diet as far back as 7000 BCE. The mass cultivation and use of red chilli have spread worldwide due to European contact with the Americas (Garces-Claver et al. 2006; Govindarajan 1985), with subsequent adaptations to various natural conditions due to the hybridisation process (Das et al. 2016; Govindarajan 1985). Chillies are utilised as spices in foods, food flavours and colouring agents (pigments), as well as in treating stomach problems and headaches (Govindarajan 1985; Govindarajan and Sathyanarayana 1991; Sganzerla et al. 2014). Chilli consumption in food also as a traditional medicine has drawn many experts to investigate the chemical constituents and its features. In the meantime, the utilisation of green or red chillies spread and gained more applications, basically as cosmetics and in pharmaceuticals and agro-food industry (Govindarajan 1985; Govindarajan and Sathyanarayana 1991). More recently, with the growth of knowledge about the genus, their attributes were credited to the bountiful existence of capsaicinoids, carotenoids as well as flavonoids (Lu et al. 2017). Researchers have identified three classes of chemical compounds as bioactive chemical constituents (Giuffrida et al. 2013; Sricharoen et al. 2017). Secondary metabolites, including carotenoids and flavonoids, are additionally responsible for the properties of chillies; along with capsaicinoids, these molecules impart a savour attribute known as sharpness which is accountable for the world fame of these pepper products (Baby and Ranganathan 2016). Acerbity causes inflammation in the cranial nerves that is perceived as a vibe of agony or torment as well as intensity, which empowers its utilisation for gastronomy purposes as a proficient structure to eliminate insipidity or lifelessness and add new tastes to meals (Hayman and Kam 2008). As of late, *Capsicum* species have been utilised for different industrial applications, including food colourants, spices, and herbs, anti-inflammatory, prevention of cardiovascular diseases, painkillers, anticancer, antitumor, and antioxidant (Luo et al. 2011; Reilly and Yost 2006). The raw substance of chillies or peppers can be used for industrial applications based on the presence of bioactive chemical constituents present in the *Capsicum* fruits or peppers or chillies. These peppers are generally used in modern development as a paprika or a *Capsicum* natural resin (oleoresin) extricate, including capsaicinoids, carotenoids, flavonoids, volatile oils, and waxes (de Aguiar et al. 2018; Yeung and Tang 2015). These pepper or chilli-based products are commonly purchased from the cultivator and further applied in various industrial purposes. Hence, the *Cap* natural resin or dry powder could be applied widely in agriculture, agro-food industry, cosmetics, and pharmaceutical industries (de Aguiar et al. 2018; Dussault et al. 2014; Monago-Marana et al. 2018; Yeung and Tang 2015). A ramification of this use is the conceivable loss of efficacy and production of the output, thus a specified chemical composition is required for each unique usage. (Haar et al. 2017; Reyes-Escogido et al. 2011). For example, the food utilisation of *Cap* natural resin needs a product with less than 1.6% of capsaicinoids; for pharmaceutical utilisation, the concentration of capsaicin should not be more than 22.64 mg, and for the colourant dry powder, the presence of the red fraction of carotenoids is needed (Kim et al. 2016a, b; Nugroho 2016). As the chemical constituents of chillies differ depending on species, ecological circumstances, climatic conditions, and, surprisingly, the life pattern of the plant (Barbero et al. 2016; Nugroho 2016; Sweat et al. 2016), it could be difficult to select the *Capsicum* genus as well as agricultural practices that are cost-effective and suitable for a particular purpose (Bae et al. 2014; Loizzo et al. 2015; Wang et al. 2018). Consequently, enumerating the chemical constituents, structures, properties, and factors influencing the chemical profile of *Cap* red chillies (fruits) might help the choice of cultivator/development methods, propelling the industrial utilisation of pepper-based substances, enhancing productivity and the improvement of postharvest technology. Since the process of hybridisation has produced a variety of cultivators with similar physical characteristics, it is challenging to distinguish or isolate each cultivar and give them a clear identification (Domínguez-Martínez et al. 2014; Martínez et al. 2006; Singh et al. 2013). Enhancing the efficacy and viability of a chilli-based product likewise requires an unambiguous recognition of the cultivar that is being utilised. The methodical identification of the family may be achieved by the identification of the chemical constituents of each variety (Patel et al. 2016; Ritota et al. 2010).

In spite of the many varieties, researchers have reported significant photochemistry, biological activity, and other industrial applications of *Cap*. This chapter mainly reviews the innovative uses of capsaicin, such as antimicrobial applications, cell reinforcement, and antioxidant properties; additionally, it can be used in food processing industries. Recognising the broad utilisation of capsaicin in the cosmetic, pharmaceutical, and food processing industries, exemplification is one of the techniques that can be utilised to reduce the challenges with the utilisation of capsaicin.

12.2 Red Pepper as a Wellspring of Nutrients and Bioactive Compounds

The red pepper consist biologically active chemical constituents. these active chemical constituents mainly depends on the red pepper fruit (grains, amnion, and endocarp), the variety, the development and storage environment, and the processing technique. (Jayaprakasha et al. 2012). Carotenoids are the most significant phytochemicals found in pepper varieties, which enhance the trade value of these peppers with regard to colour, flavour, and antioxidant activities, among other bioactive chemical constituents. In hot peppers, capsaicinoids are significantly present and are accountable for the tartness of this diversity. Peppers contain many phenolic substances, basically flavonoids and phenolic acid and its derivatives, and supplements like nutrients including vitamin A and vitamin C, as well as nutrients like manganese (Mn), calcium (Ca), and iron, which are important nutrients (Pandey and Rizvi 2009). The chemical constituents of pepper, including bioactive chemical constituents (Fig. 12.1) and micronutrients, depend upon the development stage (Table 12.1).

12.2.1 Phenolic Compounds

Generally, peppers have extremely ample natural phenolic compounds, including phenolic acids and flavonoids, which have more chemical constituents. They add to the colour, flavour, as well as savour of the pepper fruits, also provide antioxidant properties (Jayaprakasha et al. 2012; Padilha et al. 2015). Furthermore, the amount of pepper antioxidant activity is additionally impacted by the collaboration among the complete antioxidants in the sample; for instance, carotenoids and nutrients like vitamin C and vitamin E (Conforti et al. 2007). Several toxicological studies show a possible interconnection among the utilisation of flavonoids and phenolic acids and improvements in treatments for diabetes, cancer growth, coronary disturbance, malignant growth, neurodegenerative diseases, and osteoporosis (Pandey and Rizvi 2009). Most of the flavonoids, the flavonol glycosides, for instance, *O*-glycosides quercetin, are essentially initiated in pepper natural products, such as thujin. In addition, a few glycosides (GCs) and aglycones (AgCs) of apigenin, kaempferol, luteolol, and myricetol analogues exist in red pepper fruits; for example, two luteolol *O*-glycosides, five luteolol C-GCs, and two apigenin C-GCs have been identified in the endocarp of different types of sweet bell pepper (Asnin and Park 2015; Jayaprakasha et al. 2012). Further, these chemical constituents have been analysed qualitatively and quantitatively by using hyphenated techniques like HPLC, HPLC-TLC, LC-MS/MS, ^{1}H-NMR IR, as well as mass spectrometry data, which identified their concentrations in pepper assortments to be from 4 to

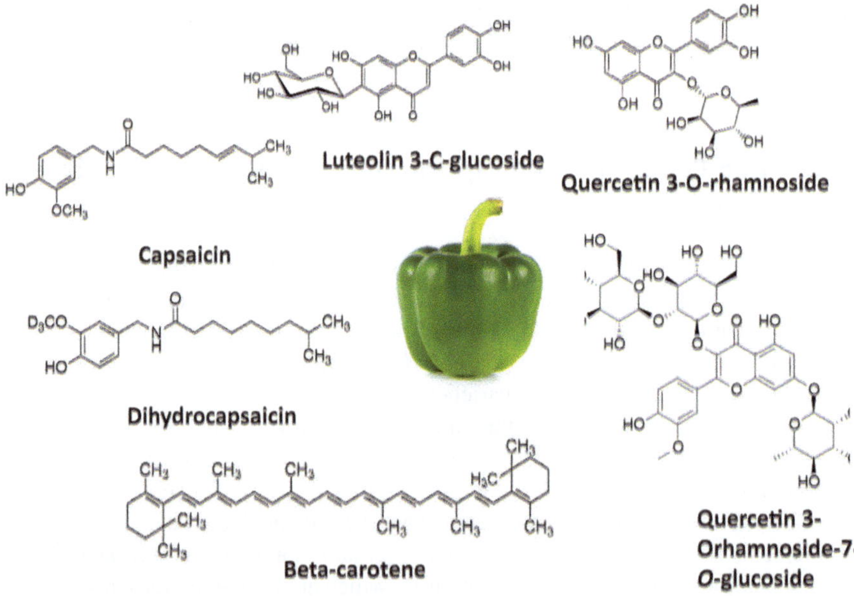

Fig. 12.1 Major chemical constituents of *Capsicum annuum* L

20 mg 100 g^{-1} fresh weight (Asnin and Park 2015). Normally, flavonoids are recognized for their potential antioxidant and antimicrobial effects, associated with the presence of a greater number of hydroxyl (–OH) functional groups at particular points as well as in between C_2 and C_3 carbons C=C bonds, as in quercetin-3-*O*-L-rhamnopyranoside in peppers (Materska and Perucka 2005). Additionally, phenolic acids, including 3,4-dihydroxycinnamic acid, 4-hydroxycinnamic acid, sinapic acid, and coniferic acid glycosides, are present in red pepper fruits and their derivatives, including *p*-sinapoyl-β-D-glycopyranoside and *p*-feruloyl-β-D-glycopyranosides, which indicate significant in vitro antioxidant activity (Materska and Perucka 2005). The placentas and pericarps of peppers consist of hydroxycinnamic acids as major chemical constituents, ranging between 50 and 500 mg 100 g^{-1} fresh weight, based upon the type of growing environmental conditions and refining stage (Sakakibara et al. 2003). The percentage of total phenols decreases during the ripening of pepper fruits as well, the chemical composition might be different, as insufficient studies have been carried out (Conforti et al. 2007; Marin et al. 2004), with the unripe green sweet peppers exhibiting more flavonols when contrasted with unripe red and red matured sweet peppers (≈ 5.0 times less). The variation of chemical constituents is not significant; for example, the ferulic acid and sinapic acid glycosides are elevated when the peppers go from a green to a red colour stage. Other researchers found anthocyanins in a few genera of purple as well as red peppers, which are defined by the fundamental flavylium cation (FLC). The quantity of anthocyanins ranges in red ripe fruits from ≈28 mg to 100 g^{-1} fresh weight and in yellow ripe natural fruits from ≈0.5 mg 100 g^{-1} fresh weight. According to different

Table 12.1 Constituents of bioactive compounds and micronutrients in (*Capsicum annuum* L.) sweet pepper (100 g fresh weight)

Type of the substance	Micronutrients/bioactive compounds	Quantity in (mg)
Vitamins	Vitamin A	0.01–0.17
	Vitamin B6	0.16–0.30
	Betaine	0.1
	Folate total	0.01–0.05
	Choline	5.5
	Vitamin C	127–300
	Niacin	0.9
	Pantothenic acid	0.315
	Riboflavin	0.02–0.1
	Thiamin	0.02–0.1
	Vitamin E (α-tocopherol)	0.6–1.7
	Vitamin K	0.002–0.015
Anthocyanins	Total anthocyanins	0.3–30
Capsaicinoids	Capsaicin	0.05–0.1
	Dihydrocapsaicin	0.05–0.1
	Total	0.1–0.4
Capsinoids	Total capsinoids	2–50
Carotenoids (provitamin A carotenoids)	α-carotene	0.02
	β-carotene	0.2–1.65
	β-cryptoxanthin	0.2–1.5
Carotenoids (non-provitamin A carotenoids)	Lutein + zeaxanthin	5–52
	Total	13–325
Phenolic compounds	Luteolin	0.6
Flavonols	Quercetin	0.5
	Total	5–22
Minerals	Boron (B)	0.02–0.20
	Calcium (Ca)	5–20
	Copper (Cu)	0.012–0.09
	Iron (Fe)	0.1–1.4
	Magnesium (Mg)	10–28
	Manganese (Mn)	0.02–0.15
	Phosphorus (P)	20–50
	Potassium (K)	210–345
	Selenium (Se)	0.0001–0.001
	Sodium (Na)	3–9
	Zinc (Zn)	0.1–0.9

studies, these peppers exhibit major anthocyanins, including delphinidin-3-*trans*-coumaroylrutinoside-5-glucoside (Padilha et al. 2015). The assessment of biologically active polyphenols is relevant material for this investigation due to the need to elucidate their potential benefits for human health. Polyphenol substances are processed in the intestinal and liver cells by sulphation, hypermethylation, and glucuronidation and are then retained within the digestion process and disseminated to various organs of the body. Among others, the solubility and biological activity strongly depend on various factors, including the intensity and conveying process in the digestive epithelium, the polyphenol fraction's stability in the gastrointestinal tract, as well as the action of the stomach microbicides in the bioconversion of these substances between their separate formats (Minatel et al. 2017). The utilisation of advanced nano-formulations with polyphenol substances might compound their harmonious impacts or join their utilisation with specific medications, which would likewise address new lines of examination, innovations, improvement, and advancement in the definition of pepper-derived products (Munin and Edwards-Lévy 2011).

12.2.2 Carotenoids

Carotenoids are mainly liquid dissolvable substances obtained from the 2-methyl 1,3-butadiene (Isoprenoid) route and deposited in the plastids in red peppers. These terpenes split the 40-carbon 2-methyl 1,3-butadiene pillar with a different type of cyclic structure at both or one end. The carbon structure is obtained from isoprenoid functional groups and carries alternating conjugated C=C bonds (Arimboor et al. 2015). Fresh peppers are decent wellsprings of carotenoids that might differ in content and composition because of hereditary contrasts and level of ripening or development; they are also affected by their growth environment and postharvest handling. In immature pepper fruits, the complete quantity of carotenoids content displays extraordinary fluctuation, going from 0.51 to 30 mg^{-1} dry mass or 10–310 mg^{-1} 100 g^{-1} crude mass (Munin and Edwards-Lévy 2011; Padilha et al. 2015); the placenta and pericarp exhibit the most comparative content of carotenoids (~0.4%). These chemical constituents are not identified in the grains. There are somewhere around 34 types of carotenoids in peppers. The red colourant of capsorubin as well as capsanthin are also important constituents in red pepper. Taking into account, capsanthin comprises between 30% and 70% of the total carotenoids. Diverse carotenoids present in orange and red chime peppers are zeaxanthin, beta-carotene, and beta-cryptoxanthin (Arimboor et al. 2015; Ha et al. 2007). The green colour of the pepper changes to red yellow or orange the presence of various compounds (Azevedo et al. 2019). The utilisation of carotenoids by the body has been connected with benefits to mental capability and cardiovascular well-being and might assist with forestalling some diseases. Additional enhanced carotenoids including xanthophyll and β,β-carotene-3,3′-diol are important for the eye health (Eggersdorfer and Wyss 2018). Regarding solubility, the greater part of the carotenoids, for instance alpha and beta-carotene, zeaxanthin, β-cryptoxanthin, and lycopene, must be integrated into micelles and blended in with mucus salts, lipoproteins, and compounds to be conveyed to the stomach intestinal cells (Fiedor and Burda 2014). This process is the main step impacting carotenoids' bioaccessibility, which is undertaken in the gut and small gastrointestinal tract.

There are a few variables impacting micellisation, which certainly influences the bioavailability and assimilation; for instance, the existence of fibre as a food constituent or the restraint of colorectal amylase utilised to medicate or handle obesity both lessen micellisation. Fernández-García et al. (2012) observed that the design of carotenoids, including the esterification of lutein, capsanthin, zeaxanthin, and β-cryptoxanthin, increases their solubility and accordingly enhances their pharmacokinetics. Bohn et al. (2017) reported that when communicating with extracellular fluid, these chemical constituents are incorporated towards different materials with enormous contrasts among the organs.

12.2.3 Capsaicinoids

Capsaicinoids are basically acid (–COOH) amines (–NH_2) of 4-hydroxy-3-methoxybenzylamine and branches of unsaturated fatty acids consisting of 8–13 carbons (Kobata et al. 2013). Generally, the capsaicinoids mainly present in peppers are capsaicin and dihydrocapsaicin. Jointly, in many cases, they are present in amounts more than 80% of aggregate capsaicinoids, while different analogues happen in much more modest amounts (Perucka and Materska 2001). Other than these two significant capsaicinoids, other capsaicinoids are separated or isolated from red peppers or hot peppers, including nonivamide, norcapsaicin, nornorcapsaicin, nordihydrocapsaicin homocapsaicin, and homodihydrocapsaicin (Barbero et al. 2006a). Capsaicinoids are the chemical constituent in pepper that is responsible for the pungency. The strength of the pungency depends on the chemical structure of capsaicinoids, which consists of an amide bond connected to the acyl chain as well as vanillyl ring (Barbero et al. 2008). The Scoville Heat Unit (SHU), a simple organoleptic test to assess the peppers' sharpness, is typically used to gauge how strong the capsaicinoids are at stinging, irritating, burning, or pungency (Dang et al. 2018). Capsaicinoids, including capsaicin and dihydrocapsaicin, show approximately 16.1×10^{-6} SHU. Capsaicinoids could be synthesised through the phenylacrylic acid route in the pepper placenta and the white rib (Topuz and Ozdemir 2007). Normally, the amount of capsaicinoids differs with genotype and repine stage (0.1–40 mg 100 g^{-1} fresh weight). For instance, capsaicin levels depend on the genotype: different types of sweet pepper exhibit extreme capsaicin variation (0.07–0.1 mg 100 g^{-1} fresh weight) (Deepa et al. 2007), with the progressions of these substances not so articulated by various ripening phases. In hot or red pepper assortments, the amounts of capsaicinoids are significantly more than in sweet pepper, which is from 0.01% to 0.5% fresh weight. In powerful chillies the amount would comprise 1–2% of fresh weight in a few Mexican species (Orellana-Escobedo et al. 2013; Perucka and Materska 2001; Topuz et al. 2011). Capsaicin and capsaicinoids exhibit significant antibacterial activity and, subsequently, potential pharmacological and clinical applications for the therapy of sensual, rheumatological, cardiovascular, respiratory, neurological, gastrointestinal, inflammatory, and oxidative stress diseases (Hayman and Kam 2008).

12.2.4 Capsinoids

Normally capsinoids are non-spicy substances; they can be identified in several genotypes of peppers with identical molecular constitution to the capsaicinoids, including 4-hydroxy-3-methoxybenzyl 8-methylnon-6-enoate and its derivatives dihydrocapsiate and 4-hydroxy-3-methoxybenzyl 7-methyloctanoate (Singh et al. 2009), which are found in sweet chilli peppers including *C. annum* L. var. (CH-19). The basic structure of capsaicin is an aromatic with unsaturated amide (Kobata et al. 2013). Although these substances are known, their systems of activity have not yet been defined. In any case, they exhibit fascinating antimicrobial activity (Bacon et al. 2017). Ohnuki et al. (2001) and Saito and Yoneshiro (2013) observed that lowered muscle-to-lipid ratio workouts in humans and enhanced energy digestion in mice as well as brown fat oxidisation. However, the particular substance has been introduced as an encouraging elective for people who avoid capsaicin-containing food varieties because of their pungency (Ludy et al. 2012).

12.2.5 Vitamins

Peppers are a rich source of vitamins, having significant amounts of vitamin A, vitamin E (α-tocopherol), vitamin C as well as vitamin B_9 (Guil-Guerrero et al. 2006; NIH 2008) represented in Table 12.1. For example, a moderate-size bell pepper (~100 g fresh weight) provides ~160% of the recommended daily consumption of vitamin C (L-ascorbic acid), which is 80 mg for male and 65 mg for female, as well as around 15% of the recommended daily allowance for previtamin A (Pre-VM-A) (900 μg for male as well as 600 μg for female individually), vitamin E (12 mg) as well as vitamin B_9 (400 μg normal female, 600 μg for expectant female) (Kantar et al. 2016; Navarro et al. 2006). Vitamin C and vitamin E have a specifically significant antioxidant activity, decreasing the degrees of unpaired electrons and controlling over-oxidation responses in humans and, thus, decreasing the risk of cardiac disease and some tumours (Navarro et al. 2006). Carotenoids, including β-carotene as well as β-cryptoxanthin, are responsible for the provitamin A (pro-VM-A) response, as they exhibit significant biological activity, as the dynamic forms of vitamin A. They are connected to decreasing the risk of age-related macular degeneration (RAMD) and lung cancer (Eggersdorfer and Wyss 2018). Toxicological studies have proposed a converse relationship between folate level and the risk of malignant growth, intellectual disability, and cardiovascular diseases, with higher levels supporting foetal and new born child mental health (Kantar et al. 2016). Peppers contain various amounts of vitamins, depending upon many factors, including genotype (assortment), ripening stage and season of harvest, and postharvest treatment and repository conditions. Generally, the studies that estimate vitamin C levels during ripening often show that red peppers have significantly more vitamin C than green peppers. (Marin et al. 2004; Martínez et al. 2007). Postharvest storage at 4 °C or less is critical to maintain the concentrations of vitamin in pepper. Howard and Wildman (2007) reported that raw red peppers stored in modified atmosphere packing (MAP) about 15 days preserved more vitamin C (~80%) when contrasted with open air system (~50%); further, the amount of vitamin C also decreased because of handling or processing procedures, including freezing, cutting, stripping, slicing, crushing, and drying, when plant tissues might be exposed to more atmospheric oxygen. However, drip-drying was shown to be a suitable technique for the protection of biologically active substances (Martínez et al. 2005).

12.2.6 Minerals

Peppers also contain necessary dietary minerals; for instance, sweet peppers contain minerals like boron (B), calcium (Ca), copper (Cu), iron (Fe), magnesium (Mg), manganese (Mn), phosphorus (P), potassium (K), selenium (Se), and sodium (Na). Their capacity present in sweet PPs is represented in Table 12.1. The amount of minerals shifts extraordinarily with assortment, environmental exposure, and maturity phase (Rubio et al. 2002). Red peppers have elevated levels of B, Cu, Mg, Mn, Fe, K, P, and Zn compared to green peppers, although Pérez-López et al. (2007) research study found remarkably less concentration levels of Cu, K, Mn as well as Fe in red peppers contrasted with green peppers. However, mineral substances normally showed a critical difference among evaluated cultivators regardless of the maturing stage. The concentration of potassium (K) significantly identified in red and green peppers (~2.5 g/100 g dry weight (DW)) contrasted with different minerals, and higher concentration (2 overlay) of calcium (~150 mg 100 DW) as well as sodium (20 mg 100 g DW) in green peppers from a store contrasted with freshly collected green peppers. Some minerals hardly varied between samples, such as Cu, Zn, and Mn. Pérez-López et al. (2007) observed that agricultural techniques, like traditional and organic farming, might also affect the amount of minerals in peppers, which overall showed significant mineral capacity as it would be grown naturally.

12.3 Explicit Extraction Techniques and Isolation of Chemical Constituents

Peppers contain a mixture of phytochemical compounds which can be isolated using various extraction strategies. These methods are mainly partitioned into two groups; physical strategies, including continuous extraction, magnetic stirring, and maceration, as well as hyphenated methods, like Ultrasound Assisted Extraction (UAE), Microwave Assisted Extraction (MAE), Enzymatic Assisted Extraction (EAE), Compressed Fluid Extraction (CFE), Pressurised Liquid Extraction (PLE), and Supercritical Fluid Extraction (SFE) (Saini and Keum 2018; Table 12.2). Traditional strategies give significant recovery of total carotenoids, although they need more time as well as solvent for the process of extraction. Hence, they may be commercially more expensive and of restricted use because of ecological and security problems (Luque de Castro and Garcıá-Ayuso 1998). Current methods have decreased extraction time and used fewer solvents like ethanol, with a lower influence on the climate. These techniques have achieved greater extract purity with the most noteworthy complete yield of the objective carotenoids contrasted with conventional techniques. Advanced methods like UAE, MAE, and SFE are the most importantly utilised, and a few models are represented in Table 12.2. The fundamental measurements of these extraction strategies, including the amount of solvent, amount of sample taken, time, and temperature, must be exactly followed as preparatory job for various analyte lattices. The key factors influencing the deviation from these standards are the concentration of the specific compounds in the extract, their consistency and fixation, as well as the kind of compartment lattices. The selection of a suitable solvent, mainly based on the extremity of the specific carotenoids, is a highly crucial stage for their systematic extraction (Saini and Keum 2018). For instance, in SFE, carbon dioxide (CO_2) is mostly used as a solvent,

Table 12.2 Extraction procedures for *Capsicum annuum* and its derived products

Extraction technique	Origin	Conditions of extraction	Isolated compounds	References
Supercritical fluid extraction of emulsion (SFEE) and ultrasound emulsification	Dry pepper	Flow rate 1.98×10^{-4} kg/s; 15 MPa; 40 °C emulsification with Hi-Cap 100 and flow rate 0.5–1 mL/min	Capsaicinoids	de Aguiar et al. (2016)
Superficial CO_2 extraction (SC-CO_2) and spray dry microencapsulation	Red pepper by-products (Piquillo)	Flow rate 2000 mL/h; 60 °C and 24 MPa; spray dried Buchi B-290 and gum Arabica; time 90–120 min; pepper weight 30 g	Carotenoids and vitamin E	Romo-Hualde et al. (2012)
Ultrasound assisted extraction (UAE)	Hot pepper	Solvent 100% methanol; time 10 min; temperature 50 °C	Capsaicinoids	Barbero et al. (2008)
Enzyme assisted extraction (EAE)	Paprika powder (red chilli)	Two-stage extraction process with ethanol and hexane. Non-commercial enzymes from *R. nigricans*	Carotenoids and capsaicinoids	Salgado-Roman et al. (2020)
Microwave assisted extraction (MAE)	Triturated peppers (fresh)	Solvent 100% ethanol 25 mL; time 5 min; temperature 125 °C; weight 0.5 g	Capsaicinoids	Barbero et al. (2006a)
Pressurised liquid extraction (PLE)	Hot pepper	Temperature 200 °C; 100 atm; extraction time 30 min	Capsaicinoids	Barbero et al. (2006b)
Superficial CO_2 extraction (SC-CO_2)	Red pepper flakes (commercial)	Particle size 0.27 mm; flow rate 0.57–1.25 mm/s; 40 °C and 540 bar	Carotenoids	Uquiche et al. (2004)

sometimes along with co-solvents, such as ethanol (Perva-Uzunalić et al. 2004), and it is usually evaluated at various temperatures and pressures for the development of the supercritical carbon dioxide extract (SC-CO_2-E), yielding more capsaicinoids and colouring substances from red chillies at 40 °C of temperature and 400 bar pressure, with the 80% of carotenoids and 96% of capsaicinoids separated from the unrefined substances. SC-CO_2-E has several advantages, including the chance of extracting thermally unstable chemical constituents and the assortment of an extract free from solvent residues. Also, it is possible to separate parts of *Capsicum* oleoresin simultaneously, for example, the assortment of one portion enhanced in capsaicinoids and a maturity division enhanced in carotenoids (de Aguiar et al. 2016; Fernández-Ronco et al. 2013). Generally, in the pepper matrices, the main extracted substances are capsaicinoids; furthermore, carotenoids are principally separated as pepper oleoresin, which is generally utilised as a characteristic food colour (Uquiche et al. 2004). *Capsicum* oleoresin is a significant result of pepper process, with use in the pharmaceutical as well as food production industries. Fernández-Ronco et al. (2013) observed that natural resin (oleoresin) is made of liquid chemical constituents, such as triglycerides and unsaturated fatty acids, pigments like carotenoids and capsaicinoids, and some other chemical constituents. Among the various procedures to concentrate and utilise these substances for well-being, de Aguiar et al. (2016) observed advancing capsaicinoids for food colouring and carotenoids for use; the utilisation of SC-CO_2-E procedures might be the best way to extract natural resin and to isolate it within improved portions of biologically active substances like carotenoids, capsaicinoids, and vitamins (Fernández-Ronco et al. 2013). In that regard, the US Food and Drug Administration (FDA 2017a) this innovation is perceived as eco-friendly, utilising small quantity of other solvents, for instance, hexane and ethanol that have been approved by the European Medication Organization (EMA 2015). The costs and benefits of SC cycles have been evaluated in order to determine their financial viability. Fernández-Ronco et al. (2013) observed an economically beneficial add-on to valued nutritional products in the food business.

12.4 Applications

12.4.1 Applications in Agro-Food Industry

The various peppers are a spices made from dried and ground fine red powder are known as paprika or red chilli powder, it is traditionally made from *C. annum*, which has been utilised as zest in meat items, tidbits, snacks, soups, and sauces for more than 70% of its manufacture. Buckenhüskes (2003) reported that the overall production of paprika substances is ≈ 60,000 tons each year, compared to the 1500 tons of paprika natural resin fabrication. In addition to the conventional purposes of paprika in production, adding to astringency, taste ascribe, and colour, different values can be obtained for food manufacturing, for instance, the antioxidant and antimicrobial agents. Table 12.3 illustrates a few of these numerous applications of *Capsicum* in the agro-food industry (AFI).

12.4.1.1 Antimicrobial Activity of Capsaicin

Many research investigations have shown the efficacy of capsaicin in antimicrobial activity against gram-negative (gram −ve) and gram-positive (gram +ve) microorganisms that cause food poisoning; however, data about dependable substances for these impacts is generally scant. Further, studies reported that *C. annuum* extracts, including morron, habanero, and serrano, show potential maximum inhibition activity against *Listeria monocytogenes* (*L. monocytogenes*), *Staphylococcus aureus* (*S. aureus*), *Salmonella enterica* (*S. enterica*), *Bacillus cereus* (*B. cereus*), and *Typhimurium*, proposing that polyphenolic acids like cinnamic acid and *m*-coumaric acid were more likely responsible for these inhibition activities, in contrast to capsaicin and dihydrocapsaicin that did not exhibit these impacts (Buckenhüskes 2003; Dorantes et al. 2000). One study described significant antimicrobial activity specifically against listeriosis from *C. annuum* from Jalapeno separated fractions high in cinnamic acids (~5 ppm). The researchers also described a particular antimicrobial activity of capsaicinoids; the derivatives of capsaicinoids exhibited potential antimicrobial activity against gram +ve organisms through the interruption of biofilm arrangement. This effect might be because of their ability to chelate calcium and other minerals, which impacts biofilm engineering advancement. In agreement with these outcomes the minimum inhibitory concentrations of *C. annuum* L., bell pepper, was reported at 1.5 mL 100 g^{-1}. These extracts prevent the growth of *S. typhimurium* in ground beef meat (Bacon et al. 2017; Careaga et al. 2003). *C. annuum* extracts exhibit significant antifungal activity against various fungi including *Debaryomyces hansenil* and *Penicillium expansum* (Nazzaro et al. 2009). Likewise, ethanolic extracts of green and cayenne pepper was used in cow's milk while manufacturing Egyptian Kareish cheese, which exhibited potential antimicrobial activity against *S. aureus*, coliforms, moulds, and microflora, being consistent in the amount of cayenne pepper (1% w/v) and green pepper (3–6% w/v) (Kurita et al. 2002; Wahba et al. 2010). Hence, pepper-determined substances might be valuable in the improvement of advanced and secure additives, which can be utilised as antimicrobial activities for food protection, to resist the food poisoning microbes in food sources as well as items' deterioration, avoiding the utilisation of other artificial additives, for instance, sodium metabisulfite ($Na_2S_2O_5$), sodium benzoate and nitrate, which have been periodically connected with hypersensitive responses and the expected arrangement of nitrosamines (Nazzaro et al. 2009). The bactericidal property of capsaicin is because of its capacity to prompt bibulous pressure, obliterate the cell film, and restrain the articulation of the nucleic acid connected with ribosomal constituents. Additionally, Marini et al. (2015) observed that cell growth and entry had detrimental impacts on the cell. The antimicrobial activity mainly depends on the concentrations of capsaicin. The antimicrobial activity of capsaicin is presented in Table 12.4.

Table 12.3 Applications of *C. annum* and its derived products in pharmaceutical, cosmetic, and agro-food industries

Bioactive substance	Source	Characteristic properties and applications	Uses	References
Capsaicinoids and carotenoids	Pungent paprika	Colour and flavour of food dishes and products	Agro-food industry	Tepic et al. (2008)
Carotenoids	Sweet red pepper	Colour enhancer, lipid oxidation inhibitor in pork meat, and nitrite replacer		Bazan-Lugo et al. (2012)
Capsaicinoids and carotenoids	Sweet red pepper	Soups, snacks, sauces. Colour stability in meat products		Fernández-López et al. (2002), Pruthi (2004)
Capsaicinoids and carotenoids	Sweet paprika, Pungent paprika	Increasing the sensory properties of food products		Tepic et al. (2008), Uquiche et al. (2004)
Carotenoids	Yellow pepper	Increase of sensory properties in wheat bread and excellent antioxidant		Danza et al. (2014)
Carotenoids	Sweet paprika	Increase of sensory and physical properties of cooked marinated chicken		Yusop et al. (2012)
Capsaicinoids	Chilli pepper	Antimicrobial activity against *P. aeruginosa*, *S. typhimurium* in raw beef meat in combination with sodium chloride		Careaga et al. (2003)
Carotenoids and capsaicinoids	Capsicum fruits	Colouring and biological activities. antioxidant activity, analgesic activity, provitamin A	Cosmetic, pharmaceutical, and agro-food industry	Fernández-Ronco et al. (2013)
Carotenoids and capsaicinoids	Chilli peppers	Increase sensor properties and biological activity including antimicrobial, anti-inflammatory, and antioxidant activity		de Aguiar et al. (2016)
o-coumaric acid, *m*-coumaric acid, ferulic acid, and cinnamic acid	Chilli peppers	Antibacterial activity against *B. cereus*, *L. monocytogenes*, *S. aureus*, and *S. typhimurium*		Dorantes et al. (2000)
Polyphenols and carotenoids	Sweet pepper	Antibacterial activity against *E. coli*, *B. cereus*, etc., antifungal activity against *D. hansenil* and *P. expansum*		Nazzaro et al. (2009)

(continued)

Table 12.3 (continued)

Bioactive substance	Source	Characteristic properties and applications	Uses	References
Capsidiol	Sweet pepper	In vitro bacteriostatic effect against *H. pylori*		De Marino et al. (2006)
Vitamin C and carotenoids	Capsicum fruits	Anti-inflammatory and antioxidant activity, protecting skin oxidative and UVA medicated damage	Cosmetic & pharmaceutical industry	Telang (2013)
Capsaicin	Chilli pepper	Medicinal agent in chronic pain syndromes and in chronic inflammatory skin diseases	Pharmaceutical industry	Căruntu et al. (2015), Lysy et al. (2003)
Capsaicinoids	Cayenne pepper	Pharmacological activities: alter appetite sensations by higher satiation		Andersen et al. (2017)

12.4.1.2 Antibacterial Activity

Wang et al. (2022) recently observed that capsaicin has potential antimicrobial activity against gram +ve and gram –ve bacteria. Several researchers reported the antimicrobial activity of capsaicin, and noticed that capsaicin significantly constrained or inhibited *Bacillus subtilis* from *E. coli* can be inhibited at concentrations of 150–300 μg/mL (Molina-Torres et al. 1999). Another study assessed the antimicrobial activity against *S. typhimurium*, *S. aureus*, *B. cereus*, and *L. monocytogenes* using different extracts from chilli peppers, including pimento, habanero, and Serrano (Dorantes et al. 2000). Further, one study reported the in vitro effects of various concentrations of capsaicin against *H. pylori* and found that capsaicin exhibited significant antibacterial activity against *H. pylori* even at a minimal concentration (25 μg/mL). Additionally, it was found that after 4 hours of incubation, capsaicin showed significant antibacterial activity at a concentration of 30 μg/mL. Furthermore, at a dosage of 100 g/mL, capsaicin strongly inhibited the synthesis of choleragen also known as cholera toxin (CTX) in *Asiatic cholera* (Chatterjee et al. 2010). Furthermore, various compounds including capsaicin from *Capsicum frutescens* showed antimicrobial activity against several microbes (Nascimento et al. 2014). In any case, their outcomes demonstrated that less minimal inhibition concentration values were required for *E. coli* (5 μg/mL), *K. pneumonia* (0.6 μg/mL), and *S. aureus* (1.2 μg/mL). Furthermore, researchers examined the antiviral and antibacterial effects of capsaicin on erycette-resistant, cell-intrusive-type *S. pyogenes*. These researchers found that the minimal inhibition concentration of capsaicin was 64–128 μg/mL, and it showed antibacterial activity. Also, they found that infective amount of capsaicin brought about an expansion in the biofilm fabrication and a decrease in the haemolysis action and cell intrusiveness. Further,

Table 12.4 Antimicrobial activity of free and blended forms of capsaicin in foodborne pathogens

Type	Microorganisms	Concentration	Results	References
Capsaicin	*V. cholera*	0.5, 1.0, 15, 50, and 100 μg/mL	Production of cholera toxin in V. cholera was inhibited by capsaicin in a dose-dependent manner.	Chatterjee et al. (2010)
	B. subtilis, *C. albicans*, *E. coli*, *E. faecalis*, *K. pneumonia*	0.02–100 μg/mL	Strong inhibition activity at lower concentrations	Nascimento et al. (2014)
	S. pyogenes	–	Moderate inhibition against *S. pyogenes*	Marini et al. (2015)
	Murine norovirus	0.1 mg/mL	Excellent inhibition against virus	Yang et al. (2016)
	Murine virus and feline calicivirus	10–200 μM	CSn potentially reduced murine virus strong inhibition against feline calicivirus	Bae et al. (2018)
	A. carbonarius and *Aspergillus* part of *Nigri* strains	–	Excellent inhibition against nigri strains	Kollia et al. (2019)
	E. coli	–	Significantly reduced	Silva et al. (2020)
	H. pylori	–	Strong inhibitory effect against *H. pylori*	Zhao et al. (2020)
	B. thermosphacta, *E. coli*, *L. plantarum*	–	Excellent inhibition effect against *B. thermosphacta*, *E. coli*, *L. plantarum*	Orlo et al. (2021)
	C. albicans, *C. parapsilosis*, *B. cereus*, *E. coli*, and *S. aureus*	–	Strong inhibition activity	Goci et al. (2021)
Chitosan/gelatin composite films having CSn loaded hallow metal organic structure	*E. coli*	–	Excellent antibacterial activity with less viable bacteria	Zhao et al. (2020)
Ethyl cellulose composite films having different amount of CSn	*E. coli* and *S. aureus*	–	Excellent antibacterial activity against *S. aureus*	Su et al. (2020)

(continued)

Table 12.4 (continued)

Type	Microorganisms	Concentration	Results	References
Albumin bovine serum-CSn-NPs	*R. mucilaginosa*	–	–	Sanchez-Arreguin et al. (2019)
CSn-loaded nanoemulsions	*E. coli* and *S. aureus*	–	–	Akbas et al. (2018)
Sodium alginate/pullulan films incorporated with CSn	*E. coli* and *S. aureus*	–	All films incorporated with CSn inhibited the bacterial development	Zhang et al. (2018)

in vitro antibacterial activity of capsaicin against *H. pylori* was investigated (Tayseer et al. 2020), and it was found that capsaicin exhibited excellent antibacterial activity at an amount of 0.0625 μg/mL. It was also assessed that the capsaicin blend with metronidazole shows potential antibacterial activity against *H. pylori*. A study investigated the antibacterial activity of phytochemicals such as capsaicin, eugenol, and vanillin against food decomposition bacteria, including *Lactobacillus plantarum* (*L. plantarum*), *Brochothrix thermosphacta* (*B. thermosphacta*), and *Shewanella putrefaciens* (*S. putrefaciens*) and foodborne microbes, including *E. coli*, *P. aeruginosa*, and *S. aureus*, utilising the microbial sensitivity test and minimal bacterial concentration (MBC) and MIC evaluations. As shown by the outcomes, capsaicin and eugenol were the main successful chemical constituents against food decomposition microbes and foodborne microorganisms. Among these microorganisms, *L. plantarum* and *S. aureus* were the maximum and minimum impacted microorganisms by these chemical constituents, individually. The study showed that the IC_{50} of capsaicin for *E. coli*, *B. thermosphacta*, *L. plantarum*, *P. aeruginosa*, *S. aureus*, and *S. putrefaciens* was 4.79, 7.77, >10, 1.21, 0.68, and 4.28 mM, respectively. Due to the lower solubility of capsaicin, it was not possible to assess higher concentrations of capsaicin to avoid the impact of the solvent (ethanol). In addition, a researcher estimated the capsaicin's antibacterial activity against *B. cereus*, *E. coli*, *S. aureus*, and *Salmonella typhi* (*S. typhi*) and established the MIC of capsaicin for *B. cereus* (5–10 μL/mL), *E. coli* (5–10 μL/mL), *S. aureus* (5–10 μL/mL), as well as *S. typhi* (>10 μL/mL). Many studies have investigated the antibacterial activity of capsaicin in epitomised structure, arranged capsaicin-stacked high-pressure homogenisation (HPH) and nano-emulsions (NEMs) by ultra-sonication, as well as evaluated the antibacterial activity of NEMs against foodborne microorganisms *E. coli* and *S. aureus*. They noticed that the NMEs prepared by the HPH technique displayed significant antibacterial activity on *E. coli* and *S. aureus* as well as decreased the populace of these microbes by up to 2.8 and 5.9 logs, individually, at pH of 7.4. They likewise identified that the antibacterial activity of NMEs against *E. coli* decreased with pH. Accordingly, they recommended that the capsaicin could have lower antibacterial activity against these pathogens' resistance in pH <6.99. Further, a researcher prepared sodium alginate $(C_6H_7NaO_6)_n$ (SG) and pulluan $(C_6H_{10}O_5)_n$ (P) blended consumable films combined with different capsaicin capacity to assess the antibacterial activity of these edible sheets against *S. aureus* and *E. coli*. As demonstrated by their outcomes, the unadulterated SG-P film displayed minimal antibacterial activity against *E. coli* and *S. aureus*, contrasted with the unadulterated bacterial development design; while the above composite films mixed with capsaicin exhibited a reasonable restraint impact against *E. coli* and *S. aureus* (Zhang et al. 2018). Therefore, the antibacterial activity of blended films increased with higher capsaicin content, so the greatest antibacterial activity, particularly against *E. coli*, was shown by the composite film with the highest amount (8%) of capsaicin content (Zhang et al. 2018). In addition, another researcher estimated the antibacterial activity of poliglusam-blended films consisting of various amounts of capsaicin (0.3, 0.6, and 1.2 mg) against to *Bacillus thuringiensis* (*B. thuringiensis*), *E. coli*, *Enterobacter aerogenes* (*E. aerogenes*), *Pseudomonas aeruginosa* (*P. aeruginosa*), *Streptococcus mutans* (*S. mutans*), and *S. typhimurium* (Akyuz et al. 2018). Akyuz et al. (2018) reported that with the increasing amount of capsaicin against almost all bacteria, the inhibitory impact also increases. Furthermore, each of their samples displayed maximum inhibition against gram –ve bacteria when contrasted with gram +ve bacteria. They observed that the composite film exhibits minimum and maximum resistance against gram-negative bacteria such as *E. aerogenes* and *P. vulgaris*. Moreover, composite films exhibited maximum resistance against *B. thuringiensis* (gram +ve bacteria). Furthermore, in the evaluation of antibacterial activity, various concentrations of capsaicin-stacked metal-organic frameworks (MOFW) (like 0%, 1%, 2%, 4%, and 8%) loaded with chitosan/gelatin composite films were prepared against *E. coli*. Researchers found that increasing capsaicin concentration resulted in superior antibacterial activity of the blended films; those carrying 4% and 8% of capsaicin-MOFW displayed excellent antibacterial activity (Zhao et al. 2020). Later, ethyl cellulose (EC) blended films were integrated with various concentrations of capsaicin (like 0.1, 0.3, and 0.5 g) to estimate antibacterial activity against *E. coli* and *S. aureus* utilising the disc diffusion method (Su et al. 2020). The study found that the zone of inhibition increases with the augmentation of capsaicin amounts of the sheets ($\rho < 0.05$); moreover, the inhibition of the sheets was very clear on *E. coli* and *S. aureus*. They additionally found that the EC composite films integrated with 0.5 g of capsaicin displayed the maximum antibacterial impact, particularly against *S. aureus*.

12.4.1.3 Antifungal Activity

Capsaicin and its analogues exhibit significant antifungal activity. Capsaicin displayed potential antifungal activity against *Saccharomyces cerevisiae* (*S. cerevisiae*). Likewise, from the beginning, the maximum amount of capsaicin (150–300 μg/mL) promoted the improvement of the cell development of *S. cerevisiae*; but a high amount of capsaicin (300 μg/mL) potentially affected the long-term (24 h) development of this yeast. Furthermore, a researcher investigated the antibacterial activity of capsaicin microcapsules (MCs) acquired from acacia tannins and gelatin against *Aspergillus niger* (*A. niger*) and *Botrytis cinerea*

(*B. cinerea*). In addition, they also assessed the impact of certain factors, for instance, pH, concentration of MCs, and microcapsule (MC) delivery behaviour. An acidic medium (pH = 5.0) was viewed as the optimum pH for the antibacterial activity of MCs (Xing et al. 2006). Furthermore, in accordance with the results, *A. niger* and *B. cinerea* displayed indistinguishable reactivity to the MCs. Additionally, the capsaicin MCs displayed minimum inhibitory impact on *A. niger* and *B. cinerea* at less than 590 μg/mL. In any case, increasing the capsaicin MCs increases the antimicrobial activity until it becomes steady at 1170 μg/mL. The researchers found that the MCs delivery method had a potential effect on antifungal activity, which is viewed as a significant factor in the storage time span of food sources. Hence, they suggested that capsaicin microcapsules could be utilised as antimicrobial agents in foods. Further, in another study, capsaicin was isolated from methyl cyanate (MeCN) extract of *Capsicum frutescens* (*C. frutescens*) to estimate the antimicrobial activity against *Candida albicans* (*C. albicans*). The study found that the MIC of capsaicin was 25 μg/mL for *C. albicans* (Nascimento et al. 2014). Further, another study assessed the inhibition of albumin bovine serum (ABS)-capsaicin nanoparticles (NPs) on the yeast *Cryptococcus rubber* (*C. rubber*). The study showed that the antimicrobial activity of these NPs was reliant on the grouping of capsaicin (Sanchez-Arreguin et al. 2019). Furthermore, researchers investigated that C. rubber cells does not grow at low concentarion of capsaicin (16.2 μg/mg of ABS), and that C. rubber cells exhibit at high concentraions of capsaicin (65.0 μg/mg) due to significant growth inhibition. Furthermore, researchers considered anti-ochratoxigenic and antimicrobial properties of capsaicin on the *Aspergillus* group Nigri strains in grapes, including *Vitis vinifera* L. (Kollia et al. 2019). The study showed capsaicin exhibited significant inhibitory impacts against parasitic development and fabrication of ochratoxin A (OTA) by the above nigri strains at almost 29–78% and especially *A. carbonarius* at 61.5% (Kollia et al. 2019). Hence, they recommended that CSn would be able to influence the biosynthesis of OTA in every examined strain and, hence, it may be useful as an agent to impede OTA pollution in grape plantations. Later, researchers examined the anti-aflatoxigenic and antifungal impacts of the ethanol concentrates of *C. chinense* consisting of capsaicin (65.51 mg/g) against *A. parasiticus*. The study showed the pure concentrates of capsaicin and *C. chinense* exhibited a significant zone of inhibition against *A. parasiticus*. Between these two, pure capsaicin showed an excellent zone of inhibition when contrasted with *C. chinense*. The development of the zone of inhibition of pure capsaicin showed 51–70% and *C. chinense* showed 24–45% at 50–200 μg/mL. Further, they found that capsaicin and *C. chinense* could impede the germination rate of spores by almost 42% and 48%, individually, at 200 μg/mL. Also, these researchers assessed the anti-aflatoxigenic activity of capsaicin and *C. chinense*; pure capsaicin exhibited less anti-aflatoxigenic activity when contrasted with *C. chinense*. Pure capsaicin had an effect at 200 μg/mL, and *C. chinense* had an effect at 150 μg/mL. The highest anti-aflatoxigenic activity appeared at 150 μg/mL. Furthermore, a researcher estimated the antifungal activity of capsaicin against *C. albicans* (YEPGA 6183 and 6379), *Candida tropicalis* (*C. tropicalis*), and *Candida parapsilosis*

(*C. parapsilosis*). Generally, capsaicin exhibited significant antifungal activity against all fungal strains (Goci et al. 2021).

12.4.1.4 Antiviral Activity

It has been described that capsaicin could have potential antiviral activity as well as antimicrobial activity. Although one study reported that capsaicin exhibited significant antiviral activity against the essential cutaneal herpes simplex virus (HSV) in *Cavia porcellus*. Stanberry et al. (1992) reported that capsaicin does not show antiviral activity against HSV in cell cultures. Further, in other investigation, capsaicin exhibited potential antiviral activity against murine norovirus (MNV). A potential inhibitory impact on MNV (24.14 ± 2.28%) was observed at 1 mg/mL of capsaicin (Yang et al. 2016). In addition, capsaicin exhibited significant antiviral activity against MNV and cat calicivirus (CCV) or feline calicivirus (FCV) in Napa cabbage kimchi (NCK). Capsaicin displayed a concentration-reliant restraint impact on MNV at 5–200 μM and decreased MNV reproduction by around 0.1 log PFU/mL at 200 μM. But capsaicin does not exhibit impeding activity against FCV. In a study of antiviral activity of red pepper capsaicinoids against MNV in the agitation of baek kimchi as well as cabbage kimchi (Lee et al. 2019), the researchers observed that capsaicin impeded MNV during fermentation. Both dihydrocapsaicin and capsaicin were shown to diminish MNV strongly (Lee et al. 2019). Additionally, it has been observed that capsaicin displayed a potential antiviral effect when compared with dihydrocapsaicin at 100 μM. Furthermore, similar research investigated the antiviral activity of capsaicin against Lassa fever virus (LFV). The findings demonstrated that capsaicin restrained the LFV lineage section at half maximum effective concentration (EC_{50}) ranging from 5.0 to 11.0 mmol/L. Moreover, in spite of displaying a particular antiviral activity as well as restraining the section of LFV strains, capsaicin does not impede the arenavirus, including LFV and benign lymphocytic. Further, a researcher examined the connection between a high-capsaicin everyday diet and coronavirus (CoV); it has been reported that people who consumed a high amount of capsaicin daily were less likely to contract CoV, so that the nations with high capsaicin diets had fewer CoV cases (Gharaibeh et al. 2020). Furthermore, a researcher examined molecular dynamics and docking methods to investigate the structure underlying the disturbance of the viral 3CL-protease of coronavirus actuated by irrevocable capsaicin. They observed that capsaicin attached to the 3CL-protease of CoV and initiated conformational changes, including flopping in its molecular arrangement. These researchers reported that additional investigations are needed to assess whether such molecular structural alteration in protease can supress CoV.

The antimicrobial activity of capsaicin has been demonstrated in many studies. The amount of capsaicin is an essential characteristic in its antibacterial, antifungal, and antiviral activity so that at less amounts, it could apparently act as an advance supplement in certain strains. Moreover, pH is also another essential feature in the antimicrobial activity of capsaicin. Regardless, the assessment of the impact of various aspects on this capsaicin needed future investigation. There is, in fact, widespread agreement about the effectiveness of capsaicin against both gram+ve and gram-ve bacteria, especially *E. coli* and *S. aureus*. Most investigations showed that capsaicin influences *E. coli* and *S. aureus*.

12.4.1.5 Antioxidant Activity

Normally, in the food processing industry, antioxidant activity is characterised as the action of a material or chemical substance to impede or postpone the decomposition of oxidisable compounds, such as fats, in very small amounts (>1%, normally 1–1000 mg/L) contrasted with how much materials that it ought to protect (Amorati et al. 2013). Generally, antioxidants are mainly divided into natural and artificial antioxidant groups. For instance, artificial antioxidants have impeding consequences for the body's well-being without giving any nutritional benefits. However, natural substances with antioxidant activity are important because of the significance of food safety to consumers and the advantages of eating food without artificial food preservatives (Hęś et al. 2019). A few investigations have looked at the antioxidant activity and its mechanism. One study described that the hydroxyl groups of capsaicin do not play a role in the free radical searching response since it stayed intact after capsaicin's response with 2,2-diphenylpicrylhydrazyl (DPPH) free extremist. As shown by their discoveries, 8-methylnon-6-eamide as well as 4-hydroxy-3-methoxybenzaldehyde would be the principal results of the capsaicin-DPPH response. In like manner and furthermore founded on the aftereffects of the quantum compound computations of various free extreme mediators acquired from the imitation substances *N*-vinylacetamide (NVA), these researchers demonstrated that at C_7-carbon was the capsaicin's scavenge free radical's location. With the aim of eliminating the decomposition of cis-9-Octadecenoic acid and producing the expected outcomes of the degenerated and disintegration of cis-9-Octadecenoic acid in the presence of capsaicin, Henderson and Henderson (1992) conducted research on the thermal decomposition of cis-9-Octadecenoic acid ($C_{18}H_{34}O_2$) and capsaicin at higher temperatures. The outcomes exhibit potential antioxidant activity. Further, research has compared antioxidant activity of capsaicin with dibutylated hydroxytoluene (dBHT) and 5-methoxytryptamine ($C_{13}H_{16}N_2O_2$) during fabrication of lipid peroxides (LPOs) of (9Z,12Z)-Octadeca-9,12-dienoic acid impeded by 2,2′-azobis(2-methylpropionitrile). Findings showed that, like bBHT, capsaicin also exhibits significant antioxidant activity. With capsaicinoids concentrations of 0.0001–0.02 M, the development of absolute lipid peroxidation (LPO) was reduced by 31–87%, and capsaicin dimers were the fundamental decomposition by-products. Hence, these researchers proposed that capsaicin's amide function group apparently does not contribute to its antioxidant activity, and subsequently, the antioxidant activity would be ascribed basically to the phenolic part of capsaicin (Henderson et al. 1999). Further, another study investigated the antioxidant activity of *C. annuum* L., including the immature (green pepper) and mature (red pepper) stages. The study looked at phenolic compounds like capsaicinoids and flavonoids of *C. annuum* L. The fundamental chemical constituents of the capsaicinoids were capsaicin and dihydrocapsaicin. They assessed the antioxidant activity in β-linoleic acid, β-carotene utilising heat-initiated deterioration and the DPPH test separately. The researchers found that phenolic acid, capsaicinoids, and flavonoid fractions displayed identical antioxidant activity. They identified that the antioxidant activity of both parts profoundly corresponded to the substance of these parts. Based on their findings, capsaicin exhibited potential antioxidant activity when compared with dihydrocapsaicin. The researchers theorised that capsaicin lipid chains consist of unsaturated double bonds that may impact the antioxidant activity, and they proposed that the ether functional group at ortho carbon to –OH might be accountable for the antioxidant activity (Materska and Perucka 2005). Furthermore, another study was carried out on the impact of antioxidant activity of various concentrations of vitamin C, such as tocopherol (TCP) (0.3–0.6%) as well as capsaicin (0.12–0.24%), on the thermal decomposition stability of *C. annuum* grain oil while searing at 135 ± 5 °C for 0, 6, 12, 24 as well as 48 h by estimating the chemiluminescence (CL) and 2-thiobarbituric acid responses (2-TBARS) of reactive substances utilising soy oil for correlation. The studies found the pepper oil with various capsaicin amounts displayed ($\rho < 0.05$) minimum CL and 2-TBARS during broiling and storage contrasted with the pepper oil without capsaicin. Hence, Yang et al. (2010) recommended capsaicin as a good antioxidant, which could impede the thermal oxidation of pepper grain oil. Another researcher investigated the extremist conversion activity (ECA) and the rate of kinetics of capsaicin in identical solutions (Okada et al. 2010). They concluded that polyphenol hydroxyl functional groups in capsaicin were overwhelmingly responsible for its ECA. It mainly depends on the solvent and the capsaicin-containing system in its antioxidant activity. They identified that one molecule of capsaicin would be able to scavenge two radicals. However, a recent study evaluated capsaicin, vanillin, and eugenol antioxidant activity using 2-TBARS, oxygen radical absorbing activity (ORAC), 2,2′-Azino-bis(3-ethylbenzothiazoline-6-sulfonic acid) (ABTS), and DPPH techniques. Capsaicin displayed excellent antioxidant activity, close to eugenol. Additionally, capsaicin displayed potential antioxidant activity regarding the ABTS and DPPH approaches. Concurring with the outcomes, IC_{50} of capsaicin for ABTS, 2-TBARS and DPPH approaches 0.014–0.017, 0.088–0.092 as well as 0.155–0.351 mM, separately (Orlo et al. 2021). Additionally, capsaicin applied maximum antiradical activity when compared with anti-lipoperoxidant activity (ALPOA). In the ORAC approach, capsaicin exhibited potential antioxidant activity when compared with trolox ($\rho < 0.05$), and its IC_{50} was 6.50–7.24 μM.

Pepper extracts and their derived products exhibit potential antioxidant activity due to phenolic acids and their derivatives, for instance, cinnamic acid and *m*-coumaric acid, etc. A few investigations have shown that antioxidant activity improves with ripening of peppers (Deepa et al. 2007), and it is primarily due to the increased content of biologically active chemical constituents like polyphenols, carotenoids, and nutrients like vitamins. Regarding vitamin C, studies by Marin et al. (2004) and MartÚnez et al. (2007) have shown a 40% increase in antioxidant activity during fruit ripening. Regarding polyphenols, immature green peppers contain a natural abundance of peppers (Marin et al. 2004), the concentrations of which vary during ripening (Deepa et al. 2007). The amounts of capsaicinoids are different in green and red ripe peppers: green peppers show less capsaicinoids compared with red ripe peppers. Hence, red ripe peppers exhibit significant antioxidant activity

(Materska and Perucka 2005). Accordingly, the distinctions in biologically active constituents found in green and ripening pepper at various ageing stages should be meticulously assessed to comprehend the compound interconnections that take place among various biologically active secondary metabolites and other decreasing chemical constituents in the extracts (Guil-Guerrero et al. 2006; Howard and Wildman 2007). Most of the secondary metabolites or phytochemicals exhibit significant antioxidant activity. For example, polyphenols, carotenoids, flavonols, etc. have been identified in antioxidant activity, as well as the pigments, affecting the colour of the pepper fruits while maturing. Apparently, Matsufuji et al. (2007) observed that the orange and red peppers exhibit significant in vitro antioxidant activity when contrasted with yellow and green peppers. Red sweet bell pepper exhibits significant antioxidant activity, which can essentially decrease the lipid disintegration as well as docosahexaenoic acid (DHA) abasement while cooking, consequently, staying away from the development of poisonous oxidation products that might produce off-flavours, crumbling of food purity, and fight cancer and ischemic heart disease (IHD) (Sun et al. 2007). Models of inexpensive red ringer pepper powders are available in various configurations, for example, crumbled or pulverised (Sun et al. 2007), for instance Jinuhua Huayang Co., Ltd. Although most investigations dealing with the antioxidant activity of *C. annuum* have focused on lipotropic substances, including carotenoids, the non-polar distillation, consisting mostly of flavonoids and phenolic acids, has been also examined for essential antioxidants, specifically forestalling DNA and in vitro deoxyribose oxidation. These antioxidants also have a potential for benefiting human well-being (Materska 2014). Sandoval-Castro et al. (2017) reported that in the modern handling of pepper, grains are regularly wasted, regardless of their biologically active chemical constituents for cancer and antioxidant prevention, including flavonols, phenolic acids, and vitamin C. In consideration, pepper grains would be taken advantage of as an antioxidant chemical constituent, and that can be additionally advertised in the business sectors (for instance, Jinuha Huayang Co., Ltd).

Based on existing research, capsaicin has exhibited significant antioxidant activity, yet the capsaicin containing system and nature of the solvent play important roles in its cancer prevention action. The methoxy group, at the C_7-carbon, the hydroxyl group at ortho position, the phenolic part, and the phenolic hydroxyl groups are functions of the capsaicin chemical constituent that has been specified as supporting its interactions with free radicals. Overall, capsaicin has been shown to have significant antioxidant activity when contrasted with dihydrocapsaicin, which may be due to the antioxidant activity of the unsaturated C=C bonds existing in the capsaicin lipoid chain. Also, capsaicin dimers, including dicapsaicin, are the fundamental degradation outcomes of capsaicin. The red pepper extracts and seeds provide excellent antioxidant activity capability that facilitates their applications as organic antioxidants in nutritional enhancements or specialised food supplements, replacement of synthetic preservatives, as well as encouraging chemical constituents for “clean label” products.

12.4.2 Applications of Capsaicin in the Food Industry

Capsaicin has different uses in diverse enterprises as an effective chemical compound for pharmaceuticals, self-deterrence, insect deterrence, cosmetics, and oral natural or additive supplements. In certain nations, the everyday intake of stinging food varieties is extensive, and chilli pepper is extensively utilised as an herb and for food flavour enhancement. For instance, some nations consume chilli pepper in the range of 2.5–20.0 g/individual daily, such as India, Mexico, Kingdom of Saudi Arabia, and Thailand (Hernandez-Perez et al. 2020). However, some people are unable or unwilling to ingest very spicy foods (Sun et al. 2016). This section on the pepper examines the different uses of capsaicin in the food business.

12.4.2.1 Food Flavouring

Capsaicin is broadly used globally as an herb and spice for food enhancing and flavouring. Generally, capsaicin is available in the form of a red chilli powder called paprika and a red pepper paste called oleoresin or natural resin. These products increase food varieties to provide a hot or tangy flavour to them. They are also utilised as additives in the articulation or formulations of different sauces, including hot ketchup and hot pepper sauce for instance (Barbero et al. 2016; Cheok et al. 2017; De Farias et al. 2020; Kim et al. 2016a, b). Capsaicin is additionally utilised in the production of spiced and flavoured Italian rice, savoury paneer cheese, instant noodles, and various types of kimchi, like cabbage kimchi and radish kimchi (Azevedo et al. 2019; Kim et al. 2018; Kwon 2021).

12.4.2.2 Food Preservation

Capsaicin, in its condensed form like capsaicin microcapsules (MCs), is utilised as an antimicrobial and antioxidant agent in food conservation or prevention (Schlossareck and Ross 2019; Xing et al. 2006). Capsaicin is likewise utilised as an additive in food varieties that are sensitive or prone to contamination with plant toxins. Buitimea-Cantua et al. (2020) reported that capsaicin has been shown to be effective against the fabrication of aflatoxin B_1 and B_2 by *A. flavus*. According to Kollia et al. (2019), capsaicin suppresses the production of ochratoxin A by *Aspergillus* group Nigri strains (≈30–79%) and *A. carbonarius* (≈62%), respectively. Further, research has been carried out on antioxidant activity of rapeseed at various temperatures showing that increasing the temperature decreased linoleic acid, because of more utilisation of oxygen (Si et al. 2012). As indicated by the outcomes, capsaicinoids displayed higher viability when compared with dBHT in impeding the decomposition of rapeseed oil at 65 °C. Likewise, the findings showed that the antioxidant activity of capsaicin was somewhat weaker than dBHT for further temperatures like 90, 120, and 180 °C (Si et al. 2021). Further, researchers concentrated on the inhibition effect of capsaicin and chilli pepper and their impact on the construction of cyclic amines in charred hamburger sandwich. It was found that the development of cyclic amines was impeded by various amounts of capsaicin as well as chilli pepper, although no remarkable alteration was noticed in the surface of the croquettes. Findings showed that all concentrations of capsaicin had an inhibitory effect on the production of all cyclic amines and 2-amino-1-methyl-6-phenylimidazo [4,5-b] pyridine (PhIP) so that the highest impediment rates were almost 81% and 97%, individually, at capsaicin of 2 mg. In addition, researchers constructed capsaicin-stacked mesoporous (MP) silica nanoparticles (SNPs) with different molecular magnitudes from 50 to 400 nm to assess their effect with slow delivery time. As demonstrated by the outcomes, the capsaicin-stacked mesoporous silica nanoparticles (MPSNPs) with a molecular magnitude of 50 nm displayed a maximum decomposition obstruction compared to unadulterated capsaicin following 120 h of capacity, which was credited to the slow delivery impact. In addition, the capsaicin's sharpness in the hamburger patty was decreased as an outcome of the slow delivery impact of MPSNPs.

12.4.2.3 Functional Foods

CSn plays an important role in functional foods. A researcher prepared capsaicin-enhanced stacked pasta by sandwiching a capsaicin-enhanced preparation with two gastro-resistant batter thicknesses to hold the capsaicin inside the pasta substructure for an extended frame and deliver it to the gastrointestinal area in a managed way. The concentration of capsaicin in the paprika flakes utilised in this research was 3.0 mg/g, which is equal to SHV, and the amount of capsaicin in the penne was ≈14 to 16 mg/100 g of the penne. The findings showed that these stacked penne helped the consumers to avoid extreme agony while eating them since many Asians tend to guzzle down the noodles instead of biting them. Further a researcher prepared a dairy product strengthened with a capsaicin extract containing monounsaturated vegetable oil, expecting to get a functional dairy product with another savour and health benefits. A dairy product was selected because capsaicin is soluble in non-polar solvents including casein, but it is insoluble in polar solvents including water. The dairy products had a better quality, spicy savour, firmer surface, and excellent sensate quality contrasted with the conventional yoghurt (Askin et al. 2021).

12.4.2.4 Packing Films

Capsaicin has been utilised in the food business for packing. A researcher prepared a progression of sodium alginate (SG)/pulluan blended consumable sheets combined with capsaicin. They found that the expansion of capsaicin to the blended sheets prompted a plain stacked arrangement of the sheets (Zhang et al. 2018, 2021). Further, when sheets were coated with apple candy with SG/pullulan/capsaicin for a while, after 72 h, the apple candy without coating and those with SG/pullulan/capsaicin −0% blended sheets covering got dim and ruined while the sheets-covered apples with various contents of capsaicin remained fresh longer. Furthermore, researchers prepared chitosan films integrated with various concentrations of capsaicin (0.3, 0.6, and 1.2 mg). The findings showed the development of the mechanical and physico-chemical attributes of the sheets coming about because of chitosan films blended with various concentrations of capsaicin. Likewise, enhancing the quantity of capsaicin enhanced the antimicrobial, flexibility, elasticity, anti-quorum sensing, antioxidant, translucency, and hydrophobicity characteristics of sheets (Akyuz et al. 2018). Consistently, the researchers proposed that mainly capsaicin-chitosan sheets could be utilised as a bundle and packing in the food industry. Furthermore, researchers prepared chitosan/gelatin composite films with capsaicin-loaded hollow metal-organic systems (Cap-Fe(III)HMOS-5) as antimicrobial biomaterials for food packaging (Zhao et al. 2020). Likewise, Cap-Fe(III) HMOS-5 was utilised as a capsaicin nano-carrier, which was aquaphobic and competently resolved the stage isolation issues (Zhao et al. 2020). These researchers likewise considered the experimental utilisation of the films for food safety and food storage. They coated high-in-water, fresh and white apple cubes with blended sheets, and they were stored at 25 °C at 50% humidity for 120 h. They found that apple cubes become black or dim with pathogens on the exterior after 24, 72, and 120 h, and those coated with 2%, 4%, and 8% CSn-Fe(III)-HMOF-5-chitosan/gelatin turned somewhat dark and exhibited a small amount of water loss. At 120 h, the 8% capsaicin-Fe(III)-HMOF-5-chitosan/gelatin sheet covered apple cubes had only a small amount of loss of moisture and decomposition without any pathogens pollution. In addition, they prepared EC-blended sheets integrated with capsaicin to assess the impacts of different capsaicin containment on physico-chemical effects of the sheets. They acquired smooth, delicate, sticky, transparent and waterproof EC-capsaicin composite films. The EC-capsaicin-blended films displayed significant antimicrobial activity against *S. aureus* and *E. coli*. Additionally, the experimental utilisation of the EC-capsaicin sheets in the protection of bell-shaped peppers was explored, the observed outcomes of which displayed that the polymeric mixed-capsaicin sheets successfully hindered the maturing procedure of bell-shaped pepper, which could assist with keeping the fruits almost fresh (Su et al. 2020). It was found that the non-pressed bell-shaped peppers shrank and spoiled, while those covered with sheets appeared new or new following multi-week storage. Findings support these sheets as a packing substance conductive to extend the time span of usability of food sources.

In view of the previously considered investigations, it could be reasoned that taking everything into account capsaicin's antimicrobial and antioxidant activity could be utilised in sheets for fabricating food packing substances to continue the food varieties fresh and extend their life span. The impacts of the expansion of capsaicin as a packing substance for the sheets are encapsulated in Fig. 12.2.

12.4.3 Pharma and Cosmetic Industrial Applications

Chilli pepper exhibited antimicrobial and antioxidant activity; hence, it could be used in pharma and cosmetic industrial applications. Due to the growth of environmental awareness, there is an extraordinary interest in looking at biologically active chemical compounds, plant extracts, or crude plant materials as natural phytochemicals for pharmaceutical and cosmetic applications. In that regard, a few attributes that could impact product performance, like biological properties such as antibacterial, antifungal, antiviral, antioxidant, and anticancer activity (Fig. 12.3) and physical properties such as flavour, variety, and penetration, have been examined (EMA 2007, 2015). The utilisation of natural chemical constituents rather than

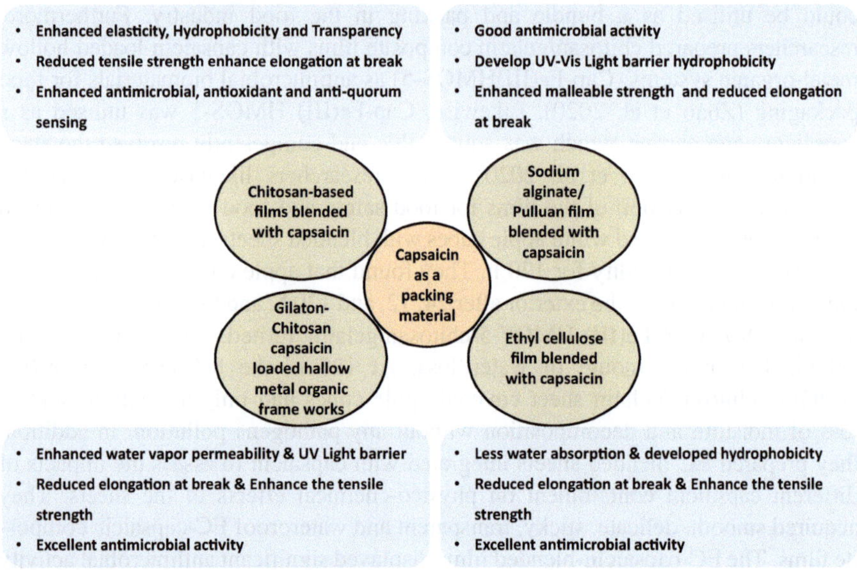

Fig. 12.2 The effect of adding capsaicin as a packing material to films

artificial preservatives can improve the benefits of pharmaceutical and cosmetic substances, avoiding the contact sensitivities produced as secondary impacts.

These plant extracts could add to the significant product assets in the restorative business, including age control products as well as the decrease of wrinkly or crinkled by battling against free extremist and UV energy. Each biologically active chemical constituent acts as a cosmeceutical. Such products have been normalised following the worldwide quality guidelines for medicine maximum pertinent group of experts like Japanese Pharma (JP), the United States Pharma (USP), and European Pharma (Ph. Eur.) (Ph. Eur. 2010). For instance, the utilisations of capsicum-determined chemical constituents in cosmetic and pharmaceutical fabrications are shown in Table 12.3.

12.4.3.1 Advancement of Physical Properties

As specified, red chilli flakes or paprika and natural resins or oleoresins are rich in capsaicinoids and carotenoids, and these biologically active chemical constituents are an outstanding source of colours in the pharmaceutical and cosmeceutical industries. As indicated by the FDA and EFSA, these natural chemical constituents do not need a particular endorsement by these associations; nonetheless, they should be secure for purchaser below standard regulatory authorities of application, with producers possessing the liability regarding the well-being of their final result and chemical constituents. In this consideration, chilli powder is usually utilised as an herb as well as a spice; hence, there is no specific cause to anticipate unwanted secondary impacts from its utilisation for improving colour, the same way already used as an edible supplement (EFSA 2015; FDA 2017a, b). For instance, paprika and

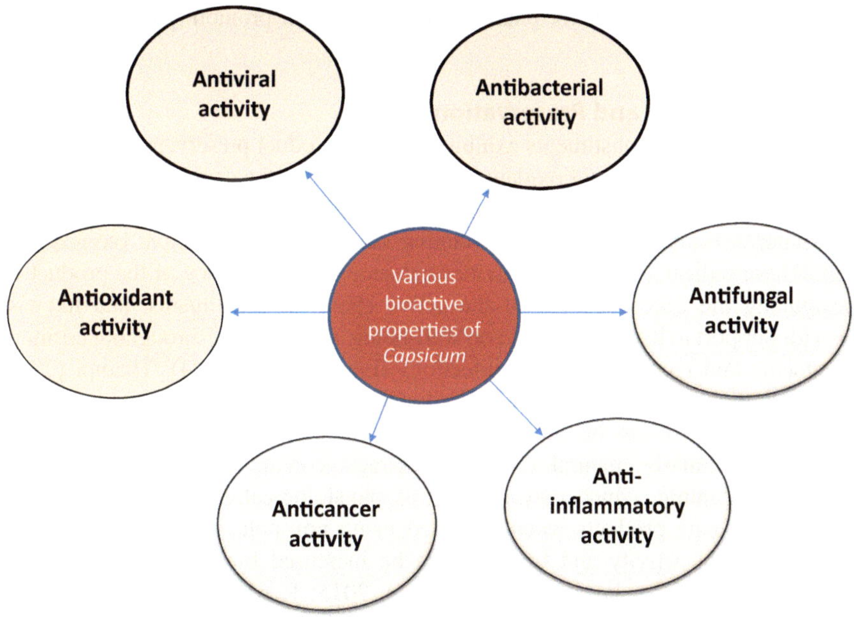

Fig. 12.3 The major biological activity of *Capsicum annuum* L. and its derivatives mainly used in agro-food, cosmetic, and pharmaceutical applications

its natural resins are utilised as soaps, shampoo, shower gels, and cosmetic colourants in bath oils by companies such as Lush Handmade Cosmetics, S.L. and also used in many herbal beauty substances such as lip gloss and eye shadow (colour marker Inc.). The biologically active chemical constituents extracted from pepper by-products could be utilised as fragrance and enhancing agents in beauty products, which might be extricated from ethanol, hexane, and cooking oil, as well as its use can be increased up to 4% (w/w) with no harmful impact (Johnson 2007). Red pepper oleoresin gives producers a significant benefit with regard to the utilisation of volatile oils; this property is known as lipophilicity, permitting this product to be soluble in lipids, oils, fats, etc., while volatile oils do not have that capacity. Generally, peppers contain natural pigments, including polyphenols and chlorophylls; mostly anthocyanins are present. Various peppers, such as violet and black peppers, are mostly composed of glycosylated and anthocyanin delphinidin, along with aglycone. Commonly these pigments are water soluble substances; hence they could be used as natural colourants, antioxidant agents, and therapeutic agents (UV protectors) in cosmetic and pharmaceutical products (Zillich et al. 2015). On a similar note, red pepper by-products were investigated with acetone-water solvent extraction and were demonstrated as a useful natural colour to be used on woollen material with satisfactory antimicrobial activity (Ksibi et al. 2015). The enhanced utilisation of *Capsicum* constituents in the business demands the normalisation of the extracts, oleoresins and pepper or powders. For instance, capsaicin content could be used as colour, flavour for pharmaceutical or health promoting products and in cosmetic applications.

12.4.3.2 Stability and Preservation

Capsicum chemical constituents exhibit utility for product preservation and product stability because of their antioxidant activity to decrease the degradation of dynamic chemical constituents in therapeutic agents. The antioxidant ability of these chemical constituents is mainly to their scavenging activity of the reactive oxygen species. These radical scavenging activities enhance the consistency of the product by postponing the decomposition of the active chemical constituents, and likewise provide support to its useful impact on well-being, for example, throughout cosmetic treatments and recuperating skin infections (Pieroni et al. 2004). Thampi (2004) observed that the green pepper (dehydrated) contains a significant amount of vitamin C, which can be utilised as protecting antioxidant in the beauty care industries. Similarly, natural *Capsicum* extracts contain lutein and carotenoids such as capsanthin, which may be used as physiologically useful ingredients in healthy skin care products when extracted using non-polar solvents. In addition, their biological activity and firmness can be increased by their consolidation to conjugated protein or layers (Arimboor et al. 2015; Kerdudo et al. 2016). Dayan and Kromidas (2011) observed that *Capsicum* could be utilised as an antimicrobial conservative to inhibit or forestall the development of fungi, pathogens, and moulds that can introduce a threat of contamination to the pharmaceutical, beauty care substances, as these normally have over 3 years of realistic usability. Normally, *p*-coumaric acid and cinnamic acid exhibit significant antimicrobial activity against *B. cereus*, *Listeria monocytogenes* (*L. monocytogenes*), *S. aureus*, and *S. typhimurium*, through several other biologically active chemical constituents present in *Capsicum* extracts, which have the possibility to reduce fermentation and infective pathogens in products. These studies show that use of natural additives for beauty care products is growing, based on purchaser interest in cosmetic compounds produced normally with plant substances.

12.4.3.3 Pharma and Cosmetic Applications in Health and Beauty Products

Pepper and its analogues are currently utilised as cosmeceuticals and pharmaceutical products, mostly as oral supplementations, analgesic therapeutics (Palombo et al. 2007), tropical applications (Telang 2013), and antioxidants. With respect to the anticipation as well as therapeutics of epidermis infections, advanced treatment alternatives are consistently needed. The effective utilisation of remedial agents is extremely well known, being utilised in small amounts every day, contingent upon the skin status. Tiwari et al. (2012) observed that the viability of effective remedial agents could be impacted by various factors, like active chemical constituents, excipients and potential interactions of chemical constituents, preparation, region impacted, condition of mucosa, and galenic properties. Capsaicin has been endorsed by the FDA as well as EU as a medication for the mucosa therapeutics of neuropathy torment, with its palliative effects being amount reliant. Despite there is no exact mechanism of action, proposed evidence shows the benefit of capsaicin through continual applications (6–8 intervals every day for 2 months) (Lysy et al. 2003). As per the pharmacokinetics results, capsaicin could be utilised for pain therapy and problems like severe excruciating diabetic neuropathy, osteoarthritis, post-herpetic neuralgia, and on-going rheumatic pain. Additionally, these compounds could be utilised in convalescence from bladder treatment to minimise incontinence. Likewise, it could decrease postoperative nausea, sore throat, and vomiting; it could assist in convalescence with irritation related to nephritic non-success as well as inpatient with cardiac infarction. Lesser amounts of capsaicin are recommended for non-prescription pain relief lotions, and a maximum amount of capsaicin has been investigated as a therapeutic for peripheral neuropathy, for example, Qutenza®; cluster headaches, for example, Civamide; and postoperative pain, for example, Adlea. Anesiva Inc. researchers have used capsaicin and lipid-recharging capsaicin lotion against cutaneous hypersensitivity and persistent irritation or pruritus (Lysy et al. 2003). Importantly, daily capsaicin utilisation must be adapted to convalescent requirements and can have notable side effects, including burning and irritation. To prevent the aftereffects, in vivo investigations have demonstrated that capsaicin can be enclosed in lipid NP transporters; further developing epidermis penetration, maintenance, and pain reliving impacts, limiting its consequences for epidermis irritation contrasted with traditional lotions (Raza et al. 2014). In any case, a blend of lipids, humectants, emollients, and water are fundamental in skin treatment; water-rich skin treatment with a hydrophilic lotion carries capsaicinoids (0.025–0.1% w/w) that should benefit a severe stage of irritation (Staubach and Metz 2013). Thapa et al. (2013) reported a straightforward use of raw *Capsicum* for vesicular formulations in the use of pain relief. In this regard, the European Medicine Agency (EMA) proposed that *Capsicum* could be utilised for the help of muscle pain, like lumbar spine pain (EMA 2015), which are typically used with a sterile bandage on the skin like creams as per the Ph. Eur. (EDQM 2016). The agency has endorsed the utilisation of various normalised chemical constituents from cayenne pepper that have been discussed, including *Cap* extracts, like soft extract (#2529), fruit (#1859), tincture (#2337), and oleoresin (#2336). These range from 31,000 to 49,000 SHU values. For instance, Qutenza® is an important system that consists of 8% of capsaicin, which is used for the administration of nerve pain due to herpes. With reference to biological availability, capsaicin is transcutaneously consumed, with experimental findings indicating a fundamental biological availability of importantly used capsaicin ranging from 27% to 34%. EMA (2015) has reported that the consumed capsaicin is metabolised essentially in the liver and eliminated as waste in the stool, excrement, and urine.

However, additionally, investigations are required to demonstrate further accurately the scope of successful capsaicin concentration in lifelong medicines. Vitamin C is a very powerful antioxidant agent identified in pepper and highly utilised in epidermis care products, which when used as a powerful antioxidant forestalls epidermis oxidative defilement, in this manner preventing the epidermis from response oxygen species (ROS) as well as additionally a significant chemical constituent for treating skin pathogens, like cancer and inflammation. The delivery of Vitamin C into the dermis is a challenging task (Telang 2013). Various carriers including liposomes are now a days used to improve vitamin C treatment (Serrano et al. 2015). Moreover, the biological availability of vitamin C (L-ascorbic acid) in the epidermis may be deficient at the point where it is directed orally; subsequently, the utilisation of effective vitamin C is promoted in dermatological practice, as well as in mixtures of various substances like zinc and tyrosine, enhancing the biological activity of vitamin C (Telang 2013). Likewise, pepper powder extract highly abundant in carotenoids, like lutein and zeaxanthin, has been also utilised. One study (Palombo et al. 2007) illustrated that an integrated therapy with lutein as well as zeaxanthin exhibited increasing skin versatility, with dermal moisture more articulated (contrasted with the individual medicinal therapy). Instances of pharmaceutics experiments include nutritional improvement with carotenoids, iron, and zinc after vitamin A deficiency in daily diet. Kana-Sop et al. (2015) review has shown superior carotenoids phases in plasma as well as retinol phases, individually. Furthermore, research conducted by Heinrich et al. (2003) revealed that an analysis of ≥24 mg carotenoids daily for around 84 days demonstrated a feasible defence against UV-induced erythema. Capsidiol extracted from *C. annuum* using methyl alcohol as a solvent exhibited in vitro effects against *H. pylori* (≥250 μg/mL), De Marino et al. (2006) demonstrating its capacity as a therapy for anti-toxin safe strains and for a convalescent who does not want to take artificial antimicrobials. Additionally, pepper extracts containing flavonol glycosides, for instance, quercetin-3-*o*-glycosides, are mainly utilised in the beauty care industry. Hence, these chemical constituents exhibit excellent anti-inflammatory and antioxidant activity when contrasted with other phenolic chemical constituents and have been connected with the treatment of various health issues; for instance, supporting skin recovery in wound mending or cell protection from UV irradiation (Hatahet et al. 2016). Quercetin might be employed utilising water/oil microemulsions (MEMs) as an excipient to expand its stability as well as solubility and, thus, improve the percutaneous conveyance of this biologically active chemical constituent (Malaj et al. 2010). For instance, it can be identified in facial skin protection (for instance, the Korres products). Phenolic chemical constituents, such as 3-caffeoylquinate and its analogues, can be utilised as antioxidant agents in the cosmetics industry by utilising thermally stable O/W microemulsions as excipients to improve their pervasion in the skin and protect it against UV instigated oxidation destruction (Kitagawa et al. 2011). For example, quercetin and coniferic acid, amidst a few biologically active chemical constituents are applied in cosmetic serum (for instance: Dr. Dennis Cross Ltd.). The red coloured dye, anthocyanins, can be utilised as a colouring specialist, antioxidants in cosmetic product preparations for UV protection, melanin synthesis, and anti-ageing chemical components. These chemical compounds might be encased in the appropriate coating to boost their biological activity and usage in effective skin-healthy treatments (Westfall and Giusti 2017). As to the utilisation of pepper powder extracts as chemical constituents in health and beauty utterance, few beauty care products can be identified with health claims against oxidation and ageing, such as facial masks and natural skin care cosmetic products (for instance, Eminence Organic Skin Care).

12.5 Conclusions and Future Directions

The growing business sector of biologically active chemical constituents and natural pharmaceuticals has arisen in light of the on-going recognition of consuming natural chemical constituents rather than artificial compounds. These natural products could increase the organoleptic properties and once blended or coordinated in equivalent edibles, they would likewise enhance human well-being. In this regard, *Capsicum* products, including paprika powder, natural resin, clean extract and portions enhanced in carotenoids as well as capsaicin, have been examined in terms of pungency or sharpness, flavour, colour, variety, and nutritional values, as potential chemical constituents for food sources and for the cosmetic and pharmaceutical industries. Besides, the use of these chemical constituents in food or drugs could be a used to further improve human health because of their biological activity, which includes antimicrobial (antibacterial, antifungal, and antiviral), antioxidant, and anti-inflammatory properties, and others. To this reality, the normalisation of *Capsicum* chemical constituents in the expression of flavour, sharpness, colour, variety, and bioactivity is expected to enhance the use of these chemical constituents for commercial purposes. For the cosmetic industry and dermatological treatment, the use of formulations with plant bioactive chemical constituents is to ensure safety, stability, and effectiveness in new natural beauty care products, for instance, transdermal patches, cosmetic materials and analgesic lotions, and oral nutri-beauty care products. Research is needed to systematisc specifications, and commercial producers must establish the security and adequacy of these nutritive enhancements before they are advertised. Thus, there is a gap in the medical sector, which must propose new regulation for the utilisation of these extracts as biologically active chemical constituents in pharmaceutical and cosmetic formulations. Further, the high popularity of natural phytochemicals as replacements for artificial additives and colours in the food and medicinal industries requires encouraging green or eco-friendly approaches that can meet these needs economically. For example, decreasing the energy necessary for extricate as well as the volume of excess solvent, alongside the growth of value of by-products, will give profitable advantages to manufacturers and agriculture businesses. Additionally, investigations are expected to illuminate the mechanism of activity for *Capsicum* chemical constituents. These investigations can also contribute fascinating possibilities for the use of pepper by-products as a wellspring of biological chemical constituents. Additionally, precise studies are needed concerning their technological benefits and ensuring food security for buyers. Finally, personal discoveries and on-going examinations are expected to more readily elucidate the well-being advancing exercises of *Capsicum* analogues, which can experimentally validate a well-being guarantee and, subsequently, advance the utilisation of these products.

Over the last few decades, capsaicin has received a lot of attention because of its functional and biological attributes. In this chapter, the practical attributes, exemplification, and applications in the food fabrication of capsaicin were examined profoundly. In line with the investigations, capsaicin demonstrates potential as an antioxidant chemical constituent for free radical scavenging, including DPPH and

ABTS. Various investigations have demonstrated its antimicrobial activity against different food waste microscopic organisms and foodborne microorganisms, particularly *E. coli* and *S. aureus*. Furthermore, studies have demonstrated capsaicin antifungal activity against *A. parasiticus* and *A. carbonarius*; these sharpness substances have also displayed antiviral activity. The inhibitory impact of capsaicin on the production of naturally occurring toxins, for instance, aflatoxin and ochratoxin, has also been accounted for in various investigations. However, these chemical constituents are delicate to various conditions including temperature treatment, light, and UV radiation. As well, capsaicin exhibits partial solubility in water, but sharpness and less bioavailability have obstructed its broad applications in food production. Among different encapsulation procedures, emulsions have been considerably designed for the adjustment of capsaicin because of the ease of their fabrication. The sharpness of the chemical constituents is broadly utilised in food sources as food flavour and spice worldwide. In addition, several researchers have concentrated on the utilisation of capsaicin as an additive in food varieties because of its antimicrobial and antioxidant properties and furthermore for delivering practical food varieties because of its biological impacts. Notwithstanding the previously mentioned applications, a few examinations have assessed the utilisation of capsaicin as in food packaging for fabricating sheets to extend the food shelf life. Moreover, further investigations could be centred on the adjustment of capsaicin and its broad use in food fabrication because of its utilitarian and biological effects; regardless, further investigations should be directed to assess the health and safety-associated effects of capsaicin to support its broad utilisation in food varieties. The consistency of capsaicin in various situations, for example, various pH and temperatures, is another significant factor that needs more exact investigation to extend its utilisation in food production.

References

Akbas E, Soyler B, Oztop MH (2018) Formation of capsaicin loaded nanoemulsions with high pressure homogenization and ultrasonication. LWT 96:266–273

Akyuz L, Kaya M, Mujtaba M, Ilk S, Sargin I, Salaberria AM, Labidi J, Cakmak YS, Islek C (2018) Supplementing capsaicin with chitosan-based films enhanced the anti-quorum sensing, antimicrobial, antioxidant, transparency, elasticity and hydrophobicity. Int J Biol Macromol 115:438–446

Amorati R, Foti MC, Valgimigli L (2013) Antioxidant activity of essential oils. J Agric Food Chem 61(46):10835–10847

Andersen BV, Byrne DV, Bredie WLP, Møller P (2017) Cayenne pepper in a meal: effect of oral heat on feelings of appetite, sensory specific desires and wellbeing. Food Qual Prefer 60:1–8

Arimboor R, Natarajan RB, Menon KR, Chandrasekhar LP, Moorkoth V (2015) Red pepper (*Capsicum annuum*) carotenoids as a source of natural food colors: analysis and stability—a review. J Food Sci Technol 52(3):1258–1271

Askin B, Ozturk D, Durusoy B, Kaya Y (2021) Properties of yogurt fortified with capsaicin extracts dissolved in high oleic sunflower oil. J Food Process Preserv 46(10):e15633

Asnin L, Park SW (2015) Isolation and analysis of bioactive compounds in Capsicum peppers. Crit Rev Food Sci Nutr 55(2):254–289

Azevedo M, Leite IB, Queiroz C, Fialho E (2019) Spiced risotto: cooking processing and simulated in vitro digestion on curcuminoids, capsaicin and piperine. J Culin Sci Technol 17(3):256–267

Baby KC, Ranganathan TV (2016) Effect of enzyme pre-treatment on yield and quality of fresh green chilli (Capsicum annuum L) oleoresin and its major capsaicinoids. Biocatal Agric Biotechnol 7:95–101

Bacon K, Boyer R, Denbow C, O'Keefe S, Neilson A, Williams R (2017) Antibacterial activity of jalapeño pepper (*Capsicum annuum* var. annuum) extract fractions against select foodborne pathogens. Food Sci Nutr 5(3):730–738

Bae H, Jayaprakasha GK, Crosby K, Yoo KS, Leskovar DI, Jifon J, Patil BS (2014) Ascorbic acid, capsaicinoid, and flavonoid aglycone concentrations as a function of fruit maturity stage in greenhouse-grown peppers. J Food Compos Anal 33:195–202

Bae G, Kim J, Kim H, Seok JH, Lee DB, Kim KH, Chung MS (2018) Inactivation of norovirus surrogates by kimchi fermentation in the presence of black raspberry. Food Control 91:390–396

Barbero GF, Palma M, Barroso CG (2006a) Determination of capsaicinoids in peppers by microwave-assisted extraction–high-performance liquid chromatography with fluorescence detection. Anal Chim Acta 578(2):227–233

Barbero GF, Palma M, Barroso CG (2006b) Pressurized liquid extraction of capsaicinoids from peppers. J Agric Food Chem 54(9):3231–3236

Barbero GF, Liazid A, Palma M, Barroso CG (2008) Ultrasound-assisted extraction of capsaicinoids from peppers. Talanta 75(5):1332–1337

Barbero GF, Liazid A, Azaroual L, Palma M, Barroso CG (2016) Capsaicinoid contents in peppers and pepper-related spicy. Int J Food Prop 19:485–493

Barceloux DG (2009) Pepper and capsaicin (*Capsicum* and Piper species). Dis Mon 55(6):380–390

Basith S, Cui M, Hong S, Choi S (2016) Harnessing the therapeutic potential of capsaicin and its analogues in pain and other diseases. Molecules 21(8):966

Bazan-Lugo E, Garcia-Martinez I, Alfaro-Rodriguez RH, Totosaus A (2012) Color compensation in nitrite-reduced meat batters incorporating paprika or tomato paste. J Sci Food Agric 92(8): 1627–1632

Bode AM, Dong Z (2011) The two faces of capsaicin. Cancer Res 71(8):2809–2814

Bohn T, Desmarchelier C, Dragsted LO, Nielsen CS, Stahl W, Rühl R, Borel P (2017) Host-related factors explaining interindividual variability of carotenoid bioavailability and tissue concentrations in humans. Mol Nutr Food Res 61(6):1600685

Buckenhüskes HJ (2003) Current requirements on paprika powder for food industry. In: Krishna A (ed) *Capsicum*: the genus *Capsicum*. Taylor and Francis Ltd, London, pp 223–230

Buitimea-Cantua GV, Velez-Haro JM, Buitimea-Cantua NE, Molina-Torres J, Rosas-Burgos EC (2020) GC-EI MS analysis, antifungal and anti-aflatoxigenic activity of *Capsicum chinense* and *Piper nigrum* fruits and their bioactive compounds capsaicin and piperine upon *Aspergillus parasiticus*. Nat Prod Res 34(10):1452–1455

Careaga M, Fernandez E, Dorantes L, Mota L, Jaramillo ME, Hernandez-Sanchez H (2003) Antibacterial activity of Capsicum extract against *Salmonella typhimurium* and *Pseudomonas aeruginosa* inoculated in raw beef meat. Int J Food Microbiol 83(3):331–335

Căruntu C, Negrei C, Ghiţă MA, Căruntu A, Bădărău AI, Buraga I, Brănişteanu D (2015) Capsaicin, a hot topic in skin pharmacology and physiology. Farmacia 63(4):487–491

Chapa-Oliver AM, Mejía-Teniente L (2016) Capsaicin: from plants to a cancer-suppressing agent. Molecules 21(8):931

Chatterjee S, Asakura M, Chowdhury N, Neogi SB, Sugimoto N, Haldar S, Awasthi SP, Hinenoya A, Aoki S, Yamasaki S (2010) Capsaicin, a potential inhibitor of cholera toxin production in Vibrio cholerae. FEMS Microbiol Lett 306(1):54–60

Cheok CY, Sobhi B, Adzahan NM, Bakar J, Rahman RA, Ab Karim MS, Ghazali Z (2017) Physicochemical properties and volatile profile of chili shrimp paste as affected by irradiation and heat. Food Chem 216:10–18

Cho H, Kwon Y (2020) Development of a database of capsaicinoids contents in foods commonly consumed in Korea. Food Science & Nutrition 8 (8):4611–4624.

Conforti F, Statti GA, Menichini F (2007) Chemical and biological variability of hot pepper fruits (*Capsicum annuum* L. var. acuminatum) in relation to maturity stage. Food Chem 102(4): 1096–1104

Costanzo MT, Yost RA, Davenport PW (2014) Standardized method for solubility and storage of capsaicin-based solutions for cough induction. Cough 25(10):6

Dang YM, Hong YS, Lee CM, Khan N, Park S, Jeong SW, Kim KS (2018) Determination of capsaicinoids in red pepper products from South Korea by high performance liquid chromatography with fluorescence detection. Anal Lett 51(9):1291–1303

Danza A, Mastromatteo M, Cozzolino F, Lecce L, Lampignano V, Laverse J, Del Nobile MA (2014) Processing and characterization of durum wheat bread enriched with antioxidant from yellow pepper flour. LWT Food Sci Technol 59(1):479–485

Das S, Teja KC, Duary B, Agrawal PK, Bhattacharya SS (2016) Impact of nutrient management, soil type and location on the accumulation of capsaicin in *Capsicum chinense* (Jacq.): one of the hottest chili in the world. Sci Hortic 213:354–366

Dayan N, Kromidas L (2011) Formulating, packaging, and marketing of natural cosmetic products. John & Wiley Sons, New York

de Aguiar AC, Silva LPS, Alves De Rezende C, Barbero GF, Martínez J (2016) Encapsulation of pepper oleoresin by supercritical fluid extraction of emulsions. J Supercrit Fluids 112:37–43

de Aguiar AC, Osorio-Tobon JF, Silva LPS, Barbero GF, Martınez J (2018) Economic analysis of oleoresin production from malagueta peppers (*Capsicum frutescens*) by supercritical fluid extraction. J Supercrit Fluids 133:86–93

De Farias VL, Da Silva Araujo IM, Da Rocha RFJ, Garruti DDS, Pinto GAS (2020) Enzymatic maceration of Tabasco pepper: effect on the yield, chemical and sensory aspects of the sauce. LWT 127:109311

De Marino S, Borbone N, Gala F, Zollo F, Fico G, Pagiotti R, Iorizzi M (2006) New constituents of sweet *Capsicum annuum* L. fruits and evaluation of their biological activity. J Agric Food Chem 54(20):7508–7516

Deepa N, Kaur C, George B, Singh B, Kapoor HC (2007) Antioxidant constituents in some sweet pepper (*Capsicum annuum* L.) genotypes during maturity. LWT Food Sci Technol 40(1): 121–129

Domínguez-Martínez I, Meza-Márquez OG, Osorio-Revilla G et al (2014) Determination of capsaicin, ascorbic acid, total phenolic compounds and antioxidant activity of *Capsicum annuum* L. var. serrano by mid infrared spectroscopy (mid-FTIR) and chemometric analysis. J Korean Soc Appl Biol Chem 57:133–142

Dong X, Li X, Ding L, Cui F, Tang Z, Liu Z (2014) LWT–stage extraction of capsaicinoids and red pigments from fresh red pepper (*Capsicum*) fruits with ethanol as solvent. Food Sci Technol 59: 396–402

Dorantes L, Colmenero R, Hernandez H, Mota L, Jaramillo ME, Fernandez E, Solano C (2000) Inhibition of growth of some foodborne pathogenic bacteria by Capsicum annum extracts. Int J Food Microbiol 57(1):125–128

Dussault D, Vu KD, Lacroix M (2014) *In vitro* evaluation of antimicrobial activities of various commercial essential oils, oleoresin and pure compounds against food pathogens and application in ham. Meat Sci 96:514–520

EDQM (2016) Reference substances. Origin of goods: capsaicin. Division of reference standards & samples (DRS), European Pharmacopoeia Reference Standards. European Directorate for the Quality of Medicine. Accessed 26 Sep 2017

EFSA (2015) Re-evaluation of paprika extract (E 160c) as a food additive. EFSA J 13(12):4320

Eggersdorfer M, Wyss A (2018) Carotenoids in human nutrition and health. Arch Biochem Biophys 652:18–26

EMA (2007) Guideline on excipients in the dossier for application for marketing authorisation of a medicinal product. Directive 2001/83/EC. Committee for medicinal products for human use (CHMP). European Medicines Agency Inspections. Accessed 26 Sep 2017

EMA (2015) EMA/HMPC/674139/2013. European Union herbal monograph on *Capsicum annuum* L. var. minimum (Miller) Heiser and small fruited varieties of *Capsicum frutescens* L., fructus. Committee on Herbal Medicinal Products (HMPC). European Medicines Agency Inspections. Accessed 27 Sep 2017

FDA (2017a) Q3C impurities: residual solvents. Guidance for industry. Food and Drug Administration. U.S. Department of Health and Human Services, Washington, DC. Accessed 8 Aug 2018

FDA (2017b) Cosmetic ingredients. FDA database. Food and Drug Administration U.S, Silver Spring. Accessed 8 Mar 2018

Feng Y, Zhu Y, Wan J, Yang X, Firempong CK, Yu J, Xu X (2018) Enhanced oral bioavailability, reduced irritation and increased hypolipidemic activity of self-assembled capsaicin prodrug nanoparticles. J Funct Foods 44:137–145

Fernández-García E, Carvajal-Lérida I, Jarén-Galán M, Garrido-Fernández J, Pérez- Gálvez A, Hornero-Méndez D (2012) Carotenoids bioavailability from foods: from plant pigments to efficient biological activities. Food Res Int 46(2):438–450

Fernández-López J, Pérez-Alvarez JA, Sayas-Barberá E, López-Santoveña F (2002) Effect of paprika (*Capsicum annum*) on color of Spanish-type sausages during the resting stage. J Food Sci 67(6):2410–2414

Fernández-Ronco MP, de Lucas A, Rodríguez JF, García MT, Gracia I (2013) New considerations in the economic evaluation of supercritical processes: separation of bioactive compounds from multicomponent mixtures. J Supercrit Fluids 79:345–355

Fiedor J, Burda K (2014) Potential role of carotenoids as antioxidants in human health and disease. Nutrients 6(2):466–488

Garces-Claver A, Arnedo-Andres MS, Abadıa J, Gil-Ortega R, Alvarez-Fernandez A (2006) Determination of capsaicin and dihydrocapsaicin in *Capsicum* fruits by liquid chromatography−electrospray/time-of-flight mass spectrometry. J Agric Food Chem 54:9303–9311

Gharaibeh A, Gharaibeh MM, Gharaibeh S (2020) Relation between Covid-19 and high capsaicin diets. Magna Sci Adv Res Rev 1(1):30–32

Giuffrida D, Dugo P, Torre G, Bignardi C, Cavazza A, Corradini C, Dugo G (2013) Characterization of 12 Capsicum varieties by evaluation of their carotenoid profile and pungency determination. Food Chem 140(4):794–802

Goci E, Haloci E, Stefano AD, Chiavaroli A, Angelini P, Miha A, Cacciatore I, Marinelli L (2021) Evaluation of in vitro capsaicin release and antimicrobial properties of topical pharmaceutical formulation. Biomol Ther 11(3):432

Govindarajan VS (1985) Capsicum production, technology, chemistry, and quality. Part 1: history, botany, cultivation, and primary processing. Crit Rev Food Sci Nutr 22(2):109–176

Govindarajan VS, Sathyanarayana MN (1991) Capsicum—production, technology, chemistry, and quality. Part V. Impact on physiology, pharmacology, nutrition, and metabolism; structure, pungency, pain, and desensitization sequences. Crit Rev Food Sci Nutr 29(6):435–474

Guil-Guerrero JL, Martínez-Guirado C, del Mar Rebolloso-Fuentes M, Carrique-Pérez A (2006) Nutrient composition and antioxidant activity of 10 pepper (*Capsicum annuun*) varieties. Eur Food Res Technol 224(1):1–9

Ha SH, Kim JB, Park JS, Lee SW, Cho KJ (2007) A comparison of the carotenoid accumulation in Capsicum varieties that show different ripening colours: deletion of the capsanthin-capsorubin synthase gene is not a prerequisite for the formation of a yellow pepper. J Exp Bot 58(12): 3135–3144

Haar RJ, Iacopino V, Ranadive N, Weiser SD, Dandu M (2017) Health impacts of chemical irritants used for crowd control: a systematic review of the injuries and deaths caused by tear gas and pepper spray. BMC Public Health 17:1–14

Hatahet T, Morille M, Hommoss A, Devoisselle JM, Müller RH, Bégu S (2016) Quercetin topical application, from conventional dosage forms to nanodosage forms. Eur J Pharm Biopharm 108 (Supplement C):41–53

Hayman M, Kam PCA (2008) Capsaicin: a review of its pharmacology and clinical applications. Curr Anaesth Crit Care 19(5):338–343

Heinrich U, Gärtner C, Wiebusch M, Eichler O, Sies H, Tronnier H, Stahl W (2003) Supplementation with beta-carotene or a similar amount of mixed carotenoids protects humans from UV-induced erythema. J Nutr 133(1):98–101

Henderson DE, Henderson SK (1992) Thermal decomposition of capsaicin. 1. Interactions with oleic acid at high temperatures. J Agric Food Chem 40(11):2263–2268

Henderson DE, Slickman AM, Henderson SK (1999) Quantitative HPLC determination of the antioxidant activity of capsaicin on the formation of lipid hydroperoxides of linoleic acid: a comparative study against BHT and melatonin. J Agric Food Chem 47(7):2563–2570

Hernandez-Perez T, Gomez-Garcia MDR, Valverde ME, Paredes-Lopez O (2020) *Capsicum annuum* (hot pepper): an ancient Latin-American crop with outstanding bioactive compounds and nutraceutical potential. A review. Compr Rev Food Sci Food Saf 19(6):2972–2993

Hęś M, Dziedzic K, Gorecka D, Jędrusek-Golińska A, Gujska E (2019) *Aloe vera* (L.) Webb.: natural sources of antioxidants—a review. Plant Foods Hum Nutr 74(3):255–265

Howard LR, Wildman REC (2007) Antioxidant vitamin and phytochemical content of fresh and processed pepper fruit (*Capsicum annuum*). In: Wildman REC (ed) Chapter 9. Handbook of nutraceuticals and functional foods. CRC Press, London

Jayaprakasha GK, Bae H, Crosby K, Jifon JL, Patil BS (2012) Bioactive compounds in peppers and their antioxidant potential. 1109 Hispanic foods: chemistry and bioactive compounds. American Chemical Society, Washington, DC, pp 43–56

Johnson W (2007) Final report on the safety assessment of *Capsicum annuum* extract, *Capsicum annuum* fruit extract, *Capsicum annuum* resin, *Capsicum annuum* fruit powder, *Capsicum frutescens* fruit, *Capsicum frutescens* fruit extract, *Capsicum frutescens* resin, and capsaicin. Int J Toxicol 26(1):103–106

Kana-Sop MM, Gouado I, Achu MB, Van Camp J, Amvam Zollo PH, Schweigert FJ, Ekoe T (2015) The influence of iron and zinc supplementation on the bioavailability of provitamin A carotenoids from papaya following consumption of a vitamin A-deficient diet. J Nutr Sci Vitaminol (Tokyo) 61(3):205–214

Kantar MB, Anderson JE, Lucht SA, Mercer K, Bernau V, Case KA, Le NC, Frederiksen MK, DeKeyser HC, Wong ZZ, Hastings JC, Baumler DJ (2016) Vitamin variation in *Capsicum* spp. provides opportunities to improve nutritional value of human diets. PLoS One 11(8):e0161464

Kerdudo A, Burger P, Merck F, Dingas A, Rolland Y, Michel T, Fernandez X (2016) Development of a natural ingredient—natural preservative: a case study. C R Chim 19(9):1077–1089

Kim JS, An CG, Park JS, Lim YP, Kim S (2016a) Carotenoid profiling from 27 types of paprika (*Capsicum annuum* L.) with different colors, shapes, and cultivation methods. Food Chem 15(201):64–71

Kim YJ, Payal AR, Daly MK (2016b) Effects of tear gases on the eye. Surv Ophthalmol 61:434–442

Kim MK, Lee MA, Lee KG (2018) Determination of compositional quality and volatile flavor characteristics of radish-based Kimchi suitable for Chinese consumers and its correlation to consumer acceptability. Food Sci Biotechnol 27(5):1265–1273

Kitagawa S, Yoshii K, Morita SY, Teraoka R (2011) Efficient topical delivery of chlorogenic acid by an oil-in-water microemulsion to protect skin against UV-induced damage. Chem Pharm Bull(Tokyo) 59(6):793–796

Kobata K, Sugawara M, MimuFigra M, Yazawa S, Watanabe T (2013) Potent production of capsaicinoids and capsinoids by capsicum peppers. J Agric Food Chem 61(46):11127–11132

Kollia E, Proestos C, Zoumpoulakis P, Markaki P (2019) Capsaicin, an inhibitor of Ochratoxin A production by *Aspergillus* section Nigri strains in grapes (*Vitis vinifera* L.). Food Addit Contam Part A Chem Anal Control Expos Risk Assess 36(11):1709–1721

Ksibi IE, Slama RB, Faidi K, Ticha MB, Mhenni MF (2015) Mixture approach for optimizing the recovery of colored phenolics from red pepper (*Capsicum annum* L.) by-products as potential source of natural dye and assessment of its antimicrobial activity. Ind Crop Prod 70:34–40

Kurita S, Kitagawa E, Kim CH, Momose Y, Iwahashi H (2002) Studies on the antimicrobial mechanisms of capsaicin using yeast DNA microarray. Biosci Biotechnol Biochem 66(3): 532–536

Kwon Y (2021) Estimation of dietary capsaicinoid exposure in Korea and assessment of its health effects. Nutrients 13(7):2461

Lee HM, Kim SJ, Lee JY, Park B, Yang JS, Ha SD, Choi C, Ha JH (2019) Capsaicinoids reduce the viability of a norovirus surrogate during kimchi fermentation. LWT 115:108460

Liu Z, Zhu P, Tao Y, Shen C, Wang S, Zhao L, Wu H, Fan F, Lin C, Chen C, Zhu Z, Wei Z, Sun L, Liu Y, Wang A, Lu Y (2015) Cancer-promoting effect of capsaicin on DMBA/TPA-induced skin tumorigenesis by modulating inflammation, Erk and p38 in mice. Food Chem Toxicol 81: 1–8

Loizzo MR, Pugliese A, Bonesi M, Menichini F, Tundis R (2015) Evaluation of chemical profile and antioxidant activity of twenty cultivars from *Capsicum annuum*, *Capsicum baccatum*, *Capsicum chacoense* and *Capsicum chinense*: a comparison between fresh and processed peppers. LWT Food Sci Technol 64:623–631

Lu M, Ho CT, Huang Q (2017) Extraction, bioavailability, and bioefficacy of capsaicinoids. J Food Drug Anal 25:27–36

Lu M, Chen C, Lan Y, Xiao J, Li R, Huang J, Huang Q, Cao Y, Ho CT (2020) Capsaicin—the major bioactive ingredient of chili peppers: bio-efficacy and delivery systems. Food Funct 11(4):2848–2860

Ludy MJ, Moore GE, Mattes RD (2012) The effects of capsaicin and capsiate on energy balance: critical review and meta-analyses of studies in humans. Chem Senses 37(2):103–121

Luo XJ, Peng J, Li YJ (2011) Recent advances in the study on capsaicinoids and capsinoids. Eur J Pharmacol 650:1–7

Luque de Castro MD, Garcıá-Ayuso LE (1998) Soxhlet extraction of solid materials: an outdated technique with a promising innovative future. Anal Chim Acta 369(1):1–10

Lysy J, Sistiery-Ittah M, Israelit Y, Shmueli A, Strauss-Liviatan N, Mindrul V, Goldin E (2003) Topical capsaicin—a novel and effective treatment for idiopathic intractable pruritus ani: a randomised, placebo controlled, crossover study. Gut 52(9):1323–1326

Malaj L, Martena V, Giovenali S, Di Martino P (2010) A new microemulsion for the transdermal delivery of quercetin. Sci Pharm 78(3):600

Marin A, Ferreres F, Tomas-Barberan FA, Gil MI (2004) Characterization and quantitation of antioxidant constituents of sweet pepper (*Capsicum annuum* L.). J Agric Food Chem 52(12): 3861–3869

Marini E, Magi G, Mingoia M, Pugnaloni A, Facinelli B (2015) Antimicrobial and anti-virulence activity of capsaicin against erythromycin-resistant, cell invasive group A streptococci. Front Microbiol 13(6):1281

Martínez S, López M, González-Raurich M, Bernardo-Alvarez A (2005) The effects of ripening stage and processing systems on vitamin C content in sweet peppers (*Capsicum annuum* L.). Int J Food Sci Nutr 56(1):45–51

Martínez L, Cilla I, Antonio Beltrán J, Roncalés P (2006) Effect of *Capsicum annuum* (red sweet and cayenne) and *Piper nigrum* (black and white) pepper powders on the shelf life of fresh pork sausages packaged in modified atmosphere. J Food Sci 71(1):S48–S53

Martínez S, Curros A, Bermúdez L, Carballo J, Franco I (2007) The composition of Arnoia peppers (*Capsicum annuum* L.) at different stages of maturity. Int J Food Sci Nutr 5(2):150–161

Materska M (2014) Bioactive phenolics of fresh and freeze-dried sweet and semi-spicy pepper fruits (*Capsicum annuum* L.). J Funct Foods 7:269–277

Materska M, Perucka I (2005) Antioxidant activity of the main phenolic compounds isolated from hot pepper fruit (*Capsicum annuum* L.). J Agric Food Chem 53(5):1750–1756

Matsufuji H, Ishikawa K, Nunomura O, Chino M, Takeda M (2007) Antioxidant content of different coloured sweet peppers, white, green, yellow, orange and red (*Capsicum annuum* L.). Int J Food Sci Technol 42(12):1482–1488

Minatel IO, Borges CV, Ferreira MI, Gomez HAG, Chen CYO, Lima GPP (2017) Phenolic compounds: functional properties, impact of processing and bioavailability. In: Soto-Hernandez M, Palma-Tenango M, Garcia-Mateos MDR (eds) Chapter 1. Phenolic compounds—biological activity. IntechOpen, London

Molina-Torres J, Garcia-Chavez A, Ramirez-Chavez E (1999) Antimicrobial properties of alkamides present in flavouring plants traditionally used in Mesoamerica: affinin and capsaicin. J Ethnopharmacol 64(3):241–248

Monago-Marana O, Guzman-Becerra M, Munoz de la Pena A, Galeano-Dıaz T (2018) Determination of pungency in spicy food by means of excitation-emission fluorescence coupled with second-order chemometric calibration. J Food Compos Anal 67:10–18

Mosqueda-Solís A, Lafuente-Ibáñez de Mendoza I, Aguirre-Urizar JM, Mosqueda-Taylor A (2021) Capsaicin intake and oral carcinogenesis: a systematic review. Med Oral Patol Oral Cir Bucal 26(2):e261–e268

Munin A, Edwards-Lévy F (2011) Encapsulation of natural polyphenolic compounds; a review. Pharmaceutics 3(4):793–829

Nascimento PL, Nascimento TC, Ramos NS, Silva GR, Gomes JE, Falcão RE, Moreira KA, Porto AL, Silva TM (2014) Quantification, antioxidant and antimicrobial activity of phenolics isolated from different extracts of *Capsicum frutescens* (*Pimenta Malagueta*). Molecules 19(4): 5434–5447

Navarro JM, Flores P, Garrido C, Martinez V (2006) Changes in the contents of antioxidant compounds in pepper fruits at different ripening stages, as affected by salinity. Food Chem 96(1):66–73

Nazzaro F, Caliendo G, Arnesi G, Veronesi A, Sarzi P, Fratianni F (2009) Comparative content of some bioactive compounds in two varieties of *Capsicum annuum* L. sweet pepper and evaluation of their antimicrobial and mutagenic activities. J Food Biochem 33(6):852–868

NIH (2008) National Institutes of Health. U.S. Department of Health & Human Services. Vitamin and mineral supplement fact sheets. Accessed 13 Aug 2022

Nugroho LH (2016) Red pepper (*Capsicum* spp.) fruit: a model for the study of secondary metabolite product distribution and its management. AIP Conf Proc 1744:020034

Ohnuki K, Haramizu S, Oki K, Watanabe T, Yazawa S, Fushiki T (2001) Administration of capsiate, a non-pungent capsaicin analog, promotes energy metabolism and suppresses body fat accumulation in mice. Biosci Biotechnol Biochem 65(12):2735–2740

Okada Y, Tanaka K, Sato E, Okajima H (2010) Kinetics and antioxidative sites of capsaicin in homogeneous solution. J Am Oil Chem Soc 87(12):1397–1405

Orellana-Escobedo L, Garcia-Amezquita LE, Olivas GI, Ornelas-Paz JJ, Sepulveda, DR (2013) Capsaicinoids content and proximate composition of Mexican chili peppers (Capsicum spp.) cultivated in the State of Chihuahua. CyTA – Journal of Food 11(2):179–184.

Orlo E, Russo C, Nugnes R, Lavorgna M, Isidori M (2021) Natural methoxyphenol compounds: antimicrobial activity against foodborne pathogens and food spoilage bacteria, and role in antioxidant processes. Foods 10(8):1807

Padilha HKM, Pereira EDS, Munhoz PC, Vizzotto M, Valgas RA, Barbieri RL (2015) Genetic variability for synthesis of bioactive compounds in peppers (*Capsicum annuum*) from Brazil. Food Sci Technol 35:516–523

Palombo P, Fabrizi G, Ruocco V, Ruocco E, Fluhr J, Roberts R, Morganti P (2007) Beneficial long-term effects of combined oral/topical antioxidant treatment with the carotenoids lutein and zeaxanthin on human skin: a double-blind, placebo-controlled study. Skin Pharmacol Physiol 20(4):199–210

Pandey KB, Rizvi SI (2009) Plant polyphenols as dietary antioxidants in human health and disease. Oxidative Med Cell Longev 2(5):270–278

Patel K, Ruiz C, Calderon R, Marcelo M, Rojas R (2016) Characterisation of volatile profiles in 50 native Peruvian chili pepper using solid phase microextraction–gas chromatography mass spectrometry (SPME–GCMS). Food Res Int 89:471–475

Pérez-López AJ, López-Nicolas JM, Núñez-Delicado E, Amor FMD, Carbonell-Barrachina ÁA (2007) Effects of agricultural practices on color, carotenoids composition, and minerals contents of sweet peppers, cv. almuden. J Agric Food Chem 55(20):8158–8164

Perucka I, Materska M (2001) Phenylalanine ammonia-lyase and antioxidant activities of lipophilic fraction of fresh pepper fruits *Capsicum annum* L. Innov Food Sci Emerg Technol 2(3):189–192

Perva-Uzunalić A, Škerget M, Weinreich B, Knez Ž (2004) Extraction of chilli pepper (var. Byedige) with supercritical CO_2: effect of pressure and temperature on capsaicinoid and colour extraction efficiency. Food Chem 87(1):51–58

Ph. Eur. (2010) European Pharmacopoeia. Working Procedures of the Pharmacopoeial Discussion Group (PDG). EDQM Council of Europe, 22(4). Accessed 29 Sep 2017

Pieroni A, Quave CL, Villanelli ML, Mangino P, Sabbatini G, Santini L, Tomasi M (2004) Ethnopharmacognostic survey on the natural ingredients used in folk cosmetics, cosmeceuticals and remedies for healing skin diseases in the inland marches, Central-Eastern Italy. J Ethnopharmacol 91(2–3):331–344

Pruthi JS (2004) Chemistry and quality control of capsicums and capsicum products. In: De AK (ed) Chapter 3. Capsicum, the genus capsicum. CRC Press, London

Raza K, Shareef MA, Singal P, Sharma G, Negi P, Katare OP (2014) Lipid based capsaicin-loaded nano-colloidal biocompatible topical carriers with enhanced analgesic potential and decreased dermal irritation. J Liposome Res 24(4):290–296

Reilly CA, Yost GS (2006) Metabolism of capsaicinoids by P450 enzymes: a review of recent findings on reaction mechanisms, bio-activation, and detoxification processes. Drug Metab Rev 38:685–706

Reyes-Escogido ML, Gonzalez-Mondragon EG, Vazquez-Tzompantzi E (2011) Chemical and pharmacological aspects of capsaicin. Molecules 16(2):1253–1270

Ritota M, Marini F, Sequi P, Valentini M (2010) Metabolomic characterization of Italian sweet pepper (*Capsicum annum* L.) by means of HRMAS-NMR spectroscopy and multivariate analysis. J Agric Food Chem 58:9675–9684

Romo-Hualde A, Yetano-Cunchillos AI, González-Ferrero C, Sáiz-Abajo MJ, González-Navarro CJ (2012) Supercritical fluid extraction and microencapsulation of bioactive compounds from red pepper (*Capsicum annum* L.) by-products. Food Chem 133(3):1045–1049

Rubio C, Hardisson A, Martín R, Báez A, Martín M, Álvarez R (2002) Mineral composition of the red and green pepper (*Capsicum annuum*) from Tenerife Island. Eur Food Res Technol 214(6): 501–504

Saini RK, Keum YS (2018) Carotenoid extraction methods: a review of recent developments. Food Chem 240:90–103

Saito M, Yoneshiro T (2013) Capsinoids and related food ingredients activating brown fat thermogenesis and reducing body fat in humans. Curr Opin Lipidol 24(1):71–77

Sakakibara H, Honda Y, Nakagawa S, Ashida H, Kanazawa K (2003) Simultaneous determination of all polyphenols in vegetables, fruits, and teas. J Agric Food Chem 51(3):571–581

Salgado-Roman M, Botello-Alvarez E, Rico-Martinez R, Jimenez-Islas H, Cardenas-Sanatombi K, Rajkumari S (2020) Effect of processing on quality of pepper: a review. Food Rev Int 36(6): 626–643

Sanatombi K, Rajkumari S (2020) Effect of processing on quality of pepper: A review. Food Reviews International 36(6):626–643.

Sanchez-Arreguin A, Carriles R, Ochoa-Alejo N, Lopez MG, Sanchez-Segura L (2019) Generation of BSA-capsaicin nanoparticles and their hormesis effect on the *Rhodotorula mucilaginosa* yeast. Molecules 24(15):2800

Sandoval-Castro CJ, Valdez-Morales M, Oomah BD, Gutierrez-Dorado R, Medina-Godoy S, Espinosa-Alonso LG (2017) Bioactive compounds and antioxidant activity in scalded Jalapeno pepper industrial by product (*Capsicum annuum*). J Food Sci Technol 54(7):1999–2010

Schlossareck C, Ross CF (2019) Electronic tongue and consumer sensory evaluation of spicy paneer cheese. J Food Sci 84(6):1563–1569

Serrano G, Almudéver P, Serrano JM, Milara J, Torrens A, Expósito I, Cortijo J (2015) Phosphatidylcholine liposomes as carriers to improve topical ascorbic acid treatment of skin disorders. Clin Cosmet Investig Dermatol 8:591–599

Sganzerla M, Coutinho JP, de Melo AMT, Godoy HT (2014) Fast method for capsaicinoids analysis from *Capsicum chinense* fruits. Food Res Int 64:718–725

Shah I, Shah MA, Nawaz MA, Pervez S, Noreen N, Vargas-De La Cruz C, Khan F, Blundell R, Briffa J, Azzopardi J (2020) Analysis of other phenolics (capsaicin, gingerol, and alkylresorcinols). In: Silva AS, Nabavi SF, Saeedi M, Nabavi SM (eds) Recent advances in natural products analysis. Elsevier, Amsterdam, pp 255–271

Sharif MK, Ejaz R, Pasha I (2018) Nutritional and therapeutic potential of spices. In: Therapeutic, probiotic, and unconventional foods. Elsevier, London, pp 181–199

Si W, Liang Y, Ma KY, Chung HY, Chen ZY (2012) Antioxidant activity of capsaicinoid in canola oil. J Agric Food Chem 60(24):6230–6234

Si W, Gao Y, Mei X, Wu C, Li J, Zhang J (2021) Mesoporous silica nanoparticles loaded with capsaicin and their oxidation resistance in meat preservation. Food Chem 344:128737

Silva HRA, De Souza GM, Fernandes JD, Constantino CJL, Winkelstroter LK (2020) Unravelling the effects of the food components ascorbic acid and capsaicin as a novel anti-biofilm agent against Escherichia coli. J Food Sci Technol 57(3):1013–1020

Singh S, Jarret R, Russo V, Majetich G, Shimkus J, Bushway R, Perkins B (2009) Determination of capsinoids by HPLC-DAD in capsicum species. J Agric Food Chem 57(9):3452–3457

Singh S, Singh DR, Balaji R, Nayak D, Singh AK, Kumar N (2013) Molecular and functional diversity in Capsicum landraces of Andaman Islands. Afr J Biotechnol 12:5729–5737

Sricharoen P, Lamaiphan N, Patthawaro P, Limchoowong N, Techawongstien S, Chanthai S (2017) Phytochemicals in *Capsicum* oleoresin from different varieties of hot chilli peppers with their antidiabetic and antioxidant activities due to some phenolic compounds. Ultrason Sonochem 38: 629–639

Srinivasan K (2016) Biological activities of red pepper (*Capsicum annuum*) and its pungent principle capsaicin: a review. Crit Rev Food Sci Nutr 56(9):1488–1500

Stanberry LR, Bourne N, Bravo FJ, Bernstein DI (1992) Capsaicin-sensitive peptidergic neurons are involved in the zosteriform spread of herpes simplex virus infection. J Med Virol 38(2): 142–146

Staubach P, Metz M (2013) Magistral formulations and pruritus therapy—what is established, what is confirmed, what is new? JDDG 11(11):1049–1055

Su X, Yang Z, Tan KB, Chen J, Huang J, Li Q (2020) Preparation and characterization of ethyl cellulose film modified with capsaicin. Carbohydr Polym 241:116259

Sun T, Xu Z, Wu CT, Janes M, Prinyawiwatkul W, No HK (2007) Antioxidant activities of different colored sweet bell peppers (*Capsicum annuum* L.). J Food Sci 72(2):98–102

Sun F, Xiong S, Zhu Z (2016) Dietary capsaicin protects cardiometabolic organs from dysfunction. Nutrients 8(5):174

Sweat KG, Broatch J, Borror C, Hagan K, Cahill TM (2016) Variability in capsaicinoid content and Scoville heat ratings of commercially grown Jalapeño, Habanero and Bhut Jolokia peppers. Food Chem 210:606–612

Tayseer I, Aburjai T, Abu-Qatouseh L, Al-Karabieh N, Ahmed W, Al-Samydai A (2020) In vitro anti-helicobacter pylori activity critical reviews in food science and nutrition 17 of capsaicin. J Pure Appl Microbiol 14(1):279–286

Telang PS (2013) Vitamin C in dermatology. Indian Dermatol Online J 4(2):143–146

Tepic AN, Dimic GR, Vujicic BL, Kevresan ZS, Varga M, Sumic ZM (2008) Quality of commercial ground paprika and its oleoresins. Acta Period Technol 39:77–83

Thampi PSS (2004) A glimpse of the world trade in capsicum. In: De AK (ed) Chapter 2. Capsicum, the genus capsicum. CRC Press, London

Thapa B, Pepic I, Vanic Z, Basnet P, Skalko-Basnet N (2013) Topical delivery system for phytochemicals: capsaicin and capsicum tincture. J Pharm Drug Dev 1(2):201

Tiwari G, Tiwari R, Sriwastawa B, Bhati L, Pandey S, Pandey P, Bannerjee SK (2012) Drug delivery systems: an updated review. Int J Pharm Investig 2(1):2–11
Topuz A, Ozdemir F (2007) Assessment of carotenoids, capsaicinoids and ascorbic acid composition of some selected pepper cultivars (*Capsicum annuum* L.) grown in Turkey. J Food Compos Anal 20(7):596–602
Topuz A, Dincer C, Ozdemir KS, Feng H, Kushad M (2011) Influence of different drying methods on carotenoids and capsaicinoids of paprika (cv., Jalapeno). Food Chem 129(3):860–865
Uquiche E, Valle JM, Ortiz J (2004) Supercritical carbon dioxide extraction of red pepper (*Capsicum annuum* L.) oleoresin. J Food Eng 65(1):55–66
Wahba NM, Ahmed AS, Ebraheim ZZ (2010) Antimicrobial effects of pepper, parsley, and dill and their roles in the microbiological quality enhancement of traditional Egyptian Kareish cheese. Foodborne Pathog Dis 7(4):411–418
Wang J, Shi T, Yang X, Han W, Zhou Y (2014) Environmental risk assessment on capsaicin used as active substance for antifouling system on ships. Chemosphere 104:85–90
Wang X, Zou C, Zhang Y, Shi X, Liu J, Fan S, Liu Y, Du Y, Zhao Q, Tan Y, Wu C, Chen X (2018) Environmental impacts of pepper (*Capsicum annuum* L.) production affected by nutrient management: a case study in southwest China. J Clean Prod 171:934–943
Wang F, Xue Y, Fu L, Wang Y, He M, Zhao L, Liao X (2022) Extraction, purification, bioactivity and pharmacological effects of capsaicin: a review. Crit Rev Food Sci Nutr 62(19):5322–5348
Westfall A, Giusti MM (2017) Color profiles and stability of acylated and nonacylated anthocyanins as novel pigment sources in a lipstick model: a viable alternative to synthetic colorants. J Cosmet Sci 68(3):233–244
Xing F, Cheng G, Yi K (2006) Study on the antimicrobial activities of the capsaicin microcapsules. J Appl Polym Sci 102(2):1318–1321
Yang XY, Du GH (2018) Capsaicin. In: Natural small molecule drugs from plants. Springer, Singapore, pp 397–402
Yang CY, Mandal PK, Han KH, Fukushima M, Choi K, Kim CJ, Lee CH (2010) Capsaicin and tocopherol in red pepper seed oil enhances the thermal oxidative stability during frying. J Food Sci Technol 47(2):162–165
Yang M, Lee G, Si J, Lee SJ, You HJ, Ko G (2016) Curcumin shows antiviral properties against norovirus. Molecules 21(10):1401
Yeung MF, Tang WYM (2015) Clinicopathological effects of pepper (oleoresin capsicum) spray. Hong Kong Med J 21:542–552
Yusop SM, O'Sullivan MG, Preuß M, Weber H, Kerry JF, Kerry JP (2012) Assessment of nanoparticle paprika oleoresin on marinating performance and sensory acceptance of poultry meat. LWT- Food Sci Technol 46(1):349–355
Zhang S, Wei F, Han X (2018) An edible film of sodium alginate/pullulan incorporated with capsaicin. New J Chem 42(21):17756–17761
Zhang R, Chen G, Yang B, Wu Y, Du M, Kan J (2021) Insights into the stability of carotenoids and capsaicinoids in water-based or oil-based chili systems at different processing treatments. Food Chem 342:128308
Zhao J, Wei F, Xu W, Han X (2020) Enhanced antibacterial performance of gelatin/chitosan film containing capsaicin loaded MOFs for food packaging. Appl Surf Sci 510:145418
Zillich OV, Schweiggert-Weisz U, Eisner P, Kerscher M (2015) Polyphenols as active ingredients for cosmetic products. Int J Cosmet Sci 37(5):455–464

Applications of Capsaicin in the Food Industry

13

Maria Vanaina Souza Gonçalves, Deepak Chandran, Lishma Nelliyaparambath, Aparna Kozhampurathu Gokul, and Luiz Everson da Silva

Abstract

The change in the eating habits of a considerable portion of the population, aiming at a better quality of life, has required the food industry to search for healthier formulations. This is especially evident in the use of natural additives that ensure adequate sensory, microbiological, chemical, and physical characteristics. In addition, increasing shelf life is an important requirement to guarantee quality to industrialized products in a highly competitive market. Chili peppers are an interesting natural source because they present differentiated characteristics that contribute to the consumer's health, besides providing color, flavor, and spiciness to the preparations. The high nutritional value also makes them a differentiated product, because they are good sources of vitamins, especially C and A, and because they have several minerals that are essential to the perfect functioning of the body. In this context, we have the capsaicin, a phytochemical found in hot pepper fruit extract, that belongs to capsaicinoids group, which shows high nutritional value. The capsaicin is accumulated in the placenta surface tissue, only released when the fruit is cut. In this chapter, we highlight the

M. V. S. Gonçalves · L. E. da Silva (✉)
Post-graduate Program in Sustainable Territorial Development, Universidade Federal do Paraná, Matinhos, Brazil

D. Chandran
Department of Veterinary Sciences and Animal Husbandry, School of Agricultural Sciences, Amrita Vishwa Vidyapeetham University, Coimbatore, Tamil Nadu, India

L. Nelliyaparambath
Regional Agricultural Research Station, Kerala Agricultural University, Wayanad, Kerala, India
e-mail: lishma.np@kau.in

A. K. Gokul
Department of Plant Pathology, Kerala Agricultural University, Wayanad, Kerala, India

M. K. Swamy, A. Kumar (eds.), *Capsaicinoids*,
https://doi.org/10.1007/978-981-99-7779-6_13

use of capsaicin as an additive and pigment, besides its antioxidant and antimicrobial properties in the food industry.

Keywords

Food industry · Additive · Pigment · Biological properties · Natural products · Antifungal and antibacterial activities

13.1 Introduction

Chili peppers are spices with a wide acceptance in the market, mainly due by their several special characteristics, which can be accentuated or reduced according to according to each variety analyzed. These alterations can occur due to some of their its coloring, antioxidant capacity and especially its pungency, fact that turns chilis into a peculiar product (Marincaş et al. 2018). This way, the chili peppers that have been on Brazilian tables for many years, have been gaining adepts all over the around the world and have become the target of researches to take full advantage of the production. Chili peppers belong to the same genus as bell peppers (Capsicum) from the tropical regions of America, more specifically, the first records date back to approximately 900 B.C. as confirmed during archaeological explorations in Tehuacán, Mexico, where they were used as medicine, a common practice among the Mayas. Other evidence was found in Peru and Bolivia, which already used some varieties as food, even before the discovery (Fan et al. 2017).

The interest of people from all over the world in chili peppers due to their high pungency, a characteristic derived from the presence of presence of capsaicin (Fig. 13.1), which is accumulated in the surface tissue of the released only when the fruit is cut and, has become a competitor to the black pepper coming from Asia and that already had great commercial potential. It became a competitor to the black pepper from Asia, which already had great commercial potential, mainly due to the Malagueta variety, which had equal or superior pungency. The Malagueta variety was nicknamed the new pepper (Akbas et al. 2018).

Alkaloid capsaicin is biosynthesized in capsicum fruit as a defense mechanism against insects and mold. As trans-8-methyl-N-vanillyl-6-nonenamide, capsaicin belongs to the class of compounds known as vanilloids. They are biologically active because vanilloids include a vanillyl (4-hydroxy-3-methoxybenzyl) moiety. Capsaicin, like other vanilloids, has a structure made up of a benzene ring and a hydrophobic carbon tail containing an amide group. Capsaicin is a crystalline, lipophilic molecule that is colorless, odorless, and extremely poorly soluble in water (10.3 mg/L at 25 °C), but is soluble in ethanol, chloroform, methanol, acetone,

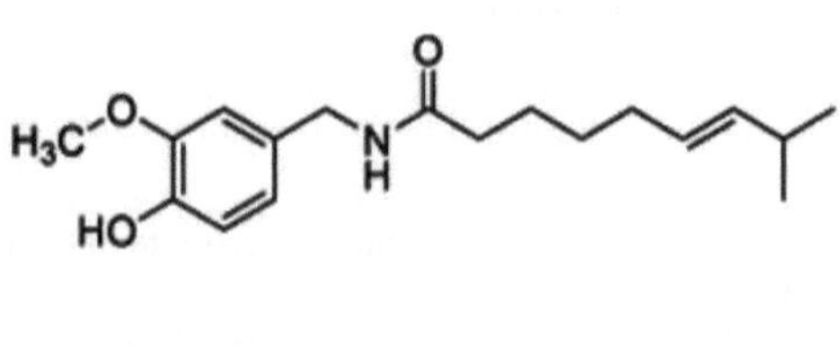

Fig. 13.1 Molecular structure of capsaicin

and alkaline aqueous solutions. At room temperature (25 °C), this chemical is reasonably stable in both acids and alkalis. In addition, it has a molecular weight of 305.4 kDa, a melting point of 62–65 °C, and an off-white color (Bode and Dong 2011; Chang et al. 2011; Saha et al. 2022).

Malagueta becomes then the main variety of American chilis, especially in Brazil, where it is cultivated in the states of Bahia, Goiás, and Minas Gerais. It is the most important chili pepper in the food industry, second to garlic and onion, with their characteristic pungency and excellent levels of vitamins A and C. We can notice that the level of pungency is what guarantees the use and commercialization of several varieties of chilis (Carvalho et al. 2014). Other characteristics that have called the attention of different research groups, chilis, mainly due to their different colors (pigments), presence of bioactive compounds, among others, that will amplify their functional properties and consequently boost the cultivation, processing, and commercialization, favoring all the productive chain and turning the pepper into a real spice (Chatterjee et al. 2010; Tyagi et al. 2022).

Capsaicin, sometimes known as chili powder, is a popular seasoning and culinary addition around the world. It can be used to preserve food by acting as a preservative because of its antibacterial and oxidation-resistance qualities. There are a variety of foods that contain capsaicin, including red pepper powder, spicy sauce, kimchi, pickle, salted seafood, red pepper paste, and instant noodles. Capsaicin has been found to have beneficial biological benefits in a number of different areas, including as an anti-inflammatory, an anti-cancer agent, a cardiovascular protector, an enhancer of digestive function, and a remedy for skin disorders (Cho and Kwon 2020). Capsaicin is used sparingly in wet foods due to its lipophilic nature and low water solubility. Capsaicin's limited water solubility also reduces its bioavailability. The inherent spiciness of capsaicin, as well as the loss of flavor and stability in the case of exposure to heat and light, are further technological challenges associated with its application. In addition, ingesting large quantities of capsaicin can result in stomach pain, nausea, vomiting, and acidic, burning stools. On the other hand, capsaicin in large doses may have a carcinogenic effect. Capsaicin, when in contact with eyes or skin, can generate a painful burning or stinging sensation. Capsaicin's toxicity and efficacy are both influenced by how much of it is consumed (Hayman and Kam 2008; Wang et al. 2020).

This book chapter gives an extensive insight into the applications of capsaicin in the food industry.

13.2 Botanical Information About *Capsicum spp*

The name pepper comes from the Latin *pigmentum* which was translated into Spanish as pimienta. It has several meanings, such as coloring matter, aromatic spice and due to its characteristics, it has conquered the world, where nowadays in Brazil dozens of varieties are produced in Brazil and the economy moves millions of people. Agronomically, chilis and peppers belong to the Plantae Kingdom, Magnoliophyta, Class Magnoliopsida, with two families: the Piperaceae family, more specifically the Piper genus, whose main representative is the bell pepper and the Solanaceae family, with the Capsicum genus, whose main variety is the malagueta. Under botanical perspective Chili peppers are shrubby, perennial plants with semi-woody stems and can exceed 120 cm in height, with wide lateral ramification. They are usually self-pollinated, but cross-pollination also occurs and with good intensity, depending on the entomological fauna existing in the region (Bode and Dong 2011; Chang et al. 2011).

Generally, chili peppers have hermaphrodite flowers (they have male and female gametes), they have calyx with 5 (in some cases, 6 to 8) sepals and corolla with corolla with (in some cases, 6 to 8) petals; they are also autogamous, that is, the pollen and the fertilized ovule belong to a single flower, thus facilitating their reproduction, although cross-fertilization can occur as already mentioned. This whole process is facilitated by possible morphological alterations of the flower and by the action of insects that exist in the region and act as pollinating agents and by diverse and effective cultural practices (Hernández-Pérez et al. 2020).

13.3 Food Additive

Nowadays, people have been searching for more and more healthy foods, which are free of certain components that can cause problems to the consumer's health. On the other hand, the food industry is looking for substances that durability of their products, so that they can reach the most diverse places and offer the most diverse places and offer quality food to all people, thus driving the use of food additives. The FAO (Food and Agriculture Organization) and WHO (Word Health Organization) treats additives as any substance that is added to food to prevent alterations, maintain, confer, or to some extent, intensify their organoleptic characteristics such as aroma, color, and flavor; they can also modify or maintain or maintain its physical state for certain periods of time or exert any action to facilitate the manufacture of the food (Aida et al. 2021).

The chilis have been used daily only in food formulations, through direct application *in natura* or through the production of sauces, extracts, jams, dried or dehydrated spices in the most diverse forms. However, in the last few years this practice has been changing that the several varieties of chilis have different properties, both culinary interests, but mainly of industrial interest in several areas. The chili peppers are already being used in different ways in the development of products for the food industry, as an increment, or to some extent, an additive in these formulations (Marincaş et al. 2018).

With the advance of the pepper production in several regions of the world, the researches were also changing and looking for more technological contours. A study conducted by Severo (2015) produced boursin cheese supplemented with biquinho pepper brans in different proportions and highlighted the low caloric value with only 4% fat obtained in the formulations. The tasting tests showed that the samples added with 3% of chili pepper bran had good acceptance and highlighted the high content of phenolic compounds that contribute to the stability of the product increasing its shelf life.

Other products have also been elaborated with the addition of chili pepper flour. Chili pepper flour, as is the case of the studies conducted by Gangabhagirathi and Joshi (2015) which elaborated and sensorially characterized savory cookies with the addition of chili pepper flour in different proportions. The authors conclude in the research that the implementation of "biquinho" pepper in cookies is a viable alternative and a new initiative for the use of chili pepper in different food industries.

Carvalho et al. (2014) in turn, determined the content of carotenoids and the antioxidant activities of peppers from the Amazon region, among them the biquinho pepper and obtained in all samples the carotenoids all-trans-lutein and zeaxanthin. The author exposes in his study that the antioxidant activity is 4905.06 g/g DPPH, which is lower than the others analyzed.

13.3.1 Natural Additives: Colorants, Preservatives and Antioxidants

Chili peppers can be classified as a functional food for having several nutrients, such as high levels of carbohydrates and dietary fiber that contribute to reduce fat levels and improve people's health. Besides, they are rich in vitamins A, E and C, folic acid, zinc, and potassium, and may also have antioxidant properties and bioflavonoids, which are plant pigments that have been linked to cancer prevention (Nedovic et al. 2011).

13.3.2 Natural Pigments and Their Antioxidant Properties

Many studies have tried to show the antioxidant potential of the several varieties of chili peppers (Bae et al. 2012; Razola-Díaz et al. 2022). These antioxidant properties evidenced for the different types of chili peppers contribute to the reduction of chronic diseases such as prostate cancer cataract and diabetes (Batiha et al. 2020). In addition, they can have a blood detoxifying effect, also contribute to prevent diseases such as thrombosis, because they are related to the impediment to blood clotting. The use of chili peppers is also related to the relief of headaches and rheumatoid arthritis. Rheumatoid arthritis and are part of the composition of plasters (patches) for muscular pain (Narang et al. 2017).

13.4 Applications of Capsaicin in Food Industry

Capsicum annum bioactive compound research has been getting a lot of attention over the past few decades. *Capsicum* sp. contains a wide variety of nutrients, including phenolic acid derivatives and flavonoids, carotenoids, pungency-imparting capsaicinoids (capsaicin and dihydrocapsaicin), non-pungent capsinoids, vitamin C, vitamin E, provitamin A, and many essential minerals such as potassium, phosphorus, magnesium, and calcium (Baenas et al. 2018). Because of its antioxidant and antibacterial characteristics, chili has several uses in the food sector beyond its more well-known culinary ones. Products developed from *Capsicum annum* are discussed briefly in this section, including their use in the food sector for purposes ranging from packaging to extending the shelf life of perishable items (Wang et al. 2020).

13.4.1 Chili Pepper as Antioxidant in Food Preservation Industry

The deleterious consequences of food autoxidation on stability and preservation, as well as on human health, are well-documented. Despite the widespread availability of synthetic antioxidants such as BHA, BHT, and TBHQ, interest in naturally occurring and new chemicals with antioxidant characteristics is rising in response to growing public health concerns (Hes et al. 2019). Capsaicin's antioxidant properties were first demonstrated in a seminal work by Henderson and Henderson (1992), which showed that the presence of capsaicin prevented the development of expected products of oleic acid oxidative breakdown. The antioxidant nature of pepper oil as compared to soybean oil during frying was studied by Yang et al. (2010). The study showed significantly lower TBARS, PV, and CL of pepper oil as compared to soy oil after frying. The addition of capsaicin to pepper oil further reduced lipid oxidation significantly and enhanced the thermal oxidative stability. In a comparative study of antioxidant potential of phytometabolities, capsaicin exhibited highest antioxidant activity second to eugenol (Orlo et al. 2021). Mesoporous silica nanoparticles (MSNs) containing capsaicin were produced by Si et al. (2021) for application in modified environment packaging of ground beef. Meat samples from MSNs had a thiobarbituric acid reactive substance (TBARS) value of 0.483 mg/kg, whereas the value for the control sample was 1.809 mg/kg, indicating significantly less lipid oxidation. Capsaicin's efficacy in long-term antioxidation preservation of meals was validated, and the lower TBARS value implies a greater antioxidant action of capsaicin loaded MSNs.

13.4.2 Antimicrobial Activity of Chili Pepper in Food Preservation Industry

Capsaicin's antimicrobial effect may stem from the compound's ability to inhibit cell growth, destroy cell membranes and ribosome components, or both. Capsaicin may be hazardous after it enters microbial cells, according to a new study (Wang et al. 2020). Using subinhibitory doses of capsaicin (7 mM), Kurita et al. (2002) investigated its effect on *Escherichia* coli biofilm development. Using a biofilm dispersion experiment, researchers found that cells treated with 7 mM capsaicin for 1 h significantly decreased in viability. Efforts were made to understand the antibacterial effect of eugenol, capsaicin, and vanillin against common foodborne pathogens and food spoilage bacteria (*E. coli*, *Pseudomonas aeruginosa*,

Staphylococcus aureus, *Shewanella putrefaciens*, *Brochothrix thermosphacta*, and *Lactobacillus plantarum*). Eugenol and capsaicin showed broader spectrum of action against these microbes. *Staphylococcus aureus* was severely affected by these phytometabolites and IC50 value of capsaicin was 0.68 mM (Orlo et al. 2021). Capsaicin microtubules are a bonus to the food business. Encapsulation technology protects the active ingredient from unstable environmental conditions, providing higher stability and aids in delayed release of the compounds. Xing et al. (2006) evaluated the inhibitory impact of capsaicin microcapsules on Botrytis cinerea and *Aspergillus niger*. The slower release rate of encapsulated capsaicin, leads to increased antibacterial action in the system.

The impact of capsaicin on mycotoxin was examined by Kindt et al. (2009). Capsaicin was found to decrease ochratoxin A (OTA) production by *Aspergillus* strains by as much as 78%. The results of the study demonstrated that capsaicin can act as a natural preservative against OTA, a nephrotoxic and carcinogenic mycotoxin. Application of capsaicin in vineyards might be done as a prophylactic step to obtain mycotoxin free grapes.

13.4.3 Beneficial Effects of Chili Pepper in Food Packaging Industry

Efficient packing materials is important for safe and long-term storage of commodities. Also, the concept of edible packaging is an emerging area of research that can progressively pave way for waste reduction in the long term. Zhang et al. (2018) created capsaicin integrated edible sodium alginate/pullulan composite films. The physico-chemical characteristics of the film were improved by the incorporation of capsaicin in the films. A decrease in transmittance, elongation at break, and moisture content and an increase in the tensile strength, water vapor permeability, and surface contact angle was found. Capsaicin also imparted antimicrobial properties to the film. They demonstrated the practical utility of these capsaicin included edible composite films by assessing their influence on prolonging the shelf life of newly sliced apple for up to 7 days at room temperature.

Capsaicin blended into chitosan gel solution at three different concentrations viz., 0.3, 0.6 and 1.2 mg resulted in betterment of physicochemical, mechanical and biological properties of the film. An improvement in anti-quorum sensing and antimicrobial activities, anti-oxidative and hydrophobicity with increasing concentration of capsaicin in the film was also observed (Akyuz et al. 2018). As nanocarriers for capsaicin, Zhao et al. (2020) integrated Fe III doped hollow metal-organic frameworks into a gelatin/chitosan matrix to create antibacterial packaging sheets. Capsaicin's hydrophilic characteristics on a nanocarrier allowed it to overcome the phase separation issue and provide a viable antibacterial packaging solution for the food market. Not only were the film's mechanical qualities improved, but so were its water vapor permeability and ultraviolet (UV) block properties.

13.4.4 Clinical and Pharmaceutical Industrial Applications of Capsaicin

Since antiquity, capsaicin has been linked to pharmacological effects, but it has not been until the past 20 years that in-depth study has been done to identify specific applications, including as in the gastrointestinal tract, for weight loss, and as an analgesic. Research into the potential protective effects of capsaicin in the gastrointestinal tract has yielded contentious findings, which amply demonstrate the need for additional clinical trials to identify appropriate dosages more precisely (Hayman and Kam 2008). In rats, capsaicin is known to prevent gastric mucous ulceration when administered at low concentrations (0.13–160 μM), but when given at high concentrations (1 or 2 mg/mL), it is believed to exacerbate damage to the gastric mucous brought on by ethanol or aspirin. Capsaicin was found to have protective properties against ethanol- and indomethacine-associated gastropathy in a study involving 84 healthy human participants. It did this by reducing base gastric acid output in a dose-dependent manner (ED50 for 400 g of capsaicin), increasing gastric emptying, and increasing gastric emptying. Increasing stomach mucosal blood flow (GMBF) and vasodilation by capsaicin, which is mediated by the release of nitric oxide and calcium gene-related peptide (CGRP) by the vanilloid type-1 transient receptor potential channel (TRPV1) bearing cells, may be one explanation for the antiulcerous activity of the spice (Reyes-Escogido et al. 2011).

Another significant pharmacological effect of capsaicin was discovered nearly 20 years ago when it was discovered that it decreased adipose tissue in rodents by improving energy and lipid metabolism. This effect may have been caused by capsaicin increasing catecholamine secretion from the adrenal medulla in response to sympathetic nervous system activation. Further study into the potential therapeutic effects and mechanisms of capsaicin in the management of obesity has been made possible by these discoveries. Although it can cause skin irritation, capsaicin has been used in topical therapies to relieve a variety of neuropathic pain problems (Papoiu and Yosipovitch 2010; Sharma et al. 2013). For the treatment of chronic pain disorders as post-herpetic neuralgia, musculoskeletal pain, diabetic neuropathy, osteoarthritis, and rheumatoid arthritis, topical creams and patches containing capsaicin and its analogues have been utilized. Additionally, it has been used to manage pain following mastectomy, psoriasis, rashes, and bladder issues. Burning, stinging, and erythema are the most common unfavorable reactions, while systemic reactions and infrequent pulmonary irritations from cream inhalation have also been noted. Due to the few participants and various definitions of pain utilized in the research, its true efficacy in pain reduction is still somewhat in question. However, whether it is given alone or in conjunction with other medications, it appears to be a suitable alternative for pain management in people who do not react to conventional therapy (Zhang et al. 2020; 2022).

Capsaicin has been given the green light by the Food and Drug Administration (FDA) and the European Union (EU) to be used as a topical analgesic for neuropathic pain. Hayman and Kam (2008) have reviewed and summarized the use of various capsaicin preparations in a variety of clinical conditions for the treatment of chronic pain, gastroprotection, reduction of overactive bladder and detrusor hyperactivity, relief of post-operative nausea, vomiting, and sore throat, assistance for patients with renal failure and myocardial ischemia, and improvement of cough and swallowing dysfunction. Obesity-related issues are being addressed with the help of a wide range of natural products. Capsaicin's effectiveness in combating obesity has recently been demonstrated in scientific studies. Capsaicin inhibited lipase activity in in-vitro research, and a higher quantity of the compound slowed lipase's ability to hydrolyze its substrate. Future research and development are likely to focus on creating capsaicin-containing functional foods that aid in weight loss (Wu et al. 2022). Wu et al. (2022) reported the effectiveness of capsaicin in dermatology using in vitro and in vivo models. Capsaicin prevented UV-induced collagen depletion in cutaneous fibroblasts. These results demonstrate capsaicin's promise as an anti-aging ingredient in medications.

There has been a lot of research into the potential cancer-fighting properties of chili pepper extracts and other compounds derived from pepper. The effects of chili pepper seed extract at varying doses on breast, colon, and gastric carcinoma cells were investigated by Jeon et al. (2012). At 500 mg/mL, capsaicin showed significant anti-cancer action due to enhanced apoptosis. Additionally, the study found that breast cell lines were especially vulnerable to growth suppression by chili pepper seed extract. Researchers looking at capsaicin's effect on pancreatic cancer found that it caused cancer cells to produce reactive oxygen species (ROS) and undergo apoptosis by disrupting their mitochondrial membrane potential. Capsaicin exhibited no adverse effects when given orally to mice, and it dramatically slowed the development of pancreatic tumors (Zhang et al. 2008). Diets high in capsaicin have been shown to stimulate the immune system and protect against the spread of the COVID-19 virus, according to research by Gharaibeh et al. (2020).

13.4.5 Beneficial Effects of Chili Pepper in Cosmetic Industry

A developing trend in the cosmetics sector is the exploration for new bioactive compounds, plant materials, or plant extracts. The eco-friendliness of this movement is a major reason for its success. In addition to increasing the health benefits for buyers, using natural chemicals can lessen the risk of any potential negative reactions that may occur with the usage of synthetic preservatives. Plant-based chemicals may also contribute to cosmetic industry product claims like anti-aging and skin healing (Baenas et al. 2018). There are currently a wide variety of homemade recipes and organic goods available in the cosmetics market that include chili pepper as an ingredient. Paprika based sheet masks by Maccaron and Eminence Organic Skin Care are few products claiming skin nourishment, hydration, skin repair and rejuvenation (Aida et al. 2021).

13.4.6 Uses as Functional Foods

To create a useful product that could store the capsaicin within the noodle matrix for a longer period and release it into the gastro-intestinal tract in a controlled-manner, Li et al. (2014) sandwiched a capsaicin-enriched dough with two gastro-resistant dough layers. The study's chili powder contained 2.0 mg/g of capsaicin, which is regarded to be hot at 31,384 Scoville Heat Units, and the study's noodles contained 15–17 mg/100 g of capsaicin. Based on the findings, the authors hypothesized that these layered noodles would assist customers avoid experiencing too much discomfort when eating because most Asians prefer to slurp their noodles rather than chew them. To create a functional yogurt with a novel flavor and a wealth of nutritional benefits, Askin et al. (2021) fortified yogurt with capsaicin extract and oleic sunflower oil. Because capsaicin is insoluble in water but soluble in non-polar casein molecules, yogurt was the product of choice. In comparison to ordinary yogurt, the prepared yogurts had a hotter flavor, greater quality, a firmer texture, and better sensory qualities.

13.5 Microcapsule Technology of Capsaicin Used in Food Industry

Capsaicin has many benefits, but there are also certain downsides that must be addressed. Direct exposure to capsaicin can induce eye irritation and pain, skin irritation, and gastrointestinal distress such as nausea, vomiting, and diarrhea if consumed in excessive amounts (Hayman and Kam 2008). Because of its lipophilic nature, capsaicin has poor water solubility, poor bioavailability, and poor chemical stability, all of which can reduce its usefulness. These obstacles can be conquered with the aid of microencapsulation, making capsaicin more useful in the food business. To encapsulate is to enclose something entirely within another substance. When it comes to encapsulation, you can choose from a number of different physical, chemical, and physicochemical methodologies and procedures (Aida et al. 2021). The benefits of encapsulated capsaicin in the food business are shown in Fig. 13.1, and some of the more frequent microencapsulation processes are discussed below.

13.5.1 Coacervation

In the coacervation method, a colloidal dispersion is broken down into two liquid phases that are inherently incompatible with one another: a colloid-rich phase and a diluted phase. The process of coacervation can be broken down into two categories: easy and difficult. The addition of dehydrating chemicals or a change in temperature can trigger simple coacervation, which only requires a single type of surfactant. Complex coacervation, on the other hand, necessitates only one type of each of two oppositely charged colloidal species in solution (Zhao and Wang 2017). Through a sophisticated coacervation process, Jincheng et al. (2010) created capsaicin encapsulated in gelatin and acacia. The study found that by encapsulating capsaicin in nanoparticles, its stability under heat was significantly increased. Bovine serum albumin (BSA)-capsaicin nanoparticles were created by Sánchez-Segura et al. (2018) utilizing a desolvation-pH-coacervation process and varying capsaicin concentrations. They found that as the concentration of capsaicin rose, the particles trapped within the BSA also rose. There was also a correlation between the concentration of capsaicin and a shift in the nanoparticles' physicochemical characteristics.

13.5.2 Emulsion Technology

Capsaicin-loaded double- and triple-layer nanoemulsions were made by Choi et al. (2013) using alginate and chitosan, two naturally occurring biopolymers. In comparison to a capsaicin control, they found that these nanoemulsions had a prolonged half-life and a bioavailability that was up to 131.7 times higher. This means that nanoemulsions can contain significantly less capsaicin than conventional formulations. Capsaicin nanoemulsions' distinguishing properties indicated that the substance could be successfully marketed as a dietary supplement. Using an ultrasonication technique, Lu et al. (2016) attempted to create capsaicin nanoemulsions produced from an organogel suitable for use in food. These had improved solubility and bioavailability compared to pure capsaicin. Moreover, these did not cause any stomach mucosal irritation in rats in in-vivo trials, although capsaicin capsules did cause ulcerative lesions on the mucosal surface 6 h after oral treatment. One possible explanation is that the capsaicin in the encapsulated form makes less direct touch with the mucosal surface. They hypothesized that in the future, multipurpose beverages with capsaicin-loaded nanoemulsions could be created.

Akbas et al. (2018) studied the effect of capsaicin loaded nanoemulsions developed by two homogenization techniques viz., high pressure homogenization and ultrasonication. Both the homogenization techniques showed difference in the encapsulation efficiencies. Efficiency was significantly higher for pressurized samples than ultrasonicated samples. Similarly, high pressure homogenized nanoemulsions showed higher antimicrobial action against the food-borne pathogens *E. coli* and *S. aureus* as compared to ultrasonicated samples. Pressurized nanoemulsions reduced *E. coli* and *S. aureus* population up to 5.89 and 2.79 log respectively at pH 7.4. The high-pressure homogenization method was used by Han et al. (2020) to prepare tocopheryl polyethylene glycol 1000 succinate (TPGS) based capsaicin emulsions. The stability of capsaicin-loaded emulsions under different temperatures (4 °C, room temperature and 60 °C), different salt concentrations and different pH conditions were studied for 45 days. Capsaicin and TPGS employed in the study had a synergistic impact, providing more protection from oxidative damage. Emulsions were also found to be effective in reducing irritations caused by capsaicin in a rat model of gastrointestinal damage.

13.5.3 Liposome Technology

Bilayer structures called liposomes are created when molecules of lipids and phospholipids come into contact with an aqueous phase. In the presence of sufficient heat or agitation, such a system would evolve into liposomes, with the latter forming in a way that seeks to reduce the amount of interaction the hydrophobic structure has with water molecules (Khorasani et al. 2018). Liposomes loaded with capsaicin were created by employing phospholipids and cholesterol and the film-dispersion technique. Liposomes loaded with capsaicin at a pH of 7.4 demonstrated somewhat sluggish release. Liposomal encapsulation of capsaicin improved its solubility and bioavailability (3.34-fold as compared to control) (Zhu et al. 2015).

13.5.4 Spray Chilling and Spray Drying

To create pungency-suppressed capsaicin microcapsules with palm oil and whey protein as the wall material, Gunel et al. (2021) used the spray chilling approach. When the intake temperature was 10 °C, the highest microencapsulation efficiency of 98.17% was achieved after studying the effects of various inlet spray chilling temperatures. In the in-vitro model systems, the study demonstrated that microencapsulation decreased the quantities of surface capsaicin and capsaicin release. To enhance the functional qualities of foods such as salad dressings, ketchup, mayonnaise, yoghurt, and cheese, the study advised adding non-pungent, non-irritating capsaicin microcapsules. Using whey protein and citrate mung bean starch in various ratios (10:0, 9:1, 7:3, 5:5, 3:7, 1:9, 0:10), Zhang et al. (2022) created capsaicin microcapsule powder, in the size range of 1–60 m, by spray drying process, and the physicochemical characteristics and stability parameters were examined. Inlet and output temperatures for spray drying were kept at 185 and 85 °C, respectively. The aqueous solubility of capsaicin microcapsules ranged from 65.97% to 96.32%. The maximum retention rate and stability were found in microcapsules produced with a 7:3 ratio of whey protein to mung bean starch citrate.

13.5.5 Other Microencapsulation Techniques

Emulsification-solvent evaporation, inclusion complexation, interfacial cross-linking, interfacial polymerization, fluidized bed coating, centrifugal extrusion, the use of supercritical fluids, and many more techniques can be used for encapsulation (Aida et al. 2021). To develop the most effective method for encapsulating capsaicin, it is necessary to conduct extensive research into the practicality of various methods. Capsaicin has some drawbacks, but encapsulation ought to be able to fix those without reducing the substance's usefulness in the food sector (Fig. 13.2).

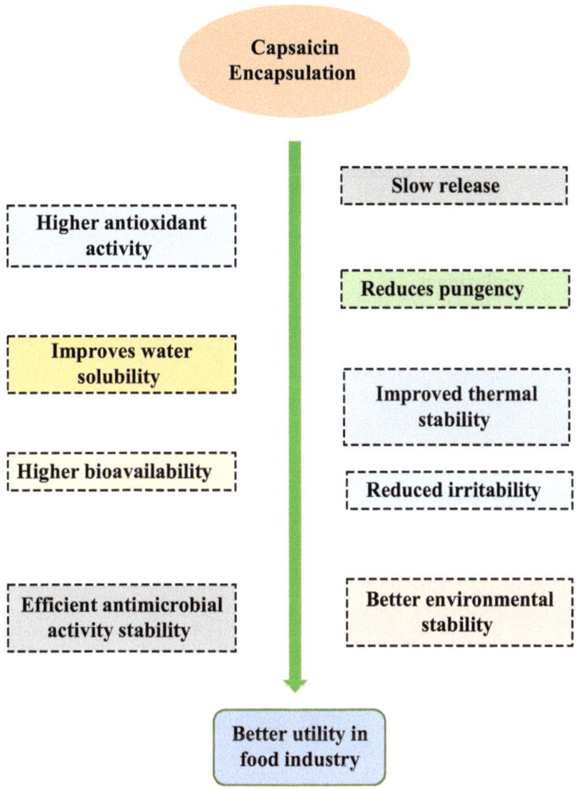

Fig. 13.2 Advantages of capsaicin microcapsule technology in the food industry

13.6 Therapeutic Properties of Capsaicin

Several chemical compounds with major therapeutic applications against human diseases have their origin in plants. *Capsicum annuum*, one such plant, has several chemicals with medicinal capabilities beyond its pharmacological properties. Capsaicin, a plant component, is present in this plant in high enough concentrations to be responsible for its pungent flavor. It is also known as N-vanillyl-8-methyl-6-(E)-noneamide (C18H27NO3, molecular weight 305.418 g/mol) and it is an alkaloid unique to the genus Capsicum in the family Solanaceae. It belongs to the category of secondary plant metabolites, collectively known as capscinoids. Aside from capsaicin, other chemicals in this group include dihydrocapsaicin, homocapsaicin, homodihydrocapsaicin, and nordihydrocapsaicin. Capsaicin has the appearance of a crystalline solid and has no discernible odor or color but is actually an off-white solid that is lipophilic. Capsaicin's widespread notoriety as a useful therapeutic agent and culinary ingredient has piqued the public's interest for as long as anybody can remember. Despite its disagreeable features, capsaicin is utilized as a fundamental ingredient in a wide variety of formulations for the treatment of many human

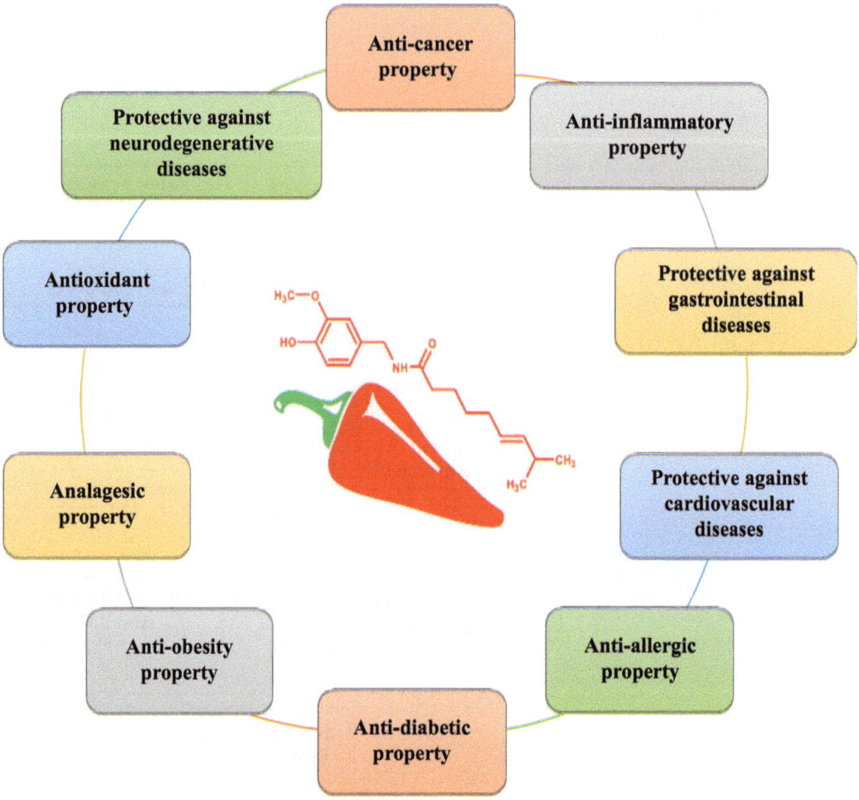

Fig. 13.3 Therapeutic properties of capsaicin

ailments, including pain, ulcers, cancer, tumors, obesity, diabetes, inflammation, neurological disorders, digestive tract diseases, cardiovascular diseases, skin disorders and CNS disorders as illustrated in Fig. 13.3.

13.6.1 Antioxidant Properties

The metabolic changes brought on by radiation, such as protein oxidation and lipid peroxidation, have been discovered to be effectively inhibited by the use of capsaicin. In an experiment conducted by Gangabhagirathi and Joshi (2015), the mitochondrial membrane of the rat liver was subjected to gamma (γ) rays from a ^{60}Co source to generate radiation induced oxidative damage and to cause damage to lipid and protein production. According to the study, capsaicin dramatically decreased dose-dependent metabolic changes brought on by radiation, such as protein and lipid oxidation and peroxidation. The loss of antioxidant enzymes and essential endogenous antioxidant glutathione activity brought on by radiation was also significantly

Table 13.1 Summary of studies on antioxidant properties of capsaicin

Model organism	Dosage of Capsaicin (μM)	Mechanism of action	Reference
Wistar rat	40 μM	Restoration of gamma radiation-induced depletion of protein thiols	Gangabhagirathi and Joshi (2015)
	5–50 μM	Protection from oxidative damage due to radiation	
Mouse	500 μM	Increased nitric oxide and scavenged free radicals	Thongin et al. (2022)

decreased. Researchers have shown that capsaicin has antioxidant effects, as shown in Table 13.1.

13.6.2 Anti-Cancer and Anti-Tumor Properties

Secondary metabolites derived from plants, such as capsaicin found in chili, have been found to actively contribute to the suppression of cancer in a number of ways, as demonstrated by multiple high-quality clinical and experimental chemoprevention investigations. Cancer's growth and metastasis from one part of the body to another are both stymied by these inhibitory actions.

13.6.2.1 Role of Capsaicin in Apoptosis

Activation of proteases and endonucleases, loss of plasma membrane phospholipid asymmetry, enzymatic cleavage of the DNA into oligonucleosomal fragments, and segmentation of the cells into membrane-bound apoptotic bodies are all characteristics of the highly organized cell death process known as apoptosis. It is now known that apoptosis plays a significant part in maintaining tissue homeostasis by carefully removing surplus cells. Carcinogenesis is expected to be largely influenced by genetic alterations that cause apoptosis to disappear or that disrupt the signaling mechanisms that trigger it. A significant method for treating cancer is also known as inducing apoptosis in cancer cells (Basith et al. 2016).

Capsaicin's chemopreventive properties stem from its ability to induce apoptosis, slow cell proliferation by blocking cell cycle progression, halt the formation of new blood vessels, and reduce the spread of cancer. Capsaicin inhibits the proliferation and metabolism of cancer cells and induces apoptosis, a potent defense mechanism against malignancy, in cancer cell lines derived from the intestinal, breast, skin, stomach, blood, nasopharynx, liver, prostate, and pancreas, among others. Human KB cells (epithelial carcinoma cells) treated with capsaicin for 24, 48, and 72 h showed a significant decrease in proliferation/viability that was dose-dependent, as shown by Lin et al. (2013). Cell cycle analysis shows that KB cells exposed to capsaicin are arrested in the G2/M phase. Capsaicin's inability to stimulate proliferation in KB cells appears to be linked to its ability to trigger apoptosis. Additionally, the potential of the mitochondrial membrane was changed by capsaicin in KB cells, and caspases 9, 3, and poly-(ADP-ribose) polymerase were activated.

13.6.2.2 Role of Capsaicin in Anti-Angiogenesis

Endothelial cell migration, differentiation, and proliferation promote angiogenesis, the formation of new blood vessels. Angiogenesis is crucial for the development of cancer because cancer cells need an oxygen-rich blood supply to continue growing above a few millimeters. Therefore, limiting angiogenesis is essential for minimizing the spread of cancer cells. Capsaicin has been proven in recent research to be beneficial in blocking angiogenesis in cancer-affected regions of the body. About 80% of lung cancers are non-small-cell lung cancers (NSCLCs), or tumors of the larger airways. Treatments like chemotherapy and radiation have never been shown to successfully treat NSCLC (Reyes-Escogido et al. 2011; Zhang et al. 2020). Tumor-associated angiogenesis is crucial for NSCLC's growth and spread, therefore blocking it may be a practical strategy for slowing the disease's progression. To counteract the angiogenesis that results from treatment-resistant NSCLC, capsaicin has been shown to effectively target the p53-SMAR1 auto-regulatory loop (Chakraborty et al. 2014).

13.6.2.3 Role of Capsaicin in Metastasis

During metastasis, cancer cells split apart from the primary tumor (primary cancer) they created. Lymph nodes and blood vessels are common pathways for cancer cells to travel when they metastasis to other regions of the body. Capsaicin's anti-invasive and anti-migratory properties are a result of its regulation of signaling pathways related with cell migration and invasion, which helps to slow the progression of cancer even in its advanced stages (Hayman and Kam 2008; Zhang et al. 2018). For example, Wu et al. (2020) reported that capsaicin alters the CDK8/PI3K/Akt/Wnt/β-catenin signaling pathway to decrease the viability of breast cancer cells. Cyclin-dependent kinases (CDK8), a serine/threonine protein kinase, can boost up the β-catenin levels in the cytoplasm of cells. Breast cancer MDA MB231 cells express CD K8 more frequently than MCF10A normal breast cells. The abnormal stimulation of the Wnt/ β-catenin signaling system causes β-catenin to accumulate in the nucleus, which persuades cancer in breast cells. Malignant melanoma cell proliferation and metastasis were severely slowed down when CD K8 was deleted. Capsaicin treatment lowered CD K8 expression in the cancer cell lines, which in turn inhibited the CDK8/PI3K/Akt/Wnt/β-catenin signaling cascade and decreased breast cancer cell viability and migration. These findings imply that capsaicin has anti-cancer properties (Table 13.2).

13.6.3 Anti-Inflammatory Properties

Chronic inflammation may result from the body being constantly stimulated by inflammatory chemicals. Numerous studies have demonstrated the anti-inflammatory characteristics of capsaicin, among other potential health benefits. A number of inflammatory conditions, such as solid tumors, refractory female urethral syndrome, post-herpetic neuralgia, and hay fever, have been treated with capsaicin (Li et al. 2021). In this experiment, mice had their peritoneal macrophages removed

Table 13.2 Impact of capsaicin on malignant cells in humans

Malignancy type	Cell line	Capsaicin concentration (μM)	Mechanism involved	Reference
Small cell lung cancer	NCI-H69	50	Anti-proliferative activity (apoptosis)	Brown et al. (2010)
	NCI-H82			
	DMS53			
	DMS114			
Colorectal cancer	SW480	≥200	Anti-proliferative activity (apoptosis)	Yang et al. (2013)
	CT-26	≥200		
	HCT 116	≥25		
Breast cancer	MCF-7	200	Mitochondrial dysfunction (apoptosis)	Chang et al. (2011)
	BT-20			
Bladder cancer	5637	150	Suppresses cell proliferation, induces cell cycle arrest and ROS production	Qian et al. (2016)
	T24	300		
Cervical cancer	HeLa	100–200	Induces premature senescence and apoptosis via mitochondrial dysfunction	Pawar et al. (2022)
Non-small cell lung cancer (NSCLC)	Hy-A549	~50	Anti-angiogenesis; increases HIF1α degradation and reduces VEGF	Chakraborty et al. (2014)

and stimulated with either lipopolysaccharide or capsaicin for 24 hours. At all tested dosages (25, 50, 75, and 100 g/mL), capsaicin greatly increased the phagocytosis of neutral red dye by macrophages. By altering the nuclear factor-kappa B (NF-B) and microtubule-associated protein kinase (MAP kinase) signaling pathways, capsaicin prevented lipopolysaccharide-induced inflammation in macrophages by reducing the production of inflammatory cytokines such interleukins and nitric oxide (Aida et al. 2021).

Physical activity-induced inflammation has been shown to be lessened by capsaicin. Several inflammatory cytokines, such as interleukin-6 (IL-6) and interleukin-1 (IL-1), are increased in the blood after exercise. They have been identified as central nervous system fatigue mediators in a number of illnesses. Specifically, capsaicin's ability to suppress the production of inflammatory cytokines and chemokines during physical activity is how it earned its reputation as an anti-inflammatory agent (Giuriato et al. 2022). Through changes in afferent signaling or the kinetics of neuromuscular relaxation, capsaicin may lessen this neuromuscular weariness after exercise. The rate of Ca^{2+} reuptake and relaxation may be accelerated as a result, possibly through the sarco-endoplasmic reticulum Ca^{2+} ATPase pumps (Askin et al. 2021).

13.6.4 Analgesic Properties

Capsaicin has long been used as an effective pain reliever (Clark and Lee 2016). According to a study report by Vianna et al. (2018), capsaicin-based painkiller balms have been shown to lessen the skeletal muscular metaboreflex in healthy individuals. Capsaicin has also been developed as a topical patch and an injectable therapy for pain management. They are currently often used for the treatment of chronic pain caused by neuropathic disorders and other sources (Arora et al. 2021).

13.6.5 Anti-Diabetic Properties

The term "diabetes" is more frequently used to describe diabetes mellitus. It refers to a metabolic condition that may result in unusually elevated blood sugar levels. Additionally, the risk of a number of adverse effects such retinopathy, blindness, peripheral vascular disease, stroke, nerve damage, renal failure, and foot injury is increased by the presence of diabetes mellitus (Jincheng et al. 2010).

Diabetes can develop in one of three ways: in people with type 1 diabetes (T1DM), in people with type 2 diabetes (T2DM), and in pregnant women with gestational diabetes (GDM) (diabetes while pregnant). Lack of insulin production in the pancreas, which normally regulates blood sugar levels, characterizes T1DM, an autoimmune disorder. Type 2 diabetes, on the other hand, develops when cells become resistant to the actions of insulin. As a result of the body's inefficiency in responding to insulin, the patient's blood sugar level inevitably rises and never falls back to a healthy range. Studies conducted over the past decade have shown that TRPV1 receptors are vital for keeping glucose metabolism stable, suggesting that these receptors may play an important role in the treatment and progression of both types of diabetes (Table 13.3).

Table 13.3 An overview of the research on the effects of capsaicin on diabetes

Diabetes type	Model	Result	Reference
T2DM	Mouse	Reduces gluconeogenesis and increases glycogen synthesis in the liver to improve glucose tolerance and insulin sensitivity.	Hui et al. (2019)
T2DM	Rat	Hypoglycemic regulation of metformin and gut microbiota profiles	Kang et al. (2022)
T1DM	Mouse	Alleviate diabetic neuropathy in T1DM patients	Saab et al. (2021)
TIDM	Rat	Ameliorates diabetic retinopathy by inhibiting poldip2-induced oxidative stress	Liu et al. (2022)
T2DM	Mouse	Stimulation of TRPV1 in leads to activation of CAMKK2 and AMPK and increased glucose oxidation which is concomitant with an elevation in intracellular ATP level	Vahidi et al. (2021)

13.6.6 Anti-Obesity Properties

Capsaicin, an alkaloid, is commonly used as a weight loss supplement due to its ability to boost core body temperature and energy expenditure and its involvement in weight control by reducing adipose tissue mass. Capsaicin combats obesity through a variety of methods, including decreased appetite, satiety induction, improved lipid metabolism and thermogenesis, inhibition of pancreatic lipase and α-amylase activity, and suppression of adipocyte differentiation (Narang et al. 2017).

Wang et al. (2020) investigated the anti-obesity effects of capsaicin and whether or not TRPV1 played a role. Mice were split into three groups according to their diets: those fed a standard lipid diet (SLD), those fed a high fat diet (HFD), and those fed both HFD and capsaicin (2 mg/kg), with each group receiving a unique treatment over the course of 12 weeks. Comparing TRPV1 knockout mice to HFD-fed animals, the HFD + CAP-fed animals showed considerably slower weight gain and lower levels of triglycerides, cholesterol, glucose, and insulin.

13.6.7 Protective Effect Against Cardiovascular Diseases

Despite medical advancements, cardiovascular diseases (CVD) remain the leading cause of death worldwide. Capsaicin has been shown to improve vascular health by stimulating the production of nitric oxide (NO) and decreasing inflammatory reactions in vascular endothelial cells, which are located within the blood vessel lumen (Thongin et al. 2022). They first showed that primary human endothelial cells of the circulatory system responded to tumor necrosis factor (TNF)-α by increased Ser536 phosphorylation of p65 NF-B, expression of adhesion molecules such vascular cell adhesion molecule (VCAM)-1 and intercellular adhesion molecule (ICAM)-1, and production of IL-6. Examining how capsaicinoids affect endothelium activity was the aim of this investigation. In this study, the researchers demonstrated that human vascular endothelial cell cultures can be shielded from inflammatory attack by capsaicinoids.

Capsaicin is an extremely potent transient receptor potential vanilloid 1 (TRPV1) channel activator. It has been hypothesized that capsaicin can reduce cardiac microvascular damage by activating the TRPV1 (Munjuluri et al. 2021). Heart microvascular damage is common in people with T2DM, especially those with hyperlipidemia and hyperglycemia. Capsaicin was studied by Li et al. (2018) to see if it may mitigate the effects of diabetes on the heart (known as "cardiac microvascular injury"). Cardiac microvascular injury is caused by T2DM in a general sense by exacerbating the TRPV1 blockage and reactive oxygen species excess. They found that by decreasing oxidative/nitrogen stress in cardiac microvascular endothelial cells, long-term therapy with capsaicin can protect against T2DM in cardiac microvascular endothelial cells (CMECs). Specifically, CMECs accomplish this by regulating the TRPV1/Ca2+/PGC-1/OPA1 pathway.

13.6.8 Protective Effect Against Neurodegenerative Disorders

Degenerative nerve disorders can have a negative impact on a wide range of body processes, including balance, mobility, speech, breathing, and cardiac function. There are hereditary components to several of these disorders. Alzheimer's disease and Parkinson's disease are two of the most common types of neurodegenerative disorders (Saha et al. 2022). Capsaicin's effect on memory and learning was studied by Shalaby et al. (2019) in a rat model of streptozotocin-induced Alzheimer's disease. One injection of 3 mg/kg into the rat's intracerebroventricular space was used to induce Alzheimer's disease in 20 animals. Some of the rats that did badly on the passive avoidance test were then given an intragastric infusion of capsaicin at a dose of 10 mg/kg of weight for 47 days. The study found that the behavioral deficiencies were reduced after capsaicin treatment was administered. Because capsaicin reverses the behavioral and biochemical abnormalities observed in streptozotocin-induced Alzheimer's disease, it may be useful as a potential element in food additives and pharmaceuticals for treating the condition.

Numerous investigations have demonstrated that capsaicin can shield neurons from a range of neurodegenerative conditions, including oxidative stress, behavioral problems, 6-hydroxydopamine-induced Parkinson's disease, pentylenetetrazol-induced seizures, global cerebral ischemia, and streptozotocin-induced Alzheimer's disease (Tyagi et al. 2022). According to research, the development of neuro-pharmacotherapeutics for the management and treatment of neurological disorders may benefit from the use of capsaicin (Table 13.4).

13.6.9 Protective Effect Against Skin Disorders

Although capsaicin is an old chemical, it is still a hot topic in medical science. The synthesis of novel capsaicinoids is a viable medical strategy for the treatment of dermatological problems associated with inflammation and severe itching of the skin (Serafini et al. 2018). An expanding number of cutaneous conditions that involve pain and inflammation are being treated topically with capsaicin. Its application is also helpful for painful or itchy cutaneous problems brought on by surgeries, wounds or tumors. Serafini et al. (2018) described the Production and testing of soft medicines containing capsaicin that are hydrolyzed by skin esterases, which results in capsaicin deactivation. Agonists and antagonists that nevertheless have the ability to modify the TRPV1 channel and are hydrolyzable are created by introducing an ester group into the lipophilic moiety of capsaicinoids using the Passerini multicomponent reaction. This demonstrates the effectiveness of capsaicin as a potential treatment for skin conditions like pruritus that include TRPV1 channel dysfunction.

Capsaicin is an ancient molecule, but it remains a focus of intense interest in the field of medicine. Creating new capsaicinoids to treat dermatological issues such as inflammation and acute itching is a practical medical approach (Serafini et al. 2018). Capsaicin is being used topically for a growing variety of painful and inflammatory skin disorders. Its use is also beneficial for relieving discomfort and itching

Table 13.4 An overview of capsaicin's role in neurodegenerative disorders

Neurodegenerative disease	Model	Result	Reference
Alzheimer's disease	Mouse	Reduces brain amyloid-β-generation and attenuates Alzheimer's disease	Wang et al. (2020)
Alzheimer's disease	Mouse	Attenuates amyloid-β-induced synapse loss and cognitive impairments	Chen et al. (2017)
Parkinson's disease	Mouse	Prevents degeneration of dopamine neurons by inhibiting glial activation and oxidative stress	Chung et al. (2017)
Stroke	Rat	Injection of capsaicin in the peri-infarct area alleviates neuronal ischemic injury	Huang et al. (2017)
Parkinson's disease	Rat	Protect the cells and reduce apoptosis rate by regulating Actg1 and Gsta2	Liu et al. (2020)
Parkinson's disease	Drosophila	Increased the glutathione and dopamine content and reduced protein carbonyl, lipid peroxidation, glutathione-S-transferase and monoamine oxidase activity	Siddique et al. (2018)
Parkinson's disease	Rat	Protects against oxidative insults and alleviates behavioral deficits	Zhao et al. (2017)

associated with surgical incisions, sores, and tumors on the skin. Hydrolysis of capsaicin by skin esterases renders the compound inactive, and this process was detailed by Serafini et al. (2018), who developed and tested a soft medication containing capsaicin. With the Passerini multicomponent reaction, the lipophilic moiety of capsaicinoids can be converted into hydrolysable agonists and antagonists without losing their ability to control the TRPV1 channel. As a result, capsaicin's promise as a treatment for pruritus and other skin conditions caused by a dysfunctional TRPV1 channel is strengthened.

13.6.10 Anti-Allergic Properties

The immune system overreacts to ordinarily harmless items such as food or pollen, dust, bee venom, etc., leading to the development of allergies. When the body's immune system overreacts to antigens that are taken in the form of food, an allergic reaction to that food results (Aida et al. 2021). Antunes et al. (2019) evaluated the impact of supplemental capsaicin on a food allergy model. Mice sensitized to ovalbumin were given an ovalbumin solution, and then fed a diet containing capsaicin for 7 days. In the group that served as the "control," no capsaicin supplements were administered. The results showed that taking capsaicin pills did not reduce Immunoglobulin synthesis or body weight, both of which are common reactions to food allergies. In contrast, IL-33 and macrophage infiltration in the proximal jejunum were reduced in groups given capsaicin supplements. Hydroperoxide levels in the gut and hepatic triglyceride levels had also dropped.

13.6.11 Protective Effect Against Gastrointestinal Diseases

Capsaicin influences digestive health and gastrointestinal disease in a number of ways, much as it does with other human systems. Both TRPV1-dependent and TRPV1-independent effects on the GI tract have been seen after ingestion of capsaicin, with effects varying with dose (Xiang et al. 2022). A TRPV1-dependent quick and transient intracellular Ca2+ raise that triggers plasma membrane depolarization, negatively influences plasma membrane permeability, and ultimately results in cell death has been demonstrated to be the mechanism by which small concentrations of capsaicin (1 μM) kill HeLa cells. However, larger capsaicin doses (10 M) cause cell death by means of TRPV1-independent mechanisms. These mechanisms entail a gradual increase in intracellular Ca^{2+} that weakens mitochondria, depolarizes the plasma membrane, disrupts the plasma membrane's integrity, and ultimately leads to apoptosis (Patcharatrakul and Gonlachanvit 2016). Table 13.5 provides a summary of the results of using capsaicin to treat various physiological and digestive disorders.

Table 13.5 Effect of capsaicin in various gastrointestinal diseases

Disease	Dose of capsaicin	Result	Reference
Heartburn (Gastroesophageal reflux)	0.84 mg	Reduced heartburn	Kindt et al. (2009)
Colorectal cancer	0–100 μM	Inhibit cell proliferation and induce apoptosis through modulating transcriptional activity of NF-κB, p53	Clark et al. (2015)
Irritable bowel syndrome	2.1 g/day	Decreased postprandial abdominal burning and increased the rectal sensory threshold	Aniwan and Gonlachanvit (2014)
Esophageal cancer	0–120 μM	Suppresses esophageal sequamous cell carcinoma's migration and invasion through the AMPK/NF-B signalling pathway	Guo et al. (2019)
Inflammation and gastric damage	100 μM	Inhibits inflammation and gastric damage during *Helicobacter pylori* infection by targeting NF-kB–miRNA Axis	Saha et al. (2022)
Gastric ulcer	0.3 g/kg	Reduced ethanol/hydrochloric acid induced gastric ulcers in mice	Kwon and Kim (2021)
Colitis	30 μM	Inhibits intestinal Cl^- secretion and promotes Na^+ absorption by blocking TRPV4 channels	Wang et al. (2020)

13.7 Conclusions and Future Prospects

The therapeutic properties of capsaicin were highlighted. The present chapter described some interesting aspect related with application of capsaicin in food industry. In addition, the use of medicinal plants in therapy, in developing countries like Brazil and India, can be seen as a therapeutic alternative. Moreover, the use of medicinal plants to cure diseases is related to human evolution, because its use can be identified in all ages, in all social classes and almost in the entire humanity. The use of herbal medicines nowadays shows the importance of this class of drugs, and, therefore, the problems related to their use must be considered. It was observed that the pepper is capable of provoking other improvements, and not only its anticancer action, such as anticoagulant action, because it helps to lower blood pressure, reduces cholesterol can prevent the formation of blood clots, avoiding thrombosis, heart attack or stroke can also help against diseases such as rheumatoid arthritis. Capsaicin, the substance that gives chili peppers their pungency, was also discussed in detail in this chapter. It is the only compound that, when applied topically to the body, causes endorphins to be produced internally and these endorphins help people feel good and boost immunity.

We can conclude that processed foods provide variability, durability and nutritional quality for consumers and has been keeping the agribusiness economic stability. Chili peppers are considered a functional food, they are powerful allies in the maintenance of quality of life and well being of the consumer. The results regarding the potential of chili peppers was highlighted due to the bioactive substances in the fresh chilis remain in the processed products, especially capsaicin.

Finally, capsaicin, as the most present substance in the fruits of capsicum chilis, presents health benefits to the consumers, when well administered by them, helping mainly in weight control, due to its good thermogenic action, besides being an antioxidant. Therefore, it was possible to observe the versatility of capsaicin, which allows adding value to several from capsicum chilis, consolidating economically the food industry.

Capsaicin is sensitive to a variety of factors, including heat, light, and UV radiation. Its limited use in the food sector has been hampered by its poor water solubility, pungency, and low bioavailability. Emulsions have been actively researched for the stabilization of capsaicin among the many encapsulation approaches because of how simple they are to produce. Around the world, this pungent ingredient is frequently used as a spice, food flavoring, and seasoning in meals. Additionally, because of its anti-inflammatory and antibacterial qualities as well as its biological impacts, numerous studies have looked at the use of capsaicin as a food preservative. Several research have examined the use of capsaicin as a food packaging material for creating films to increase the shelf life of the foods, in addition to the aforementioned applications. Future research can also concentrate on preserving capsaicin and expanding its application in the food business because of its biological and functional attributes. To lay the foundation for capsaicin's widespread use in foods, additional research must be done to assess its safety and health-related qualities. Another crucial factor that needs more thorough research to increase the usage of capsaicin in the food sector is its stability under various conditions, such as various temperatures and pH levels.

References

Aida R, Hamed H, Ali E et al (2021) Applications of capsaicin in food industry: functionality, utilization and stabilization. Crit Rev Food Sci Nutr. https://doi.org/10.1080/10408398.2021.1997904

Akbas E, Soyler B, Oztop MH (2018) Formation of capsaicin loaded nanoemulsions with high pressure homogenization and ultrasonication. LWT 96:266–273. https://doi.org/10.1016/j.lwt.2018.05.043

Akyuz L, Kaya M, Mujtaba M et al (2018) Supplementing capsaicin with chitosan-based films enhanced the anti-quorum sensing, antimicrobial, antioxidant, transparency, elasticity and hydrophobicity. Int J Biol Macromol 115:438–446. https://doi.org/10.1016/j.ijbiomac.2018.04.040

Aniwan S, Gonlachanvit S (2014) Effects of chili treatment on gastrointestinal and rectal sensation in diarrhea-predominant irritable bowel syndrome: a randomized, double-blinded, crossover study. J Neurogastroenterol Motil 20(3):400

Antunes MM, Coelho BSL, Vichi TM et al (2019) Oral supplementation with capsaicin reduces oxidative stress and IL-33 on a food allergy murine model. World Allergy Organ J 12(7):100045

Arora V, Campbell JN, Chung MK (2021) Fight fire with fire: neurobiology of capsaicin-induced analgesia for chronic pain. Pharmacol Ther 220:107743

Askin B, Ozturk D, Durusoy B et al (2021) Properties of yogurt fortified with capsaicin extracts dissolved in high oleic sunflower oil. J Food Process Preserv. https://doi.org/10.1111/jfpp.15633

Bae H, Jayaprakasha GK, Jifon J et al (2012) Variation of antioxidant activity and the levels of bioactive compounds in lipophilic and hydrophilic extracts from hot pepper (*Capsicum* spp.) cultivars. Food Chem 134:1912–1918. https://doi.org/10.1016/j.foodchem.2012.03.108

Baenas N, Belović M, Ilic N et al (2018) Industrial use of pepper (*Capsicum annum* L.) derived products: technological benefits and biological advantages. Food Chem. https://doi.org/10.1016/j.foodchem.2018.09.047

Basith S, Cui M, Hong S, Choi S (2016) Harnessing the therapeutic potential of capsaicin and its analogues in pain and other diseases. Molecules 21(8):966

Batiha GE, Alqahtani A, Ojo OA, Shaheen HM, Wasef L, Elzeiny M, Ismail M, Shalaby M, Murata T, Zaragoza-Bastida A et al (2020) Biological properties, bioactive constituents, and pharmacokinetics of some Capsicum spp. and capsaicinoids. Int J Mol Sci 21:5179. https://doi.org/10.3390/ijms21155179

Bode AM, Dong Z (2011) The two faces of capsaicin. Cancer Res 71(8):2809–2814. https://doi.org/10.1158/0008-5472.CAN-10-3756

Brown KC, Witte TR, Hardman WE et al (2010) Capsaicin displays anti-proliferative activity against human small cell lung cancer in cell culture and nude mice models via the E2F pathway. PLoS One 5(4):10243

Carvalho AV, Mattietto RA, Rios AO et al (2014) Mudanças nos compostos bioativos e atividade antioxidante de pimentas da região amazônica. Pesq Agropec Trop 44:4

Chakraborty S, Adhikary A, Mazumdar M et al (2014) Capsaicin-induced activation of p53-SMAR1 auto-regulatory loop down-regulates VEGF in non-small cell lung cancer to restrain angiogenesis. PLoS One 9(6):e99743

Chang HC, Chen ST, Chien SY et al (2011) Capsaicin may induce breast cancer cell death through apoptosis-inducing factor involving mitochondrial dysfunction. Hum Exp Toxicol 30(10):1657–1665

Chatterjee S, Asakura M, Chowdhury N et al (2010) Capsaicin, a potential inhibitor of cholera toxin production in Vibrio cholerae. FEMS Microbiol Lett 306(1):54–60. https://doi.org/10.1111/j.1574-6968.2010.01931.x

Chen L, Huang Z, Du Y et al (2017) Capsaicin attenuates amyloid-β-induced synapse loss and cognitive impairments in mice. J Alzheimers Dis 59(2):683–694

Cho H, Kwon Y (2020) Development of a database of capsaicinoid contents in foods commonly consumed in Korea. Food Sci Nutr 8(8):4611–4624. https://doi.org/10.1002/fsn3.1785

Choi AY, Kim CT, Park HY et al (2013) Pharmacokinetic characteristics of capsaicin-loaded nanoemulsions fabricated with alginate and chitosan. J Agric Food Chem 61(9):2096–2102. https://doi.org/10.1021/jf3052708

Chung YC, Baek JY, Kim SR et al (2017) Capsaicin prevents degeneration of dopamine neurons by inhibiting glial activation and oxidative stress in the MPTP model of Parkinson's disease. Exp Mol Med 49(3):e298

Clark R, Lee J, Lee SH (2015) Synergistic anticancer activity of capsaicin and 3, 3′-diindolylmethane in human colorectal cancer. J Agric Food Chem 63(17):4297–4304

Clark R, Lee SH (2016) Anticancer properties of capsaicin against human cancer. Anticancer Res 36(3):837–843

Fan Y, Lu YM, Yu B et al (2017) Extraction and purification of capsaicin from capsicum oleoresin using an aqueous two-phase system combined with chromatography. J Chromatogr B Analyt Technol Biomed Life Sci 1063:11–17

Gangabhagirathi R, Joshi R (2015) Antioxidant activity of capsaicin on radiation-induced oxidation of murine hepatic mitochondrial membrane preparation. Res Rep Biochem 5:163

Gharaibeh A, Gharaibeh MM, Gharaibeh S (2020) Relation between Covid-19 and high capsaicin diets. Magna Scientia Adv Res Rev 1(1):30–32. https://doi.org/10.30574/msarr.2020.1.1.0029

Giuriato G, Venturelli M, Matias A et al (2022) Capsaicin and its effect on exercise performance, fatigue and inflammation after exercise. Nutrients 14(2):232

Gunel Z, Varhan E, Koç M et al (2021) Production of pungency-suppressed capsaicin microcapsules by spray chilling. Food Biosci 40:100918. https://doi.org/10.1016/j.fbio.2021.100918

Guo Y, Liu N, Liu K et al (2019) Capsaicin inhibits the migration and invasion via the AMPK/NF-κB signaling pathway in esophagus squamous cell carcinoma by decreasing matrix metalloproteinase-9 expression. Biosci Rep. https://doi.org/10.1042/BSR20190819

Han J, Zhang S, Liu X et al (2020) Fabrication of capsaicin emulsions: improving the stability of the system and relieving the irritation to the gastrointestinal tract of rats. J Sci Food Agr 100(1):129–138. https://doi.org/10.1002/jsfa.10002

Hayman M, Kam PC (2008) Capsaicin: a review of its pharmacology and clinical applications. Curr Anaesth Crit Care 19(5–6):338–343. https://doi.org/10.1016/j.cacc.2008.07.003

Henderson DE, Henderson SK (1992) Thermal decomposition of capsaicin. 1. Interactions with oleic acid at high temperatures. J Agric Food Chem 40(11):2263–2268. https://doi.org/10.1021/jf00023a042

Hernández-Pérez T, Gómez-García MDR, Valverde ME et al (2020) *Capsicum annuum* (hot pepper): an ancient Latin-American crop with outstanding bioactive compounds and nutraceutical potential. A review. Compr Rev Food Sci Food Saf 19(6):2972–2993. https://doi.org/10.1111/1541-4337.12634

Hes M, Dziedzic K, Górecka D et al (2019) *Aloe vera* (L.) Webb.: natural sources of antioxidants—a review. Plant Foods Hum Nutr 74(3):255–265. https://doi.org/10.1007/s11130-019-00747-5

Huang M, Cheng G, Tan H et al (2017) Capsaicin protects cortical neurons against ischemia/reperfusion injury via down-regulating NMDA receptors. Exp Neurol 295:66–76

Hui S, Liu Y, Chen M et al (2019) Capsaicin improves glucose tolerance and insulin sensitivity through modulation of the gut microbiota-bile acid-FXR axis in type 2 diabetic db/db mice. Mol Nutr Food Res 63(23):1900608

Jeon G, Choi Y, Lee SM et al (2012) Antioxidant and antiproliferative properties of hot pepper (*Capsicum annuum* L.) seeds. J Food Biochem 36(5):595–603

Jincheng W, Xiaoyu Z, Sihao C (2010) Preparation and properties of nanocapsulated capsaicin by complex coacervation method. Chem Eng Comm 197(7):919–933. https://doi.org/10.1080/00986440903249700

Kang ZQ, Hu JL, Chen MY et al (2022) Effects of capsaicin on the hypoglycemic regulation of metformin and gut microbiota profiles in type 2 diabetic rats. Am J Chin Med 50(3):839–861

Khorasani S, Danaei M, Mozafari M (2018) Nanoliposome technology for the food and nutraceutical industries. Trends Food Sci Technol 79:106–115. https://doi.org/10.1016/j.tifs.2018.07.009

Kindt S, Vos R, Blondeau K et al (2009) Influence of intra-oesophageal capsaicin instillation on heartburn induction and oesophageal sensitivity in man. Neurogastroenterol Motil 21(10): 1032–1082

Kurita S, Kitagawa E, Kim CH (2002) Studies on the antimicrobial mechanisms of capsaicin using yeast DNA microarray. Biosci Biotechnol Biochem 66(3):532–536. https://doi.org/10.1271/bbb.66.532

Kwon SC, Kim JH (2021) Gastroprotective effects of irsogladine maleate on ethanol/hydrtochloric acid induced gastric ulcers in mice. Korean J Int Med 36(1):67

Li J, Wang H, Zhang L et al (2021) Capsaicin affects macrophage anti-inflammatory activity via the MAPK and NF-κB signaling pathways. Int J Vitam Nutr Res. https://doi.org/10.1024/0300-9831/a000721

Li LY, Tan TC, Liong MT et al (2014) Preparation and evaluation of chili powder-enriched layered noodles. LWT 59:566–571. https://doi.org/10.1016/j.lwt.2014.04.040

Li X, Hou J, Du J et al (2018) Potential protective mechanism in the cardiac microvascular injury. Hypertension 72(1):116–127

Lin CH, Lu WC, Wang CW et al (2013) Capsaicin induces cell cycle arrest and apoptosis in human KB cancer cells. BMC Complement Altern Med 13:46. https://doi.org/10.1186/1472-6882-13-46

Liu J, Liu H, Zhao Z et al (2020) Regulation of Actg1 and Gsta2 is possible mechanism by which capsaicin alleviates apoptosis in cell model of 6-OHDA-induced Parkinson's disease. Biosci Rep. https://doi.org/10.1042/BSR20191796

Liu K, Gao X, Hu C et al (2022) Capsaicin ameliorates diabetic retinopathy by inhibiting poldip2-induced oxidative stress. Redox Biol 56:102460

Lu M, Cao Y, Ho CT et al (2016) Development of organogel-derived capsaicin nanoemulsion with improved bioaccessibility and reduced gastric mucosa irritation. J Agric Food Chem 64(23): 4735–4741. https://doi.org/10.1021/acs.jafc.6b01095

Marincaş O, Feher I, Magdas DA et al (2018) Optimized and validated method for simultaneous extraction, identification and quantification of flavonoids and capsaicin, along with isotopic composition, in hot peppers from different regions. Food Chem 267:255–262. https://doi.org/10.1016/j.foodchem.2017.10.031

Munjuluri S, Wilkerson DA, Sooch G et al (2021) Capsaicin and TRPV1 channels in the cardiovascular system: the role of inflammation. Cell 11(1):18

Narang N, Jiraungkoorskul W, Jamrus P (2017) Current understanding of antiobesity and property of capsaicin. Pharmacogn Rev 11(21):23–26. https://doi.org/10.4103/phrev.phrev_48_16

Nedovic V, Kalusevic A, Manojlovic V et al (2011) An overview of encapsulation technologies for food applications. Procedia Food Sci 1:1806–1815. https://doi.org/10.1016/j.profoo.2011.09.265

Orlo E, Russo C, Nugnes R et al (2021) Natural methoxyphenol compounds: antimicrobial activity against foodborne pathogens and food spoilage bacteria, and role in antioxidant processes. Foods 10(8):1807. https://doi.org/10.3390/foods10081807

Papoiu AD, Yosipovitch G (2010) Topical capsaicin. The fire of a 'hot' medicine is reignited. Expert Opin Pharm 11(8):1359–1371

Patcharatrakul T, Gonlachanvit S (2016) Chili peppers, curcumins, and prebiotics in gastrointestinal health and disease. Curr Gastroenterol Rep 18(4):1–11

Pawar JS, Mustafa S, Ghosh I (2022) Chrysin and Capsaicin induces premature senescence and apoptosis via mitochondrial dysfunction and p53 elevation in cervical cancer cells. Saudi J Biol Sci 29(5):3838–3847

Qian K, Wang G, Cao R, Liu T et al (2016) Capsaicin suppresses cell proliferation, induces cell cycle arrest and ROS production in bladder cancer cells through FOXO3a-mediated pathways. Molecules 21(10):1406

Razola-Díaz MDC, Gómez-Caravaca AM, López de Andrés J et al (2022) Evaluation of phenolic compounds and pigments content in yellow bell pepper wastes. Antioxidants (Basel) 11(3):557. https://doi.org/10.3390/antiox11030557

Reyes-Escogido MD, Gonzalez-Mondragon EG, Vazquez-Tzompantzi E (2011) Chemical and pharmacological aspects of capsaicin. Molecules 16(2):1253–1270

Saab M, Raafat K, El-Maradny H (2021) Transdermal delivery of capsaicin nanoemulgel: optimization, skin permeation and in vivo activity against diabetic neuropathy. Adv Pharm Bull 12(4): 780–790

Saha K, Sarkar D, Khan U et al (2022) Capsaicin inhibits inflammation and gastric damage during *Helicobacter pylori* infection by targeting NF-kB–miRNA axis. Pathogens 11(6):641

Sánchez-Segura L, Ochoa-Alejo N, Carriles R et al (2018) Development of bovine serum albumin–capsaicin nanoparticles for biotechnological applications. Appl Nanosci 8(8):1877–1886. https://doi.org/10.1007/s13204-018-0874-x

Serafini M, Griglio A, Aprile S et al (2018) Targeting transient receptor potential vanilloid 1 (TRPV1) channel softly: the discovery of passerini adducts as a topical treatment for inflammatory skin disorders. J Med Chem 61(10):4436–4455

Severo DS (2015) Farelo de pimenta biquinho (Capsicum chinense) aplicado na produção de Boursin. Trabalho de conclusão do curso de Engenharia de Alimentos, Universidade Federal de Campina Grande, Setembro

Shalaby MA, Nounou HA, Deif MM (2019) The potential value of capsaicin in modulating cognitive functions in a rat model of streptozotocin-induced Alzheimer's disease. Egypt J Neurol Psychiatr Neurosurg 55(1):1–13

Sharma SK, Vij AS, Sharma M (2013) Mechanisms and clinical uses of capsaicin. Euro J Pharmacol 720(1-3):55–62

Si W, Gao Y, Mei X et al (2021) Mesoporous silica nanoparticles loaded with capsaicin and their oxidation resistance in meat preservation. Food Chem 344:128737. https://doi.org/10.1016/j.foodchem.2020.128737

Siddique YH, Naz F, Jyoti S (2018) Effect of capsaicin on the oxidative stress and dopamine content in the transgenic Drosophila model of Parkinson's disease. Acta Biol Hung 69(2): 115–124

Thongin S, Den-Udom T, Uppakara K et al (2022) Beneficial effects of capsaicin and dihydrocapsaicin on endothelial inflammation, nitric oxide production and antioxidant activity. Biomed Pharmacother 154:113521

Tyagi S, Shekhar N, Thakur AK (2022) Protective role of capsaicin in neurological disorders: an overview. Neurochem Res 47:1513–1531. https://doi.org/10.1007/s11064-022-03549-5

Vahidi FP, Ahuja KD, Beckett JM et al (2021) TRPV1 activation by capsaicin mediates glucose oxidation and ATP production independent of insulin signalling in mouse skeletal muscle cells. Cell 10(6):1560

Vianna LC, Fernandes IA, Barbosa TC et al (2018) Capsaicin-based analgesic balm attenuates the skeletal muscle metaboreflex in healthy humans. J Appl Physiol 125(2):362–368

Wang J, Sun BL, Xiang Y et al (2020) Capsaicin consumption reduces brain amyloid-beta generation and attenuates Alzheimer's disease-type pathology and cognitive deficits in APP/PS1 mice. Transl Psychiatry 10(1):1–12

Wu D, Duan R, Tang L et al (2022) *In-vitro* binding analysis and inhibitory effect of capsaicin on lipase. LWT 154:112674

Wu D, Jia H, Zhang Z et al (2020) Capsaicin suppresses breast cancer cell viability by regulating the CDK8/PI3K/Akt/Wnt/β-catenin signaling pathway. Mol Med Rep 22(6):4868–4876

Xiang Y, Xu X, Zhang T et al (2022) Beneficial effects of dietary capsaicin in gastrointestinal health and disease. Exp Cell Res. https://doi.org/10.1016/j.yexcr.2022.113227

Xing F, Cheng G, Yi K (2006) Study on the antimicrobial activities of the capsaicin microcapsules. J Appl Pol Sci 102(2):1318–1321. https://doi.org/10.1002/app.23766

Yang CY, Mandal PK, Han KH et al (2010) Capsaicin and tocopherol in red pepper seed oil enhances the thermal oxidative stability during frying. J Food Sci Technol 47(2):162–165. https://doi.org/10.1007/s13197-010-0032-2

Yang J, Li TZ, Xu GH et al (2013) Low-concentration capsaicin promotes colorectal cancer metastasis by triggering ROS production and modulating Akt/mTOR and STAT-3 pathways. Neoplasma 60(4):364–372

Zhang R, Humphreys I, Sahu RP et al (2008) In vitro and in vivo induction of apoptosis by capsaicin in pancreatic cancer cells is mediated through ROS generation and mitochondrial death pathway. Apoptosis 13(12):1465–1478

Zhang S, Wang D, Huang J et al (2020) Application of capsaicin as a potential new therapeutic drug in human cancers. J Clin Pharm Ther 45(1):16–28

Zhang S, Wei F, Han X (2018) An edible film of sodium alginate/pullulan incorporated with capsaicin. New J Chem 42(21):17756–17761. https://doi.org/10.1039/C8NJ04249G

Zhang X, Zhang B, Ge X et al (2022) Fabrication and characterization of whey protein—citrate mung bean starch—capsaicin microcapsules by spray drying with improved stability and solubility. Foods 11:1049. https://doi.org/10.3390/foods11071049

Zhao J, Wei W, Xu X et al (2020) Enhanced antibacterial performance of gelatin/chitosan film containing capsaicin loaded MOFs for food packaging. Appl Surf Sci 510:145418. https://doi.org/10.1016/j.apsusc.2020.145418

Zhao W, Wang Y (2017) Coacervation with surfactants: from single-chain surfactants to gemini surfactants. Adv Colloid Interf Sci 239:199–212. https://doi.org/10.1016/j.cis.2016.04.005

Zhao Z, Wang J, Wang L et al (2017) Capsaicin protects against oxidative insults and alleviates behavioral deficits in rats with 6-OHDA-induced Parkinson's disease via activation of TRPV1. Neurochem Res 42(12):3431–3438

Zhu Y, Wang M, Zhang J et al (2015) Improved oral bioavailability of capsaicin via liposomal nanoformulation: preparation, in vitro drug release and pharmacokinetics in rats. Arch Pharm Res 38(4):512–521. https://doi.org/10.1007/s12272-014-0481-7

Capsaicin Receptor TRPV1: Biophysics, Physiology, and Pharmacology

14

Shisheng Li and Jie Zheng

Abstract

The animal receptor for capsaicin, an important metabolite found in chili peppers that bestows the signature spiciness, is TRPV1 (Transient Receptor Potential Vanilloid 1). It is a cation channel that allows cations to permeate through cell membranes to initiate action potentials in nociceptive neurons. In addition to sensing pungent chemicals in plants, TRPV1 responds to extracellular protons, animal peptide toxins and membrane depolarization. It also serves as a thermosensor involved in controlling body core high-temperature protection from noxious high temperature hazards by initialing pain signals. Rapid developments of pharmacology, electrophysiology and structural biology yield high-resolution TRPV1 structures bound with different ligands and evolving mechanistic understanding of the activation mechanisms of TRPV1. In this chapter, we review the biophysics, physiology, and pharmacology of TRPV1.

Keywords

TRPV1 · Capsaicin · Ion channel · Nociceptor · Biophysics · Physiology · Pharmacology

S. Li · J. Zheng (✉)
Department of Physiology and Membrane Biology, University of California at Davis, School of Medicine, Davis, CA, USA
e-mail: jzheng@ucdavis.edu

M. K. Swamy, A. Kumar (eds.), *Capsaicinoids*,
https://doi.org/10.1007/978-981-99-7779-6_14

14.1 Introduction

Spicy food is appreciated all over the world and is popular in many regions of south-western part of China, India, and Mexico. The history of humans cultivating and consuming chili pepper can be traced back to 6000 years ago in east-central Mexico (Kraft et al. 2014). Mirroring the enthusiasm for spicy food, scientists have put in great efforts to reveal the mechanism underlying spiciness sensory physiology. Early research focused on identification of the principal ingredients in chili peppers that are responsible for spiciness, and later work identified the receptor in animals that detect spiciness and elucidated its activation mechanism (Szallasi and Blumberg 1999).

When chili pepper plants were first transported from the Americas in the Columbian exchange, the Europeans were attracted to both their exotic appearance and pungent taste. John Clough Thresh, who is best known for his efforts advocating hygiene of public water supply, was among the first to try isolating the material in chili peppers that yielded the signature spiciness. In 1876, he succeeded in obtaining an almost pure sample of the spicy compound, apparently from tens of pounds of chili peppers; the pure compound would form beautiful crystals in the form of pearly white leaflets (Thresh 1846). It was Thresh who gave the compound its name, capsaicin. Once pure capsaicin was obtained, the logical next step was to find out its chemical nature. Achieving this goal took nearly another 50 years. E.K. Nelson, who worked at the US Department of Agriculture, pointed out in 1910 that the proposed molecular formula for capsaicin ($C_{18}H_{28}NO_3$), by a German Karl Micho, was chemically impossible. In 1923, Nelson finally solved the molecular structure of capsaicin (Fig. 14.1a) (Nelson and Dawson 1923). The final step of the decade-long project was determining the location of a C=C double-bond. Nelson proved that this double-bond locates between the sixth and seventh carbon in capsaicin's tail and, by doing so, putting down the last piece of the puzzle. In the process of this work, he also showed that this double-bond does not contribute to the spiciness of capsaicin. This corrected another misconception at the time. Capsaicin is a small lipophilic molecule with a vanillyl head group, an amide neck, and a fatty acid tail (Fig. 14.1a).

Since the discovery of capsaicin, it has been used widely in biomedical research. It is particularly useful for the research of pain, as intraplantar injection of capsaicin has served as a robust animal pain model. By 1990s, capsaicin was shown to activate primary sensory neurons, and the existence of a membrane ion channel as the capsaicin receptor was predicted (Bevan and Szolcsányi 1990). In 1997, this capsaicin receptor was cloned in David Julius' laboratory, from a cDNA library from the rat dorsal root ganglia sensory neurons (Caterina et al. 1997). Originally named vanilloid receptor member 1 (VR1), it was later changed to Transient Receptor Potential Vanilloid 1 (TRPV1) (Caterina et al. 1997; Montell et al. 2002). This landmark study started a flourishing field of research that has already yielded important findings concerning many aspects of human physiology (Yang and Zheng 2017; Cheng 2022). In this background, this chapter focuses on the biophysics, physiology, and pharmacology of TRPV1.

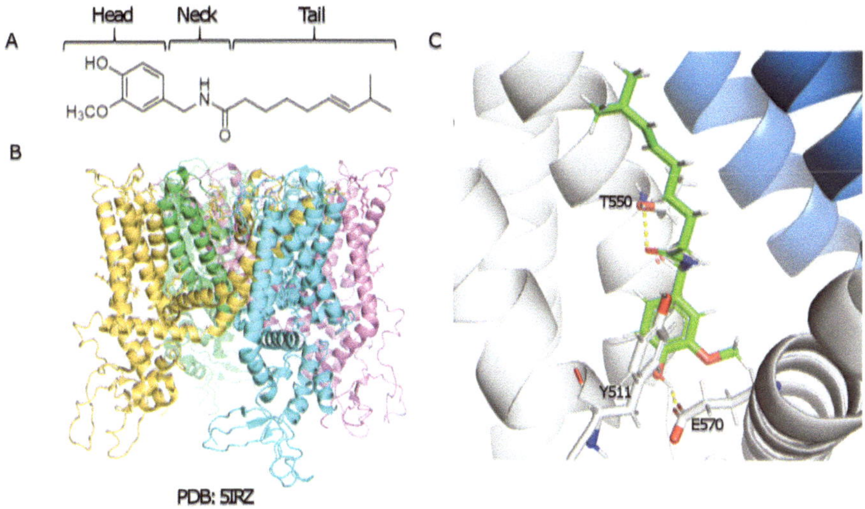

Fig. 14.1 Molecular interactions between capsaicin and TRPV1. (**a**) Molecular structure of capsaicin. (**b**) Cryo-EM structure of rat TRPV1 determined in lipid nanodisc (PDB entry code 5IRZ). Each subunit is colored differently. (**c**) The vanilloid binding pocket with a bound capsaicin. Key interacting residues are highlighted. Hydrogen bonds are indicated by a yellow dash line

14.2 TRPV1 Biophysics

14.2.1 Overall Structural Features of TRPV1

TRPV1 is a non-selective cation channel belonging to the voltage-gated potassium channel superfamily (Yu and Catterall 2004). It is a homotetramer with each subunit containing six transmembrane helixes and intracellularly located N- and C- termini (Cao et al. 2013b; Zhang et al. 2021; Gao et al. 2016; Kwon et al. 2021) (Fig. 14.1b). The S1-S4 helixes form a voltage sensor-like domain surrounding the channel pore formed by the S5-S6 helixes and the P-loop. The channel pore has two constricted regions: the upper constriction is formed by the selectivity filter which determines the ion types that can pass through the pore when it opens; the lower constriction, also called the lower gate, is formed by the S6 helixes which move during channel activation (Zheng and Ma 2014). Within the transmembrane domain of TRPV1—as well as other ion channels of the same TRPV subfamily—there is a ligand binding pocket formed by the S3 and S4 helixes and the S4-S5 linker from one subunit and the S6 helix from a neighbor subunit (Fig. 14.1c). This is the vanilloid binding pocket for pungent capsaicinoids such as capsaicin. Binding of ligands causes a global conformational rearrangement in TRPV1 that opens the channel pore to initiate ion conduction.

14.2.2 Capsaicin Activation

Capsaicin is the most abundant capsaicinoids in chili peppers (Scoville 1912). Being a lipophilic compound, capsaicin has a vanilloid head and flexible hydrophobic tail, linked by a covalent amide bond neck (Fig. 14.1a). It is membrane permeable, an assessment supported by observations that capsaicin can activate TRPV1 from both intracellular and extracellular sides. Combining observations from CryoEM structures, computational modeling and mutagenesis tests, it was proposed that capsaicin takes a "head-down tail-up" configuration in the vanilloid binding pocket, which was confirmed in the latest TRPV1 structure containing capsaicin (Yang and Zheng 2017; Kwon et al. 2021) (Fig. 14.1c). The head of capsaicin forms a hydrogen bond with E570 (rat and human TRPV1, equals to E571 in mouse TRPV1); the neck of capsaicin forms another hydrogen bond with T550 (rat and human TRPV1, equals to T551 in mouse TRPV1); the hydrophobic tail of capsaicin is flexible and makes non-specific interactions with residues in the upper part of the vanilloid binding pocket. In addition, Y511 (rat and human TRPV1, equals to Y512 in mouse TRPV1) is found to interact with a bound capsaicin to stabilize its binding; mutating this residue to an alanine substantially right-shifts the concentration dependent curve of capsaicin (Hazan et al. 2015). Fluorescence resonance energy transfer (FRET) experiments combined with computational modeling methods showed that binding of capsaicin inside the vanilloid binding pocket triggers a conformational wave that propagates through the S4-S5 linker to the S6 bundle and further induces a conformational change in the selectivity filter (Yang et al. 2018, 2015a).

14.2.3 TRPV1 as a Polymodal Receptor

While it was first identified as the capsaicin receptor, TRPV1 also functions as a sensor for other noxious stimuli, such as high temperature, low pH, toxins from venomous animals and poisonous plants (Caterina et al. 1997; Julius 2013). It also exhibits weak voltage sensitivity (Fig. 14.2a), which makes TRPV1's response to a stimulus variable according to the state of the host sensory neuron (Matta and Ahern 2007; Yang et al. 2020). Voltage activation of TRPV1 does not occur within the physiologically relevant range (the half-activation voltage, V_{half}, being above +60 mV) (Fig. 14.2b). Nonetheless, it is anticipated that under physiological conditions, transmembrane voltage changes may modulate TRPV1 activity induced by other stimuli.

The activation threshold temperature of TRPV1 is around 40 °C, with a Q_{10} (a parameter for temperature dependence defined as folds change in activity upon a 10 °C rise in temperature) value over 20 (Cao et al. 2013a). This high Q_{10} value and the threshold temperature suggest that TRPV1 can serve as a body temperature sensor involved in both sensing noxious high temperatures and regulating the body core temperature. The structural mechanism for temperature sensing in TRPV1 is still not resolved. Lowering extracellular pH (a.k.a., increasing extracellular proton concentration) below pH 6.5 activates TRPV1, indicating protons can be an

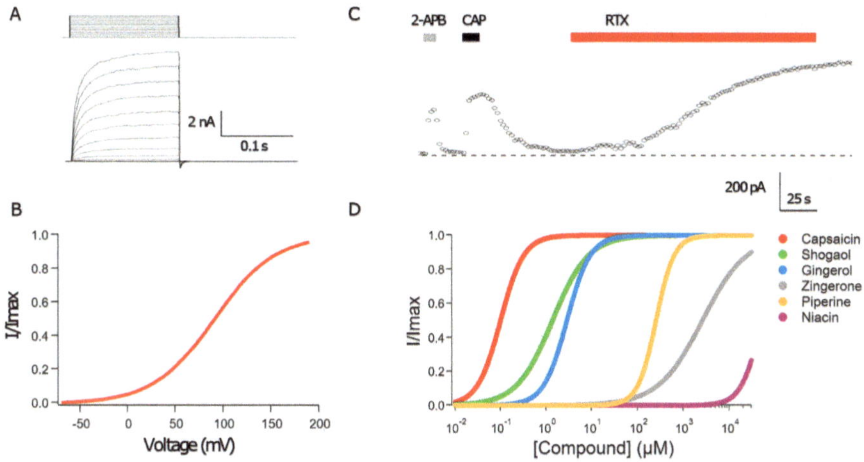

Fig. 14.2 Ligand and voltage activation of TRPV1. (**a**) Whole-cell recording of mouse TRPV1 activated by depolarization from a −80 mV holding potential up to +200 mV in 20 mV steps. (**b**) G-V curve of voltage-dependent activation. (**c**) An inside-out patch-clamp recording of mouse TRPV1 activated by 30 mM 2-APB, 10 μM capsaicin, and 100 nM RTX, at +80 mV. Current induced by 2-APB and capsaicin are fully reversible, whereas RTX-induced current is irreversible. Dash line indicates the zero current level. (**d**) Concentration dependence of several TRPV1 agonists from natural products

endogenous activator for TRPV1. Protonation of two key residues, E600 and E648 (of rat TRPV1), have been suggested as proton-sensing residues (Jordt et al. 2000; Zhang et al. 2021).

14.2.4 Ion Selectivity

When TRPV1 is activated, it allows both mono- and di-valent cations to pass through the open pore. As a non-selective cation channel, TRPV1 causes membrane depolarization when it is activated, which in turn activates voltage-gated sodium channels to initiate action potential firing (Goldstein et al. 2017). The permeabilities of TRPV1 to Li^+, Na^+, K^+, Rb^+, Cs^+ are nearly equal (Bevan et al. 2014), whereas the permeabilities to divalent cations such as Ca^{2+} and Mg^{2+} are higher compared with that for Na^+ (with a P_{Ca}/P_{Na} value of about 5 and a P_{Mg}/P_{Na} value of about 10). TRPV1 also exhibits permeability to larger cations including N-Methyl-D-glucamine (NMDG), Tetraethylammonium (TEA), and polyamines. It is reported that different stimuli or extended activation may cause changes in the ion selectivity property of TRPV1 (Chung et al. 2008). It is found that the outer pore domain contributes to the dynamic process of ion selectivity (Munns et al. 2015).

14.2.5 Desensitization

TRPV1 utilizes multiple desensitization mechanisms to prevent overloading cells with influx cations when a lasting stimulus is present. One desensitization mechanism is Ca^{2+} dependent. Repeated or extended activation of TRPV1 by capsaicin or other stimuli without chelating the intracellular calcium leads to gradually reduced responses (Lishko et al. 2007). Desensitized TRPV1 in the presence of intracellular Ca^{2+} exhibits a reduced apparent binding affinity for capsaicin (Vyklický et al. 2008). It is also reported that extended ligand activation and calcium influx can induce channel internalization, leaving less TRPV1 channels in the plasma membrane (Sanz-Salvador et al. 2012). Substantial calcium influx may produce cytotoxicity and degeneration of nerve termini (Chung and Campbell 2016). Desensitization and internalization of TRPV1 as well as degeneration of TRPV1-expressing nerve termini may help explain that capsaicin patch can relieve pain in muscles and joints. Interestingly, TRPV1 also shows heat dependent desensitization following heat activation even in the absence of calcium, suggesting a distinct mechanism (Cao et al. 2014; Yang et al. 2014). This type of desensitization can also be channel species dependent; for example, platypus TRPV1 channel lacks this heat dependent desensitization (Luo et al. 2019).

14.3 TRPV1 Physiology

14.3.1 Expression and Splice Variants

TRPV1 is mainly expressed in primary sensory neurons, located in both the cell body and the nerve endings, for sensing noxious stimuli. However, expression of TRPV1 is not limited to sensory neurons. Its expression has been reported in central nervous system (both neurons and glia in the brain), muscles, skin, endothelium cells, epithelium cells, immune cells, and adipose tissues (Shuba 2020). Expression of TRPV1 in the brain appears to be rather restricted to hypothalamus (Cavanaugh et al. 2011).

As a eukaryotic membrane protein, TRPV1 is controlled by alternative splicing. There are multiple human *Trpv1* gene splice variants, four of which generate identical proteins with 839 amino acids. This is the canonical TRPV1 channel (Dorfman et al. 2010; Schumacher and Eilers 2010). A TRPV1b variant (lack exon 7) has been reported which can also form a functional channel. The TRPV1b channel is resistant to capsaicin and protons but is responsive to heat, with a higher temperature threshold (>47 °C) compared to the canonical form. When co-expressed with the canonical TRPV1, TRPV1b exhibits a dominant negative effect, suggesting co-assembly of the canonical and variant subunits into heteromeric channels (Lu et al. 2005). It is known that TRPV1 subunits are capable of co-assembling with subunits of other TRPV types (Cheng et al. 2012, 2007), further expanding TRPV1 functional diversity (Cheng et al. 2010).

At least three murine TRPV1 variants have been reported. One of them is TRPV1beta, equivalent to human TRPV1b, with exon 7 being recognized as an intron (Wang et al. 2004). A second variant is VR.5'sv, which lacks most of the N terminal. A third variant, named TRPV1var, contains only a portion of N terminal intracellular part. The last two variants do not form functional channels by themselves, but are reported to show a dominant negative effect over the canonical form when co-expressed (Eilers et al. 2007; Tian et al. 2006).

14.3.2 Temperature Sensation and Regulation

One major physiological role of TRPV1 is sensing noxious high temperatures (threshold 40–42 °C). Strong heat activation of TRPV1 generates pain sensation by depolarizing the membrane potential in primary sensory neurons in the dorsal root ganglia and trigeminal ganglia. It should be noted that, TRPV1 probably is not the only high temperature sensor. Indeed, TRPV1 knock out mice still exhibit robust heat responses (Caterina et al. 2000). Transient receptor potential cation channel, subfamily A member 1 (TRPA1), Transient receptor potential cation channel subfamily M member 2 (TRPM2) and Transient receptor potential cation channel subfamily M member 3 (TRPM3) are also reported to be involved in heat sensing (Vilar et al. 2020; Vandewauw et al. 2018).

Besides sensing noxious high temperatures, TRPV1 is also involved in body temperature regulation. TRPV1 knock out mice experience prolonged hyperthermia when exposed to ambient temperatures above 35 °C; in comparison, wildtype mice only exhibit transient hyperthermia (Yonghak et al. 2020). Subcutaneous injection of capsaicin induces acute hypothermia in rat; the effect exhibits a clear dose-dependent manner (Hori 1984); similar effects have been reported for other TRPV1 agonists such as resiniferatoxin. On the other hand, systemic administration of TRPV1 antagonists can induce hyperthermia in laboratory mammals (Garami et al. 2020).

14.3.3 Target for Plant and Animal Toxins

Since activation of TRPV1 can induce strong pain sensation, TRPV1 has been the target for many plant and animal toxins. Capsaicin and resiniferatoxin (produced by resin spurge *Euphorbia resinifera*) are good examples of the protective role in plants against fungal infection (Tewksbury et al. 2008) and animal herbivory (Han et al. 2018). Humans appear to be the only animal enjoying the spiciness sensation elicited by consuming spicy food, which is clearly an acquired behavior through learning. For most animals the spicy sensory from consuming chili peppers is clearly repulsive. For these animals, a spicy diet appears to elicit a hot and painful sensation. Certain animals, such as tree shrew and birds, eat plants containing capsaicinoids. It is found that in these animals, the TRPV1 channels are much less sensitive to capsaicin, due to point mutations to the key hydrogen bond forming residues mentioned earlier (Han et al. 2018; Jordt and Julius 2002). It is hypothesized that the genetic changes may be driven by the beneficial outcome: birds get food whereas the chili pepper seeds get spread around in the process (together with fertilizer!). Like plants, certain venomous animals generate peptide toxins in their venoms that target TRPV1, for either hunting or protecting themselves from predators. Double-knot toxin (DkTx) from Chinese bird spider *Ornithoctonus huwena* and red head toxin (RhTx) from Chinese red-headed centipede *Scolopendra subspinipes mutilans* are good examples of peptide toxins targeting TRPV1 as potent agonists (Yang et al. 2015b; Cao et al. 2013b). These toxins bind to the outer pore region of TRPV1 to induce strong activation. It is anticipated that the pain produced by strong TRPV1 activation would send a deterrent signal.

14.3.4 Regulation of Blood Vessel Tone

TRPV1 is expressed in smooth muscle cells, endothelial cells, and peri-vascular nerves where the channel participates in the regulation of vascular tone and blood pressure (Baylie and Brayden 2011; Storozhuk et al. 2019). In endothelia cells, TRPV1 can be activated by endocannabinoids and contribute to vasodilation. Capsaicin has been reported to evoke constriction in skeleton muscle arteries and carotid artery (Tóth et al. 2013). Long-term use of capsaicin has been shown to have the effect of reducing hypertension induced by long-term high-salt diet (Hao et al. 2011). The mechanism for regulating vasoconstriction and blood pressure may be linked to downstream effects of TRPV1 activation, such as activating substance P (SP) and calcitonin gene related peptide (CGRP) release, nitrogen oxide (NO) generation (Du et al. 2019).

14.4 TRPV1 Pharmacology

14.4.1 Ligand Regulation of TRPV1

As a nociceptor, TRPV1 has attracted substantial attention as a potential target of new drugs for pain relief and anesthesia. In support of these efforts, vanilloids have been thoroughly tested for their effects on TRPV1 (Table 14.1). Capsaicin activates TRPV1 with an EC50 value of about 700 nM at room temperature (Fig. 14.2c, d). Resiniferatoxin (RTX) activates TRPV1 with a reported EC50 value of about 40 nM (Raisinghani et al. 2005), although the slow and irreversible binding kinetics of RTX makes estimation of the EC50 value challenging. Dermal patches containing 8% capsaicin have been approved by FDA for topical pain relief. RTX is undergoing clinical trials to treat severe pain associated with advanced cancer and osteoarthritic knee. The mechanism for capsaicin and RTX acting as pain relief agents is likely related to defunctionalization of afferent neurons and loss of unmyelinated as well as myelinated nerve fibers through cytotoxicity associated with strong activation of TRPV1.

Table 14.1 Commonly used TRPV1 agonists, antagonists, and pore blockers

Category	Name	EC50 OR IC50 range
Agonist	Capsaicin	500–700 nM
	Resiniferatoxin (RTX)	10 pM–40 nM
	2-Aminoethoxydiphenyl Borate (2-APB)	110 μM
	Proton	>0.3 μM (<pH 6.5)
	Allicin	50 μM
Antagonist	Capsazepine	400 nM
	5'-Iodoresiniferatoxin	4 nM
Pore blocker	Ruthenium Red	–
	Barrium	0.89 mM

Vanilloids produced by other plants are found to also activate TRPV1 (Fig. 14.2d). Gingers, for example, produce a group of vanilloids that vary in molecular structures and potencies in activating TRPV1 (Yin et al. 2019). Even non-vanilloid molecules are found to activate TRPV1 by binding to the same ligand binding pocket when used at high enough concentrations; an example is piperine, the pungent compound found in black peppers (Dong et al. 2019). Activation of TRPV1 by non-vanilloids may help explain certain flushing reactions (cutaneous vasodilation and symptoms of hot flashes and burning) in humans, for example, when patients with dyslipidemias are treated with niacin (Ma et al. 2014; Clifton et al. 2015).

Vanilloids can also be strong antagonists for TRPV1 (Table 14.1). Capsazepine is a strong inhibitor with an IC50 value of about 400 nM (Phillips et al. 2004). 5′-iodoresiniferatoxin is also a strong inhibitor with an IC50 of about 4 nM (Wahl et al. 2001); it has been used in animal models to inhibit TRPV1 activity. Many non-vanilloid chemicals have been explored to regulate TRPV1. Allicin can covalently modify a cystine in the N terminus of TRPV1 to activate the channel with an EC50 value of about 50 μM (Macpherson et al. 2005). Allyl isothiocyanate (AITC), the pungent component in wasabi, can also activate TRPV1 as a partial agonist.

14.4.2 Challenges for Targeting TRPV1 for Analgesic Purposes

TRPV1 plays an important role in heat sensing and body temperature regulation. While there are plenty of chemicals which can either potently activate TRPV1 or strongly inhibit its function, attempts for using TRPV1 agonists or antagonists has been hindered by the hypothermia or hyperthermia caused by these chemicals. Global administration of TRPV1 agonists such as capsaicin and RTX induces hypothermia in animal models; on the other hand, global application of TRPV1 inhibitors would induce hyperthermia. A selective TRPV1 blockade AMG517 induced hyperthermia in rats, dogs, monkeys, and human trials which prevented it from systemic usage to treat pain and other diseases (Gavva et al. 2008). New strategies need to be developed to reduce the sensitivity of TRPV1 to a specific stimulus without affecting the temperature sensing. This is indeed the goal of many ongoing promising research projects.

14.5 Conclusions and Future Perspectives

Being a temperature sensor and nociceptor, TRPV1 has been the subject of intense biophysical and pharmacological investigation as an attractive potential target for novel analgesics. With rapidly developing techniques and newly designed ligands, we are expecting to see more detailed research works to elucidate the mechanism of TRPV1 activation and regulation, which will guide the pharmaceutical efforts. For structural investigation, TRPV1 structures representing different functional states are eagerly anticipated. While in the last two decades much has been learned about how capsaicin binds and activates TRPV1, how the receptor distinguishes partial and full agonists as well as antagonists is still not well understood. Much less is known about how TRPV1 responds to heat and various other stimuli. Future studies will also help to elucidate TRPV1's contributions to a wide range of physiological processes.

References

Baylie RL, Brayden JE (2011) TRPV channels and vascular function. Acta Physiol (Oxf) 203(1): 99–116. https://doi.org/10.1111/j.1748-1716.2010.02217.x

Bevan S, Quallo T, Andersson DA (2014) TRPV1. In: Nilius B, Flockerzi V (eds) Mammalian transient receptor potential (TRP) cation channels, vol I. Springer, Berlin Heidelberg, Berlin, Heidelberg, pp 207–245. https://doi.org/10.1007/978-3-642-54215-2_9

Bevan S, Szolcsányi J (1990) Sensory neuron-specific actions of capsaicin: mechanisms and applications. Trends Pharmacol Sci 11(8):330–333. https://doi.org/10.1016/0165-6147(90)90237-3

Cao E, Cordero-Morales JF, Liu B, Qin F, Julius D (2013a) TRPV1 channels are intrinsically heat sensitive and negatively regulated by phosphoinositide lipids. Neuron 77(4):667–679

Cao E, Liao M, Cheng Y, Julius D (2013b) TRPV1 structures in distinct conformations reveal activation mechanisms. Nature 504(7478):113–118

Cao X, Ma L, Yang F, Wang K, Zheng J (2014) Divalent cations potentiate TRPV1 channel by lowering the heat activation threshold. J Gen Physiol 143(1):75–90

Caterina MJ, Leffler A, Malmberg AB, Martin WJ, Trafton J, Petersen-Zeitz KR, Koltzenburg M, Basbaum AI, Julius D (2000) Impaired nociception and pain sensation in mice lacking the capsaicin receptor. Science 288(5464):306–313. https://doi.org/10.1126/science.288.5464.306

Caterina MJ, Schumacher MA, Tominaga M, Rosen TA, Levine JD, Julius D (1997) The capsaicin receptor: a heat-activated ion channel in the pain pathway. Nature 389(6653):816–824

Cavanaugh DJ, Chesler AT, Jackson AC, Sigal YM, Yamanaka H, Grant R, O'Donnell D, Nicoll RA, Shah NM, Julius D (2011) Trpv1 reporter mice reveal highly restricted brain distribution and functional expression in arteriolar smooth muscle cells. J Neurosci 31(13):5067–5077

Cheng W, Sun C, Zheng J (2010) Heteromerization of TRP channel subunits: extending functional diversity. Protein Cell 1(9):802–810. https://doi.org/10.1007/s13238-010-0108-9

Cheng W, Yang F, Liu S, Colton CK, Wang C, Cui Y, Cao X, Zhu MX, Sun C, Wang K (2012) Heteromeric heat-sensitive transient receptor potential channels exhibit distinct temperature and chemical response. J Biol Chem 287(10):7279–7288

Cheng W, Yang F, Takanishi CL, Zheng J (2007) Thermosensitive TRPV channel subunits coassemble into heteromeric channels with intermediate conductance and gating properties. J Gen Physiol 129(3):191–207. https://doi.org/10.1085/jgp.200709731

Cheng Y (2022) TRPV1 and piezo: the 2021 Nobel prize in physiology or medicine. IUCrJ 9(Pt 1): 4–5. https://doi.org/10.1107/s2052252521013488

Chung M-K, Campbell JN (2016) Use of capsaicin to treat pain: mechanistic and therapeutic considerations. Pharmaceuticals 9(4):66

Chung M-K, Güler AD, Caterina MJ (2008) TRPV1 shows dynamic ionic selectivity during agonist stimulation. Nat Neurosci 11(5):555–564. https://doi.org/10.1038/nn.2102

Clifton HL, Inceoglu B, Linlin M, Zheng J, Schaefer S (2015) TRPV1 channels are involved in niacin-induced cutaneous vasodilation in mice. J Cardiovasc Pharmacol 65(2):184

Dong Y, Yin Y, Vu S, Yang F, Yarov-Yarovoy V, Tian Y, Zheng J (2019) A distinct structural mechanism underlies TRPV1 activation by piperine. Biochem Biophys Res Commun 516(2): 365–372. https://doi.org/10.1016/j.bbrc.2019.06.039

Dorfman R, Tsui H, Salter MW, Dosch H-M (2010) TRPV1 genetics. J Wiley & Sons, Hoboken, NJ

Du Q, Liao Q, Chen C, Yang X, Xie R, Xu J (2019) The role of transient receptor potential vanilloid 1 in common diseases of the digestive tract and the cardiovascular and respiratory system. Front Physiol 10. https://doi.org/10.3389/fphys.2019.01064

Eilers H, Lee SY, Hau CW, Logvinova A, Schumacher MA (2007) The rat vanilloid receptor splice variant VR.5'sv blocks TRPV1 activation. Neuroreport 18(10):969–973. https://doi.org/10.1097/WNR.0b013e328165d1a2

Gao Y, Cao E, Julius D, Cheng Y (2016) TRPV1 structures in nanodiscs reveal mechanisms of ligand and lipid action. Nature 534(7607):347–351. https://doi.org/10.1038/nature17964

Garami A, Shimansky YP, Rumbus Z, Vizin RCL, Farkas N, Hegyi J, Szakacs Z, Solymar M, Csenkey A, Chiche DA, Kapil R, Kyle DJ, Van Horn WD, Hegyi P, Romanovsky AA (2020) Hyperthermia induced by transient receptor potential vanilloid-1 (TRPV1) antagonists in human clinical trials: insights from mathematical modeling and meta-analysis. Pharmacol Ther 208: 107474. https://doi.org/10.1016/j.pharmthera.2020.107474

Gavva NR, Treanor JJ, Garami A, Fang L, Surapaneni S, Akrami A, Alvarez F, Bak A, Darling M, Gore A (2008) Pharmacological blockade of the vanilloid receptor TRPV1 elicits marked hyperthermia in humans. Pain 136(1–2):202–210

Goldstein RH, Katz B, Lev S, Binshtok AM (2017) Ultrafast optical recording reveals distinct capsaicin-induced ion dynamics along single nociceptive neurite terminals in vitro. J Biomed Opt 22(7):76010. https://doi.org/10.1117/1.Jbo.22.7.076010

Han Y, Li B, Yin TT, Xu C, Ombati R, Luo L, Xia Y, Xu L, Zheng J, Zhang Y, Yang F, Wang GD, Yang S, Lai R (2018) Molecular mechanism of the tree shrew's insensitivity to spiciness. PLoS Biol 16(7):e2004921. https://doi.org/10.1371/journal.pbio.2004921

Hao X, Chen J, Luo Z, He H, Yu H, Ma L, Ma S, Zhu T, Liu D, Zhu Z (2011) TRPV1 activation prevents high-salt diet-induced nocturnal hypertension in mice. Pflugers Arch - Eur J Physiol 461(3):345–353. https://doi.org/10.1007/s00424-011-0921-x

Hazan A, Kumar R, Matzner H, Priel A (2015) The pain receptor TRPV1 displays agonist-dependent activation stoichiometry. Sci Rep 5(1):1–13

Hori T (1984) Capsaicin and central control of thermoregulation. Pharmacol Ther 26(3):389–416. https://doi.org/10.1016/0163-7258(84)90041-x

Jordt SE, Julius D (2002) Molecular basis for species-specific sensitivity to "hot" chili peppers. Cell 108(3):421–430. https://doi.org/10.1016/s0092-8674(02)00637-2

Jordt S-E, Tominaga M, Julius D (2000) Acid potentiation of the capsaicin receptor determined by a key extracellular site. Proc Natl Acad Sci 97(14):8134–8139. https://doi.org/10.1073/pnas.100129497

Julius D (2013) TRP channels and pain. Annu Rev Cell Dev Biol 29:355–384

Kraft KH, Brown CH, Nabhan GP, Luedeling E, Luna Ruiz JJ, Coppens d'Eeckenbrugge G, Hijmans RJ, Gepts P (2014) Multiple lines of evidence for the origin of domesticated chili pepper, *Capsicum annuum*, in Mexico. Proc Natl Acad Sci 111(17):6165–6170. https://doi.org/10.1073/pnas.1308933111

Kwon DH, Zhang F, Suo Y, Bouvette J, Borgnia MJ, Lee S-Y (2021) Heat-dependent opening of TRPV1 in the presence of capsaicin. Nat Struct Mol Biol 28(7):554–563. https://doi.org/10.1038/s41594-021-00616-3

Lishko PV, Procko E, Jin X, Phelps CB, Gaudet R (2007) The ankyrin repeats of TRPV1 bind multiple ligands and modulate channel sensitivity. Neuron 54(6):905–918

Lu G, Henderson D, Liu L, Reinhart PH, Simon SA (2005) TRPV1b, a functional human vanilloid receptor splice variant. Mol Pharmacol 67(4):1119–1127. https://doi.org/10.1124/mol.104.009852

Luo L, Wang Y, Li B, Xu L, Kamau PM, Zheng J, Yang F, Yang S, Lai R (2019) Molecular basis for heat desensitization of TRPV1 ion channels. Nat Commun 10(1):2134. https://doi.org/10.1038/s41467-019-09965-6

Ma L, Lee BH, Mao R, Cai A, Jia Y, Clifton H, Schaefer S, Xu L, Zheng J (2014) Nicotinic acid activates the capsaicin receptor TRPV1: potential mechanism for cutaneous flushing. Arterioscler Thromb Vasc Biol 34(6):1272–1280

Macpherson LJ, Geierstanger BH, Viswanath V, Bandell M, Eid SR, Hwang S, Patapoutian A (2005) The pungency of garlic: activation of TRPA1 and TRPV1 in response to allicin. Curr Biol 15(10):929–934. https://doi.org/10.1016/j.cub.2005.04.018

Matta JA, Ahern GP (2007) Voltage is a partial activator of rat thermosensitive TRP channels. J Physiol 585(2):469–482

Montell C, Birnbaumer L, Flockerzi V, Bindels RJ, Bruford EA, Caterina MJ, Clapham DE, Harteneck C, Heller S, Julius D, Kojima I, Mori Y, Penner R, Prawitt D, Scharenberg AM, Schultz G, Shimizu N, Zhu MX (2002) A unified nomenclature for the superfamily of TRP cation channels. Mol Cell 9(2):229–231. https://doi.org/10.1016/s1097-2765(02)00448-3

Munns CH, Chung M-K, Sanchez YE, Amzel LM, Caterina MJ (2015) Role of the outer pore domain in transient receptor potential vanilloid 1 dynamic permeability to large cations. J Biol Chem 290(9):5707–5724

Nelson EK, Dawson LE (1923) The constitution of Capsaicin, the pungent principle of Capsicum. III. J Am Chem Soc 45(9):2179–2181. https://doi.org/10.1021/ja01662a023

Phillips E, Reeve A, Bevan S, McIntyre P (2004) Identification of species-specific determinants of the action of the antagonist capsazepine and the agonist PPAHV on TRPV1. J Biol Chem 279(17):17165–17172

Raisinghani M, Pabbidi RM, Premkumar LS (2005) Activation of transient receptor potential vanilloid 1 (TRPV1) by resiniferatoxin. J Physiol 567(Pt 3):771–786. https://doi.org/10.1113/jphysiol.2005.087874

Sanz-Salvador L, Andrés-Borderia A, Ferrer-Montiel A, Planells-Cases R (2012) Agonist- and Ca2+−dependent desensitization of TRPV1 channel targets the receptor to lysosomes for degradation. J Biol Chem 287(23):19462–19471. https://doi.org/10.1074/jbc.M111.289751

Schumacher MA, Eilers H (2010) TRPV1 splice variants: structure and function. Front Biosci (Landmark Ed) 15(3):872–882. https://doi.org/10.2741/3651

Scoville WL (1912) Note on capsicums. J Am Pharm Assoc 1(5):453–454. https://doi.org/10.1002/jps.3080010520

Shuba YM (2020) Beyond neuronal heat sensing: diversity of TRPV1 heat-capsaicin receptor-channel functions. Front Cell Neurosci 14:612480. https://doi.org/10.3389/fncel.2020.612480

Storozhuk MV, Moroz OF, Zholos AV (2019) Multifunctional TRPV1 ion channels in physiology and pathology with focus on the brain, vasculature, and some visceral systems. Biomed Res Int 2019:5806321. https://doi.org/10.1155/2019/5806321

Szallasi A, Blumberg PM (1999) Vanilloid (Capsaicin) receptors and mechanisms. Pharmacol Rev 51(2):159–212

Tewksbury JJ, Reagan KM, Machnicki NJ, Carlo TA, Haak DC, Peñaloza ALC, Levey DJ (2008) Evolutionary ecology of pungency in wild chilies. Proc Natl Acad Sci 105(33):11808–11811. https://doi.org/10.1073/pnas.0802691105

Thresh L (1846) Isolation of capsaicin. Pharm J 6:941

Tian W, Fu Y, Wang DH, Cohen DM (2006) Regulation of TRPV1 by a novel renally expressed rat TRPV1 splice variant. Am J Physiol Renal Physiol 290(1):F117–F126. https://doi.org/10.1152/ajprenal.00143.2005

Tóth A, Czikora Á, Pásztor ET, Dienes B, Bai P, Csernoch L, Rutkai I, Csató V, Mányiné IS, Pórszász R, Édes I, Papp Z, Boczán J (2013) Vanilloid Receptor-1 (TRPV1) expression and function in the vasculature of the rat. J Histochem Cytochem 62(2):129–144. https://doi.org/10.1369/0022155413513589

Vandewauw I, De Clercq K, Mulier M, Held K, Pinto S, Van Ranst N, Segal A, Voet T, Vennekens R, Zimmermann K, Vriens J, Voets T (2018) A TRP channel trio mediates acute noxious heat sensing. Nature 555(7698):662–666. https://doi.org/10.1038/nature26137

Vilar B, Tan C-H, McNaughton PA (2020) Heat detection by the TRPM2 ion channel. Nature 584(7820):E5–E12. https://doi.org/10.1038/s41586-020-2510-7

Vyklický L, Nováková-Tousová K, Benedikt J, Samad A, Touska F, Vlachová V (2008) Calcium-dependent desensitization of vanilloid receptor TRPV1: a mechanism possibly involved in analgesia induced by topical application of capsaicin. Physiol Res 57(Suppl 3):S59–S68. https://doi.org/10.33549/physiolres.931478

Wahl P, Foged C, Tullin S, Thomsen C (2001) Iodo-resiniferatoxin, a new potent vanilloid receptor antagonist. Mol Pharmacol 59(1):9–15

Wang C, Hu HZ, Colton CK, Wood JD, Zhu MX (2004) An alternative splicing product of the murine trpv1 gene dominant negatively modulates the activity of TRPV1 channels. J Biol Chem 279(36):37423–37430. https://doi.org/10.1074/jbc.M407205200

Yang F, Ma L, Cao X, Wang K, Zheng J (2014) Divalent cations activate TRPV1 through promoting conformational change of the extracellular region. J Gen Physiol 143(1):91–103

Yang F, Xiao X, Cheng W, Yang W, Yu P, Song Z, Yarov-Yarovoy V, Zheng J (2015a) Structural mechanism underlying capsaicin binding and activation of the TRPV1 ion channel. Nat Chem Biol 11(7):518–524. https://doi.org/10.1038/nchembio.1835

Yang F, Xiao X, Lee BH, Vu S, Yang W, Yarov-Yarovoy V, Zheng J (2018) The conformational wave in capsaicin activation of transient receptor potential vanilloid 1 ion channel. Nat Commun 9(1):1–9

Yang F, Xu L, Lee BH, Xiao X, Yarov-Yarovoy V, Zheng J (2020) An unorthodox mechanism underlying voltage sensitivity of TRPV1 ion channel. Adv Sci 7(20):2000575

Yang F, Zheng J (2017) Understand spiciness: mechanism of TRPV1 channel activation by capsaicin. Protein Cell 8(3):169–177

Yang S, Yang F, Wei N, Hong J, Li B, Luo L, Rong M, Yarov-Yarovoy V, Zheng J, Wang K, Lai R (2015b) A pain-inducing centipede toxin targets the heat activation machinery of nociceptor TRPV1. Nat Commun 6:8297. https://doi.org/10.1038/ncomms9297

Yin Y, Dong Y, Vu S, Yang F, Yarov-Yarovoy V, Tian Y, Zheng J (2019) Structural mechanisms underlying activation of TRPV1 channels by pungent compounds in gingers. Br J Pharmacol 176(17):3364–3377. https://doi.org/10.1111/bph.14766

Yonghak P, Miyata S, Kurganov E (2020) TRPV1 is crucial for thermal homeostasis in the mouse by heat loss behaviors under warm ambient temperature. Sci Rep 10(1):8799. https://doi.org/10.1038/s41598-020-65703-9

Yu FH, Catterall WA (2004) The VGL-Chanome: a protein superfamily specialized for electrical Signaling and ionic homeostasis. Sci STKE 253:re15. https://doi.org/10.1126/stke.2532004re15

Zhang K, Julius D, Cheng Y (2021) Structural snapshots of TRPV1 reveal mechanism of polymodal functionality. Cell 184(20):5138–5150. e5112

Zheng J, Ma L (2014) Chapter nine - structure and function of the ThermoTRP channel pore. In: Islas LD, Qin F (eds) Current topics in membranes, vol 74. Academic Press, pp 233–257. https://doi.org/10.1016/B978-0-12-800181-3.00009-9

Pharmacological Applications of Capsaicin in the Treatment of Diseases

15

Juliana Pimenta Salles and Robson Xavier Faria

Abstract

Capsaicin is the bioactive substance found in Capsicum plants. Pungency is a particular characteristic of such substance. The therapeutic use of chili extracts was made more than 200 years ago, and the properties of capsaicin are still being studied. It is recognized by cells through binding to the transient receptor potential vanilloid 1 (TRPV1) ion channel, being the main receptor agonist. Previous studies described TRV1 expression in afferent neurons and highlighted the essential role of channel activation in the cellular transduction of signs of pain. After interacting with its receptor, capsaicin may modulate substance P release. Interestingly, after prolonged exposure to the agonist, TRPV1 desensitizes and does not respond to capsaicin stimuli and thus promotes pain relief. Additionally, capsaicin modulates the gastrointestinal tract by regulating microbiota that favor short-chain fatty acid production and regulate ghrelin hormone and proinflammatory cytokines. The anti-cancer activity of capsaicin occurs through different pathways. On breast malignant cells capsaicin promote cell death through the cyclin-dependent kinase 8 pathway. In prostate cancer cells the substance impairs androgen receptor (AR) activity while in enterocytes cancer cells it triggers oxidative stress, which will lead to apoptosis. Prevention of

J. P. Salles
Laboratory of Studies in Experimental Pharmacology, Federal University of Rio de Janeiro, Rio de Janeiro, Brazil

R. X. Faria (✉)
Evaluation and Promotion of the Ambiental Health Laboratory, Oswaldo Cruz Institute, Oswaldo Cruz Foundation, Rio de Janeiro, Brazil

Postgraduate Program in Sciences and Biotechnology, Institute of Biology, Federal Fluminense University, Niterói, Brazil
e-mail: salvador@ioc.fiocruz.br

M. K. Swamy, A. Kumar (eds.), *Capsaicinoids*,
https://doi.org/10.1007/978-981-99-7779-6_15

oxidative stress and endothelial dysfunction was reported, impairing atherosclerosis development. Currently, researchers are developing new techniques to deliver capsaicin with few side effects. Conventional formulations and nanotechnological strategies have been studied to enhance capsaicin bioavailability. It is noteworthy that capsaicin exerts significant beneficial effects and thus is considered a new strategy for the management of pathological conditions.

Keywords

Capsaicin · Capsaicinoid · Vanilloid receptor · TRPV1 · Pungency · Disease

15.1 Introduction

Found as the major capsaicinoid compound in the *Capsicum* plant, capsaicin gives the pungency characteristic of the fruit. Nilius and Appendino (2013) describe capsaicin's effect as a chemesthetic sensation, in which the compound can activate gustatory receptors causing effects other than taste, such as pain, thermal, and touch. The authors suggest that the consumption of spicy food was maintained throughout the evolutionary process due to its biological relevance (Nilius and Appendino 2013).

Written records of chili's medicinal uses are described beginning in the twelfth century and are derived from the Aztec culture (Schumacher 2010). In 1850, Turnbull formally characterized the pain-reducing property of chili extract for the treatment of burning and itching extremities (Turnbull 1850). Despite its use since ancient times, the capsaicin receptor was indeed identified in the 1990s, when Caterina and coworkers cloned the receptor from rat dorsal root ganglia (DRG) neurons, initially called vanilloid receptor 1 (VR1) (Caterina et al. 1997). After this event, the mechanistic explanation of the capsaicin effect began to be elucidated. Capsaicin-induced desensitization is an important characteristic reported about its mechanism of action. Reports describe that neurons previously activated turn unresponsive to various noxious stimuli (Dray 1992). Given this, researchers turned their attention to the pharmacological applicability of capsaicin, since this molecule is capable of reducing neuronal responses. The involvement of capsaicin in the pathophysiology of pain is one of the main elucidated pathways. The neurobiological analgesic effect of capsaicin was characterized by its ability to desensitize neuronal afferents (Fattori et al. 2016). It is important to note that capsaicin-induced neuronal desensitization depends on the capsaicin concentration. When applied in lower doses, it elicits a reversible response, while higher concentrations lead to long-lasting desensitization and are not limited to capsaicin stimuli. Different pathways are activated in such events. Low concentrations of capsaicin trigger TRPV1 activation, induces Ca^{2+} mobility inside the cell, and inhibit voltage-dependent sodium channels (VCSCs) and voltage-dependent calcium channels (VDCCs) (Arora et al. 2021).

Blocking these receptors, in turn, will lead to the inhibition of signal transduction in the conduction of action potentials. On the other hand, high concentrations lead to a Ca^{2+} increase through TRPV1 activation that will stimulate Ca^{2+}-dependent protease calpain, resulting in prolonged analgesia due to terminal axonal ablation. An additional study by the same group suggests that such ablation is due to the depolymerization of axonal microtubules induced by capsaicin (Arora et al. 2022).

Extensive studies have shown capsaicin modulation of inflammation and cancer. Additionally, it may trigger antiobesity and antioxidant effects and is involved in metabolic regulation. Spicy-food consumption has presented controversial effects. The World Health Organization (WHO) reports an elevated incidence of cases of oral and gastrointestinal cancers in countries with high spicy-food ingestion. Southern Asian countries present 7.4% of oral cancer incidence, followed by 6.2%, 4.3%, and 4.2% for colorectum, stomach, and esophagus, respectively (World Health Organization - Cancer Country Profiles 2020). Nevertheless, the protective effects of capsaicin have overcome these findings. The same countries have demonstrated the smallest incidence of prostate cancer (Braga Ferreira et al. 2020).

Not only in the context of diseases, but clinical results also show capsaicin may help improve resistance and exercise performance, as well as attenuate fatigue during and after the activity (Giuriato et al. 2022; de Freitas et al. 2022; da Silva et al. 2022).

Capsaicin exerts its effects through distinct pathways, and so leads to different outcomes that contribute to homeostasis both in vitro, in vivo, and human experimental models. Thus, this work aims to summarize the main uses of capsaicin and its application in pathological conditions.

Due to its high lipophilicity, capsaicin penetrates cellular membranes and easily reaches its target in organisms, showing great bioavailability. Conventional administration routes are explored by researchers planning new formulations to deliver capsaicin the better way. Capsaicin patches with low and high concentrations are well described in the literature for pain treatment in general. An injectable formulation is also reported in a phase III study. Recent advances in technology aimed to find new ways to deliver capsaicin without compromising the stomach and oral cavity and enhancing bioavailability. Capsaicin liposomes, nanoemulsions, and nanoparticles showed good stability, good gastric tolerance, and enhanced bioavailability of the compound (Zhu et al. 2015; de Freitas et al. 2018; Nigam et al. 2019; Han et al. 2020). Figure 15.1 shows the different ways to deliver capsaicin.

15.2 Pain

Capsaicin can evoke pain-like behavior through the activation of TRPV1 (Caterina et al. 1997). Curiously, repeated doses of the compound cause analgesia when the receptor loses sensitivity to painful stimuli caused by the desensitization of C-fibers (Palazzo et al. 2002; Szallasi and Blumberg 1999).

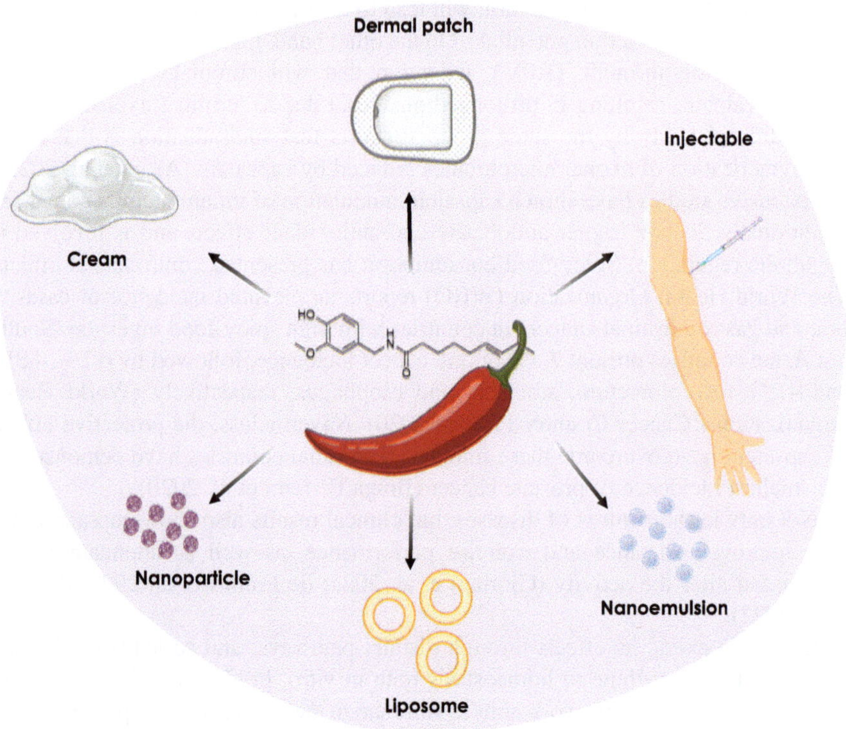

Fig. 15.1 Capsaicin delivery systems

Thus, capsaicin began to be applied as a treatment option for pain relief. The use of capsaicin is well-described in chronic pain and represents the main studied disease to its potential analgesic activity.

15.2.1 Neuropathic Pain

Special Interest Group on Neuropathic Pain (NeuPSIG) suggested neuropathic pain treatment with a high concentration of capsaicin, 8%, and considered it as a second treatment choice (Finnerup et al. 2015).

In 2009, the Food and Drug Administration (FDA) approved the use of capsaicin in neuropathic pain conditions with the release of Qutenza®, a topical capsaicin 8% patch (D'Souza et al. 2022). Neuropathies such as diabetic neuropathy, HIV-associated neuropathy, and postherpetic neuralgia showed improvement in pain when treated with capsaicin (Peppin and Pappagallo 2014).

15.2.1.1 Postherpetic Neuralgia

Postherpetic neuralgia is a recurrent and localized pain following herpes zoster infection. Its treatment consists of systemic agents such as tricyclic antidepressants, anticonvulsants, and opioids (Dworkin et al. 2007; Baron and Wasner 2006). However, tolerability and safety issues compromise its benefits. Therefore, topical treatments provide an alternative with no or few adverse effects.

Due to its analgesic effect, capsaicin is extensively studied in this condition in a topical formulation. Bernstein et al. developed a preliminary study with 14 patients, varying from 59 to 91 years old, reporting severe postherpetic neuralgia of greater than 6 months. The 0.025% capsaicin cream was applied five times a day for 7 days and three times a day for the lasting 21 days. At the end of the treatment period, it presented 75% of pain relief, in which patients declared that pain was completely gone or much better. Only one reported intermittent burning sensation following capsaicin application (Bernstein et al. 1987).

In a vehicle-controlled study developed by the same group over 6 weeks, it was observed that 0.075% capsaicin cream treatment was able to reduce pain by 46%, 40%, and 29% compared with the vehicle cream in the physician's global rating, categoric pain scale, and visual analog pain scale, respectively (Bernstein et al. 1989).

A sustained, up to 2 years, and amplified of approximately 86% response was suggested as a clinical benefit of 0.075% capsaicin therapy during a period of 6 weeks (Watson et al. 1993). Low-concentration capsaicin-containing creams (0.025–0.075%), however, present limitations due to the high frequency of application. The burning sensation is the main complaint after topical capsaicin application. NGX-4010 (commercially called Qutenza®, NeurogesX), is a high-concentration (8%) capsaicin dermal patch, that can deliver the dose to promote the analgesic effect in one short application. Backonja et al. examined patients with postherpetic neuralgia treated with a single dose of NGX-4010 for 12 weeks. The groups were divided into a high-concentration capsaicin-treated group and a low-concentration capsaicin-treated group, taken as the control group to ensure blinding due to the burning effect and erythema caused by topical capsaicin. The subjects reported significant pain reduction of 29.6% in the numeric pain rating scale (NPRS) score for the NGX-4010 group. In addition, NGX-4010 promoted pain improvement with a mean change of 29.9% in the NPRS score. Similarly, patients using concomitant neuropathic pain medication reported such improvement in pain, so the authors suggested an additive benefit of the systemic medications used for neuropathic pain and the topical treatment with the 8% capsaicin patch (Backonja et al. 2008). Nonetheless, Irving and colleagues showed that subjects who were treated with NGX-4010 alone had an improvement in pain of about 52%, while patients who used drugs for neuropathic pain had an improvement of only 36%. In addition, for patients who were not under pain therapeutic treatment a greater improvement was observed in about 45%, than in those who used systemic analgesics, 29% (Irving et al. 2012).

An open-label extension study developed by Backonja and colleagues reported that NGX-4010 could maintain reduced NPRS scores during 12 weeks and for subsequent retreatment cycles over a 48-week (Backonja et al. 2010).

15.2.1.2 Diabetic Neuropathy

Low concentrations of capsaicin have been studied since the 1990s. The Capsaicin Study Group demonstrated that 0.075% capsaicin could decrease pain intensity by 38.1% and improve pain relief by 58.4% (Capsaicin Study Group 1991). The group also observed an amelioration in working, 18.3%; sleeping, 29.5%; and leisure time, 22.8% (Capsaicin Study Group 1992). In addition, approximately 50% of the treated group presented improvement in pain regulation or was completely restored (Tandan et al. 1992). Despite being well tolerated and safe, 0.025% capsaicin did not cause significant pain reduction (Kulkantrakorn et al. 2013).

Treatment with 0.05% capsaicin cream, can decrease the pain, dysesthesia, hypesthesia, and muscle weakness scores. There is a slight decrease in the neurovascular response in the capsaicin-treated group and an increase in substance P levels in the first 4 weeks. Interestingly, during the course of the study, it decreased to levels compared to the baseline, suggesting a desensitization of substance P containing C-fibers (T Forst et al. 2002)

A phase III study showed that 30 minutes of high-concentration capsaicin (8%) treatment not only improved pain control but also sleep quality. Moreover, the treated group showed a shorter onset of response than the nontreated group (19 vs. 72 days) (Simpson et al. 2017). Compared to oral systemic medications, capsaicin treatment in diabetic neuropathic patients is more effective than oral agents such as pregabalin and gabapentin and presented efficacy compared with duloxetine and amitriptyline (Biesbroeck et al. 1995; Kiani et al. 2015).

15.2.1.3 HIV-Induced Neuropathy

HIV-associated sensory neuropathy is not an uncommon debilitating condition that increases as the use of anti-retroviral treatment increases. It is characterized by pain and often presents as distal symmetric polyneuropathy (DSPN). There is no consensus for the treatment, so general neuropathic pain medication is prescribed (e.g., gabapentin) (Jazebi et al. 2021). The good tolerance observed for capsaicin is an important characteristic that turns it into a topical adjuvant choice for the treatment of HIV-associated neuropathy (Penfold and Clark 1992; Simpson and Olney 1992). NGX-4010 evoked pain relief during the first week of the treatment and persisted for up to 12 weeks. Subjects reported a 41% pain reduction, with a sustained response in "average pain for past 24 hours" NPRS scores, which was observed similarly for "worst pain for past 24 hours" and "pain now" scores. Additionally, 67% of patients described a 30% or greater pain decrease, and 33% reported a 50% or greater in pain reduction. Similar results were observed after 30, 60, or 90 min of capsaicin treatment. One-third of treated patients presented 30% or greater pain reduction (Simpson et al. 2008a, b). Low-concentration capsaicin (0.075%), however, failed to elicit significant pain reduction scores (Paice et al. 2000).

In a 52-week open-label study, subjects received up to three retreatments of capsaicin 8% dermal patch. The study showed a similar mean percentage NPRS score reduction up to 12 weeks for all groups and demonstrated benefits regardless of the number of treatments (Simpson et al. 2014).

Katz and colleagues identified that the female sex could present better outcomes for capsaicin treatment than male patients (Katz et al. 2015).

15.2.2 Arthritis

Pain is a common symptom observed in arthritis, but its management is limited due to the inefficiency in pain relief either with pharmacological treatment or with knee joint replacement surgery. Preclinical studies have implicated TRPV1 as a nociceptive sensitizer in osteoarthritis models (Szabó et al. 2005; Barton et al. 2006; Kelly et al. 2015).

Injectable capsaicin, CNTX-4975, has a rapid onset of action and provides long-term analgesia. Stevens and colleagues investigated the efficacy of a single dose of 0.5 and 1.0 mg of CNTX-4975 in 172 patients in a randomized, double-blind, study over 6 months. A meaningful reduction in pain scores was observed during walking on flat surfaces that persisted for up to 24 weeks. The result was significant for CNTX-4975 1.0 mg presenting a rapid onset in the first week after the treatment, while CNTX-4975 0.5 mg had an intermediate effect. The study supports the idea that capsaicin treatment is not dependent on exposure time, since the elimination half-life is <4 h (Stevens et al. 2019). Persson and collaborators (2020) suggested that topical 0.025% capsaicin was more effective than 5% ibuprofen gel for knee osteoarthritis, in which 41% of patients beneficiated from capsaicin treatment than from ibuprofen (18%) (Persson et al. 2021).

15.3 Gastrointestinal Tract

Healthy intestines present bacterial homeostasis between commensal and potentially pathogenic bacteria. Such colonization protects foreign microbiomes, interacts with the host immune system, and presents metabolic functions that can generate beneficial substrates. The disruption of homeostasis will lead to gut vulnerability to pathogenic insults. This dysbiosis, in turn, may favor the development of pathological conditions, such as autoimmune and inflammatory diseases (Lathrop et al. 2011; Larsen et al. 2010; Turnbaugh et al. 2009; Sartor 2008; Frank et al. 2007; Tang and Hazen 2017).

Evidence suggests that capsaicin can regulate the gut microenvironment (Lakes et al. 2020; Kang et al. 2016). It was demonstrated that capsaicin at low and high concentrations led to an increase in the *Roseburia* population and decreased *Bacteroides* and *Parabacteroides* contents in diabetic obese mice. Additionally, an antihyperglycemic effect was observed. The group suggested an association of microbiota regulation that improved short-chain fatty acid production and regulated GLP-1, ghrelin levels, and proinflammatory cytokines (Song et al. 2017).

Dysbiosis may promote a Gram-negative bacterial population that secretes lipopolysaccharide (LPS) and favors gut permeability. LPS may increase in plasma, leading to chronic inflammation, which will contribute to the development of obesity and diabetes. Kang et al. showed that capsaicin intake (0.01%) enhanced butyrate-producing bacteria, *Ruminococcaceae* and *Lachnospiraceae*, which strengthen the gut barrier. In addition, capsaicin decreased the levels of systemic chronic low-grade inflammation markers and LPS-producing bacteria. Ultimately, capsaicin was demonstrated to ameliorate insulin resistance and obesity markers (Kang et al. 2017).

The connection between capsaicin-sensitive sensory nerves and defense mechanisms against peptic ulcers has been described (J Szolcsányi and L Barthó 1981). Abdel-Salam and collaborators showed that capsaicin in low concentrations or RTX, a capsaicin analog, provided gastric protection against noxious agents such as acidified and alcoholic compounds, also identified by Mózsik and colleagues (Abdel-Salam et al. 1999; Mózsik et al. 2005). Evidence suggests that reduction in blood flow may aggravate gastric damage by ulcerogenic agents or by physiological volume of luminal acid. The group observed that topical capsaicin treatment evoked a significant increase in microcirculation, which is considered to be the main mechanism for the anti-ulcer effect. Indeed, capsaicin triggers the release of tachykinins and calcitonin gene-related peptide, from sensory nerve terminals, enhancing local microcirculation and forming a plexus around gastric submucosal arterioles (Sternini et al. 1987; Sharkey et al. 1984).

Moreover, capsaicin induces somatostatin release that has antinociceptive and anti-inflammatory properties and suppresses gastrointestinal hormone secretion, thus contributing to the antiemetic effect of the pungent compound (Pintér et al. 2006).

A case report described the success of capsaicin treatment of a man diagnosed with cannabinoid hyperemesis syndrome (CHS) due to prolonged use of marijuana. The patient reported severe periumbilical pain, nausea, and vomiting, which were completely abolished after the second application of 0.075% capsaicin cream (Moon et al. 2018).

A study conducted with 13 cases of suspected cannabis hyperemesis syndrome in two medical centers described positive outcomes and symptom relief in all patients treated with capsaicin (Dezieck et al. 2017).

15.4 Cancer

Capsaicin displays robust anti-cancer activity in several cancer types. The mechanism by which capsaicin underlies the anti-cancer effect is multiple, but it was reported that it may occur in a TRPV1-independent manner in some cases (Braga Ferreira et al. 2020).

15.4.1 Breast Cancer

One of the most common tumors, breast cancer is responsible for the highest number of cancer deaths in the world (Bray et al. 2018). Several reports demonstrate that capsaicin may elicit its anti-tumoral effect by distinct mechanisms. Wu et al. noted high expression of cyclin-dependent kinase 8 (CDK8) activity, a critical regulator during the cell cycle and proliferation that can promote tumor development, and mRNA in the breast cancer cell lineage. High CDK8 expression led to reduced cell viability and increased cell migration, which was abolished after capsaicin treatment. Downregulation of the CDK-8 pathway was proposed to be the cause of cell viability reduction. Wnt/β-catenin signaling is increased in human cancers and can promote cell proliferation. The study showed that capsaicin inhibited such a pathway that has an intimal correlation with CDK8. In addition, capsaicin induced G_2/M cell cycle arrest (Wu et al. 2020).

In vitro and in vivo studies pointed out that capsaicin promotes caspase 3-induced apoptosis by downregulating the proto-oncogene FBI-1, which is mediated by the NF-κB pathway (Chen et al. 2021). Capsaicin also inhibited breast cells proliferation in immunodeficient mice by approximately 50% and impaired preneoplastic breast lesion progression by up to 80% (Thoennissen et al. 2010).

15.4.2 Prostate Cancer

Prostate cancer represents the second most common type of cancer worldwide (Bray et al. 2018). Cell proliferation in prostate cancer depends on androgen levels and androgen receptors (AR). Zheng et al. (2015) reported capsaicin prevented cell proliferation by AR inactivation. In addition, capsaicin restored miR-449a, an important tumor suppressor, and caused AR inhibition (Zheng et al. 2015).

A recent study demonstrated that capsaicin inhibited prostate cancer cell proliferation significantly by activating AMP-activated kinase (AMPK) and induced apoptosis in liver kinase B1 (LKB1)-expressing cells but not in cells that were not expressed, revealing the involvement of LKB1 in cancer progression regulation (Sánchez et al. 2022).

15.4.3 Gastric Cancer

Stomach cancer was classified as one of the major cancer types diagnosed, reaching approximately 1.1 million cases reported in 2020 (Ilic and Ilic 2022). It represents the fourth leading cause of cancer deaths to date. Gastric lesions, *Helicobacter pylori* infection, and diet can influence the development of the disease. Park and coworkers (2013) observed that capsaicin-treated gastric cancer cells exhibited decreased proliferation with cleaved caspase 3 enhancement in a dose-dependently. A reduction in Bcl-2 protein, an important regulator of cell-programmed death, was shown. In addition, capsaicin reduced ERK1/2, p38 MAPK, and Janus Kinase activity in cancerous enterocytes (Park et al. 2014).

Wang and colleagues reported inhibition of the tumor-associated NADH oxidase tNOX (ENOX2) after capsaicin treatment. The pungent compound enhanced oxidative stress associated with mitochondria-dependent apoptosis (Wang et al. 2011).

Table 15.1 Other cancer types that capsaicin can regulate

Cancer type	Cell type	Effect	Reference
Brain cancer	U87-MG U251	Cell viability reduction; redox imbalance and ferroptosis	Hacioglu and Kar (2022)
Oral cancer	NPC cell line	Cell proliferation, mobility inhibition; apoptosis enhancement; inhibition of MKK3-induced p38	Chiang et al. (2020)
Pancreatic cancer	AsPC-1 and BxPC-3 cells	Cell death—apoptosis; ROS generation; Bax expression induction; downregulation of Bcl-2 protein; cytochrome c release; JNK induction	Zhang et al. (2008)
	PANC-1	G0/G1 phase cell cycle arrest and apoptosis induction; phospho-PI3 Kinase p85 and phospho-Akt downregulation; growth inhibition of athymic nude mice	Zhang et al. (2013)
Liver cancer	HPCs	HPCs activation inhibition; SIRT1 downregulation; SOX2 degradation	Xie et al. (2022)
Lung cancer	SCLC cells	Contribution to camptothecin-induced apoptosis; rise in intracellular calcium and calpain signaling activity	Friedman et al. (2017)

Capsaicin also regulates other cancer types by different pathways. Table 15.1 summarizes its activity in other cancer types.

15.5 Cardiovascular Diseases

Cardiovascular diseases, in general, are conditions that disturb the structures and/or function of the heart and blood vessels. The development of cardiovascular diseases may be associated with several behavioral factors, such as smoking, alcoholism, physical inactivity, and inappropriate dietary intake.

Reports pointed TRPV1 channels are expressed in peptidergic primary sensory neurons that innervate the heart and in cardiomyocytes, endothelial and vascular smooth muscle cells (Randhawa and Jaggi 2017). Considered a chronic inflammatory disease, anti-inflammatory therapy has emerged, with TRPV1 being identified as a new treatment alternative. TRPV1 activation by capsaicin leads to reduced lipid accumulation on vascular smooth muscle cells in wild-type animals It increased the expression of ATP-binding cassette transporter A1 (ABCA1) and inhibited low-density lipoprotein-related protein 1 (LRP1), which suggests cell cholesterol efflux and reduced cholesterol uptake. Capsaicin treatment during 24 weeks reduced lipid reserves and atherosclerotic lesions from apolipoprotein E knockout (ApoE (−/−)) mice. However, on ApoE(−/−) and TRPV1 knockout mice in a high-fat diet, it was not observed, suggesting the cardioprotective effect of capsaicin (Ma et al. 2011).

Development of hyperlipidemia and atherosclerosis were investigated in pigs treated with capsaicin in a high-fat diet. The groups were differentiated by diet

Table 15.2 Capsaicin mechanisms in other pathological conditions

Disease	Study Model	Effect	Reference
Nephrotoxicity	In vivo	Nephroprotective effect; Serum creatinine urea, NO, and renal MDA reduction; SOD and GPx activities increase	Aldossary et al. (2021)
Psoriasis	In vivo	IL-23/IL-17 pathway activation inhibition; psoriatic core cytokines reduction	Chan et al. (2021)
Itch	In vivo and human	Itch and heat pain sensation decrease	Malewicz et al. (2022)
Inflammation	In vitro	TLR4/NF-κB signaling pathway-induced cytokines reduction; gene and protein expression of tight junctions and trans-epithelial electrical resistance induction; SGLT1 upregulation	Zhao et al. (2021)
Nonallergic rhinitis	Human	Blocked nose, sinus inflammation symptoms, and headache improvement	Bernstein et al. (2011)
			Ciabatti and D'Ascanio (2009)
Migraine	Human	Arterial pain reduction of over 50%	Cianchetti (2010)

(normal or high-fat) and treatment (low (2.5 mg/kg), moderate (5 mg/kg), or high-dose (10 mg/kg) capsaicin). Serum evaluation showed impaired levels of total cholesterol (TC), triglycerides (TGs), low-density lipoprotein cholesterol (LDL-C), high-density lipoprotein cholesterol (HDL-C), and apolipoprotein B100 (ApoB100). Capsaicin, however, prevented the increase in such serum protein levels and the decrease in HDL-C. Histological analysis demonstrated capsaicin inhibited atherosclerosis progression. Lipid peroxidation and antioxidant potential were impaired in the high-fat diet group, but capsaicin alleviated this imbalance in a dose-dependent manner by decreasing malondialdehyde (MDA), inducible nitric oxide synthase (iNOS) and endothelin-1 (ET-1), and increasing superoxide dismutase (SOD), endothelial nitric oxide synthase (eNOS) and nitric oxide (NO). These data suggest that capsaicin has a beneficial effect on preventing atherosclerosis by decreasing hyperlipidemia, oxidative stress, and endothelial dysfunction (Yang et al. 2019). The capsaicin involvement in the conditions cited in this work is well described, but evidence shows capsaicin effects in other pathological conditions, as summarized in Table 15.2.

15.6 Conclusions

Although the therapeutic activity of capsaicin has been known for a long period, the mechanistic understanding of capsaicin is recent. The study of capsaicin has provided us with rich knowledge about the physiology and existence of its receptor, TRPV1, which plays an essential role in the response induced by capsaicin. Figure 15.2 summarizes the pharmacological effects of capsaicin discussed in this paper. It is important to point out that the studies do not agree on the capsaicin concentration used for the treatment of the various dysfunctions discussed here, but its high efficacy is highlighted. Also, capsaicin proves to be a great alternative for the treatment of such diseases due to its less discomfort and greater effectiveness over conventional drugs. Thus, the development of new formulations or new analogs that improve capsaicin response are promising for the future of the pharmaceutical market.

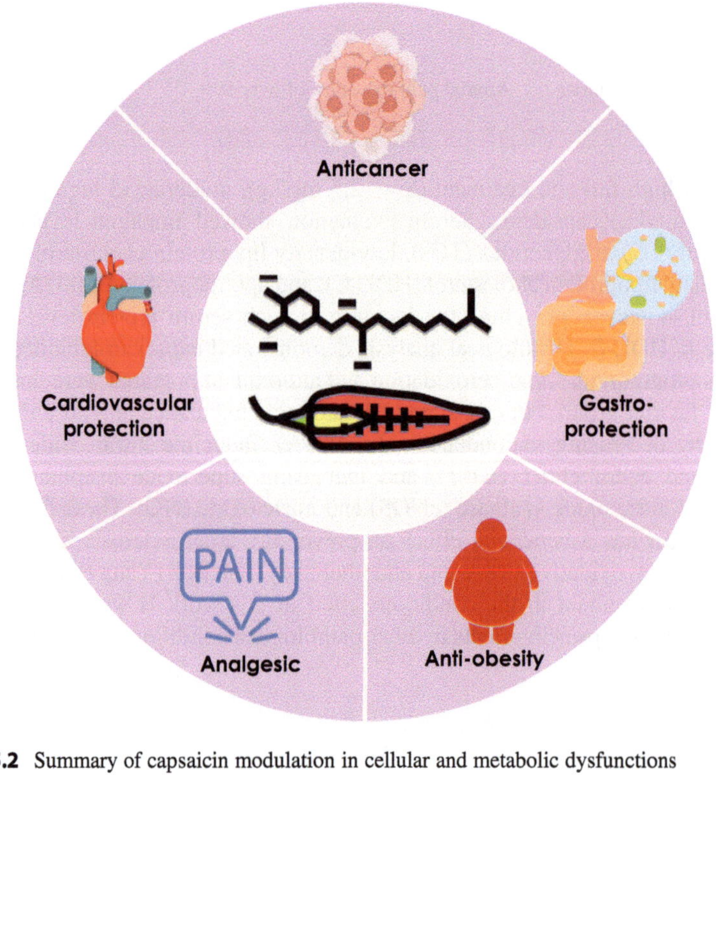

Fig. 15.2 Summary of capsaicin modulation in cellular and metabolic dysfunctions

References

Abdel-Salam OME, Debreceni A, Mózsik G, Szolcsányi J (1999) Capsaicin-sensitive afferent sensory nerves in modulating gastric mucosal defense against noxious agents. J Physiol Paris 93:443–454. https://doi.org/10.1016/S0928-4257(99)00115-1

Aldossary SA, Chohan MS, Mohaini MA, Tasleem Rasool S (2021) Capsaicin ameliorate the nephrotoxicity induced by methotrexate. Pak J Pharm Sci 34:2191–2195

Arora V, Campbell JN, Chung M-K (2021) Fight fire with fire: neurobiology of capsaicin-induced analgesia for chronic pain. Pharmacol Ther 220:107743. https://doi.org/10.1016/j.pharmthera.2020.107743

Arora V, Li T, Kumari S, Wang S, Asgar J, Chung M-K (2022) Capsaicin-induced depolymerization of axonal microtubules mediates analgesia for trigeminal neuropathic pain. Pain 163:1479–1488. https://doi.org/10.1097/j.pain.0000000000002529

Backonja M, Wallace MS, Blonsky ER, Cutler BJ, Malan P, Rauck R, Tobias J (2008) NGX-4010, a high-concentration capsaicin patch, for the treatment of postherpetic neuralgia: a randomised, double-blind study. Lancet Neurol 7:1106–1112. https://doi.org/10.1016/S1474-4422(08)70228-X

Backonja MM, Malan TP, Vanhove GF, Tobias JK (2010) NGX-4010, a high-concentration capsaicin patch, for the treatment of postherpetic neuralgia: a randomized, double-blind, controlled study with an open-label extension. Pain Med 11:600–608. https://doi.org/10.1111/j.1526-4637.2009.00793.x

Baron R, Wasner G (2006) Prevention and treatment of postherpetic neuralgia. Lancet 367:186–188. https://doi.org/10.1016/S0140-6736(06)68010-0

Barton NJ, McQueen DS, Thomson D, Gauldie SD, Wilson AW, Salter DM, Chessell IP (2006) Attenuation of experimental arthritis in TRPV1R knockout mice. Exp Mol Pathol 81:166–170. https://doi.org/10.1016/j.yexmp.2006.04.007

Bernstein JA, Davis BP, Picard JK, Cooper JP, Zheng S, Levin LS (2011) A randomized, double-blind, parallel trial comparing capsaicin nasal spray with placebo in subjects with a significant component of nonallergic rhinitis. Ann Allergy Asthma Immunol 107:171–178. https://doi.org/10.1016/j.anai.2011.05.016

Bernstein JE, Bickers DR, Dahl MV, Roshal JY (1987) Treatment of chronic postherpetic neuralgia with topical capsaicin. J Am Acad Dermatol 17:93–96. https://doi.org/10.1016/S0190-9622(87)70178-9

Bernstein JE, Korman NJ, Bickers DR, Dahl MV, Millikan LE (1989) Topical capsaicin treatment of chronic postherpetic neuralgia. J Am Acad Dermatol 21:265–270. https://doi.org/10.1016/S0190-9622(89)70171-7

Biesbroeck R, Bril V, Hollander P, Kabadi U, Schwartz S, Singh SP, Ward WK, Bernstein JE (1995) A double-blind comparison of topical capsaicin and oral amitriptyline in painful diabetic neuropathy. Adv Ther 12:111–120

Braga Ferreira LG, Faria JV, dos Santos JPS, Faria RX (2020) Capsaicin: TRPV1-independent mechanisms and novel therapeutic possibilities. Eur J Pharmacol 887:173356. https://doi.org/10.1016/j.ejphar.2020.173356

Bray F, Ferlay J, Soerjomataram I, Siegel RL, Torre LA, Jemal A (2018) Global cancer statistics 2018: GLOBOCAN estimates of incidence and mortality worldwide for 36 cancers in 185 countries. CA Cancer J Clin 68:394–424. https://doi.org/10.3322/caac.21492

Capsaicin Study Group (1991) Treatment of painful diabetic neuropathy with topical capsaicin: a Multicenter, double-blind, vehicle-controlled study. Arch Intern Med 151:2225. https://doi.org/10.1001/archinte.1991.00400110079017

Capsaicin Study Group (1992) Effect of treatment with capsaicin on daily activities of patients with painful diabetic neuropathy. Diabetes Care 15:159–165. https://doi.org/10.2337/diacare.15.2.159

Caterina MJ, Schumacher MA, Tominaga M, Rosen TA, Levine JD, Julius D (1997) The capsaicin receptor: a heat-activated ion channel in the pain pathway. Nature 389:816–824. https://doi.org/10.1038/39807

Chan TC, Lee M-S, Huang W-C, Chang W-Y, Krueger JG, Tsai T-F (2021) Capsaicin attenuates imiquimod-induced epidermal hyperplasia and cutaneous inflammation in a murine model of psoriasis. Biomed Pharmacother 141:111950. https://doi.org/10.1016/j.biopha.2021.111950

Chen M, Xiao C, Jiang W, Yang W, Qin Q, Tan Q, Lian B, Liang Z, Wei C (2021) Capsaicin inhibits proliferation and induces apoptosis in breast cancer by down-regulating FBI-1-mediated NF-κB pathway. DDDT 15:125–140. https://doi.org/10.2147/DDDT.S269901

Chiang C, Zhang M, Wang D, Xiao T, Zhu L, Chen K, Huang J, Huang J, Zhu J, Li L, Chen C, Chen Y, Hu H, Jiang W, Zou Y, Wang T, Zheng D (2020) Therapeutic potential of targeting MKK3-p38 axis with capsaicin for nasopharyngeal carcinoma. Theranostics 10:7906–7920. https://doi.org/10.7150/thno.45191

Ciabatti PG, D'Ascanio L (2009) Intranasal capsicum spray in idiopathic rhinitis: a randomized prospective application regimen trial. Acta Otolaryngol 129:367–371. https://doi.org/10.1080/00016480802510186

Cianchetti C (2010) Capsaicin jelly against migraine pain: capsaicin jelly in migraine. Int J Clin Pract 64:457–459. https://doi.org/10.1111/j.1742-1241.2009.02294.x

D'Souza RS, Barman R, Joseph A, Abd-Elsayed A (2022) Evidence-based treatment of painful diabetic neuropathy: a systematic review. Curr Pain Headache Rep 26:583–594. https://doi.org/10.1007/s11916-022-01061-7

De Freitas GBL, De Almeida DJ, Carraro E, Kerppers II, Martins GAG, Mainardes RM, Khalil NM, Messias-Reason IJT (2018) Formulation, characterization, and in vitro/in vivo studies of capsaicin-loaded albumin nanoparticles. Mater Sci Eng C 93:70–79. https://doi.org/10.1016/j.msec.2018.07.064

Dezieck L, Hafez Z, Conicella A, Blohm E, O'Connor MJ, Schwarz ES, Mullins ME (2017) Resolution of cannabis hyperemesis syndrome with topical capsaicin in the emergency department: a case series. Clin Toxicol 55:908–913. https://doi.org/10.1080/15563650.2017.1324166

Dray A (1992) Mechanism of action of capsaicin-like molecules on sensory neurons. Life Sci 51: 1759–1765. https://doi.org/10.1016/0024-3205(92)90045-Q

Dworkin RH, O'Connor AB, Backonja M, Farrar JT, Finnerup NB, Jensen TS, Kalso EA, Loeser JD, Miaskowski C, Nurmikko TJ, Portenoy RK, Rice ASC, Stacey BR, Treede R-D, Turk DC, Wallace MS (2007) Pharmacologic management of neuropathic pain: evidence-based recommendations. Pain 132:237–251. https://doi.org/10.1016/j.pain.2007.08.033

Fattori V, Hohmann M, Rossaneis A, Pinho-Ribeiro F, Verri W (2016) Capsaicin: current understanding of its mechanisms and therapy of pain and other pre-clinical and clinical uses. Molecules 21:844. https://doi.org/10.3390/molecules21070844

Finnerup NB, Attal N, Haroutounian S, McNicol E, Baron R, Dworkin RH, Gilron I, Haanpää M, Hansson P, Jensen TS, Kamerman PR, Lund K, Moore A, Raja SN, Rice ASC, Rowbotham M, Sena E, Siddall P, Smith BH, Wallace M (2015) Pharmacotherapy for neuropathic pain in adults: a systematic review and meta-analysis. Lancet Neurol 14:162–173. https://doi.org/10.1016/S1474-4422(14)70251-0

Frank DN, St. Amand AL, Feldman RA, Boedeker EC, Harpaz N, Pace NR (2007) Molecular-phylogenetic characterization of microbial community imbalances in human inflammatory bowel diseases. Proc Natl Acad Sci U S A 104:13780–13785. https://doi.org/10.1073/pnas.0706625104

de Freitas MC, Cholewa JM, Panissa VLG, Toloi GG, Netto HC, Zanini de Freitas C, Freire RV, Lira FS, Rossi FE (2022) Acute capsaicin supplementation improved resistance exercise performance performed after a high-intensity intermittent running in resistance-trained men. J Strength Cond Res 36:130–134. https://doi.org/10.1519/JSC.0000000000003431

Friedman JR, Perry HE, Brown KC, Gao Y, Lin J, Stevenson CD, Hurley JD, Nolan NA, Akers AT, Chen YC, Denning KL, Brown LG, Dasgupta P (2017) Capsaicin synergizes with camptothecin to induce increased apoptosis in human small cell lung cancers via the calpain pathway. Biochem Pharmacol 129:54–66. https://doi.org/10.1016/j.bcp.2017.01.004

Forst T, Pohlmann T, Kunt T, Goitom K, Schulz G, Löbig M, Engelbach M, Beyer J, Pfützner A (2002) The influence of local capsaicin treatment on small nerve fibre function and neurovascular control in symptomatic diabetic neuropathy. Acta Diabetol 39(1):1–6. https://doi.org/10.1007/s005920200005

Giuriato G, Venturelli M, Matias A, Soares EMKVK, Gaetgens J, Frederick KA, Ives SJ (2022) Capsaicin and its effect on exercise performance, fatigue and inflammation after exercise. Nutrients 14:232. https://doi.org/10.3390/nu14020232

Hacioglu C, Kar F (2022) Capsaicin induces redox imbalance and ferroptosis through ACSL4/GPx4 signaling pathways in U87-MG and U251 glioblastoma cells. Metab Brain Dis. https://doi.org/10.1007/s11011-022-00983-w

Han J, Zhang S, Liu X, Xiao C (2020) Fabrication of capsaicin emulsions: improving the stability of the system and relieving the irritation to the gastrointestinal tract of rats. J Sci Food Agric 100:129–138. https://doi.org/10.1002/jsfa.10002

Ilic M, Ilic I (2022) Epidemiology of stomach cancer. WJG 28:1187–1203. https://doi.org/10.3748/wjg.v28.i12.1187

Irving GA, Backonja M, Rauck R, Webster LR, Tobias JK, Vanhove GF (2012) NGX-4010, a capsaicin 8% dermal patch, administered alone or in combination with systemic neuropathic pain medications, reduces pain in patients with postherpetic neuralgia. Clin J Pain 28:101–107. https://doi.org/10.1097/AJP.0b013e318227403d

Jazebi N, Evans C, Kadaru HS, Kompella D, Raji M, Fang F, Pappolla M, Tang S-J, Chung JM, Hammock B, Fang X (2021) HIV-related neuropathy: pathophysiology, treatment and challenges. J Neurol Exp Neurosci 7:15–24. https://doi.org/10.17756/jnen.2021-082

Kang C, Wang B, Kaliannan K, Wang X, Lang H, Hui S, Huang L, Zhang Y, Zhou M, Chen M, Mi M (2017) Gut microbiota mediates the protective effects of dietary capsaicin against chronic low-grade inflammation and associated obesity induced by high-fat diet. MBio 8:e00470-17. https://doi.org/10.1128/mBio.00470-17

Kang C, Zhang Y, Zhu X, Liu K, Wang X, Chen M, Wang J, Chen H, Hui S, Huang L, Zhang Q, Zhu J, Wang B, Mi M (2016) Healthy subjects differentially respond to dietary capsaicin correlating with specific gut enterotypes. J Clin Endocrinol Metabol 101:4681–4689. https://doi.org/10.1210/jc.2016-2786

Katz NP, Mou J, Paillard FC, Turnbull B, Trudeau J, Stoker M (2015) Predictors of response in patients with postherpetic neuralgia and HIV-associated neuropathy treated with the 8% capsaicin patch (Qutenza). Clin J Pain 31:859–866. https://doi.org/10.1097/AJP.0000000000000186

Kelly S, Chapman RJ, Woodhams S, Sagar DR, Turner J, Burston JJ, Bullock C, Paton K, Huang J, Wong A, McWilliams DF, Okine BN, Barrett DA, Hathway GJ, Walsh DA, Chapman V (2015) Increased function of pronociceptive TRPV1 at the level of the joint in a rat model of osteoarthritis pain. Ann Rheum Dis 74:252–259. https://doi.org/10.1136/annrheumdis-2013-203413

Kiani J, Ahmad Nasrollahi S, Esna-Ashari F, Fallah P, Sajedi F (2015) Amitriptyline 2% cream vs. capsaicin 0.75% cream in the treatment of painful diabetic neuropathy (double blind, randomized clinical trial of efficacy and safety). Iran J Pharm Res 14:1263–1268

Kulkantrakorn K, Lorsuwansiri C, Meesawatsom P (2013) 0.025% capsaicin gel for the treatment of painful diabetic neuropathy: a randomized, double-blind, crossover, placebo-controlled trial. Pain Pract 13:497–503. https://doi.org/10.1111/papr.12013

Lakes JE, Richards CI, Flythe MD (2020) Inhibition of bacteroidetes and firmicutes by select phytochemicals. Anaerobe 61:102145. https://doi.org/10.1016/j.anaerobe.2019.102145

Larsen N, Vogensen FK, van den Berg FWJ, Nielsen DS, Andreasen AS, Pedersen BK, Al-Soud WA, Sørensen SJ, Hansen LH, Jakobsen M (2010) Gut microbiota in human adults with type 2 diabetes differs from non-diabetic adults. PLoS One 5:e9085. https://doi.org/10.1371/journal.pone.0009085

Lathrop SK, Bloom SM, Rao SM, Nutsch K, Lio C-W, Santacruz N, Peterson DA, Stappenbeck TS, Hsieh C-S (2011) Peripheral education of the immune system by colonic commensal microbiota. Nature 478:250–254. https://doi.org/10.1038/nature10434

Ma L, Zhong J, Zhao Z, Luo Z, Ma S, Sun J, He H, Zhu T, Liu D, Zhu Z, Tepel M (2011) Activation of TRPV1 reduces vascular lipid accumulation and attenuates atherosclerosis. Cardiovasc Res 92:504–513. https://doi.org/10.1093/cvr/cvr245

Malewicz NM, Rattray Z, Oeck S, Jung S, Escamilla-Rivera V, Chen Z, Tang X, Zhou J, LaMotte RH (2022) Topical capsaicin in poly(lactic-co-glycolic)acid (PLGA) nanoparticles decreases acute itch and heat pain. Int J Mol Sci 23:5275. https://doi.org/10.3390/ijms23095275

Moon AM, Buckley SA, Mark NM (2018) Successful treatment of cannabinoid hyperemesis syndrome with topical capsaicin. ACG Case Rep J 5:e3. https://doi.org/10.14309/crj.2018.3

Mózsik G, Szolcsányi J, Rácz I (2005) Gastroprotection induced by capsaicin in healthy human subjects. World J Gastroenterol 11:5180–5184. https://doi.org/10.3748/wjg.v11.i33.5180

Nigam K, Gabrani R, Dang S (2019) Nano-emulsion from capsaicin: formulation and characterization. Mater Today Proc 18:869–878. https://doi.org/10.1016/j.matpr.2019.06.517

Nilius B, Appendino G (2013) Spices: the Savory and beneficial science of pungency. In: Nilius B, Amara SG, Lill R, Offermanns S, Gudermann T, Petersen OH, Jahn R (eds) Reviews of physiology, biochemistry and pharmacology, vol 164. Springer International Publishing, Cham, pp 1–76

Paice JA, Ferrans CE, Lashley FR, Shott S, Vizgirda V, Pitrak D (2000) Topical capsaicin in the management of HIV-associated peripheral neuropathy. J Pain Symptom Manag 19:45–52. https://doi.org/10.1016/S0885-3924(99)00139-6

Palazzo E, de Novellis V, Marabese I, Cuomo D, Rossi F, Berrino L, Rossi F, Maione S (2002) Interaction between vanilloid and glutamate receptors in the central modulation of nociception. Eur J Pharmacol 439:69–75. https://doi.org/10.1016/S0014-2999(02)01367-5

Park S-Y, Kim J-Y, Lee S-M, Jun C-H, Cho S-B, Park C-H, Joo Y-E, Kim H-S, Choi S-K, Rew J-S (2014) Capsaicin induces apoptosis and modulates MAPK signaling in human gastric cancer cells. Mol Med Rep 9:499–502. https://doi.org/10.3892/mmr.2013.1849

Penfold J, Clark AJM (1992) Pain syndromes in HIV infection. Can J Anaesth 39:724–730. https://doi.org/10.1007/BF03008237

Peppin JF, Pappagallo M (2014) Capsaicinoids in the treatment of neuropathic pain: a review. Ther Adv Neurol Disord 7:22–32. https://doi.org/10.1177/1756285613501576

Persson MSM, Stocks J, Sarmanova A, Fernandes G, Walsh DA, Doherty M, Zhang W (2021) Individual responses to topical ibuprofen gel or capsaicin cream for painful knee osteoarthritis: a series of n-of-1 trials. Rheumatology 60:2231–2237. https://doi.org/10.1093/rheumatology/keaa561

Pintér E, Helyes Z, Szolcsányi J (2006) Inhibitory effect of somatostatin on inflammation and nociception. Pharmacol Ther 112:440–456. https://doi.org/10.1016/j.pharmthera.2006.04.010

Randhawa PK, Jaggi AS (2017) TRPV1 channels in cardiovascular system: a double edged sword? Int J Cardiol 228:103–113. https://doi.org/10.1016/j.ijcard.2016.11.205

Sánchez BG, Bort A, Mora-Rodríguez JM, Díaz-Laviada I (2022) The natural chemotherapeutic capsaicin activates AMPK through LKB1 kinase and TRPV1 receptors in prostate cancer cells. Pharmaceutics 14:329. https://doi.org/10.3390/pharmaceutics14020329

Sartor RB (2008) Microbial influences in inflammatory bowel diseases. Gastroenterology 134:577–594. https://doi.org/10.1053/j.gastro.2007.11.059

Schumacher MA (2010) Transient receptor potential channels in pain and inflammation: therapeutic opportunities: TRP channels in pain and inflammation. Pain Pract 10:185–200. https://doi.org/10.1111/j.1533-2500.2010.00358.x

Sharkey KA, Williams RG, Dockray GJ (1984) Sensory substance P innervation of the stomach and pancreas. Demonstration of capsaicin-sensitive sensory neurons in the rat by combined immunohistochemistry and retrograde tracing. Gastroenterology 87:914–921 da Silva BVC, Mota GR, Marocolo M, Martin JS, Prado LS (2022) Acute supplementation with capsaicin enhances upper-limb performance in male jiu-jitsu athletes. Sports 10:120. https://doi.org/10.3390/sports10080120

Simpson DM, Brown S, Tobias J, For the NGX-4010 C107 Study Group (2008a) Controlled trial of high-concentration capsaicin patch for treatment of painful HIV neuropathy. Neurology 70: 2305–2313. https://doi.org/10.1212/01.wnl.0000314647.35825.9c

Simpson DM, Brown S, Tobias JK, Vanhove GF (2014) NGX-4010, a capsaicin 8% dermal patch, for the treatment of painful HIV-associated distal sensory polyneuropathy: results of a 52-week open-label study. Clin J Pain 30:134–142. https://doi.org/10.1097/AJP.0b013e318287a32f

Simpson DM, Estanislao L, Brown SJ, Sampson J (2008b) An open-label pilot study of high-concentration capsaicin patch in painful HIV neuropathy. J Pain Symptom Manag 35:299–306. https://doi.org/10.1016/j.jpainsymman.2007.04.015

Simpson DM, Olney RK (1992) Peripheral neuropathies associated with human immunodeficiency virus infection. Neurol Clin 10:685–711

Simpson DM, Robinson-Papp J, Van J, Stoker M, Jacobs H, Snijder RJ, Schregardus DS, Long SK, Lambourg B, Katz N (2017) Capsaicin 8% patch in painful diabetic peripheral neuropathy: a randomized, double-blind, placebo-controlled study. J Pain 18:42–53. https://doi.org/10.1016/j.jpain.2016.09.008

Song J-X, Ren H, Gao Y-F, Lee C-Y, Li S-F, Zhang F, Li L, Chen H (2017) Dietary capsaicin improves glucose homeostasis and alters the gut microbiota in obese diabetic ob/ob mice. Front Physiol 8:602. https://doi.org/10.3389/fphys.2017.00602

Sternini C, Reeve JR, Brecha N (1987) Distribution and characterization of calcitonin gene-related peptide immunoreactivity in the digestive system of normal and capsaicin-treated rats. Gastroenterology 93:852–862. https://doi.org/10.1016/0016-5085(87)90450-1

Stevens RM, Ervin J, Nezzer J, Nieves Y, Guedes K, Burges R, Hanson PD, Campbell JN (2019) Randomized, double-blind, placebo-controlled trial of intraarticular trans—capsaicin for pain associated with osteoarthritis of the knee. Arthritis Rheumatol 71:1524–1533. https://doi.org/10.1002/art.40894

Szabó Á, Helyes Z, Sándor K, Bite A, Pintér E, Németh J, Bánvölgyi Á, Bölcskei K, Elekes K, Szolcsányi J (2005) Role of transient receptor potential vanilloid 1 receptors in adjuvant-induced chronic arthritis: in vivo study using gene-deficient mice. J Pharmacol Exp Ther 314: 111–119. https://doi.org/10.1124/jpet.104.082487

Szallasi A, Blumberg PM (1999) Vanilloid (capsaicin) receptors and mechanisms. Pharmacol Rev 51:159–212

J., Szolcsányi L., Barthó (1981) Gastrointestinal Defence Mechanisms IMPAIRED DEFENSE MECHANISM TO PEPTIC ULCER IN THE CAPSACIN-DESENSITIZED RAT Elsevier 39-51

Tandan R, Lewis GA, Krusinski PB, Badger GB, Fries TJ (1992) Topical capsaicin in painful diabetic neuropathy: controlled study with Long-term follow-up. Diabetes Care 15:8–14. https://doi.org/10.2337/diacare.15.1.8

Tang WHW, Hazen SL (2017) The gut microbiome and its role in cardiovascular diseases. Circulation 135:1008–1010. https://doi.org/10.1161/CIRCULATIONAHA.116.024251

Thoennissen NH, O'Kelly J, Lu D, Iwanski GB, La DT, Abbassi S, Leiter A, Karlan B, Mehta R, Koeffler HP (2010) Capsaicin causes cell-cycle arrest and apoptosis in ER-positive and -negative breast cancer cells by modulating the EGFR/HER-2 pathway. Oncogene 29:285–296. https://doi.org/10.1038/onc.2009.335

Turnbaugh PJ, Hamady M, Yatsunenko T, Cantarel BL, Duncan A, Ley RE, Sogin ML, Jones WJ, Roe BA, Affourtit JP, Egholm M, Henrissat B, Heath AC, Knight R, Gordon JI (2009) A core gut microbiome in obese and lean twins. Nature 457:480–484. https://doi.org/10.1038/nature07540

Turnbull A (1850) Tincture of capsaicin as a remedy for chilblains and toothache. Dublin Free Press 1:95–96

Wang H-M, Chuang S-M, Su Y-C, Li Y-H, Chueh PJ (2011) Down-regulation of tumor-associated NADH oxidase, tNOX (ENOX2), enhances capsaicin-induced inhibition of gastric cancer cell growth. Cell Biochem Biophys 61:355–366. https://doi.org/10.1007/s12013-011-9218-0

Watson CP, Tyler KL, Bickers DR, Millikan LE, Smith S, Coleman E (1993) A randomized vehicle-controlled trial of topical capsaicin in the treatment of postherpetic neuralgia. Clin Ther 15:510–526

World Health Oragnization. Cancer Country Profiles 2020. https://www.who.int/teams/noncommunicable-diseases/surveillance/data/cancer-profiles

Wu D, Jia H, Zhang Z, Li S (2020) Capsaicin suppresses breast cancer cell viability by regulating the CDK8/PI3K/Akt/Wnt/β-catenin signaling pathway. Mol Med Rep 22:4868–4876. https://doi.org/10.3892/mmr.2020.11585

Xie Z, Li H, Hou X, Huang M, Zhu Z, Wei L, Tang C (2022) Capsaicin suppresses hepatocarcinogenesis by inhibiting the stemness of hepatic progenitor cells via SIRT1 / SOX2 signaling pathway. Cancer Med:cam4.4777. https://doi.org/10.1002/cam4.4777

Yang S, Liu L, Meng L, Hu X (2019) Capsaicin is beneficial to hyperlipidemia, oxidative stress, endothelial dysfunction, and atherosclerosis in Guinea pigs fed on a high-fat diet. Chem Biol Interact 297:1–7. https://doi.org/10.1016/j.cbi.2018.10.006

Zhang J-H, Lai F-J, Chen H, Luo J, Zhang R-Y, Bu H-Q, Wang Z-H, Lin H-H, Lin S-Z (2013) Involvement of the phosphoinositide 3-kinase/Akt pathway in apoptosis induced by capsaicin in the human pancreatic cancer cell line PANC-1. Oncol Lett 5:43–48. https://doi.org/10.3892/ol.2012.991

Zhang R, Humphreys I, Sahu RP, Shi Y, Srivastava SK (2008) In vitro and in vivo induction of apoptosis by capsaicin in pancreatic cancer cells is mediated through ROS generation and mitochondrial death pathway. Apoptosis 13:1465–1478. https://doi.org/10.1007/s10495-008-0278-6

Zhao X, Dong B, Friesen M, Liu S, Zhu C, Yang C (2021) Capsaicin attenuates lipopolysaccharide-induced inflammation and barrier dysfunction in intestinal porcine epithelial cell line-J2. Front Physiol 12:715469. https://doi.org/10.3389/fphys.2021.715469

Zheng L, Chen J, Ma Z, Liu W, Yang F, Yang Z, Wang K, Wang X, He D, Li L (2015) Capsaicin causes inactivation and degradation of the androgen receptor by inducing the restoration of miR-449a in prostate cancer. Oncol Rep 34:1027–1034. https://doi.org/10.3892/or.2015.4055

Zhu Y, Wang M, Zhang J, Peng W, Firempong CK, Deng W, Wang Q, Wang S, Shi F, Yu J, Xu X, Zhang W (2015) Improved oral bioavailability of capsaicin via liposomal nanoformulation: preparation, in vitro drug release and pharmacokinetics in rats. Arch Pharm Res 38:512–521. https://doi.org/10.1007/s12272-014-0481-7

Clinical Pharmacology of Capsaicin, Its Synthetic and Semisynthetic Analogues

16

Appala Raju Nemala and Anitha Nandagopal

Abstract

The spicy flavour of chilli pepper fruit comes from chemicals, called capsaicinoids. Capsaicin, dihydrocapsaicin, nordihydrocapsaicin, homodihydrocapsaicin, and homocapsaicin are the most important capsaicinoids in chilli pepper. Over 92% of the capsaicinoids in chilli pepper fruit are capsaicin and dihydrocapsaicin. These are the most powerful capsaicinoids, and the only difference between them is the saturation of the acyl group. Amides of vanillyl-amine plus long-chain fatty acids yield capsaicin and all other capsaicinoid derivatives. Analogues of capsaicin that keep the homovanillyl group safe are good options for blocking TRPV1 (a membrane channel). Because vanilloids function on sensory nerve fibres, they are anti-inflammatory and painkilling. Non-spicy capsaicin analogues synthesised from n-3 polyunsaturated fatty acid were studied for antioxidant and carbohydrate-hydrolyzing enzyme-blocking activities. Both two capsaicin analogues, N-eicosapentaenoyl vanillylamine and N-docosahexaenoyl vanillylamine, non-spicy forms of capsaicin can inhibit amylase and glucosidase. Capsaicinoids and capsaicin have antioxidant, anticarcinogenic, energy metabolism, lipid profile, and inflammation-reducing properties, according to study. Still, these molecules' fragrance can make people irritated, limiting their utilisation. 'CH-19-Sweet pepper', a mild red pepper, contains capsinoids similar to capsaicin. Capsiate, dihydrocapsiate, and nordihydrocapsiate are capsinoids. Capsinoids offer consistent biopotency to capsaicin devoid of pungency or spiciness. The anti-cancer effect of capsaicin epoxide is the same as capsaicin. Resiniferatoxin is a thousand times

A. R. Nemala
Department of Pharmacognosy, Sultan-ul-Uloom College of Pharmacy, Hyderabad, Telangana, India

A. Nandagopal (✉)
Department of Pharmacology, Sultan-ul-Uloom College of Pharmacy, Hyderabad, Telangana, India

M. K. Swamy, A. Kumar (eds.), *Capsaicinoids*,
https://doi.org/10.1007/978-981-99-7779-6_16

stronger than capsaicin as an agonist of TRPV1. Zucapsaicin, the cis-isomer of capsaicin, treats cluster migraines, episodic collection headaches, and neuropathic pain. The first antagonist was capsazepine, a synthetic form of capsaicin produced by modifying its molecules. Despite its negative consequences, capsaicin is employed in many treatments to treat human ailments. Several capsaicin equivalents from nature or the laboratory were employed to create novel medications with fewer or no lethal side effects. This chapter emphasises the necessity for Structure Activity Relationships (SAR) investigations to develop novel capsaicin analogues with minimal or no harmful effects, as well as bioisostere analogues as a prospective scaffold.

Keywords

Capsaicin · *Capsicum* · Capsaicin derivatives · Pharmacology

16.1 Introduction

Capsaicinoids are also the name for the chemicals dihydrocapsaicin, nordihydrocapsaicin, homodihydrocapsaicin, and homocapsaicinis. One of these compounds is called capsaicin (8-methyl-vanilyl-none-6-amide). Capsaicin and dihydrocapsaicin are common capsaicinoids. They are mostly made in red pepper fruits of *Capsicum* genre. The main place where capsaicinoid is made is in the epidermal cell of the placenta. Capsaicinoids are bioactive vanillylamides with 9 to 11 carbons that give the *Capsicum* fruit its spicy taste. It is thought that capsaicin and dihydrocapsaicin make up more than 80% of all capsaicinoids, and their amounts are mostly based on how hot the pepper is (Hayman and Kam 2008). Capsaicin was one of the more common of the two main capsaicinoids that were measured and listed above. It was found in many of the *Capsicum* species that were studied. All of the capsaicinoids have been looked at, but clinical and lab research has focused most on capsaicin. Capsaicin has good qualities that could help treat problems with the nervous system like arthritis, cystitis, and the human immunodeficiency virus. This is especially true for newer drugs. It has a lot of antibacterial properties that can be used to stop pathogenic microbes from growing in food.

A small quantity of nordihydrocapsaicin, another capsaicinoid, was also found in these peppers. According to Menichini et al. (2009), the *C. chinense* species is one of the spiciest on the planet, like the Habanero pepper, which can be measured by its acidity. Sarpras et al. (2016) looked at how spicy *C. chinense* is compared to two other types of pepper, *C. annuum* and *C. frutescens,* and found that *C. chinense* is the hottest. *C. chinense*, on the other hand had more flavour. This chapter emphasises the development of novel capsaicin analogues with minimal or no harmful effects.

16.1.1 Chemistry of Capsaicin and Design of Analogues

Nelson and Dawson determined the molecular structure of capsaicin in 1919, after Tresh refined and named it in 1876. Capsaicin is a crystalline alkaloid that is odourless and has no colour. Its empirical formula is $C_{18}H_{27}NO_3$. It has a Mol.Wt. of 305.40 gm-mol^{-1} and can be dissolved in fat, alcohol, or oil. Dividing capsaicin into its three main domains has previously been used to elucidate the structure-activity relationships (SAR) of capsaicin agonist. The possible modifications of the capsaicin structure were examined in detail in this chapter (Fig. 16.1), which was neatly divided into three sections: the Amide-bridge-'B' region (Walpole et al. 1993a, b, c) or the amide bond, the Aromatic-'A' region and the structure-based activity profiles, and the hydrophobic-'C'-moiety (Walpole et al. 1993a, b, c).

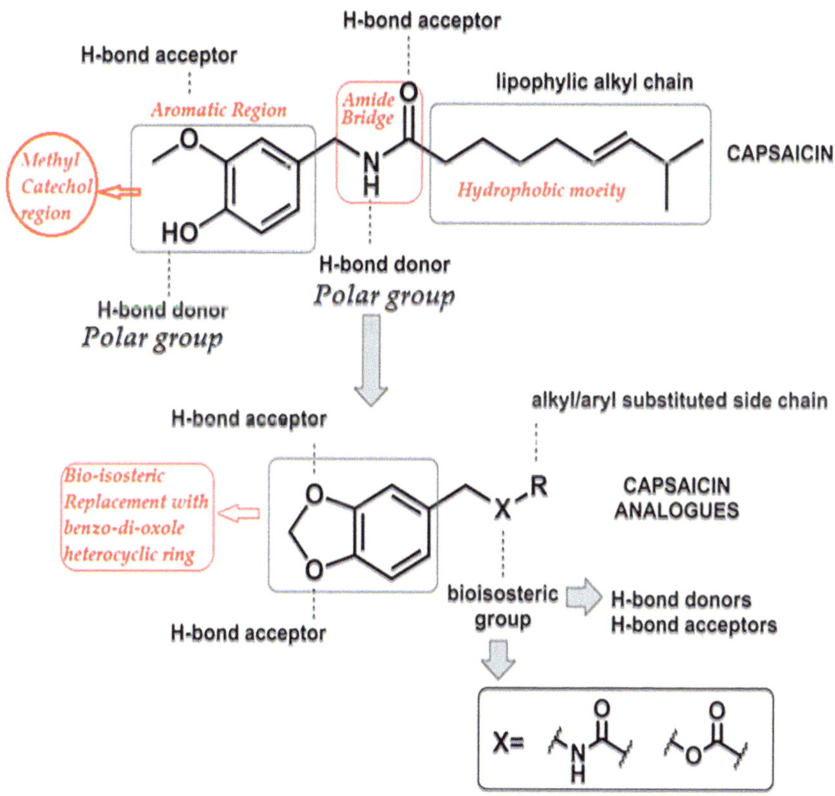

Fig. 16.1 Structural features of capsaicin analogues

16.1.2 Structural Features and Functional Groups: Aromatic Region

An in vitro study found that the most powerful analogues are those that replace the 'parent' 3-methoxy-4-hydroxy in the aromatic ring (A-region) of the capsaicin molecule with a benzyl group (Kosuge and Furuta 1970). Replacing the second, fifth and sixth positions on Ring-A by rather themselves or in any substitute led to ineffective substances. When the 4-OH substituent is removed or alkylated, the activity decreases or ceases (Rosa et al. 2009). It is interesting to note that the nitro substituent still works in other forms where the 4-substituent has been removed or reduced, like SH or NH_2. Also examined was the influence of Iodine-position in the Ring-A-region (9, 10, and 11) on the antioxidant activity of capsaicinoids, with nonivamide >5-I > 6-I > 2-I. Generally, adding an iodine atom to non-vanillyl ivamide's moiety decreases its antioxidant activity. 5th-Iodine retains most of ivamide's non-ability to protect cholesterol and linoleic fatty acids against radicalisation due to iodine's aromatic ring position.

3-&-4-Substitutions at A-ring positions are important for agonistic property. The capsaicin 4-OH moiety is pivotal; due to the phenol group's ability to give and take H-bonds; is important for agonist activity. Some studies go into a lot of detail about the SAR of C- moiety in capsaicin analogues. High efficacy requires the presence of a lipophilic-group, such as an 8-carbon chain or a substituted benzyl group (Walpole et al. 1993a, b, c). Small hydrophobic groups could best replace these aralkyl groups at the para position. According to Barbero et al. (2010), the potency of capsaicinoids is maximum between 8 and 9 carbon atoms. Geometric isomers of capsaicin are existed due to restricted rotation by C=C bond. Trans-capsaicin is prevalent because the -CH-$(CH_3)_2$ and alkyl side chain are across to each other with reference to the double bond are near-by but the same in the cis-capsaicin, which makes them repel each other. Trans- isomer has no steric hindrance. Increased strain makes cis-isomer less stable than trans.

16.1.3 Bioisosteres Analogues

Bioisosterism is used to reconfigure lead compounds to improve potency, selectivity, pharmacokinetic properties, and side effects (Zhou et al. 2009). Numerous clinically recognised pharmaceuticals and other biomolecules utilise the amide functional group for modification. These compounds were made by changing the structure of capsaicin, which was split into tri-segments: region-A with the methyl catechol, region-B with amide link, and region-C has alkyl side chain (Choi and Yoon 1999).

The process of cyclisation, which turned the methyl catechol group into the 1,3-benzodioxole bicyclic motif, was studied in area A. This switch may be more advantageous because the benzodioxol nucleus is preferred for molecules with anti-tumour effects. In the first set of analogs, the amide-linking bond in the B-region was either held or bioisosterically replaced with an ester. Various alkyls- and aryl-branches were inserted to modify alkyl-chain in the C-moiety to keep it hydrophobic.

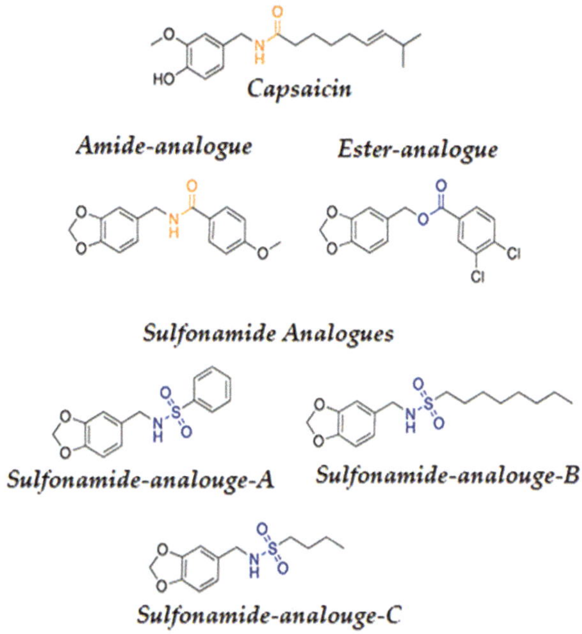

Fig. 16.2 Chemical structures of bioisosteres of capsaicin derivatives

To maintain the hydrophobic properties of the alkyl-chain in the C-moiety, the amide bridge in moiety-B was retained (or) bioisosterically substituted by an ester in the first analogue series. Several amides-and/or-ester analogues were as lethal to Mus musculus cutaneous melanoma cells (B16F10) and Anderson metastatic breast cancer cells as capsaicin, as depicted in Fig. 16.2 (MDA-MB-231). The sulphonamide group was used as a bioisosteric alternative to the amide and ester linkers to make sulphonamide analogues A, B, and C. This is because it is very resistant to chemical hydrolysis and other physiochemical processes that break it down.

The amide functional group is used in a lot of well-known medicines and other biomolecules. A 1,2,3-triazole group was looked at as a bioisosteric factor in a study that tried to find a way to replace an amide bridge. Capsaicin and other capsaicinoids were compared to synthetic versions of vanilloid and cannabinoid (CB) profiles. To sum up, the bioisostere derivative was altered by adding a 1,2,3-triazole ring to its pharmacophore (Fig. 16.3). This gave it strong selective antagonist activity against TRPV1 and cannabinoid receptor 1.

16.1.4 Pungent Analogs of Capsaicin

Capsaicinoid compounds have an amide-link connecting an acyl chain to a vanillyl moiety are responsible for their spicy flavour (Srinivasan 2016). Capsaicin and dihydrocapsaicin are the most pungent capsaicinoid compounds. They are followed

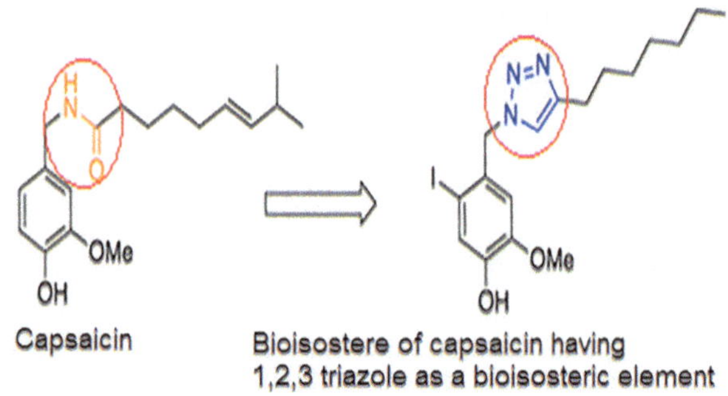

Fig. 16.3 Bioisosterism in capasiacin molecule

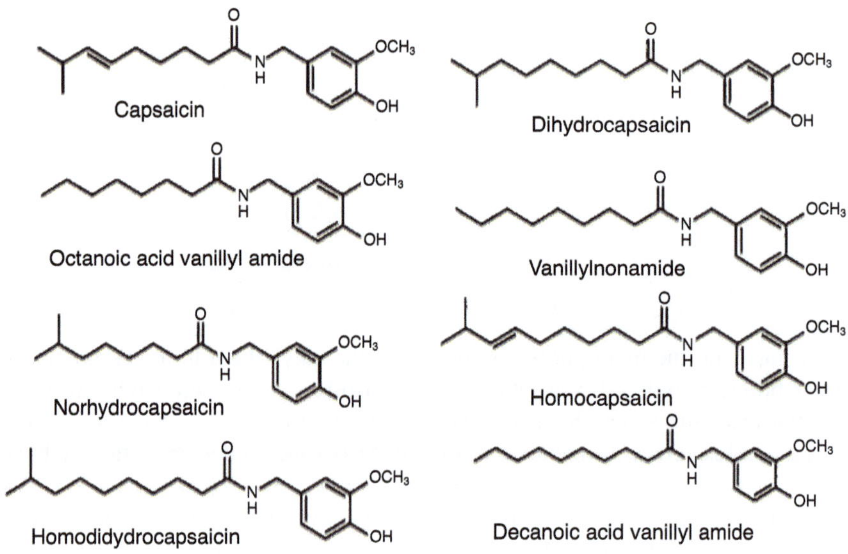

Fig. 16.4 Pungent analogues of capsaicin derived compounds

by Octanoic-acid vanillyl-amide, vanillylnonamide, nor-hydrocapsaicin, and homodihydrocapsaicin. Saturation of the acyl group distinguishes capsaicin from dihydrocapsaicin. In capsaicin derivatives, the substitution of the vanillyl portion and the acyl-chain-length will determine their pungent nature. C8-C11 acyl chain was pungent, but C-6-C-12 was not (Menichini et al. 2009). However, the spiciness can also be altered by modifying the vanillyl ring. For instance, benzylamides and 3-methoxy benzylamides lacked any pungent properties.

Research on the structure-activity relationship (SAR) between acyl chain length and capsaicinoids' pungency reveals that a nine-carbon chain is the most pungent. As with nordihydrocapsaicin, homocapsaicin, and homodihydrocapsaicin (Fig. 16.4), losing or gaining a single carbon atom reduces the heat by about half (Appendino et al. 2002). Capsaicinoids with an acyl chain with more-(or)-less than

C-9 are low-pungent, whereas those with a chain length of C18 or longer do not exhibit this characteristic.

16.1.5 Non-Pungent Capsaicin Derivatives

Capsaicin has pharmacological properties, as demonstrated by numerous studies; however, its pungent taste limits its application. As a result, the mild CH-19 sweet pepper, which is prime source for natural capsinoids, is a good substitute. On the other hand, capsinoids like capsiate, dihydrocapsiate, and nor-dihydro capsiate (Fig. 16.5) are not spicy-components. Capsinoids are chemically similar to capsaicin, which is a highly concentrated component of sweet peppers called 'CH-19 Sweet' (Ohnuki K et al. 2001). However, instead of an amide bond between the vanillyl groups, capsinoids have an ester link. Capsiate is a carboxylic ester that is made by combining the COO− group of (6E)-8-methylnon-6-enoic acid with the OH-benzyl moiety of vanillyl alcohol. This substance consists of phenol, monomethoxybenzene, and a carboxylic ester (Barbero et al. 2010). The structure of the fatty acid chain and capsaicin is similar like that of vanillyl alcohol. It is a replacement for capsaicin that is biologically the same and has no pungency (Torsten and Einar 2006).

The body temperature and oxygen utility were enhanced by giving a single-dose of CH-19-Sweet pepper while regular consumption reduces body weight and promotes metabolism. Like capsaicin, capsiate makes the body use more energy and raises its temperature. It is a plant metabolite, an angiogenesis inhibitor, an antioxidant, an anti-allergy agent, an anti-inflammatory agent, and an agonist for the capsaicin receptor. Along with capsaicin and similar PUFAs, non-spicy n-3

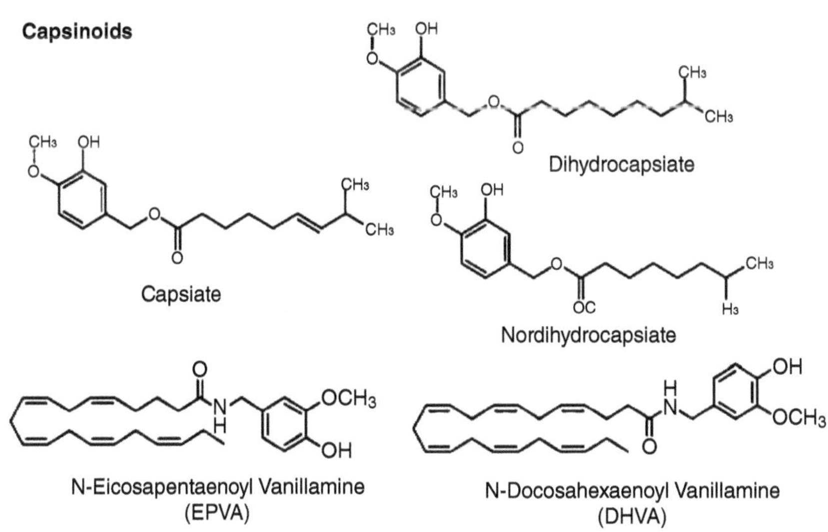

Fig. 16.5 Ester derivative of capsaicin derivatives

polyunsaturated fatty acid (PUFA)-derived analogues of capsaicin were studied for their antioxidant and carbohydrate-hydrolysing enzyme-blocking properties. Researchers found that the activities of capsaicin, EPVA, and DHVA were all similar. In addition, the novel compounds are spice-free, which facilitates their application.

16.1.6 Pungency and Structural Features

The placement of -OH and -OCH_3 in the mono-methoxy-benzene was one of the most important factors that affected the correlations between structure and level of spiciness. Modifying either of these two features will definitely change the pungent aroma. This conclusion was derived from the observation that the pungency degree of N-vanillyl nonanamide was 1,550,000 SHU (Scoville Heat Units, SHU), whereas the level of spiciness of N-(3-methoxy benzyl)-nonanamide came down to 600,000 SHU owing to removal of -OH- in the benzene ring and N-(4-hydroxybenzyl) nonanamid's pungency. Consequently, the -OH and -OCH3 bonds had a substantial effect on the level of pungency. The level of spiciness was also affected by acyl chain-length on right side of the amide plane. Dihydrocapsaicin had a pungency level of 15,000,000 SHU, while nor-dihydrocapsaicin only had a level of 9,100,000 SHU.

With a single methylene (CH_2) change, the pungency level decreased by over 5,400,000 SHU number. Furthermore, the impact of the branched-CH_3 group on the pungency level was unlikely to be overlooked. Both dihydrocapsaicin and capsaicin had 15,000,000 SHU, making them the two substances with the highest levels of pungency. N-vanillylnonanamide) was only one-tenth as pungent as dihydrocapsaicin owing to different branching on methyl at 8-C.

Moreover, the effect of the acyl-chain unsaturation on the level-of-spiciness should be carefully considered. Capsaicin and dihydrocapsaicin had the same level of heat, which was 15,000,000 SHU. The only difference between them was a double bond at 6-C which is the acyl-chain unsaturation. The level of pungency is directly related to the location of acyl-double bond. Suppose that, if its location changed from C-6 to C-4 (N-(4-OH--3-CH_3O-benzyl) -8-methylene-trans-4-enamide), the pungency level reduced from 15,000,000 SHU to 12,000,000 SHU. Lastly, the methylene that is between the acyl chain and the benzene ring deserves special attention. If the number went from 1 (N-vanillyl nonanamide) to (N-[2-(4-OH--3-CH_3O-phenyl) ethyl] nonanamide), the level of spiciness would drop by 58%, from 1,550,000 SHU to 650,000 SHU.

16.1.7 Esters and Amides of Capsaicin Derivatives

Xuan Wang et al. have made six capsaicin derivatives with phenol groups using the Freidel–Crafts (F–C) alkylation reaction. Their antioxidant and antibacterial activities were measured and analysed. The antioxidant and antimicrobial properties were disclosed by all six capsaicin derivatives, which share similar structural elements such as the phenolic hydroxyl, benzene ring, and amide groups. Capsaicin amide derivative, which has two phenolic-OH-groups, had the best reducing power, scavenging effect, and antimicrobial effect. The compound which has a R-S-R (thioether) moiety had the best-reducing power, scavenging effect, and antimicrobial effect. Such a study of effective capsaicin derivatives paves the way for future food and medicine uses.

16.1.8 Biosynthesis of Capsaicin

From valine (Val) and phenylalanine, the placentas of hot *Capsicum* genotypes produce capsaicinoids (CAP) (Phe). Habanero pepper placentas can generate the essential amino acids required for CAP biosynthesis by processing inorganic nitrogen into amino acids (Bennet and Kirby 1968). Plants make capsaicin through two processes: phenylpropanoid, and fatty-acid biosynthetic pathway, which controls the fatty acids in the molecule. The condensation reaction catalysed by capsaicin synthase generates capsaicinoids, as depicted in Fig. 16.3. Capsaicin synthase merges vanillylamine and branched fatty acid residues from two diverging pathways. The phenylpropanoid pathway takes phenylalanine and turns it into vanillylamine and its building blocks. Leucine or valine are deaminated and lengthened by fatty acid synthase using acetyl-CoA. Thioesterase (FAT) controls the length of the chain and also has a role in elongation cycles. This leads to capsaicinoids with an even number of branching carbons in the acyl moieties (Castillo et al. 2007). Capsaicinoids get a distinct trans-structure, which must be desaturated before the thio-esterase FAT will release the free fatty acids (Aza-Gonzalez et al. 2011).

Like capsaicinoid, capsinoid is made from valine and phenylalanine (Leete and Louden 1968). Phenyl-acrylic acid, 4-hydroxy-cinnmic acid, 3,4-dihydoxy-Cinnamic acid, and 3-methoxy,4-dihydoxy-trans-Cinnamic acid from the phenylpropanoid pathway are involved in capsinoid synthesis. In addition, the metabolic pathway between capsinoid and capsaicinoid includes vanillin, a key precursor. Capsaicin content reaches its maximum level after 40 to 50 days of fruit development, and thereafter prefers to breakdown into secondary compounds due to peroxidase activity (Reyes-Duarte et al. 2002). Researchers who were looking for ways to increase or improve the production of spicy chemicals found that a lack of water affects the phenylpropanoid pathway and makes capsaicinoid levels raise. Hydric-stress elevates capsaicin content by increasing the enzymatic functions involved in its production, phenylalanine ammonia-lyase (PAL), cinnamic acid-4-hydroxylase (C4H), and capsaicin synthase (CS) (Appendino et al. 2003).

16.2 Clinical Pharmacology of Capsaicinoids and Related Compounds

In the past decade, much study has been done to determine how capsaicin works. Capsaicin binds to the vanilloid receptor, commonly known as TRPV1 or transient receptor potential vanilloid 1, which is mostly evinces in sensory neurons. TRPV1 has a Mol.wt of 95 kDa, 838 AA, and six transmembrane domains. In between 5th- and 6th- covering proteins, it has very small pore-developing space. It has been identified in both humans and rodents. The capsaicin burn is prompted by a heat-sensitive TRPV1 protein part. Capsaicin couples with TRPV1 and increases intra-cellular Ca^{2+} levels producing neuropeptides like substance-P and calcium gene-related peptide (CGRP). When capsaicin comes in contact with sensory neurons, it causes pain, inflammation, and a localised feeling of heat. Owing to desensitisation of sensory neurons caused by depletion of substance P, it produces an analgesic effect when applied topically to the skin. This mechanism is used to develop new synthetic TRPV1 ligands in structure-activity studies.

16.2.1 Role of Capsaicin in CVS-Disorders

Capsaicin-sensitive sensory neurons are numerous in the cardiovascular system and affect heart function by releasing neurotransmitters including CGRP and Substance P (Yang et al. 2019). According to many studies by different researchers, there is evidence that capsaicinoids may be good for the circulatory system and that CGRP protects cardiovascular function. In addition to having positive effects on the cardiovascular system, capsaicinoids have antioxidant properties. LDL oxidation is thought to cause atherosclerosis. In vitro studies show that capsaicin, dihydrocapsaicin, and chilli extracts can make LDLs more resistant to β-oxidation either by lagging the start of and/or slows the rate of reaction. In high-fat-fed rats, capsaicin treatment decreased lipid peroxidation and blood total cholesterol (Renaud et al. 1995). The resistance of blood lipoproteins to oxidation is found to be higher in both adult men and women who regularly eat chilli for 4 weeks. According to study, capsaicin may improve cardiovascular function by activating TRPV1 to produce CGRP. In fact, capsaicin can mimic the way ischemia preconditioning protects the heart and capsaicinoids may help treat circulatory problems like atherosclerosis and coronary disease (Hachiya et al. 2007). Experts believe that capsinoids may be better for the heart than capsaicinoids because they are not as spicy but have similar qualities. More research should be done to find a solution to this problem. Angiotensin-converting enzyme (ACE) inhibition, long-term activation of TRPV1, release of N=O- and calcitonin gene-dependent peptide from endothelial cells, and inhibition of L-Class-Ca^{2+} channels in smooth muscle cells are all possibilities that nutritional capsaicin lowers blood pressure.

16.2.2 Effect of Capsaicin-Analogues as Anti-Cancer Compounds

Chemoprevention slows or stops tumour growth before tumour cells invade healthy cells. Capsaicin has shown promise as a cancer-prevention drug. Capsaicin may reduce lung carcinogenesis in a mouse model by preventing benzo(a)pyrene from binding to DNA. Studies show that capsaicin protects against chemical carcinogens like N-nitroso dimethylamine, vinyl carbamate, and 4-(methyl nitrosamino)-1-(3-pyridyl)-1-butanone (Clark and Lee 2016). Capsaicin activates 2+/calmodulin (CaM)-dependent protein kinase (CaMK) and CCAAT/enhancer-binding protein (C/EBP), which limits CYP1A1 mRNA. Capsaicin can prevent cancer by stopping the production of the CYP enzyme and its upstream modulator.

Capsaicin fights cancer in two ways: directly by getting rid of free radicals and indirectly by creating more antioxidant enzymes. Pure capsaicin's antiradical activity was very good at eliminating of the 2,2′-and-bis-(3-ethyl-benzo-thiazoline-6-sulphonic acid (ABTS) radical, with an IC50 of 187.69 M. (Zhang et al. 2008). Similarly, there was a correlation between the redox potential of other species belonging to the *Capsicum* genre and the concentrations of capsaicin and its analogs. Also, it was found that capsaicin can stop Fe2+- and autooxidation from oxidising linoleic acid. This substance's ability to scavenge free radicals suggests it may reduce oxidative stress, which is related to cellular malfunction and cancer.

Researchers used a variety of model systems to study how capsaicin affects angiogenesis. Capsaicin stopped angiogenesis mediated by VEGF in Matrigel matrix, in-vivo Matrigel plug tests, ex-vivo rat aortic ring models, and chicken chorioallantoic membrane (CAM) models. The effects of VEGF on endothelial cell invasion and growth were greatly reduced by capsaicin (Lu et al. 2010). Non-pungent capsinoids, capsiate, and dihydrocapsiate produced similar results in animal models and cell cultures. Capsiate and dihydrocapsiate blocked VEGF-initiated cell-to-cell and endothelial permeability (Mateos et al. 2013). Both capsiate and dihydrocapsiate have anti-angiogenic effects independent of TRPV1. Evodiamine, a capsaicin like substance, activates TRPV1 and induces angiogenesis in the Matrigel Plug model.

Auto-dock research reveals that capsaicin and capsiate can coupled with Src-kinase near the ATP site. Another mechanism for capsaicin's anti-tumour potential is its interaction with AMP-dependent protein kinase (AMPK). Capsaicin suppresses lipid production by activating AMPK and inhibiting acetyl-CoA carboxylase (ACC) in HT29 human-colorectal adenocarcinoma cells. The apoptosis caused by capsaicin analogues was linked to the strategies listed above. According to studies, capsaicin's anti-cancer activity targets the p53 gene. Capsaicin increases p53 acetylation by sirtuin-1 downregulation and phosphorylates site Ser-15 to induce apoptosis (Thoennissen et al. 2010).

Capsaicin has synergistic anti-cancer effects with other substances. Capsaicin increased nitric oxide in a p53-dependent manner when combined with resveratrol. TRPV1 activation in bladder cancer synergises with pirarubicin. Capsaicin and genistein work synergistically to control AMPK and cyclooxygenase-2 in breast cancer cells. Both capsaicin and 3,3′-methylene-bis-1H-indole, an endogenous intermediate of 3-indole-methyl alcohol prevalent in cruciferous herbs, induce cell death in colorectal cancer cells, as shown by investigations (Gunthorpe and Szallasi 2008). Brassin, a cruciferous vegetable indole, and capsaicin inhibited prostate carcinoma cell invasion, migration, and enzymatic activities.

Most studies focus on capsaicin as an anti-proliferative substance, but clinical development is difficult due to side effects. Capsaicin causes acute gut discomfort, hyperalgesia, stomach cramps, and nausea in humans. These annoying adverse properties have led patients to stop using oral capsaicin, according to clinical research. This limitation can be overcome by discovering capsaicin analogues, which preserve capsaicin's anti-tumour efficacy but do not cause burning sensation.

In human SCLC cell lines, Arvanil® and Olvanil® were more effective than capsaicin at preventing invasion. Arvanil® and Olvanil® anti-invasive property was mediated by the AMPK pathway and is independent of TRPV1. Lee et al. (2017) tested the anti-invasive properties of a capsazepine using prostate tumour cells. They found that capsazepine blocked JAK/STAT3 pathway, which ceased DU145 human prostate cancer cells from spreading. Capsaicin and its related compounds may be useful in future treatments if we learn more about how they combat cancer. This will also help us understand how they work to treat and prevent cancer.

16.2.3 Capsaicin-Analogues as Potential Antioxidants

C. annuum, C. chinense, and *C. frutescens* possess antioxidant secondary metabolites. Carotenoids, capsaicinoids, flavonoids, quercetin, and luteolin are some examples. Capsiate, dihydrocapsiate, and similar chemicals have been shown to have a great deal of antioxidant activity. The antioxidant ability of capsaicinoids aids in cardiovascular disease management. Consuming *C. annuum* for 4 weeks steadily increases oxidation resistance in adult men and women. Redox potential of tetra-terpenoids isolated from *C. annuum* was studied. All capsinoids and their analogues are effective chain-breaking antioxidants in scavenging lipid peroxyl radicals at non-cytotoxic concentrations for vanillyl nonanoate (Murakami et al. 2001). This means that these molecules primarily donate hydrogen atoms. The simple capsinoids mimic vanillyl nonanoate, which we discovered in recent research, was able to boost the activity of superoxide dismutase in an ethanol-induced mucosal lesion in a rat's model. This result was only partially reversed by capsazepine, the TRPV1 antagonist, which implies that capsinoids may directly affect antioxidant enzyme activity, which may partially describe how capsinoids are an antioxidant. Overall, capsiate and dihydrocapsiate have similar antioxidant effects to capsaicinoids (capsaicin and dihydrocapsaicin). Capsiate and dihydrocapsiate are promising antioxidants because they lack capsaicinoids' pungency (Ochi et al. 2003).

16.2.4 Pharmacological Action of Capsaicin Derivatives on Obesity

Several studies show that capsaicin lowers lipid levels by modulating lipid metabolism. This molecule can treat diabetes, obesity, and other diseases (Ludy et al. 2012). Gut tissue changes can disrupt metabolic balance. The intestinal microbiota has an important effect on metabolism. Several studies have found that the gut microbiota directly affects or controls energy metabolism (Hsu and Yen 2007). Studies have shown a clear link between gut microbiota and the onset of obesity. Studies link obesity and metabolic syndrome to abnormal gut microbiota (Magied et al. 2014). The gut microbiota of healthy people is more diverse than those of obese people.

Obese people have less diverse gut microbiota, and fast-growing harmful bacteria cause dysbacteriosis. When mice are fed a high-fat diet, capsaicin lowers their body fat and diversifies their gut microbiota. Capsaicin may reverse gut microbiota dysbiosis and reduce lipid levels by killing harmful bacteria. Researchers Wang et al. found in 2009 that mice fed a high-fat diet had a lot more *Bacteroides*, *Coprococcus, Prevotella,* and *Akkermansia* bacteria in their digestive tracts (Panchal et al. 2018).

16.2.5 Effect of Capsaicin in Metabolic Syndrome

Metabolic syndrome is mainly composed of obesity, insulin resistance, high blood pressure, poor glucose tolerance, and abnormal cholesterol levels. These changes in the body render heart ailments, NIDDM, and non-alcoholic fatty liver disease (NAFLD) more likely to happen. Damage from oxidation and low-level inflammation that persists for a long time are major causes of several number of metabolic problems. Adipose tissue inflammation in obese people creates reactive oxygen species. This adipokine instability can lead to metabolic syndrome diseases. Capsaicin makes insulin and key molecules that send messages work better. It also minimises on factors that cause inflammation, which helps keep glucose levels stable. In C2C12 cells, capsaicin makes glucose uptake go up without IRS-1 or AKT phosphorylation, which are insulin signalling molecules.

16.2.6 Effect of Capsaicinoids on Neuropathic Pain and Dermatological Disorders

Inappropriate care may lead to neuropathic pain and itch, which are common dermatologist visits. Capsaicin may help those with neuropathic pain and itching who do not respond to analgesics, antidepressants, antiepileptics, and other sedatives. Although capsaicin has been used therapeutically for many years, there is little evidence of its use as a topical analgesic for neuropathic pain or itching. As early as 1850, it was reported in the Dublin Free Press that capsaicin relieved dental discomfort (Lotti et al. 1994).

Post-Herpetic Neuralgia Qutenza® is a pain patch containing loading dose of capsaicin, an extract from chilli peppers. The patch is applied by qualified professionals after numbing the affected area using medication. The treatment takes at least 2 h since you must be monitored after receiving the high-concentration patch, yet a single application can reduce pain for up to 3 months. Successful applications can be resubmitted every 3 months.

Notalgia Paresthetica Many people with Notalgia Paresthetica are adults. The most common sign is a feeling of itching, burning, or tingling that goes up the spine and down the inside of the shoulder blade. Capsaicin cream can be used three times a day for 3–6 weeks after being used five times a day for 1 week. At first, the application makes the skin burn, but this should go away over time. A recent study suggests that an 8% capsaicin patch may be helpful (Goodless and Eaglstein 1993).

Brachioradial Pruritus Solar (brachioradialis) pruritus is a confusing condition that affects the sides of the arm(s). It is more common in Caucasians who live in tropical or subtropical climates. An 8% capsaicin patch once was used as a single treatment for brachioradial pruritus (Knight and Hayashi 1994). To confirm these results, more research needs to be done. A high-concentration capsaicin patch gave fast and long-lasting relief from itch and pain. Patients consistently said they felt much better the day after therapy. This promotes an 8% capsaicin patch as a viable, albeit expensive, new BRP treatment option (€340/patch).

Pruritus from Hemodialysis Itching is the most common and troubling skin problem for people on continuous hemodialysis (Ellis et al. 1993). Topical capsaicin 0.05% significantly reduced itching in hemodialysis patients compared to healthy controls. People who get chronic hemodialysis may feel less itchy if they take capsaicin before their treatment.

Psoriasis The cause of psoriasis and itching has been attributed to Substance P. Capsaicin, a substance-P depletion, was studied as a therapy for itching psoriasis. In a double-blind study, 44 people with psoriatic lesions on both sides of their bodies used capsaicin four times a day on one side and a placebo on the other side for 6 weeks. This study showed that putting capsaicin on itchy psoriasis helped. Several clinical studies have shown that capsaicin helps people with psoriasis. When applied to the skin, capsaicin is effective at treating psoriasis that causes itching, which suggests that substance P is involved. These two studies reported growing evidence that Substance-P plays a key role in the pathophysiology of (Lotti et al. 1994). Capsaicin is also an effective treatment for a number of conditions, such as alopecia areata, lipodermatosclerosis, and apocrine chromhidrosis. At higher doses, practically every patient develops cutaneous irritation, making compliance difficult. But this can be kept to a minimum with enough training and increasing the dose.

16.2.7 Effect of Capsaicin and Related Compounds as Pain-Relieving Agents

Capsaicin is applied to the skin to treat post-herpetic neuralgia. Pain can be treated with topical capsaicin products. Capsaicin is most often used as a topical pain reliever. It comes in different concentrations of cream, liquid, and patches. Qutenza®, a patch containing 8% capsaicin, has recently gained approval in the US and the EU. Patients with neuropathic pain got significant pain relief from a single 60-min treatment that lasted up to 12-weeks. Benefits of the capsaicin patch include reduced risk of adverse effects and drug interactions, more patient assent, and reduced pain intensity over a longer period of time. Substance P depletion is thought to be the way that topical capsaicin works (Derry et al. 2013). The capsaicin 8% patch is suggested for treating post-herpetic neuralgia. Capsaicin creams and solutions are available for temporary analgesia in cases of muscular and joint pain, as well as neuropathic pain (Iida et al. 2003). Capsaicin is a neuropeptide-activating drug that regulates the synthesis, storage, transport, and release of substance P, the principal active mediator of pain impulses from the PNS to the CNS (He et al. 2009). Moreover, substance P has been identified to be secreted into joint tissues, where it activates inflammatory intermediates implicated in the pathogenesis of rheumatoid arthritis.

16.2.8 Emphasis on Novel Capsaicin Derived Compounds

Capsaicin's naturally occurring cis isomer is called zucapsaicin (civamide®). Compared to capsaicin, zucapsaicin is much less spicy. This substance is thought to both block calcium channels in neurons and stimulate the vanilloid receptor VR1 (TRPV1). Zucapsaicin was also tested on guinea pigs for its efficacy in treating herpes simplex virus infections (Luo et al. 2011). It has been shown that zucapsaicin is better tolerated than the topical application of capsaicin. Importantly, this drug's topical application is better tolerated than its trans isomer in humans (capsaicin). The pharmacology of zucapsaicin is unknown.

The plant Euphorbia resinifera, which looks like a cactus, is used to get resiniferatoxin, which is a very powerful form of capsaicin. It stimulates the transient receptor vanilloid TRPV1 receptor as an agonist. When painful heat is applied to C fibres, the TRPV1 receptor transmits the heat. Resiniferatoxin-induced prolonged calcium influx kills only TRPV1-expressing sensory neuronal cell bodies (Caterina et al. 2000). Resiniferatoxin represents a potential target for analgesia (Holzer 2008).

Pelargonic acid vanillyl amide is a synthetic version of capsaicin. It is also called PAVA or nonivamide. Constant and Cordell (1996) found nonivamide in *Capsicum* annum peppers; however, most PAVA is synthesised rather than taken from plants. There are two commercial versions of PAVA named Captor I and Captor II. Captor I is dissolved in water and ethanol in the same amount, and it has 0.3% PAVA. Captor II is made mostly of propylene glycol, water, ethanol, and 0.3% PAVA.

Anesiva Inc. is developing Adlea® (ALRGX-4975), a parenteral form of capsaicin, a TRPV1-agonist, for the prospective management of osteoarthritis, tendonitis, post-surgical conditions, and neuropathic pain caused by nerve damage. Adlea® desensitises neurons that cause throbbing, persistent pain.

Adlea® might be an innovative method for treating chronic neuropathic pain. Tennis elbow is caused by overusing arm muscles and repetitive arm motion. ALGRX 4975 has the potential to relieve pain for weeks or months after just one injection. This trial will evaluate ALGRX 4975's effectiveness and safety in treating tennis elbow.

Capsiate is a non-spicy version of capsaicin that is found in C. annuum. It has several biological functions. Capsiate forms when (6E)-8-methylnon-6-enoic acid's carboxy group interacts with vanillyl alcohol's benzylic hydroxy group. Biochemically similar to capsaicin, but without heat. It is a plant metabolite, an agonist of the capsaicin receptor, hypo-glycemic, anti-allergic, antioxidant, and angiogenesis inhibitor. It contains monomethoxy benzene, phenol, and carboxylic ester. It prevents Src and VEGF-mediated tube development and multiplication in human umbilical vein endothelial cells at 5 to 25 M. (HUVECs).

Capsazepine is the first synthetic derivative and acts as a capsaicin receptor antagonist (Walpole et al. 1994). Since capsazepine was discovered as a TRPV1 antagonist, other compounds have been identified (Saper et al. 2002)—it shows anticancer and anti-proliferative effects in oral squamous cell carcinoma, NSCLC, breast cancer, and prostate cancer cell lines. Capsazepine improves the deteriorating pulmonary dynamics of endotoxemia and may be an effective treatment for acute lung injury (ALI).

Abbott's A-425619 is a very effective TRPV1 competitive-blocker at the capsaicin binding site. It has shown dose-dependent effectiveness in inflammatory models and neuropathic pain without affecting motor or memory/learning functions. AZD1386 is an orally administered TRPV1 transient receptor potential channel antagonist. It works as an inhibitor at the capsaicin-binding site of the channel (Trbovich and Yang 2015). AZD1386 was well tolerated, the study was done without early termination or major side effects, and there were no hyperthermic effects.

16.3 Conclusions

Capsaicin's pungency prevents it from being used in clinical trials, despite its medical importance. In response, non-pungent analogues were created. Despite having many potential medical uses, a lack of clinical trials prevents its widespread use. Capsaicin's pharmacological and therapeutic uses show its efficacy. Capsaicin has been studied extensively and may have numerous medical uses. In the coming years, novel pharmaceutical formulations, novel analogues, or targeting the capsaicin-activated receptor TRPV1 will be interesting pharmacological techniques. Future problems include determining the best dosage of capsaicin and how to administer it. Comprehensive SAR is needed to develop capsaicin analogues with fewer or no side effects (Chan et al. 2020). Furthermore, the on-going debate over topical vs. oral administration is productive. More regulated studies are needed to support the safety of capsaicin and its derivatives.

References

Appendino G, Harrison L, De Petrocellis N et al (2003) Halogenation of a capsaicin analogue leads to novel vanilloid TRPV1 receptor antagonists. Br J Pharmacol 139:1417–1424

Appendino G, Minassi A, Morello AS et al (2002) N-acylvanillamides: development of an expeditious synthesis and discovery of new acyl templates for powerful activation of the vanilloid receptor. J Med Chem 45:3739–3745

Aza-González C, Núñez-Palenius H, Ochoa-Alejo N (2011) Molecular biology of capsaicinoid biosynthesis in chili pepper (*Capsicum* spp.). Plant Cell Rep 30:695–706

Barbero GF, Molinillo JMG, Varela RM, Palma M, Macías FA, Barroso CG (2010) Application of Hansch's model to capsaicinoids and capsinoids: a study using the quantitative structure activity relationship. A novel method for the synthesis of capsinoids. J Agric Food Chem 58:3342–3349

Bennet DJ, Kirby GW (1968) Constitution and biosynthesis of capsaicin. J Chem Soc 4:442–446

Castillo E, López-González I, Regil-Hernández D et al (2007) Enzymatic synthesis of capsaicin analogs and their effect on the T-type Ca2+ channels. Biochem Biophys Res Commun 356:424–430

Caterina MJ, Leffler A, Malmberg AB et al (2000) Impaired nociception and pain sensation in mice lacking the capsaicin receptor. Science 288:306–313

Chan SH, Azlan A, Ismail A (2020) Capsaicin: current understanding in therapeutic effects, drug interaction, and bioavailability. Mal J of Med Health Sci 16:216–224

Choi HY, Yoon SH (1999) Bioisoster of capsaicin: synthesis of 1-hydroxy-2-pyridone analogue. Bull Kor Chem Soc 20:857–859

Clark R, Lee SH (2016) Anticancer properties of capsaicin against human cancer. Anticancer Res 36:837–843

Constant HL, Cordell GA (1996) Nonivamide, a constituent of *Capsicum* oleoresin. J Nat Prod 59: 425–426

Derry S, Sven-Rice A, Cole P et al (2013) Topical capsaicin (high concentration) for chronic neuropathic pain in adults. Cochrane Database Syst:1–52

Ellis CN, Berberian B, Sulica VI (1993) A double-blind evaluation of topical capsaicin in pruritic psoriasis. J Am Acad Dermatol 29:438–442

Goodless DR, Eaglstein WH (1993) Brachioradial pruritus: treatment with topical capsaicin. J Am Acad Dermatol 29:783–784

Gunthorpe MJ, Szallasi A (2008) Peripheral TRPV1 receptors as targets for drug development: new molecules and mechanisms. Curr Pharm Des 14:32–41

Hachiya S, Kawabata F, Ohnuki K et al (2007) Effects of CH-19 sweet, a non-pungent cultivar of red pepper, on sympathetic nervous activity, body temperature, heart rate, and blood pressure in humans. Biosci Biotechnol Biochem 71:671–676

Hayman M, Kam PCA (2008) Capsaicin: a review of its pharmacology and clinical applications. Curr Anaesth Crit Care 19:338–343

He GJ, Ye XL, Mou X (2009) Synthesis and antinociceptive activity of capsinoid derivatives. Eur J Med Chem 44:3345–3349

Holzer P (2008) The pharmacological challenge to tame the transient receptor potential vanilloid-1 (TRPV1) nocisensor. Br J Pharmacol 155:1145–1162

Hsu CL, Yen GC (2007) Effects of capsaicin on induction of apoptosis and inhibition of adipogenesis in 3T3-L1 cells. J Agric Food Chem 55:1730–1736

Iida T, Moriyama T, Kobata K et al (2003) TRPV1 activation and induction of nociceptive response by a non-pungent capsaicin-like compound, capsiate. Neuropharmacology 44:958–967

Knight TE, Hayashi T (1994) Solar (brachioradial) pruritus–response to capsaicin cream. Int J Dermatol 33:206–209
Kosuge S, Furuta M (1970) Study on pungent principle of *Capsicum* part chemical constitution of the pungent principle. Agric Boil Chem 34:248–256
Lee JH, Kim C, Baek SH et al (2017) Capsazepine inhibits JAK/STAT3 signaling, tumor growth, and cell survival in prostate cancer. Oncotarget 8:17700–17711
Leete E, Louden MCL (1968) Biosynthesis of capsaicin and dihydrocapsaicin in *Capsicum* frutescens. J Am Chem Soc 90:6837–6841
Lotti T, Teofoli P, Tsampau D (1994) Treatment of aquagenic pruritus with topical capsaicin cream. J Am Acad Dermatol 30:232–235
Lu HF, Chen YL, Yang JS (2010) Antitumor activity of capsaicin on human colon cancer cells in vitro and colo 205 tumor xenografts in vivo. J Agric Food Chem 58:12999–13005
Ludy M-J, Moore GE, Mattes RD (2012) The effects of capsaicin and capsiate on energy balance: critical review and meta- analyses of studies in humans. Chem Senses 37:103–121
Luo XJ, Peng J, Li YJ (2011) Recent advances in the study on capsaicinoids and capsinoids. Eur J Pharmacol 650:1–7
Magied MMA, Salama NAR, Ali MR (2014) Hypoglycemic and hypocholesterolemia effects of intragastric administration of dried red chili pepper (*capsicum* annum) in alloxan-induced diabetic male albino rats fed with high-fat-diet. J Food Nutr Res 2:850–856
Mateos RM, Jiménez A, Román P (2013) Antioxidant systems from pepper (*Capsicum* annuum L): Involvement in the response to temperature changes in ripe fruits. Int. J. Mol. Sci 14:9556–9580
Menichini F, Tundis R, Bonesi M et al (2009) The influence of fruit ripening on the phytochemical content and biological activity of *Capsicum* chinense Jacq. cv Habanero. Food Chem 114:553–560
Murakami K, Ito M, Htay HH et al (2001) Antioxidant effect of capsaicinoids on the metal-catalyzed lipid peroxidation. Biomed Res 22:15–17
Ochi T, Takaishi Y, Kogure K (2003) Antioxidant activity of a new capsaicin derivative from *Capsicum* annuum. J Nat Prod 66:1094–1096
Ohnuki K et al (2001) Administration of capsiate, a non-pungent capsaicin analog, promotes energy metabolism and suppresses body fat accumulation in mice. Biosci Biotechnol Biochem 65: 2735–2740
Panchal SK, Bliss E, Brown L (2018) Capsaicin in metabolic syndrome. Nutrients 10:630
Renaud S, de Lorgeril M, Delaye J (1995) Cretan Mediterranean diet for prevention of coronary heart disease. Am J Clin Nutr 61:1360S–1367S
Reyes-Duarte D, Castillo E, Martínez R (2002) Lipase-catalysed synthesis of olvanil in organic solvents. Biotechnol Lett 24:2057–2061
Rosa A, Appendino G, Melis MP et al (2009) Protective effect and relation structure-activity of nonivamide and iododerivatives in several models of lipid oxidation. Chemico-Bio Int 180: 183–192
Saper JR, Klapper J, Mathew NT et al (2002) Intranasal civamide for the treatment of episodic cluster headaches. Arch Neurol 59:990–994
Sarpras M, Gaur R, Sharma V et al (2016) Comparative analysis of fruit metabolites and pungency candidate genes expression between Bhut Jolokia and other *Capsicum* species. PLoS One 11:1–19
Srinivasan K (2016) Biological activities of red pepper (*Capsicum* annuum) and its pungent principle capsaicin: a review. Crit Rev Food Sci Nutr 56:1488–1500
Thoennissen NH, O'Kelly J, Lu D et al (2010) Capsaicin causes cell-cycle arrest and apoptosis in ER-positive and negative breast cancer cells by modulating the EGFR/HER-2 pathway. Oncogene 29:285–296
Torsten H, Einar B (2006) Capsaicin derivatives and their synthetic methods and applications. CN patent 1849068A

Trbovich M, Yang H (2015) Capsaicin 8% patch for central and peripheral neuropathic pain of persons with incomplete spinal cord injury: two case reports. Am J Phys Med Rehabil 94:e66–e72

Walpole CSJ, Bevan S, Bovermannm G et al (1994) The discovery of capsazepine, the first competitive antagonist of the sensory neuron excitants capsaicin and resiniferatoxin. J Med Chem 37:1942–1954

Walpole CSJ, Wrigglesworth R, Bevan S et al (1993a) Analogues of capsaicin with agonist activity as novel analgesic agents; structure-activity studies. 1. The aromatic “A-region”. J Med Chem 36:2362–2372

Walpole CSJ, Wrigglesworth R, Bevan S et al (1993b) Analogues of capsaicin with agonist activity as novel analgesic agents; structure-activity studies. 2. The amide bond B-region. J Med Chem 36:2373–2380

Walpole CSJ, Wrigglesworth R, Bevan S et al (1993c) Analogues of capsaicin with agonist activity as novel analgesic agents; structure-activity studies. 3. The hydrophobic side-chain C-region. J Med Chem 36:2381–2389

Wang BF, Yang YF, Shan WW et al (2009) Highly efficient synthesis of capsaicin analogues by condensation of vanillylamine and acyl chlorides in a biphase H2O/CHCl3 system. Tetrahedron 65:5409–5412

Yang S, Liu L, Meng L, Hu X (2019) Capsaicin is beneficial to hyperlipidemia, oxidative stress, endothelial dysfunction, and atherosclerosis in Guinea pigs fed on a high-fat diet. Chem Biol Interact 297:1–7

Zhang R, Humphreys I, Sahu RP et al (2008) In vitro and in vivo induction of apoptosis by capsaicin in pancreatic cancer cells is mediated through ROS generation and mitochondrial death pathway. Apoptosis 13:1465–1478

Zhou SZ, Zhou XF, Peng BX (2009) Synthetic methods of capsaicin analogues. Application No. 200910087

Role of Capsaicin in Cardiovascular Diseases 17

Anitha Nandagopal and Appala Raju Nemala

Abstract

Experimental evidence supports the use of dietary phytochemicals to lower the risk of cardio-related ailments. Phytochemicals found in fruits, leafy greens, whole grains, seasonings, and teas have been popular in the field of cardiovascular diseases management since from many years ago. Furthermore, phytochemicals are widely available, inexpensive, and generally non-toxic, which differentiate them from existing treatment methods that are indiscriminately harmful and result in numerous undesirable side effects as well as medication resistance in patients who already received therapy. Capsicum or red pepper plants are commonly used in foods all over the world for their poignant flavor, color, and to delay food deterioration. The primary spicy component of chili peppers is the dietary compound capsaicin. Researchers have found that capsaicin has pain-relieving, weight reducing, inflammation reducing, anti-diabetes, and antitumor properties. Capsaicin's scientific validity is restricted because of troublesome side effects; which results in substantial investigation on second-generation capsaicin analogs with greater therapeutic potential. Capsaicin can be used in atherosclerosis, cardiomyopathy, hypertension, and stroke. It also inhibits platelet aggregation via a TRPV1-dependent and independent mechanism. Capsaicin promotes platelet activation by augmenting ADP and thrombin-mediated calcium release from cellular platelet reserves. Capsaicin enhances ATP-binding cassette transporter A1 and lessens LDL-related protein 1, restricting atherosclerosis. Capsaicin-sensitive nerves are in the heart. By

A. Nandagopal (✉)
Department of Pharmacology, Sultan-ul-Uloom College of Pharmacy, Hyderabad, Telangana, India

A. R. Nemala
Department of Pharmacognosy, Sultan-ul-Uloom College of Pharmacy, Hyderabad, Telangana, India

M. K. Swamy, A. Kumar (eds.), *Capsaicinoids*,
https://doi.org/10.1007/978-981-99-7779-6_17

stimulating TRPV1 and substance P, these nerves govern the liberation of the peptide linked to calcitonin gene. Capsaicin causes vasodilation by releasing nitric oxide from vascular endothelial cells. It also acts as an antihypertensive agent by blocking the L-type calcium channels in vascular smooth muscles. This chapter highlights the multiple mechanisms of capsaicin in various cardiovascular diseases such as coronary heart disease, hypertension, cardiac hypertrophy, arrhythmias, stroke, and myocardial infarction.

Keywords

Cardiovascular diseases · Hypertension · Capsaicin · Stroke · Atherosclerosis · TRPV1 · Nitric oxide

17.1 Introduction

Coronary vascular disease (CVD) has been the main reason of morbidity-&-mortality in many industrialized and developing nations for decades (Menees and Bates 2010). The most typical underlying factor in many CVDs is atherosclerosis (AS). As lipids build up inside the arterial wall, plaque forms and the lumen finally becomes narrowed, causing AS. Increased serum cholesterol, especially LDL-C, is a major risk factor for AS (Libby et al. 2011). According to the INTERHEART study, major cardiovascular events occurred at higher rates in low-income nations despite having fewer risk factors than in high-income countries. The high-fat and high-salt diets of the West are well-known contributors to CVD risk (Hu 2008). There are many medications in clinical practice for morbid lipid profiles, but their effectiveness is limited and they are linked to negative effects (Wiklund et al. 2013).

Evidence have suggested that dietary phytochemicals lower the risk of cardiovascular disease. To cure the diseases linked to AS, several scientists are interested in using plant-derived chemicals. These include extracts from blueberry, grape seed, fenugreek, bitter melon, cinnamon, Russian tarragon, and other plants (Graf et al. 2010). Because of their redox-potential, anti-inflammatory, and apoptosis features, phytochemicals found in fruits, leafy greens, whole grains, seasonings, and teas have been popular in the field of cardiovascular disease for many years. Furthermore, phytochemicals are widely available, inexpensive, and generally non-toxic, which differentiate them from existing treatment methods that are indiscriminately harmful and result in numerous undesirable side effects as well as medication resistance in patients who already received therapy. The fundamental cure for mankind's illness is powerful biomolecules derived from natural items such as plants, animals, and minerals. *Capsicum* or red pepper plants were classified as members of the Plantae kingdom and the Solanaceae family. It is commonly used in foods all over the world for their poignant flavor, color, and to delay food deterioration. Chili peppers are the source of capsaicin. Studies have demonstrated the remarkable analgesic, anti-inflammatory, anticancer, and redox-potential properties of capsaicin (Jaganathan et al. 2020). Capsaicin's beneficial effects on the heart may be grouped into three categories: a decline in the heart rate, an increase in myocardial tissue blood circulation, and an elevated antioxidant potential of plasma lipoprotein. Capsaicin was already demonstrated to help with a variety of health problems including AS, metabolic syndrome, high-BP, and stroke. Modulating blood lipid levels, capsaicin may have cardioprotective effects. Additionally, capsaicin may inhibit the production of oxidized LDL particles and the subsequent buildup of these particles in foam cells.

This chapter highlights the multiple mechanisms of capsaicin in various cardiovascular diseases such as coronary heart disease, hypertension, cardiac hypertrophy, arrhythmias, stroke, and myocardial infarction.

17.2 Capsaicin Receptors and Their Second Messenger System

A naturally occurring, pungent phytotoxin, known as capsaicin is found in the fruits of plants from the genus *Capsicum*, which is part of the *Solanaceae* family. Foods containing capsaicin, notably chili peppers, have apparently been consumed by humans since 9000–7000 B.P. according to archaeological findings from Mexico (Govindarajan 1985; Kraft et al. 2014). After discovering America by European explorers in the sixteenth century, these hot fruits were first only known to those who lived in the Latin America. Thereafter, they spread throughout the world. Since ancient times, people have used chili peppers to taste, color, and preserve food, as well as for medicinal purposes. Chilies have since become an essential component of culinary traditions all around the world. Some people are intolerant of capsaicin's pungency due to its mouth and digestive side effects, but there are still numerous dedicated users of this South American plant. Chili peppers are becoming increasingly popular in food around the world (Tapsell et al. 2006).

The vanilloid subtype 1 transient receptor potential is the name of the receptor for capsaicin (TRPV1). This receptor belongs to Transient Receptor Potential (TRP) family. The TRPV1 ion channel is found in sensory neurons, and is sensitive to vanilloids and capsaicin (Caterina et al. 1997).

On structure similarity, TRP channels are classified into six subfamilies: TRP canonical (TRPC1–7), TRP vanilloid (TRPV1–6), TRP melastatin (TRPM1–8), TRP mucolipin (TRPML1–3), TRP ankyrin (TRPA1), and TRP polycystin (TRPP) (Nilius and Owsianik 2010). It is generally known that TRP channels are involved in the signaling pathways that transmit pain, taste, hearing, and visual information. An increasing number of studies report that TRP channels are involved in the control of cellular growth & survival, mineral assimilation, body fluid balance, gastrointestinal movement, and cardiac performance (Nilius et al. 2007). Much study has been conducted on the TRPV1 subfamily member. TRPV1 is vital in maintaining physiological homeostasis along with typical functions in the nervous system. Absorption of capsaicin is passive in stomach and git tract and bound to albumin in the blood (Kawada et al. 1984); hence, it enables the $TRPV_1$-channels and triggers physiological effects.

TRPV1 is a receptor for vanilloids such as capsaicin, acid with a pH of 6.5, and heat. Ca^{2+}, Na^{+}, and other monovalent cations can pass through TRPV1 channels. It

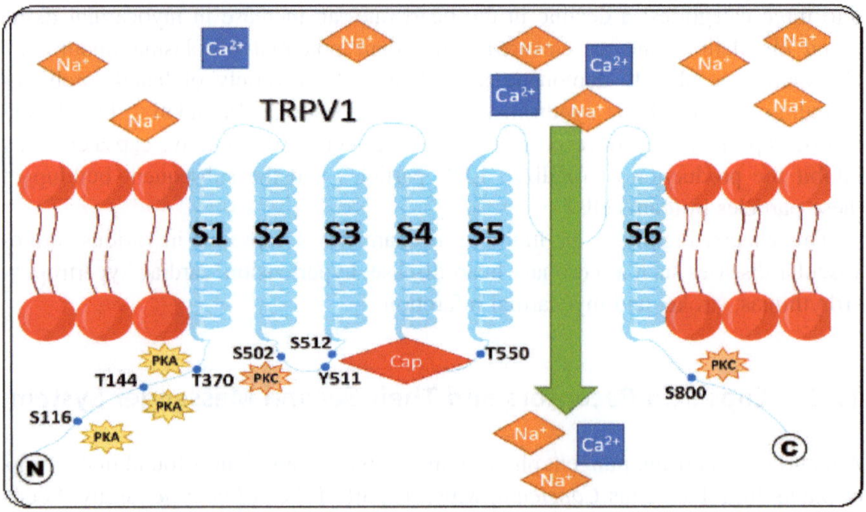

Fig. 17.1 Topology of TRPV1 in rats. TRPV1 is sensitized when PKA or PKC phosphorylates its intracellular N- and C-termini. TRPV1 sensitization increases cation upsurge through the channel when stimulated by heat, H+, or capsaicin, especially in contrast to nonresponsive form. Caps = capsaicin (Munjuluri et al. 2022)

is noteworthy that prolonged capsaicin-induced TRPV1 activation results in channel desensitization. TRPV1 is activated at 43 °C (Caterina et al. 2000). Numerous additional naturally occurring hydrophobic substances, such as 20-OH-eicosatetraenoic acid (20-HETE), N-arachidonoyl-ethanolamine (AEA), and phospholipids like LPA, can also activate TRPV1 channels in addition to heat, acid, and capsaicin. The TRPV1 protein, which mostly present in plasma membrane, can also be found in the ER and SR (Olah et al. 2001; Xin et al. 2005), which enables pathway for the efflux of intra-cellular Ca^{2+} ions.

According to Munjuluri et al. (2022), each TRPV1 subunit includes two lengthy intra-cellular N- and C-terminal and six trans-membrane domains which is shown in Fig. 17.1. It was discovered that the S3 and S5 trans-membrane moiety of the rat TRPV1 protein includes capsaicin binding site. Residues Y511, S512, and T550 were found to be crucial for capsaicin surface adherence with the rat TRPV1-protein. The temperature at which a channel is activated by heat is lowered when capsaicin binds to the TRPV1 protein. According to Mózsik (2014), TRPV1 has a significant role in pain perception, oxidative stress, and inflammation. It is essential for the regulation of metabolic balance and cardiovascular function, according to recent findings. According to Yang et al. (2010), stimulation of TPRV1 by its ligand like capsaicin encourages EDRF-mediated dilation of blood vessels, which in turn helps to reduce blood pressure.

17.3 Role of Capsaicin in Various Cardiovascular Diseases

Capsaicin is a secondary metabolite found in plants. It is an ingredient found in chili peppers of the Capsicum genus of plants and has been used for medicinal and culinary purposes (Sun et al. 2016). It is a dietary component with a distinctive impact on the pain sensory system. Capsaicin also has therapeutic promise in the management of a number of clinical disorders. Dietary phytochemicals have been proven to have health benefits, and the development of natural foods and phytonutrients has drawn increasing interest. However, one of the difficult paths in finding viable chemotherapeutic agents is to research and design a medicine with the fewest side effects.

17.3.1 Role of Capsaicin in Coronary Heart Disease

Atherosclerosis is marked by plaque build-up in large vascular bed. Atherosclerotic plaques contain oxidized LDL, cholesterol crystals, fat-depots, $CaPO_4$, macrophages, and foam cells. The artery lumen becomes more constrained when atherosclerotic plaque advances. A blood vessel may become entirely blocked by atherosclerotic plaque in extreme circumstances, causing an interruption in blood flow and possibly persistent ischemia. When the fibrous layer of plaque ruptures in the coronary or cerebral circulation, it is a life-threatening emergency that could lead to myocardial infarction or stroke. High cholesterol and dyslipidemia are main concerns for the onset and propagation of atherosclerosis. Endothelial malfunction and smooth muscle cell multiplication in vessels are also important phases in the pathology of atherosclerosis (Davignon and Ganz 2004; Tabas et al. 2015; Li et al. 2017). TRPV1 activation is reported in studies to monitor fat-metabolism, protect endothelial cells, restrict vascular tissue growth, and prevents inflammation and oxidative stress (Harris et al. 1991; Kunieda et al. 2005; Wang et al. 2008a, b; Zhang et al. 2020). Following capsaicin treatment, lipoprotein lipase activity was increased in fat-tissues (Tani et al. 2004). Stimulation of TRPV1 receptors in nerves plays a vital part in atherosclerosis decrease and provides major cardio protective role against ischemia injury (Szabados et al. 2020).

TRPV1 activation greatly decreases foam cell formation, according to Ma et al. (2011), by enhancing ATP-binding cassette transporter A1 precipitation and decreasing LDL-dependent-Protein-1 expression. Capsaicin forbids high-fat diet complications such as glucose intolerance and hypercholesterolemia. It yields the above effects by upregulating thermogenic genes that regulate metabolisms, such as UCP-1, BMP-8b, SIRT-1, PGC-1a, and PRDM-16 (Baskaran et al. 2017). Capsaicin can enhance sympathetic BAT thermogenesis and lower body fat (Joo et al. 2010).

Adiponectin is upregulated at the protein level by capsaicin, which also inhibits the appearance of leptin, CCAAT-enhancer-binding protein-α (C/EBP-α), peroxisome proliferator-activated receptor-γ (PPAR γ), and other genes. Adipocytes and 3 T3-L1 pre-adipocytes can therefore be efficiently brought to death, and adipogenesis can be inhibited in a lab setting. In reducing hyperlipidemia in turkey fed a diet high in cholesterol, dihydrocapsaicin was more efficient than capsaicin. Their results show that LDL oxidation pertains to atherosclerosis. Hence, researchers conclude that capsaicinoids inhibit serum lipid oxidation in a concentration-dependent fashion.

Harper et al. (2009) said that vanilloid channels are found in thrombocytes and that clustering of these platelets may cause cardiac injury in people with CAD. Scientists concluded that TRPV1 might be required for 5-HT and ADP release. They observed that TRPV1 may have an effect on human platelets by mediating the correlation among platelet activation during atherogenesis and pro-inflammatory cytokines like 12- and 15-hydroxyeicosatetraenoic acid. As a result, patients with atherosclerosis may experience negative effects due to increases in intracellular Ca^{2+} concentration of platelets caused by capsaicin, which is necessary for their degranulation.

According to Xiong et al. (2016), capsaicin in food may reduce endothelial dysfunction, which is a leading cause of atherosclerosis. ApoE −/− mice were used in the investigation. ApoE −/− mice fed a high-lipid meal showed reduced coronary vasodilation and a lower survival rate in an in vivo examination. ApoE −/− mice with genetically altered TRPV1 and UCP2 exhibited an aggravation of this trait. According to the research, capsaicin lowered mitochondrial reactive oxygen species production while increasing UCP-2 release in endothelium was dependent on protein kinase-A and TRPV1. Capsaicin-added high-fat diets gave the mice a better chance of surviving. So, supplementing with dietary capsaicin may be helpful for reducing the risk of CAD. Figure 17.2 shows the summary of role of capsaicin in coronary artery disease.

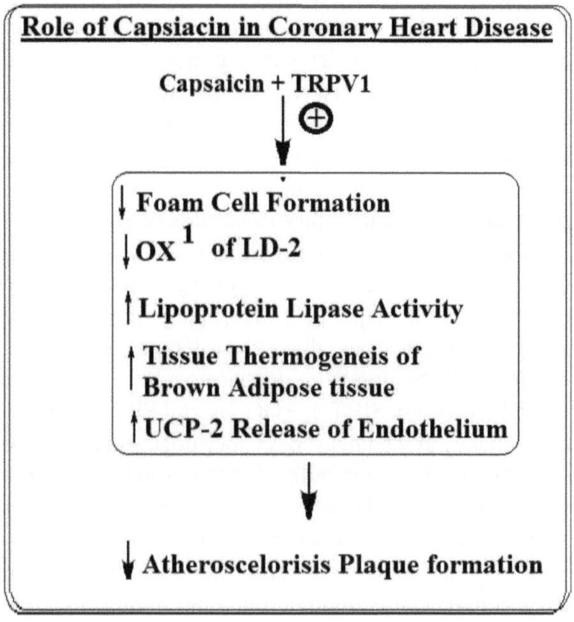

Fig. 17.2 Role of capsaicin in coronary artery disease

17.3.2 Role of Capsaicin in Hypertension

The force the heart exerts as it pumps blood throughout the body is measured as blood pressure. The amount and force of blood pumped, as well as the size and flexibility of the arteries, influence whether the blood pressure is normal or abnormal. The etiology of hypertension is defined as an imbalance between the vasoconstriction and vasodilatation properties of blood vessels. Vascular function and blood pressure regulation require intracellular Ca^{2+} equilibrium (Firth et al. 2007). An imbalance in Ca^{2+} levels in blood leads to increased vascular resistance (Liu et al. 2009; Thilo et al. 2009). Capsaicin showed its antihypertensive action by many mechanisms, including the release of vasodilator neuropeptides via TRPV1 activation (Jang et al. 2020), stimulation of natriuresis and diuresis (Zhang et al. 2015), angiotensin-converting enzyme (ACE) inhibitory activity (Jeon et al. 2010; Adefegha and Oboh 2013), and L-type Ca^{2+} channel (Sun et al. 2016; Shuba 2021). By preventing the COX-2 gene from being expressed, capsaicin also enhances endothelial function.

Vascular tone integrity, release of substance-P (SP), neurokinin A, and calcitonin gene-related protein (CGRP) may all play a role in the potential antihypertensive processes of capsaicin (Franco-Cereceda 1991; Sessa 2010). Aside from oxidation of fatty acids and glucose, other processes, such as brown adipose tissue (BAT) thermogenesis reduces fat. According to Li et al. (2014), stimulation of TRPVI by capsaicin reduces blood pressure by encouraging sodium excretion in the urine and prevents the beginning of stroke (Xu et al. 2011).

N=O is released from endothelial cells and CGRP is liberated from capsaicin responsive neurons when TRPV1 is activated by the substance. When capsaicin was given acutely, CGRP surged and BP dropped. Yang et al. (2010) found that dietary capsaicin enabled TRPV1, raising PKA and eNOS phosphorylation, and NO bio-distribution in endothelial cells. Capsaicin therapy reduced BP in genetically hypertensive rats by getting better endothelium-dependent relaxation. The hypotensive impact of prolonged capsaicin ingestion was also aided, if somewhat, by CGRP release.

Mode of vascular smooth muscle relaxation initiated by capsaicin in rat aorta was discovered by blockage of L-type Ca^{2+} channels (Rocha and Bendhack 2009). Dahl salt-resistant rats with TRPV1 activated were protected from salt's effects on blood pressure, although the TRPV1-expression and function in Dahl salt-sensitive lab rats were impaired, making them vulnerable to the consequences of salt load on BP control (Wang and Wang 2006). Regular intake of dietary capsaicin lessened vascular diseases and night time BP brought on by excessive salt consumption by limiting vascular oxidative stress in a TRPV1-mediated path (Hao et al. 2011) Mice fed a high-salt diet with capsaicin intake had much higher urinary sodium excretion than mice fed a high-salt diet alone. TRPV1-induced salt aversion low salt intake (Ruiz et al. 2006). Capsaicin's TRPV1 activation minimizes high blood pressure due to excessive salt intake and genetic origin. Figure 17.3 shows the summary of role of capsaicin in hypertension.

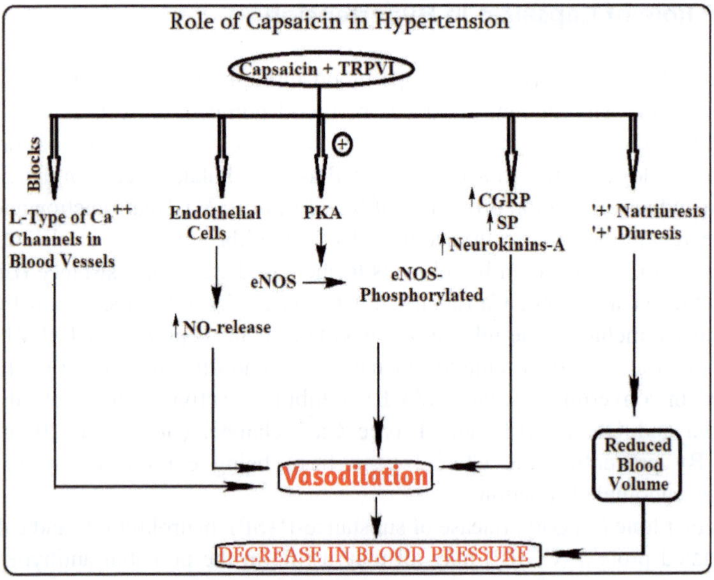

Fig. 17.3 Role of capsaicin in hypertension

17.3.3 Capsaicin and Cardiac Hypertrophy

Cardiac hypertrophy is a response to pressure overload, fluid pressure, and ischemic stroke, and is a risk factor for heart failure. Myocardial enlargement, cardiomyocyte apoptosis, inflammatory reactions, cardiac fibrosis, and ultimately ventricular failure result from overload pressure (Lorell and Carabello 2000). The development of cardiac hypertrophy and remodeling is heavily dependent on inflammatory cytokines such tumor necrosis factor (TNF), interleukin 6 (IL-6), and nuclear factor-B (NF-B) activation (Uozumi et al. 2001; Opie et al. 2006). Additionally, hypertrophic cardiomyopathy is also developed when TNF is overexpressed (Kubota et al. 1997). As a result, declining inflammation and cytokine release may benefit cardiac hypertrophy.

Diastolic stiffness, systolic dysfunction, and ventricular enlargement are all caused by the reactive vascular and granular fibrosis. As such, they contribute to the fibrosis of the heart, a trait of cardiac hypertrophy (Barasch et al. 2009; Ho et al. 2012; Lijnen et al. 2012). Myocardial fibrosis is triggered due to inflammation (Gupta et al. 2012; Doltra et al. 2013). Eventually, inflammatory cells migrate into the perivascular domain, where fibroblasts create collagen I and build perivascular fibrosis, and then into the interstitial zone, where fibrosis spreads between the cardiomyocytes and causes interstitial fibrosis (Nicoletti and Michel 1999).

An endogenous ligand called N-arachidonoyl-dopamine (NADA) or N-oleoyl-dopamine (OLDA) as well as protons, capsaicin, and lipid metabolites can all activate the TRPV1 receptor (Caterina et al. 1997; Huang et al. 2002; Chu et al. 2003). Immunosuppressive properties are found in capsaicin and endo-vanilloids such as NADA and OLDA. NADA inhibits NFAT, AP1, and NF-B mediated biochemical routes (Singh et al. 1996; Sancho et al. 2004). TRPV1 excitation diminishes TNF and IL-6 while raising IL-10 (Demirbilek et al. 2004). Therefore, TRPV1 may prevent heart remodeling and hypertrophy through its anti-inflammatory properties. TRPV1's anti-inflammatory properties have been linked to regulating vascular tone and cardiovascular function in a number of diseases, including hypertension, endotoxic shock, and myocardial injury (Clark et al. 2007; Wang 2008; Wang et al. 2008a, b; Miller et al. 2021).

SP and CGRP, which may mediate sensory neuropeptides, lead to vasodilation and be implicated with the regulation of inflammation, are released by TRPV1 to protect the heart (Levite et al. 1998; Okajima and Harada 2006). TRPV1 expression and substantial capsaicin-sensitive sensory nerve innervation in the myocardial and coronary systems suggest that TRPV1 may significantly affect cardiac reflexes, adaptive response, and heart function (Pan and Chen 2004). TRPV1 has been proven to be prophylactic against myocardial injury (Zhong and Wang 2007) and doxorubicin-induced cardio toxicity (Katona et al. 2004). Furthermore, cardiac sensory neurons have a profound influence on gene expression levels in the heart (Zvara et al. 2006).

Signs of cardiomegaly and fibrosis caused by excess salt consumption were an upsurge in heart weight, a decline in heart function, and more collagen accumulation. These changes were connected to decreased PPAR and UCP2 expression, increased iNOS synthesis, and elevated oxidative/nitrated tyrosine stress (Feng et al. 2014; Gao et al. 2014). Mice fed a high-salt diet or lacking TRPV1 exhibited decreased oxidative phosphorylation and mitochondrial complex I enzyme activity. Chronic capsaicin administration did indeed, in a TRPV1-dependent way, attenuate these negative consequences of long-term high-salt ingestion (Hongmei et al. 2014). Similarly, diets enriched with capsaicin reduced pressure overload- and angiotensin II-induced heart fibrosis and hypertrophy (Qiang et al. 2014).

Collagen I, collagen III, and fibronectin protein expression were used to determine how much more collagen was produced as a result of aortic banding (AB). Collagen and fibronectin were dramatically downregulated by dietary capsaicin. TGF-1 and CTGF expression, as well as Smad2/3 phosphorylation, were greatly enhanced in pressure-loaded cardiac hypertrophy and significantly decreased by capsaicin treatment. Angiotensin II-induced cardiac fibroblast growth was dramatically reduced by capsaicin. Thus, it is evident that capsaicin protects the cardiac function from pressure overload through the TRPV1 channel.

17.3.4 Capsaicin and Myocardial Infarction

A primary reason for mortality in humans is ischemic heart disease, defined by lack of blood flow to some areas of the myocardium, which results in tissue necrosis (infarction). As a result of factors such as diabetes, dyslipidemia, atherosclerosis, hypertension, and other conditions, ischemic heart disease can develop. Reperfusion has entered a new phase in the treatment of this ailment, allowing for a decrease in mortality. Reperfusion, on the other hand, could cause more trouble, such as the heart not being able to contract as well (shocking) and arrhythmias (Braunwald 1985). To increase myocardial function, reduce arrhythmia risk, minimize necrotic tissue, and delay necrosis during ischemia-reperfusion, cardioprotective therapies are required. The heart adjusts to ischemic stress by means of preconditioning and postconditioning (Murry et al. 1986).

An adaptive response known as "ischemic preconditioning" describes how a brief period of ischemia significantly improves the heart's capacity to resist a subsequent ischemic insult. According to Przyklenk and Kloner (1998) and Boengler et al. (2020) preconditioning gives exceptional cardio protection in several in vivo models. Ageing, hyperlipidemia, and diabetes may reduce the cardio-beneficial effects of ischemia preconditioning (Ferdinandy et al. 1998; Ferdinandy 2003; Ferdinandy et al. 2007). Preconditioning can be mediated by ischemia, hypoxia, fast electrical pacing, heat stress, or bacterial endotoxin. Preconditioning's cardio protective impact includes two separate phases. The early phase has a manifestation time of less than 3 h and begins shortly after the preconditioning stimulus. The late phase has a slower (20 h) onset and can last for up to 72 h (Baxter & Ferdinandy 2001). Preconditioning involves reducing the bulk of necrotic tissue (infarct size), restoring heart function, and lessening the frequency of arrhythmias after ischemia and reperfusion in both stages.

The investigation into the molecular mechanisms behind this extraordinary adaptability was sparked by the identification of this endogenous cardio protective mechanism. The precise cellular process of pre and postconditioning, however, is still unknown. Adenosine, bradykinin, NO, superoxide, peroxy-nitrite, CGRP, cGMP, protein kinases, norepinephrine, and K^{+}ATP are involved in ischemia-reperfusion injury and the formation of the cardio protective effect of preconditioning (Bolli et al. 1998; Schulz et al. 2001). It was hypothesized that cardiac sensory neurons, with their ability to sense chemical signals and emit NO and CGRP, are pivotal in process of cardiac stress adaptation.

Capsaicin, a highly selective sensory neurotoxin, specifically blocks the function of a group of 1^{0} sensory neurons and/or kills them off (Jancso 1968; Jancso et al. 1977). It serves as a critical probe for understanding the morphology, physiology, pathogenesis, and pharmacology of the sensory neural system (Holzer 1991; Jancso 1992). Capsaicin-sensitive cardiac C-fiber afferents include CGRP co-receptors, SP, neurokinin A, and NO. Capsaicin-sensitive sensory neurons promote heart performance and adaptive activity because of their NO and vasoactive peptide levels (Takaki et al. 1994; Sosunov et al. 1996; Yaoita et al. 1994).

Capsaicin-sensitive thin afferent fibers have nerve endpoints that are activated by cardiac ischemia due to a decrease in pH, spike in potassium ions and lactate levels, and discharge of bradykinin and adenosine (Bevan and Geppetti 1994). Because of the CGRP, NO, and SP concentration in these afferent C-fibers, they exhibit local efferent activities. NO and CGRP were discovered to be key mediators of preconditioning (Ferdinandy et al. 1996; Li et al. 2000). All afferent axons in the heart contain CGRP, as is widely known. Myocytes, vascular, and endocardial endothelium, and few cardiac neurons, are the heart's biological sources of NO.

Additionally, it appears that capsaicin-sensitive neurons control cardiac NO-cGMP transmission at the basal level. This implies that a substantial proportion of the overall amount of cardiac NO may come from capsaicin-sensitive afferent nerve end-points. Another finding that sensory ganglion, sensitive to capsaicin, produce NO synthase lends credence to this (NOS). This hypothesis is also strongly supported by Pabla and Curtis (1996) who suggested a neurological origin for cardiac NO efflux in rats. There is strong link between CGRP and NO in the rabbit coronary blood stream and a person's ductal artery, according to various studies (Mitchell et al. 1997; Raddino et al. 1997). According to a recent DNA-microarray study, capsaicin-sensitive sensory neurons control the rat heart's gene expression pattern, which includes the capsaicin receptor gene (Zvara et al. 2006).

CGRP is involved in neuro-inflammation, migraines, gestation, menopause and cardiac-arrest (Brain et al. 2002). Cardioprotective CGRP interacts with unknown G-protein-coupled receptors (Doggrell 2001).

CGRP receptors in heart and distal conducting system suggest that CGRP may affect the rate of impulse transmission and cardio-modulation by parasympathetic nervous system (Chang et al. 2001). When severe hypoxia and simulated reperfusion are present, cardiac cells are directly protected by CGRP. Changes in membrane fluidity were linked to CGRP's ability to protect cardiac cells. It is proven that CGRP protects against myocardial ischemia in humans (Kallner 1998). An in-situ "efferent" release of CGRP in myocardium leads to myocardial ischemia, which has as its clinical symptoms such as angina pectoris and infarction. The release of CGRP slows the progression of myocardial infarction and leads to coronary vasodilatation. Thus, CGRP may constitute an endogenous local cardiac protective molecule with intriguing therapeutic implications (Franco-Cereceda and Liska 2000). The CGRP efflux owing to ischemic precondition in cardiac surgery suggests that this peptide's vasodilating and cardio protective characteristics may be significant in human myocardial ischemia (Kallner et al. 1999).

Myocardial cells, vascular endothelium, and endocardial endothelial cells, as well as specialized myocardial neurons, all create NO via Ca^{2+} – mediated NOSs (NOS I). By improving coronary vasodilatation, decreasing platelet and neutrophil actions, modifying myocardial contractile function, diminishing myocardial oxygen consumption, and limiting free radical activities, it plays a vital role in the regulation of cardiac function in the healthy heart. It is unclear if exact fraction of cardiac NO is formed by cardiac sensory fibers. At least in rat hearts, undamaged cardiac sensory nerves correspond with basal NO release (Bencsik et al. 2008). NO protects the myocardium via (a) activating soluble guanylate cyclase and lowering [Ca^{2+}], (b) inhibiting chain propagating lipid radical processes generated by oxidative stress (Rubbo et al. 1994), and (c) suppressing activation of platelet and neutrophil and attachment to the surface of endothelium (Radomski et al. 1987; Kubes et al. 1991). Exogenous ONOO- damages coronary circulation, platelets, and liposomes (Villa et al. 1994). NO (from NOS or NO donors) prevents these effects.

As a result of producing cardioprotective mediators including NO and CGRP during ischemia and reperfusion, cardiac capsaicin-sensitive nerves are engaged in early and delayed preconditioning. Activation of cardiac sensory nerve terminals can

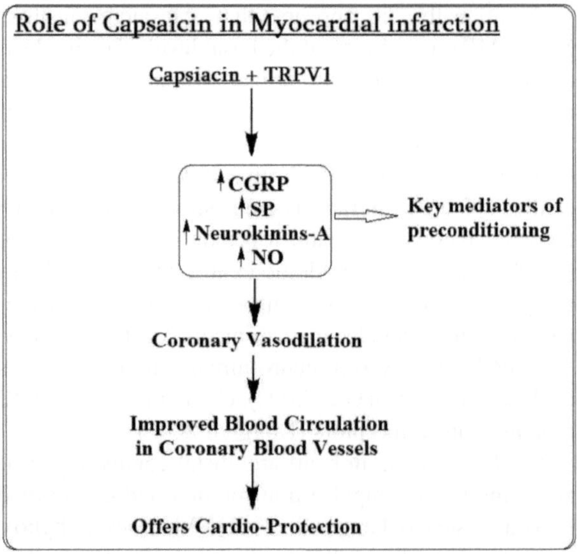

Fig. 17.4 Role of capsaicin in myocardial infarction

minimize injury due to ischemic stress and protect endogenous stress adaptation in specific illness circumstances. Figure 17.4 shows the summary of role of capsaicin in myocardial infarction.

17.3.5 Capsaicin and Stroke

Stroke is a main factor in death and disability in many nations, considerably increasing the economic and social cost. The management of stroke patients is still far from ideal, despite the positive effects of intravascular thrombolysis, medication therapy, and endovascular techniques. Furthermore, research evidence indicates that higher fruit and vegetable diet is linked to a lower incidence of stroke (He et al. 2006). Vascular NO and eNOS shows a critical part in preserving cerebral blood flow and preventing neuronal damage, according to several experimental and clinical investigations (Li and Forstermann 2000; Endres et al. 2004). After blockage of the middle cerebral artery, eNOS-lacking mice have huge cerebral infarctions (Baumbach et al. 2004). In animals, further reduction of eNOS activity lowers cerebral blood circulation and enhances the extent of cerebral infarct. Genetic variation at the eNOS locus is an inherited detrimental feature for human ischemic stroke (Carr et al. 2002). Capsaicin links TRPV1 channels. Eight TRPV1 channels have been detected in sensory neurons, in addition to blood arteries and the brain. Dietary capsaicin opens TRPV1 channels, which speeds up protein kinase A and eNOS phosphorylation. This increases endothelial NO.

Capsaicin dilates basilar arteries and decreases intracranial arterial hypertrophy in spontaneously hypertensive rats through phosphorylating eNOS (SHRsp). Chronic dietary capsaicin decreases stroke risk and increases SHRsp lifespan via activating

TRPV1 channels in cerebral arteries. Effective stroke prevention is still the excellent manner to reduce stroke burden, despite advances in stroke treatment (Goldstein et al. 2006). Over 70% of strokes each year are primary strokes, making its avoidance difficult worldwide (Ohira et al. 2006). For the sake of public health, it is especially crucial to find vulnerable people to stroke (Bogousslavsky and Kaste 2000). To control many precipitating factors which leads to primary stroke, numerous preventative measures are available (Hatano 2006). Strokes are mostly caused by high BP, which has significant consequences on brain (Wong 2009). High blood pressure alters how the endothelium relaxes cerebral blood vessels and causes cerebrovascular hypertrophy that makes the brain more susceptible to ischemic injury (Leonardi-Bee et al. 2002). Medications that lower blood pressure make a big difference in how many people die from strokes.

A stroke was associated with higher levels of endocannabinoids and capsaicin receptors. Capsaicin is a very effective TRPV1 channel agonist. Sensory neurons and blood arteries both express the TRPV1. In perennial hypertensive rats, feeding capsaicin makes the endothelium relax more and lowers arterial pressure. The ability of cerebral blood arteries to relax over time is significantly improved by chronic dietary capsaicin. The importance of eNOS and vascular NO in sustaining cerebral blood flow and preventing neuronal damage is amply supported by the available data (Rikitake et al. 2005). Larger brain infarctions occur in eNOS-deficient mice due to reduced blood flow in the ischemic border zone. In animals lacking neuronal NO synthase, additional suppression of eNOS activity declines in blood-brain-circulation and an expansion of the cerebral infarct (Limbourg et al. 2002). In contrast, increasing NO synthesis by giving eNOS substrate L-arginine enhances local blood circulation in cerebral ischemic location, offering prophylaxis against stroke (Morikawa et al. 1994). Human ischemic stroke vulnerability is increased, according to clinical investigations, by genetic variation at the eNOS gene locus.

The phosphorylation of eNOS in mesenteric arteries and circulating NO derivatives can both rise as a result of capsaicin's long-term activation of TRPV1. It also increases eNOS expression in carotid arteries, connected to the growing relaxation of cerebral vasculature. Another intriguing finding was that chronic capsaicin treatment dramatically reduced cerebral vascular enlargement. Mouse cerebral arterioles grew larger when eNOS was blocked using NG-nitro-L-arginine methyl ester. Furthermore, eNOS-deficient mice experienced cerebral arteriole hypertrophy. Consuming dietary capsaicin could be a positive lifestyle change for people who are prone to stroke. Figure 17.5 shows the summary of role of capsaicin in stroke.

17.3.6 Capsaicin and Arrhythmias

Cardiac arrhythmias cause at least half of unexpected cardiac arrests and are a leading cause of cardiovascular deaths. Arrhythmias or dysrhythmias are disturbances in the heart's electrical activity, such as tachycardia or bradycardia (Wong et al. 2019). Anti-arrhythmic agents (AADs) are used to treat cardiac

Fig. 17.5 Role of capsaicin in stroke

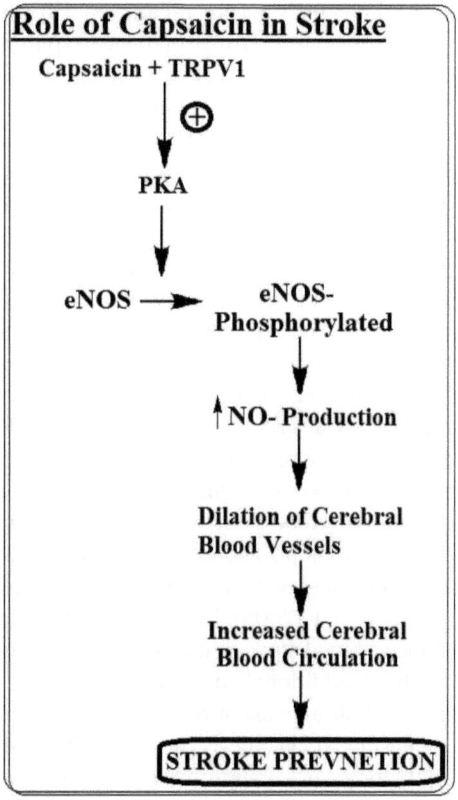

arrhythmias. The primary effects of these medications are to modify electrical conduction velocity, impede trigger action, or impact membrane ionic currents.

Capsaicin triggers cardiac nerve terminals to secrete substance P and CGRP (Franco-Cereceda et al. 1989; Manzini et al. 1989). In isolated rat cardiac myocytes, it suppresses K^+ channel conduction. Capsaicin blocks transient outward (Ito), delayed rectifier (IK), and inward rectifier (IK1) currents with IC50 values as low as 6.4, 11.5 and 46.9 M. (Castle 1992). K^+ current inhibition may have antiarrhythmic effects (Hondeghem 1991; Hondeghem and Snyders 1990).

In myocardium, capsaicin shortens the time between action potentials. A multitude of actions, such as inhibiting an inward current (such as Na^+ or Ca^{2+}) or boosting an outward K^+ current, could explain for the decrease in action potential duration (i.e., IKr, IKs, Ito, IK1, or IKATP). According to single microelectrode tests, capsaicin concentrations up to 100 M had no effect on Vmax or amplitude, showing that fast Na^+ channels were not inhibited. Capsaicin's capacity to shorten action potential duration was thus unrelated to Na^+ channel blocking. IK1 involvement is also unlikely because resting potentials and action potential amplitudes were not depolarized.

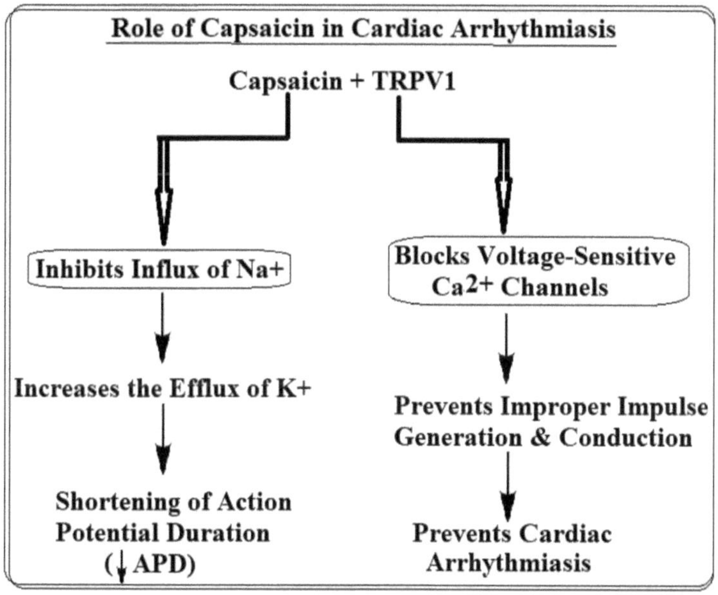

Fig. 17.6 Role of capsaicin in cardiac arrhythmias

Since capsaicin blocks IK in rats (it is uncertain if rats have IKr or IKs or if these channels are pertinent to other species), IK activation cannot be ruled out, but it is dubious (Castle 1992). However rare it may be, capsaicin may open K^+ATP channels. Diltiazem shortened action potential duration values to support Ca^{2+} channel modulation with capsaicin. Capsaicin enhanced Vmax and action potential at huge doses (30 and 100 IzM), but diltiazem diminished both, indicating blockade of fast inward Na^+ channels. Based on the results, it does seem probable that capsaicin can block Ca^{2+} channels, directly via voltage-gated channels or indirectly through one or more mechanisms. Castle (1992) did show that 10 mM of capsaicin restricted some Ca^{2+} channel current, but he also concluded that most of capsaicin's implications are triggered by changes in K^+ channels. Capsaicin's ability to prevent ischemia is another sign that the Ca^{2+} channel is involved. Capsaicin and diltiazem were found to increase by 36% and 50%, respectively, the amount of time it took for isolated, perfused rat hearts that were exposed to global ischemia to contract. The fact that glyburide had no effect on reviving the anti-ischemic effects of these medications suggests that K^+ATP channels are not important to the mechanism by which they work. Researchers found that glyburide did not counteract the actions of capsaicin on action potential length. They concluded that capsaicin's antiarrhythmic and anti-ischemic properties in guinea pig entailed actions on other ionic components, probably voltage-gated Ca^{2+} channels rather than K^+ channels. Figure 17.6 shows the summary of role of capsaicin in cardiac arrhythmias.

17.4 Conclusion and Future Aspects

Capsaicin's uses are no longer limited to the kitchen. It is a bioactive food ingredient that comes from nature. At the moment, it is used with the newest medical technologies. Capsaicin is involved in preventing cancer, oxidative stress, inflammation, lipid metabolism, and thermogenesis. TRPV1 may be a possible target for the therapy of cardiovascular problems since capsaicin can prevent cardiovascular diseases by activating TRPV1 in target organs or tissues. TRPV1 releases SP, CGRP, and neurokinin-A to protect the cardiovascular system. It makes blood vessels extend by decreasing the amount of calcium inside cells, causing more nitric oxide to be released, and lowering blood pressure. Brown adipose tissue (BAT) may also be involved in glucose and fatty acid oxidation. Li et al. (2014) found that capsaicin activates TRPV1 and lowers blood pressure by making more sodium leave the body through the urine. This also delays the start of a stroke. Dietary capsaicin or its analogs need to be reformulated into a more pleasant dose form to be efficient in alleviating cardiovascular disease.

References

Adefegha SA, Oboh G (2013) Phytochemistry and mode of action of some tropical spices in the management of type-2 diabetes and hypertension. African J Pharm Pharmacol 7:332–346

Barasch E, Gottdiener JS, Aurigemma G et al (2009) Association between elevated fibrosis markers and heart failure in the elderly: the cardiovascular health study. Circ Heart Fail 2:303–310

Baskaran P, Krishnan V, Gao F et al (2017) TRPV1 activation counters diet-induced obesity through sirtuin-1 activation and PRDM-16 deacetylation in brown adipose tissue. Int J Obese (Lond) 41:739–749

Baumbach GL, Sigmund CD, Faraci FM (2004) Structure of cerebral arterioles in mice deficient in expression of the gene for endothelial nitric oxide synthase. Circ Res 95:822–829

Baxter GF, Ferdinandy P (2001) Delayed preconditioning of myocardium: current perspectives. Basic Res Cardiol 96:329–344

Bencsik P, Kupai K, Giricz Z (2008) Cardiac capsaicin-sensitive sensory nerves regulate myocardial relaxation via S-nitrosylation of SERCA: role of peroxynitrite. Br J Pharmacol 153:488–496

Bevan S, Geppetti P (1994) Protons: small stimulants of capsaicin-sensitive sensory nerves. TINS 17:509–512

Boengler K, Schlüter KD, Schermuly RT et al (2020) Cardio protection in right heart failure. Br J Pharmacol 177:5413–5431

Bogousslavsky J, Kaste M (2000) Risk factors and stroke prevention. European Stroke Initiative (EUSI). Cerebrovasc Dis 10:12–21

Bolli R, Dawn B, Tang XL (1998) The nitric oxide hypothesis of late preconditioning [review]. Basic Res Cardiol 93:325–338

Brain SD, Poyner DR, Hill RG (2002) CGRP receptors: a headache to study, but will antagonists prove therapeutic in migraine? Trends Pharmacol Sci 23:51–53

Braunwald EK (1985) Myocardial reperfusion: a double-edged sword? J Clin Invest 76:1713–1719

Carr FJ, McBride MW, Carswell HV et al (2002) Genetic aspects of stroke: human and experimental studies. J Cereb Blood Flow Metab 22:767–773

Castle N (1992) Differential inhibition of potassium currents in rat ventricular myocytes by capsaicin. Cardiovasc Res 26:1137

Caterina MJ, Leffler A, Malmberg AB (2000) Impaired nociception and pain sensation in mice lacking the capsaicin receptor. Science 288:306–313

Caterina MJ, Schumacher MA, Tominaga M (1997) The capsaicin receptor: a heat-activated ion channel in the pain pathway. Nature 389:816–824

Chang Y, Stover SR, Hoover DB (2001) Regional localization and abundance of calcitonin generelated peptide receptors in Guinea pig heart. J Mol Cell Cardiol 33:745–754

Chu CJ, Huang SM, De Petrocellis L (2003) N-oleoyldopamine, a novel endogenous capsaicin-like lipid that produces hyperalgesia. J Biol Chem 278:13633–13639

Clark N, Keeble J, Fernandes ES (2007) The transient receptor potential vanilloid 1 (TRPV1) receptor protects against the onset of sepsis after endotoxin. FASEB J 21:3747–3755

Davignon J, Ganz P (2004) Role of endothelial dysfunction in atherosclerosis. Circulation 109: III27–III32

Demirbilek S, Ersoy MO, Demirbilek S (2004) Small-dose capsaicin reduces systemic inflammatory responses in septic rats. Anesth Analg 99:1501–1507

Doggrell SA (2001) Migraine and beyond: cardiovascular therapeutic potential for CGRP modulators. Expert Opin Investig Drugs 10:1131–1138

Doltra A, Stawowy P, Dietrich T (2013) Magnetic resonance imaging of cardiovascular fibrosis and inflammation: from clinical practice to animal studies and back. Biomed Res Int 2013:1–10

Endres M, Laufs U, Liao JK et al (2004) Targeting eNOS for stroke protection. Trends Neurosci 27: 283–289

Feng Z, Hai-ning Y, Xiao-man C, Zun-chen W, Sheng-rong S, Das UN (2014) Effect of yellow *Capsicum* extract on proliferation and differentiation of 3T3-L1 preadipocytes. Nutrition 30: 319–325

Ferdinandy P (2003) Myocardial ischaemia/reperfusion injury and preconditioning: effects of hypercholesterolaemia/hyperlipidaemia. Br J Pharmacol 138:283–285

Ferdinandy P, Schulz R, Baxter GF (2007) Interaction of cardiovascular risk factors with myocardial ischemia/reperfusion injury, preconditioning, and postconditioning. Pharmacol Rev 59:418–458

Ferdinandy P, Szilvassy Z, Balogh N et al (1996) Nitric oxide is involved in active preconditioning in isolated working rat hearts. Ann N Y Acad Sci 793:489–493

Ferdinandy P, Szilvassy Z, Baxter GF (1998) Adaptation to myocardial stress in disease states: is preconditioning a healthy heart phenomenon? Trends Pharmacol Sci 19:223–229

Firth AL, Remillard CV, Yuan XJ (2007) TRP channels in hypertension. Biochim Biophys Acta 1772:895–906

Franco-Cereceda A (1991) Calcitonon gene related peptide and human epicardial coronary arteries: presence, release and vasodilator effects. Br J Pharmacol 102:506–510

Franco-Cereceda A, Liska J (2000) Potential of calcitonin gene-related peptide in coronary heart disease. Pharmacology 60:1–8

Franco-Cereceda A, Saria A, Lundberg JM (1989) Differential release of calcitonin gene-related peptide and neuropeptide Y from the isolated heart by capsaicin, ischaemia, nicotine, bradykinin and ouabain. Acta Physiol Scand 135:173–187

Gao F, Liang Y, Wang X et al (2014) TRPV1 activation attenuates high-salt diet-induced cardiac hypertrophy and fibrosis through PPAR-δ up regulation. PPAR Res 2014:1–12

Goldstein LB, Adams R, Alberts MJ (2006) Primary prevention of ischemic stroke: a guideline from the American Heart Association/American Stroke Association Stroke Council: cosponsored by the atherosclerotic peripheral vascular disease interdisciplinary working group; cardiovascular nursing council; clinical cardiology council; nutrition, physical activity, and metabolism council; and the quality of care and outcomes research interdisciplinary working group: the American Academy of Neurology affirms the value of this guideline. Stroke 37:1583–1633

Govindarajan VS (1985) Capsicum production, technology, chemistry, and quality. Part 1: history, botany, cultivation, and primary processing. Crit Rev Food Sci Nutr 22:109–176

Graf BL, Raskin I, Cefalu WT (2010) Plant-derived therapeutics for the treatment of metabolic syndrome. Curr Opin Investig Drugs 11:1107–1115

Gupta S, Kumar S, Sopko N et al (2012) Thymosin beta-4 and cardiac protection: implication in inflammation and fibrosis. Ann N Y Acad Sci 1269:84–91

Hao X, Chen J, Luo Z et al (2011) TRPV1 activation prevents high-salt diet-induced nocturnal hypertension in mice. Pflugers Arch 461:345–353

Harper AG, Brownlow SL, Sage SO (2009) A role for TRPV1 in agonist-evoked activation of human platelets. J Thromb Haemost 7:330–338

Harris DW, Buckley C, Dunn EC et al (1991) Briquet's syndrome—An unusual case of urticaria. J R Soc Med 84:167

Hatano S (2006) The WHO collaborative study on the control of stroke in the community. WHO Bull 54:541–553

He FJ, Nowson CA, MacGregor GA (2006) Fruit and vegetable consumption and stroke: meta-analysis of cohort studies. Lancet 367:320–326

Ho JE, Liu C, Lyass A (2012) Galectin-3 A marker of cardiac fibrosis, predicts incident heart failure in the community. J Am Coll Cardiol 60:1249–1256

Holzer P (1991) Capsaicin: cellular targets, mechanisms of action, and selectivity for thin sensory neurons. Pharmacol Rev 43:143–201

Hondeghem LM (1991) Ideal antiarrhythmic agents: chemical defibrillators. J Cardiovasc Electrophysiol 2:S169

Hondeghem LM, Snyders DJ (1990) Class III antiarrhythmic agents have a lot of potential but a long way to go. Circulation 81:686

Hongmei L, Qiang L, Hao Y (2014) Activation of TRPV1 attenuates high salt induced cardiac hypertrophy through improvement of mitochondrial function. Br J Pharmacol 172:5548–5558

Hu FB (2008) Globalization of food patterns and cardiovascular disease risk. Circulation 118:1913–1914

Huang SM, Bisogno T, Trevisani M et al (2002) An endogenous capsaicin-like substance with high potency at recombinant and native vanilloid VR1 receptors. Proc Natl Acad Sci U S A 99:8400–8405

Jaganathan R, Purushotham B, Radhakrishnan N, Swamy MK (2020) Capsaicin and its potential anticancer mechanisms of action. In: Plant-derived bioactives, pp. 301–321. Springer, Singapore

Jancso G (1992) Pathobiological reactions of C-fibre primary sensory neurones to peripheral nerve injury. Exp Physiol 77:405–431

Jancso G, Kiraly E, Jancso-Gabor A (1977) Pharmacologically induced selective degeneration of chemosensitive primary sensory neurones. Nature 270:741–743

Jancso N (1968) Desensitization with capsaicin as a tool for studying the function of pain receptors. In: Lim RKS (ed) Pharmacology of pain. Pergamon Press, Oxford, pp 33–55

Jang HH, Lee J, Lee SH et al (2020) Effects of *Capsicum annuum* supplementation on the components of metabolic syndrome: a systematic review and meta-analysis. Sci Rep 10:20912

Jeon G, Choi Y, Lee SM et al (2010) Antiobesity activity of methanol extract from hot pepper (Capsicum annuum L.) seeds in 3T3-L1 adipocyte. Food Sci Biotechnol 19:1123–1127

Joo JI, Kim DH, Choi JW et al (2010) Proteomic analysis for antiobesity potential of capsaicin on white adipose tissue in rats fed with a high fat diet. J Proteome Res 9:2977–2987

Kallner G (1998) Release and effects of calcitonin gene-related peptide in myocardial ischaemia. Scand Cardiovasc J Suppl 49:1–35

Kallner G, Owall A, Franco-Cereceda A (1999) Myocardial outflow of calcitonin gene-related peptide in relation to metabolic stress during coronary artery bypass grafting without cardiopulmonary bypass. J Thorac Cardiovasc Surg 117:447–453

Katona M, Boros K, Santha P et al (2004) Selective sensory denervation by capsaicin aggravates adriamycin-induced cardiomyopathy in rats. Naunyn Schmiedeberg's Arch Pharmacol 370: 436–443

Kawada T, Suzuki T, Takahashi M et al (1984) Gastrointestinal absorption and metabolism of capsaicin and dihydrocapsaicin in rats. Toxicol Appl Pharmacol 72:449–456

Kraft KH, Brown CH, Nabhan (2014) Multiple lines of evidence for the origin of domesticated chili pepper, Capsicum annuum, in Mexico. Proc Natl Acad Sci U S A 111:6165–6170

Kubes P, Suzuki M, Granger DN (1991) Nitric oxide: an endogenous modulator of leukocyte adhesion. Proc Natl Acad Sci U S A 88:4651–4655

Kubota T, McTiernan CF, Frye CS et al (1997) Dilated cardiomyopathy in transgenic mice with cardiac-specific overexpression of tumor necrosis factor-alpha. Circ Res 81:627–635

Kunieda K, Someya A, Horie S (2005) Lafutidine-induced increase in intracellular Ca^{2+} concentrations in PC12 and endothelial cells. J Pharmacol Sci 97:67–74

Leonardi-Bee J, Bath PM, Phillips SJ (2002) Blood pressure and clinical outcomes in the International Stroke Trial. Stroke 33:1315–1320

Levite M, Cahalon L, Hershkoviz R (1998) Neuropeptides, via specific receptors, regulate T cell adhesion to fibronectin. J Immunol 160:993–1000

Li H, Forstermann U (2000) Nitric oxide in the pathogenesis of vascular disease. J Pathol 190:244–254

Li L, Wang F, Wei X et al (2014) Transient receptor potential vanilloid 1 activation by dietary capsaicin promotes urinary sodium excretion by inhibiting epithelial sodium channel α subunit-mediated sodium reabsorption. Hypertension 64:397–404

Li W, Chen X, Riley AM (2017) Long-term spironolactone treatment reduces coronary TRPC expression, vasoconstriction, and atherosclerosis in metabolic syndrome pigs. Basic Res Cardiol 112:54

Li YJ, Song QJ, Xiao J (2000) Calcitonin gene-related peptide: an endogenous mediator of preconditioning. Acta Pharmacol Sin 21:865–869

Libby P, Ridker PM, Hansson GK (2011) Progress and challenges in translating the biology of atherosclerosis. Nature 473:317–325

Lijnen PJ, Maharani T, Finahari N (2012) Serum collagen markers and heart failure. Cardiovasc Hematol Disord Drug Targets 12:51–55

Limbourg FP, Huang Z, Plumier JC (2002) Rapid non-transcriptional activation of endothelial nitric oxide synthase mediates increased cerebral blood flow and stroke protection by corticosteroids. J Clin Invest 110:1729–1738

Liu D, Yang D, He H (2009) Increased transient receptor potential canonical type 3 channels in vasculature from hypertensive rats. Hypertension 53:70–76

Lorell BH, Carabello BA (2000) Left ventricular hypertrophy: pathogenesis, detection, and prognosis. Circulation 102:470–479

Ma L, Zhong J, Zhao Z (2011) Activation of TRPV1 reduces vascular lipid accumulation and attenuates atherosclerosis. Cardiovasc Res 92:504–513

Manzini AF, Perretti L, DeBenedetti P et al (1989) A comparison of bradykinin- and capsaicin-induced myocardial and coronary effects in isolated perfused heart of Guinea-pig: involvement of substance P and calcitonin gene-related peptide release. Br J Pharmacol 97:303

Menees DS, Bates ER (2010) Evaluation of patients with suspected coronary artery disease. Coron Artery Dis 21:386–390

Miller M, Koch SE, Veteto A et al (2021) Role of known transient receptor potential vanilloid channels in modulating cardiac mechanobiology. Front Physiol 12:1–14

Mitchell JA, Williams FM, Williams TJ (1997) Role of nitric oxide in the dilator actions of capsaicin-sensitive nerves in the rabbit coronary circulation. Neuropeptides 31:333–338

Morikawa E, Moskowitz MA, Huang Z (1994) L-arginine infusion promotes nitric oxide-dependent vasodilation, increases regional cerebral blood flow, and reduces infarction volume in the rat. Stroke 25:429–435

Mózsik G (2014) Capsaicin as new orally applicable gastroprotective and therapeutic drug alone or in combination with nonsteroidal anti-inflammatory drugs in healthy human subjects and in patients. Prog Drug Res 68:209–258

Munjuluri S, Wilkerson DA, Sooch G, Chen X, White FA, Obukhov AG (2022) Capsaicin and TRPV1 channels in the cardiovascular system: the role of inflammation. Cell 11:1–21

Murry CE, Jennings RB, Reimer KA (1986) Preconditioning with ischemia: a delay of lethal cell injury in ischemic myocardium. Circulation 74:1124–1136

Nicoletti A, Michel JB (1999) Cardiac fibrosis and inflammation: interaction with hemodynamic and hormonal factors. Cardiovasc Res 41:532–543

Nilius B, Owsianik G (2010) Transient receptor potential channelopathies. Pflügers Arch 460:437–450

Nilius B, Owsianik G, Voets T (2007) Transient receptor potential cation channels in disease. Physiol Rev 87:165–217

Ohira T, Shahar E, Chambless LE (2006) Risk factors for ischemic stroke subtypes: the atherosclerosis risk in communities study. Stroke 37:2493–2498

Okajima K, Harada N (2006) Regulation of inflammatory responses by sensory neurons: molecular mechanism(s) and possible therapeutic applications. Curr Med Chem 13:2241–2251

Olah Z, Szabo T, Karai L (2001) Ligand-induced dynamic membrane changes and cell deletion conferred by vanilloid receptor 1. J Biol Chem 276:11021–11030

Opie LH, Commerford PJ, Gersh BJ et al (2006) Controversies in ventricular remodeling. Lancet 367:356–367

Pabla R, Curtis MJ (1996) Endogenous protection against reperfusion-induced ventricular fibrillation: role of neuronal versus non-neuronal sources of nitric oxide and species dependence in the rat versus rabbit isolated heart. J Mol Cell Cardiol 28:2097–2110

Pan HL, Chen SR (2004) Sensing tissue ischemia: another new function for capsaicin receptors? Circulation 110:1826–1831

Przyklenk K, Kloner RA (1998) Ischemic preconditioning: exploring the paradox. Prog Cardiovasc Dis 40:517–547

Qiang W, Shuangtao M, De L (2014) Dietary capsaicin ameliorates pressure overload-induced cardiac hypertrophy and fibrosis through the transient receptor potential vanilloid type 1. Am J Hypertens 27:1521–1529

Raddino R, Pelr G, Manca C et al (1997) Mechanism of action of human calcitonin gene-related peptide in rabbit heart and in human mammary arteries. J Cardiovasc Pharmacol 29:463–470

Radomski MW, Palmer RMJ, Moncada S (1987) The anti-aggregating properties of vascular endothelium: interactions between prostacyclin and nitric oxide. Br J Pharmacol 92:639–646

Rikitake Y, Kim HH, Huang Z (2005) Inhibition of Rho kinase (ROCK) leads to increased cerebral blood flow and stroke protection. Stroke 36:2251–2257

Rocha M, Bendhack L (2009) Relaxation evoked by extracellular Ca^{2+} in rat aorta is nerve-independent and involves sarcoplasmic reticulum and l-type Ca^{2+} channel. Vasc Pharmacol 50:98–103

Rubbo H, Radi R, Trujillo M et al (1994) Nitric oxide regulation of superoxide and peroxynitrite-dependent lipid peroxidation. Formation of novel nitrogen-containing oxidized lipid derivatives. J Biol Chem 269:26066–26075

Ruiz C, Gutknecht S, Delay E et al (2006) Detection of Nacl and Kcl in TRPV1 knockout mice. Chem Senses 31:813–820

Sancho R, Macho A, de La Vega L (2004) Immunosuppressive activity of endovanilloids: N-arachidonoyl-dopamine inhibits activation of the NF-kappa B, NFAT, and activator protein 1 signaling pathways. J Immunol 172:2341–2351

Schulz R, Cohen MV, Behrends M (2001) Signal transduction of ischemic preconditioning. Cardiovasc Res 52:181–198

Sessa WC (2010) A new way to lower blood pressure: pass the chili peppers please! Cell Metab 12:109–110

Shuba YM (2021) Beyond neuronal heat sensing: diversity of TRPV1 heat-capsaicin receptor-channel functions. Front Cell Neurosci 14:1–17

Singh S, Natarajan K, Aggarwal BB (1996) Capsaicin (8-methyl-N-vanillyl-6-nonenamide) is a potent inhibitor of nuclear transcription factor-kappa B activation by diverse agents. J Immunol 157:4412–4420

Sosunov AA, Hassall CJS, Loesch A (1996) Nitric oxide synthase-containing neurons and nerve fibers within cardiac ganglia of rat and Guinea-pig: an electron- microscopic immune cytochemical study. Cell Tissue Res 284:19–28

Sun F, Xiong S, Zhu Z (2016) Dietary capsaicin protects cardio metabolic organs from dysfunction. Nutrients 8:174–187

Szabados T, Gömöri K, Pálvölgyi L et al (2020) Capsaicin-sensitive sensory nerves and the TRPV1 ion channel in cardiac physiology and pathologies. Int J Mol Sci 21:1–23

Tabas I, Garcia-Cardena G, Owens GK (2015) Recent insights into the cellular biology of atherosclerosis. J Cell Biol 209:13–22

Takaki M, Akashi T, Ishioka K et al (1994) Effects of capsaicin on mechano energetics of excised cross-circulated canine left ventricle and coronary artery. J Mol Cell Cardiol 26:1227–1239

Tani Y, Fujioka T, Sumioka M (2004) Effects of capsinoid on serum and liver lipids in hyperlipidemic rats. J Nutr Sci Vitaminol 50:351–355

Tapsell LC, Hemphill I, Cobiac L (2006) Health benefits of herbs and spices: the past, the present, the future. Med J Aust 185:S4–S24

Thilo F, Loddenkemper C, Berg E et al (2009) Increased trpc3 expression in vascular endothelium of patients with malignant hypertension. Mod Pathol 22:426–430

Uozumi H, Hiroi Y, Zou Y (2001) Gp130 plays a critical role in pressure overload induced cardiac hypertrophy. J Biol Chem 276:23115–23119

Villa LM, Salas E, Darley-Usmar VM (1994) Peroxynitrite induces both vasodilatation and impaired vascular relaxation in the isolated perfused rat heart. Proc Natl Acad Sci U S A 91: 12383–12387

Wang DH (2008) Transient receptor potential vanilloid channels in hypertension, inflammation, and end organ damage: an imminent target of therapy for cardiovascular disease? Curr Opin Cardiol 23:356–363

Wang Y, Babankova D, Huang J (2008a) Deletion of transient receptor potential vanilloid type 1 receptors exaggerates renal damage in deoxy corticosterone acetate-salt hypertension. Hypertension 52:264–270

Wang Y, Wang DH (2006) A novel mechanism contributing to development of dahl salt-sensitive hypertension: role of the transient receptor potential vanilloid type 1. Hypertension 47:609–614

Wang YX, Wang J, Wang C (2008b) Functional expression of transient receptor potential vanilloid related channels in chronically hypoxic human pulmonary arterial smooth muscle cells. J Membr Biol 223:151–159

Wiklund O, Pirazzi C, Romeo S (2013) Monitoring of lipids, enzymes, and creatine kinase in patients on lipid-lowering drug therapy. Curr Cardiol Rep 15(390):397

Wong CX, Brown A, Lau DH (2019) Epidemiology of sudden cardiac death: global and regional perspectives. Heart Lung Circ 28:6–14

Wong KSL (2009) Hemodynamic changes with cerebral augmentation. J Neurol Sci 285:S10–S10

Xin H, Tanaka H, Yamaguchi M et al (2005) Vanilloid receptor expressed in the sarcoplasmic reticulum of rat skeletal muscle. Biochem Biophys Res Commun 332:756–762

Xiong S, Wang P, Ma L et al (2016) Ameliorating endothelial mitochondrial dysfunction restores coronary function via transient receptor potential vanilloid 1-mediated protein kinase A/uncoupling protein 2 pathway. Hypertension 67:451–460

Xu X, Wang P, Zhao Z (2011) Activation of transient receptor potential vanilloid 1 by dietary capsaicin delays the onset of stroke in stroke-prone spontaneously hypertensive rats. Stroke 42: 3245–3251

Yang D, Luo Z, Ma S et al (2010) Activation of TRPV1 by dietary capsaicin improves endothelium-dependent vasorelaxation and prevents hypertension. Cell Metab 12:130–141

Yaoita H, Sato E, Kawaguchi M et al (1994) Nonadrenergic noncholinergic nerves regulate basal coronary flow via release of capsaicin-sensitive neuropeptides in the rat heart. Circ Res 75:780–788

Zhang C, Ye L, Zhang Q et al (2020) The role of TRPV1 channels in atherosclerosis. Channels 14: 141–150

Zhang MJ, Yin YW, Li BH et al (2015) The role of TRPV1 in improving VSMC function and attenuating hypertension. Prog Biophys Mol Biol 117:212–216

Zhong B, Wang DH (2007) TRPV1 gene knockout impairs preconditioning protection against myocardial injury in isolated perfused hearts in mice. Am J Physiol Heart Circ Physiol 293: H1791–H1798

Zvara A, Bencsik P, Fodor G et al (2006) Capsaicin-sensitive sensory neurons regulate myocardial function and gene expression pattern of rat hearts: a DNA microarray study. FASEB J 20:160–162

Therapeutic Implications of Capsaicin in Diabetes

18

Arshpreet Kaur, Manasi Verma, Ranjana Bhandari, and Anurag Kuhad

Abstract

Capsaicin, chemically known as trans-8-methyl-N-vanillyl-6-noneamide, is a primary active principle found in members of chili peppers (*Capsicum annum* and *Capsicum frutescens*). First isolated in the nineteenth century, it is known to belong to the capsaicinoid alkaloid family, and carries a vanilloid moiety, which plays a role in its bioactivity. Through extensive research over the years, capsaicin has shown evidence of multiple biological effects, leading to it being considered a "miracle molecule" for decades and even centuries. This is mainly due to its therapeutic potential in ailments ranging from pain and obesity to urinary complications and even cancer. From the multitude of studies that have been performed to evaluate the effectiveness of capsaicin and its analogs in various medical conditions, compelling evidence for their anti-inflammatory, analgesic, anti-pruritic, antioxidant, anti-cancer, anti-apoptotic, and neuroprotective functions have been accumulated. Among these myriads of potential applications, the efficacy of capsaicin in tackling diabetes has been extensively studied, and the results have brought forth encouraging hints about its therapeutic role in the same. While the exact mechanisms remain largely unknown, it has been established that this molecule can lower blood glucose by processes inducing high levels of insulin, glycogen in diabetic rats induced by streptozotocin. Moreover, one report says that dietary intake of capsaicin rich foods has a favorable impact on glycemic control in obese (ob/ob) diabetic mice, despite causing any significant change in the incidence of an increase in body weight. More recently, capsiate, a non-pungent analog of capsaicin, has shown potential as an insulin sensitizer,

A. Kaur · M. Verma · R. Bhandari · A. Kuhad (✉)
Pharmacological Research Laboratory, University Institute of Pharmaceutical Sciences, UGC-CAS, Panjab University, Chandigarh, India
e-mail: anurag.kuhad@pu.ac.in

M. K. Swamy, A. Kumar (eds.), *Capsaicinoids*,
https://doi.org/10.1007/978-981-99-7779-6_18

sparking interest in its activity. These studies, along with numerous others, have warranted further research to explore the efficacy of capsaicinoids in metabolic syndromes such as diabetes mellitus, and this article aims to review and discuss the varied therapeutic avenues for capsaicin and its analogs in diabetes, and for evaluating its effectiveness in the same.

Keywords

Capsaicin · Diabetes mellitus · GLP-1 · Gut microbiota · TRPV-1

18.1 Introduction

Blood glucose balance is essential to conserve our body up with regular biological functions. The loss of glucose balance gives rise to a number of metabolic syndromes and their associated diseases (Liang et al. 2023). "Metabolic syndrome" refers to a group of symptoms, such as poor glucose tolerance, central obesity, hypertension, insulin resistance, and dyslipidemia that frequently co-occur. The likelihood of heart related problem, type II diabetes mellitus (DM), and fatty liver disease (non-alcoholic) is increased by these physiological alterations (Huang 2009). The development of these metabolic problems is significantly influenced by oxidative stress and persistent low-grade inflammation (Nishida and Otsu 2017). There are numerous ways to prevent and regulate glucose metabolic problems, including exercising and using medications, but some of these methods call for perseverance and may have undesirable side effects (Marín-Peñalver et al. 2016).

A metabolic condition known as diabetes mellitus (DM) is caused by improper carbohydrate metabolism and its utilization (especially glucose). As per the causative reason, it might be classified into two main types—type I DM is caused by pancreatic dysfunction and a consequent reduction in insulin release, and hence named as "insulin-dependent diabetes." Type II DM, on the other hand, arises mainly owing to desensitization of insulin receptors, and therefore is referred to as "non-insulin-dependent diabetes mellitus." The WHO estimates that 5–10% of persons worldwide have type I DM, and the multitude of population detected with the condition has been gradually rising up (Zhang et al. 2017).

A growing body of research reveals that natural bioactive phytochemicals such as resveratrol, epigallocatechin gallate (EGCG), anthocyanins, and curcumin may be able to alleviate glucose metabolic dysfunction and its negative effect. These natural substances can be consumed as supplements to enhance glucose homeostasis because of their low toxicity (Gowd et al. 2017; Hewlings and Kalman 2017; Legeay et al. 2015; Springer and Moco 2019). One of the most popular natural substances is capsaicin, which is effective in treating many illnesses, is being reviewed in the present article for potential pharmacological activity in treating diabetes mellitus.

Capsaicin (trans-8-methyl-N-vanillyl-6-nonenamide), a phenolic compound, is primary hot and pungent compound found in capsicum fruit species, including

Capsicum annum, *C. frutescens*, *C. longum*, and others, belonging to the family Solanaceae (Szolcsányi 2014). With its consumption dating back to between 7000 and 6000 BCE, Capsicum is today considered to be one of the oldest-domesticated plants (Mózsik et al. 2009). A naturally occurring irritant, capsaicin was isolated as the first member of the capsaicinoid class of compounds in 1816 by Christian Friedrich Bucholz, who named it "capsicin." Then, in 1876, the pure form was extracted by Thresh and renamed "capsaicin." Several other compounds belonging to the capsaicinoid category have been discovered since then, and are reported to be defensive substances synthesized by plants against infections (Basith et al. 2016).

Capsiate (4-hydroxy-3-methoxybenzyl (E)-8-methyl-6-nonenoate), the non-pungent analog of capsaicin, is a fatty acid ester of vanillyl alcohol and a member of the family of capsinoids. According to reports, every chemical in these two families-capsaicin and capsiate is odorless and is what gives them their distinct pungent characteristics (Gupta et al. 2022).

While its culinary uses have been well-documented over the course of centuries, the therapeutic applications of capsaicin have also been the target of significant scientific research, and have made a niche for themselves amid the myriad of potential uses of the same, including self-defense products, insects, and animal repellents, among others. Some of the oldest records of capsaicin's use as a therapeutic highlighted its effectiveness in enhancing tolerability to hot weather by improving skin vasodilation, perspiration, and heat loss. Other folk applications include its use in the treatment of sore throat, cough, tonsillitis, gastric ulcers, cholera, backache, gout, and others, along with an anti-inflammatory, analgesic, anti-pruritic, antioxidant, and anti-diabetic agent (Conway 2008).

18.2 Mechanism of Action

Other vanilloids, including capsaicin, interact with the TRPV1 neuronal membrane receptor (previously known as vanilloid receptor, VR1). Endovanilloids, or ligands, that either directly or indirectly activate TRPV1 include capsaicin, protons, heat (greater than 43 °C), and endogenous lipid molecules (Clapham 2003).

TRPV1 belongs to the transient receptor potential family of non-specific cation channels (TRP). Mammalian transient receptor potential channels can be categorized into six subfamilies based on their structural similarity, viz., the TRP canonical (TRPC; TRPC1–7), melastatin (TRPM; TRPM1–8), vanilloid (TRPV; TRPV1–6) mucolipin (TRPML; TRPML1–3), ankyrin (TRPA; TRPA1) and polycystin (TRPP; TRPP2, TRPP3, TRPP5) (Nilius and Owsianik 2010). Sun et al. (2016) showed that capsaicin stimulated inflow of cations, activation of TRPV1, and several additional physiological processes (Fig. 18.1).

The intracellular attachment of capsaicin to TRPV1's vanilloid-binding pocket changes its characteristics allosterically, opening the pore and allowing Ca^{2+} inflow (Kumar et al. 2016; Patowary et al. 2017). Neurotransmitters released like calcitonin gene-related protein (CGRP) as well as substance P are made possible by this Ca^{2+} influx, which also changes the electrical characteristics of the cell (Devesa et al.

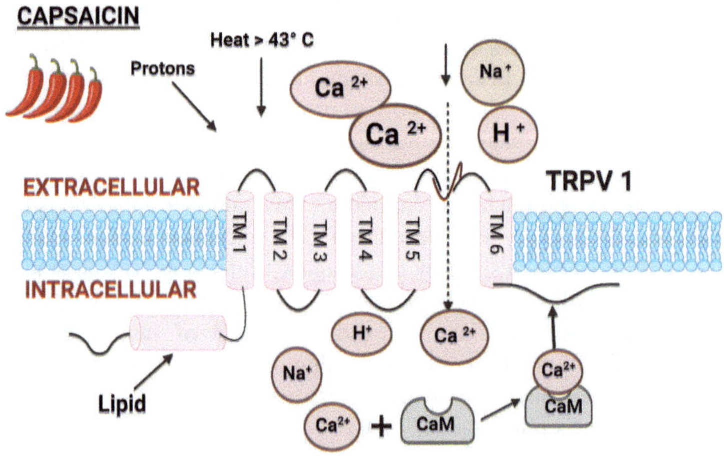

Fig. 18.1 Structure and function of Transient Receptor Potential Vanilloid 1. There are six transmembrane domains in TRPV1. Between fifth and sixth transmembrane domains, there is a brief hydrophobic stretch that can generate pores. Voltage, acidity (pH 5.9), toxic heat (>43 °C), and a variety of lipids all activate TRPV1. Additionally, capsaicin stimulates cation inflow, TRPV1 activation, and several additional physiological processes. (Adapted from Sun et al. 2016)

2014; Smutzer and Devassy 2016). A probable relationship between TRPV1 and GLP-1 is further suggested by the fact that capsaicin boosted GLP-1 and decreased ghrelin production (Smeets et al. 2009). On intestinal cells that express GLP-1, TRPV1 was found, and when activated, it induced the release of GLP-1 (Wang et al. 2012). An incretin hormone called GLP-1 stimulates the growth of insulin-secreting beta cells, which increases insulin production in regards to glucose (Macdonald et al. 2002). Additionally, GLP-1 increased both human and rodent insulin sensitivity (Parlevliet et al. 2010). Pancreatic-cells have been found to contain TRPV1, and channel activation has been linked to increased insulin release (Zsombok 2013).

Several prior investigations suggested that the stimulation of the Transient Receptor Potential Vanilloid 1 and its overexpression in the pancreas, liver was accountable for hypoglycemic effect of capsaicin (Zhang et al. 2018).

Because TRPV1 is involved in pain perception, pain management has received the greatest clinical attention as a therapeutic response to capsaicin through TRPV1 binding (Basith et al. 2016; Frias and Merighi 2016). Capsaicin initially causes pain by sensitizing TRPV1, and then inflammation causes pain as a result. Capsaicin exposure at high or repeated dosages causes TRPV1 to become desensitized, resulting in analgesia (Fattori et al. 2016). In addition to treating pain, capsaicin may also be helpful in the management of other illnesses, including cardiovascular disease and obesity. Additionally, capsaicin has antipruritic, anti-inflammatory, anti-apoptotic, anti-tumor, antioxidant, and neuroprotective properties. TRPV1 receptors allow capsaicin to deliver these health advantages (Basith et al. 2016). Lately, Zhang et al. (2021) showed that capsaicin blocked the TRPV1 channel in the glycometabolism of streptozotocin (STZ)-induced diabetic rats (Fig. 18.2).

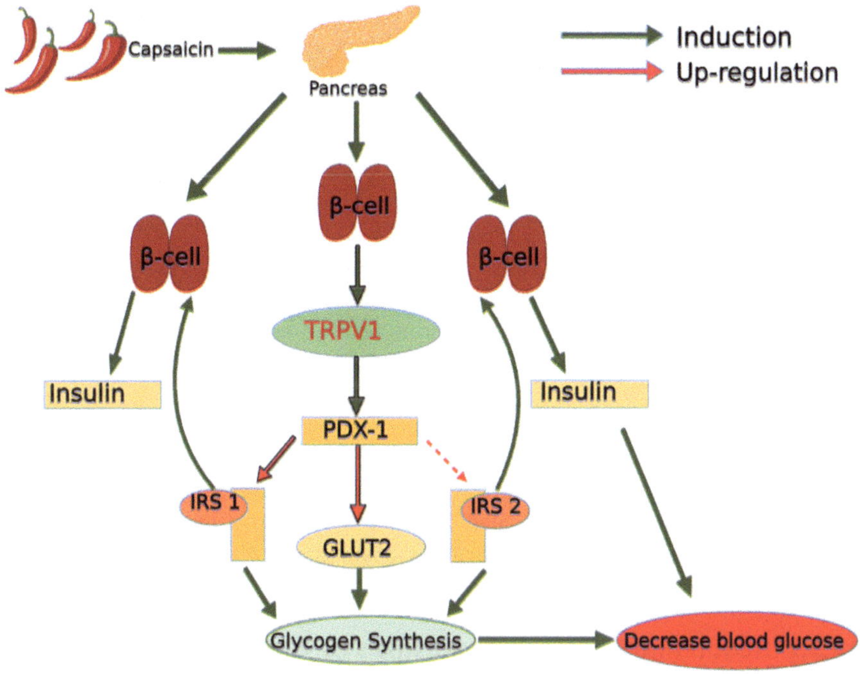

Fig. 18.2 In diabetic rats caused by streptozotocin, a particular pathway in pancreas signals the hypoglycemia-inducing effects of capsaicin. (Adapted from Zhang et al. 2021)

18.3 Capsaicin as Anti-diabetic Agent

Research on the impact of capsaicin on blood glucose levels has given researchers a thorough understanding of the compound's potential as an anti-diabetic, and they have discovered multiple methods by which it enhances carbohydrate metabolism. According to one study, the *C. annuum* acetone fraction (CAAF) fruit has significant anti-T2D effect that is mediated by improvement in insulin resistance and enhanced-cell function, both of which are linked to the fruit's active components (Mohammed et al. 2017).

18.3.1 Inhibiting Enzymes: Glucosidase and Amylase

One of these strategies entails inhibiting the action of the enzymes glucosidase and amylase, which can hydrolyze polysaccharides into glucose. The digestive enzyme α-amylase, which is present in the pancreas as well as saliva, hydrolyzes the α-1-4-glycosidic bond on starch (polysaccharide), generating tiny sugar molecules such as glucose and maltose in the process. Salivary α-amylase initiates starch digestion in the mouth, while pancreatic α-amylase continues it to the small intestine. The results of the process, glucose units, are taken up by the intestinal wall. The blood then transports the glucose to the intended cells all over the body.

Similar to this, the enzyme found in the small intestine, α-glucosidase, may hydrolyze polysaccharides at the final non-reducing stage into glucose α-1-4-glycosidic linkages. Monosaccharides like glucose, which are produced by α-glucosidase, are absorbed into the intestinal wall. Therefore, by delaying carbohydrate breakdown and absorption of glucose, inhibition of α-glucosidase as well as α-amylase can manage diabetes.

The research on these enzymes α-glucosidase as well as α-amylase in chili peppers revealed that all of the peppers might prevent the enzymes to varying degrees, and the inhibition was discovered to be dose-dependent. Compared to the other species of the capsicum family, C. annum extracts (5 mg/mL) had the strongest anti—glucosidase activity (66% inhibition). Additionally, the examined chili peppers' α-amylase inhibitory properties showed that each extract could inactivate the enzyme to varying degrees (5 mg/mL).

These results suggest that chili peppers may be used to treat diabetes by preventing the activity of two vital enzymes, glucosidase and amylase, which transform carbs into sugar (Watcharachaisoponsiri et al. 2016).

18.3.2 Regulating Glucose Homeostasis by GLP-1 Release via TRPV1

According to another study, dietary capsaicin controls glucose balance by triggering the manufacturing of hormone (GLP-1) glucagon-like peptide-1 through intestinal tissues and cells via potential transient receptor vanilloid 1 (Wang et al. 2012). It is present in neurons, rat pancreas, rat RIN and INS1 cell lines, islet, and other cell types (Akiba et al. 2004). Through activating TRPV1, treatment with capsaicin increase GLP-1 production in calcium-dependent behavior, from STC-1 cells. The first-ever data show that gastric gavage of capsaicin abruptly boosted insulin and glucagon-like-peptide-1 secretion in in-vivo WT mice, but had no impact on $TRPV1^{-/-}$ animals. Furthermore, prolonged dietary capsaicin enhanced glucose tolerance as well as decreased blood glucose daily profiles in WT mice while decreasing levels of plasma insulin as well as GLP-1, whereas such results were absent in $TRPV1^{-}/^{-}$ animals. In db/db mice, dietary capsaicin activated Transient Receptor Potential Vanilloid 1, enhancing abnormal glucose balance and increasing levels of glucagon-like peptide-1 in blood as well as terminal ileum (Wang et al. 2012).

18.3.3 Improved Glucose Metabolism

According to (Kwon et al. 2013), consuming capsaicin and capsiate orally may enhance metabolism of glucose by boosting βeta cell mass and its activity while reducing resistance by insulin in type II Px diabetes model. It was discovered that 0.025% capsaicin and 0.025% capsiate raised β-cell mass and enhancement in insulin release stimulated by glucose via enhancing β-cell proliferation and lowering apoptosis by heightening insulin/IGF-1 signaling.

But whereas capsiate, a non-pungent drug, caused a significant reduction in hepatic insulin resistance through improving hepatic insulin signaling, capsaicin did not reduce it. During a euglycemic hyperinsulinemic clamp, capsiate improved hepatic insulin sensitivity. It improves glycogen storage while decreasing hepatic glucose production and lipid buildup in the hyperinsulinemic state. This had to do with improved pAkt→PEPCK and pAMPK signaling.

18.3.4 Improving Glucose Homeostasis by Controlling Gut Microbiota

Numerous reports have revealed that by controlling the gut microbiota, a variety of dietary elements can prevent obesity and enhance glucose homeostasis. The microbiota presents in our gut, a diverse and intricate population of commensal microorganisms, which inhabits the mammalian digestive tract. Numerous pathological and physiological occurrences in the hosts are significantly influenced by the gut microbiota. Increased body fat content and insulin resistance were observed after microbe-free adult mice were conventionalized with typical microbiota collected through healthy mice's cecum. This resulted in the identification of gut microbiota's role in obesity. The ingestion of capsaicin markedly enhanced the homeostasis of glucose in ob/ob obese diabetic mice, despite not possessing any inhibitory activity on phenotypes related to obesity, according to the study. Through the stimulation of short-chain FA, the control of GI hormones, inhibiting pro-inflammatory cytokines, capsaicin in diet enhances glucose balance. This is due to high levels of *Roseburia* and decreased levels of *Parabacteroides* and B*acteroides* at genus level (Drew et al. 2017).

18.3.5 By Modulating Microbiota in Gut-Bile Acid-FXR Axis

To help in uptake of lipid-soluble vitamins, cholesterol, triglycerides, duodenum receives BA (bile acid), which is produced through cholesterol present in hepatocytes (Wahlström et al. 2016). Besides its function in digestion, BA is thought to be a signaling molecule that controls the production of its own, insulin sensitivity, and glucose metabolism (Fiorucci and Distrutti 2015). By potentiating small heterodimer partner (SHP) expression, a rate-limiting enzyme for bile acid production, cholesterol seven-hydroxylase (CYP7A1) is suppressed in liver by activating bile acid of FXR (SHP) (Goodwin et al. 2000; Lu et al. 2000). Additionally, BA production is influenced by the quantity of fibroblast growth factor 15 (FGF15, also known as FGF19 in humans). FGF receptor 4 binds to the postprandial hormone FGF15, which is triggered by FXR in the enterocytes of ileum and enters the portal circulation, inhibiting the synthesis of CYP7A1 (FGFR4) (Inagaki et al. 2006; Potthoff et al. 2012). Also, via controlling the expression of enzymes phosphoenolpyruvate carboxykinase (PEPCK), glycogen synthase kinase three (GSK3) and glucose-six-phosphatase (G6Pase), respectively, pathways related to FXR are linked to the control of glycogen formation and hepatic gluconeogenesis (Fiorucci and Distrutti 2015; Li and Chiang 2014).

According to the research, altering the gut microbiota may influence FXR activity by modifying BA profiles, which would then enhance host metabolism and make this an effective approach to treat metabolic diseases. One study suggests that by boosting hepatic glycogen synthesis and suppressing its gluconeogenesis, capsaicin helped diabetic db/db mice maintain normal blood sugar levels and reduced insulin resistance. These beneficial outcomes were also related with altered gut microbiota, as evidenced by lower number of Lactobacillus genus species and their BSH activity. This resulted in buildup of TβMCA, a natural FXR antagonist. Finally, enterohepatic FXR-FGF15 signaling was suppressed by Capsaicin therapy, which improved hepatic BA production at least in part. These findings point to the potential for future type II DM therapies that focus on FXR-related signals, particularly the gut microbiota's ability to convert BA (Hui et al. 2019).

18.3.6 In-Vivo Studies

According to one study, rats induced with diabetes via alloxan were fed a high-fat diet underwent a 4-week examination to find out how the serum levels of cholesterol, triglycerides, and glucose changed after receiving 0.015% capsaicin treatment (Mohamed Abdel Magied et al. 2014). To examine the hypolipidemic and hypoglycemic effects of red pepper in insulin-dependent diabetes, four groups of 36 male Wistar rats were randomly assigned to the control, pepper-treated, diabetic, and pepper-treated diabetic groups. Such findings revealed no appreciable variation in levels of serum glucose between the control and pepper-treated control groups. Red pepper therapy in diabetic rats reduced serum glucose levels in the short term, but it could only drop TG levels over time. This was demonstrated at 4 weeks, when the diabetic group receiving pepper treatment had significant TG serum levels, and at 2 weeks, when this group had significantly lower serum glucose levels than the diabetic group (Roghani et al. 2004).

According to Kang et al., mice fed on high-fatty diet for 10 weeks were given capsaicin (0.015% in food for 10 weeks), which reduced leptin, insulin, fasting glucose, obesity, hepatic triglycerides, markers of inflammation in both liver and adipose tissue. The effects of the treatment enhanced PPAR/PGC-1 and mRNA/ protein adiponectin in the liver and adipose tissue, respectively. Therefore, better inflammatory and metabolic effects in adipose tissue and liver point to ability of dietary capsaicin to lessen insulin resistance (Kang et al. 2010).

18.3.7 In-Vitro Studies

Skeletal muscle accounts for around 75% of the post-meal blood glucose clearance as well as glucose transport is mediated by two different pathways: the insulin signaling route as well as activation of AMP-activated protein kinase (AMPK). Type II diabetes, most common type of diabetes owing to insulin resistance, reduces glucose absorption even when there is insulin present (Björnholm and Zierath 2005). However, type II diabetes may benefit from the use of AMPK treatment since it promotes glucose uptake without the need for insulin (Hilder et al. 2005).

By activating AMPK, capsaicin boosted glucose absorption in C2C12 muscle cells while having little effect on IRS-1 or Akt, two key insulin signaling molecules. Further research revealed that capsaicin significantly increases p38 MAPK phosphorylation and SB203580, a particular inhibitor of p38 MAPK, blocked capsaicin-mediated glucose absorption, demonstrating that capsaicin-induced glucose transport is also regulated by p38 MAPK and occurs downstream of AMPK (Kim et al. 2013). In C2C12 muscle cells, it was shown to greatly increase the generation of ROS. The finding that effects caused by capsaicin were blocked as ROS were bound by the non-specific antioxidant NAC provided additional proof that these ROS are necessary for capsaicin-induced activation of AMPK and glucose absorption as they function as an upstream signal of AMPK (Fryer et al. 2002).

These findings imply that capsaicin promotes glucose absorption independently of insulin via generating ROS and then activating p38 MAPK and AMPK (Kim et al. 2013).

18.3.8 Human Studies

Capsaicin-containing chilies were used as a 4-week intervention in pregnant women having gestational diabetes mellitus to lower postprandial hyperinsulinemia and hyperglycemia and ameliorate fasting lipid metabolic abnormalities. The frequency of large-for-gestational-age babies was further decreased by these reactions. An enhanced release of CGRP was linked to these effects of capsaicin (Yuan et al. 2016).

Reduced inflammatory adipokines and insulin resistance were seen in another randomized, placebo-controlled, double-blind, 8-week experiment using a combination of supplements including capsaicin, indicating improved metabolic condition. According to the conclusions drawn from the assessment of homeostasis model, respiratory quotient, leptin/adiponectin ratio, and low-density level cholesterol levels, dietary supplement use was linked to a much larger reduction in insulin resistance. Although there was no statistically significant variation between groups, adiponectin caused a significant decrease in the placebo group while no change in supplemented group. Compared to placebo group, supplemented group had high resting energy expenditure and low levels of leptin, ghrelin, and C-reactive protein. Both groups experienced a significant increase in resting energy expenditure and free fat mass, but had low body mass index, fat mass (as determined by DXA), and VEGF levels (Rondanelli et al. 2012).

18.4 Conclusions

Considering how frequently it is used in both food and medicine, capsaicin has always garnered a lot of interest. Despite its negative consequences, capsaicin is nevertheless a common active ingredient in medicines used to treat a number of human diseases (Rios and Olivo 2014). Additionally, more recent studies have shown that capsaicin plays a larger role than previously believed. Numerous experimental and clinical studies have shown that it is important for other important pathological conditions, such as cancer, obesity, skin conditions, cardiovascular diseases, etc. (Jaganathan et al. 2020). These studies demonstrate its importance in other important clinical disorders although it is clearly described in the domain of nociception and pain. It has also been connected to numerous other processes, such as adaptive and innate immune response regulation, upper respiratory reflexes treatment, adipogenesis prevention, the stimulation of metabolism, and the treatment of adipogenesis.

The rising interest in capsaicin's potential benefits in treating a range of illnesses has allowed the pharmaceutical industry to suggest a brand-new chapter of the substance's use in treating diabetes (Basith et al. 2016). Diabetes can be successfully treated with chili peppers that contain the therapeutically effective compound capsaicin. The studies show that it can help fight diabetic mellitus through a variety of processes. As was already said, this review study discusses the functions and methods of treating diabetes by either inhibiting or activating the two key enzymes, amylase and glucosidase. Additionally, several research suggested that capsiate, a non-pungent component of chili peppers, works more efficiently than capsaicin to increase insulin sensitivity and glucose metabolism. Another study concluded that by enhancing beta-cell mass and activity and lowering insulin resistance, capsaicin and capsiate can both be beneficial in diabetes. Studies have also suggested that enterohepatic FXR-FGF15 signaling, which is suppressed by capsaicin, is a mechanism by which gut microbiota aids in the treatment of diabetes. This in turn boosts bile acid synthesis, which is further involved in the metabolism of glucose. Additionally, numerous human and in vitro studies point to capsaicin's potential relevance in the treatment of diabetes. Therefore, more research on capsaicin may be beneficial given its important role in metabolic syndrome and cardiovascular illnesses.

References

Akiba Y, Kato S, Katsube KI, Nakamura M, Takeuchi K, Ishii H, Hibi T (2004) Transient receptor potential vanilloid subfamily 1 expressed in pancreatic islet β cells modulates insulin secretion in rats. Biochem Biophys Res Commun 321(1):219–225. https://doi.org/10.1016/j.bbrc.2004.06.149

Basith S, Cui M, Hong S, Choi S (2016) Harnessing the therapeutic potential of capsaicin and its analogues in pain and other diseases. Molecules 21(8):966. https://doi.org/10.3390/molecules21080966

Björnholm M, Zierath JR (2005) Insulin signal transduction in human skeletal muscle: identifying the defects in Type II diabetes. Biochem Soc Trans 33(2):354–357

Clapham DE (2003) TRP channels as cellular sensors. www.nature.com/nature

Conway SJ (2008) TRPing the switch on pain: an introduction to the chemistry and biology of capsaicin and TRPV1. Chem Soc Rev 37(8):1530–1545. https://doi.org/10.1039/b610226n

Devesa I, Ferrándiz-Huertas C, Mathivanan S, Wolf C, Luján R, Changeux JP, Ferrer-Montiel A (2014) αCGRP is essential for algesic exocytotic mobilization of TRPV1 channels in peptidergic nociceptors. Proc Natl Acad Sci U S A 111(51):18345–18350. https://doi.org/10.1073/pnas.1420252111

Drew BG, Coughlan M, Yepuri G, Chen H, Song JX, Ren H, Gao YF, Lee CY, Li SF, Zhang F, Li L (2017) Dietary capsaicin improves glucose homeostasis and alters the gut microbiota in obese diabetic Ob/Ob mice. Front Physiol 1:602. https://doi.org/10.3389/fphys.2017.00602

Fattori V, Hohmann MSN, Rossaneis AC, Pinho-Ribeiro FA, Verri WA (2016) Molecules capsaicin: current understanding of its mechanisms and therapy of pain and other pre-clinical and clinical uses. Molecules 21(7):844. https://doi.org/10.3390/molecules21070844

Fiorucci S, Distrutti E (2015) Bile acid-activated receptors, intestinal microbiota, and the treatment of metabolic disorders. Trends Mol Med 21(11):702–714. https://doi.org/10.1016/J.MOLMED.2015.09.001

Frias B, Merighi A (2016) Molecules capsaicin, nociception and pain. Molecules 21(6):797. https://doi.org/10.3390/molecules21060797

Fryer LGD, Parbu-Patel A, Carling D (2002) The anti-diabetic drugs rosiglitazone and metformin stimulate AMP-activated protein kinase through distinct signaling pathways. J Biol Chem 277(28):25226–25232. https://doi.org/10.1074/JBC.M202489200

Goodwin B, Jones SA, Price RR, Watson MA, McKee DD, Moore LB, Galardi C, Wilson JG, Lewis MC, Roth ME, Maloney PR, Willson TM, Kliewer SA (2000) A regulatory cascade of the nuclear receptors FXR, SHP-1, and LRH-1 represses bile acid biosynthesis. Mol Cell 6(3):517–526. https://doi.org/10.1016/S1097-2765(00)00051-4

Gowd V, Jia Z, Chen W (2017) Anthocyanins as promising molecules and dietary bioactive components against diabetes—a review of recent advances. Trends Food Sci Technol 68:1–13. https://doi.org/10.1016/J.TIFS.2017.07.015

Gupta R, Kapoor B, Gulati M, Kumar B, Gupta M, Singh SK, Awasthi A (2022) Sweet pepper and its principle constituent capsiate: functional properties and health benefits. Crit Rev Food Sci Nutr 62(26):7370–7394. https://doi.org/10.1080/10408398.2021.1913989

Hewlings SJ, Kalman DS (2017) Curcumin: a review of its' effects on human health. Foods 6(10):92. https://doi.org/10.3390/foods6100092

Hilder TL, Baer LA, Fuller PM, Fuller CA, Grindeland RE, Wade CE, Graves LM (2005) Insulin-independent pathways mediating glucose uptake in hindlimb-suspended skeletal muscle. J Appl Physiol 99:2181–2188. https://doi.org/10.1152/japplphysiol.00743.2005.-In

Huang PL (2009) A comprehensive definition for metabolic syndrome. Dis Model Mech 2(5–6):231–237. https://doi.org/10.1242/dmm.001180

Hui S, Liu Y, Chen M, Wang X, Lang H, Zhou M, Yi L, Mi M (2019) Capsaicin improves glucose tolerance and insulin sensitivity through modulation of the gut microbiota-bile acid-FXR axis in type 2 diabetic db/db mice. Mol Nutr Food Res 63(23):e1900608. https://doi.org/10.1002/mnfr.201900608

Inagaki T, Moschetta A, Lee YK, Peng L, Zhao G, Downes M, Yu RT, Shelton JM, Richardson JA, Repa JJ, Mangelsdorf DJ, Kliewer SA, O'malley BW (2006) Regulation of antibacterial defense in the small intestine by the nuclear bile acid receptor. Proc Natl Acad Sci U S A 103(10):3920–3925

Jaganathan R, Purushotham B, Radhakrishnan N, Swamy MK (2020) Capsaicin and its potential anticancer mechanisms of action. In: Swamy M (ed) Plant-derived bioactives. Springer, Singapore, pp 301–321

Kang JH, Tsuyoshi G, Han IS, Kawada T, Min Kim Y, Yu R (2010) Dietary capsaicin reduces obesity-induced insulin resistance and hepatic steatosis in obese mice fed a high-fat diet. Obesity (Silver Spring) 18(4):780–787. https://doi.org/10.1038/oby.2009.301

Kim SH, Hwang JT, Park HS, Kwon DY, Kim MS (2013) Capsaicin stimulates glucose uptake in C2C12 muscle cells via the reactive oxygen species (ROS)/AMPK/p38 MAPK pathway. Biochem Biophys Res Commun 439(1):66–70. https://doi.org/10.1016/j.bbrc.2013.08.027

Kumar R, Hazan A, Basu A, Zalcman N, Matzner H, Priel A (2016) Tyrosine residue in the TRPV1 vanilloid binding pocket regulates deactivation kinetics. J Biol Chem 291(26):13855–13863. https://doi.org/10.1074/JBC.M116.726372

Kwon DY, Kim YS, Ryu SY, Cha MR, Yon GH, Yang HJ, Kim MJ, Kang S, Park S (2013) Capsiate improves glucose metabolism by improving insulin sensitivity better than capsaicin in diabetic rats. J Nutr Biochem 24(6):1078–1085. https://doi.org/10.1016/j.jnutbio.2012.08.006

Legeay S, Rodier M, Fillon L, Faure S, Clere N (2015) Epigallocatechin gallate: a review of its beneficial properties to prevent metabolic syndrome. Nutrients 7:5443–5468. https://doi.org/10.3390/nu7075230

Li T, Chiang JYL (2014) Bile acid signaling in metabolic disease and drug therapy. Pharmacol Rev 66:948–983. https://doi.org/10.1124/pr.113.008201

Liang W, Lan Y, Chen C, Song M, Xiao J, Huang Q, Cao Y, Ho CT, Lu M (2023) Modulating effects of capsaicin on glucose homeostasis and the underlying mechanism. Crit Rev Food Sci Nutr 63(19):3634–3652. https://doi.org/10.1080/10408398.2021.1991883

Lu TT, Makishima M, Repa JJ, Schoonjans K, Kerr TA, Auwerx J, Mangelsdorf DJ (2000) Molecular basis for feedback regulation of bile acid synthesis by nuclear receptors. Mol Cell 6(3):507–515. https://doi.org/10.1016/S1097-2765(00)00050-2

Macdonald PE, El-Kholy W, Riedel MJ, Marie A, Salapatek F, Light PE, Wheeler MB (2002) The multiple actions of GLP-1 on the process of glucose-stimulated insulin secretion. Diabetes 51 (Suppl 3):S434–S442. https://doi.org/10.2337/diabetes.51.2007.s434

Marín-Peñalver JJ, Martín-Timón I, Sevillano-Collantes C, Javier Del Cañizo-Gómez F (2016) Update on the treatment of type 2 diabetes mellitus. World J Diabetes 7(17):354–395. https://doi.org/10.4239/wjd.v7.i17.354

Mohamed Abdel Magied M, Abdel Rahman Salama N, Rashad Ali M (2014) Hypoglycemic and hypocholesterolemia effects of intragastric administration of dried red chili pepper (capsicum annum) in alloxan-induced diabetic male albino rats fed with high-fat-diet. J Food Nutr Res 2(11):850–856. https://doi.org/10.12691/jfnr-2-11-15

Mohammed A, Koorbanally N, Islam MS (2017) Anti-diabetic effect of Capsicum annuum L. fruit acetone fraction in a type 2 diabetes model of rats

Mózsik G, Past T, Abdel Salam OME, Kuzma M, Perjési P (2009) Interdisciplinary review for correlation between the plant origin capsaicinoids, non-steroidal antiinflammatory drugs, gastrointestinal mucosal damage and prevention in animals and human beings. Inflammopharmacology 17(3):113–150. https://doi.org/10.1007/s10787-009-0002-3

Nilius B, Owsianik G (2010) Transient receptor potential channelopathies. Pflugers Arch 460(2): 437–450. https://doi.org/10.1007/s00424-010-0788-2

Nishida K, Otsu K (2017) Inflammation and metabolic cardiomyopathy. Cardiovasc Res 113(4): 389–398. https://doi.org/10.1093/cvr/cvx012

Parlevliet ET, de Leeuw Van Weenen JE, Romijn JA, Pijl H (2010) GLP-1 treatment reduces endogenous insulin resistance via activation of central GLP-1 receptors in mice fed a high-fat diet. Am J Physiol Endocrinol Metab 299:318–324. https://doi.org/10.1152/ajpendo.00191.2010.-Glucagon-like

Patowary P, Pathak MP, Zaman K, Raju PS, Chattopadhyay P (2017) Research progress of capsaicin responses to various pharmacological challenges. Biomed Pharmacother 96:1501–1512. https://doi.org/10.1016/J.BIOPHA.2017.11.124

Potthoff MJ, Kliewer SA, Mangelsdorf DJ (2012) Endocrine fibroblast growth factors 15/19 and 21: from feast to famine. Genes Dev 26(4):312–324. https://doi.org/10.1101/gad.184788.111
Rios MY, Olivo HF (2014) Natural and synthetic alkamides: applications in pain therapy. Stud Nat Prod Chem 43:79–121. https://doi.org/10.1016/B978-0-444-63430-6.00003-5
Roghani M, Baluchnejadmojarad T, Sohrabi Z, Sadeghi M (2004) Anti-hyperlycemic and hypolipidemic effect of oral administration of capsicum frutescens in male STZ-diabetic rats. J Med Plants 3(10):47–52
Rondanelli M, Opizzi A, Perna S, Faliva M, Solerte SB, Fioravanti M, Klersy C, Cava E, Paolini M, Scavone L, Ceccarelli P, Castellaneta E, Savina C, Donini LM (2012) Improvement in insulin resistance and favourable changes in plasma inflammatory adipokines after weight loss associated with two months' consumption of a combination of bioactive food ingredients in overweight subjects. Endocrine 44(2):391–401. https://doi.org/10.1007/s12020-012-9863-0
Smeets AJ, Margriet AE, Westerterp-Plantenga S (2009) The acute effects of a lunch containing capsaicin on energy and substrate utilisation, hormones, and satiety. Eur J Nutr 48(4):229–234. https://doi.org/10.1007/s00394-009-0006-1
Smutzer G, Devassy RK (2016) Integrating TRPV1 receptor function with capsaicin psychophysics. Adv Pharmacol Sci 2016:1512457. https://doi.org/10.1155/2016/1512457
Springer M, Moco S (2019) Resveratrol and its human metabolites-effects on metabolic health and obesity. Nutrients 11(1):143. https://doi.org/10.3390/nu11010143
Sun F, Xiong S, Zhu Z (2016) Dietary capsaicin protects cardiometabolic organs from dysfunction. Nutrients 8(5):174. https://doi.org/10.3390/nu8050174
Szolcsányi J (2014) Capsaicin and sensory neurones: a historical perspective. Prog Drug Res 68:1–37. https://doi.org/10.1007/978-3-0348-0828-6_1
Wahlström A, Sayin SI, Marschall HU, Bäckhed F (2016) Intestinal crosstalk between bile acids and microbiota and its impact on host metabolism. Cell Metab 24(1):41–50. https://doi.org/10.1016/J.CMET.2016.05.005
Wang P, Yan Z, Zhong J, Chen J, Ni Y, Li L, Ma L, Zhao Z, Liu D, Zhu Z (2012) Transient receptor potential vanilloid 1 activation enhances gut glucagon-like peptide-1 secretion and improves glucose homeostasis. Diabetes 61(8):2155–2165. https://doi.org/10.2337/db11-1503
Watcharachaisoponsiri T, Sornchan P, Charoenkiatkul S, Suttisansanee U (2016) The α-glucosidase and α-amylase inhibitory activity from different chili pepper extracts. Int Food Res J 23(4):1439–1445
Yuan LJ, Qin Y, Wang L, Zeng Y, Chang H, Wang J, Wang B, Wan J, Chen SH, Zhang QY, Zhu JD, Zhou Y, Mi MT (2016) Capsaicin-containing chili improved postprandial hyperglycemia, hyperinsulinemia, and fasting lipid disorders in women with gestational diabetes mellitus and lowered the incidence of large-for-gestational-age newborns. Clin Nutr 35(2):388–393. https://doi.org/10.1016/J.CLNU.2015.02.011
Zhang S, Ma X, Zhang L, Sun H, Liu X (2017) Capsaicin reduces blood glucose by increasing insulin levels and glycogen content better than capsiate in streptozotocin-induced diabetic rats. J Agric Food Chem 65(11):2323–2330. https://doi.org/10.1021/acs.jafc.7b00132
Zhang S, Tang L, Xu F, Hui Y, Lu H, Liu X (2021) TRPV1 receptor-mediated hypoglycemic mechanism of capsaicin in streptozotocin-induced diabetic rats. Front Nutr 8:750355. https://doi.org/10.3389/fnut.2021.750355
Zhang S, You Y, Liu J, Qin X, Liu X (2018) Hypoglycaemic effect of capsaicinoids via elevation of insulin level and inhibition of glucose absorption in streptozotocin-induced diabetic rats. J Funct Foods 51:94–103. https://doi.org/10.1016/J.JFF.2018.10.015
Zsombok A (2013) Vanilloid receptors—do they have a role in whole body metabolism? Evidence from TRPV1. J Diabetes Complicat 27(3):287–292. https://doi.org/10.1016/J.JDIACOMP.2012.11.006

Anti-obesity Effects of Capsaicin via Gut Microbiota 19

Cemre Elmas and Ceren Gezer

Abstract

Bioactive components obtained from plants have been used for the prevention, treatment and reduction of diseases' symptoms. Capsaicin is the most frequently used capsaicinoids. It is the important bioactive component found in the *Capsicum* species. Capsaicin has an antioxidant, anti-inflammatory and antiproliferative metabolic effects thus it is related to numerous lifestyle related chronic diseases (obesity, diabetes and different type cancer, etc.). In animal and human researches, it has been shown that capsaicin provides intestinal integrity by changing the gut microbiota profile, prevents chronic low-grade inflammation, regulates hunger-satiety hormones by affecting the gut-brain axis, and has obesity preventive effects through energy metabolism. Understanding the relation between capsaicin, obesity, gut microbiota is suggested as an effective strategy for preventing obesity and co-morbidities. However, to take advantage of the pathways in which capsaicin is effective in preventing and treating obesity, more clinical studies are needed on what dose, how, and when it should be taken. This chapter summarises the anti-obesity effects of capsaicin through gut microbiota and potential mechanisms using updated literature.

Keywords

Capsaicin · Obesity · Gut microbiota · Peppers · Nutrition

C. Elmas
Department of Nutrition and Dietetics, Faculty of Health Sciences, European University of Lefke, Lefke, North Cyprus, Turkey

C. Gezer (✉)
Department of Nutrition and Dietetics, Faculty of Health Sciences, Eastern Mediterranean University, Famagusta, North Cyprus, Turkey
e-mail: ceren.gezer@emu.edu.tr

M. K. Swamy, A. Kumar (eds.), *Capsaicinoids*,
https://doi.org/10.1007/978-981-99-7779-6_19

19.1 Introduction

Globally, there are differences in the quality of life of individuals because of socio-economic differences. Although this fact has been in existence virtually since the first days of humanity, it is seen that these differences between societies have recently increased dramatically. While some parts of the world are struggling with hunger and the problems brought on by hunger, which is the result of negative energy balance, a large proportion of the world struggle with obesity and its complications, the primary cause of which is known as excessive food intake. It is known that the incidence of obesity has increased gradually, especially in the passing 40 years, and it seriously affects people's health negatively. As a result, obesity has been begun to be handled in different ways by researchers (Crovesy et al. 2020; Ballini et al. 2020; Loos and Yeo 2022). Imbalance in energy (between intake—expenditure), unhealthy life-style and heredity variables are among the general factors known to cause obesity problem. However, to better define the etiology of obesity, instead of focusing on a single factor among these factors, it was thought that multidisciplinary approaches would be more effective and multi-disciplinary approaches have become popular (Loos and Yeo 2022). One of the popular approaches in the prevention and treatment of obesity is related to the microbiota which is known to have a role in the formation of many diseases. In addition, the factors such as lifestyle and genetics not only affect microbiota but also affect formation of obesity. Because of these similar interactions, the relationship between obesity and microbiota has gained importance in recent years (Crovesy et al. 2020). The gut microbiota is recognised as an important potential factor for obesity and its comorbidities (Muscogiuri et al. 2019). The microbiota is thought to be a significant indicator of both the environmental factors and the genetic material of humans (Abenavoli et al. 2019; Ayariga et al. 2023). One of the significant factors affecting the gut microbiota is dietary consumption. Different components of the consumed foods become a source of nutrients for the gut microbiota. In addition, it is seen that bioactive components, especially those found in some foods as natural, have important functions in metabolism, energy pathways, and regulatory signal pathways (Ballini et al. 2020; Ramírez-Aragón et al. 2023). One of the bioactive components that has a modulatory effect on the gut microbiota is capsaicin. It is known that capsaicin can change the gut microbiota by increasing the beneficial bacteria profile by supporting the energy sources of the microbiota (Rosca et al. 2020; Mahalak et al. 2022). Furthermore, capsaicin has functions that affect body weight via various ways such as activating lipolysis in adipose tissue, formation affecting hunger-satiety hormones hence feeling of satiety, and stimulating energy expenditure (Elmas and Gezer 2022). Depending on these effects, capsaicin is thought to have an important role on energy homeostasis, obesity, and its complications. Therefore, microbiota-centred treatments supported by bioactive components have become important in obesity (Loos and Yeo 2022; Crovesy et al. 2020; Abenavoli et al. 2019). It has been aimed to examine the obesity prevention effects of capsaicin via microbiota in this chapter.

19.2 Capsaicin

Capsaicin (trans-8-methyl-N-vanilyl-6-nonenamide) is the main bioactive component of the *Capsicum* genus hot pepper, which is widely used as a pungent flavouring agent (Wu et al. 2022a, b). When hot pepper is consumed, a burning sensation occurs. It is the capsaicin, a group of alkaloids called capsaicinoids that causes this burning sensation. There are five different type capsaicinoids (capsaicin, dihydrocapsaicin, nordihydrocapsaicin, homodihydrocapsaicin, homocapsaicin) (Lu et al. 2020). The ratio of capsaicinoids that provide hotness in the analyses made using the chromatogram is shown as follow: Capsaicin 69%, dihydrocapsaicin 26% and other detected capsaicinoids provide 5% effect. Therefore, the major hotness component is considered to be capsaicin (Shafiee et al. 2022; Elmas and Gezer 2022).

The chemical structure of capsaicin consists of vanilloid ring, amide bond, and fatty acid side chain. The main factor that enables the detection of capsaicin in body is the 'transient receptor potential vanilloid receptor-1 (TRPV1)' (Kwon et al. 2021; Xiang et al. 2022b). TRPV1 is a family of receptors stimulated by severe stimuli such as heat and vanilloid, and has important roles in body composition and, regulation of glucose-lipid metabolism and cardio-vascular system. While there exist limited data on how this family of receptors works, it is accepted that our ability to sense heat is mediated by TRPV1, a calcium-permeable ion channel located in the primary-sensory-nerve-terminals (Munjuluri et al. 2021; Xiang et al. 2022b). Absorption of capsaicin in the body takes place in the stomach and duodenum. The absorption model of capsaicin is passive absorption. After capsaicin is absorbed, it binds to TRPV1 and activates this channel. Thus, a chemically induced heat sensation occurs (Kwon et al. 2021; Xiang et al. 2022b). The first TRPV1 agonist discovered is capsaicin, and it is known that most of its functions are TRPV1-mediated, although it has some effects independent of TRPV1 depending on the dose. Most researches on capsaicin consider the TRPV1 related mechanism of capsaicin (Wu et al. 2022a, b; Xiang et al. 2022b). For example, capsaicin exerts its analgesic effect, known as one of the most common effects, through TRPV1. Capsaicin can bind to the TRPV1 receptor on the nociceptive neurons of the peripheral nervous system hence can show an analgesic effect (Lu et al. 2020). Besides its analgesic effect, capsaicin appears to have antioxidant, anti-inflammatory, anti-carcinogen, anti-obesity effects; cardioprotection, blood lipids and blood glucose regulation and metabolic modulation. Because of these properties, it has been used in medicine for centuries (Lu et al. 2020; Wu et al. 2022a, b; Kwon et al. 2021). In in vitro studies with different cell types, capsaicin was found to inhibit lipid peroxidation, and it was found to be a stronger antioxidant than alpha-tocopherol (Chung and Campbell 2016; Tanrıkulu-Küçük et al. 2019; Zheng et al. 2017). In studies investigating its anti-inflammatory effect, it was observed that capsaicin may be effective in relation to release of substances as pro-inflammatory mediators that activate the TRPV1 channel (Wang et al. 2017). In many studies, capsaicin also shows potent anti-cancer activities (inhibition of proliferation, angiogenesis and metastasis; activation of apoptosis and autophagy) in various cancer types. In addition, researches have shown that it can rise the activity of chemotherapy and radiotherapy medicines used in cancer treatment and reduce side effects (Zhang et al. 2020b; Lan et al. 2019; Popescu et al. 2020; Wang et al. 2018; Parashar et al. 2019). It is also seen that capsaicin may have anti-obesity effects with its effects on the adipose tissue, and the sympathetic nervous system mediated gastrointestinal system (hunger-satiety hormones, digestive enzymes, and microbiota) (Karri et al. 2019; Lu et al. 2020; Elmas and Gezer 2022; Wu et al. 2022a, b).

19.3 Obesity and Microbiota

19.3.1 Obesity

Excessive increase in body fat rate has become one of the biggest health problems of today. This condition, called obesity, has affected almost every age group (Müller et al. 2022; Klein et al. 2022). It is known that approximately 20% of children and adolescents and approximately 30% of adults are obese worldwide (Müller et al. 2022). Obesity can develop due to genetic factors and other health problems (Loos and Yeo 2022). However, the ease of access to foods rich in energy and the decrease in physical activity due to living conditions are shown as the main reason for this global prevalence of obesity caused by positive energy balance (Yasmin et al. 2021). It is accepted that obesity is a health problem that affects quality of life. It is seen that obesity, especially in the early stages of life, brings along many psychological, neurological, endocrine, musculoskeletal and gastrointestinal diseases. It is stated that obesity can rise the incidence of cardiovascular diseases, diabetes, cancer later in life (Müller et al. 2022; Klein et al. 2022; Rohm et al. 2022).

It is indicated that capsaicin can positively affect obesity and insulin resistance, diabetes, hyperlipidemia, some types of cancer as comorbidities of obesity. Capsaicin can control obesity by preventing lipid oxidation and lipid storage in the body, regulating hunger-satiety through the hypothalamus (Karri et al. 2019; Lu et al. 2020; Elmas and Gezer 2022). It is also stated that capsaicin acts as a lipase inhibitor (Wu et al. 2022a, b). Although it is seen that these effects of capsaicin on obesity are mostly mediated by the sympathetic nervous system and adipose tissue (brown and white) associated with hunger-satiety metabolism, recent studies have shown that it may have an effect on the intestines microbiota profile as well as the impacts of hormones and enzymes in the gastrointestinal system. With the effect of capsaicin, the amount of bacterial species such as *Akkermansia*, *Allobaculum*, *Bacteroides*, *Coprococcus*, *Odoribacter* and *Prevotella* in the gut microbiota increase, and a decrease occurs in some other bacteria species such as *Desulfovibrio*, *Escherichia*, *Helicobacter* and *Sutterella* (Karri et al. 2019; Wu et al. 2022a, b; Elmas and Gezer 2022; Wang et al. 2020a). In the absence of capsaicin, when the relationship between body fat rate and gut microbiota is examined, it is seen that gut microbiota may differ according to body fat rate. In the gut microbiota of obese individuals, the *Firmicutes/Bacteroidetes ratio and Firmicutes*, *Fusobacteria*, *Proteobacteria*, *Mollicutes*, *Lactobacillus reuteri* counts are higher, *Bacteroidetes*, *Akkermansia muciniphila*,

Faecalibacterium prausnitzii, *Methanobrevibacter smithii*, *Lactobacillus plantarum* and *Lactobacillus paracasei* numbers are lower (Crovesy et al. 2020; Newman et al. 2021; Amabebe et al. 2020). This bacterial imbalance in the microbiota is positively regulated by capsaicin (Karri et al. 2019; Wu et al. 2022a, b; Elmas and Gezer 2022; Wang et al. 2020a).

19.3.2 Microbiota

Microbial communities constitute an important section of the impact of environmental factors on human optimal health and diseases. These microorganisms, which are generally called microbiota, coexist on all body surfaces and cavities. These microorganisms can consist of a large size of bacteria, archaea, bacteriophages, eukaryotic viruses and fungi (Fan and Pedersen 2021). Most of the microorganisms living in humans are found in the intestine. The bacterial density in the colon is estimated to be between 10^{11} and 10^{12} per mL.

Bacteria are classified taxonomically. This classification is made as follows: phyla, class, order, family, genera and species. The predominant gut microbial phyla are *Actinobacteria*, *Bacteroidetes*, *Firmicutes*, *Fusobacteria*, *Proteobacteria*, and *Verrucomicrobia*. *Bacteroidetes* and *Firmicutes* constitute 90% of gut microbiota (Rinninella et al. 2019). More than 200 different genera such as *Bacillus*, *Clostridium*, *Enterococcus*, *Lactobacillus* and *Ruminococcus* are composed of *Firmicutes* phylum. The dominant genera of *Bacteroidetes* phylum are *Bacteroides* and *Prevotella*. Another gut microbiota phylum, *Actinobacteria*, mainly includes the *Bifidobacterium* genera (Fan and Pedersen 2021). However, since the intestinal microbiota is impacted by many agents, the estimated number and profile of microorganisms show individual differences. Mode of birth, life-long nutritional model from infancy (such as breast milk consumption, etc.), lifestyle, drugs, and genetics of the host are factors that affect the intestinal microbiota (Rinninella et al. 2019; Lynch and Pedersen 2016; Gacesa et al. 2022).

The gut microbiota is an important component of digestion, and is effective in the minor degradation of macro-nutrients that arrive the lower digestive system (Oliphant and Allen-Vercoe 2019). A large number of microbial metabolites are formed end of this process which have various activities in different parts of the body. Some of these microbial metabolites are beneficial and some are toxic. At the same time, the impact on human health is complex, as the effects of microbial metabolites depend on metabolite concentration or organ location (Wu et al. 2022a, b; Wang et al. 2021a). Gut microbiota performs biotransformation of bioactive components such as phytochemicals and micro-nutrients can show antioxidant, anti-inflammatory, anti-cancer and anti-obesity effects, as well as macro-nutrients, enabling the host to benefit from these substances (Oliphant and Allen-Vercoe 2019). Consequently, the intestinal microbiota appears to have important roles in enhancing host immunity, digestive and endocrine system functions, regulating neurological signals, altering the effect of drugs, removing toxins, and the formation of certain compounds that influence the host (Fan and Pedersen 2021; Lynch and

Pedersen 2016). According to the most recent data, it is stated that the intestinal microbiota may affect the mental health, metabolic health of the human host, and in case of deterioration, it may be effective in the pathophysiological formation of various health problems (obesity, malnutrition, cardiovascular diseases, diabetes and non-alcoholic fatty liver disease) (Fan and Pedersen 2021; Vijay and Valdes 2022). In spite of the great variation in the pathophysiology of widespread chronic diseases, it is known that they might be associated to the content and function of the intestinal microbiota. It is widely accepted that the intestinal microbiota is particularly associated with the development of obesity (Fan and Pedersen 2021; Lynch and Pedersen 2016; Isacco et al. 2021; Vijay and Valdes 2022). The relationship between the intestinal microbiota and the formation of obesity and other chronic disorders is thought to be because of low-grade inflammation associated with endotoxemia and energy provided by toxic metabolites from the microbiota in the intestinal microbiota (Cao et al. 2019). The control of energy metabolism and the change of preferred nutrients depending on different bacterial compositions are shown as the main reason for the relationship between microbiota and obesity (Ai et al. 2017; Mameli et al. 2019; Abenavoli et al. 2019; Cao et al. 2019).

It is known that plants rich in bioactive components (hot pepper, grapes, berries, apples, turmeric, and soy) can be influence in body composition via intestinal microbiota. For this reason, capsaicin, which is especially contained in hot pepper, is a bioactive component that has been widely used in the therapy and prevention of obesity recently via its effects on the intestinal microbiota (Cao et al. 2019; Yin et al. 2019; Sudheer et al. 2022). By increasing the number of beneficial bacteria, capsaicin positively affects the intestinal microbiota diversity. The most important result of bacterial diversity is the increase in the production of short-chain fatty acids (SCFA), along with many metabolic processes. It also provides the necessary energy for gut bacteria. Thus, by supporting intestinal permeability, it prevents inflammation that supports the formation of obesity. Capsaicin supports the gut microbiota profile by increasing the number of beneficial bacteria (Wu et al. 2022a, b; Wang et al. 2020a; Coppola et al. 2021).

19.3.3 Capsaicin, Microbiota and Obesity

Capsaicin exerts its anti-obesity effects through one of its important components microbiota. In the realisation of this effect, there are three basic conditions: changing the profile of microorganisms that make up the intestinal microbiota, supporting the intestinal permeability by rising the formation of short-chain fatty acids (SCFA), and stimulating the microbiota-gut-brain axis (Xiang et al. 2021; Coppola et al. 2021; Song et al. 2019) (Fig. 19.1). Capsaicin increases butyrate-producing bacteria (*Roseburia* spp., *Ruminococcaceae*, etc.) and mucin-producing bacteria (*Akkermansia muciniphila*, etc.) and supports the formation of butyrate, which is a significant energy source of gut cells. Capsaicin can also stimulate the expression of ZO-1 and occludin which are tight junction proteins. With these effects, intestinal barrier integrity can be ensured and bacterial endotoxins can be prevented from

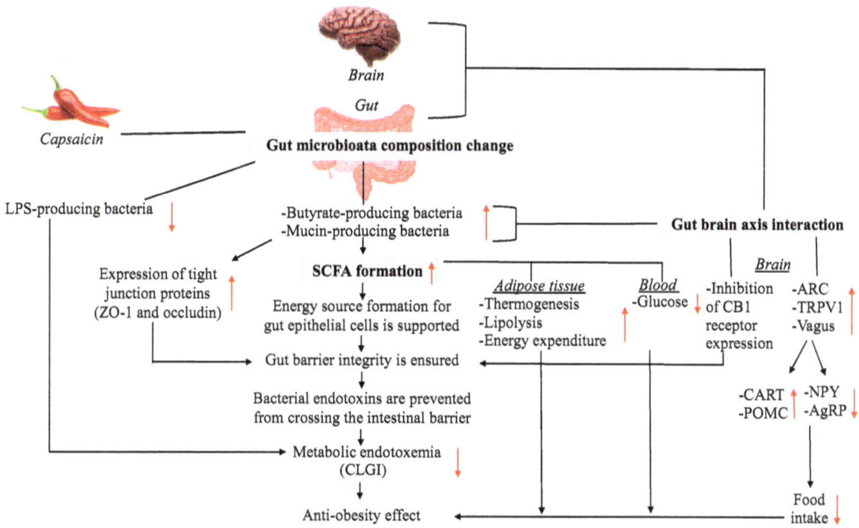

Fig. 19.1 Summary of the effect of capsaicin on microbiota-mediated obesity. *LPS* lipopolysaccharide, *SCFA* short-chain fatty acids, *CLGI* chronic low-grade inflammation, *CB1* cannabinoid receptor type 1, *ARC* arcuate nucleus, *TRPV1* transient receptor potential vanilloid receptor-1, *CART* amphetamine-regulated transcript, *POMC* proopiomelanocortin, *NPY* neuropeptide Y, *AgRP* agouti-related peptide, *(red)* — ↓ decrease, *(red)* — ↑ increase

crossing the intestinal barrier. Inhibition of metabolic endotoxemia creates anti-inflammatory (inhibition of chronic low-grade inflammation) and anti-obesity effects. The rise in the amount of butyrate due to such changes in the intestinal microbiota profile may also affect the body fat (Weng et al. 2021; Roshanravan et al. 2021; Kang et al. 2017; Wang et al. 2021b). Oxidation of fatty acids and glucose increases (increased thermogenesis) due to the increase in brown uncouple proteins-1 under the influence of capsaicin. As a result, energy expenditure also increases. In white adipose tissue, with the effect of capsaicin lipolysis and adiponectin increase and body weight decreases as a result of the increase in energy expenditure, thus obesity can be prevented. In addition, capsaicin can prevent obesity by reducing blood glucose levels (Elmas and Gezer 2022; Weng et al. 2021).

Changes in the gut microbiota due to different reasons (increase in lipopolysaccharide producing bacteria and/or change in SCFA production) may also affect intestinal barrier integrity (Coppola et al. 2021; Song et al. 2019; Scheithauer et al. 2020). Intestinal integrity is impaired, especially due to the change in SCFA production. As a result of the impaired intestinal barrier, intestinal bacteria producing gram-negative lipopolysaccharide directly enter the systemic circulation, causing an increase in bacterial LPS levels in plasma and triggering chronic low-grade inflammation due to metabolic endotoxemia. Thus, metabolic endotoxemia resulting from the impaired microbiota profile is accepted as the main mediator in starting CLGI responsible for the formation of obesity, and it is related with up levels of circulating interleukin-1beta (IL-1β), interleukin-6 (IL-6) and tumour necrosis factor alpha (TNF-α). It is also seen that chronic low-grade inflammation disrupts pancreatic β-cells, impairs insulin secretion, and causes glucose intolerance in obese. While capsaicin causes an increase in butyrate-producing bacteria (*Roseburia* spp., *Ruminococcaceae*, etc.), it reduces the number of lipopolysaccharide-producing bacteria (*S24_7 bacteria family*, *Proteobacteria*, etc.) and lipopolysaccharide biosynthesis. Thus, metabolic endotoxemia and systemic chronic low-grade inflammation prevention can be resulted as prevention of obesity (Gasmi et al. 2021; Scheithauer et al. 2020; Kang et al. 2017).

In addition to its effect on hunger and satiety control, it is accepted that capsaicin also provides the connection between the intestine and the brain through the intestinal microbiota. One of these pathways is mediated by the cannabinoid receptor type 1 (CB1). CB1 is one of the factors affecting intestinal integrity. Cannabinoids modulate intestinal permeability through expression of CB1. By inhibiting the expression of the CB1 receptor by capsaicin, intestinal barrier integrity is preserved and metabolic endotoxemia is prevented (Kang et al. 2017; Wang et al. 2021b). The major bacteria responsible for this effect of capsaicin are *Akkermansia* and other SCFA-producing bacteria. TRPV1 channels in hypothalamus and arcuate nucleus (ARC) are stimulated as a result of the rise in the size of these microorganism with the effect of capsaicin. While neuropeptides amphetamine-regulated transcript (CART) and proopiomelanocortin (POMC) suppressing appetite in hypothalamus are activated by the vagus nerve activation, agouti-related peptide (AgRP) and neuropeptide Y (NPY) are inactivated. Thus, food intake is reduced and body weight loss can be achieved (Wang et al. 2020a, 2021b; Tobita et al. 2021). Figure 19.1 summarises the effect of capsaicin on microbiota-mediated obesity.

19.3.3.1 Animal Researches

In the study with C57BL/6 mice, the mice were separated into groups. Two groups were formed and the diets of both groups were adjusted to contain high amount fat (45%). For 9 weeks, capsaicin was not added to the diet of one group while the other group received it (0.01%), and a significant reduction in weight gain and a decrease in glucose intolerance were seen in the mice in the capsaicin-treated group (Shen et al. 2017). In addition, it was determined that the ratio of *Proteobacteria* decreased and the number of *Akkermansia muciniphila* increased in the mice in the capsaicin-treated group. Therefore, it was stated that capsaicin may have an anti-obesity effect (Shen et al. 2017). In a 4-week study, male Sprague-Dawley (SD) rats were separated into four categories: (1) a low-fat diet, (2) a high amount fat diet, (3) a high amount fat diet with 0.01% capsaicin, (4) a high amount fat diet with 0.05% capsaicin. At the end of the study, it was seen that the number of *Firmicutes* decreased and the number of *Akkermansia* increased in the groups given capsaicin. In addition, capsaicin was shown to affect SFCA production and reduce weight gain by difference the intestinal microbiota components (Gong et al. 2022). In a similar research, a group of mice divided into four groups. A diet containing 70% carbohydrates, 20% protein, 10% fat was given to one group and a different diet containing 35% carbohydrates, 20% protein, 45% fat was given to another group. In the other two groups, capsaicin was added to these two diets (0.01 g capsaicin/100 g diet) and given to the other two groups of mice. As a result of the research, it was determined that the number of butyrate-producing *Lachnospiraceae* and *Ruminococcaceae* increased in the intestinal microbiota of mice with capsaicin added to their diet, and the number of LPS-producing *S24_7 bacteria family* decreased. In addition, it was determined that capsaicin can prevent the deterioration of the intestinal barrier that can be caused by a high amount fat diet (Kang et al. 2017). In another study, it was established that the level of glucagon-like peptide 1 (GLP-1) and the number of *Bacteroides* increased in mice consuming high amount fat and capsaicin (Hui et al. 2020). In a study with obese mice, lower dose (0.01%) capsaicin was used to the diet of first group of mice and higher dose (0.02%) to the other group for 6 weeks. As a result, it was determined that *Roseburia* number increased significantly, *Firmicutes/Bacteroidetes* ratio changed, *Bacteroides* and *Parabacteroides* numbers decreased in both groups; a significant improvement in glucose homeostasis was observed but no effect was observed on obesity-related phenotypes (Song et al. 2017). In the study of Xiang et al. (2022a) in which 40, 60, and 80 mg/kg doses of capsaicin were given to mice, higher doses of capsaicin (60 and 80 mg/kg) were found to cause inflammation in the jejunum, ileum, and colon, while 40 mg/kg capsaicin did not cause any adverse effects in the gastrointestinal system. However, it was also observed that 80 mg/kg capsaicin significantly increased the amount of SCFA (Xiang et al. 2022a). In another research, capsaicin was given intragastrically to mice and it was found that *Akkermansia* bacteria in the gut microbiota increased and body weight reduced (Wang et al. 2020b). In a research conducted with mice with type 2 diabetes, two different diets were applied to the mice for 8 weeks: a normal diet and a diet containing capsaicin. It was found that capsaicin decreased the amount of *Lactobacillus* and showed anti-hyperglycemic effect in these mice. It was determined that the proliferation of *Firmicutes* bacteria was prevented at phylum grade. In this study, although there was no evaluation showing the effect of capsaicin on obesity directly, it was stated that it may also have positive effects on obesity (Hui et al. 2019). When the results of these animal researches are examined, although there are exceptions, in general, capsaicin seems to be affecting SFCA production by decreasing the number of *Proteobacteria*, *Firmicutes*, *Parabacteroides*, LPS-producing S24_7 bacteria families and increasing the number of *Akkermansia muciniphila*, *Ruminococcaceae*, *Lachnospiraceae*, *Bacteroides*, *Roseburia* and it seems to be effective in anti-obesity (Shen et al. 2017; Gong et al. 2022; Kang et al. 2017; Hui et al. 2020; Song et al. 2017; Xiang et al. 2022a; Hui et al. 2019). The animal researches cited in this chapter are summarised in Table 19.1.

19.3.3.2 Human Researches

In a research that studied the human intestinal microbiota in vitro, it was determined that capsaicin caused a rise in the amount of SFCA-producing-bacteria and especially in the amount of butanoic acid (Mahalak et al. 2022). In a study conducted with 12 adult individuals aged 18–40 years and investigating the preventive effect of capsaicin via microbiota, powdered pepper was added to their diets for 6 weeks (0 mg/day in the first week; 5 mg/dL (weeks 2 and 3); 0 mg/dL (week 4); 10 mg/dL

Table 19.1 Researches on capsaicin-gut microbiota-anti-obesity

Animal researches			
Materials/ participants	Method	Results	Reference
C57BL/ 6 mice	– Number of group: 2	About capsaicin:	Shen et al. (2017)
	– Duration: 9 weeks	– Reduce in weight gain	
	– Diet type:	– Increase in blood glucose tolerance	
	• High amount fat (45%)	– Decreased amount of *Proteobacteria*	
	• High amount fat (45%) with capsaicin (0.01%)	– Increased number of *Akkermansia muciniphila*	
	– Anthropometric measurements and nutritional status were evaluated	– Anti-obesity effect	
	– Feces were collected and analysed		
	– Oral glucose tolerance test assessed		
Male Sprague-Dawley (SD) rats	– Four groups for 4 weeks:	In relation to capsaicin:	Gong et al. (2022)
	– Diet type:	– Reduce in the amount of *Firmicutes*	
	• Low amount fat	– Rise in the amount of *Akkermansia*	
	• High amount fat	– Increase in SFCA production	
	• High amount fat +0.01% capsaicin	– Decrease in weight gain	
	• High amount fat +0.05% capsaicin		
	– Biochemical analysis		
	– Gut microbiota analysis		
Male wild-type (WT) mice of the C57BL/6 J	– Four groups for 12 weeks:	In relation to capsaicin:	Kang et al. (2017)
	– Diet type:	– Increase in the number of butyrate-producing *Ruminococcaceae* and *Lachnospiraceae*	
	• 70% carbohydrate, 20% protein, 10% fat	– Decrease in the number of LPS-producing *S24_7 family* of bacteria	
	• 35% carbohydrate, 20% protein, 45% fat	– Prevent intestinal barrier disruption associated with high amount fat diet	
	• 70% carbohydrate, 20% protein, 10% fat +0.01 g capsaicin/100 g		
	• 35% carbohydrate, 20% protein, 45% fat +0.01 g capsaicin/100 g		

(continued)

Table 19.1 (continued)

Animal researches			
Materials/ participants	Method	Results	Reference
	– Fecal samples were collected		
	– Markers of CLGI, metabolic endotoxemia, and obesity related parameters were examined		
Male C57BL/ 6 J mice	– Three groups for 8 weeks:	About capsaicin:	Hui et al. (2020)
	• Chow diet (70% carbohydrate, 20% protein, 10% fat)	– Rise in GLP-1 level	
	• High amount fat diet (35% carbohydrate, 20% protein, 45% fat)	– Increase in the number of *Bacteroides*	
	• High amount fat diet +0.01% capsaicin		
	– Fecal samples were collected		
	– Blood samples were taken		
Ob/Ob mice	– Number of group: 2	In both groups:	Song et al. (2017)
	– Duration: 9 weeks	– Increase in the number of *Roseburia*	
	– Diet type:	– Change in *Firmicutes/ Bacteroidetes* ratio	
	• Normal	– Decrease in *Bacteroides* and *Parabacteroides* numbers	
	• Low amount capsaicin (0.01%) • High amount capsaicin (0.02%)	– Improvement in glucose homeostasis – No change in obesity-related phenotypes	
	– Obesity phenotypes		
	– Glucose homeostasis		
	– Intestinal microbiota composition		
	– Gastrointestinal hormones		
	– Cytokines (pro-inflammatory) were evaluated		
Male C57BL/ 6 J mice	– Number of group: 4	– No adverse effects on gastrointestinal system (40 mg/kg capsaicin)	Xiang et al. (2022a)
	– Duration:	– Inflammation increased in jejunum, ileum and colon (60 and 80 mg/kg capsaicin)	

(continued)

Table 19.1 (continued)

Animal researches			
Materials/ participants	Method	Results	Reference
	– Diet type:	– Change in gut microbiota (60 and 80 mg/kg capsaicin)	
	• Control group.	– Increase the amount of SCFA (80 mg/kg capsaicin)	
	• 40 mg/kg capsaicin		
	• 60 mg/kg capsaicin		
	• 80 mg/kg capsaicin administered intragastrically		
	– Blood samples		
	– Fecal content were analyzed		
	– Tissues were examined (stomach, jejunum, ileum, and colon)		
Male and female ICR clean-grade mice	– 2 groups:	In relation to capsaicin:	Wang et al. (2020b)
	• Control group	– Increase in *Akkermansia* bacteria	
	• Capsaicin group (8 mg/kg/day mice body weight)	– Decrease in body weight	
	– Capsaicin was administered as gastric perfusion		
	– Body weight was measured		
	– Gut composition was analyzed		
Male mice C57BL/KsJ (db/db)	– Two groups for 8 weeks:	In relation to capsaicin:	Hui et al. (2019)
	• Chow diet	– Decreased amount of *lactobacillus*	
	• Chow diet + capsaicin (0.01%)	– Anti-hyperglycemic effect	
	– Body weight, food intake, and fasting glucose were measured	– Prevented increase in the number of *Firmicutes*	
	– Cecal content was analyzed	– May have positive effects on obesity	
Human researches			
Participants	*Method*	*Results*	*Reference*
Age: 18–40 years	– Participants' diets were adjusted to meet the recommended normal requirement for 6 weeks	In relation to chili powder:	Kang et al. (2016)
$n = 12$	– Added chili powder to the participants' diets	– Change in Firmicutes/ Bacteroidetes ratio	
	• 1 week: 0 mg/dL	– Increased number of Faecalibacterium	
	• 2 and 3 week: 5 mg/dL	– Increase in GLP-1	
	• 4 week: 0 mg/dL	– Increase in GIP levels	

(continued)

Table 19.1 (continued)

Animal researches			
Materials/ participants	Method	Results	Reference
	• 5 and 6 week: 10 mg/dL	– Decrease in ghrelin	
	– Blood, stool samples, anthropometric and basal metabolism measurements, appetite evaluation (at the end of each period)		
Age: 18–50 years	– Two groups for 12 weeks:	In relation to capsaicinoid powder:	Manca et al. (2021)
$n = 61$	• Nutrition program with energy restriction of 500 kcal + capsule with capsaicinoid powder	– Change in gut microbiota profile	
Gender: Female	• Nutrition program with energy restriction of 500 kcal + placebo	– Rise in the amount of bacteria with anti-obesity effect	
Body Mass Index: 25–35 kg/m^2	– Blood and stool samples were collected		

(week 5 and 6)). At the end of the study, it was established that dietary capsaicin changed the ratio of *Firmicutes/Bacteroidetes* and the amount of *Faecalibacterium* increased (Kang et al. 2016). It was also found that there was an increase in gastric inhibitory polypeptide (GIP) and GLP-1 levels, and a decrease in ghrelin levels (Kang et al. 2016). In a study conducted with 61 people (gender: women; age: 18–50 years) with 25–35 kg/m^2 body mass index, individuals were given capsules containing capsaicinoid powder in addition to a nutrition program with 500 kcal energy restriction for 12 weeks. As a result of this research, it was seen that the gut microbiota profile changed and the number of bacteria can show anti-obesity effect increased (Manca et al. 2021). The human researches cited in this chapter are summarised in Table 19.1.

As a result, gut microbiota profile can be effective on obesity development in different ways. Metabolites formed in the intestinal microbiota may affect the regulation of body weight in different ways depending on the regulation of metabolism of fatty acids, alleviation of intestinal inflammation, decreased appetite, regulation of genes involved in lipogenesis, decrease in serum lipid and glucose levels, and changes in white adipose and brown adipose tissues. It is seen that the effects of gut microbiota on obesity may vary depending on the species/strain (Cao et al. 2019). In contrast to the effect of the intestinal microbiota on body weight, the gut microbiota may differ depending on the body fat ratio. It is known that as body fat rate increases, intestinal diversity decreases, *Firmicutes* rate increases, and *Akkermansia* bacteria generally decreases. When the intestinal microbiota of obese people with high fat content is examined in more detail, the *Firmicutes/Bacteroidetes* ratio and

Firmicutes, Fusobacteria, Lactobacillus reuteri, Mollicutes, Proteobacteria numbers are higher, and *Akkermansia muciniphila, Bacteroidetes, Faecalibacterium prausnitzii, Lactobacillus paracasei, Lactobacillus plantarum* and *Methanobrevibacter smithii* numbers are lower (Crovesy et al. 2020; Newman et al. 2021; Amabebe et al. 2020). Therefore, it is thought that the adding of capsaicin as an alternative method in the remedy of obesity can correct this dysbacteriosis by different mechanisms (Szallasi 2022; Kumar et al. 2022; Roshanravan et al. 2021). For instance, capsaicin is known to be effective in obesity by promoting the proliferation of *Akkermansia* and stimulating intestinal mucus secretion (Zhang et al. 2020a; Neyrinck et al. 2021). Similarly, capsaicin is considered to be effective on obesity by providing a substantial decrease in the number of some microorganism *(S24_7 families)* (Kang et al. 2017). In addition, it is stated that capsaicin can provide its anti-obesity effect by directly eliminating different pathogens and improving bacterial diversity (Wang et al. 2020a; Füchtbauer et al. 2021).

19.4 Conclusions

Capsaicin exerts its effect on body weight control through different metabolic pathways. One of them is its effect on the intestinal microbiota. In researches (animal and human), it has been shown that capsaicin provides intestinal integrity by changing the gut microbiota profile, prevents chronic low-grade inflammation, regulates hunger-satiety hormones by affecting the gut-brain axis, and has obesity preventive effects through energy metabolism. Capsaicin is seen as an effective factor in reducing and treating the risk of many chronic diseases, especially obesity, through its effects on the gut microbiota. However, to take advantage of the effects of capsaicin in preventing and treating obesity, more clinical studies are needed on what dose, how, and when it should be taken.

It is important to carry out studies in different age groups, especially on people, where this missing information can be eliminated. These studies will shed light on the prevention and treatment of obesity, a serious global health problem. It will be a very important, up-to-date and new approach in the treatment of obesity that the goal is to improve gut health, which is now accepted as one of the most important organs of human beings. For this, it is thought that the use of bioactive substances especially those found in foods as natural such as capsaicin can be an economical, feasible and effective method.

References

Abenavoli L, Scarpellini E, Colica C, Boccuto L, Salehi B, Sharifi-Rad J, Capasso R (2019) Gut microbiota and obesity: a role for probiotics. Nutrients 11:2690. https://doi.org/10.3390/nu11112690

Ai D, Huang R, Wen J, Li C, Zhu J, Xia LC (2017) Integrated metagenomic data analysis demonstrates that a loss of diversity in oral microbiota is associated with periodontitis. BMC Genomics 18:1–15. https://doi.org/10.1186/s12864-016-3254-5

Amabebe E, Robert FO, Agbalalah T, Orubu ES (2020) Microbial dysbiosis-induced obesity: role of gut microbiota in homoeostasis of energy metabolism. Br J Nutr 123:1127–1137. https://doi.org/10.1017/S0007114520000380

Ayariga JA, Ibrahim I, Gildea L, Abugri J, Villafane R (2023) Microbiota in a long survival discourse with the human host. Arch Microbiol 205:5. https://doi.org/10.1007/s00203-022-03342-6

Ballini A, Scacco S, Boccellino M, Santacroce L, Arrigoni R (2020) Microbiota and obesity: where are we now? Biology (Basel) 9:415. https://doi.org/10.3390/biology9120415

Cao SY, Zhao CN, Xu XY, Tang GY, Corke H, Gan RY, Li HB (2019) Dietary plants, gut microbiota, and obesity: effects and mechanisms. Trends Food Sci Technol 92:194–204. https://doi.org/10.1016/j.tifs.2019.08.004

Chung MK, Campbell JN (2016) Use of capsaicin to treat pain: mechanistic and therapeutic considerations. Pharmaceuticals (Basel) 9:66. https://doi.org/10.3390/ph9040066

Coppola S, Avagliano C, Calignano A, Berni Canani R (2021) The protective role of butyrate against obesity and obesity-related diseases. Molecules 26:682. https://doi.org/10.3390/molecules26030682

Crovesy L, Masterson D, Rosado EL (2020) Profile of the gut microbiota of adults with obesity: a systematic review. Eur J Clin Nutr 74:1251–1262. https://doi.org/10.1038/s41430-020-0607-6

Elmas C, Gezer C (2022) Capsaicin and its effects on body weight. J Am Nutr Assoc 41:831–839. https://doi.org/10.1080/07315724.2021.1962771

Fan Y, Pedersen O (2021) Gut microbiota in human metabolic health and disease. Nat Rev Microbiol 19:55–71. https://doi.org/10.1038/s41579-020-0433-9

Füchtbauer S, Mousavi S, Bereswill S, Heimesaat MM (2021) Antibacterial properties of capsaicin and its derivatives and their potential to fight antibiotic resistance—a literature survey. Eur J Microbiol Immunol 11:10–17. https://doi.org/10.1556/1886.2021.00003

Gacesa R, Kurilshikov A, Vich Vila A, Sinha T, Klaassen MAY, Bolte LA, Weersma RK (2022) Environmental factors shaping the gut microbiome in a Dutch population. Nature 604:732–739. https://doi.org/10.5281/zenodo.5910709

Gasmi A, Noor S, Menzel A, Pivina L, Bjørklund G (2021) Obesity and insulin resistance: associations with chronic inflammation, genetic and epigenetic factors. Curr Med Chem 28: 800–826. https://doi.org/10.2174/0929867327666200824112056

Gong T, Wang H, Liu S, Zhang M, Xie Y, Liu X (2022) Capsaicin regulates lipid metabolism through modulation of bile acid/gut microbiota metabolism in high-fat-fed SD rats. Food Nutr Res 66. https://doi.org/10.29219/fnr.v66.8289

Hui S, Liu Y, Chen M, Wang X, Lang H, Zhou M, Mi M (2019) Capsaicin improves glucose tolerance and insulin sensitivity through modulation of the gut microbiota-bile acid-FXR axis in type 2 diabetic db/db mice. Mol Nutr Food Res 63:1900608. https://doi.org/10.1002/mnfr.201900608

Hui S, Huang L, Wang X, Zhu X, Zhou M, Chen M, Mi M (2020) Capsaicin improves glucose homeostasis by enhancing glucagon-like peptide-1 secretion through the regulation of bile acid metabolism via the remodeling of the gut microbiota in male mice. FASEB J 34:8558–8573. https://doi.org/10.1096/fj.201902618RR

Isacco CG, Ballini A, De Vito D, Nguyen KC, Cantore S, Bottalico L, Inchingolo F (2021) Rebalancing the oral microbiota as an efficient tool in endocrine, metabolic and immune disorders. Endocr Metab Immune Disord Drug Targets 21:777–784. https://doi.org/10.2174/1871530320666200729142504

Kang C, Zhang Y, Zhu X, Liu K, Wang X, Chen M, Mi M (2016) Healthy subjects differentially respond to dietary capsaicin correlating with specific gut enterotypes. J Clin Endocrinol Metab 101:4681–4689. https://doi.org/10.1210/jc.2016-2786

Kang C, Wang B, Kaliannan K, Wang X, Lang H, Hui S, Mi M (2017) Gut microbiota mediates the protective effects of dietary capsaicin against chronic low-grade inflammation and associated obesity induced by high-fat diet. MBio 8:e00470–e00417. https://doi.org/10.1210/jc.2016-2786

Karri S, Sharma S, Hatware K, Patil K (2019) Natural anti-obesity agents and their therapeutic role in management of obesity: a future trend perspective. Biomed Pharmacother 110:224–238. https://doi.org/10.1016/j.biopha.2018.11.076

Klein S, Gastaldelli A, Yki-Järvinen H, Scherer PE (2022) Why does obesity cause diabetes? Cell Metab 34:11–20. https://doi.org/10.1016/j.cmet.2021.12.012

Kumar V, Kumar V, Mahajan N, Kaur J, Devi K, Dharavath RN, Bishnoi M (2022) Mucin secretory action of capsaicin prevents high fat diet-induced gut barrier dysfunction in C57BL/6 mice colon. Biomed Pharmacother 145:112452. https://doi.org/10.1016/j.biopha.2021.112452

Kwon DH, Zhang F, Suo Y, Bouvette J, Borgnia MJ, Lee SY (2021) Heat-dependent opening of TRPV1 in the presence of capsaicin. Nat Struct Mol Biol 28:554–563. https://doi.org/10.1038/s41594-021-00616-3

Lan Y, Sun Y, Yang T, Ma X, Cao M, Liu L, Liu Y (2019) Co-delivery of paclitaxel by a capsaicin prodrug micelle facilitating for combination therapy on breast cancer. Mol Pharm 16:3430–3440. https://doi.org/10.1021/acs.molpharmaceut.9b00209

Loos RJ, Yeo GS (2022) The genetics of obesity: from discovery to biology. Nat Rev Genet 23:120–133. https://doi.org/10.1038/s41576-021-00414-z

Lu M, Chen C, Lan Y, Xiao J, Li R, Huang J, Ho CT (2020) Capsaicin—the major bioactive ingredient of chili peppers: bio-efficacy and delivery systems. Food Funct 11:2848–2860. https://doi.org/10.1039/D0FO00351D

Lynch SV, Pedersen O (2016) The human intestinal microbiome in health and disease. N Engl J Med 375:2369–2379

Mahalak KK, Bobokalonov J, Firrman J, Williams R, Evans B, Fanelli B, Liu L (2022) Analysis of the ability of capsaicin to modulate the human gut microbiota in vitro. Nutrients 14:1283. https://doi.org/10.3390/nu14061283

Mameli C, Cattaneo C, Panelli S, Comandatore F, Sangiorgio A, Bedogni G, Pagliarini E (2019) Taste perception and oral microbiota are associated with obesity in children and adolescents. PLoS One 14:e0221656. https://doi.org/10.1371/journal.pone.0221656

Manca C, Lacroix S, Pérusse F, Flamand N, Chagnon Y, Drapeau V, Silvestri C (2021) Oral capsaicinoid administration alters the plasma endocannabinoidome and fecal microbiota of reproductive-aged women living with overweight and obesity. Biomedicines 9:1246. https://doi.org/10.3390/biomedicines9091246

Müller TD, Blüher M, Tschöp MH, DiMarchi RD (2022) Anti-obesity drug discovery: advances and challenges. Nat Rev Drug Discov 21:201–223. https://doi.org/10.1038/s41573-021-00337-8

Munjuluri S, Wilkerson DA, Sooch G, Chen X, White FA, Obukhov AG (2021) Capsaicin and TRPV1 channels in the cardiovascular system: the role of inflammation. Cell 11:18. https://doi.org/10.3390/cells11010018

Muscogiuri G, Cantone E, Cassarano S, Tuccinardi D, Barrea L, Savastano S, Colao A (2019) Gut microbiota: a new path to treat obesity. Int J Obes Suppl 9:10–19. https://doi.org/10.1038/s41367-019-0011-7

Newman TM, Shively CA, Register TC, Appt SE, Yadav H, Colwell RR, Cook KL (2021) Diet, obesity, and the gut microbiome as determinants modulating metabolic outcomes in a non-human primate model. Microbiome 9:1–17. https://doi.org/10.1186/s40168-021-01069-y

Neyrinck AM, Sánchez CR, Rodriguez J, Cani PD, Bindels LB, Delzenne NM (2021) Prebiotic effect of berberine and curcumin is associated with the improvement of obesity in mice. Nutrients 13:1436. https://doi.org/10.3390/nu13051436

Oliphant K, Allen-Vercoe E (2019) Macronutrient metabolism by the human gut microbiome: major fermentation by-products and their impact on host health. Microbiome 7:1–15. https://doi.org/10.1186/s40168-019-0704-8

Parashar P, Tripathi CB, Arya M, Kanoujia J, Singh M, Yadav A, Saraf SA (2019) A facile approach for fabricating CD44-targeted delivery of hyaluronic acid-functionalized PCL nanoparticles in urethane-induced lung cancer: Bcl-2, MMP-9, caspase-9, and BAX as potential markers. Drug Deliv Transl Res 9:37–52. https://doi.org/10.1007/s13346-018-0575-8

Popescu GDA, Scheau C, Badarau IA, Dumitrache MD, Caruntu A, Scheau AE, Caruntu C (2020) The effects of capsaicin on gastrointestinal cancers. Molecules 26:94. https://doi.org/10.3390/molecules26010094

Ramírez-Aragón MG, Troyo-Diéguez E, Preciado-Rangel P, Borroel-García VJ, García-Carrillo EM, García-Hernández JL (2023) Antioxidant profile of hot and sweet pepper cultivars by two extraction methods. Hortic Bras 40:411–417. https://doi.org/10.1590/s0102-0536-20220409

Rinninella E, Raoul P, Cintoni M, Franceschi F, Miggiano GAD, Gasbarrini A, Mele MC (2019) What is the healthy gut microbiota composition? A changing ecosystem across age, environment, diet, and diseases. Microorganisms 7:14. https://doi.org/10.3390/microorganisms7010014

Rohm TV, Meier DT, Olefsky JM, Donath MY (2022) Inflammation in obesity, diabetes, and related disorders. Immunity 55:31–55. https://doi.org/10.1016/j.immuni.2021.12.013

Rosca AE, Iesanu MI, Zahiu CDM, Voiculescu SE, Paslaru AC, Zagrean AM (2020) Capsaicin and gut microbiota in health and disease. Molecules 25:5681. https://doi.org/10.3390/molecules25235681

Roshanravan N, Bastani S, Tutunchi H, Kafil B, Nikpayam O, Mesri Alamdari N, Ostadrahimi A (2021) A comprehensive systematic review of the effectiveness of Akkermansia muciniphila, a member of the gut microbiome, for the management of obesity and associated metabolic disorders. Arch Physiol Biochem 129:741–751. https://doi.org/10.1080/13813455.2021.1871760

Scheithauer TP, Rampanelli E, Nieuwdorp M, Vallance BA, Verchere CB, Van Raalte DH, Herrema H (2020) Gut microbiota as a trigger for metabolic inflammation in obesity and type 2 diabetes. Front Immunol 2546:571731. https://doi.org/10.3389/fimmu.2020.571731

Shafiee SA, Danial WH, Perry SC, Ali ZI, Mohamed Huri MA, Mohmad Sabere AS (2022) Qualitative and quantitative methods of capsaicinoids: a mini-review. Food Anal Methods 15:2424–2435. https://doi.org/10.1007/s12161-022-02306-2

Shen W, Shen M, Zhao X, Zhu H, Yang Y, Lu S, Le S (2017) Anti-obesity effect of capsaicin in mice fed with high-fat diet is associated with an increase in population of the gut bacterium Akkermansia muciniphila. Front Immunol 8:272. https://doi.org/10.3389/fmicb.2017.00272

Song JX, Ren H, Gao YF, Lee CY, Li SF, Zhang F, Chen H (2017) Dietary capsaicin improves glucose homeostasis and alters the gut microbiota in obese diabetic Ob/Ob mice. Front Physiol 8:602. https://doi.org/10.3389/fphys.2017.00602

Song B, Zhong YZ, Zheng CB, Li FN, Duan YH, Deng JP (2019) Propionate alleviates high-fat diet-induced lipid dysmetabolism by modulating gut microbiota in mice. J Appl Microbiol 127:1546–1555. https://doi.org/10.1111/jam.14389

Sudheer S, Gangwar P, Usmani Z, Sharma M, Sharma VK, Sana SS, Nabavi SM (2022) Shaping the gut microbiota by bioactive phytochemicals: an emerging approach for the prevention and treatment of human diseases. Biochimie 193:38–63. https://doi.org/10.1016/j.biochi.2021.10.010

Szallasi A (2022) Capsaicin for weight control: "exercise in a pill" (or just another fad)? Pharmaceuticals (Basel) 15:851. https://doi.org/10.3390/ph15070851

Tanrıkulu-Küçük S, Başaran-Küçükgergin C, Seyithanoğlu M, Doğru-Abbasoğlu S, Koçak H, Beyhan-Özdaş Ş, Öner-İyidoğan Y (2019) Effect of dietary curcumin and capsaicin on testicular and hepatic oxidant–antioxidant status in rats fed a high-fat diet. Appl Physiol Nutr Metab 44:774–782. https://doi.org/10.1139/apnm-2018-0622

Tobita N, Tsuneto K, Ito S, Yamamoto T (2021) Human TRPV1 and TRPA1 are receptors for bacterial quorum sensing molecules. J Biochem 170:775–785. https://doi.org/10.1093/jb/mvab099

Vijay A, Valdes AM (2022) Role of the gut microbiome in chronic diseases: a narrative review. Eur J Clin Nutr 76:489–501. https://doi.org/10.1038/s41430-021-00991-6
Wang J, Tian W, Wang S, Wei W, Wu D, Wang H, Li Y (2017) Anti-inflammatory and retinal protective effects of capsaicin on ischaemia-induced injuries through the release of endogenous somatostatin. Clin Exp Pharmacol Physiol 44:803–814. https://doi.org/10.1111/1440-1681.12769
Wang Y, Deng X, Yu C, Zhao G, Zhou J, Zhang G, Zhang Y (2018) Synergistic inhibitory effects of capsaicin combined with cisplatin on human osteosarcoma in culture and in xenografts. J Exp Clin Cancer Res 37:1–17. https://doi.org/10.1186/s13046-018-0922-0
Wang Y, Tang C, Tang Y, Yin H, Liu X (2020a) Capsaicin has an anti-obesity effect through alterations in gut microbiota populations and short-chain fatty acid concentrations. Food Nutr Res 64:3525. https://doi.org/10.29219/fnr.v64.3525
Wang F, Huang X, Chen Y, Zhang D, Chen D, Chen L, Lin J (2020b) Study on the effect of capsaicin on the intestinal flora through high-throughput sequencing. ACS Omega 5:1246–1253. https://doi.org/10.1021/acsomega.9b03798
Wang HJ, Battousse O, Ramadas A (2021a) Modulation of gut microbiota by dietary macronutrients in type 2 diabetes: a review. Prog Mic Mol Biol 4:a0000182
Wang Y, Zhou Y, Fu J (2021b) Advances in antiobesity mechanisms of capsaicin. Curr Opin Pharmacol 61:1–5. https://doi.org/10.1016/j.coph.2021.08.012
Weng G, Duan Y, Zhong Y, Song B, Zheng J, Zhang S, Deng J (2021) Plant extracts in obesity: a role of gut microbiota. Front Nutr 612:727951. https://doi.org/10.3389/fnut.2021.727951
Wu S, Bhat ZF, Gounder RS, Mohamed Ahmed IA, Al-Juhaimi FY, Ding Y, Bekhit AED (2022a) Effect of dietary protein and processing on gut microbiota—a systematic review. Nutrients 14: 453. https://doi.org/10.3390/nu14030453
Wu D, Duan R, Tang L, Zhou D, Zeng Z, Wu W, Sun Q (2022b) In-vitro binding analysis and inhibitory effect of capsaicin on lipase. LWT 154:112674. https://doi.org/10.1016/j.lwt.2021.112674
Xiang Q, Guo W, Tang X, Cui S, Zhang F, Liu X, Chen W (2021) Capsaicin—the spicy ingredient of chili peppers: a review of the gastrointestinal effects and mechanisms. Trends Food Sci Technol 116:755–765. https://doi.org/10.1016/j.tifs.2021.08.034
Xiang Q, Tang X, Cui S, Zhang Q, Liu X, Zhao J, Chen W (2022a) Capsaicin, the spicy ingredient of chili peppers: effects on gastrointestinal tract and composition of gut microbiota at various dosages. Foods 11:686. https://doi.org/10.3390/foods11050686
Xiang Y, Xu X, Zhang T, Wu X, Fan D, Hu Y, Xie R (2022b) Beneficial effects of dietary capsaicin in gastrointestinal health and disease. Exp Cell Res 417:113227. https://doi.org/10.1016/j.yexcr.2022.113227
Yasmin I, Khan WA, Naz S, Iqbal MW, Awuchi CG, Egbuna C, Uche CZ (2021) Etiology of obesity, cancer, and diabetes. In: Egbuna C, Hassan S (eds) Dietary phytochemicals. Springer, Cham, pp 1–27
Yin R, Kuo HC, Hudlikar R, Sargsyan D, Li S, Wang L, Kong AN (2019) Gut microbiota, dietary phytochemicals, and benefits to human health. Curr Pharmacol Rep 5:332–344. https://doi.org/10.1007/s40495-019-00196-3
Zhang M, Zhu J, Zhang X, Zhao DG, Ma YY, Li D, Huang Q (2020a) Aged citrus peel (chenpi) extract causes dynamic alteration of colonic microbiota in high-fat diet induced obese mice. Food Funct 11:2667–2678. https://doi.org/10.1039/C9FO02907A
Zhang S, Wang D, Huang J, Hu Y, Xu Y (2020b) Application of capsaicin as a potential new therapeutic drug in human cancers. J Clin Pharm Ther 45:16–28. https://doi.org/10.1111/jcpt.13039
Zheng J, Zheng S, Feng Q, Zhang Q, Xiao X (2017) Dietary capsaicin and its anti-obesity potency: from mechanism to clinical implications. Biosci Rep 37:BSR20170286. https://doi.org/10.1042/BSR20170286

Exploring the Potential of Capsaicin in Cancer Therapy: In Vitro and In Vivo Effects and Their Mechanism of Action

20

Mandira Basumatary and Amit Talukdar

Abstract

Cancer is the second most major cause of human mortality globally, and according to the International Agency for Research on Cancer (IARC), more than 19 million cancer cases are prevalent worldwide and approximately 10 million cases of death are reported in 2020. Cancer prevention and treatment is very challenging due to the involvement of multiple systems, genetic factors and pathways which manifest over a long period. The ineffectiveness and side effects of modern cancer treatment have motivated researchers to explore phytochemicals for alternative or supportive cancer therapy. The phytochemical capsaicin is a well-known naturally occurring plant secondary metabolite, which is responsible for imparting pungency or hotness in chilli peppers of the *Capsicum* genus. The medicinal properties of capsaicin have been explored by researchers since its discovery in the nineteenth century to the modern day. The discovery of the capsaicin receptor, transient receptor potential subfamily V type 1 (TRPV1), has shed light on the mechanisms by which capsaicin induces some of its pharmacological effects if not all. Capsaicin has been explored as an anti cancer agent and has shown wide applications against various types of cancer because of its availability, affordability, non-toxicity and fewer side effects. This chapter highlights the in vitro and in vivo effects of capsaicin in different types of cancers, namely breast, lung, colorectal, prostate, gastric, liver, pancreatic, bladder, skin, bone, and leukemia, and helps to understand the mechanism of action of capsaicin, which elucidates the potential of capsaicin as an anti-cancer agent. Capsaicin has been reported to display a chemopreventive role by regulating the crucial stages of carcinogenesis, i.e., initiation, promotion and malignancy

M. Basumatary (✉) · A. Talukdar
Department of Molecular Biology and Biotechnology, School of Sciences, Tezpur University, Tezpur, Assam, India

M. K. Swamy, A. Kumar (eds.), *Capsaicinoids*,
https://doi.org/10.1007/978-981-99-7779-6_20

(metastasis). The mechanism of action of capsaicin involves inhibition of cell viability, cell cycle arrest at different phases, induction of angiogenesis, and apoptosis, via modulation of the levels of various downstream intermediaries. Newer approaches using nanotechnology are explored to improve the bioavailability of capsaicin, and synthetic non-pungent analogues are also developed to overcome the shortcomings of capsaicin. Co-treatment of capsaicin along with chemotherapy or radiation therapy are also explored. Further studies involving robust models and cancer patients may help in better understanding and utilisation of this compound.

Keywords

Capsaicin · TRPV1 · Anti-cancer · Chemoprevention · Carcinogenesis · Cell cycle arrest · Apoptosis

20.1 Introduction

Cancer ranks second among the major causes of human mortality globally. The International Agency for Research on Cancer (IARC) has estimated that globally one out of every five human beings develop cancer during their lifetime and 1 out of 8 males and 1 out of 11 females succumb to the disease. IARC has reported in its online database GLOBOCAN 2020 that the burden of cancer has reached 19.3 million cases worldwide and nearly 10 million cases of death are recorded in 2020. Breast cancer (11.7%) ranks as the most frequently diagnosed cancer among all cancer cases combined. However, lung cancer is responsible for the highest number of deaths among both the sexes at around 1.8 million (18.0%) in 2020. The prevention and treatment of cancer is a very challenging task despite the progress and developments made in the field of medical science (Akhtar and Swamy 2018; Swamy and Akhtar 2018a). The incidence and progression of cancer involve multiple steps which develop over a relatively long period. Cancer cells develop unique properties such as resistance to growth-regulatory signals, non-dependence on growth-stimulatory factors, uncontrolled replication, resistance to apoptosis, invasion of healthy tissues and angiogenesis (Hanahan and Weinberg 2000; Santosh et al. 2017; Narayanaswamy and Swamy 2018; Ravichandra et al. 2018). Numerous genetic factors and signalling pathways lead to the onset and development of cancer. The ineffectiveness along with associated side-effects of modern treatment approaches have motivated scientists to explore natural compounds such as phytochemicals in the area of chemoprevention (Akhtar and Swamy 2018; Swamy 2020). Phytochemicals have played a critical role in limiting the initiation and progression of cancer to severe stages such as malignancy (Aggarwal and Shishodia 2006). Besides, they are easily available, non-toxic, affordable, and without many side effects or development of drug resistance in patients (Clark and Lee 2016; Swamy and Akhtar 2018b; Swamy 2020).

Capsaicin (*trans*-8-methyl-*N*-vanillyl-6-noneamide) is the most biologically active constituent of the *Capsicum* genus. The *Capsicum* commonly known as chilli peppers or just chillis is used as a spice and it is capsaicin which is responsible for its distinct pungent flavour. It is an alkyloid which occurs naturally and belongs to a group of compounds called capsaicinoids believed to protect plants from plant-eating animals and microbes. Under controlled stress conditions, the capsaicin levels may be increased in peppers (Sung et al. 2005). The concentration of capsaicin is highest in the ovary and fruit tip whereas the lowest concentration is in the seeds (Supalkova et al. 2007). Capsaicin has been extensively explored in the last few decades to understand its beneficial effects in treating various human diseases. The molecular mass of capsaicin is ~305.4 kDa and the melting temperature ranges from 62 to 65 °C. It is a hydrophobic, highly unstable, whitish powder lacking any colour or odour and has a crystal structure (Sharma et al. 2013). Initially isolated in 1846, the structure of capsaicin was chemically resolved in 1919 (Hayman and Kam 2008) which was followed by its chemical synthesis in 1930 (Adetunji et al. 2022). The head group has a vanillyl ring (methyl catechol group) which is responsible for its spicy nature and a fatty-acid aliphatic tail which are connected by an amide bond (Zheng et al. 2016). Numerous structural analogues have existed for the compound and analysis of their structure-function relationship has revealed that two regions, namely the vanillyl domain and the regions containing the amide bonds are necessary for its activity involving the TRPV1 receptor and hence it excites the sensory nerve cells to exert its pungency (Walpole et al. 1996).

Capsaicin exerts its pharmacological activities by binding to its receptor, i.e., transient receptor potential vanilloid 1 (TRPV1). This receptor is a member of the ion channel family, the transient release potential (TRP) which is necessary for heat sensation in animals (Ramsey et al. 2006). The TRPV1 is a ligand-gated channel and it prefers Ca^{2+} ions over Na^{+} ions. This membrane protein has four subunits which are identical in nature surrounding an aqueous pore at the centre. Each of the subunits is composed of one N-terminal end consisting of six repeats of ankyrin proteins, six transmembrane domains, and a C-terminal end (Díaz-Laviada and Rodríguez-Henche 2014). Capsaicin is an exogenous activator of TRPV1 which binds to its S2–S4 region and results in the influx of cations and this leads to a variety of physiological effects (Jordt and Julius 2002; Sun et al. 2016). Besides capsaicin, TRPV1 is also activated by high temperature (>43 °C), low pH (below 5.9), voltage and several lipids, and capsaicin stimulated by heat may work in synergy to initiate channel gating (Cui et al. 2012). In numerous studies concerning capsaicin-induced anti-cancer effects, it appears that TRPV1 receptors are not responsible for the mediation of these effects, since capsazepine, which is a competitive inhibitor of TRPV1 receptor, could not inhibit the apoptosis mediated by capsaicin (Kim et al. 2004; Raisinghani et al. 2005; Athanasiou et al. 2007; Zhang et al. 2008; Jaganathan et al. 2020). However, some reports also suggest an active role of TRPV1 receptors in the increase of intracellular calcium ions which initiates the apoptotic process (Ziglioli et al. 2009). The level of TRPV1 expression is also related to different stages of tumour progression and prognosis. For example, high TRPV1 levels have been reported in epithelial cells of prostate cancer and metastasis in the liver (Czifra et al. 2009; Rychkov and Barritt 2011), and a decrease in TRPV1 levels have been observed in transitional cell carcinoma of the human bladder (Santoni et al. 2012).

In this chapter, we have tried to comprehensively explore the in vitro and in vivo role of the phytochemical capsaicin to understand how it affects different cancers, and also tried to understand its mechanism of action at different stages of cancer progression.

20.2 Anti-cancer Role of Capsaicin Identified via In Vitro and In Vivo Studies

Capsaicin has been widely adapted into numerous cuisines of the world right from Mexico and South America to Africa and Asia. Even if extensively consumed by people of numerous countries, there is no consensus among researchers regarding the safety related to its consumption and topical application. There are conflicting reports on capsaicin, some of which indicate an anti-cancer role while others point towards a procarcinogenic one (Bode and Dong 2011). The chemopreventive and chemotherapeutic role of capsaicin has been widely explored in numerous studies of cancer types (Zhang et al. 2008; Amantini et al. 2009). However, the mechanism of action which is responsible for these effects of capsaicin is not completely known. The role of capsaicin in reducing cancer cell proliferation, apoptosis, arrest of the cell cycle (G_2/M phase), inhibition of angiogenesis as well as metastasis have been reported previously by various authors (Zhang et al. 2008; Chen et al. 2012). Capsaicin and its synthetic analogues have also been explored as a supporting therapy along with chemotherapeutic drugs or radiation therapy (Zhang et al. 2020). The effect of capsaicin in vitro on cancer cell lines and in vivo on experimental models have been summarised in Tables 20.1 and 20.2 respectively and briefly described as follows.

20.2.1 Breast Cancer

The most frequently diagnosed type of cancer globally is breast cancer, with one in every eight women in the USA having a lifetime risk of the disease. More than 1.67 million cases are reported worldwide in 2012, and the number increased to 2.2 million in 2020 (Momenimovahed and Salehiniya 2019; Lukasiewicz et al. 2021). It is also the fourth leading cause of cancer-related deaths worldwide and annually around 684,996 deaths are reported globally (Sung et al. 2021).

Capsaicin has been reported to inhibit the viability of numerous breast cancer cell lines such as MDA-MB-231 and MCF-7 cell lines (Wu et al. 2020). Capsaicin regulated necrotic cell death in the MCF-7 cell line with overexpression of TRPV1 receptors (Wu et al. 2014). Chang et al. reported the inhibition of cell lines MCF-7 and BT-20 dose-dependently by induction of apoptosis and the arrest of cell cycle (S-phase) (Chang et al. 2011). Similarly, Chou et al. have also reported that capsaicin induced apoptosis in MCF-7 cell lines (Chou et al. 2009). Capsaicin targets the

Table 20.1 Anti-cancer roles of capsaicin in vitro

Cancer types	Cell lines utilised	Effective dose (μM)	Anti-cancer effect	References
Breast cancer	MDA-MB-231	200	Inhibition of cell viability as well as migration via arrest of cell cycle at G2/M phase	Wu et al. (2020)
	MCF-7	3	Regulation of necrotic cell death with overexpression of TRPV1	Wu et al. (2014)
	MCF-7, BT-20	200	Apoptotic cell death and cell cycle arrest in S-phase	Chang et al. (2011)
	T47D, BT-474, SKBR-3, MDA-MB-231, MCF-7	200	Inhibition of growth, cell cycle arrest in G0/G1 phase followed by increased apoptosis	Thoennissen et al. (2010)
	MCF-7, MDA-MB-231	250	Induction of autophagy regulated via p38 and ERK	Choi et al. (2010)
	MCF-7	5–20	Apoptosis in cancer cell lines as well as cancer stem cells	Shim and Song (2015)
Lung cancer	DMS 53, DMS 114	10–100	Anti-invasive activity regulated by AMPK pathway	Hurley et al. (2017)
	NCI-H82, NCI-H69, DMS 53, DMS 114	50	TRPV6 receptor dependent apoptosis	Brown et al. (2010); Lau et al. (2014)
	A549	37.5	Cell migration inhibition and network formation, and restricted angiogenesis	Chakraborty et al. (2014)
	H1299	200	Degradation of mutant p53, restoration of wild-type p53, and contribution to cell death	Garufi et al. (2016)
Colorectal cancer	HCT 116, LoVo	300	Cell cycle arrest followed by apoptosis	Jin et al. (2014)
	HCT 116	100	Growth inhibition, stimulation of p53 and Bax, followed by apoptosis	Kim et al. (2009)
	Colo 205	150–500	Cytotoxicity due to increased p53 and Ca^{2+} and decreased mitochondrial membrane potential	Lu et al. (2010)
	Colo320DM, LoVo	100–500	Apoptosis induction, ROS generation and disruption of mitochondrial membrane potential	Yang et al. (2009)
	HT-29, RKO	2.5–200	Inhibition of cancer cell proliferation, stimulation of IL-6 cytokine	Bessler and Djaldetti (2017)
	HT-29	100–300	Induction of apoptosis, activation of PPARγ and AMPK	Kim et al. (2004)

(continued)

Table 20.1 (continued)

Cancer types	Cell lines utilised	Effective dose (μM)	Anti-cancer effect	References
	SW480, LoVo, HCT 116	50–100	Inhibition of cell proliferation, decrease in expression and interaction of β-catenin gene	Lee et al. (2012)
Prostate cancer	LNCaP, PC-3, DU-125	200	Inhibition of cell proliferation, increase of p53, p21 and Bax	Mori et al. (2006)
	C4-2, LNCaP	100	Inactivate androgen receptors by restoring miR-449a	Zheng et al. (2015)
	PC-3	20	Apoptosis of cells	Sánchez et al. (2006)
	PC-3	10–20	Synthesis and secretion of IL-6 increased	Malagarie-Cazenave et al. (2011)
	LNCaP, PC-3	20–80	Block autophagy in cell lines which leads to cell death	Ramos-Torres et al. (2016)
	PC-3, DU145	1–10	Downregulate cancer stem cell markers which could be mediated by the Wnt/β-catenin pathway	Zhu et al. (2020)
	LNCaP	1–20	Pro-carcinogenic activity, increased viability of cell lines, reversed by using TRPV1 antagonists	Malagarie-Cazenave et al. (2009)
Gastric cancer	AGS	10–200	Anti-proliferative effects on cell lines, increase of apoptotic bodies after 24 h of incubation	Lo et al. (2005)
	AGS	10–300	Increase of cleaved caspase-3 and decrease in Bcl-2 population	Park et al. (2014)
	SNU-668	300	Apoptotic cell death	Huh et al. (2011)
	SCM-1	100–250	Inhibition of growth-related protein tNOX	Wang et al. (2009)
	SGC-7901	16–65	Inhibited growth of cells, reduced the activity of hMOF	Wang et al. (2016)
Liver cancer	HepG2	50–200	Apoptosis of cells by increase in Ca^{2+} and ROS, regulation of Bcl-2 and increase in caspase-3	Huang et al. (2009)
	HepG2, Hep3B	200	TRIAL induced apoptosis in cells via upregulation of DR-5 cell surface receptor	Moon et al. (2012)
	HepG2	50–200	Induction of autophagy in cells via ROS generation	Chen et al. (2016)
	HepG2	200	Induction of apoptosis due to inhibition of the enzyme fatty acid synthase	Impheng et al. (2014)

(continued)

Table 20.1 (continued)

Cancer types	Cell lines utilised	Effective dose (μM)	Anti-cancer effect	References
	SMMC-7721	150–300	Inhibition of cell-viability and apoptosis via phosphorylation of JNK and p38 MAPK	Bu et al. (2015)
Pancreatic cancer	BxPC-3, AsPC-1	25–250	Mitochondrial damage in cells due to reduced antioxidants followed by apoptosis	Zhang et al. (2008)
	PANC-1	50–300	Reduced cell viability via cell cycle arrest at G0/G1 phase and subsequent apoptosis	Zhang et al. (2012a)
	AsPC-1, BxPC-3	150	Induction of apoptosis via activation of ASK1 and other downstream effectors	Pramanik and Srivastava (2012)
Bladder cancer	5637	50–200	Induction of cell death via enhanced ROS and lowered mitochondrial membrane potential	Chen et al. (2012)
	5637, T24	300	Suppress cell proliferation and trigger autophagy	Amantini et al. (2016)
	T24	100–200	Cytotoxicity and impaired cell migration via downregulation of SIRT 1 deacetylase	Islam et al. (2019b)
Leukemia	NB4	300	Induction of apoptosis via modulation of mitochondrial membrane protein Prohibitin 2	Kuramori et al. (2009)
	HL-60	70	Decreased viability of cells via cell cycle arrest at G0/G1 phase and induction of apoptosis	Tsou et al. (2006)
	HPB-ATL-T, HPB-CTL-1, HUT-102	100–200	Decreased viability of cells via cell cycle arrest at G0/G1 phase and induction of apoptosis	Zhang et al. (2003)
	K562	130	Induction of apoptosis through probable interaction of miR-520a-5p/STAT3	Kaymaz et al. (2014)
Melanoma of skin	A-375, SK-MEL-28	100	Inhibitory effect on NADH oxidase and cell-growth	Morré et al. (1996)
	A-375, C8161	50–200	Induced PARP and caspase-3 dependent apoptosis as well as autophagy	Chu et al. (2019)
	A375P	300	Time and dose-dependent inhibition of IL-8 expression	Patel et al. (2002)
Bone cancer	MG63	200	Increased phagocytosis by dendritic cells leading to immunogenic cell death	Jin et al. (2016)

(continued)

Table 20.1 (continued)

Cancer types	Cell lines utilised	Effective dose (μM)	Anti-cancer effect	References
	MG63	20–40	Induction of apoptosis via TRPV1 dependent as well as TRPV1 independent pathways	Bao et al. (2019)
	U2OS, MG63	20	Induction of apoptosis via AMPK-dependent as well as independent pathways	Ying et al. (2013)

growth of cell lines containing (MCF-7, T47D and BT-474) as well as lacking (SKBR-3 and MDA-MB-231) the estrogen receptor, which is followed by cell cycle arrest in the G0/G1 stage and subsequent rise in apoptosis (Thoennissen et al. 2010; Dou et al. 2011). It also induces autophagy in malignant MCF-7 and MDA-MB-231 cells (Choi et al. 2010). Shim and Song have reported that capsaicin induced apoptotic cell death in MCF-7 breast cancer cells and also in stem cells of breast cancer (Shim and Song 2015). Capsaicin combined with genistein inhibited the proliferation of MCF-7 cell lines and exhibited anti-carcinogenic and anti-inflammatory properties (Hwang et al. 2009).

The in vivo effects of capsaicin in 5-week-old female BNX nu/nu mice have been studied and it was found to inhibit the formation of lesions in breasts by nearly 80% without any side effects (Thoennissen et al. 2010). In female athymic mice (4–5 weeks old), capsaicin has been reported to inhibit the proliferation of breast cancer cells and promote cell death (Chen et al. 2021).

20.2.2 Lung Cancer

It is responsible for the highest number of deaths in men and women combined and ranks second in terms of frequency. According to GLOBOCAN 2020, about 2.2 million new cases were reported worldwide, along with 1.79 million deaths (Sung et al. 2021). Lung cancer is responsible for nearly 18.4% of total cancer deaths, and its incidence is expected to increase in developing countries as the prevalence of tobacco use and smoking is increasing in these countries (Bade and Dela Cruz 2020). From a treatment perspective, lung cancer cases are classified into two types, small cell lung cancer (SCLC) or non-small cell lung cancer (NSCLC).

Capsaicin exhibited anti-invasive activity and apoptosis in SCLC cell lines such as DMS 53 and DMS 114 (Hurley et al. 2017). In other SCLC cell lines NCI-H82 and NCI-H69, capsaicin induced apoptosis (Lau et al. 2014). Brown et al. also observed the anti-proliferative effect of capsaicin on these SCLC cell lines (Brown et al. 2010). In case of NSCLC cell line A549, capsaicin restricted angiogenesis in these cell lines (Chakraborty et al. 2014). Lee et al. reported that capsaicin induces apoptosis in A549 NSCLC cell lines (Lee et al. 2015). Capsaicin inhibited human H1299 NSCLC and further contributed to apoptosis (Garufi et al. 2016).

Table 20.2 Anti-cancer roles of capsaicin in vivo

Cancer types	Animal models	Effective dose	Results	References
Breast cancer	Female BNX nu/nu mice	5 mg/kg	Inhibition of formation of pre-neoplastic lesions of breasts by up to 80%	Thoennissen et al. (2010)
	Female BALB/c athymic mice	10 mg/kg	Inhibition of proliferation cancer cells and promotion of apoptosis	Chen et al. (2021)
Lung cancer	Male nude mice	10 mg/kg	Inhibition of cancer cell proliferation	Brown et al. (2010)
	Male Swiss albino mice	10 mg/kg	Countered alteration of extracellular matrix components due to benzo(a)pyrene induced lung cancer	Anandakumar et al. (2015)
Colorectal cancer	Male F244 rats	40–400 mg/kg	Significant reduction in the initiation of colon tumourigenesis	Yoshitani et al. (2001)
	nu/nu nude mice	1–3 mg/kg	Significant decrease in tumour volumes and weights	Lu et al. (2010)
Prostate cancer	Athymic male nude mice	5 mg/kg	Inhibited growth of tumour xenografts and induced apoptosis	Sánchez et al. (2006)
	BNX nu/nu male mice	5 mg/kg	Slowed the growth progression of tumour xenografts with reduction in size and weight	Mori et al. (2006)
Gastric cancer	Adult Sprague-Dawley rats	0.13–160 μM	Protection of gastric mucosa against ethanol or aspirin induced ulceration	Mózsik et al. (2005)
	Adult Sprague-Dawley rats	1–2 mg/mL	Ethanol or aspirin induced damage to the gastric mucosa is worsened	Mózsik et al. (2005)
Liver cancer	Male F344 rats	4.4 mg/kg	Decrease of liver cancer marker GST-P+ as well as tumour growth	Jang et al. (1991)
	Male Wistar rats	1–10 mg/kg	Increased activity of chlorzoxazone 6-hydrolase as well as expression of CYP2E1 enzyme	Zhang et al. (2012b)
Pancreatic cancer	Female athymic	5 mg/kg	Inhibited growth of pancreatic tumour xenografts without any side-effects	Zhang et al. (2008)

(continued)

Table 20.2 (continued)

Cancer types	Animal models	Effective dose	Results	References
	BALB/c mice			
	LSL-Kras^{G12D}/Pdxi-Cre mice	10–20 ppm	Reduction in the severity of chronic pancreatitis with reduction of acini, inflammatory cells and fibrosis	Bai et al. (2011)
	Female athymic nude mice	2.5 mg/kg	Reduction in tumour superoxide dismutase by 60% along with 1.8-fold decrease in GSSG/GSH levels	Pramanik et al. (2011)
Bladder cancer	Athymic male nude mice	5 mg/kg	Reduction in the size of T24 bladder cancer xenografts	Yang et al. (2010)
Leukemia	NOD/SCID mice	50 mg/kg	Inhibition of tumour growth and induction of apoptosis	Ito et al. (2004)
Melanoma of skin	Female ICR mice	2.5 mg/kg	Lowering of vinyl carbamate induced skin tumour by 62%	Surh et al. (1995)
	Female ICR mice	10 mg/kg	Moderate inhibition of skin tumour papilloma formation	Park and Surh (1997)
Bone cancer	Male nude mice	20 mg/kg	Significant reduction in tumour volume	Wang et al. (2018)

The effect of capsaicin in vivo was observed in 4-week-old male nude mice given solid diets (10 mg/kg body weight) which exhibited non-proliferation of cells (Brown et al. 2010). Anandakumar et al. have exhibited that intraperitoneal administration of capsaicin (10 mg/kg body weight) in male Swiss albino mice effectively countered the alteration of extracellular matrix components due to the development of benzo(a)pyrene-induced lung cancer (Anandakumar et al. 2015).

20.2.3 Colorectal Cancer

Colorectal cancer (CRC) ranks as the second-most frequent cause of cancer-related deaths globally. Every year, about 1.88 million humans are diagnosed with colorectal cancer and around 915,000 deaths occur (Sung et al. 2021). During diagnosis, around 20% of people have the metastatic disease and fewer than 20% of them survive beyond 5 years (Biller and Schrag 2021). Capsaicin has exhibited anti-tumoural properties in numerous cell lines of CRC and negatively affects their growth, differentiation and proliferation.

Capsaicin has been reported to inhibit the proliferation of human colon cancer cell lines HCT 116 and LoVo via arrest of the cell cycle followed by apoptosis (Jin et al.

2014). Treatment of human colon carcinoma HCT 116 cell lines by capsaicin and resveratrol inhibited the growth of cells which was followed by apoptosis (Kim et al. 2009). Lu et al. have exhibited the cytotoxicity of capsaicin in colon cancer Colo 205 cell line from humans (Lu et al. 2010). Yang et al. reported capsaicin induced cell-death in colon cancer cell line Colo320DM and LoVo (Yang et al. 2009). Capsaicin induced concentration-dependent inhibition of colon cancer carcinoma in HT-29 and RKO cell lines (Bessler and Djaldetti 2017). Treatment of human colon cancer cells (HT-29) by capsaicin exhibited apoptosis. Capsaicin suppressed colon cancer cells SW480, LoVo and HCT 116 from proliferating in vitro (Lee et al. 2012). Studies involving in vivo effects of capsaicin have indicated a chemopreventive role. Yoshitani et al. reported that in male F344 rats, dietary exposure to capsaicin during a 4-week initiation phase of carcinogen exposure (Azoxymethane) significantly reduced the initiation of colon tumourigenesis (Yoshitani et al. 2001). In immunodeficient nude mice (nu/nu) carrying xenografts of human Colo 205 cells, capsaicin treatment exhibited a significant reduction in tumour volumes and weights when compared to DMSO-treated control groups (Lu et al. 2010).

However, some authors have reported a carcinogenic role of capsaicin as well. Yang et al. have reported that a high concentration of capsaicin inhibits the proliferation of colon cancer cell lines (CT-26, SW480 and HCT 116), however, at a lower concentration, the proliferative and invasive nature of these cells was enhanced (Yang et al. 2013). Liu et al. have reported that elevated levels of Tumour-associated NADH oxidase expression in HCT116 cell line was enhanced by capsaicin treatment followed by increased cell migration (Liu et al. 2012).

20.2.4 Prostate Cancer

It is the second-most frequent cancer type occurring in males as well as the ninth major mortality causing cancer worldwide, with a higher prevalence being reported from developed countries. According to GLOBOCAN 2020 data, more than 1.4 million incidences of prostate cancer were reported globally and the highest number of cases were diagnosed in elderly men above the age of 65 years (Sung et al. 2021).

Capsaicin has demonstrated a tumour-suppressive role in cancer of the prostate both in vitro and in vivo, which correlates to the induction of oxidative stress as well as apoptosis (Diaz-Laviada 2010). Capsaicin profoundly halted the proliferation of androgen receptor (AR)-containing prostate cancer cell line (LNCaP) and also AR-negative (PC-3 and DU145) cell lines (Mori et al. 2006). Capsaicin is reported to inactivate the androgen receptor in C4-2 and LNCaP prostate cancer cell lines (Zheng et al. 2015) and induce apoptosis on androgen-resistant PC-3 cell lines (Sánchez et al. 2006, 2007, 2008). Interleukin-6 (IL-6) synthesis and secretion were increased in capsaicin-treated PC-3 cells (Malagarie-Cazenave et al. 2011). Capsaicin has been suggested to block autophagy in LNCaP and PC-3 cells resulting in a dose and time-dependent cell death (Ramos-Torres et al. 2016). Co-treatment of capsaicin with brassinin, which is an indole compound obtained from cruciferous vegetables, exhibited apoptotic and anti-metastatic effects synergistically in PC-3 cells (Kim et al. 2015). Zhu et al. have reported that capsaicin suppressed cancer stem cell (CSC) growth in PC-3 as well as in DU145 cell lines (Zhu et al. 2020). There are also some conflicting studies which have reported a pro-carcinogenic role of capsaicin on prostate cancer cells. Malagarie-Cazenave et al. have reported that capsaicin enhanced the viability of LNCaP cell line which could be reversed by using antagonists of TRPV1 (capsazepine, I-RTx and SB 366791) (Malagarie-Cazenave et al. 2009).

The effects of capsaicin in vivo for prostate cancer were exhibited in cell xenografts of PC-3 cells. Subcutaneous injection of capsaicin in male athymic nude mice every 2 days for 14 days (5 mg/kg body weight) inhibited tumour cell proliferation and induced apoptosis (Sánchez et al. 2006). Oral administration of capsaicin in BNX nu/nu male mice slowed the growth progression of PC-3 cell xenografts with reduced dimension and weight (Mori et al. 2006). Co-treatment of capsaicin along with radiotherapy in LNCaP xenograft model in male nu/nu athymic nude mice exhibited a reduction in 'greater than additive' growth, and modulated tumour growth at a higher rate than capsaicin or radiotherapy alone (Venier et al. 2015a). In a case study involving human patients of prostate cancer, who had biochemical recurrence after radiation treatment, was given a daily diet of capsaicin (2.5–15 mL). The patient exhibited stable prostate specific antigen (PSA) levels for 4–5 months with no symptoms of recurrence (Jankovic et al. 2010).

20.2.5 Gastric Cancer

Gastric cancer or stomach cancer has been the fourth most frequent and deadly form of cancer with higher than 768,000 deaths worldwide in 2020. It has the sixth-highest incidence rate with about 1.08 million new detections in 2020 (Sung et al. 2021). The incidence and mortality of gastric cancer vary largely by region and diet preferences, and also by *Helicobacter pylori* infection (Rawla and Barsouk 2019). Capsaicin has exhibited protection of the gastric mucosa and healed gastric injury by inhibiting the secretion of acids and increasing blood circulation (Satyanarayana 2006).

Sokolova and Naumann have reported that capsaicin partially inhibits *H. pylori* (ATCC 43504) in MKN-45 cell lines, and hence, it could be further explored as a potential treatment against infection of *H. pylori* (Hyoung and Lee 2006). Capsaicin has been demonstrated to induce dose-dependent anti-proliferative effects on gastric cancer cells AGS after 24 h of incubation and apoptotic bodies are also significantly increased (Lo et al. 2005). Korean human gastric cancer cells SNU-668 treated with capsaicin along with cisplatin resulted in higher apoptotic deaths than when cells were treated with only capsaicin or cisplatin (Huh et al. 2011). Capsaicin treatment inhibited growth-related protein tNOX in SCM-1 gastric carcinoma cell lines (Wang et al. 2009), and also inhibited the gastric cancer SGC-7901 cell line from growing (Wang et al. 2016). Administration of capsaicin along with 5-fluorouracil (5-FU) as a chemotherapeutic in HGC-27 gastric cancer cell line caused a multi-fold decrease in the IC_{50} value of 5-FU (Meral et al. 2014).

Conversely, some studies have produced results which present a rather carcinogenic role of capsaicin at a high dose. For instance, a low dosage (0.13–160 μM) of capsaicin protected the gastric mucous of adult Sprague-Dawley rats against ethanol-induced ulceration, however, it is reported that when capsaicin is administered at a high dosage (1–2 mg/mL), it worsens the damage to gastric mucous (Mózsik et al. 2005; Reyes-Escogido et al. 2011). In a hospital-based case study, Lopez-Carrillo et al. reported that a moderate to high intake of capsaicin elevated the risk of developing gastric cancer in genetically vulnerable human populations of Mexico who were also suffering from *Helicobacter pylori* (CagA+) infection (López-Carrillo et al. 2012).

20.2.6 Liver Cancer

Liver cancer has been the third most frequent type of cancer and the incidence rate globally was more than 905,000 in 2020 (Sung et al. 2021). The factors which are responsible for increasing liver cancer incidence include viral hepatitis C infection, diabetes, obesity, aflatoxin intake, tobacco smoking and fatty liver disease. The most common form of liver cancer is hepatocellular carcinoma (HCC) and the risk factors include chronic infections of hepatitis B or C, and liver cirrhosis due to alcohol abuse (Arem et al. 2018).

Capsaicin is previously reported to induce apoptotic cell death in the HepG2 hepatoma cell lines (Huang et al. 2009). Capsaicin has also been exhibited to activate tumour necrosis factor-related apoptosis-inducing ligand (TRIAL)-induced apoptosis in the HepG2 and Hep3B cells (Moon et al. 2012). Chen et al. have reported that capsaicin induced autophagy in HepG2 cell line and inhibition of this autophagy could enhance the apoptosis induced by capsaicin (Chen et al. 2016). Capsaicin also inhibited the synthesis of saturated long-chain fatty acids in HepG2 cell line which then ultimately led to apoptosis (Impheng et al. 2014). It has been suggested that ROS generation might be essential to capsaicin mediated apoptosis of the human hepatoblastoma cell line HepG2 (Lee et al. 2004). Capsaicin is well-known for inhibiting cell viability and inducing apoptosis in SMMC-7721 cells dose-dependently (Bu et al. 2015). Co-treatment of capsaicin, along with multi-kinase inhibitor sorafenib, exerted synergistic inhibition of LM3, Hep3B and HuH7 cell growth, and also inhibited cell invasion and metastasis, and induced autophagy (Dai et al. 2018). The inhibitory function of capsaicin studied on cytochrome P450 (CYP450) enzyme superfamily in human liver microsomes is metabolically linked to the activation of carcinogens. The results exhibited moderate inhibition of five isoenzymes namely, CYP1A2, CYP2C9, CYP2C19, CYP2D6 and CYP3A4/5, except for CYP2E1 (Zhang et al. 2012b).

In vivo treatment of capsaicin (4.4 mg/kg body weight) for 20 weeks decreased the liver cancer marker GST-P+ and tumour growth in Male F344 rats (Jang et al. 1991). The effect of capsaicin in vivo on CYP450 enzymes in liver microsomes of male Wistar rats revealed increased activity of chlorzoxazone 6-hydrolase as well as elevated levels of the enzyme CYP2E1 (Zhang et al. 2012b).

20.2.7 Pancreatic Cancer

Pancreatic cancer has been the seventh leading cause of cancer-related mortality globally, while in the USA it ranks third in terms of frequency of prevalence. One of the leading causes of such lethality is that patients rarely exhibit disease symptoms until they reach an advanced stage of the malignant neoplasm. According to estimates by GLOBOCAN 2020, the total number of deaths due to pancreatic cancer was more than 466,000 and the current 5-year survival rate is only a meagre 9% (Rawla et al. 2019; Sung et al. 2021).

Capsaicin has exhibited inhibition of complex-I and III activity in BxPC-3 as well as in AsPC-1 cell lines which finally lead to apoptosis (Zhang et al. 2008; Pramanik et al. 2011). It also significantly reduced PANC-1 pancreatic cancer cell line viability in a dose-dependent way, by inducing arrest of the cell cycle at the G0/G1 stage, and subsequent apoptotic cell death which indicated the involvement of the phosphoinositide 3-kinase/Akt pathway (Zhang et al. 2012a). Pramanik et al. have exhibited that in AsPC-1 and BxPC-3 cells, capsaicin reduces Thioredoxin (Trx) expression and dissociates Trx-ASK1 (apoptosis signal-regulating kinase 1) complex which results in the activation of ASK1 and other downstream molecules inducing apoptosis (Pramanik and Srivastava 2012). A combination treatment of nano-liposome encapsulated capsaicin exhibited enhanced anti-cancer activity and selectivity compared to capsaicin alone in different cancer cell lines, which include PANC-1 pancreatic cancer cell lines (Al-Samydai et al. 2021).

In vivo effects of capsaicin in pancreatic cancer have been widely studied by various authors. Zhang et al. reported that capsaicin when administered orally (5 mg/kg body weight) inhibited the enlargement and proliferation of tumour xenografts (AsPC-1) in female athymic BALB/c mice and did not exhibit any adverse effects (Zhang et al. 2008). In another study involving 45 *LSL-Kras*^{G12D}*/Pdxi-Cre* mice, oral feeding of capsaicin (10–20 ppm) ameliorated the pathogenesis of chronic pancreatitis via significant reduction of the acini, inflammatory cell infiltration as well as stromal fibrosis. It also prevented pancreatic intraepithelial neoplasia phase-1 (PanIN-1) to pass on to the more severe grades of PanIN-2 and PanIN-3 (Bai et al. 2011). Oral feeding of female athymic nude mice with capsaicin (2.5 mg/kg body weight) reduced the growth of tumour along with a 60% reduction in tumour superoxide dismutase (SOD) and an increase in GSSG/GSH content by 1.8 times (Pramanik et al. 2011).

20.2.8 Other Cancers

20.2.8.1 Bladder Cancer

Capsaicin has been investigated for the management of bladder cancer by various authors (Zhang et al. 2020; Wang et al. 2022). Chen et al. exhibited that this compound induces death in cell lines of human bladder cancer 5637 dose-dependently (Chen et al. 2012). A high dose of capsaicin (300 μM) is reported to suppress cell proliferation and triggers autophagy in 5637 and T24 bladder cancer cells (Yang et al. 2010; Amantini et al. 2016; Qian et al. 2016). Along with cytotoxicity, capsaicin is also reported to impair cell migration in T24 cells (Islam et al. 2019b). The in vivo effects of capsaicin in athymic male nude mice have been studied previously, where T24 bladder cancer cells were grown as xenografts and an injection of (5 mg/kg body weight) of capsaicin subcutaneously decreased the size of T24 xenografts (Yang et al. 2010).

20.2.8.2 Leukemia

Leukemia generally refers to cancer of the white blood cells (WBCs). Capsaicin induced suppression of leukemic cell growth and promoted apoptosis (Ito et al. 2004). Kuramori et al. have reported the compound induced apoptotic cell death in NB4 myeloid leukemia cells via binding and translocation of inner mitochondrial membrane protein prohibitin 2 (Kuramori et al. 2009). Capsaicin has also been reported to reduce the percentage viability of leukemic HL-60 cells via G0/G1 cell cycle arrest followed by apoptosis (Tsou et al. 2006). Comparable effects of capsaicin were also observed in three T-cell leukemic (ATL) cell lines (HPB-ATL-T, HPB-CTL-1 and HUT-102) from adults in a dose and time-dependent mode (Zhang et al. 2003). It also inhibits the proliferation of K562 leukemic cell lines by the induction of apoptotic cell death (Kaymaz et al. 2014). On the other hand, there are reports of capsaicin increasing the differentiation of human promyelocytic leukemia HT-60 cell lines to monocytes, when co-treated with 1,25-dihydroxyvitamin D3, and to granulocytes, when co-treated with all *trans* retinoic acid (Kang et al. 2001). Capsaicin was also observed to inhibit the growth of tumour and induce apoptotic cell death in NOD/SCID mice (Ito et al. 2004).

20.2.8.3 Melanoma of Skin

The effect of capsaicin in melanoma of the skin was previously studied by various authors (Bode and Dong 2011; Ng et al. 2018; Iqbal et al. 2019). Capsaicin exhibited an inhibitory effect on A-375 and SK-MEL-28 melanoma cell lines (Morré et al. 1996), as well as the activation of apoptosis and autophagy in A-375 and C8161 human melanoma cell lines (Chu et al. 2019). Capsaicin is reported to inhibit IL-8 expression time and dose-dependently in the low metastatic human melanoma cell lines A375P (Patel et al. 2002). In female ICR mice, pre-treatment with capsaicin applied on the skin lowered skin tumours by 62% (Surh et al. 1995). Also, when capsaicin was applied before topical application in female ICR mice, it inhibited the formation of skin tumour papilloma moderately (Park and Surh 1997). Application of *trans*-capsaicin in Tg.AC mice were found non-carcinogenic as they did not enhance the incidence of skin lesions (preneoplastic or neoplastic) compared to control groups (Chanda et al. 2007).

20.2.8.4 Bone Cancer

Osteosarcoma is one of the most frequently observed types of bone cancer. Jin et al. reported that capsaicin increased phagocytosis of the human osteosarcoma MG63 cell line by dendritic cells, leading to immunogenic cell death (Jin et al. 2016). Capsaicin has been reported to exhibit apoptotic cell death in MG63 and U2OS cell lines (Ying et al. 2013; Bao et al. 2019). Combination treatment of capsaicin along with cisplatin, both in vitro (MG63, 143B and HOS cell lines) and in vivo (male nude mice models), exhibited strong inhibition of osteosarcoma cells compared to capsaicin or cisplatin alone, or to control groups (Wang et al. 2018).

20.3 Mechanism of Action of Capsaicin on Carcinogenesis

In the last few decades, numerous research has emerged to validate the anti-cancer effects of capsaicin and various theories have also been proposed by different authors to explain its mechanism (Sharma et al. 2013). Capsaicin exhibits a chemopreventive role in different cancer types by modulating the sequential stages in carcinogenesis, i.e., initiation, promotion and metastasis (malignancy). During the first stage of carcinogenesis, capsaicin blocks carcinogens from being activated by inhibiting various metabolic enzymes and activating detoxification pathways, which ultimately helps in the elimination of carcinogens, thus protecting cellular macromolecules (DNA, RNA and proteins) from reactive oxygen species (ROS)-mediated damage. In the second stage of carcinogenesis, i.e., tumour promotion (pre-malignancy), capsaicin exhibits various anti-cancer activities such as modulation of the aberrant inflammatory signalling pathways, cell cycle arrest, cell death (by apoptosis, autophagy, necrosis, paraptosis or necroptosis), inhibition of angiogenesis, and increased sensitivity of chemoresistant cells to chemotherapy. The final stage of carcinogenesis is metastasis, and capsaicin has also exhibited concentration-dependent anti-migration and anti-invasive properties in some metastatic cancer cells, however, some conflicting studies have also reported that capsaicin promotes metastasis at certain concentrations in cancer cells (Cho et al. 2017; Adetunji et al. 2022). Here, we have described the anti-cancer roles as well as some of the pro-carcinogenic roles which capsaicin has demonstrated in various stages of cancer, both in cell lines as well as in experimental animal models (Fig. 20.1).

20.3.1 Anti-cancer Effect of Capsaicin in Initiation Phase of Carcinogenesis

Capsaicin blocks the bioactivation of procarcinogens by inhibition of phase I xenobiotic metabolising enzymes, which would otherwise lead to covalent bond formation between the pro-carcinogens and DNA, which could lead to DNA damage

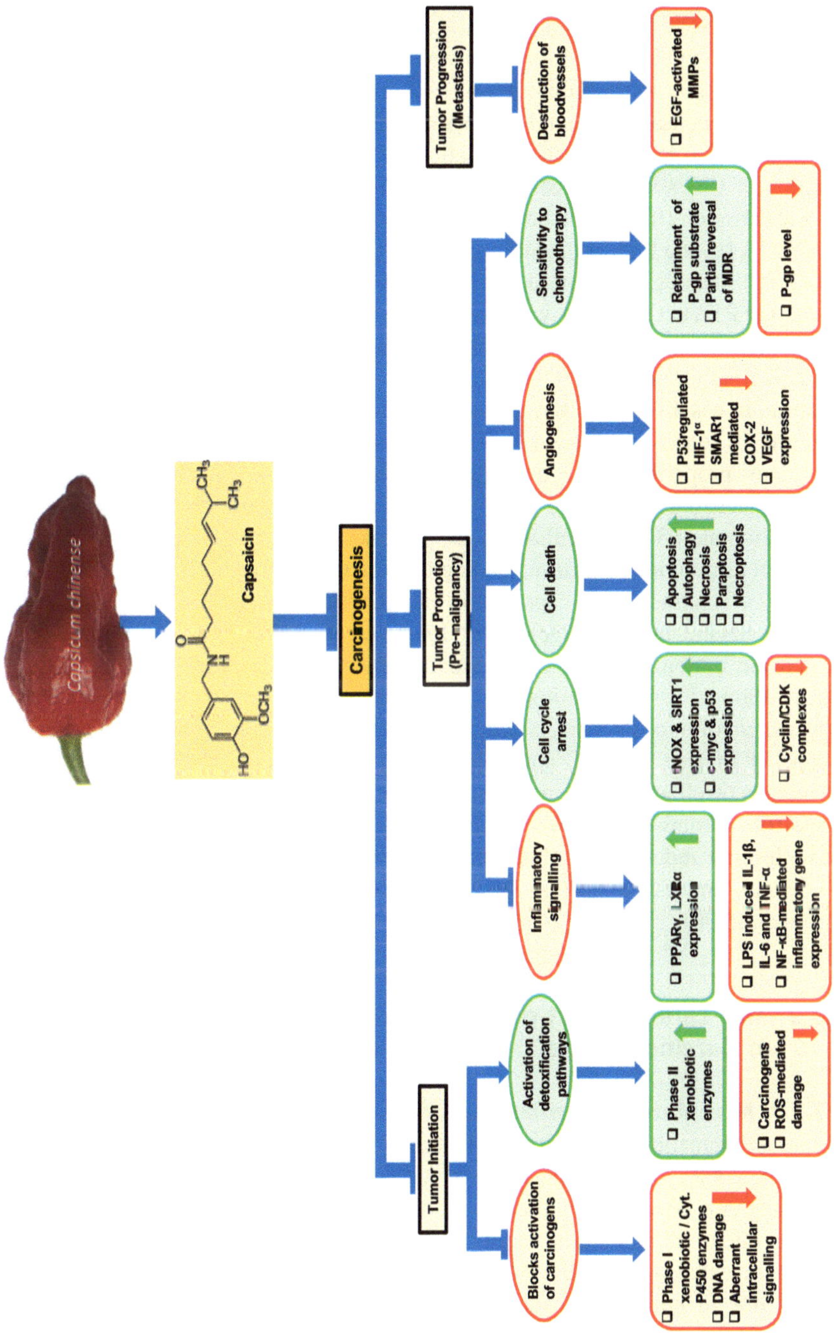

Fig. 20.1 Capsaicin mediated anti-cancer effects and their underlying mediators

and aberrant intracellular signalling (Cho et al. 2017). Capsaicin has been exhibited to inhibit the benzo(a)pyrene (a carcinogenic hydrocarbon) activation in human and mouse keratinocytes, preventing its interaction with the DNA (Modly et al. 1986). NNK (a potent carcinogen in tobacco) was also regulated from the formation of all its metabolites due to the inhibition of α-hydroxylation by capsaicin (Miller et al. 1993). Capsaicin has been found to block cytochrome P450 enzymes (CYPs), which play a critical role during the activation of numerous pro-carcinogens. Direct inhibition of CYPs is observed in vitro for CYP1A2, CYP2C9 and CYP2C19 at a low concentration (IC_{50} value between 2.0 and 4.0 μM), and for CYP2B6, CYP2D6 and CYP3A4/5 at a high concentration (IC_{50} value between 12 and 38 μM) (Babbar et al. 2010). Han et al. reported the inhibitory effect of capsaicin in another CYP enzyme CYP1A1. Capsaicin when incubated with murine hepatoma cells (Hepa-1c1c7) alone, weakly induced the activity of CYP1A1, but inhibited 3-methylcholanthrene induced CYP1A1 levels significantly (Han et al. 2011).

Capsaicin has also been shown to increase the expression and activity of numerous phase II xenobiotic enzymes which perform detoxification of cells and have antioxidant properties, and thus provide protection to macromolecules from ROS-mediated damages (Cho et al. 2017). Oral administration of capsaicin and rotenone in male F344 rats enhanced the enzyme activity of glutathione-S-transferase as well as quinone reductase, leading to reduction in tongue carcinoma (Tanaka et al. 2002). Treatment with capsaicin in HepG2 cells elevated the expression of the phase II enzyme heme oxygenase-1 (Joung et al. 2007). Various studies have reported the inhibition of lipid peroxidation by capsaicin which is modulated by the activation of various antioxidant enzymes. Anandakumar et al. reported that pre-treatment of mice using capsaicin in benzo(a)pyrene-induced lung cancer acts as a quencher of free radicals, and helps in the maintenance of normal balance of enzymatic as well as some non-enzymatic antioxidants (Anandakumar et al. 2008). Intragastric administration of capsaicin in Sprague-Dawley rats suppressed the gastric lesions induced by absolute alcohol, and significantly suppressed lipid peroxidation and expression of myeloperoxidase and cyclooxygenase-2 (COX-2) (Park et al. 2000).

20.3.2 Anti-cancer Effect of Capsaicin in Promotion Phase of Carcinogenesis

20.3.2.1 Modulation of Inflammatory Signalling

Chronic inflammation has a critical role to play in the progression of cancer (Kundu and Surh 2008), and capsaicin acts as an anti-inflammatory compound by modulating the aberrant inflammatory signalling pathways. It is observed that capsaicin elicits inflammation due to the topical application on human skin, however, on repeated application, the inflammatory response and burning sensation subsides due to inhibition of the release of a neuropeptide, Substance P (Tafler et al. 1993; Inoue et al. 1995). Capsaicin when injected in the hind paw of rats significantly inhibited paw swelling and inflammation, which is comparable to the anti-inflammatory drug diclofenac (Jolayemi and Ojewole 2013). It was earlier reported that the anti-inflammatory responses are mediated by the capsaicin receptor TRPV1, however, recent studies have suggested that these responses are independent of the TRPV1 receptor (Kim et al. 2003; Shin et al. 2013; Tang et al. 2015). Capsaicin exhibits its anti-inflammatory effect by a time and dose-dependent inhibition of pro-inflammatory cytokines and production of TNF-α. Capsaicin increases target gene (LXRα) expression which signifies the activation of the pathway involving PPARγ (Tang et al. 2015). It is been reported that in lipopolysaccharide-stimulated murine peritoneal macrophages, capsaicin exhibits an anti-inflammatory response by modulating the expression and activity of proinflammatory mediators, such as prostaglandin E2 (PGE2), COX-2 as well as inducible nitric oxide synthase (iNOS). COX-2 and iNOS inhibition is associated with the inactivation of NF-κB and stabilisation of IkB-a protein (Kim et al. 2003). Another study has explored the mechanism of signal transduction mediated by capsaicin in RAW264.7 macrophages, where capsaicin inhibited the production of nitric oxide along with iNOS, COX-2 expression, as well as PGE2 production. It also inhibited NF-κB and AP-1 activation, activation of STAT1, and protein kinases (ERK, JNK and IKK) (Chen et al. 2003).

20.3.2.2 Regulation of Cell Cycle by Capsaicin

The eukaryotic cells have developed inbuilt mechanisms which restrict damaged cells or cells having DNA mutations from progressing through the cell cycle due to the presence of 'cell cycle checkpoints' at G0/G1, S, G2/M as well as M/Anaphase shift, which are regulated by cyclins, CDKs and various CDK inhibitors. Cancer cells have developed mechanisms to escape these checkpoints and ensure uncontrolled proliferation. Numerous dietary agents extracted from fruits and vegetables have shown the capability to repair the deregulated cell cycle checkpoints in cancer cells (Meeran and Katiyar 2008). Capsaicin has been reported in numerous studies to mediate arrest of cell cycle by modulating key signal transducers, such as cyclins and CDKs, and thus halt the progression of cancer. Capsaicin exhibited arrest of cell cycle at the G1-phase in ORL-48 oral cancer cell lines in a time-dependent manner (Kamaruddin et al. 2019). Capsaicin also exhibited a significant arrest of cell cycle at G0/G1 phase in bladder cancer (5637 cell line) cells, along with downregulation of the proteins involved (cyclin D1 and CDK2/4/6) in the shift from G0/G1 to S phase (Qian et al. 2016). Arrest of cell cycle at the G2/M phase, followed by capsaicin treatment, has also been observed in MDA-MB-231 breast cancer cells along with a lowering in the expression levels of CDK8 (Wu et al. 2020). Besides regulating CDKs, capsaicin has also been reported to degrade tNOX, which then results in the inhibition of SIRT1 deacetylase. This results in acetylation of c-myc and p53 followed by inhibition of cyclin/CDK complexes, finally leading to cell cycle arrest (Jin et al. 2014; Islam et al. 2019a). Capsaicin suppressed the p53/MDM2 interaction in human colon cancer cells due to which the p53 ubiquitination by MDM2 was blocked. This inhibited the degradation of p53, and resulted in its accumulation in the cells and subsequent G0/G1 cell cycle arrest (Jin et al. 2014).

20.3.2.3 Induction of Cell Death by Capsaicin

Capsaicin Induced Apoptotic Cell Death

It has been reported that capsaicin has cytotoxic effects towards a number of cancer cells such as prostrate, leukemic, colon, lung, pancreas, liver, breast and skin in both in vitro and in vivo studies. The primary pathway via which capsaicin induces cell death involves apoptosis (Adetunji et al. 2022). Capsaicin induces apoptosis in more than 40 cancer cell lines without affecting normal healthy cells (Bley et al. 2012). Apoptosis involves programmed cell death which acts as a critical barrier against development and progression of cancer, and inhibition of this signalling in cells leads to malignancy. Apoptosis is carried out involving two major signalling pathways: the intrinsic mitochondrial pathway and the extrinsic pathway. The first pathway stimulates mitochondrial release of cytochrome c and apoptosis inducing factor (AIP), which is triggered by the increasing levels of Bax and decreasing levels of Bcl-2. The intrinsic mitochondrial pathway followed by apoptosis has been reported in cell lines of gastric cancer (AGS), epidermoid carcinoma (KB), lung cancer (FaDu), colon cancer (HCT116), lymphoblastic leukemia, and in experimental animals of lung cancer as well. The extrinsic pathway is mediated by activated caspase-8 due to the stimulation by activated death receptor (DR4 or DR5). DR4 mediated extrinsic signalling, leading to activation of caspase-8, is reported in HCT116 and A375 cell lines. Other important signalling mechanisms of apoptosis include intracellular increase in calcium influx, ROS generation, p53 phosphorylation/acetylation, alteration of transcription factor β-catenin, and activation of NFκB and STATS (Chapa-Oliver and Mejía-Teniente 2016; Cho et al. 2017) (Fig. 20.2).

TRPV receptors are also critically involved in mediating apoptosis by altering the structure and function of mitochondrial membrane and depolarising the membrane of intact cells (Gunthorpe et al. 2002). Capsaicin mediates pro-apoptotic activity via TRPV1 or TRPV6 in various cancers types (Clark and Lee 2016). Reports suggest that Ca^{2+} influx due to stimulation of TRPV1 induces cell death in urothelial and glioma cells involving the p38 signalling pathway (Amantini et al. 2007). Various studies have reported that capsaicin triggers TRPV1 dependent intrinsic and extrinsic apoptotic pathways, via ATM-dependent manner, in urothelial cancer cell lines (Amantini et al. 2009). Capsaicin has similar effects against anaplastic thyroid cancer cells where it induces mitochondrial dysfunction via TRPV1-mediated mitochondrial Ca^{2+} overload (Xu et al. 2020). Capsaicin effectively induces apoptosis in human osteosarcoma cells (MG63) via TRPV1-dependent pathways, involving impairment of mitochondria, ROS and JNK production, whereas, pathways independent of TRPV1 include activation of AMPK-p53 (Bao et al. 2019). One of the reports suggested that capsaicin mediated activation of TRPV1 induced apoptosis in bladder cancer 5637 cell line by enhancing efficacy of pirarubicin (main drug used in urinary bladder chemotherapy) (Zheng et al. 2016). Capsaicin also induces apoptosis by increasing TRPV6 expression in human small cell lung cancer by activating the calpain pathway (Lau et al. 2014). TRPV6 activation following capsaicin treatment in gastric cancer cells increases membrane permeability by activating Bax and p53 via JNK activation (Chow et al. 2007).

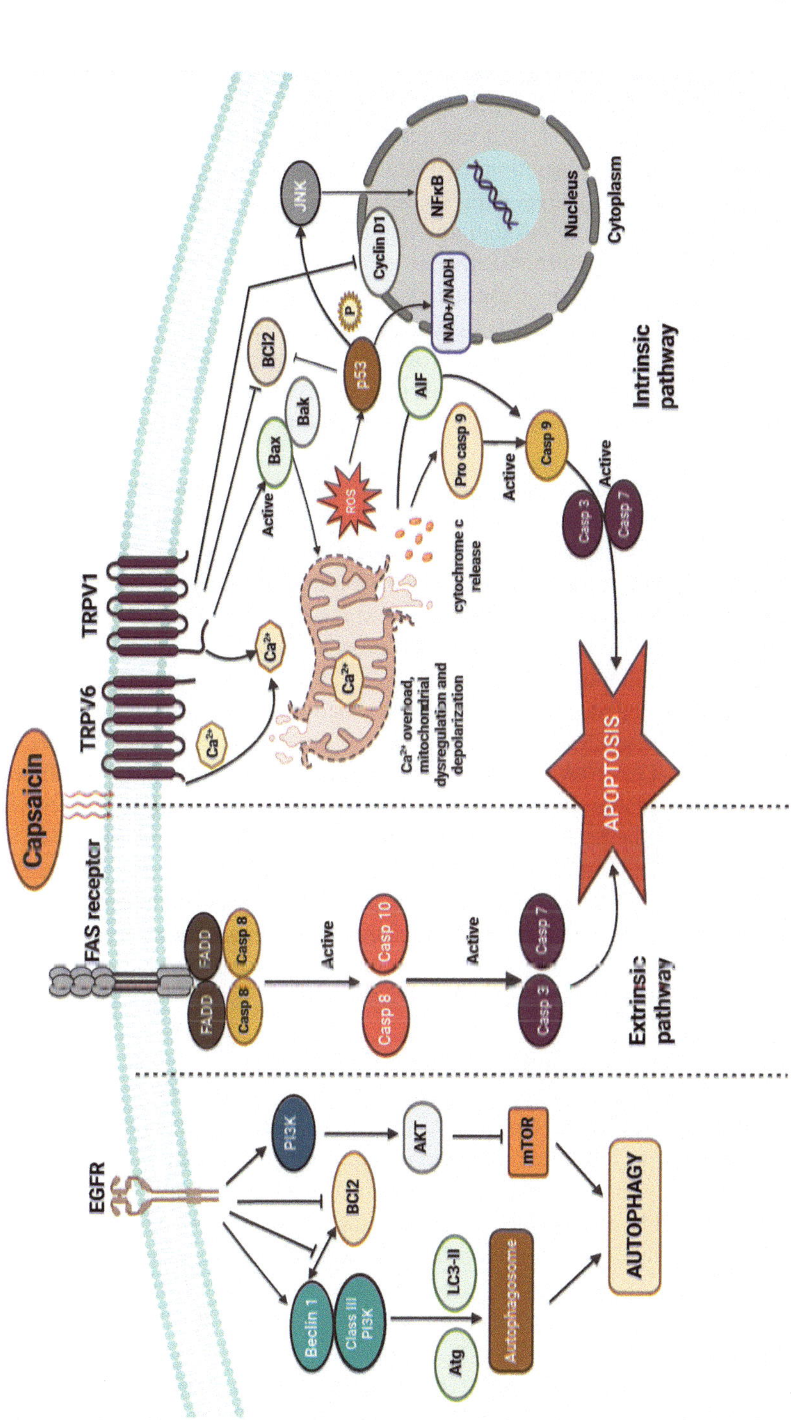

Fig. 20.2 Mechanism of capsaicin mediated autophagy and apoptotic cell death

One of the main targets of anti-cancer agents involves the activation of the tumour suppressor protein p53 whose activation promotes cell death, suggesting that it plays a critical role in apoptosis (Vogelstein et al. 2000; Michael and Oren 2002). Capsaicin induces apoptosis in leukemic cells via phosphorylation of p53 at Ser-15 residue by ROS generation, and it also effectively inhibits tumour growth and induces apoptosis in vivo with no side effects (Ito et al. 2004). Capsaicin increases acetylation of p53 through downregulation of sirtuin 1 and decreases the NAD+/NADH ratio, which then activates apoptosis in lung cancer cell line A549 (Lee et al. 2015). Capsaicin treatment led to enhanced expression of p53 and phosphorylation at Ser-15, Ser-20 as well as in Ser-392 which regulates intrinsic and extrinsic apoptosis pathway in cells of urothelial cancer (Amantini et al. 2009). Similar treatment in colorectal cancer cells (HCT116) using this compound activated TRPV1 followed by induction of apoptosis via calcineurin-NFAT2-p53 signalling pathway (Hou et al. 2019).

Another major signalling pathway which regulates cancer progression involves the deregulation of β-catenin-dependent signalling. Capsaicin has been reported to downregulate the β-catenin mediated transcription in human colorectal cancer cells (Lee et al. 2012). Phosphorylation of C/EBPb through glycogen synthase kinase 3b in human colorectal cancer cells mediates expression of anti-inflammatory drug-activated gene-1 which induces apoptosis (Lee et al. 2010).

Capsaicin also induces apoptosis through other pro-apoptotic mechanisms including AMPK mediated apoptosis, which was observed following the treatment in osteosarcoma cells U2OS and MG63 (Ying et al. 2013) and in colon cancer cell line HT29 (Kim et al. 2007). A study suggested that capsaicin treatment inhibited p65 nuclear translocation through suppression of NF-κB activation in both in vivo models as well as in breast cancer cell lines (MDA-MB-231 and MCF-7). The pro-apoptotic activities might be related to downregulation of NF-κB signalling pathway (Chen et al. 2021). Reports also suggested that Capsaicin induces apoptosis by downregulating the PI3K/AKT/mTOR pathway in human nasopharyngeal carcinoma cell lines (Lin et al. 2017). The compound was also able to induce apoptotic cell death in human tongue cancer cells SCC-4 (Ip et al. 2012).

Capsaicin Induced Autophagy Regulated Cell Death

Autophagy is a process which is conserved in organisms and transfers dysfunctional or redundant constituents of the cytoplasm for lysosomal degradation and can be used subsequently to fulfil the metabolic requirements of the cell during starvation periods. Previously thought of as a cell-survival mechanism, autophagy-mediated cell death is a highly context-based role as some studies have shown autophagy to promote the survival of cancer cells, whereas, a high degree of autophagy stimulates cell death (Denton and Kumar 2019; Elekofehinti et al. 2021). Due to this dichotomous nature of autophagy, the role of cancer therapeutics to inhibit or promote autophagy remains debatable (Levy et al. 2017). The role capsaicin plays in autophagic cell death needs more understanding as it exhibits inhibition of autophagy in some cancer cells whereas promotion in others. Capsaicin induces autophagy followed by apoptosis in human melanoma cells (A375 and C8161) (Chu et al. 2019), human glioma cells (U251) (Liu et al. 2016) and nasopharyngeal carcinoma cells (NPC-TW01) of humans (Lin et al. 2017). However, capsaicin is also reported to inhibit autophagy induced by 5-FU in cholangiocarcinoma cells and also in in vivo xenograft models via the activation of PI3K/AKT/mTOR signalling route (Hong et al. 2015) (Fig. 20.2). Numerous studies involving autophagy induced by capsaicin in cancer cells indicates that it is a pro-survival process, as inhibition of autophagy by using inhibitors like 3-methyladenine or chloroquine enhanced apoptosis (Chen et al. 2016; Chu et al. 2019). In breast cancer (MCF-7 and MDA-MB-231) cell lines, capsaicin mediated autophagy is mediated by p38 and ERK which ultimately retards endoplasmic reticulum stress-mediated apoptosis (Choi et al. 2010).

Capsaicin Induced Cell Death via Necrosis, Paraptosis and Necroptosis

Necrosis is a type of premature cell death which is generally unregulated and may occur due to exposure to extreme environmental factors, injury or deleterious mutations and lacks the characteristics of both apoptosis and autophagy (Syntichaki and Tavernarakis 2002; Golstein and Kroemer 2007). Capsaicin induced necrosis in 5637 and T24 bladder cancer cells time-dependently, which was enhanced when a combination treatment with lysosomal inhibitor bafilomycin A was provided (Amantini et al. 2016). Another study, which explored the role of capsaicin in cell death of MCF-7 cells observed necrotic cell death (Wu et al. 2014). Similarly, a high concentration (>10 μM) of capsaicin mediated cell death in HeLa cells via necrosis or apoptosis (Ramírez-Barrantes et al. 2018).

Another type of programmed cell death, named paraptosis, generally occurs due to endoplasmic reticulum or mitochondrial dysfunction. It is mediated by disruption in proteostasis or alterations in redox/ion homeostasis (Fontana et al. 2020). Jambrina et al. have reported that in Jurkat cells, capsaicin triggered a Ca^{2+} overload in the cytoplasm which resulted in the loss of mitochondrial membrane potential by paraptosis (Jambrina et al. 2003). Necroptosis is another cell death program which represents a caspase-independent subtype of necrosis, although the exact mechanism is not well understood. Huang et al. have reported that capsaicin induced a decrease in cell viability through necroptosis, when co-treated with four anticancer agents in human oral squamous carcinoma cell lines (HSC3 and SAS) (Huang et al. 2021).

20.3.2.4 Capsaicin Mediated Suppression of Angiogenesis

Angiogenesis is the physiological formation of blood vessels from pre-existing ones and viewed as an essential step for tumour growth. Capsaicin induced anti-angiogenic effects in vitro are mediated by inhibition of VEGF/VEGFR signalling, DNA synthesis, cell migration and formation of capillary tubes. The in vitro effect of capsaicin in chick chorioallantoic membrane assay is the inhibition of tumour-induced angiogenesis, and in in vivo experiments capsaicin also selectively inhibited the genesis of newer blood vessels without any side effect on the pre-existing ones (Min et al. 2004). Capsaicin has exhibited down-regulation of the expression of the angiogenic gene product VEGF via inhibition of the STAT3 pathway in multiple myeloma cell lines in a dose and time-dependent mode (Bhutani et al. 2007).

Chakraborty et al. have observed that capsaicin down-regulates expression of VEGF in NSCLC cell lines under hypoxic conditions. Capsaicin dosage activated p53-SMAR1 positive feedback loop followed by degradation of p53-regulated HIF-1α as well as SMAR1-mediated Cox-2 expression, which restrained nuclear localisation of HIF-1α. This leads to VEGF down-regulation, thwarting cell migration and network formation finally inhibiting angiogenesis (Chakraborty et al. 2014).

20.3.2.5 Enhanced Sensitivity to Chemotherapy by Capsaicin

Out of many problems encountered in cancer treatment one important problem is the non-responsiveness of cancer cells due to multi-drug resistance (MDR) to available chemotherapeutic drugs and radiotherapy (Takara et al. 2005). MDR is mediated by several mechanisms among which the drug transporter P-glycoprotein (P-gp) was the first mediator to be characterised (Johnson 2004). Nabekura et al. reported that capsaicin exhibited inhibition of P-gp in MDR carcinoma cells KB-C2 as evidenced by the accumulation and retainment of P-gp substrate vinblastine, which mediates cytotoxicity in the carcinoma cells leading to a partial reversal of MDR in these cells (Nabekura et al. 2005). Capsaicin also inhibited the efflux of P-gp substrate rhodamine123 in human CaCo-2 cells via modulation of intestinal TJ barrier at low non-cytotoxic doses (Tsukura et al. 2007). Okura et al. described the capsaicin mediated inhibition of P-gp-controlled efflux of rhodamine 123 in colon carcinoma cell lines (LS-180V) and also significantly enhanced the mRNA expression of MDR-associated protein-1 (MDR1) and MDR3. Thus, exposure to dietary ingredients like capsaicin can cause food-drug interactions which could further regulate the expression of mRNA and the functioning intestinal carrier molecules (Okura et al. 2010). Similarly, Han et al. reported that capsaicin treatment for a short duration and lower concentration inhibited the P-gp-induced efflux of [^{3}H]-digoxin from CaCo-2 cell monolayers, however, incubation for a longer duration (24–48 h) with higher concentration increased the activity of P-gp and upregulated mRNA levels of MDR1 (Han et al. 2006). The combination treatment of chemotherapeutic agents along with capsaicin can be a promising approach in different studies. Hong et al. reported that capsaicin treatment along with 5-FU demonstrated improved sensitivity in CCA cells and in an in vivo xenograft model, indicating a useful role in adjunct therapy for cancer (Hong et al. 2015).

20.3.3 Role of Capsaicin in Metastasis Phase of Carcinogenesis

In certain cases, the cancer cells can migrate to other distant sites of the body via the blood or lymphatic systems which is known as metastasis and this complex process is liable for most of the cancer-related mortality (Adetunji et al. 2022). Capsaicin has been reported by various authors to modulate critical pathways preventing cancer metastasis. For example, the matrix metalloproteases (MMPs) are important group of proteolytic enzymes which play a critical role in destruction of the basement membrane of arteries and veins and facilitate metastasis to different organs. Capsaicin has exhibited inhibition of the epidermal growth factor activated MMP9, MMP2 and MT1-MMP in human fibrosarcoma (HT-1080) cell lines (Hwang et al. 2011). It has been reported in in vitro and in vivo experiments that capsaicin has demonstrated anti-invasive and anti-metastatic property. In PC-3 cells capsaicin exhibited an inhibition of cancer cell migration dose-dependently, further in vivo treatment of capsaicin in TRAMP x C57BL/6 mice model exhibited a significant reduction of metastasis along with a reduction in p27KIP1 expression (Venier et al. 2015b). Angiogenesis is another important process for secondary tumour growth followed by metastasis and capsaicin regulates this metastasis by inhibiting angiogenesis which has been described in previous sections. However, some reports have also suggested a pro-metastatic role of capsaicin at certain concentrations. For example, Yang et al. have demonstrated that capsaicin at a high concentration inhibited the CRC cell proliferation (SW480, CT-26 and HCT116), however, a lower concentration of the compound triggered ROS production involving the Akt/mTOR and STAT-3 signalling routes enhancing the migratory and invasive nature of the CRC cells (Yang et al. 2013). Erin et al. have reported that capsaicin treatment resulted in systemic denervation of sensory nerves and enhanced metastasis in lung and cardiac cancer in adult mice (Erin et al. 2004).

20.4 Conclusions and Future Perspectives

One of the approaches of cancer research which is extensively explored in recent times is the search for natural and effective anti-cancer molecules which can target the molecular pathways responsible for the progression of cancer. Capsaicin has emerged as a viable candidate over the years which has exhibited strong anti-cancer effects in numerous cancer types both in vitro and in vivo. Capsaicin mediates these effects by targeting the multiple signalling pathways and genes associated with different cancer stages. The mechanism of action of capsaicin activity includes cell cycle arrest, and modulation of apoptosis, angiogenesis and metastasis. However, the molecular mechanisms known so far for the manifestation of these effects need more study and illustration. Further research in this field will shed light on the various receptors, pathways and intermediaries involved in these activities. Some of the challenges which have restricted the use of capsaicin include pungency, limited aqueous solubility, gastric and skin irritation, poor oral bioavailability as well as pro-carcinogenic properties at high concentrations in certain cancers. The development of synthetic analogues of capsaicin which are non-pungent or slightly pungent have been explored in recent years. The effective concentration of capsaicin in different cancers vary considerably and hence one aspect of research should be focused on improving the bioavailability of capsaicin. Nanotechnology-based approaches such as nano-liposome or solid-lipid-nanoparticle mediated targeted delivery of capsaicin to improve bioavailability are currently being explored by some researchers. Clinical trials of capsaicin in cancer patients have mostly focused on employing capsaicin for supporting therapy such as pain relief and in chemotherapy-induced neuropathy such as regeneration and restoration of nerve fibres of the skin. Co-treatment of capsaicin with other chemotherapy drugs or alongside radiotherapy has emerged as another promising approach to reduce the dependency on conventional anti-cancer drugs. Further, well-controlled studies involving diverse cell lines, animal models and clinical trials could help in better understanding of the anti-cancer potential of this compound.

References

Adetunji TL, Olawale F, Olisah C et al (2022) Capsaicin: a two-decade systematic review of global research output and recent advances against human cancer. Front Oncol 12:908487. https://doi.org/10.3389/fonc.2022.908487

Aggarwal BB, Shishodia S (2006) Molecular targets of dietary agents for prevention and therapy of cancer. Biochem Pharmacol 71:1397–1421

Akhtar MS, Swamy MK (2018) Anticancer plants: properties and application. Springer, Singapore

Al-Samydai A, Alshaer W, Al-Dujaili EAS et al (2021) Preparation, characterization, and anticancer effects of capsaicin-loaded nanoliposomes. Nutrients 13(11):3995. https://doi.org/10.3390/nu13113995

Amantini C, Mosca M, Nabissi M et al (2007) Capsaicin-induced apoptosis of glioma cells is mediated by TRPV1 vanilloid receptor and requires p38 MAPK activation. J Neurochem 102:977–990. https://doi.org/10.1111/j.1471-4159.2007.04582.x

Amantini C, Ballarini P, Caprodossi S et al (2009) Triggering of transient receptor potential vanilloid type 1 (TRPV1) by capsaicin induces Fas/CD95-mediated apoptosis of urothelial cancer cells in an ATM-dependent manner. Carcinogenesis 30:1320–1329. https://doi.org/10.1093/carcin/bgp138

Amantini C, Morelli MB, Nabissi M et al (2016) Capsaicin triggers autophagic cell survival which drives epithelial mesenchymal transition and chemoresistance in bladder cancer cells in an Hedgehog-dependent manner. Oncotarget 7:50180–50194. https://doi.org/10.18632/oncotarget.10326

Anandakumar P, Kamaraj S, Jagan S et al (2008) Capsaicin modulates pulmonary antioxidant defense system during benzo(a)pyrene-induced lung cancer in Swiss albino mice. Phytother Res 22:529–533. https://doi.org/10.1002/ptr.2393

Anandakumar P, Kamaraj S, Jagan S et al (2015) The anticancer role of capsaicin in experimentally induced lung carcinogenesis. J Pharmacopuncture 18:19–25. https://doi.org/10.3831/kpi.2015.18.011

Arem H, Loftfield E, Saint-Maurice PF et al (2018) Physical activity across the lifespan and liver cancer incidence in the NIH-AARP Diet and Health Study cohort. Cancer Med 7:1450–1457. https://doi.org/10.1002/cam4.1343

Athanasiou A, Smith PA, Vakilpour S et al (2007) Vanilloid receptor agonists and antagonists are mitochondrial inhibitors: how vanilloids cause non-vanilloid receptor mediated cell death. Biochem Biophys Res Commun 354:50–55. https://doi.org/10.1016/j.bbrc.2006.12.179

Babbar S, Chanda S, Bley K (2010) Inhibition and induction of human cytochrome P450 enzymes in vitro by capsaicin. Xenobiotica 40:807–816. https://doi.org/10.3109/00498254.2010.520044

Bade BC, Dela Cruz CS (2020) Lung cancer 2020. Clin Chest Med 41:1–24. https://doi.org/10.1016/j.ccm.2019.10.001

Bai H, Li H, Zhang W et al (2011) Inhibition of chronic pancreatitis and pancreatic intraepithelial neoplasia (panin) by capsaicin in lsl-kras g12d/pdx1-cre mice. Carcinogenesis 32:1689–1696. https://doi.org/10.1093/carcin/bgr191

Bao Z, Dai X, Wang P et al (2019) Capsaicin induces cytotoxicity in human osteosarcoma MG63 cells through TRPV1-dependent and -independent pathways. Cell Cycle 18:1379–1392. https://doi.org/10.1080/15384101.2019.1618119

Bessler H, Djaldetti M (2017) Capsaicin modulates the immune cross talk between human mononuclears and cells from two colon carcinoma lines. Nutr Cancer 69:14–20. https://doi.org/10.1080/01635581.2017.1247893

Bhutani M, Pathak AK, Nair AS et al (2007) Capsaicin is a novel blocker of constitutive and interleukin-6—inducible STAT3 activation. Clin Cancer Res 13:3024–3032. https://doi.org/10.1158/1078-0432.CCR-06-2575

Biller LH, Schrag D (2021) Diagnosis and treatment of metastatic colorectal cancer. JAMA 325: 669. https://doi.org/10.1001/jama.2021.0106

Bley K, Boorman G, Mohammad B et al (2012) A comprehensive review of the carcinogenic and anticarcinogenic potential of capsaicin. Toxicol Pathol 40:847–873. https://doi.org/10.1177/0192623312444471

Bode AM, Dong Z (2011) The two faces of capsaicin. Cancer Res 71:2809–2814

Brown KC, Witte TR, Hardman WE et al (2010) Capsaicin displays anti-proliferative activity against human small cell lung cancer in cell culture and nude mice models via the E2F pathway. PLoS One 5:e10243. https://doi.org/10.1371/journal.pone.0010243

Bu HQ, Cai K, Shen F et al (2015) Induction of apoptosis by capsaicin in hepatocellular cancer cell line SMMC-7721 is mediated through ROS generation and activation of JNK and p38 MAPK pathways. Neoplasma 62:582–591. https://doi.org/10.4149/neo_2015_070

Chakraborty S, Adhikary A, Mazumdar M et al (2014) Capsaicin-induced activation of p53-SMAR1 auto-regulatory loop down-regulates VEGF in non-small cell lung cancer to restrain angiogenesis. PLoS One 9:e99743. https://doi.org/10.1371/journal.pone.0099743

Chanda S, Erexson G, Frost D et al (2007) 26-Week dermal oncogenicity study evaluating pure trans-capsaicin in Tg.AC hemizygous mice (FBV/N). Int J Toxicol 26:123–133. https://doi.org/10.1080/10915810701225281

Chang HC, Chen ST, Chien SY et al (2011) Capsaicin may induce breast cancer cell death through apoptosis-inducing factor involving mitochondrial dysfunction. Hum Exp Toxicol 30:1657–1665. https://doi.org/10.1177/0960327110396530

Chapa-Oliver AM, Mejía-Teniente L (2016) Capsaicin: from plants to a cancer-suppressing agent. Molecules 21:931

Chen CW, Lee ST, Wu WT et al (2003) Signal transduction for inhibition of inducible nitric oxide synthase and cyclooxygenase-2 induction by capsaicin and related analogs in macrophages. Br J Pharmacol 140:1077–1087. https://doi.org/10.1038/sj.bjp.0705533

Chen D, Yang Z, Wang Y et al (2012) Capsaicin induces cycle arrest by inhibiting cyclin-dependent-kinase in bladder carcinoma cells. Int J Urol 19:662–668. https://doi.org/10.1111/j.1442-2042.2012.02981.x

Chen X, Tan M, Xie Z et al (2016) Inhibiting ROS-STAT3-dependent autophagy enhanced capsaicin–induced apoptosis in human hepatocellular carcinoma cells. Free Radic Res 50: 744–755. https://doi.org/10.3109/10715762.2016.1173689

Chen M, Xiao C, Jiang W et al (2021) Capsaicin inhibits proliferation and induces apoptosis in breast cancer by down-regulating FBI-1-mediated NF-κB pathway. Drug Des Devel Ther 15: 125–140. https://doi.org/10.2147/DDDT.S269901

Cho SC, Lee H, Choi BY (2017) An updated review on molecular mechanisms underlying the anticancer effects of capsaicin. Food Sci Biotechnol 26:1–13

Choi CH, Jung YK, Oh SH (2010) Autophagy induction by capsaicin in malignant human breast cells is modulated by p38 and extracellular signal-regulated mitogen-activated protein kinases and retards cell death by suppressing endoplasmic reticulum stress-mediated apoptosis. Mol Pharmacol 78:114–125. https://doi.org/10.1124/mol.110.063495

Chou CC, Wu YC, Wang YF et al (2009) Capsaicin-induced apoptosis in human breast cancer MCF-7 cells through caspase-independent pathway. Oncol Rep 21:665–671. https://doi.org/10.3892/or_00000269

Chow J, Norng M, Zhang J, Chai J (2007) TRPV6 mediates capsaicin-induced apoptosis in gastric cancer cells-mechanisms behind a possible new “hot” cancer treatment. Biochim Biophys Acta, Mol Cell Res 1773:565–576. https://doi.org/10.1016/j.bbamcr.2007.01.001

Chu H, Li M, Wang X (2019) Capsaicin induces apoptosis and autophagy in human melanoma cells. Oncol Lett 17:4827–4834. https://doi.org/10.3892/ol.2019.10206
Clark R, Lee SH (2016) Anticancer properties of capsaicin against human cancer. Anticancer Res 36:837–844
Cui Y, Yang F, Cao X et al (2012) Selective disruption of high sensitivity heat activation but not capsaicin activation of TRPV1 channels by pore turret mutations. J Gen Physiol 139:273–283. https://doi.org/10.1085/jgp.201110724
Czifra G, Varga A, Nyeste K et al (2009) Increased expressions of cannabinoid receptor-1 and transient receptor potential vanilloid-1 in human prostate carcinoma. J Cancer Res Clin Oncol 135:507–514. https://doi.org/10.1007/s00432-008-0482-3
Dai N, Ye R, He Q et al (2018) Capsaicin and sorafenib combination treatment exerts synergistic anti-hepatocellular carcinoma activity by suppressing EGFR and PI3K/Akt/mTOR signaling. Oncol Rep 40:3235–3248. https://doi.org/10.3892/or.2018.6754
Denton D, Kumar S (2019) Autophagy-dependent cell death. Cell Death Differ 26:605–616
Diaz-Laviada I (2010) Effect of capsaicin on prostate cancer cells. Future Oncol 6:1545–1550. https://doi.org/10.2217/fon.10.117
Díaz-Laviada I, Rodríguez-Henche N (2014) The potential antitumor effects of capsaicin. Prog Drug Res 68:181–208
Dou D, Ahmad A, Yang H, Sarkar FH (2011) Tumor cell growth inhibition is correlated with levels of capsaicin present in hot peppers. Nutr Cancer 63:272–281. https://doi.org/10.1080/01635581.2011.523497
Elekofehinti OO, Iwaloye O, Olawale F, Ariyo EO (2021) Saponins in cancer treatment: current progress and future prospects. Pathophysiology 28:250–272
Erin N, Boyer PJ, Bonneau RH et al (2004) Capsaicin-mediated denervation of sensory neurons promotes mammary tumor metastasis to lung and heart. Anticancer Res 24:1003–1009
Fontana F, Raimondi M, Marzagalli M et al (2020) The emerging role of paraptosis in tumor cell biology: perspectives for cancer prevention and therapy with natural compounds. Biochim Biophys Acta Rev Cancer 1873:188338
Garufi A, Pistritto G, Cirone M, D'Orazi G (2016) Reactivation of mutant p53 by capsaicin, the major constituent of peppers. J Exp Clin Cancer Res 35:136. https://doi.org/10.1186/s13046-016-0417-9
Golstein P, Kroemer G (2007) Cell death by necrosis: towards a molecular definition. Trends Biochem Sci 32:37–43
Gunthorpe MJ, Benham CD, Randall A, Davis JB (2002) The diversity in the vanilloid (TRPV) receptor family of ion channels. Trends Pharmacol Sci 23:183–191
Han Y, Tan TMC, Lim LY (2006) Effects of capsaicin on P-gp function and expression in Caco-2 cells. Biochem Pharmacol 71:1727–1734. https://doi.org/10.1016/j.bcp.2006.03.024
Han EH, Hwang YP, Kim HG et al (2011) CCAAT/enhancer-binding protein β activation by capsaicin contributes to the regulation of CYP1A1 expression, mediated by the aryl hydrocarbon receptor. Br J Pharmacol 164:1600–1613. https://doi.org/10.1111/j.1476-5381.2011.01232.x
Hanahan D, Weinberg RA (2000) The hallmarks of cancer. Cell 100:57–70
Hayman M, Kam PCA (2008) Capsaicin: a review of its pharmacology and clinical applications. Curr Anaesth Crit Care 19:338–343. https://doi.org/10.1016/j.cacc.2008.07.003
Hong ZF, Zhao WX, Yin ZY et al (2015) Capsaicin enhances the drug sensitivity of cholangiocarcinoma through the inhibition of chemotherapeutic-induced autophagy. PLoS One 10:e0121538. https://doi.org/10.1371/journal.pone.0121538
Hou N, He X, Yang Y et al (2019) TRPV1 induced apoptosis of colorectal cancer cells by activating calcineurin-NFAT2-p53 signaling pathway. Biomed Res Int 2019:6712536. https://doi.org/10.1155/2019/6712536
Huang SP, Chen JC, Wu CC et al (2009) Capsaicin-induced apoptosis in human hepatoma HepG2 cells. Anticancer Res 29:165–174

Huang YC, Yuan TM, Liu BH et al (2021) Capsaicin potentiates anticancer drug efficacy through autophagy-mediated ribophorin II downregulation and necroptosis in oral squamous cell carcinoma cells. Front Pharmacol 12:676813. https://doi.org/10.3389/fphar.2021.676813

Huh HC, Lee SY, Lee SK et al (2011) Capsaicin induces apoptosis of cisplatin-resistant stomach cancer cells by causing degradation of cisplatin-inducible aurora-A protein. Nutr Cancer 63: 1095–1103. https://doi.org/10.1080/01635581.2011.607548

Hurley JD, Akers AT, Friedman JR et al (2017) Non-pungent long chain capsaicin-analogs arvanil and olvanil display better anti-invasive activity than capsaicin in human small cell lung cancers. Cell Adhes Migr 11:80–97. https://doi.org/10.1080/19336918.2016.1187368

Hwang JT, Lee YK, Shin JI, Park OJ (2009) Anti-inflammatory and anticarcinogenic effect of genistein alone or in combination with capsaicin in TPA-treated rat mammary glands or mammary cancer cell line. Ann N Y Acad Sci 1171:415–420

Hwang YP, Yun HJ, Choi JH et al (2011) Suppression of EGF-induced tumor cell migration and matrix metalloproteinase-9 expression by capsaicin via the inhibition of EGFR-mediated FAK/Akt, PKC/Raf/ERK, p38 MAPK, and AP-1 signaling. Mol Nutr Food Res 55:594–605. https://doi.org/10.1002/mnfr.201000292

Hyoung LK, Lee YC, Kim T II et al (2006) Inhibitory effect of capsaicin on interleukin-8 production by Helicobacter pylori-infected MKN-45 cells. J Microbiol Biotechnol 16:1084–1089

Impheng H, Pongcharoen S, Richert L et al (2014) The selective target of capsaicin on FASN expression and de novo fatty acid synthesis mediated through ROS generation triggers apoptosis in HepG2 cells. PLoS One 9:e107842. https://doi.org/10.1371/journal.pone.0107842

Inoue H, Nagata N, Koshihara Y (1995) Involvement of substance P as a mediator in capsaicin-induced mouse ear oedema. Inflamm Res 44:470–474. https://doi.org/10.1007/BF01837912

Ip SW, Lan SH, Huang AC et al (2012) Capsaicin induces apoptosis in SCC-4 human tongue cancer cells through mitochondria-dependent and -independent pathways. Environ Toxicol 27:332–341. https://doi.org/10.1002/tox.20646

Iqbal J, Abbasi BA, Ahmad R et al (2019) Potential phytochemicals in the fight against skin cancer: current landscape and future perspectives. Biomed Pharmacother 109:1381–1393

Islam A, Su AJ, Zeng ZM et al (2019a) Capsaicin targets tNOX (ENOX2) to inhibit G1 Cyclin/CDK complex, as assessed by the cellular thermal shift assay (CETSA). Cells 8:1275. https://doi.org/10.3390/cells8101275

Islam A, Yang Y-T, Wu W-H et al (2019b) Capsaicin attenuates cell migration via SIRT1 targeting and inhibition to enhance cortactin and β-catenin acetylation in bladder cancer cells. Am J Cancer Res 9:1172–1182

Ito K, Nakazato T, Yamato K et al (2004) Induction of apoptosis in leukemic cells by homovanillic acid derivative, capsaicin, through oxidative stress: implication of phosphorylation of p53 at Ser-15 residue by reactive oxygen species. Cancer Res 64:1071–1078. https://doi.org/10.1158/0008-5472.CAN-03-1670

Jaganathan R, Purushotham B, Radhakrishnan N, Swamy MK (2020) Capsaicin and its potential anticancer mechanisms of action. In: Swamy MK (ed) Plant-derived bioactives: production, properties and therapeutic applications. Springer, Singapore, pp 301–321

Jambrina E, Alonso R, Alcalde M et al (2003) Calcium influx through receptor-operated channel induces mitochondria-triggered paraptotic cell death. J Biol Chem 278:14134–14145. https://doi.org/10.1074/jbc.M211388200

Jang JJ, Cho KJ, Lee YS, Bae JH (1991) Different modifying responses of capsaicin in a wide-spectrum initiation model of F344 rat. J Korean Med Sci 6:31–36. https://doi.org/10.3346/jkms.1991.6.1.31

Jankovic B, Loblaw DA, Nam R (2010) Capsaicin may slow PSA doubling time: case report and literature review. Can Urol Assoc J 4:E9–E11. https://doi.org/10.5489/cuaj.784

Jin J, Lin G, Huang H et al (2014) Capsaicin mediates cell cycle arrest and apoptosis in human colon cancer cells via stabilizing and activating p53. Int J Biol Sci 10:285–295. https://doi.org/10.7150/ijbs.7730

Jin T, Wu H, Wang Y, Peng H (2016) Capsaicin induces immunogenic cell death in human osteosarcoma cells. Exp Ther Med 12:765–770. https://doi.org/10.3892/etm.2016.3368
Johnson WW (2004) ABC transporter proteins and cellular drug resistance. In: Lu DR, Øie S (eds) Cellular drug delivery. Humana Press, Totowa, pp 129–162
Jolayemi AT, Ojewole JAO (2013) Comparative anti-inflammatory properties of Capsaicin and ethyl-aAcetate extract of Capsicum frutescens linn [Solanaceae] in rats. Afr Health Sci 13:357–361. https://doi.org/10.4314/ahs.v13i2.23
Jordt SE, Julius D (2002) Molecular basis for species-specific sensitivity to "hot" chili peppers. Cell 108:421–430. https://doi.org/10.1016/S0092-8674(02)00637-2
Joung EJ, Li MH, Lee HG et al (2007) Capsaicin induces heme oxygenase-1 expression in HepG2 cells via activation of PI3K-Nrf2 signaling: NAD(P)H:quinone oxidoreductase as a potential target. Antioxid Redox Signal 9:2087–2098. https://doi.org/10.1089/ars.2007.1827
Kamaruddin MF, Hossain MZ, Alabsi AM, Bakri MM (2019) The antiproliferative and apoptotic effects of capsaicin on an oral squamous cancer cell line of asian origin, ORL-48. Medicina (Kaunas) 55:322. https://doi.org/10.3390/MEDICINA55070322
Kang SN, Chung SW, Kim TS (2001) Capsaicin potentiates 1,25-dihydoxyvitamin D3- and all-trans retinoic acid-induced differentiation of human promyelocytic leukemia HL-60 cells. Eur J Pharmacol 420:83–90. https://doi.org/10.1016/S0014-2999(01)00994-3
Kaymaz BT, Çetintaş VB, Aktan Ç, Kosova B (2014) MicroRNA-520a-5p displays a therapeutic effect upon chronic myelogenous leukemia cells by targeting STAT3 and enhances the anticarcinogenic role of capsaicin. Tumor Biol 35:8733–8742. https://doi.org/10.1007/s13277-014-2138-z
Kim CS, Kawada T, Kim BS et al (2003) Capsaicin exhibits anti-inflammatory property by inhibiting IkB-a degradation in LPS-stimulated peritoneal macrophages. Cell Signal 15:299–306. https://doi.org/10.1016/S0898-6568(02)00086-4
Kim CS, Park WH, Park JY et al (2004) Capsaicin, a spicy component of hot pepper, induces apoptosis by activation of the peroxisome proliferator-activated receptor γ in HT-29 human colon cancer cells. J Med Food 7:267–273. https://doi.org/10.1089/jmf.2004.7.267
Kim YM, Hwang J-T, Kwak DW et al (2007) Involvement of AMPK signaling cascade in capsaicin-induced apoptosis of HT-29 colon cancer cells. Ann N Y Acad Sci 1095:496–503. https://doi.org/10.1196/annals.1397.053
Kim MY, Trudel LJ, Wogan GN (2009) Apoptosis induced by capsaicin and resveratrol in colon carcinoma cells requires nitric oxide production and caspase activation. Anticancer Res 29: 3733–3740
Kim SM, Oh EY, Lee JH et al (2015) Brassinin combined with capsaicin enhances apoptotic and anti-metastatic effects in PC-3 human prostate cancer cells. Phytother Res 29:1828–1836. https://doi.org/10.1002/ptr.5478
Kundu JK, Surh YJ (2008) Inflammation: gearing the journey to cancer. Mutat Res 659:15–30
Kuramori C, Azuma M, Kume K et al (2009) Capsaicin binds to prohibitin 2 and displaces it from the mitochondria to the nucleus. Biochem Biophys Res Commun 379:519–525. https://doi.org/10.1016/j.bbrc.2008.12.103
Lau JK, Brown KC, Dom AM et al (2014) Capsaicin induces apoptosis in human small cell lung cancer via the TRPV6 receptor and the calpain pathway. Apoptosis 19:1190–1201. https://doi.org/10.1007/s10495-014-1007-y
Lee YS, Kang YS, Lee JS et al (2004) Involvement of NADPH oxidase-mediated generation of reactive oxygen species in the apototic cell death by capsaicin in HepG2 human hepatoma cells. Free Radic Res 38:405–412. https://doi.org/10.1080/10715760410001665262
Lee SH, Krisanapun C, Baek SJ (2010) NSAID-activated gene-1 as a molecular target for capsaicin-induced apoptosis through a novel molecular mechanism involving GSK3β, C/EBPβ and ATF3. Carcinogenesis 31:719–728. https://doi.org/10.1093/carcin/bgq016
Lee SH, Richardson RL, Dashwood RH, Baek SJ (2012) Capsaicin represses transcriptional activity of β-catenin in human colorectal cancer cells. J Nutr Biochem 23:646–655. https://doi.org/10.1016/j.jnutbio.2011.03.009

Lee YH, Chen HY, Su LJ, Chueh PJ (2015) Sirtuin 1 (SIRT1) deacetylase activity and NAD+/NADH ratio are imperative for capsaicin-mediated programmed cell death. J Agric Food Chem 63:7361–7370. https://doi.org/10.1021/acs.jafc.5b02876

Levy JMM, Towers CG, Thorburn A (2017) Targeting autophagy in cancer. Nat Rev Cancer 17: 528–542. https://doi.org/10.1038/nrc.2017.53

Lin YT, Wang HC, Hsu YC et al (2017) Capsaicin induces autophagy and apoptosis in human nasopharyngeal carcinoma cells by downregulating the PI3K/AKT/mTOR pathway. Int J Mol Sci 18:1343. https://doi.org/10.3390/ijms18071343

Liu NC, Hsieh PF, Hsieh MK et al (2012) Capsaicin-mediated tNOX (ENOX2) up-regulation enhances cell proliferation and migration in vitro and in vivo. J Agric Food Chem 60:2758–2765. https://doi.org/10.1021/jf204869w

Liu YP, Dong FX, Chai X et al (2016) Role of autophagy in capsaicin-induced apoptosis in U251 glioma cells. Cell Mol Neurobiol 36:737–743. https://doi.org/10.1007/s10571-015-0254-y

Lo YC, Yang YC, Wu IC et al (2005) Capsaicin-induced cell death in a human gastric adenocarcinoma cell line. World J Gastroenterol 11:6254–6257. https://doi.org/10.3748/wjg.v11.i40.6254

López-Carrillo L, Camargo MC, Schneider BG et al (2012) Capsaicin consumption, Helicobacter pylori CagA status and IL1B-31C>T genotypes: a host and environment interaction in gastric cancer. Food Chem Toxicol 50:2118–2122. https://doi.org/10.1016/j.fct.2012.02.043

Lu H-F, Chen Y-L, Yang J-S et al (2010) Antitumor activity of capsaicin on human colon cancer cells in vitro and colo 205 tumor xenografts in vivo. J Agric Food Chem 58:12999–13005. https://doi.org/10.1021/jf103335w

Lukasiewicz S, Czeczelewski M, Forma A et al (2021) Breast cancer—epidemiology, risk factors, classification, prognostic markers, and current treatment strategies—an updated review. Cancers (Basel) 13:4287

Malagarie-Cazenave S, Olea-Herrero N, Vara D, Díaz-Laviada I (2009) Capsaicin, a component of red peppers, induces expression of androgen receptor via PI3K and MAPK pathways in prostate LNCaP cells. FEBS Lett 583:141–147. https://doi.org/10.1016/j.febslet.2008.11.038

Malagarie-Cazenave S, Olea-Herrero N, Vara D et al (2011) The vanilloid capsaicin induces IL-6 secretion in prostate PC-3 cancer cells. Cytokine 54:330–337. https://doi.org/10.1016/j.cyto.2011.03.010

Meeran SM, Katiyar SK (2008) Cell cycle control as a basis for cancer chemoprevention through dietary agents. Front Biosci 13:2191–2202

Meral O, Alpay M, Kismali G et al (2014) Capsaicin inhibits cell proliferation by cytochrome c release in gastric cancer cells. Tumor Biol 35:6485–6492. https://doi.org/10.1007/s13277-014-1864-6

Michael D, Oren M (2002) The p53 and Mdm2 families in cancer. Curr Opin Genet Dev 12:53–59

Miller CH, Zhang Z, Hamilton SM, Teel RW (1993) Effects of capsaicin on liver microsomal metabolism of the tobacco-specific nitrosamine NNK. Cancer Lett 75:45–52. https://doi.org/10.1016/0304-3835(93)90206-O

Min JK, Han KY, Kim EC et al (2004) Capsaicin inhibits in vitro and in vivo angiogenesis. Cancer Res 64:644–651. https://doi.org/10.1158/0008-5472.CAN-03-3250

Modly CE, Das M, Con PSC (1986) Capsaicin as an in vitro inhibitor of benzo(a)pyrene metabolism and its DNA binding in human and murine keratinocytes. Drug Metab Dispos 14:413–416

Momenimovahed Z, Salehiniya H (2019) Epidemiological characteristics of and risk factors for breast cancer in the world. Breast Cancer Targets Ther 11:151–164

Moon DO, Kang CH, Kang SH et al (2012) Capsaicin sensitizes TRAIL-induced apoptosis through Sp1-mediated DR5 up-regulation: involvement of Ca^{2+} influx. Toxicol Appl Pharmacol 259: 87–95. https://doi.org/10.1016/j.taap.2011.12.010

Mori A, Lehmann S, O'Kelly J, Kumagai T, Desmond JC, Pervan M, McBride WH, Kizaki M, Koeffler HP (2006) Capsaicin, a component of red peppers, inhibits the growth of androgen-independent, p53 mutant prostate cancer cells. Cancer Res 66(6):3222–3229

Morré DJ, Sun E, Geilen C et al (1996) Capsaicin inhibits plasma membrane NADH oxidase and growth of human and mouse melanoma lines. Eur J Cancer 32A:1995–2003. https://doi.org/10.1016/0959-8049(96)00234-1

Mózsik G, Szolcsányi J, Rácz I (2005) Gastroprotection induced by capsaicin in healthy human subjects. World J Gastroenterol 11:5180–5184. https://doi.org/10.3748/wjg.v11.i33.5180

Nabekura T, Kamiyama S, Kitagawa S (2005) Effects of dietary chemopreventive phytochemicals on P-glycoprotein function. Biochem Biophys Res Commun 327:866–870. https://doi.org/10.1016/j.bbrc.2004.12.081

Narayanaswamy R, Swamy MK (2018) Elucidation of mechanisms of anticancer plant compounds against the tumor cells. In: Anticancer plants: mechanisms and molecular interactions, vol 4. Springer, Singapore, pp 99–130

Ng CY, Yen H, Hsiao HY, Su SC (2018) Phytochemicals in skin cancer prevention and treatment: an updated review. Int J Mol Sci 19:941

Okura T, Ibe M, Umegaki K et al (2010) Effects of dietary ingredients on function and expression of P-glycoprotein in human intestinal epithelial cells. Biol Pharm Bull 33:255–259. https://doi.org/10.1248/bpb.33.255

Park KK, Surh YJ (1997) Effects of capsaicin on chemically-induced two-stage mouse skin carcinogenesis. Cancer Lett 114:183–184

Park JS, Choi MA, Kim BS et al (2000) Capsaicin protects against ethanol-induced oxidative injury in the gastric mucosa of rats. Life Sci 67:3087–3093. https://doi.org/10.1016/S0024-3205(00)00890-0

Park SY, Kim JY, Lee SM, Jun CH, Cho SB, Park CH, Joo YE, Kim HS, Choi SK, Rew JS (2014) Capsaicin induces apoptosis and modulates MAPK signaling in human gastric cancer cells. Mol Med Rep 9(2):499–502

Patel PS, Varney ML, Dave BJ, Singh RK (2002) Regulation of constitutive and induced NF-κB activation in malignant melanoma cells by capsaicin modulates interleukin-8 production and cell proliferation. J Interf Cytokine Res 22:427–435. https://doi.org/10.1089/10799900252952217

Pramanik KC, Srivastava SK (2012) Apoptosis signal-regulating kinase 1-thioredoxin complex dissociation by capsaicin causes pancreatic tumor growth suppression by inducing apoptosis. Antioxid Redox Signal 17:1417–1432. https://doi.org/10.1089/ars.2011.4369

Pramanik KC, Boreddy SR, Srivastava SK (2011) Role of mitochondrial electron transport chain complexes in capsaicin mediated oxidative stress leading to apoptosis in pancreatic cancer cells. PLoS One 6:e20151. https://doi.org/10.1371/journal.pone.0020151

Qian K, Wang G, Cao R et al (2016) Capsaicin suppresses cell proliferation, induces cell cycle arrest and ROS production in bladder cancer cells through FOXO3a-mediated pathways. Molecules 21:1406. https://doi.org/10.3390/molecules21101406

Raisinghani M, Pabbidi RM, Premkumar LS (2005) Activation of transient receptor potential vanilloid 1 (TRPV1) by resiniferatoxin. J Physiol 567:771–786. https://doi.org/10.1113/jphysiol.2005.087874

Ramírez-Barrantes R, Córdova C, Gatica S et al (2018) Transient receptor potential vanilloid 1 expression mediates capsaicin-induced cell death. Front Physiol 9:682. https://doi.org/10.3389/fphys.2018.00682

Ramos-Torres Á, Bort A, Morell C et al (2016) The pepper's natural ingredient capsaicin induces autophagy blockage in prostate cancer cells. Oncotarget 7:1569–1583. https://doi.org/10.18632/oncotarget.6415

Ramsey IS, Delling M, Clapham DE (2006) An introduction to TRP channels. Annu Rev Physiol 68:619–647

Ravichandra VD, Ramesh C, Swamy MK, Purushotham B, Rudramurthy GR (2018) Anticancer plants: chemistry, pharmacology, and potential applications. In: Akhtar MS, Swamy MK (eds) Anticancer plants: properties and application, vol 1. Springer, Singapore, pp 485–515

Rawla P, Barsouk A (2019) Epidemiology of gastric cancer: global trends, risk factors and prevention. Prz Gastroenterol 14:26–38

Rawla P, Sunkara T, Gaduputi V (2019) Epidemiology of pancreatic cancer: global trends, etiology and risk factors. World J Oncol 10:10–27. https://doi.org/10.14740/wjon1166

Reyes-Escogido M, Gonzalez-Mondragon EG, Vazquez-Tzompantzi E (2011) Chemical and pharmacological aspects of capsaicin. Molecules 16:1253–1270. https://doi.org/10.3390/molecules16021253

Rychkov GY, Barritt GJ (2011) Expression and function of TRP channels in liver cells. Adv Exp Med Biol 704:667–686

Sánchez AM, Sánchez MG, Malagarie-Cazenave S et al (2006) Induction of apoptosis in prostate tumor PC-3 cells and inhibition of xenograft prostate tumor growth by the vanilloid capsaicin. Apoptosis 11:89–99. https://doi.org/10.1007/s10495-005-3275-z

Sánchez AM, Malagarie-Cazenave S, Olea N et al (2007) Apoptosis induced by capsaicin in prostate PC-3 cells involves ceramide accumulation, neutral sphingomyelinase, and JNK activation. Apoptosis 12:2013–2024. https://doi.org/10.1007/S10495-007-0119-Z

Sánchez AM, Martínez-Botas J, Malagarie-Cazenave S et al (2008) Induction of the endoplasmic reticulum stress protein GADD153/CHOP by capsaicin in prostate PC-3 cells: a microarray study. Biochem Biophys Res Commun 372:785–791. https://doi.org/10.1016/j.bbrc.2008.05.138

Santoni G, Caprodossi S, Farfariello V et al (2012) Antioncogenic effects of transient receptor potential vanilloid 1 in the progression of transitional urothelial cancer of human bladder. ISRN Urol 2012:1–11. https://doi.org/10.5402/2012/458238

Santosh S, Rajagopalan MD, Pallavi BA, Rudramurthy GR, Rajashekar V, Sridhar KA, Swamy MK (2017) Cancer therapies: current scenario, management, and safety aspects. In: Akhtar M, Swamy M (eds) Anticancer plants: clinical trials and nanotechnology, vol 3. Springer, Singapore, pp 1–25

Satyanarayana MN (2006) Capsaicin and gastric ulcers. Crit Rev Food Sci Nutr 46:275–328. https://doi.org/10.1080/10408390591379236

Sharma SK, Vij AS, Sharma M (2013) Mechanisms and clinical uses of capsaicin. Eur J Pharmacol 720:55–62. https://doi.org/10.1016/j.ejphar.2013.10.053

Shim Y, Song JM (2015) Quantum dot nanoprobe-based high-content monitoring of notch pathway inhibition of breast cancer stem cell by capsaicin. Mol Cell Probes 29:376–381. https://doi.org/10.1016/j.mcp.2015.09.004

Shin YH, Namkoong E, Choi S et al (2013) Capsaicin regulates the NF-κB pathway in salivary gland inflammation. J Dent Res 92:547–552. https://doi.org/10.1177/0022034513487376

Sun F, Xiong S, Zhu Z (2016) Dietary capsaicin protects cardiometabolic organs from dysfunction. Nutrients 8:174

Sung Y, Chang Y, Ting N (2005) Capsaicin biosynthesis in water-stressed hot pepper fruits. Bot Bull Acad Sin 46:35–42

Sung H, Ferlay J, Siegel RL et al (2021) Global cancer statistics 2020: GLOBOCAN estimates of incidence and mortality worldwide for 36 cancers in 185 countries. CA Cancer J Clin 71:209–249. https://doi.org/10.3322/caac.21660

Supalkova V, Stavelikova H, Krizkova S et al (2007) Study of capsaicin content in various parts of pepper fruit by liquid chromatography with electrochemical detection. Acta Chim Slov 54:55–59

Surh YJ, Lee RCJ, Park KK, Mayne ST, Liem A, Miller JA (1995) Chemoprotective effects of capsaicin and diallyl sulfide against mutagenesis or tumorigenesis by vinyl carbamate and N-nitrosodiinethylamine. Carcinogenesis 16(10):2467–2471

Swamy MK (2020) Plant-derived bioactives: production, properties and therapeutic applications. Springer Nature, Singapore

Swamy MK, Akhtar MS (2018a) Anticancer plants: mechanisms and molecular interactions. Springer, Singapore

Swamy MK, Akhtar MS (2018b) Natural bio-active compounds. Springer, Singapore

Syntichaki P, Tavernarakis N (2002) Death by necrosis. Uncontrollable catastrophe, or is there order behind the chaos? EMBO Rep 3:604–609. https://doi.org/10.1093/embo-reports/kvf138

Tafler R, Herbert MK, Schmidt RF, Weis KH (1993) Small reduction of capsaicin-induced neurogenic inflammation in human forearm skin by the glucocorticoid prednicarbate. Agents Actions 38:C31–C34. https://doi.org/10.1007/BF01991128

Takara K, Sakaeda T, Okumura K (2005) An update on overcoming MDR1-mediated multidrug resistance in cancer chemotherapy. Curr Pharm Des 12:273–286. https://doi.org/10.2174/138161206775201965

Tanaka T, Kohno H, Sakata K et al (2002) Modifying effects of dietary capsaicin and rotenone on 4-nitroquinoline 1-oxide-induced rat tongue carcinogenesis. Carcinogenesis 23:1361–1367. https://doi.org/10.1093/carcin/23.8.1361

Tang J, Luo K, Li Y et al (2015) Capsaicin attenuates LPS-induced inflammatory cytokine production by upregulation of LXRα. Int Immunopharmacol 28:264–269. https://doi.org/10.1016/j.intimp.2015.06.007

Thoennissen NH, O'Kelly J, Lu D et al (2010) Capsaicin causes cell-cycle arrest and apoptosis in ER-positive and-negative breast cancer cells by modulating the EGFR/HER-2 pathway. Oncogene 29:285–296. https://doi.org/10.1038/onc.2009.335

Tsou MF, Lu HF, Chen SC et al (2006) Involvement of Bax, Bcl-2, Ca^{2+} and caspase-3 in capsaicin-induced apoptosis of human leukemia HL-60 cells. Anticancer Res 26:1965–1971

Tsukura Y, Mori M, Hirotani Y et al (2007) Effects of capsaicin on cellular damage and monolayer permeability in human intestinal Caco-2 cells. Biol Pharm Bull 30:1982–1986. https://doi.org/10.1248/bpb.30.1982

Venier NA, Colquhoun AJ, Sasaki H et al (2015a) Capsaicin: a novel radio-sensitizing agent for prostate cancer. Prostate 75:113–125. https://doi.org/10.1002/pros.22896

Venier NA, Yamamoto T, Sugar LM et al (2015b) Capsaicin reduces the metastatic burden in the transgenic adenocarcinoma of the mouse prostate model. Prostate 75:1300–1311. https://doi.org/10.1002/pros.23013

Vogelstein B, Lane D, Levine AJ (2000) Surfing the p53 network. Nature 408:307–310. https://doi.org/10.1038/35042675

Walpole CSJ, Bevan S, Bloomfield G et al (1996) Similarities and differences in the structure-activity relationships of capsaicin and resiniferatoxin analogues. J Med Chem 39:2939–2952. https://doi.org/10.1021/jm960139d

Wang H-M, Chueh PJ, Chang S-P et al (2009) Effect of Ccapsaicin on tNOX (ENOX2) protein expression in stomach cancer cells. Biofactors 34:209–217. https://doi.org/10.1002/biof.5520340305

Wang F, Zhao J, Liu D et al (2016) Capsaicin reactivates hMOF in gastric cancer cells and induces cell growth inhibition. Cancer Biol Ther 17:1117–1125. https://doi.org/10.1080/15384047.2016.1235654

Wang Y, Deng X, Yu C et al (2018) Synergistic inhibitory effects of capsaicin combined with cisplatin on human osteosarcoma in culture and in xenografts 11 medical and health sciences 1112 oncology and carcinogenesis. J Exp Clin Cancer Res 37:251. https://doi.org/10.1186/s13046-018-0922-0

Wang F, Xue Y, Fu L et al (2022) Extraction, purification, bioactivity and pharmacological effects of capsaicin: a review. Crit Rev Food Sci Nutr 62:5322–5348

Wu TTL, Peters AA, Tan PT et al (2014) Consequences of activating the calcium-permeable ion channel TRPV1 in breast cancer cells with regulated TRPV1 expression. Cell Calcium 56:59–67. https://doi.org/10.1016/j.ceca.2014.04.006

Wu D, Jia H, Zhang Z, Li S (2020) Capsaicin suppresses breast cancer cell viability by regulating the CDK8/PI3K/Akt/Wnt/β-catenin signaling pathway. Mol Med Rep 22:4868–4876. https://doi.org/10.3892/mmr.2020.11585

Xu S, Cheng X, Wu L et al (2020) Capsaicin induces mitochondrial dysfunction and apoptosis in anaplastic thyroid carcinoma cells via TRPV1-mediated mitochondrial calcium overload. Cell Signal 75:109733. https://doi.org/10.1016/j.cellsig.2020.109733

Yang KM, Pyo JO, Kim G-Y et al (2009) Capsaicin induces apoptosis by generating reactive oxygen species and disrupting mitochondrial transmembrane potential in human colon cancer cell lines. Cell Mol Biol Lett 14:497–510. https://doi.org/10.2478/s11658-009-0016-2

Yang ZH, Wang XH, Wang HP et al (2010) Capsaicin mediates cell death in bladder cancer T24 cells through reactive oxygen species production and mitochondrial depolarization. Urology 75: 735–741. https://doi.org/10.1016/j.urology.2009.03.042

Yang J, Li TZ, Xu GH et al (2013) Low-concentration capsaicin promotes colorectal cancer metastasis by triggering ROS production and modulating Akt/mTOR and STAT-3 pathways. Neoplasma 60:364–372. https://doi.org/10.4149/neo_2013_048

Ying H, Wang Z, Zhang Y et al (2013) Capsaicin induces apoptosis in human osteosarcoma cells through AMPK-dependent and AMPK-independent signaling pathways. Mol Cell Biochem 384:229–237. https://doi.org/10.1007/s11010-013-1802-8

Yoshitani SI, Tanaka T, Kohno H, Takashima S (2001) Chemoprevention of azoxymethane-induced rat colon carcinogenesis by dietary capsaicin and rotenone. Int J Oncol 19:929–939. https://doi.org/10.3892/ijo.19.5.929

Zhang J, Nagasaki M, Tanaka Y, Morikawa S (2003) Capsaicin inhibits growth of adult T-cell leukemia cells. Leuk Res 27:275–283. https://doi.org/10.1016/S0145-2126(02)00164-9

Zhang R, Humphreys I, Sahu RP et al (2008) In vitro and in vivo induction of apoptosis by capsaicin in pancreatic cancer cells is mediated through ROS generation and mitochondrial death pathway. Apoptosis 13:1465–1478. https://doi.org/10.1007/s10495-008-0278-6

Zhang JH, Lai FJ, Chen H et al (2012a) Involvement of the phosphoinositide 3-kinase/Akt pathway in apoptosis induced by capsaicin in the human pancreatic cancer cell line PANC-1. Oncol Lett 5:43–48. https://doi.org/10.3892/ol.2012.991

Zhang Q-H, Hu J-P, Wang B-L, Li Y (2012b) Effects of capsaicin and dihydrocapsaicin on human and rat liver microsomal CYP450 enzyme activities in vitro and in vivo. J Asian Nat Prod Res 14:382–395. https://doi.org/10.1080/10286020.2012.656605

Zhang S, Wang D, Huang J et al (2020) Application of capsaicin as a potential new therapeutic drug in human cancers. J Clin Pharm Ther 45:16–28

Zheng L, Chen J, Ma Z et al (2015) Capsaicin causes inactivation and degradation of the androgen receptor by inducing the restoration of miR-449a in prostate cancer. Oncol Rep 34:1027–1034. https://doi.org/10.3892/or.2015.4055

Zheng J, Zhou Y, Li Y et al (2016) Spices for prevention and treatment of cancers. Nutrients 8:495. https://doi.org/10.3390/nu8080495

Zhu M, Yu X, Zheng Z et al (2020) Capsaicin suppressed activity of prostate cancer stem cells by inhibition of Wnt/β-catenin pathway. Phyther Res 34:817–824. https://doi.org/10.1002/ptr.6563

Ziglioli F, Frattini A, Maestroni U et al (2009) Vanilloid-mediated apoptosis in prostate cancer cells through a TRPV-1 dependent and a TRPV-1-independent mechanism. Acta Biomed 80:13–20

21 The Inhibitory Mechanisms of Capsaicin Against Colorectal Cancer

Saabira Banu Sahubar Sadik and Ashok Kumar Pandurangan

Abstract

Colorectal cancer (CRC) is the one of the leading cancers and causes mortality worldwide. There is a steady increase in the incidences of CRC in India due to the change of Western food style. Despite having lower incidence rates than many Western nations, there have been notable improvements in recent years. The two most widely used clinical treatments for CRC are surgery and chemotherapy. To combat the growth of drug resistance and side effects, new and effective medications are however urgently needed. Since many natural substances are well tolerated by patients and don't have hazardous side effects even at high dosages, they are the target drug for cancer treatment. In this chapter we discuss the effect of capsaicin on regulating the CRC. Capsaicin ($C_{18}H_{27}NO_3$, 8-methyl-N-vanillyl-6-nonenamide), a bioactive component of red chili, is used widely as a food additive and an analgesic. Capsaicin exhibits antiproliferative properties against lung, prostate, pancreatic, and breast cancer. This chapter summarizes the recent findings on capsaicin pharmacological activity, which shed light on the mechanism underlying the apoptotic effects of capsaicin, suggesting that capsai cin may present interesting new options for improving CRC therapy.

Keywords

Colorectal cancer · Capsaicin · Apoptosis · B-catenin · Capsicum · Vegetable

S. B. S. Sadik · A. K. Pandurangan (✉)
School of Life Sciences, B.S. Abdur Rahman Crescent Institute of Science and Technology, Chennai, Tamil Nadu, India

M. K. Swamy, A. Kumar (eds.), *Capsaicinoids*,
https://doi.org/10.1007/978-981-99-7779-6_21

21.1 Introduction

The third most prevalent cancer type and the main cause of mortality globally is colorectal cancer (CRC). According to projections from the World Health Organization (WHO) GLOBOCAN for 2015, there were 746,000 new cases of CRC in both men and women, or up to 9.7% of all incident malignancies (Torre et al. 2015, 17; Thimmarayan et al. 2020). According to forecasts for 2020, colon (9.4%), liver (8.3%), stomach (7.7%), and female breast (6.9%) cancers were the next most common malignancies to cause mortality, each accounting for an expected 1.8 million fatalities. Globally, there is a wide range in the incidence and fatality rates of CRC (Ferlay et al. 2015; Chandraprasad et al. 2022). The incidence of CRC is higher in industrialized nations, showing a link to Western food practices and lifestyle patterns like smoking, drinking alcohol, being overweight, and being sedentary (Kamangar et al. 2006). In 2010, it is anticipated that there would be 142,570 new cases of CRC and 51,370 deaths from CRC (Gingras and Béliveau 2011). India has had rapid economic growth over the past 20 years, and the country's changing lifestyle-related behaviors are in part to blame for the rising cancer burden. Despite having lower incidence rates than many Western nations, there have been notable improvements in recent years (Mathew et al. 2020; Chandraprasad et al. 2022).

The two most widely used clinical treatments for CRC are surgery and chemotherapy (Engblom et al. 2016). Though this treatment may occasionally be successful, a complete recovery cannot be guaranteed. Therefore, a better choice that can be used for many cancers safely and successfully needs to be devised (Kim 2015). Fatigue, GI toxicity, radiation therapy's carcinogenic effects, nausea, hair loss, loss of appetite, and vomiting are among the side effects of the traditional treatment technique (Qi et al. 2015; Ravichandra et al. 2018). Many people with CRC still have poor prognoses since the disease is typically only found in advanced stages. An improved prognosis for CRC may result from a deeper knowledge of the molecular mechanisms underlying CRC progression (Ogretmen 2017, 17; Thimmarayan et al. 2020, 14; Vaiyapuri et al. 2020a, b).

To combat the growth of drug resistance and side effects, new and effective medications are however urgently needed (Pulido-Moran et al. 2016). Since many natural substances are well tolerated by animals and patients and don't have hazardous side effects even at high dosages, they are the target as a drug for cancer treatment (Pandurangan et al. 2013; Mohebali et al. 2020). Also, natural products have apoptogenic properties and anticancer actions against CRC. The objective of the ongoing research is to demonstrate the anti-CRC properties of natural items based on their chemical makeup, such as alkaloids, terpenoids, polysaccharides, polyphenols, and unsaturated fatty acids (Gezici and Şekeroğlu 2019; Vaiyapuri et al. 2020a, b). The natural products have already been used as therapeutic drugs and proven to be efficient in treating human ailments, for example, as antidiabetic (Aziz et al. 2017), cardioprotective (Hemmati et al. 2018), antimicrobial (Pandurangan et al. 2016), and anti-inflammatory agents (Pandurangan and Esa 2014).

A key pungent constituent of red pepper and the active ingredient in chili peppers, capsaicin ($C_{18}H_{27}NO_3$, 8-methyl-N-vanillyl-6-nonenamide), is used widely as a food additive and an analgesic (Clark and Lee 2016). The chili pepper is a fruit that grows on herbaceous, *Solanaceae* family plants of the genus *Capsicum*, which are indigenous to the Americas (Bode and Dong 2011). Capsaicin is known for its antioxidant activity. Numerous in vitro models, including rat liver mitochondria, soybean phosphatidylcholine liposomal membrane, and human erythrocyte, have demonstrated the inhibitory impact of capsaicin on lipid peroxidation (Ngoc et al. 2008). By interacting with ferrous and ferric ions, capsaicin was discovered to be able to stop the redox cycling of iron in rat brain homogenate and to suppress the lipid peroxidation that ferrous causes (Dairam et al. 2008). The use of capsaicin and curcumin supplements alleviated the testicular and hepatic oxidative stress brought on by high-fat diet, according to in vivo studies employing obese rats (Tanrıkulu-Küçük et al. 2019). According to investigations on the capsaicin's mechanism, the C7-benzyl carbon served as the location of the radical reaction, and the compound both prevented the production of active oxygen species and scavenged 1,1′--diphenyl-2-picrylhydrazyl (DPPH) radicals (Kogure et al. 2022).

The topical gels and patches containing capsaicin, an effective anti-inflammatory drug, have long been used to alleviate pain (Lee et al. 2007). By suppressing the activity of the Cyclooxygenase (COX)-2 enzyme and the expression of the inducible nitric oxide synthase (iNOS), they discovered that capsaicin reduced the generation of prostaglandin E_2 (PGE_2). The results suggested the use of capsaicin for treating inflammatory diseases and prostaglandin overproduction since prostaglandins, COX-2, and iNOS were known to be important pro-inflammatory mediators in the development of inflammatory diseases (Moita et al. 2013). These reports showcase the anti-inflammatory effects of capsaicin.

Gastric, breast, lung, prostate, colorectal, and pancreatic cancers are only a few of the human cancers that capsaicin has shown to be highly anti-carcinogenic (Cao et al. 2015). Through enhancing the intracellular Ca^{2+} concentration and calpain activity, capsaicin and camptothecin have been shown to work together synergistically to promote cell death in small cell lung cancer (Friedman et al. 2017). The 20S proteasome's caspase-like activity and cyclin D1 expression levels were both observed to be reduced by capsaicin, demonstrating the antitumor effect of capsaicin on colorectal cancer. Capsaicin enhanced the activation of p38 MAP kinase and downregulated the activation of AKT in MCF- 7 cells indicating the anticancer effect of capsaicin (Vendrely et al. 2017).

Suppressing in vitro platelet aggregation and thromboxane production, capsaicin and dihydrocapsaicin showed a combined impact that decreased the possibility of cardiovascular disease (Almaghrabi et al. 2018). Apoptosis in cardiomyocytes was also suppressed, ROS levels were lowered, and Bcl-2 and Bax expression and activity were controlled. It has been hypothesized that the 14-3-3 pathway and the preservation of mitochondrial activities are implicated in how capsaicin exerts its cardioprotective effects (Liu et al. 2014). Ameliorating myocardial I/R damage in animal models, protecting rats from doxorubicin-induced cardiotoxicity, and shielding cardiometabolic organs from dysfunction are some other results that suggest the cardioprotective effect of capsaicin (Dow et al. 2014; Lu et al. 2020). Capsaicin exhibits anti-obesity property as it prevented adipogenesis, enhances lipid oxidation in adipocytes, and by modulating on ghrelin induces appetite suppression as well as inhibits weight gain by upregulating UCP2 and UCP3 (Karri et al. 2019).

A multitude of human disorders, including chronic pain, obesity, diabetes (Rosca et al. 2020), cardiovascular conditions (Dow et al. 2014), respiratory diseases, and cancer, have been identified as possible therapeutic targets for capsaicin (Merritt et al. 2022). According to reports, capsaicin exhibits antiproliferative properties against lung, prostate (Zhu et al. 2020), pancreatic, and breast cancers (Chen et al. 2021). Thus, the recent findings on capsaicin pharmacological activity shed light on the mechanism underlying the apoptotic effects of capsaicin suggesting that capsaicin may present interesting new options for improving CRC therapy efficiency along with reduced or no side effects. This chapter summarizes the recent findings on capsaicin pharmacological activity, which shed light on the mechanism underlying the apoptotic effects of capsaicin, suggesting that capsaicin may present interesting new options for improving CRC therapy.

21.2 Capsaicin as Anticancer Agent Against Colorectal Cancer (CRC)

21.2.1 Capsaicin Represses Transcriptional Activity of β-Catenin

Embryonic development, epithelial-mesenchymal transition (EMT) signaling, and cell fate are all significantly influenced by Wnt/beta-catenin signaling. Carcinogenesis, among other illnesses, is influenced by its malfunction (Brabletz et al. 2005). A recent genetic analysis of human colon and rectal cancers shows that at least 90% of patient tumors exhibit a disrupted Wnt signaling pathway, further emphasizing the crucial significance of Wnt signaling (Pate et al. 2014). It has been demonstrated that stimulation of the canonical Wnt/β-catenin pathway in colon tumors results in a partial reduction of oxidative metabolism in the TCA cycle and boosts the development of cells. MCT-1, a lactate transporter, and Phosphoinositide-dependent kinase (PDK)-1, a target of Wnt/beta-catenin, both become expressed excessively in tumor cells (Van de Wetering et al. 2022). The Wnt pathway also stimulates the transcription of the *c-Myc* gene, which is implicated in growth of cells. These results suggest a possible role for Wnt/beta-catenin signaling inhibitors as new CRC therapeutics (Niehrs and Acebron 2012).

Lee et al. (2012) reported that capsaicin reduced the growth of SW480, LoVo, and HCT-116 CRC cells in a dose- and time-dependent manner. The result of the luciferase reporter assay indicated that capsaicin treatment at doses 50 and 100 μM after 48 h reduced the β-catenin/TCF-dependent transcriptional activity. It was also observed that capsaicin reduced the accumulation of β-catenin in the nucleus and cytosol. Capsaicin treatment suppressed cyclin D1 transcription. The capsaicin treatment also reduced the TCF-4 expression and also effectively inhibited the binding of β-catenin to TCF4. The capsaicin induced reduction in the intracellular β-catenin levels and transcripts from CTNNB1 (from a *β-catenin* gene). Capsaicin also induced proteasomal-dependent degradation, thereby resulting in degradation of *β-catenin*. Through effective inhibition of the β-catenin pathway, capsaicin can act as a potent anti-CRC agent.

21.2.2 Capsaicin Reduces Genotoxicity

For inducing colorectal cancer, in rodents prominently two chemical agents as azoxymethane and 1, 2-dimethylhydrazine (DMH) were used by the researchers (Shakib Khoob et al. 2022; Ashokkumar and Sudhandiran, 2011). Caetano et al. (2018) reported that DMH-stimulated genotoxicity in leukocytes and fecal water genotoxicity in CaCo-2 cells were suppressed by capsaicin at the concentration of 50 mg/kg. Capsaicin also increased the Ki-67 proliferation index and induced caspase 3 activity. Capsaicin also enhanced the expression of genes including *Casp4*, *Sp1*, *Aifm1*, and *Dffb* in colon mucosa that is involved in apoptosis and tissue development. It also reduced the aberrant crypt foci (ACF) number and multiplicity of CRC in a dose-dependent manner. Capsaicin increased the expression of *Foxa1* and *Cdh1* genes, which is important for cell differentiation. Thus, these results suggested that capsaicin can effectively reduce cell proliferation and act as a protective agent against DMH-induced CRC.

21.2.3 Synergistic Anticancer Activity of Capsaicin and 3, 3′-Diindolylmethane

NF-κB is a transcription factor and considered as a master regulator of key signaling pathways. Many reports have stated that NF-κB activation has helped in the progression of CRC (Patel et al. 2018; Peng et al. 2020). Clark et al. (2015) reported that capsaicin along with 3,3′ diindolylmethane (DIM) isolated from cruciferous vegetables exhibited antiproliferative effect and induced apoptosis in HCT116, SW480, LoVo, Caco-2, and HT-29 CRC cells. It also synergistically increased the transcriptional activity of NF-κB and p53. They also enhanced the level of *p21* and *Bak* mRNA. They also enhanced the expression level of pro-apoptotic proteins including Bak, BID, Bax, Puma, and Fas-L and downregulated anti-apoptotic proteins. These results suggested that capsaicin along with DIM can act as a potent anti-colon cancer agent by inducing apoptosis and suppressing cancer cell growth.

21.2.4 Capsaicin Induces Apoptosis by Generating Reactive Oxygen Species (ROS)

Apoptosis is a cell death mechanism, and natural products have the tendency to induce apoptosis by inducing the hallmarks of apoptosis (Sriram et al. 2008; Kalsoom et al. 2022; Rahman et al. 2022). Yang et al. (2009) reported that capsaicin exhibited its anticancer effect by decreasing the viability of Colo320DM and LoVo CRC cells in a dose-dependent manner. It induced cell shrinkage and chromatin condensation in a cancer cell. It also effectively stimulated DNA fragmentation, morphological changes, and phosphatidylserine translocation, which results in apoptotic cell death. Capsaicin treatment increased reactive oxygen species (ROS) production and decreased mitochondrial transmembrane potential in both cancer cells in a dose-dependent manner. The study also suggests that capsaicin-induced apoptosis is highly correlated to the activation of caspase 3 in a dose- and time-dependent manner. Thus, the results suggest that capsaicin effectively inhibits colon cancer cell growth by mediating caspase activity and ROS production (Yang et al. 2009).

21.2.5 Capsaicin Modulates PI3K/Akt Signaling

A group of transcription factors called signal transducers and activators of transcription (STAT) are triggered through a number of cytokines and growth factors. The production of homo- or heterodimers and transportation to the nucleus are caused by the receptor's ligand-dependent tyrosine phosphorylation of STATs (Li et al. 2022). Leading to several downstream targets in the nucleus will be activated, which will advance the cell cycle, inhibit apoptosis, and encourage tumorigenesis. Oncogenic transformation of STAT3 occurs both in vivo and in vitro when it is constitutively activated (Zhang et al. 2020). It was observed that the inhibition of cell growth and induces cell cycle arrest in breast cancer cell by downregulating STAT3 pathway. Numerous studies state that CRC patients who have STAT3 activation on a constant basis develop tumors and proliferate their cells to high levels (Xue et al. 2019).

With poor prognosis and survival, PI3K/Akt/mTOR pathway increase is shown in a variety of malignancies, including colon, ovarian, pancreatic, glioma, melanoma, lymphatic, and breast cancers. mTOR is one of Akt's many downstream targets (Zhang et al. 2018). The translation of proteins, growth, metabolism, and angiogenesis are all regulated by the mammalian target of rapamycin (mTOR). Immunity to both standard treatments and targeted treatments is induced by the PI3K/Akt/mTOR pathway. Increased cell growth and the ability to resist apoptosis are the outcomes of uncontrolled Akt phosphorylation, which is seen in human colorectal cancer. As a result, blocking the PI3K/Akt/mTOR pathway may reduce the growth of cancer cells, trigger apoptosis, and stop the formation of new blood vessels (Padmavathi et al. 2021).

Yang et al. (2013) reported that capsaicin inhibited the growth of SW480, CT-26, and HCT116CRC cells in a concentration-dependent manner and exhibited an antiproliferative effect. The result of the transwell assay indicated that capsaicin at low concentrations, i.e., 100 μM, increased cancer cell motility after 48 h of treatment, while capsaicin inhibited the growth of cancer cells at 800 μM after 48 h of treatment. It was also found that capsaicin at lower concentrations induced epithelial-mesenchymal transition (EMT) in SW480 cells. Similarly, low levels of capsaicin treatment activated matrix metalloproteinase (MMP)-2 and MMP-9, activated Akt/mTOR and STAT-3 pathways in SW480 cells, and also enhanced ROS production. These results can be utilized to better understand capsaicin-mediated CRC treatment.

21.2.6 Capsaicin Induces Cell Cycle Arrest in CRC Cells

Using the G0/G1, S, and G2/M phases of the cell cycle, which also includes DNA checkpoints to ensure the accuracy of DNA replication, cells reproduce. Cellular responses to DNA damage are made possible by the checkpoint and repair pathways. The chance of developing cancer can rise with any change to these pathways (Kastan and Bartek 2004). Cyclins, cyclin-dependent kinases (CDKs), and CDK inhibitors are crucial components of the cell cycle mechanism. The CDKs provide the cells a push to transition from one stage to another soon after they are active. In the majority of cancer types, CDKs and cyclins are significantly active. Therefore, cell cycle arrest happens when CDK activity is suppressed to control the cell proliferation (Chen et al. 2012). Thus targeting cell cycle is developing strategy to prevent cancer occurrence.

Lu et al. (2010) reported that capsaicin suppressed the Colo 205 colon cancer cells' viability up to 97% at the concentration of 500 μM. It was also found that capsaicin-treated cells after 24 h of incubation induced 60% apoptosis through the cell cycle arrest G0/G1 phase. Capsaicin also increased the ROS and Ca^{2+} levels while suppressing the levels of mitochondrial membrane potential in colon cancer cells. Caspase activity assay and flow cytometric analysis results indicated that capsaicin stimulated apoptosis by regulating the activation of caspases 3, 8, and 9 in Colo 205 cells. At a concentration of 150 μM, increased gene expression levels of mRNA of *caspases 3*, *8*, and *9* and *apoptosis-inducing factor (AIF)* were observed in capsaicin-treated cells. Levels of Fas and cytochrome c were found to be increased in capsaicin-treated cells. It also increased the level of pro-apoptotic proteins such as Bcl-2 while suppressing the level of anti-apoptotic proteins including Bax. It also effectively exhibited an antiproliferative effect in mice bearing Colo 205 tumor xenografts by suppressing the growth of cancer cells. Thus the in vivo and in vitro studies suggest that capsaicin can be an effective alternative and protective method to treat CRC.

21.2.7 Involvement of AMPK Signaling Cascade in Capsaicin-Induced Apoptosis

The serine/threonine kinase family includes AMP-activated protein kinase (AMPK). The AMP/ATP ratio that is responsible for cellular homeostasis is monitored by AMPK pathway. The AMPK pathway is activated as a result of numerous metabolic stressors, such as a lack of oxygen, physical activity, and famine (Hwang et al. 2007). In contrast to enhancing ATP producing catabolic pathways like lipid oxidation and metabolism of glucose, AMPK phosphorylation suppresses ATP-consuming activities like the production of proteins, lipids, and cholesterol as well (Kan et al. 2016). By activating signaling pathways like COX-2, Akt, and mTOR that lead in the increased levels of p53 and p21 proteins, stimulation of caspases, and suppression of compounds involved in the development and invasion of tumor cells, the stimulation of AMPK controls cellular death in a variety of tumor cell varieties (Lee et al. 2009).

Kim et al. (2007) reported that the results of the MTT assay indicated that capsaicin exhibited an antiproliferative effect after 24 h of treatment in HT-29 cells. Capsaicin also induced chromatin condensation and fragmentation of cancer cells. It also increased the expressions of phosphorylated a AMP-activated protein kinase (AMPK) and phosphorylated acetyl-CoA carboxylase (ACC) indicating that capsaicin-induced apoptosis through the regulation of AMPK pathway. The activation of AMPK due to capsaicin treatment enhanced the expression of the inactive form of ACC. Thus, the study on capsaicin-induced AMPK activation can be used as an effective treatment against RC progression.

21.2.8 NSAID-Activated Gene 1 as a Molecular Target for Capsaicin

Lee et al. (2010) reported that capsaicin at a concentration of 50 and 100 μM induced changes in colon cancer cell morphology and reduced the cell number, thereby suppressing cell growth after 24 h of treatment. Capsaicin induced apoptosis in a dose-dependent manner after 24 h of treatment. Capsaicin treatment also increased the expression level of nonsteroidal anti-inflammatory drug-activated gene-1 (NAG-1) at a concentration of 50 μM after 24 and 48 h of treatment. Capsaicin-stimulated NAG-1 promoter activity, further increased expression of CCAAT/enhancer binding protein b (C/EBPb). An increase of phosphorylated serine/threonine residues was seen after capsaicin treatment and immunoprecipitation research revealed that capsaicin improved C/EBPb's affinity for GSK3β and activating transcription factor 3 (ATF3). The suppression of the GSK-3β and protein kinase C pathways reduces the phosphorylation and interaction of C/EBPb with GSK-3β and ATF3. According to these results, ATF3 and C/EBPb may work together in human colorectal cancer cells to mediate NAG-1 production and apoptosis caused by capsaicin.

21.2.9 Capsaicin Modulates the Immune Cross Talk

Bessler and Djaldetti (2017) reported that about 200 μM concentration of capsaicin treatment for 24 h to HT-29 and RKO CRC cells suppressed the cell proliferation by 69% and 92%, respectively. Capsaicin treatment suppressed the production of tumor necrosis factor (TNF)-α, at the concentrations of 200 mM by 68% and 88% in HT-29 and RKO cells, respectively. Capsaicin suppressed the expression level of interleukin (IL)-1β, interferon (IFN)-γ, IL-1ra, and IL-10 while enhancing the expression of IL-6 in a concentration-dependent manner. These findings suggest the presence of a further immunomodulatory mechanism through which capsaicin may inhibit the growth of colon tumors.

21.2.10 Capsaicin and Resveratrol Induce Apoptosis in CRC

Inducible nitric oxide synthase (iNOS) is a key regulator of cancer development and progression in many types of cancer (Gochman et al. 2012). Kim et al. (2009) reported that capsaicin reduced the viability of CRC cells after 24 h of treatment in a dose- and time-dependent manner. It was also observed that capsaicin along with resveratrol exhibited an antiproliferative effect more efficiently than as a single component. The results of Western blot analysis indicated that capsaicin and resveratrol elevated the level of iNOS and eNOS proteins. It was observed NO generation increased 10- to 40-fold in colon cancer cells when treated with capsaicin and the level of NO increased further when treated with both capsaicin and resveratrol together. They also synergistically induced apoptosis via increased NO production. Capsaicin increased NO• generation after treatment uniformly boosted p53 and Bax expression through Mdm2 downregulation in p53-WT cells. Regardless of the p53 status, the cells recorded a decrease in the anti-apoptotic Bcl-2 protein level, cytochrome c loss from the mitochondria, and activation of caspases 9 and 3 in tandem. It was noticed that the activation of DR4, Fas(CD95), and caspase 8 simultaneously, resulted to activate both the mitochondrial and death receptor pathways, that cooperate to induce apoptosis. Based on p53 status, these findings shed light on the mechanism underlying the apoptotic effects of capsaicin and resveratrol and suggest that altering NO• may present interesting new options for improving colon cancer therapy efficiency.

21.3 Conclusion

Colorectal cancer is the third most common and leading cause of death in the Western population. In the Asian region, there is a steady increase in the incidence. Natural products are always a better candidate for treating cancer by modulating the key signaling and less/no toxicity. Capsaicin has a variety of pharmacological benefits. It has been proven that lower concentrations in CRC cells can induce apoptosis via inducing ROS. It modulated many key signaling molecules such as β-catenin, MMP-9 and 2, AIF, and other apoptotic mediators. In a nutshell, capsaicin is an effective drug at the preclinical level and it can be transformed in a clinical setting.

References

Almaghrabi S, Adams M, Geraghty D, Ahuja K (2018) Synergistic inhibitory effect of capsaicin and dihydrocapsaicin on in-vitro platelet aggregation and thromboxane formation. Blood Coagul Fibrinolysis 29(4):351–355

Ashokkumar P, Sudhandiran G (2011) Luteolin inhibits cell proliferation during azoxymethane-induced experimental colon carcinogenesis via Wnt/β-catenin pathway. Investig New Drugs 29(2):273–284

Aziz MSA, Giribabu N, Rao PV, Salleh N (2017) Pancreatoprotective effects of *Geniotrigona thoracica* stingless bee honey in streptozotocin-nicotinamide-induced male diabetic rats. Biomed Pharmacother 89:135–145

Bessler H, Djaldetti M (2017) Capsaicin modulates the immune cross talk between human mononuclears and cells from two colon carcinoma lines. Nutr Cancer 69(1):14–20

Bode AM, Dong Z (2011) The two faces of capsaicin. Cancer Res 71(8):2809–2814

Brabletz T, Hlubek F, Spaderna S, Schmalhofer O, Hiendlmeyer E, Jung A, Kirchner T (2005) Invasion and metastasis in colorectal cancer: epithelial-mesenchymal transition, mesenchymal-epithelial transition, stem cells and β-catenin. Cells Tissues Organs 179(1–2):56–65

Caetano BFR, Tablas MB, Pereira NEF, de Moura NA, Carvalho RF, Rodrigues MAM, Barbisan LF (2018) Capsaicin reduces genotoxicity, colonic cell proliferation and preneoplastic lesions induced by 1,2-dimethylhydrazine in rats. Toxicol Appl Pharmacol 338:93–102

Cao S, Chen H, Xiang S, Hong J, Weng L, Zhu H, Liu Q (2015) Anti-cancer effects and mechanisms of capsaicin in chili peppers. Am J Plant Sci 6(19):3075

Chandraprasad MS, Dey A, Swamy MK (2022) Introduction to cancer and treatment approaches. In: Swamy MK, Pullaiah T, Chen Z-S (eds) Paclitaxel: sources, chemistry, anticancer actions, and current biotechnology. Academic Press, Cambridge, MA, pp 1–27

Chen D, Yang Z, Wang Y, Zhu G, Wang X (2012) Capsaicin induces cycle arrest by inhibiting cyclin-dependent-kinase in bladder carcinoma cells. Int J Urol 19(7):662–668

Chen M, Xiao C, Jiang W, Yang W, Qin Q, Tan Q, Lian B, Liang Z, Wei C (2021) Capsaicin inhibits proliferation and induces apoptosis in breast cancer by down-regulating FBI-1-mediated NF-κB pathway. Drug Des Devel Ther 15:125–140

Clark R, Lee SH (2016) Anticancer properties of capsaicin against human cancer. Anticancer Res 36(3):837–843

Clark R, Lee J, Lee SH (2015) Synergistic anticancer activity of capsaicin and 3,3′--diindolylmethane in human colorectal cancer. J Agric Food Chem 63(17):4297–4304

Dairam A, Fogel R, Daya S, Limson JL (2008) Antioxidant and iron-binding properties of curcumin, capsaicin, and S-allylcysteine reduce oxidative stress in rat brain homogenate. J Agri Food Chem 56(9):3350–3356

Dow J, Simkhovich BZ, Hale SL, Kay G, Kloner RA (2014) Capsaicin-induced cardioprotection. Is hypothermia or the salvage kinase pathway involved? Cardiovasc Drugs Ther 28(4):295–301

Engblom C, Pfirschke C, Pittet MJ (2016) The role of myeloid cells in cancer therapies. Nat Rev Cancer 16(7):447–462

Ferlay J, Soerjomataram I, Dikshit R, Eser S, Mathers C, Rebelo M, Parkin DM, Forman D, Bray F (2015) Cancer incidence and mortality worldwide: sources, methods and major patterns in GLOBOCAN 2012. Int J Cancer 136(5):E359–E386

Friedman JR, Perry HE, Brown KC, Gao Y, Lin J, Stevenson CD, Hurley JD, Nolan NA, Akers AT, Chen YC, Denning KL (2017) Capsaicin synergizes with camptothecin to induce increased apoptosis in human small cell lung cancers via the calpain pathway. Biochem Pharmacol 129:54–66

Gezici S, Şekeroğlu N (2019) Current perspectives in the application of medicinal plants against cancer: novel therapeutic agents. Anti Cancer Agents Med Chem 19(1):101–111

Gingras D, Béliveau R (2011) Colorectal cancer prevention through dietary and lifestyle modifications. Cancer Microenviron 4(2):133–139

Gochman E, Mahajna J, Shenzer P, Dahan A, Blatt A, Elyakim R, Reznick AZ (2012) The expression of iNOS and nitrotyrosine in colitis and colon cancer in humans. Acta Histochem 114(8):827–835

Hemmati AA, Olapour S, Varzi HN, Khodayar MJ, Dianat M, Mohammadian B, Yaghooti H (2018) Ellagic acid protects against arsenic trioxide-induced cardiotoxicity in rat. Hum Exp Toxicol 37(4):412–419

Hwang JT, Kwak DW, Lin SK, Kim HM, Kim YM, Park OJ (2007) Resveratrol induces apoptosis in chemoresistant cancer cells via modulation of AMPK signaling pathway. Ann N Y Acad Sci 1095(1):441–448

Kalsoom A, Altaf A, Ashraf M, Ali MM, Aftab S, Sattar H, Sajjad M, Aqib AI, Maqbool T (2022) *In vitro* evaluation of cytotoxic potential of Caladium lindenii extracts on human hepatocarcinoma HepG2 and normal HEK293T cell lines. Biomed Res Int 2022:1279961

Kamangar F, Dores GM, Anderson WF (2006) Patterns of cancer incidence, mortality, and prevalence across five continents: defining priorities to reduce cancer disparities in different geographic regions of the world. J Clin Oncol 24(14):2137–2150

Kan JY, Yen MC, Wang JY, Wu DC, Chiu YJ, Ho YW, Kuo PL (2016) Nesfatin-1/Nucleobindin-2 enhances cell migration, invasion, and epithelial-mesenchymal transition via LKB1/AMPK/TORC1/ZEB1 pathways in colon cancer. Oncotarget 7(21):31336

Karri S, Sharma S, Hatware K, Patil K (2019) Natural anti-obesity agents and their therapeutic role in management of obesity: a future trend perspective. Biomed Pharmacother 110:224–238

Kastan MB, Bartek J (2004) Cell-cycle checkpoints and cancer. Nature 432(7015):316–323

Kim JH (2015) Chemotherapy for colorectal cancer in the elderly. World J Gastroenterol 21(17):5158–5166

Kim YM, Hwang JT, Kwak DW, Lee YK, Park OJ (2007) Involvement of AMPK signaling cascade in capsaicin-induced apoptosis of HT-29 colon cancer cells. Ann N Y Acad Sci 1095:496–503

Kim MY, Trudel LJ, Wogan GN (2009) Apoptosis induced by capsaicin and resveratrol in colon carcinoma cells requires nitric oxide production and caspase activation. Anticancer Res 29(10):3733–3740

Kogure K, Goto S, Nishimura M, Yasumoto M, Abe K, Ohiwa C, Sassa H, Kusumi T, Terada H (2022) Mechanism of potent antiperoxidative effect of capsaicin. Biochim Biophys Acta 1573(1):84–92

Lee IO, Lee KH, Pyo JH, Kim JH, Choi YJ, Lee YC (2007) Anti-inflammatory effect of capsaicin in *Helicobacter pylori*-infected gastric epithelial cells. Helicobacter 12(5):510–517

Lee YK, Park SY, Kim YM, Park OJ (2009) Regulatory effect of the AMPK–COX-2 signaling pathway in curcumin-induced apoptosis in HT-29 colon cancer cells. Ann N Y Acad Sci 1171(1):489–494

Lee SH, Krisanapun C, Baek SJ (2010) NSAID-activated gene-1 as a molecular target for capsaicin-induced apoptosis through a novel molecular mechanism involving GSK3beta, C/EBPbeta and ATF3. Carcinogenesis 31(4):719–728

Lee SH, Richardson RL, Dashwood RH, Baek SJ (2012) Capsaicin represses transcriptional activity of β-catenin in human colorectal cancer cells. J Nutr Biochem 23(6):646–655

Li JK, Sun HT, Jiang XL, Chen YF, Zhang Z, Wang Y, Chen WQ, Zhang Z, Sze SC, Zhu PL, Yung KK (2022) Polyphyllin II induces protective autophagy and apoptosis via inhibiting PI3K/AKT/mTOR and STAT3 signaling in colorectal cancer cells. Int J Mol Sci 23(19):11890

Liu D, Yi B, Liao Z, Tang L, Yin D, Zeng S, Yao J, He M (2014) 14-3-3γ protein attenuates lipopolysaccharide-induced cardiomyocytes injury through the Bcl-2 family/mitochondria pathway. Int Immunopharmacol 21(2):509–515

Lu HF, Chen YL, Yang JS, Yang YY, Liu JY, Hsu SC, Lai KC, Chung JG (2010) Antitumor activity of capsaicin on human colon cancer cells *in vitro* and colo 205 tumor xenografts *in vivo*. J Agric Food Chem 58(24):12999–13005

Lu M, Chen C, Lan Y, Xiao J, Li R, Huang J, Huang Q, Cao Y, Ho CT (2020) Capsaicin—the major bioactive ingredient of chili peppers: bio-efficacy and delivery systems. Food Funct 11(4):2848–2860

Mathew A, Baby B, Wang K, Sirohi B, Lei F, Chen Q, Huang B (2020) Colorectal cancer incidence in younger adults in India. Gut 69(10):1899–1900

Merritt JC, Richbart SD, Moles EG, Cox AJ, Brown KC, Miles SL, Finch PT, Hess JA, Tirona MT, Valentovic MA, Dasgupta P (2022) Anti-cancer activity of sustained release capsaicin formulations. Pharmacol Ther 238:108177

Mohebali N, Pandurangan AK, Mustafa MR, Anandasadagopan SK, Alagumuthu T (2020) Vernodalin induces apoptosis through the activation of ROS/JNK pathway in human colon cancer cells. J Biochem Mol Toxicol 34(12):e22577

Moita E, Gil-Izquierdo A, Sousa C, Ferreres F, Silva LR, Valentao P, Dominguez-Perles R, Baenas N, Andrade PB (2013) Integrated analysis of COX-2 and iNOS derived inflammatory mediators in LPS-stimulated RAW macrophages pre-exposed to Echium plantagineum L. bee pollen extract. PLoS One 8(3):e59131

Ngoc TM, Hung TM, Thuong PT, Na M, Kim H, Do TH, Min BS, Minh PT, Bae K (2008) Inhibition of human low density lipoprotein and high density lipoprotein oxidation by oligostilbenes from rhubarb. Biol Pharm Bull 31(9):1809–1812

Niehrs C, Acebron SP (2012) Mitotic and mitogenic Wnt signalling. EMBO J 31(12):2705–2713

Ogretmen B (2017) Sphingolipid metabolism in cancer signalling and therapy. Nat Rev Cancer 18: 33–50

Padmavathi G, Monisha J, Bordoloi D, Banik K, Roy NK, Girisa S, Singh AK, Longkumer I, Baruah MN, Kunnumakkara AB (2021) Tumor necrosis factor-α induced protein 8 (TNFAIP8/TIPE) family is differentially expressed in oral cancer and regulates tumorigenesis through Akt/mTOR/STAT3 signaling cascade. Life Sci 287:120118

Pandurangan AK, Esa NM (2014) Luteolin, a bioflavonoid inhibits colorectal cancer through modulation of multiple signaling pathways: a review. Asian Pac J Cancer Prev 15(14): 5501–5508

Pandurangan AK, Dharmalingam P, Anandasadagopan SK, Ganapasam S (2013) Inhibitory effect of Luteolin on the status of membrane bound ATPases against Azoxymethane-induced colorectal cancer. J Chem Pharm Res 5(6):123–127

Pandurangan AK, Mohebali N, Hasanpourghadi M, Looi CY, Mustafa MR, Mohd Esa N (2016) Boldine suppresses dextran sulfate sodium-induced mouse experimental colitis: NF-κB and IL-6/STAT3 as potential targets. Biofactors 42(3):247–258

Pate KT, Stringari C, Sprowl-Tanio S, Wang K, TeSlaa T, Hoverter NP, McQuade MM, Garner C, Digman MA, Teitell MA, Edwards RA (2014) Wnt signaling directs a metabolic program of glycolysis and angiogenesis in colon cancer. EMBO J 33(13):1454–1473

Patel M, Horgan PG, McMillan DC, Edwards J (2018) NF-κB pathways in the development and progression of colorectal cancer. Transl Res 197:43–56

Peng C, Ouyang Y, Lu N, Li N (2020) The NF-κB signaling pathway, the microbiota, and gastrointestinal tumorigenesis: recent advances. Front Immunol 11:1387

Pulido-Moran M, Moreno-Fernandez J, Ramirez-Tortosa C, Ramirez-Tortosa M (2016) Curcumin and health. Molecules 21(3):264

Qi F, Zhao L, Zhou A, Zhang B, Li A, Wang Z, Han J (2015) The advantages of using traditional Chinese medicine as an adjunctive therapy in the whole course of cancer treatment instead of only terminal stage of cancer. Biosci Trends 9(1):16–34

Rahman MM, Sarker MT, Alam Tumpa MA, Yamin M, Islam T, Park MN, Islam MR, Rauf A, Sharma R, Cavalu S, Kim B (2022) Exploring the recent trends in perturbing the cellular signaling pathways in cancer by natural products. Front Pharmacol 13:950109

Ravichandra VD, Ramesh C, Swamy MK, Purushotham B, Rudramurthy GR (2018) Anticancer plants: chemistry, pharmacology, and potential applications. In: Akhtar M, Swamy M (eds) Anticancer plants: properties and application, vol 1. Springer, Singapore, pp 485–515

Rosca AE, Iesanu MI, Zahiu CDM, Voiculescu SE, Paslaru AC, Zagrean AM (2020) Capsaicin and gut microbiota in health and disease. Molecules 25(23):5681

Shakib Khoob M, Hosseini SM, Kazemi S (2022) *In vitro* and *in vivo* antioxidant and anticancer potentials of royal jelly for dimethylhydrazine-induced colorectal cancer in wistar rats. Oxid Med Cell Longev 2022:9506026

Sriram N, Kalayarasan S, Ashokkumar P, Sureshkumar A, Sudhandiran G (2008) Diallyl sulfide induces apoptosis in Colo 320 DM human colon cancer cells: involvement of caspase-3, NF-kappaB, and ERK-2. Mol Cell Biochem 311(1–2):157–165

Tanrıkulu-Küçük S, Başaran-Küçükgergin C, Seyithanoğlu M, Doğru-Abbasoğlu S, Koçak H, Beyhan-Özdaş Ş, Öner-İyidoğan Y (2019) Effect of dietary curcumin and capsaicin on testicular and hepatic oxidant–antioxidant status in rats fed a high-fat diet. Appl Physiol Nutr Metab 44(7):774–782

Thimmarayan S, Manickam M, Mekapogu M, Vasamsetti BM, Swamy MK, Natesan K, Vaiyapuri M (2020) Plant flavonoids against colorectal cancer and mechanisms of action. In: Swamy MK (ed) Plant-derived bioactives: chemistry and mode of action. Springer Nature, Singapore, pp 21–34

Torre LA, Bray F, Siegel RL, Ferlay J, Lortet-Tieulent J, Jemal A (2015) Global cancer statistics, 2012. CA Cancer J Clin 65(2):87–108

Vaiyapuri M, Thimmarayan S, Dhupal M, Rallabandi HR, Mekapogu M, Vasamsetti BM, Swamy MK, Natesan K (2020a) Pelargonidin, a dietary anthocyanidin in the prevention of colorectal cancer and its chemoprotective mechanisms. In: Swamy MK (ed) Plant-derived bioactives: chemistry and mode of action. Springer Nature, Singapore, pp 19–35

Vaiyapuri M, Natesan K, Vasamsetti BMK, Mekapogu M, Swamy MK, Thangaraj K (2020b) Orientin: a C-glycosyl flavonoid that mitigates colorectal cancer. In: Swamy MK (ed) Plant-derived bioactives: chemistry and mode of action. Springer Nature, Singapore, pp 1–19

Van de Wetering M, Sancho E, Verweij C, De Lau W, Oving I, Hurlstone A, Van Der Horn K, Batlle E, Coudreuse D, Haramis AP, Tjon-Pon-Fong M (2022) The β-catenin/TCF-4 complex imposes a crypt progenitor phenotype on colorectal cancer cells. Cell 111(2):241–250

Vendrely V, Peuchant E, Buscail E, Moranvillier I, Rousseau B, Bedel A, Brillac A, de Verneuil H, Moreau-Gaudry F, Dabernat S (2017) Resveratrol and capsaicin used together as food complements reduce tumor growth and rescue full efficiency of low dose gemcitabine in a pancreatic cancer model. Cancer Lett 390:91–102

Xue J, Ge X, Zhao W, Xue L, Dai C, Lin F, Peng W (2019) PIPKIγ regulates CCL2 expression in colorectal cancer by activating AKT-STAT3 signaling. J Immunol Res 2019:3690561

Yang KM, Pyo JO, Kim GY, Yu R, Han IS, Ju SA, Kim WH, Kim BS (2009) Capsaicin induces apoptosis by generating reactive oxygen species and disrupting mitochondrial transmembrane potential in human colon cancer cell lines. Cell Mol Biol Lett 14(3):497–510

Yang J, Li TZ, Xu GH, Luo BB, Chen YX, Zhang T (2013) Low-concentration capsaicin promotes colorectal cancer metastasis by triggering ROS production and modulating Akt/mTOR and STAT-3 pathways. Neoplasma 60(4):364–372

Zhang L, Chen C, Duanmu J, Wu Y, Tao J, Yang A, Yin X, Xiong B, Gu J, Li C, Liu Z (2018) Cryptotanshinone inhibits the growth and invasion of colon cancer by suppressing inflammation and tumor angiogenesis through modulating MMP/TIMP system, PI3K/Akt/mTOR signaling and HIF-1α nuclear translocation. Int Immunopharmacol 65:429–437

Zhang ZH, Li MY, Wang Z, Zuo HX, Wang JY, Xing Y, Jin C, Xu G, Piao L, Piao H, Ma J (2020) Convallatoxin promotes apoptosis and inhibits proliferation and angiogenesis through crosstalk between JAK2/STAT3 (T705) and mTOR/STAT3 (S727) signaling pathways in colorectal cancer. Phytomedicine 68:153172

Zhu M, Yu X, Zheng Z, Huang J, Yang X, Shi H (2020) Capsaicin suppressed activity of prostate cancer stem cells by inhibition of Wnt/β-catenin pathway. Phytother Res 34(4):817–824

Elucidating the Molecular Mechanisms Involved in Regulating Alzheimer's Disease by Capsaicin

22

Akila Muthuramalingam, Mohd Kashif, Poornima D Vijendra, Mohammed Waseem, and Ashok Kumar Pandurangan

Abstract

More than 40 million people worldwide are presently afflicted by Alzheimer's disease (AD), a neurodegenerative condition that is expected to spread rapidly over the ensuing decades. Since amyloid was isolated and identified from post-mortem brain analyses of AD patients, the field of AD study has grown exponentially over the past few decades. Because there is no cure, researching the pathophysiology of the disease and testing new drugs is essential. There have been several unsuccessful efforts to treat AD using an anti-amyloid strategy. As demonstrated in both autopsy and imaging investigations, the "amyloid cascade hypothesis" does not adequately explain neuronal damage in AD. It concentrates on the disease's potential mechanisms as well as how systemic and local factors influence disease development. Capsaicinoids are phytochemicals, and capsaicin is primarily responsible for the spiciness of chili peppers. It is biosynthesized through two pathways involved in the metabolism of phenylpropanoids and fatty acids. Capsaicin levels in plants are primarily influenced by genetic, environmental, and plant management factors. However, that composition can be improved using triggers. Capsaicin is used as a food additive and in pharmaceutical applications. In addition, capsaicin has been shown to prevent a variety of

A. Muthuramalingam · M. Kashif · A. K. Pandurangan (✉)
School of Life Sciences, B.S. Abdur Rahman Crescent Institute of Science and Technology, Chennai, Tamil Nadu, India

P. D. Vijendra
Department of Food Technology and Biochemistry, Davangere University, Davangere, Karnataka, India

M. Waseem
School of Pharmacy, University of Maryland Eastern Shore, Princess Anne, MD, USA

M. K. Swamy, A. Kumar (eds.), *Capsaicinoids*,
https://doi.org/10.1007/978-981-99-7779-6_22

cancers. This chapter highlights the role of capsaicin against AD and the molecular mechanisms involved.

Keywords

Alzheimer's disease · Capsaicin · Reactive oxidative species · Transient receptor potential vanilloid type 1

22.1 Introduction

Neurodegenerative disorders are caused majorly due to autophagy dysfunction that complicates the pathogenesis of several diseases like Alzheimer's disease (AD) (Metaxakis et al. 2018; Adil et al. 2020). Autophagy is a natural condition that is related to aging that removes the dead cells and replaces them with new cells. This process is focused on removing the toxic molecules by intracellular autophagosomes that are eventually taken to lysosomes and degraded (Menzies et al. 2015). Improper functioning of this may lead to the formation of aggregated toxic molecules and organelles in the autophagosomes, and these can be in the form of plaques like amyloid β(Aβ) and neurofibrillary tangles of hyperphosphorylated tau proteins (Tanzi 2012; Gao et al. 2018). Both are associated with neuronal loss and cognitive functions, eventually leading to severe pathogenesis of the disease (Kivipelto et al. 2018).

In the cytosol and extracellular region of brain cells, the accumulation of Aβ plaques was prominent, and that contributes to the pathogenesis of AD. It deteriorates the structure of the cell membrane and also disrupts permeability (Wang et al. 2004). One of the adverse effects includes improper functioning of the ion channels. Due to the impairment of the membrane, the Ca^{2+} influx increases, resulting in the formation of reactive oxidative species (ROS). Excessive Ca^{2+} in the mitochondria can increase the level of ROS and lead to apoptosis and triggers the caspase. Antioxidants that clear the ROS and ion channel blockers can be attenuated to prevent dysfunction and regulate normal metabolism (Holubiec et al. 2022).

Learning, memory, and motor skills depend on the central nervous system, which is safeguarded by immune cells known as microglia that stop the formation of aggregates in the early stages of neurodegenerative disease. These macrophages tend to engulf the aggregates like plaques to maintain homeostasis in the nervous system (Hool 2008). It was found that autophagy affects the functions of innate and adaptive immune responses, phagocytosis, and other antigen presentation process in the peripheral nervous system. Microglia cells were observed around the plaques both in AD patients and in AD mice models. These cells contribute to destroying the plaques while there are chances of them getting reactive and helping in AD pathogenesis. Although the role of these cells in the pathogenesis is confirmed, the mechanism is still unknown. These microglia cells are the major pathological feature in AD studies.

Normally the activation of the peripheral immune cells activates the oxidative phosphorylation process followed by the glycolysis for the energy-intensive process like phagocytosis and migration, but in the case of AD, Aβ intimates the process to continue from cellular metabolism to aerobic glycolysis (Hanisch and Kettenmann 2007; Baik et al. 2019). Since it is a known fact that microglial dysfunction causes defective metabolic processes during AD, researchers assume that this may be one of the effector functions of these cells.

22.2 Epidemiology of AD

Almost many cases of AD are prescribed to consult a psychiatrist in the beginning since it requires treatment for the behavioral and psychological problems commonly known as dementia, with an estimated 24.3 million cases worldwide (Sharma et al. 2021). Nearly all psychiatrists around the world would have faced an AD patient with giving a beginning referral regarding the status of the disease progression. In South Asia, the prevalence of dementia was 1.9% in 2005, and it is predicted that it will increase to 3.6 million by 2020 and 7.5 million by 2040. According to several epidemiological studies, dementia prevalence in India ranges from 2 per 1000 to 35 per 1000 people (Choudhary et al. 2021). Australia and New Zealand account for almost 50–70% of AD cases (Eratne et al. 2018). Nearly 30% of the younger people living in Australia suffer from dementia. The onset of the people with the disease is prevalent in people in the age category of 65 years and below. According to studies, dementia is Australia's second leading cause of mortality, and among persons in their 60s, the death rate doubles every year by 10%. Due to the increase in population and aging, the rate of incidence might increase in 2050 largely (Nichols et al. 2022). Moreover, there is no complete cure for the disease with severe symptoms and there should be efforts taken to emphasize reducing the symptoms and pain.

22.3 Markers in AD

Since this disease is without any potent drug or treatment, researchers have thrown light on possible markers that can be studied to find a treatment. This disease has hallmark histopathology history like plaque formation and tau protein aggregation, which are imparting a dysfunction in the synaptic pathway and other cognitive traits.

Figure 22.1 represents the amyloidogenic and non-amyloidogenic pathways that were involved in the processing of Aβ. In the amyloidogenic pathway, numerous enzymes, including α-secretase and β-secretase, participate in the proteolytic breakdown of an amyloid precursor protein (APP). APP is a type 1 transmembrane protein that is essential for the proteolytic processing of the Aβ protein. These APPs are shredded by the β-secretase and form sAPP-β and the c-terminal tethered fragment β-CTF. The c-terminal is further cleaved by the γ-secretase, a heterotetrameric protein for the Aβ peptide. The Aβ peptide can also be processed by α-secretase that prevents the release of the toxic Aβ peptides. Neprilysin (NEP) and insulin-

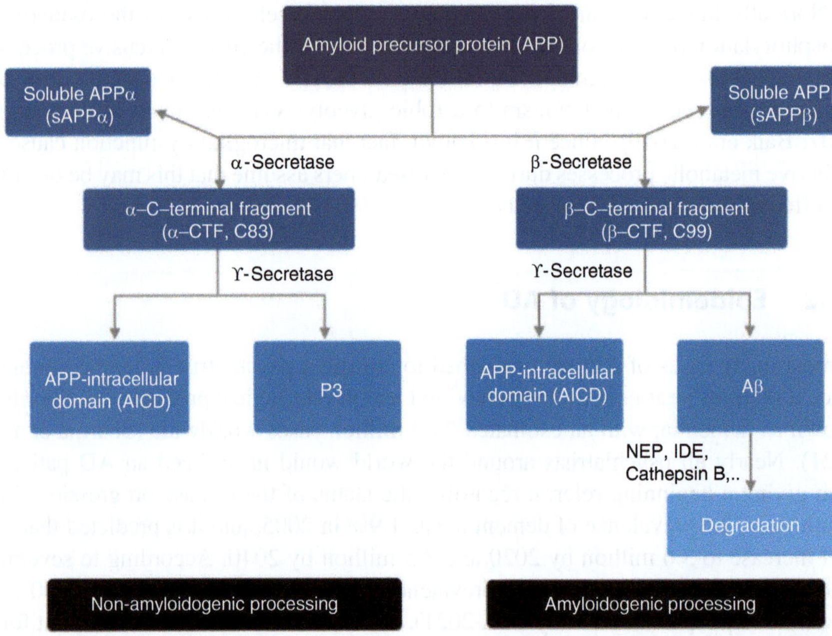

Fig. 22.1 Amyloid precursor protein (APP) processing by amyloidogenic and non-amyloidogenic pathway

degrading enzyme (IDE) are some of the major reasons for the total A β levels in the body along with the anabolism pathway.

A remethylation mechanism transforms the amino acid homocysteine (Hcy), which contains sulfur, into methionine. One of the variables contributing to AD may be Hcy accumulation, which may be brought on by Hcy buildup, which is ultimately brought on by a cystathionine synthase deficit. The deficiency is called homocysteinemia, which is one of the major threats in neurological diseases, including AD (Cankurtaran et al. 2013). TRPV1 is also activated due to the oxidative alterations made in multiple cysteine residues. Experimental animals and patients with AD have shown prominent changes in Ca^{+} homeostasis by the activation of receptors like TRPA 1, TRPM2, and TRPV1. An increase in the concentration of ca^{+} ions induces oxidative stress and causes apoptosis via activation of the receptor channels that were found in the hippocampal neurons.

Transient receptor potential vanilloid type 1 (TRPV1) is an ion channel formerly addressed as vanilloid receptor 1. It is a selective channel that is sensitive to proton and capsaicin. It is an essential component of the brain as it modulates synaptic transmission and helps in the communication between microglial cells and neurons and also regulates membrane potential (Anstötz et al. 2018). It also plays a major role in the nervous system as it is responsible for activities like cross talk and the functioning of the gamma-aminobutyric acid receptors in the brain (Ji et al. 2002). Capsaicin, the TRPV1 agonist, is found to activate the receptor to ameliorate the cognitive and synaptic functions in the AD model by inhibiting certain receptors that contain GluA2. Reports suggest that the activation of this receptor increases the Ca^+ levels in the intramembrane, thereby leading to mitochondrial depolarization resulting in the movement of the cells in response to the chemical stimuli. TRPV1 is considered as a potent marker since it is said to reduce the aggregates of proteins like amyloids and hyperphosphorylation of tau proteins by mediating autophagy in microglial cells. It is also evident in regulating the oxidative phosphorylation and glycolysis pathway by administered agonist of TRPV1; all these were the finding were demonstrated in the 3xTg mice model (Wang et al. 2022).

Memantine (MEM) is solely utilized in the treatment of AD and MEM is an agonist to N-methyl-D-aspartate (NMDA) receptor; all three Hcy, NMDA, and MEM are connected. NMDA is said to participate in the Hcy-induced neuronal damage, while MEM act as an agonist to NMDA receptors, and their activation role in TRPA1 and TRPV1 is still unknown (Raghavendra Rao et al. 2001; Lipton 2005). These possible regions are potential markers that can be studied and all are still in research to prevent the onset or to reduce dementia-related neural symptoms.

22.4 Capsaicin

Capsicum, the most common ingredient in world cuisine, comprises a therapeutic effect on various diseases. There are more than 20 different types of capsicum with a large number of bioactive compounds. The pungent-smelling bioactive compound capsaicinoids are responsible for the sharp taste. Capsaicin and dihydrocapsaicin, the two main capsaicinoids, are responsible for over 90% of the pepper's pungent flavor (Kosuge and Furuta 1970). Both capsaicinoids can dissolve in the solvent and non-solvent solutions. These compounds are found mostly in the east part of the world; however, America is known to be the birthplace of these peppers. Many kinds of capsaicinoids differ from each other according to bond (Nava-Ochoa et al. 2021). They are widely used in the cosmetic, food, and drug industry. Moreover, they are used to treat several clinical conditions since it possesses many pharmacological benefits such as antioxidant, anti-inflammatory, and anti-analgesic properties. When capsaicin binds to the TRPV1 a cationic channel, it induces the secretion of IL-10 (interleukin-10 cytokines) and suppression of IL-8 and IL-1β (interleukin-8 and interleukin-1β pro-inflammatory cytokines) using nuclear factor kappa B (NF-κB). The pathway is explained in Fig. 22.2. Capsaicin, the major compound in capsaicinoids, is primarily known for its analgesic activity, and it reduces pain caused by certain chronic diseases and was found to be more effective when applied tropically.

The urge for capsaicin in the therapeutic world is because of its property like anti-inflammatory and anti-analgesic; it was proven to reduce pain in neuropathic diseases like pain after stroke and muscular pain. The topical application showed great results, while certain commercial products indicated secondary effects like irritation and burning sensation. To avoid such conditions, researchers are working out certain formulations to vary the concentration and add some other vital

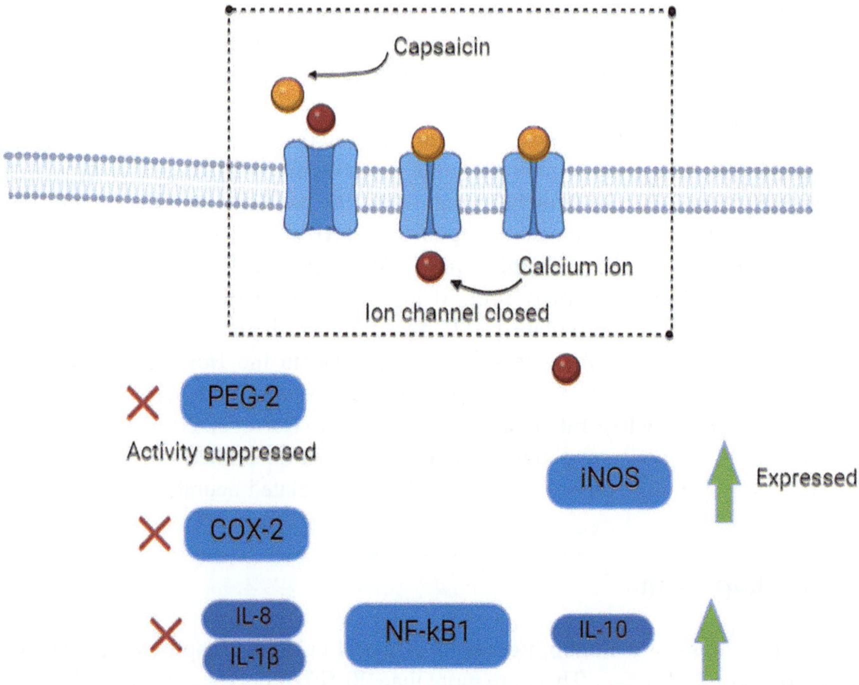

Fig. 22.2 Capsaicin as an anti-inflammatory component and the mechanism involved in it. IL-10 and iNOS are expressed, while the pro-inflammatory interleukin and prostaglandin E2 are suppressed by inhibiting cyclooxygenase 2

compounds. Apart from being anti-inflammatory and anti-analgesic, it is also a neuroprotective compound.

AD is a neurological condition that manifests in persons over the age of 40. The formation of neurofibrillary tangles and extracellular plaques in the brain is the major cause of the disease leading to the inability to speak, dementia, and improper voluntary movements (Jiang et al. 2013). Environmental factors are one of the factors that trigger these activities in the brain. There were several research conducted to ensure a proper cure, but there were no intended results. It is a disease without a cure and is believed to be controlled through a healthy diet and lifestyle. Failure of several series of clinical trials led to a change in the strategy of study to prevent the progression instead of a complete cure.

In this chapter, we have elucidated the properties of capsaicin and targeted specifically AD. It includes the pathway and tests recruited to analyze the reducing pathological and synaptic impairment conditions. Capsaicin acts as an agonist in the TRPV1 receptor and behaves like a neuroprotectant; the complete study of capsaicin's role in AD along with the future perspective of the study is explained.

22.5 Capsaicinoids

Capsicum is a fruit commonly used in cuisine and well known for the pungency that improves the spice quality of food (Caterina et al. 2000). It is one of the members of the Solanaceae family. Almost everyone around the world uses capsicum daily, which consists of active compounds that possess certain attributes, and they can officially be approved by the US formulary. The parent compound called capsaicin is responsible for the pungent taste of capsicum, while there are other compounds accounting for 0.1–1% in the whole fruit. Commonly known capsaicinoids are capsaicin and dihydrocapsaicin, and they are very similar other than the structure; the double bond is present in capsaicin, while it is not in the dihydrocapsaicin (DHC). Capsaicin is commonly used as a cream in the treatment of arthritis, diabetic neuropathy, and cluster headache. Commonly when there is a pain symptom, a kind of substance P is released by the neurons. This compound activates several cascades of events in the nervous system, which attenuates the pain in the skin, brain stem, and spinal cord.

With all these effective remedial properties of capsaicin, scientists found that this can be used to inhibit certain receptor functions like TRPV1 in AD. Capsaicin involves the central nervous system and helps in preventing ischemic pain after stress. A recent study on AD proved that the incidence rate is less in China compared to other regions since Chinese cuisine in the east had strong pungency pepper than west part of China.

22.6 Clinical Applications of Capsaicin in AD

22.6.1 Capsaicin Reduces the Hyperphosphorylated Tau Proteins

The effects of capsaicin in APP/PS1 mice were tested. The research included administering dietary capsaicin to the mice even before the onset of the disease to find the effectiveness of capsaicin in the mice. A 0.01% of capsaicin was administered almost 3 months and was observed for 9 months. The learning and spatial identification ability were better in the capsaicin-administered mice than in the control. This indicated that capsaicin can prevent the onset to a greater extent. His study regarding the ability of capsaicin to inhibit or reduce the phosphorylated tau proteins showed a significant result. Capsaicin group mice had very less phosphor tau-positive neurons compared to the control group, which showed an increase in the level of the phosphorylated protein in both neocortex and hippocampal neurons.

Neuroinflammation was also evident in the brain of capsaicin-treated mice with a decrease in the activated microglial cells, astrocytes, and pro-inflammatory factors. Postsynaptic density protein (PSD) 95, synapsin (SYN)-1, and synaptosomal-associated protein (SNAP)25 are immune proteins that help the synapse assembly and function were elevated in the test groups. Altogether it indicates that capsaicin can protect the brain from neuroinflammation, neurodegradation, and hyperphosphorylation in APP/PS1 mice.

Tau phosphorylation sites were present in both type 2 diabetic patients and rats (Liu et al. 2011; Yang et al. 2012). Reports from Weijie Xu et al. and his colleagues stated that the three most phosphorylated sites in the AD ser^{199}, ser^{202}, and Ser^{396} are present in the hippocampus cells; this was evident in the T2D groups, but in his recent study, it was proved that the T2D + capsaicin group was showed predominantly less phosphorylated sites and the level of tau1 a monoclonal antibody that recognizes the sites was merely found. All these were found using the western blotting technique and the samples were the hippocampal cells from the brain of rats. However, T2D + PF rats did not show a respective reduction in phosphorylation compared to the T2D group. These consolidated results on the use of dietary capsaicin provide a better knowledge of their effects in both normal and diabetic mice models (Xu et al. 2017).

22.6.2 Effect of Capsaicinoids in Decreasing the Degradation and Aggregation of Aβ Peptides

Capsaicinoids are known to be effective in APP processing to analyze Marcus ow Grimm used capsaicin and dihydrocapsaicin (DHC) (Cordell and Araujo 1993). Both the components are almost the same with aliphatic side chains of the same length, but only capsaicin has a double bond in it. SH-SY5Y is a Human neuroblastoma cell line and normal non-neuronal cells were used to analyzed the increased Aβ generation; this is due to the expression of *BACE1* gene eventually leading to the increase the proteins such as BEC1 and PS1; these are the components of the complex γ-secretase. Capsaicinoids exert an increase in Aβ. The total level of Aβ depends both on the anabolism and catabolism by the degrading enzymes (Iwata et al. 2001). Capsaicin, a common capsaicinoid compound, is used to examine the degradation of Aβ using the mouse neuroblastoma cells along with synthetic peptides. The process was monitored by the immunoblotting method to find the presence of human Aβ peptide instead of a mouse. Although both anabolism and catabolism of the Aβ peptide aggregation is the major cause of AD pathogenesis aggregation. Thioflavin T-based assay found that in the presence of the capsaicin, the level of Aβ aggregation was reduced notably (Reis et al. 2018).

22.6.3 Capsaicin Rescues the Microglial Metabolism from Aβ Tolerated Microglia

It was observed that the increase in the expression of AKT/mTOR phosphorylation and HIE-1α also boosted oxidative metabolism and glycolysis. Capsaicin was administered to the cells and the results showed a reduction in the ROS in microglial cells, and the mitochondrial membrane potential (MMP) was analyzed using a membrane function indicator (Louis-Gray et al. 2022). Capsaicin, the agonist of TRPV1, was tested on APP/PS1 mice and the Aβ plaque deposition and soluble Aβ levels were analyzed. Levels of $A\beta_{1-40}$ and $A\beta_{1-42}$ were analyzed by ELISA method, but they were almost similar and in less concentration in the hippocampal cells of the test group (APP/PS1+ capsaicin).

22.6.4 Capsaicin Activates the Autophagy Pathway in APP/PS 1 Mice

AD is associated with dysfunction in the synaptic pathway and the impairment of microglial cells. Capsaicin was found to induce autophagy in the APP/PS 1 mice with upregulation of microglia/macrophages (iba-1) and astrocytes [glial fibrillary acidic protein (GFAP)]; these are immunoreactive compounds found around the plaques, and they were largely found in test groups with capsaicin than the control group (de la Monte and Wands 2006; Wang et al. 2015). The arrival of the glial cells is activated by the TRPV1 agonist that proceeds with phagocytosis in the plaque region to eliminate them from the vicinity. Additionally, pro-inflammatory factors like IL-6 and TNF-α were merely found due to the increase in IL-10, an anti-inflammatory factor in the brain cells.

22.7 Future Perspectives of Capsaicin

The TRPV1 agonist capsaicin is a component present in the chili and is said to be protective against cognitive impairment. Several studies suggest that the consumption of capsicum reduces the worse conditions of pain and symptoms of AD in an animal model, and they hypothesized that it can prevent the onset of the disease. Since all these elucidations were self-reported and there was no consistent study to prove the hypothesis, the research continued, and in the present study regarding the capsaicinoids, it is proved that the consumption of chili could cure cognition. However, all these findings were not confirmed to be the case in wild-type and healthy humans, and further study is essential to confirm the effect of capsaicin on cognitive performance. Through several studies, it is found that capsaicin is capable of reducing the Aβ plaque formation and reduction in the tau levels in mouse models like 3Xtg and APP/PS 1. There are multiple targets to focus on AD pathogenesis and capsaicin was found to protect against all these conditions. Almost in every study capsaicin was administered orally or through intragastric infusion to improve the treatment in long-term research. A different result was observed for a different mode of administration. Future research can find the most suitable and effective dosage of capsaicin.

TRPV1 receptor mediation reduces the level of tau and amyloid phosphorylation and several studies have found that capsaicin is a potent agonist to activate these receptors to induce autophagy and eliminate the plaques from the brain. TRPV1 activation was found to improve the synaptic plasticity and other cognitive impairment through the mice model although there are several controversies in these findings (Du et al. 2020; Balleza-Tapia et al. 2018). Autophagy improves the availability of cells' nutrients during the deprivation and maintains homeostasis.

Metabolisms like glycolysis and mitochondrial respiration were also a threat to AD onset but activation of TRPV1 receptor. In several studies, it is reported that capsaicin is effective as dose-dependent and independent in neuronal cells of AD. Reports suggested that TRPV1 activation leading to autophagy that dependents on capsaicin (Shibata et al. 2020). Further research in these areas can bring about changes in the perspectives of research studies and come up with an effective and efficient treatment for AD.

22.8 Conclusion

Alzheimer's disease progressively gets worse over longer period of diagnosis and researchers across the globe are developing drug targets and novel drugs to control Alzheimer's disease. In this chapter, the components present in capsicum, the chili peppers a common ingredient of cuisine and its potential advantages in Alzheimer's disease are discussed along with certain clinical studies of capsaicin in vitro and in vivo studies. In conclusion, capsaicin can be a potent drug or be used in the treatment of Alzheimer's disease in order to reduce the incidence of the disease throughout the world.

References

Adil M, Ameenudeen S, Ali N, Ahmad N, Pandurangan AK, Hemalatha S, Kashif M, Waseem M (2020) Emerging role of mitophagy in inflammatory diseases: cellular and molecular episodes. Curr Pharm Design 26:1–7

Anstötz M, Lee SK, Maccaferri G (2018) Expression of TRPV1 channels by Cajal-Retzius cells and layer-specific modulation of synaptic transmission by capsaicin in the mouse hippocampus: TRPV1 in hippocampal Cajal-Retzius cells. J Physiol 596:3739–3758

Baik SH, Kang S, Lee W, Choi H, Chung S, Kim JI, Mook-Jung I (2019) A breakdown in metabolic reprogramming causes microglia dysfunction in Alzheimer's disease. Cell Metab 30:493–507

Balleza-Tapia H, Crux S, Andrade-Talavera Y, Dolz-Gaiton P, Papadia D, Chen G, Johansson J, Fisahn A (2018) TrpV1 receptor activation rescues neuronal function and network gamma oscillations from Aβ-induced impairment in mouse hippocampus *in vitro*. elife 7:e37703

Cankurtaran M, Yesil Y, Kuyumcu ME, Oztürk ZA, Yavuz BB, Halil M, Ulger Z, Cankurtaran ES, Arıoğul S (2013) Altered levels of homocysteine and serum natural antioxidants links oxidative damage to Alzheimer's disease. J Alzheimers Dis 33:1051–1058

Caterina MJ, Leffler A, Malmberg AB, Martin WJ, Trafton J, Petersen-Zeitz KR, Koltzenburg M, Basbaum AI, Julius D (2000) Impaired nociception and pain sensation in mice lacking the capsaicin receptor. Science 288:306–313

Choudhary A, Ranjan JK, Asthana HS (2021) Prevalence of dementia in India: a systematic review and meta-analysis. Indian J Public Health 65:152–158

Cordell GA, Araujo OE (1993) Capsaicin: identification, nomenclature, and pharmacotherapy. Ann Pharmacother 27:330–336

de la Monte SM, Wands JR (2006) Molecular indices of oxidative stress and mitochondrial dysfunction occur early and often progress with severity of Alzheimer's disease. J Alzheimers Dis 9:167–181

Du Y, Fu M, Huang Z, Tian X, Li J, Pang Y, Song W, Tian Wang Y, Dong Z (2020) TRPV1 activation alleviates cognitive and synaptic plasticity impairments through inhibiting AMPAR endocytosis in APP23/PS45 mouse model of Alzheimer's disease. Aging Cell 19:e13113
Eratne D, Loi SM, Farrand S, Kelso W, Velakoulis D, Looi JC (2018) Alzheimer's disease: clinical update on epidemiology, pathophysiology and diagnosis. Australas Psychiatry 26(4):347–357
Gao Y, Tan L, Yu J-T, Tan L (2018) Tau in Alzheimer's disease: mechanisms and therapeutic strategies. Curr Alzheimer Res 15:283–300
Hanisch U-K, Kettenmann H (2007) Microglia: active sensor and versatile effector cells in the normal and pathologic brain. Nat Neurosci 10:1387–1394
Holubiec MI, Gellert M, Hanschmann EM (2022) Redox signaling and metabolism in Alzheimer's disease. Front Aging Neurosci 14:1003721
Hool LC (2008) Evidence for the regulation of L-type Ca^{2+} channels in the heart by reactive oxygen species: mechanism for mediating pathology. Clin Exp Pharmacol Physiol 35(2):229–234
Iwata N, Tsubuki S, Takaki Y, Shirotani K, Lu B, Gerard NP, Gerard C, Hama E, Lee HJ, Saido TC (2001) Metabolic regulation of brain Aβ by neprilysin. Science 292:1550–1552
Ji R-RA, Jin S-X, Schmoll R, Woolf CJ (2002) p38 MAPK activation by NGF in primary sensory neurons after inflammation increases TRPV1 levels and maintains heat hyperalgesia. Neuron 36:57–68
Jiang X, Jia LW, Li XH, Cheng XS, Xie JZ, Ma ZW, Xu WJ, Liu Y, Yao Y, Du LL, Zhou XW (2013) Capsaicin ameliorates stress-induced Alzheimer's disease-like pathological and cognitive impairments in rats. J Alzheimers Dis 35:91–105
Kivipelto M, Mangialasche F, Ngandu T (2018) Lifestyle interventions to prevent cognitive impairment, dementia and Alzheimer disease. Nat Rev Neurol 14:653–666
Kosuge S, Furuta M (1970) Studies on the pungent principle of *Capsicum*: part XIV chemical constitution of the pungent principle. Agric Biol Chem 34:248–256
Lipton S (2005) The molecular basis of memantine action in Alzheimer's disease and other neurologic disorders: low-affinity, uncompetitive antagonism. Curr Alzheimer Res 2:155–165
Liu Y, Liu F, Grundke-Iqbal I, Iqbal K, Gong C-X (2011) Deficient brain insulin signalling pathway in Alzheimer's disease and diabetes. J Pathol 225:54–62
Louis-Gray K, Tupal S, Premkumar LS (2022) TRPV1: a common denominator mediating antinociceptive and antiemetic effects of cannabinoids. Int J Mol Sci 23(17):10016
Menzies FM, Fleming A, Rubinsztein DC (2015) Compromised autophagy and neurodegenerative diseases. Nat Rev Neurosci 16:345–357
Metaxakis A, Ploumi C, Tavernarakis N (2018) Autophagy in age-associated neurodegeneration. Cell 7:37
Nava-Ochoa AE, Antunes-Ricardo M, Guajardo-Flores D (2021) Nano-sized carriers for capsaicinoids with topic analgesic and anti-inflammatory effects. J Biotechnol 333:77–85
Nichols E et al (2022) Estimation of the global prevalence of dementia in 2019 and forecasted prevalence in 2050: an analysis for the Global Burden of Disease Study 2019. Lancet Public Health 7:e105–e125
Raghavendra Rao VL, Dogan A, Todd KG, Bowen KK, Dempsey RJ (2001) Neuroprotection by memantine, a non-competitive NMDA receptor antagonist after traumatic brain injury in rats. Brain Res 911:96–100
Reis AB, Arain HA, Stecker MM, Siegart NM, Kasselman LJ (2018) Amyloid toxicity in Alzheimer's disease. Rev Neurosci 29:613–627
Sharma VK, Singh TG, Singh S, Garg N, Dhiman S (2021) Apoptotic pathways and Alzheimer's disease: probing therapeutic potential. Neurochem Res 46(12):3103–3122
Shibata M, Kayama Y, Takizawa T, Ibata K, Shimizu T, Yuzaki M, Suzuki N, Nakahara J (2020) Resilience to capsaicin-induced mitochondrial damage in trigeminal ganglion neurons. Mol Pain 16:174480692096085
Tanzi RE (2012) A brief history of Alzheimer's disease gene discovery. J Alzheimers Dis 33:S5–S13

Wang Y-F, Fan L-M, Zhang W-Z, Zhang W, Wu W-H (2004) Ca^{2+}-permeable channels in the plasma membrane of Arabidopsis pollen are regulated by actin microfilaments. Plant Physiol 136:3892–3904

Wang Y, Cella M, Mallinson K, Ulrich JD, Young KL, Robinette ML, Gilfillan S, Krishnan GM, Sudhakar S, Zinselmeyer BH, Holtzman DM, Cirrito JR, Colonna M (2015) TREM2 lipid sensing sustains the microglial response in an Alzheimer's disease model. Cell 160:1061–1071

Wang C, Huang W, Lu J, Chen H, Yu Z (2022) TRPV1-mediated microglial autophagy attenuates Alzheimer's disease-associated pathology and cognitive decline. Front Pharmacol 12:763866

Xu W, Liu J, Ma D, Yuan G, Lu Y, Yang Y (2017) Capsaicin reduces Alzheimer-associated tau changes in the hippocampus of type 2 diabetes rats. PLoS One 12:e0172477

Yang Y, Ma D, Wang Y, Jiang T, Hu S, Zhang M, Yu X, Gong CX (2012) Intranasal insulin ameliorates tau hyperphosphorylation in a rat model of type 2 diabetes. J Alzheimers Dis 33: 329–338

Therapeutic Effects of Capsaicin on Central Nervous Disorders with Special Emphasis on Parkinson's and Alzheimer's Diseases

23

Debdarshan Dutta, Dipti Tripathi, Prakhar Asthana, Kratik Rana, Nami Jain, Radhika Sharma, Urshila Naithani, Devatman Jauhari, and Rachana

Abstract

The two major neurodegenerative diseases, viz., Parkinson's disease (PD) and Alzheimer's disease (AD), are incurable so far. These diseases are even difficult to manage as their exact pathogenesis is not clear, and currently the drugs that are being used for their management are just for symptomatic relief. At present, pathological pathways for these diseases are being explored at the molecular level, so that effective drugs can be discovered for the specific molecular targets. Recently, capsaicin (CAP) has been found to be effective in ameliorating some of the root causes of AD and PD, targeting some similar and some different mechanisms of action. In case of PD, dopaminergic neurons are lost and it has been found that CAP being a transient receptor potential vanilloid subtype 1 (TRPV1) agonist can help in the treatment of PD as it can further reduce oxidative stress and metabolic impairments of α-synuclein and inhibit glial activation and inflammation, protecting dopaminergic neurons in the striatum and substantia nigra. On the other hand, in the case of AD, amyloid-β (Aβ) accumulation is majorly reported to be responsible for the disease. It is suggested that inducing autophagy of Aβ can improve neural signaling in the brain. The CAP molecule has been explored for this action and has been found to attenuate Aβ amyloid, to improve locomotion, alleviating cognitive impairment. Research indicates that capsaicin, also has potential in combating one of the precursors to Alzheimer's disease, ischemia. This suggests that it might contribute in reducing the symptoms associated with the condition. It was also found to decrease proinflammatory responses and oxidative stress in various AD models. Thus, it

D. Dutta · D. Tripathi · P. Asthana · K. Rana · N. Jain · R. Sharma · U. Naithani · D. Jauhari · Rachana (✉)
Department of Biotechnology, Jaypee Institute of Information Technology, Noida, Uttar Pradesh, India
e-mail: rachana.dr@iitbombay.org

M. K. Swamy, A. Kumar (eds.), *Capsaicinoids*,
https://doi.org/10.1007/978-981-99-7779-6_23

can be suggested that CAP may be utilized as a potential drug molecule against the two neurodegenerative disorders. The present chapter discusses the basic mechanisms behind these two diseases and further details on the role of capsaicin in their amelioration. In addition, certain limitations related to the therapeutic uses of CAP have been discussed.

Keywords

Capsaicin · Trpv1 · Parkinson's disease · Alzheimer's disease · Neurodegenerative disorders · Microglia · Oxidative stress

23.1 Introduction

23.1.1 Chemical Structure, History, and Current Application of Capsaicin

Capsaicin (CAP) is an active component extracted from chili peppers that possess analgesic properties (Arora et al. 2021; Lu et al. 2020), having IUPAC name as 8-methyl-N-vanillyl-trans-6-nonenamide. Capsaicin (CAP) produces pain along with a sensation of heat and therefore has been crucial in studying pain. The chemical structure of CAP is shown in Fig. 23.1. CAP is hydrophobic, highly pungent, colorless, and a crystalline to waxy solid compound (Amruthraj et al. 2014).

CAP was isolated first by Christian Bucholz in 1816 (Chang et al. 2022). In 1873 and 1878, respectively, Rudolf Buchheim, a German pharmacologist, and Endre Hőgyes, a Hungarian doctor, both independently concluded that "capsicol" (partially purified capsaicin) leads to the burning of mucous membranes that line various body

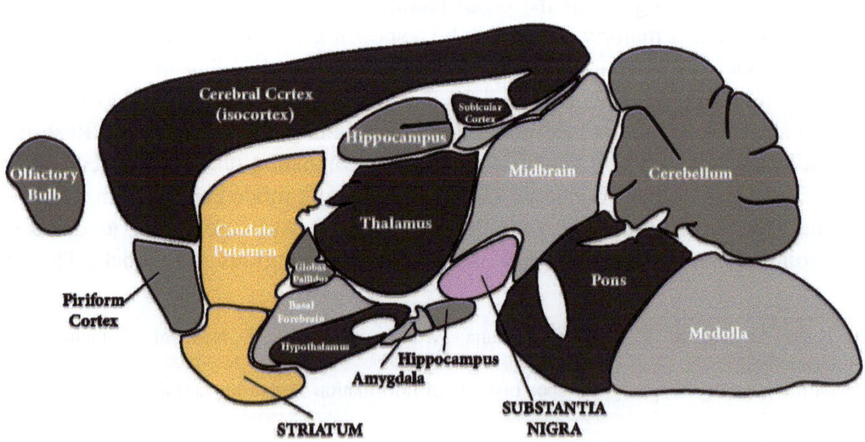

Fig. 23.1 Brain of rat showing different parts of the brain including the most affected substantia nigra and striatum

cavities and increases gastric acid secretion when exposed to mucous membranes (Basith et al. 2016). The chemical structure of CAP was determined in 1919, and its biosynthetic pathway was explained in 1960 (Chang et al. 2022).

The flowering plants of the genus *Capsicum* produce varying quantities of capsaicin and have been used by humans for over 6000 years, as spice ingredients (Fattori et al. 2016). CAP was earlier used to induce Type I and Type II hypersensitivity to innocuous and noxious stimuli by injecting it intradermally, in monkeys and rats (Lipshetz et al. 2018). In traditional medicine, people residing in tropical countries consumed capsicum to increase perspiration and skin vasodilation through capsaicin-induced skin vasodilation, to deal with hot weather (Basith et al. 2016). Other applications of capsaicin in folk medicine include treatments of pharyngitis, rheumatism, cough, tonsillitis, appetite stimulation, backache, cholera, gout, gastric ulcers, water retention, cramps, diarrhea, dyspepsia and toothache, and hair growth restoration (Basith et al. 2016).

Currently, CAP has been used to treat neuropathic pain (Ilie et al. 2019) postherpetic neuralgia, diabetic and/or HIV-related neuropathy, burning mouth syndrome, temporomandibular joint problem (Campbell et al. 2016), chemotherapy-induced peripheral neuropathy (Filipczak-Bryniarska et al. 2017), and fibromyalgia due to its analgesic and antinociceptive effects. Topical application of capsaicin has demonstrated its efficacy and safety as a substitute to systemic analgesics, which commonly result in major adverse effects in individuals with rheumatoid arthritis and osteoarthritis.

23.1.2 Neurodegenerative Disorders and Suggested Role of Capsaicin

Neurodegenerative disorders deliver high damage to the quality of life, rendering a person incapable of carrying out day-to-day activities. Due to highly decreased neuromotor and mental capacity, the patient loses all connections from the outside world and is forced to be confined in a restricted place while being dependent on others for the most basic of needs. Diseases like Alzheimer's, Parkinson's, Huntington, and amyotrophic lateral sclerosis are only a few examples of the important class of neurological conditions that are distinguished by an irreversible loss of certain kinds of neurons. Once neurodegeneration starts, it can only be slowed down but not completely halted; hence this class of illnesses presents a significant clinical challenge. Instead of treating the underlying pathological cause, current clinical treatments have a more symptom-based approach (Chen and Pan 2015). There are several US Food and Drug Administration (US-FDA)-approved therapies suggested for various neurodegenerative illnesses. However, these medications have adverse effects that include nausea, sleepiness, headaches, stiffness, tremors, hives, and gait instability (Durães et al. 2018) and provide symptomatic relief. This demonstrates that, despite the fact that modern pharmaceuticals can lessen the symptoms, there are no long-term fixes to save the neuronal population that has survived (Muddapu et al. 2020). Hence, there is an utmost need to discover a novel therapeutic that can not only slow down the neurodegeneration but also help the body regenerate the lost neurons. In the last few decades, CAP has been shown to be effective against neurodegenerative disorders like Parkinson's disease and Alzheimer's disease in their respective animal models, which shall be discussed in detail further in this review. The following section first discusses some of the basic features and causes known so far for AD and PD.

23.2 Parkinson's Disease

23.2.1 Brief Introduction to Parkinson's Disease

Parkinson's disease (PD) is a disease that involves progressive neurodegeneration of cells of the central nervous system (CNS). Loss in nigrostriatal dopamine (DA) neurons, in the pars compacta area of the substantia nigra (SNpc) in the midbrain (Fig. 23.1), is the primary feature of Parkinson's disease (PD) (Baluchnejadmojarad et al. 2009). The progressive neuronal depletion results in motor dysfunctions such as stiffness in limbs and joints, resting tremors, slowness of movement, etc. (DeMaagd and Philip 2015). The following section briefly describes the major pathways reported to be responsible for PD.

23.2.2 Major Pathological Pathways Responsible for Parkinson's Disease

PD majorly constitutes symptoms including a reduction of automatic movements (bradykinesia), stiffness of limbs, weakness, and resting tremors (Váradi 2020). These are majorly caused when tyrosine hydroxylase-positive DA neurons disintegrate in the substantia nigra (SN), which are involved in dopamine production (Charlton and Mack 1994). As a result, lesser dopamine content is present in the striatum as neurons of the substantia nigra (SN) are the ones that project into the striatum (Menegas et al. 2015).

Lewy bodies and Lewy neurites, the distinctive proteinaceous deposits in midbrain nigral neurons/striatum, are the hallmarks of Parkinson's disease (PD). They are primarily composed of α-synuclein (Gómez-Benito et al. 2020; Spillantini et al. 1997; Trojanowski et al. 1998). There is a mechanism proposed for aggregation of α-synuclein in the brain based on the "gut-brain axis" concept that says that the changes in the gut microbiota may promote α-Syn aggregation and cause an inflammatory response such as elevated cytokine levels and activated T cells, in the peripheral nervous system (Klann et al. 2022). α-Synuclein-mediated pathway is the most discussed pathological pathway for PD and both the central and peripheral inflammations are another next mostly discussed root causes for Parkinson's disease (PD) and, many a times, are shown to be linked to each other (Pajares et al. 2020). Also, human α-synuclein (h-αS) when overexpressed attaches itself to the end of nerve terminals and prevents neurotransmitter release (Stefanis 2012). Once within the brain, h-αS proteinopathy perpetuates central inflammation in a vicious cycle involving glial cells, activated endothelium, and dying dopaminergic neurons. Further peripheral immune cells penetrating the brain make the situation worse (Kouli et al. 2018). It has been suggested that α-synuclein regulates dopamine synthesis as well and, thus, further contributes to PD pathogenesis (Mavroeidi and Xilouri 2021).

The rodent models, which have been used for studying PD, have shown that LPS, MPTP, and 6-OHDA can lead to microglial activation aggravating neurotoxicity of DA neurons (Peterson and Flood 2012). When activated, the microglia are reported to stimulate the secretion of various oxidants and proinflammatory cytokines that further promote the loss of neuronal cells in SN (Miraglia and Colla 2019).

Another recently discussed mechanism of PD is related to mitochondrial dysfunction, which ultimately leads to raised oxidative stress, which is proposed to be a major cause for the death of nigral DA neurons in PD (Moon and Paek 2015). Several factors such as aging, genetic mutations, and environmental toxins or their combinations are reported to cause mitochondrial defects, which further can disturb many other cellular pathways raising oxidative stress, damaging intracellular components and cell death. Thus, various cellular mechanisms together with aggregation of α-synuclein, activation of microglia, inflammation, dysfunction of mitochondria, and increase in oxidative stress lead to neurodegeneration of the substantia nigra (SN), which ultimately results in the development of PD as described above.

On the basis of the basic information available about the pathogenesis of PD, the most commonly used PD animal models are animals treated with MPTP and MPP+ (Stilling et al. 2014). Toxic chemical compounds MPTP (1-methyl-4-phenyl-1,2,3,6-tetrahydropyridine) and its active form MPP+ (1-methyl-4-phenylpyridinium) induce SN neuronal degeneration (Spillantini et al. 1997; Trojanowski et al. 1998). Moreover, various models such as *Drosophila* models, comprising transgenes and mutants, have also been used to understand the pathogenesis of different degenerative diseases such as PD (Lu et al. 2009). The following section discusses mechanism action of capsaicin, which is reported to help in alleviating various root causes responsible for PD.

23.2.3 Mechanism of Action of Capsaicin on Parkinson's Disease

There are various mechanisms suggested for capsaicin ameliorating PD. Figure 23.2 describes the reported mechanism of action for capsaicin against PD.

23.2.3.1 Reduction in Oxidative Stress

Capsaicin is reported to reduce oxidative stress in animal models of Parkinson's disease. Zhao et al. (2017) showed that CAP-mediated reduction in oxidative stress inhibits the progression of PD and relieves behavioral deficiencies in Wister rats, treated with 6-hydroxydopamine (6-OHDA), having lesions in the brain. Four categories of rats were used by them for their experiment – the control, the SOG

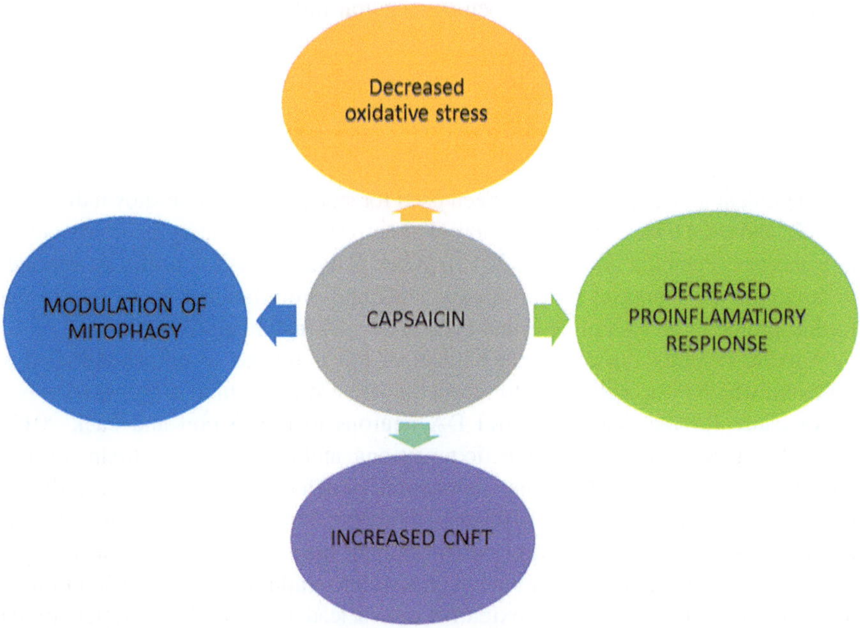

Fig. 23.2 Various modes of amelioration of Parkinson's disease by capsaicin reported in various animal models

group (injection of saline by stereotaxic method and treatment of saline for 7 days after the stereotaxic operation), the 6-OHDA group (induced 6-OHDA with no CAP treatment), and the CAP group (induced 6-OHDA with treatment of capsaicin for 7 days post-stereotaxic operation). An increase in malonaldehyde (MDA), an indicator of oxidative stress, was observed in 6-OHDA groups. CAT and SOD activities in the right striatum and SN were shown to be increased in the CAP group and there was a decrease in MDA levels. No differences were observed in the left SN and left striatum of the CAP group and the 6-OHDA group. In this open-field experiment with rats, it was observed that the rats of the CAP group showed capsaicin-ameliorated behavioral deficits and raised motor abilities. The 6-OHDA-induced rat, which was given capsaicin, displayed a heightened TRPV1 expression in the striatum and SN, along with antioxidant system activation in the ipsilateral (right) hemisphere.

Liu et al. (2020) had shown downregulation of glutathione S-transferase (GSTA2) as one of the crucial mechanisms of PD progression. GSTA2 is an antioxidant response element, which is a part of the body's defense against oxidative stress, and they found it to be upregulated in capsaicin-treated rats. Normally, antioxidant and oxidative stress responses are required to be in an equilibrium. High oxidative stress and a parallel low antioxidant response lead to disturb this equilibrium and thus cause neural cell death. CAP was found to upregulate GSTA2 and lead to an antioxidant response to oxidative stress, resulting in a neuroprotective effect. This study concluded that regulation of apoptosis and oxidative stress, by regulating ACTG1 and GSTA2, could be a possible mechanism to combat PD.

An experiment by Chung et al. (2017) showed that CAP reduces oxidative stress via decreasing the activity of microglial NADPH oxidase in the MPTP-lesioned SN. NADPH oxidase is highly expressed in microglia and is an enzyme that produces superoxides, a type of reactive oxygen species (Hou et al. 2018). Generation of these highly reactive compounds causes neuronal damage as it induces prolonged neuroinflammatory responses (Wang et al. 2015b; Che et al. 2018; Zhang et al. 2018a). Inhibition or genetic deletion of this enzyme led to the mitigation of DA neuron degeneration and activation of microglia in PD models (Hou et al. 2018). The experiment showed that MPTP led to the upregulation of p47phox–gp91phox complexes and p47phox phosphorylation, indicating NADPH oxidase activation. On the other hand, CAP-treated rats showed a significant decrease in this trend in the SN of MPTP-treated mice. It was proposed that TRPV1 activation by CAP leads to downregulation of NADPH oxidase and subsequent neuroprotection in the PD model.

In a separate study by Bok et al. (2018), a decrease in protein tyrosine nitration, an oxidative stress marker, mediated by oxidants derived by iNOS (inducible nitric oxide synthase), was observed. SN of rats treated with LPS showed sizably higher positive nitrotyrosine cells in comparison to the control group. CAP treatment led to a decrease in this effect, which was reversed by TRPV1/capsaicin receptor antagonist CZP (capsazepine) showing that CAP was involved in neuroprotection from nitrosative damage. Also, visualization of O_2^- was done by treating it with hydroethidine and observing the hydroethidine-oxidized fluorescent product (ethidium). One day after the LPS treatment, an increase in ethidium was observed, and on the other hand, CAP treatment led to a decrease in ethidium levels, indicating decreased oxidation levels, which was shown to be reversed by CZP. Treatment with CAP was found to prevent ROS production induced by LPS and the damage by oxidation, leading to neuroprotection.

A study by Siddique et al. (2018) showed that dietary capsaicin is protective against PD in flies that were genetically mutated to express the human α-synuclein (h-αS). CAP induced an increase in free radical scavenging potential, the rise of dopamine content, and lowered the oxidative stress markers.

Another experiment was conducted in the transgenic *Drosophila* models expressing h-αS, to study the effect of CAP on dopamine content and oxidative stress level. The flies were fed 20, 40, 80, and 100 μM of CAP for 24 days. Lipid peroxidation (LPO), protein carbonyl (PC), glutathione S-transferase (GST), glutathione (GSH), dopamine contents, and monoamine oxidase (MAO) activity were estimated in the fly homogenate. Substantial increase in the superoxide anions and free radical scavenging potential was observed in CAP-treated flies, in comparison to unexposed PD flies ($p < 0.05$). It was observed that the exposure to CAP increased dopamine (1.56-fold) and GSH (1.37-fold) content and noticeably reduced the levels of GST (1.26-fold), LPO (1.8-fold), PC content (1.95-fold), and MAO activities (1.60-fold). This highlights that CAP can play an effective role protective against PD symptoms (Siddique et al. 2018).

23.2.3.2 Effect of Capsaicin on Proinflammatory Action/Phagocytosis

Microglial activation is a major mechanism of progression of PD via proinflammation (Lecours et al. 2018). Studies have shown that CAP treatment in PD model rats leads to inhibition of microglial activation and thus its proinflammatory responses (Chung et al. 2017; Wi et al. 2020). Experiments conducted by Chung et al. (2017) showed that CAP inhibits microglial activation. MPTP model of rats showed an increase in anti-macrophage Ag complex-1 (MAC-1)-labeled (reactive) microglia in their substantia nigra, thus indicating increased activation of microglia. After treatment with CAP, attenuation of these microglia was observed, which was reversed by capsazepine (CZK). Moreover, CAP attenuated the MPTP-induced mRNA expression of cytotoxic and inflammation-promoting factors that are responsible for nigrostriatal dopaminergic (DA) neuronal degeneration. In addition, this was reversed by CZK treatment, which suggested TRPV1 involvement. Further, CAP treatment led to a reduction in those cells that are MPO positive in SN of MPTP rat model. MPO is a heme-containing peroxidase that can cause host cell damage due to generation of reactive nitrating and halogenating agents. Studies have shown that MPO deficiency led to decreased inflammation (Aratani 2018). The reduction of MPO-positive cells was reversed by CZK.

In the LPS-lesioned substantia nigra, CAP administration was shown to suppress proinflammatory cytokine levels and prevent DA neuronal degradation in SN (Bok et al. 2018). CAP was responsible for shifting the M1 microglia/macrophage population (proinflammatory) to an M2 state (anti-inflammatory). It was observed that CAP could restrict microglia activation and IL-6 and IL-1β expression, which are markers of the M1 phenotype.

23.2.3.3 Effect of Capsaicin on Induced Ciliary Neurotrophic Factor (CNTF) Upregulation

The study by Baek et al. (2018) reported that ciliary neurotrophic factor (CNTF) is produced by CAP-activated astrocytes through TRPV1, endogenously, and led to inhibition of activation of microglia and microglia-derived oxidative stress causing neuroprotection. There was an increased expression of CNTF receptor-α (receptor for CNTF) on microglial cells in MPP^+-induced PD rat models, and as CNTF is involved in neuroprotection by inhibiting microglial activation, this increase becomes highly relevant for PD studies. It was observed that neutralizing CNTF receptor α (CNTFRα) by neutralizing antibody (CNTFRαNAb) reversed the neuroprotective effects of CAP as well as the motor recovery.

Another experiment by Kim et al. (2019) showed that prolonged TRPV1 expression on astrocytes may lead to beneficial upregulation of endogenous production of CNTF, by delayed CAP treatment, resulting in raised activity of TH and levels of DA neurons and its associated metabolites, which led to behavioral recovery, in the complete MPP+-lesioned rats. It was shown that CAP enhanced CNTF expression in the SN pars compacta (SNpc) at 10 weeks post-PD induction by MPP^+ in rats. It was seen that in the case of delayed CAP treatment, if the treatment is terminated then, the reversal of CAP amelioration of PD was observed and rats came back to a similar state to the control groups not treated with CAP.

Experiments by Nam et al. (2015) showed that in MPP^+ rat model of PD, astrocytic CNTF was upregulated in CAP-treated rats, in comparison to the control group, and further, the inhibition of TRPV1 led to a decrease in this upregulation. Also, inhibiting CNTFRα led to decrease in neuroprotective results like decrease in the number of TH^+ and $Nissl^+$ in the substantia nigra pars compacta (SNpc) and density of TH+ density in the striatum in the CAP-treated animal.

23.2.3.4 Effect of Capsaicin on Mitophagy

Mitophagy is the process responsible for removal of defective mitochondria. In PD, this has been seen to be compromised, thus accelerating neurodegeneration. Mutated genes related to PD such as PINK1 and Parkin (Ge et al. 2020) cause mitochondrial defects and neurotoxicity due to disrupted mitophagy (Liu et al. 2019). In vitro and in vivo experiments by Lu et al. (2022) on h-αS preformed fibril (PFF) exposed mice and indicated CAP-mediated restoration of mitophagy in PFF-tolerant microglial cells. Reports show activation of microglia on release of α-synuclein from neurons leading to engulfment and degradation of α-synuclein through autophagy. The inability of this process to remove α-synuclein can lead to its accumulation, which as discussed above is a major pathological pathway for PD. Capsaicin was seen to ameliorate metabolic impairments in microglia tolerant to h-αS preformed fibril 2k, and it was also seen to mitigate mitophagy defects due to the disruption of the AKT–mTOR–HIF-1α pathway. CAP increased phagocytosis in microglia tolerant to PFF in vitro, which reduced the phosphorylation of h-αS in primary neurons.

In another study 6-OHDA-induced PD rats also showed an upregulation of ACTG1 (actin γ1) and downregulation of glutathione S-transferase α 2 (GSTA2) (Liu et al. 2020). ACTG1 is a component of the autophagy pathway and in their experiment, its upregulation led to a higher apoptosis rate and further observations showed its downregulation in capsaicin-treated rats.

The discussion above describes application of capsaicin in effectively combating PD in various animal models and provides an insight into the possibilities of using CAP as a possible drug for treating PD as it can decrease oxidative stress and inflammation and also can elevate CNTF and mitophagy.

23.3 Alzheimer's Disease

23.3.1 Brief Introduction to Alzheimer's Disease

Alzheimer's disease (AD) is a neurodegenerative condition that leads to neuronal deterioration and shrinkage of the brain (Sanabria-Castro et al. 2017). AD is the leading root of dementia, characterized by a persistent decline in behavioral, mental, and social abilities (Emmady and Tadi 2022). Symptoms of the late onset of AD often begin to manifest in the majority of individuals in their mid-60s. Early onset of Alzheimer's disease can also be seen between the age of 30 and 60, in rare cases (Sanabria-Castro et al. 2017). Memory loss, verbal difficulty, and unpredictable behavior are the major symptoms associated with AD. The following sections briefly describe the major pathological pathways responsible for AD.

23.3.2 Major Pathological Pathway for AD

Examinations of the AD brain indicate the presence of plaques of amyloid and bundles of twisted fiber called neurofibrillary or tau tangles (Pasierski and Szulczyk 2022). Losing connections between nerve cells, or neurons in the brain, is another hallmark of AD. There are numerous alternative explanations for AD pathogenesis, which often conflict each other. However, deposition of amyloid-β (Aβ) and hyperphosphorylated tau protein remain the two main characteristics of AD that cause neurodegeneration (Leeuw and Tackenberg 2019).

According to the amyloid hypothesis, various enzymes such as β-secretase (BACE1) and γ-secretase are responsible for the production of Aβ from the cleavage of amyloid precursor protein (APP) (Sanabria-Castro et al. 2017). A reduction in the breakdown of Aβ also leads to its accumulation in neurons and the amyloid plaques that cause neurodegeneration (Tabaton and Piccini 2005). Aβ42 peptide (42 amino acid form of amyloid-β) is a major constituent of these plaques and is often considered a key biomarker of AD pathogenesis (Teunissen et al. 2018). Aβ appears to specifically affect the parasympathetic neurons, either directly or via nicotinic and muscarinic cholinergic receptors (Ferreira-Vieira et al. 2016). It has long been recognized that the cholinergic neurons of the basal forebrain are more likely to get degenerated during the early AD stages (Geula et al. 2008). Aβ may alter cholinergic neurotransmission at a number of levels; for instance, Aβ peptides reduce the synthesis and release of acetylcholine, alter K+ currents, and inhibit potentiation for the long term, all of which contribute to cognitive deficits. Also, in later stages of AD, Aβ peptides lead to the death of cholinergic neurons (Ferreira-Vieira et al. 2016). Microglia contribute to the decrease of Aβ accumulation by promoting phagocytosis of Aβ and synthesis of IL-1β. However, overexpression of IL-1β increases tau phosphorylation and formation of NFTs, as well as hinders Aβ removal (Wang et al. 2015a). Microglia is a major source of inflammasomes (Hoffmann and Beyer 2020) and its activation is attributed to the inflammation of AD brain (Augusto-Oliveira et al. 2019). Activated microglia can mediate loss of synapse via engulfment of synapses (Hong et al. 2016) in a complement-dependent manner. Microglial cells can also exaggerate tau pathology via secretion of factors responsible for inflammation, injuring the neurons directly and even via activating the astrocytes secreting neurotoxins (Španić et al. 2019). The factors responsible for such a contrast in the behavior of microglial cells can be explained via its gene expression profiles, which indicate that microglial activation has various states during the pathogenesis of neurodegenerative disease, and so microglia may have contrasting roles in AD development and progression (Muzio et al. 2021) as well.

Some in vitro studies involving different types of neurons have suggested that Aβ42 can also cause alteration in the tau protein (Wang et al. 2003; Götz et al. 2001). Tau protein is a phosphoprotein that is connected to microtubules. In AD, tau protein has a pathogenic involvement mainly because of two reasons – posttranslational modifications (Bakota and Brandt 2016) and abnormal phosphorylation of tau (Sanabria-Castro et al. 2017). Hyperphosphorylated tau forms due to aberrant phosphorylation of tau. These molecules then accumulate as neurofibrillary tangles (NFTs) in neuronal cells. Subsequently, the cytoskeleton is broken down as the microtubules associated with it are lost due to accumulation of NFTs (Breijyeh and Karaman 2020).

TRPV1 is often found to be downregulated in AD brains as it gets reduced during the aging process (Du et al. 2020). Downregulation of TRPV1 results in the deposition of Aβ and the subsequent AMPAR (α-amino-3-hydroxy-5-methyl-4-isoxazolepropionic acid receptor) endocytosis through its GluA2 subunit. AMPAR is primarily involved in synaptic transmission through CNS and its endocytosis impairs cognitive functions (Du et al. 2020). Ubiquitination of AMPAR plays an important role in AD pathogenesis as it results in AMPARs loss and therefore in decreased synaptic plasticity (Qu et al. 2020). When Aβ plaques induce ubiquitination of AMPAR, DUB enzymes "USP46" and E3 ligase "Nedd4" are often upregulated and downregulated, respectively. This suggests that dysregulation of these ubiquitination-associated enzymes is responsible for reduction of AMPAR by Aβ (Zhang et al. 2018b).

Neuroinflammation caused by various factors including the above pathways destroys neurons and microglial cells by various pathways and can lead to AD. In the brain, certain NADPH oxidases like NOX2 and NOX4 are major sources of ROS. NOX4 is primarily expressed in neurons, whereas NOX2, one of the best-studied ROS-generating systems, is mainly expressed in CNS in the microglia and lesser in neurons (Eastman et al. 2020) and causes neuroinflammation if overexpressed.

In contrast to this, AD-related inflammation has also been understood to be a secondary response to the amyloid-oligomer-generated progressive lesion as it has the potential to attract neuroprotective T cells or promote microglia to absorb and remove Aβ (Fisher et al. 2010).

23.3.3 Therapeutic Roles of Capsaicin in Alzheimer's Disease

Figure 23.3 summarizes various such modes of action of capsaicin against AD. Amyloid precursor protein (APP) is considered to be essential in AD etiology (Leeuw and Tackenberg 2019), and so several transgenic mice with human APP gene mutations are used as AD models to study the basic mechanism and for the development of therapeutics. These models have been experimented upon with various molecules including capsaicin, to study their impact on APP and further effects and symptoms.

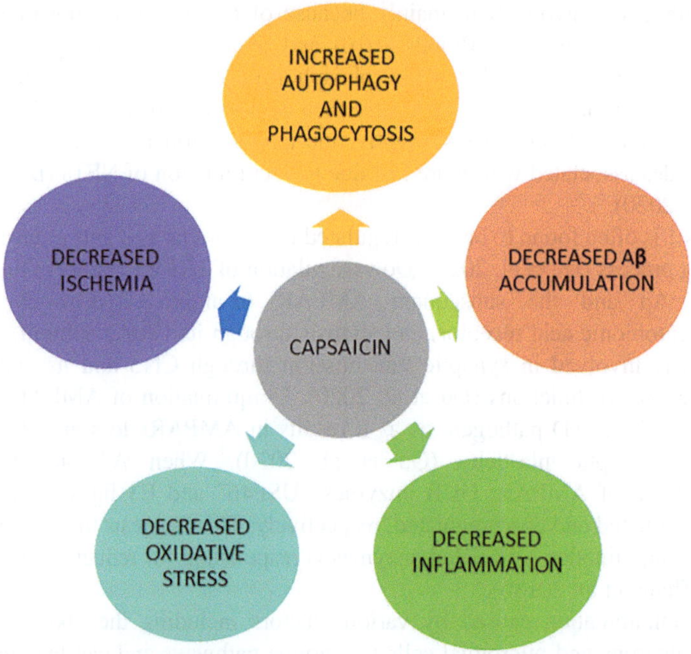

Fig. 23.3 Modes of amelioration of Alzheimer's disease by capsaicin reported in various animal models

23.3.3.1 Effect of Capsaicin on Autophagy, Phagocytosis, and aβ Accumulation

Microglia are found in the CNS where they elicit an immune response. These cells are essential for the phagocytosis of Aβ plaques in the early stages of Alzheimer's disease. One study found that giving microglial cultures 10 μM of CAP induced autophagy and increased phagocytic activity. The result showed that it was effective in removing Aβ from the culture media and TRPV1 receptors produced by microglia are the one that caused this impact. CAP therapy reduces synaptic loss and increases cortical pyramidal neuron density. Moreover, it also prevented the development of Aβ in the extracellular areas. It has been suggested that exposure to CAP enhances the ability of microglial cells to phagocytose Aβ plaques (Fig. 23.4) as CAP binds TRPV1 receptors expressed on the surface of microglial cells (Pasierski and Szulczyk 2022).

In a study, an increase in the concentration of administered CAP was found to be directly proportional to the suppression of β-secretase enzyme-1 (BACE-1). BACE-1 is responsible for APP cleavage and hence is essential for Aβ generation. In this study, inhibition of BACE-1 led to negligible production of Aβ, suggesting that this event could be correlated with the activation of TRPV1 receptor (Perumal and Hua Chong 2021). However, the association between TRPV1 receptor and BACE had been previously refuted in another study (Wang et al. 2014).

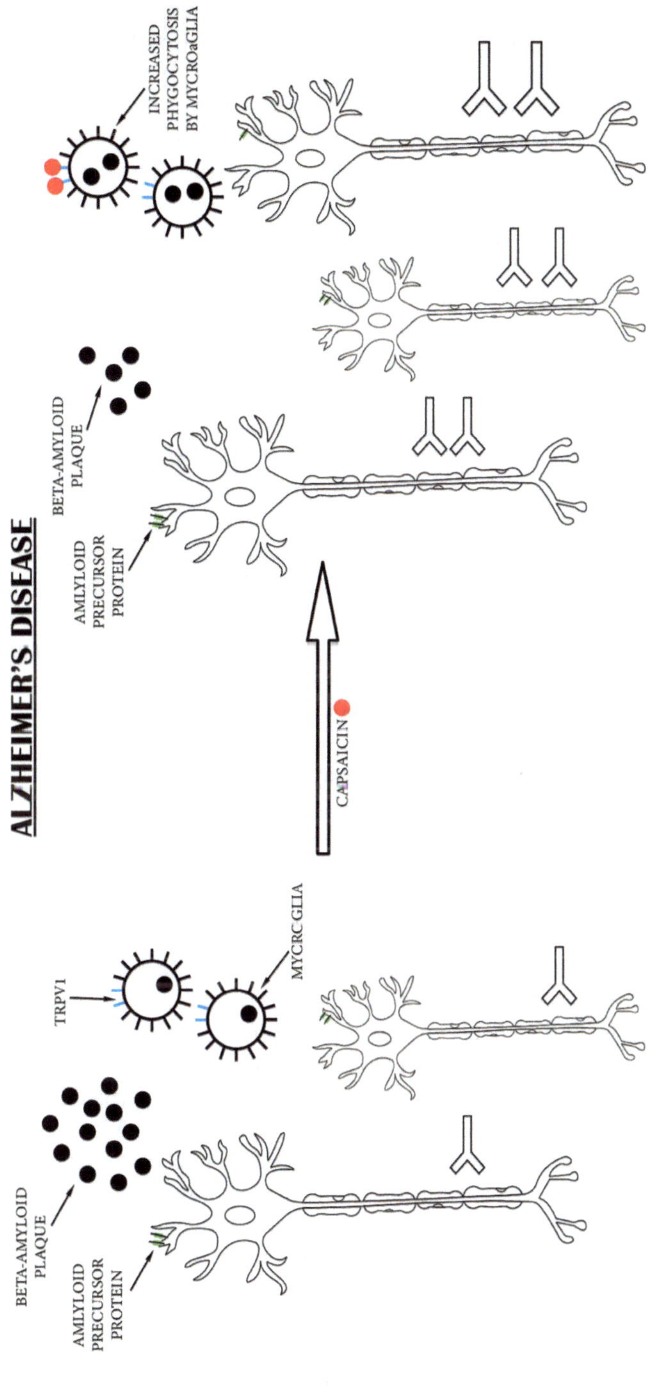

Fig. 23.4 Diagrammatic representation of beneficial effects of capsaicin autophagy, phagocytosis, and Aβ accumulation in Alzheimer's disease

In AD transgenic mice, intraperitoneal injection of capsaicin every day for 1 month decreased Aβ plaques, which resulted in enhanced cognitive ability. CAP reduced the disease's signs and symptoms by boosting microglial cells' capacity for phagocytosis and activating autophagy (Wang et al. 2020). In a different study, it was shown that Aβ affected anabolism, catabolism, and microglia's capacity to phagocytose; this impairment was reversed when microglia cultures were given CAP (10 μM). Microglial cells were thus more effective in phagocytosing Aβ after capsaicin treatment and required the TRPV1 channels (Pasierski and Szulczyk 2022).

Dietary CAP has also been proposed to be protective against cognitive deterioration in APP/PS1 AD mice as it attenuates Aβ pathologies as well as the neurodegeneration and associated immunologic responses, by increasing non-amyloidogenic processing of APP (Wang et al. 2020). In another study, APP/PS1 mice were provided either 0.01% CAP-rich food or a conventional diet at an early age when Aβ pathologies are not present. Later, mice of 9 months were examined, when severe Aβ pathologies had started to develop, followed by initiation of cognitive impairment. Mice in the control group had an impaired capacity to learn and memorize, while the mice treated with CAP indicated a massive enhancement of memory as there was a higher number of platform area crossings. These mice demonstrated an extensive traveling distance and a greater number of rearing, as well as elevated locomotor activity and improved spatial learning ability. CAP-treated mice have also displayed significant reductions in Aβ plaque density in hippocampus and neocortex regions. Moreover, Aβ accumulation in vessel walls was found to decrease in these mice (Wang et al. 2020).

23.3.3.2 Effect of Capsaicin on Ischemia

The event of ischemia has been suggested as one of the precursors for Alzheimer's disease (Pluta 2022). There is convincing evidence that ischemia leads to increased production of Aβ in the brain. Studies have highlighted the positive effects of CAP in other animal models for ischemia, where it lowered the rate of neurodegeneration and decline in cognitive function. Pegorini et al. (2005) studied the in vivo effect of CAP on transient global cerebral ischemia models, in Mongolian gerbil. It was shown that CAP is a TRPV1 agonist and completely protects the gerbil from global cerebral ischemia. Full recovery of memory performance, spectral power, spontaneous motor activity, and hippocampal CA1 neuronal density – the traits known to be impacted by global ischemia – were used to measure the neuroprotection. Except for spontaneous motor activity, a dose-dependent influence on behavioral measures was found. This exception developed as a result of reduced hyperlocomotion induced by ischemia at all dosages. Even on the first posttreatment day, CAP alone showed high locomotor activity via anandamide release and its production enhancement. Previously, Di Marzo et al. (2001) used a comparable dose range with similar results.

23.3.3.3 Decrease in Oxidative Stress by Capsaicin

During Alzheimer's disease, an increase in oxidative stress has been observed, indicated by oxidation of protein, lipid, DNA, glycoxidation, etc. (Meng-Zhen et al. 2022). CAP is proposed to help in such situations as it is known to be a good antioxidative agent as indicated by a study by Dairam et al. (2008) and Sirin et al. (2023). They analyzed the DPPH (1,1-diphenyl-2-picrylhyrazyl) radical scavenging activity of CAP, and it was shown that CAP was highly potent in scavenging DPPH free radicals. The study also showed CAP to be more effective in scavenging than well-known antioxidants glutathione and melatonin.

The study by Ali et al. (2018) showed the effect of capsaicin on heat-induced oxidative stress in 48 growing Animal Production Research Institute (APRI) rabbits. The oxidative stress was measured for 7 weeks and it was found that the neurochemistry of the brain tissue did get significantly affected. But in rabbits that underwent capsaicin treatment, lipid peroxidation contents drastically decreased and glutathione and activities of antioxidant enzymes significantly increased. Also, neurotransmitter contents seemed to be decreased in the control group, whereas because of capsaicin, it was increased in the treated group. So it can be suggested that, when introduced in the right amount and dosage, CAP can act as a protector against oxidative stress induced by heat in the brain tissues.

23.3.3.4 Decrease in Inflammation by Capsaicin

Alzheimer's is primarily caused by inflammation in the CNS, mostly induced by Aβ, oxidative stress, or glial activation. It further results in loss of control involving a network of neuronal cells (Tamagno et al. 2021). Wang et al. (2020) demonstrated that neuroinflammation was significantly attenuated in the APP/PS1 rat model. They had shown that CAP-treated rats showed lower microglia and astrocyte activation and there was depreciation in IFN-γ, TNF-α, and IL-6 levels. The significant inhibition of neuroinflammatory responses in stimulated microglial cells indicated that capsaicin is a potential therapeutic agent that may benefit novel drug development to prevent and treat neuroinflammatory diseases.

Overproduction of ROS in brain tissue raises oxidative stress and inflammation, which in turn leads to the nuclear factor kappa-B (NF-κB) signaling pathway activation to produce cytokines that are proinflammatory such as nitric oxide synthase, tumor necrosis factor-alpha (TNF-α), and interleukin 6 (IL-6) (Sivandzade et al. 2019). Neuronal inflammation is promoted by these cytokines and matrix metalloproteinase-9 (MMP-9), which lead to the extracellular matrix breakdown and direct degradation of tight junction (TJ) proteins, hence leading to damage of blood-brain barrier (BBB) (Rempe et al. 2016).

The anti-inflammatory nuclear-related factor-2 (Nrf2) pathway is associated with antioxidant-associated inhibition of oxidative stress and shows anti-inflammation response (Saha et al. 2020). CAP and dihydrocapsaicin (DHP) have been shown to induce the Nrf2 pathway (Janyou et al. 2017). Janyou et al. showed that CAP and DHP prevent morphology changes of neurons that occur due to neuroinflammation and reduce apoptotic cell death. It was also shown that DHC protects BBB from disruption by increasing expression of TJ proteins (Janyou et al. 2017).

23.4 Disadvantages of Using Capsaicin as a Therapeutic Agent

Capsaicin is known to cause burn and irritation when eaten or applied as it induces calcitonin gene-related peptide (CGRP) and P(SP) release from peripheral and central nerve terminals that contribute to neurogenic inflammation (Fattori et al. 2016). It can induce burning sensations that are not pleasant for the patient. Studies on patients with neuropathic pain have shown that in topical treatment of purine nucleoside phosphorylase (PNP), CAP shows moderate efficacy (Blair 2018). It has been shown that there is a requirement for repetitive application of CAP several times per day and the initial burning sensation caused at each application was found to be unbearable by the patient (Bonezzi et al. 2020). Experiments by Kim et al. (2019) showed that on delay of application of CAP in rat models of PD, the neuroprotective effects were reversed (Kim et al. 2019). This is another disadvantage of capsaicin that is unable to produce long-lasting effects and is needed to be administered at regular intervals to maintain positive results. In addition to the known advantages of CAP in AD treatment, the pro-amyloidogenic characteristics of CAP have also been reported in a study by Grimm et al. (2020). The study involved treatment of cancerous neurons with 10 μM CAP and increased Aβ levels were recorded. Further analysis revealed that BACE1 was upregulated, which resulted in enhanced activity of β-secretase. Similarly, activity of γ-secretase was also found to be increased (Grimm et al. 2020). This contradictory study suggests that certain properties of CAP may limit its role as potential therapeutics in AD.

23.5 Conclusions

Various studies on animal models have suggested that capsaicin can be used for the amelioration of neurodegenerative disorders such as PD, AD, and many other neural diseases (Pasierski and Szulczyk 2022). In the case of PD, capsaicin leads to various positive effects in animal models broadly in the form of decreased oxidative stress, decreased proinflammatory responses, and increased CNTF. CAP leads to a decrease in activation of microglia, which helped counter oxidative stress and inflammation associated with microglial-derived proinflammatory and oxidative stress-inducing factors. CAP also showed amelioration of α-syn-induced PD symptoms through decreased microglial α-syn tolerance, thus improving defected energy metabolism. Cap has also been shown to restore mitophagy in microglial cells that are affected during PD in a rat model. All this helps to prevent DA neuron degradation in the striatum and SN and alleviates PD symptoms by improving motor abilities in the models. Mitophagy is disrupted in the case of PD, which was shown to be ameliorated by CAP in animal models.

AD is characterized by Aβ plaque deposition, which in the body leads to the pathogenesis of AD. CAP could induce autophagy and increased phagocytosis in microglial culture, which was effective in removing Aβ from the culture. Autophagy effectively decreased Aβ after CAP treatment leading to enhanced cognitive ability in mice. Similar to the case of PD, CAP showed antioxidant and anti-inflammatory properties in AD animal models also improving AD symptoms. CAP demonstrated reduced astrocyte and microglial activation, as well as a drop in proinflammatory factors like TNF-α, IFN-γ, and IL-6. As CAP treatment showed considerable reduction of neuroinflammatory responses in activated microglial cells, it can be considered effective in developing drugs for preventing and alleviating neuroinflammation. As can be understood from the above discussion, there have been extensive studies on capsaicin with in vivo and in vitro models of PD and AD and have demonstrated major success in ameliorating the symptoms, but in case of its application for humans, the research is still scarce and requires major investigations. Besides this, application of capsaicin in various forms causes unbearable pain and burns and requires multiple applications in a day. This makes the regular use of CAP inconvenient at this point of time. The aforementioned facts suggest that it is imperative to find a safe way to apply/use capsaicin for various treatments, without causing pain and burns.

References

Ali K, Hassan W, El-Gamal A, Hassanien M, El-Aasar T, Mosalm G (2018) Protective effect of capsaicin against brain oxidative stress induced by heat stress in APRI rabbits. Egypt J Rabbit Sci 28:1–22. https://doi.org/10.21608/ejrs.2018.44321

Amruthraj NJ, Preetam Raj JP, Lebel A (2014) Capsaicin-capped silver nanoparticles: its kinetics, characterization and biocompatibility assay. Appl Nanosci 5:403–409. https://doi.org/10.1007/s13204-014-0330-5

Aratani Y (2018) Myeloperoxidase: its role for host defense, inflammation, and neutrophil function. Arch Biochem Biophys 640:47–52. https://doi.org/10.1016/j.abb.2018.01.004

Arora V, Campbell JN, Chung M-K (2021) Fight fire with fire: neurobiology of capsaicin-induced analgesia for chronic pain. Pharmacol Ther 220:107743. https://doi.org/10.1016/j.pharmthera.2020.107743

Augusto-Oliveira M, Arrifano GP, Lopes-Araújo A, Santos-Sacramento L, Takeda PY, Anthony DC, Malva JO, Crespo-Lopez ME (2019) What do microglia really do in a healthy adult brain? Cell 8:1293. https://doi.org/10.3390/cells8101293

Baek JY, Jeong JY, Kim KI, Won SY, Chung YC, Nam JH, Cho EJ, Ahn TB, Bok E, Shin WH, Jin BK (2018) Inhibition of microglia-derived oxidative stress by ciliary neurotrophic factor protects dopamine neurons in vivo from MPP^+ neurotoxicity. Int J Mol Sci 19:3543. https://doi.org/10.3390/ijms19113543

Bakota L, Brandt R (2016) Tau biology and tau-directed therapies for Alzheimer's disease. Drugs 76:301–313. https://doi.org/10.1007/s40265-015-0529-0

Baluchnejadmojarad T, Roghani M, Nadoushan MRJ, Bagheri M (2009) Neuroprotective effect of genistein in 6-hydroxydopamine hemi-parkinsonian rat model. Phytother Res 23:132–135. https://doi.org/10.1002/ptr.2564

Basith S, Cui M, Hong S, Choi S (2016) Harnessing the therapeutic potential of capsaicin and its analogues in pain and other diseases. Molecules 21:966

Blair HA (2018) Capsaicin 8% dermal patch: a review in peripheral neuropathic pain. Drugs 78: 1489–1500. https://doi.org/10.1007/s40265-018-0982-7

Bok E, Chung YC, Kim KS, Baik HH, Shin WH, Jin BK (2018) Modulation of M1/M2 polarization by capsaicin contributes to the survival of dopaminergic neurons in the lipopolysaccharide-lesioned substantia nigra in vivo. Exp Mol Med 50:1–14. https://doi.org/10.1038/s12276-018-0111-4

Bonezzi C, Costantini A, Cruccu G, Fornasari DMM, Guardamagna V, Palmieri V, Polati E, Zini P, Dickenson AH (2020) Capsaicin 8% dermal patch in clinical practice: an expert opinion. Expert Opin Pharmacother 21:1377–1387

Breijyeh Z, Karaman R (2020) Comprehensive review on Alzheimer's disease: causes and treatment. Molecules 25:5789. https://doi.org/10.3390/molecules25245789

Campbell BK, Fillingim RB, Lee S, Brao R, Price DD, Neubert JK (2016) Effects of high-dose capsaicin on TMD subjects. JDR Clin Trans Res 2:58–65. https://doi.org/10.1177/2380084416675837

Chang A, Rosani A, Quick J (2022) Capsaicin. In: StatPearls. StatPearls Publishing. Available via NCBI: https://www.ncbi.nlm.nih.gov/books/NBK459168/. Accessed Jan 2023

Charlton CG, Mack J (1994) Substantia nigra degeneration and tyrosine hydroxylase depletion caused by excess S-adenosylmethionine in the rat brain. Support for an excess methylation hypothesis for parkinsonism. Mol Neurobiol 9(1–3):149–161. https://doi.org/10.1007/BF02816115

Che Y, Hou L, Sun F, Zhang C, Liu X, Piao F, Zhang D, Li H, Wang Q (2018) Taurine protects dopaminergic neurons in a mouse Parkinson's disease model through inhibition of microglial M1 polarization. Cell Death Dis 9:35. https://doi.org/10.1038/s41419-018-0468-2

Chen X, Pan W (2015) The treatment strategies for neurodegenerative diseases by integrative medicine. Integr Med Int 1:223–225. https://doi.org/10.1159/000381546

Chung YC, Baek JY, Kim SR, Ko HW, Bok E, Shin WH, Won SY, Jin BK (2017) Capsaicin prevents degeneration of dopamine neurons by inhibiting glial activation and oxidative stress in the MPTP model of Parkinson's disease. Exp Mol Med 49:e298. https://doi.org/10.1038/emm.2016.159

Dairam A, Fogel R, Daya S, Limson JL (2008) Antioxidant and iron-binding properties of curcumin, capsaicin, and *S*-allylcysteine reduce oxidative stress in rat brain homogenate. J Agric Food Chem 56:3350–3356. https://doi.org/10.1021/jf0734931

DeMaagd G, Philip A (2015) Parkinson's disease and its management: part 1: disease entity, risk factors, pathophysiology, clinical presentation, and diagnosis. P T 40:504–532

Di Marzo V, Lastres-Becker I, Bisogno T, De Petrocellis L, Milone A, Davis JB, Fernandez-Ruiz JJ (2001) Hypolocomotor effects in rats of capsaicin and two long chain capsaicin homologues. Eur J Pharmacol 420:123–131. https://doi.org/10.1016/s0014-2999(01)01012-3

Du Y, Fu M, Huang Z, Tian X, Li J, Pang Y, Song W, Tian Wang Y, Dong Z (2020) TRPV1 activation alleviates cognitive and synaptic plasticity impairments through inhibiting AMPAR endocytosis in APP23/PS45 mouse model of Alzheimer's disease. Aging Cell 19:e13113. https://doi.org/10.1111/acel.13113

Durães F, Pinto M, Sousa E (2018) Old drugs as new treatments for neurodegenerative diseases. Pharmaceuticals (Basel) 11(2):44. https://doi.org/10.3390/ph11020044

Eastman CL, D'Ambrosio R, Ganesh T (2020) Modulating neuroinflammation and oxidative stress to prevent epilepsy and improve outcomes after traumatic brain injury. Neuropharmacology 172:107907. https://doi.org/10.1016/j.neuropharm.2019.107907

Emmady PD, Tadi P (2022) Dementia. In: StatPearls. StatPearls Publishing. Available via NCBI: https://www.ncbi.nlm.nih.gov/books/NBK557444/. Accessed Jan 2023

Fattori V, Hohmann MSN, Rossaneis AC, Pinho-Ribeiro FA, Verri WA (2016) Capsaicin: current understanding of its mechanisms and therapy of pain and other pre-clinical and clinical uses. Molecules 21:844. https://doi.org/10.3390/molecules21070844

Ferreira-Vieira TH, Guimaraes IM, Silva FR, Ribeiro FM (2016) Alzheimer's disease: targeting the cholinergic system. Curr Neuropharmacol 14:101–115. https://doi.org/10.2174/1570159X13666150716165726

Filipczak-Bryniarska I, Krzyzewski RM, Kucharz J, Michalowska-Kaczmarczyk A, Kleja J, Woron J, Strzepek K, Kazior L, Wordliczek J, Grodzicki T, Krzemieniecki K (2017) High-dose 8% capsaicin patch in treatment of chemotherapy-induced peripheral neuropathy: single-center experience. Med Oncol 34:162. https://doi.org/10.1007/s12032-017-1015-1

Fisher Y, Nemirovsky A, Baron R, Monsonego A (2010) T cells specifically targeted to amyloid plaques enhance plaque clearance in a mouse model of Alzheimer's disease. PLoS One 5: e10830. https://doi.org/10.1371/journal.pone.0010830

Ge P, Dawson VL, Dawson TM (2020) PINK1 and Parkin mitochondrial quality control: a source of regional vulnerability in Parkinson's disease. Mol Neurodegener 15:1–11. https://doi.org/10.1186/s13024-020-00367-7

Geula C, Nagykery N, Nicholas A, Wu C-K (2008) Cholinergic neuronal and axonal abnormalities are present early in aging and in Alzheimer disease. J Neuropathol Exp 67:309–318. https://doi.org/10.1097/NEN.0b013e31816a1df3

Gómez-Benito M, Granado N, García-Sanz P, Michel A, Dumoulin M, Moratalla R (2020) Modeling Parkinson's disease with the alpha-synuclein protein. Front Pharmacol 11:356. https://doi.org/10.3389/fphar.2020.00356

Götz J, Chen F, van Dorpe J, Nitsch RM (2001) Formation of neurofibrillary tangles in P301l tau transgenic mice induced by Abeta 42 fibrils. Science 293:1491–1495. https://doi.org/10.1126/science.1062097

Grimm MOW, Blümel T, Lauer AA, Janitschke D, Stahlmann C, Mett J, Haupenthal VJ, Miederer A-M, Niemeyer BA, Grimm HS, Hartmann T (2020) The impact of capsaicinoids on APP processing in Alzheimer's disease in SH-SY5Y cells. Sci Rep 10:9164. https://doi.org/10.1038/s41598-020-66009-6

Hoffmann S, Beyer C (2020) A fatal alliance between microglia, inflammasomes, and central pain. Int J Mol Sci 21:3764. https://doi.org/10.3390/ijms21113764

Hong S, Beja-Glasser VF, Nfonoyim BM, Frouin A, Li S, Ramakrishnan S, Merry KM, Shi Q, Rosenthal A, Barres BA, Lemere CA, Selkoe DJ, Stevens B (2016) Complement and microglia mediate early synapse loss in Alzheimer mouse models. Science 352:712–716. https://doi.org/10.1126/science.aad8373

Hou L, Wang K, Zhang C, Sun F, Che Y, Zhao X, Zhang D, Li H, Wang Q (2018) Complement receptor 3 mediates NADPH oxidase activation and dopaminergic neurodegeneration through a Src-Erk-dependent pathway. Redox Biol 14:250–260. https://doi.org/10.1016/j.redox.2017.09.017

Ilie MA, Caruntu C, Tampa M, Georgescu S, Matei C, Negrei C, Ion RM, Constantin C, Neagu M, Boda D (2019) Capsaicin: physicochemical properties, cutaneous reactions and potential applications in painful and inflammatory conditions (review). Exp Ther Med 18:916–925. https://doi.org/10.3892/etm.2019.7513

Janyou A, Wicha P, Jittiwat J, Suksamrarn A, Tocharus C, Tocharus J (2017) Dihydrocapsaicin attenuates blood brain barrier and cerebral damage in focal cerebral ischemia/reperfusion via oxidative stress and inflammatory. Sci Rep 7:10556. https://doi.org/10.1038/s41598-017-11181-5

Kim KI, Baek JY, Jeong JY, Nam JH, Park ES, Bok E, Shin W-H, Chung YC, Jin BK (2019) Delayed treatment of capsaicin produces partial motor recovery by enhancing dopamine function in MPP+-lesioned rats via ciliary neurotrophic factor. Exp Neurobiol 28:289–299. https://doi.org/10.5607/en.2019.28.2.289

Klann EM, Dissanayake U, Gurrala A, Farrer M, Shukla AW, Ramirez-Zamora A, Mai V, Vedam-Mai V (2022) The gut-brain axis and its relation to Parkinson's disease: a review. Front Aging Neurosci 13:782082. https://doi.org/10.3389/fnagi.2021.782082

Kouli A, Torsney KM, Kuan W-L (2018) Parkinson's disease: etiology, neuropathology, and pathogenesis. Exon Publications, Brisbane. https://doi.org/10.15586/codonpublications.parkinsonsdisease.2018.ch1

Lecours C, Bordeleau M, Cantin L, Parent M, Paolo TD, Tremblay M-È (2018) Microglial implication in Parkinson's disease: loss of beneficial physiological roles or gain of inflammatory functions? Front Cell Neurosci 12:340. https://doi.org/10.3389/fncel.2018.00282

Leeuw S, Tackenberg C (2019) Alzheimer's in a dish—induced pluripotent stem cell-based disease modeling. Transl Neurodegener 8:1–13. https://doi.org/10.1186/s40035-019-0161-0

Lipshetz B, Khasabov SG, Truong H, Netoff TI, Simone DA, Giesler GJ (2018) Responses of thalamic neurons to itch- and pain-producing stimuli in rats. J Neurophysiol 120:1119–1134. https://doi.org/10.1152/jn.00264.2018

Liu J, Liu W, Li R, Yang H (2019) Mitophagy in Parkinson's disease: from pathogenesis to treatment. Cell 8:712. https://doi.org/10.3390/cells8070712

Liu J, Liu H, Zhao Z, Wang J, Guo D, Liu Y (2020) Regulation of Actg1 and Gsta2 is possible mechanism by which capsaicin alleviates apoptosis in cell model of 6-OHDA-induced Parkinson's disease. Biosci Rep 40:BSR20191796. https://doi.org/10.1042/BSR20191796

Lu H, Leung H-T, Wang N, Pak WL, Shieh B-H (2009) Role of Ca^{2+}/calmodulin-dependent protein kinase II in drosophila photoreceptors. J Biol Chem 284:11100–11109. https://doi.org/10.1074/jbc.M806956200

Lu M, Chen C, Lan Y, Xiao J, Li R, Huang J, Huang Q, Cao Y, Ho C-T (2020) Capsaicin—the major bioactive ingredient of chili peppers: bio-efficacy and delivery systems. Food Funct 11: 2848–2860. https://doi.org/10.1039/d0fo00351d

Lu J, Wang C, Cheng X, Wang R, Yan X, He P, Chen H, Yu Z (2022) A breakdown in microglial metabolic reprogramming causes internalization dysfunction of α-synuclein in a mouse model of Parkinson's disease. J Neuroinflammation 19:1–21. https://doi.org/10.1186/s12974-022-02484-0

Mavroeidi P, Xilouri M (2021) Neurons and glia interplay in α-synucleinopathies. Int J Mol Sci 22: 4994. https://doi.org/10.3390/ijms22094994

Menegas W, Bergan JF, Ogawa SK, Isogai Y, Umadevi Venkataraju K, Osten P, Uchida N, Watabe-Uchida M (2015) Dopamine neurons projecting to the posterior striatum form an anatomically distinct subclass. Elife 4:e10032. https://doi.org/10.7554/eLife.10032

Meng-Zhen S, Ju L, Lan-Chun Z, Cai-Feng D, Shu-da Y, Hao-Fei Y, Wei-Yan H (2022) Potential therapeutic use of plant flavonoids in AD and PD. Heliyon 8:e11440. https://doi.org/10.1016/j.heliyon.2022.e11440

Miraglia F, Colla E (2019) Microbiome, Parkinson's disease and molecular mimicry. Cell 8:222. (1–16). https://doi.org/10.3390/cells8030222

Moon HE, Paek SH (2015) Mitochondrial dysfunction in Parkinson's disease. Exp Neurobiol 24: 103–116. https://doi.org/10.5607/en.2015.24.2.103

Muddapu VR, Dharshini SAP, Chakravarthy VS, Gromiha MM (2020) Neurodegenerative diseases—is metabolic deficiency the root cause? Front Neurosci 31:14:213. https://doi.org/10.3389/fnins.2020.00213

Muzio L, Viotti A, Martino G (2021) Microglia in neuroinflammation and neurodegeneration: from understanding to therapy. Front Neurosci 15:742065. https://doi.org/10.3389/fnins.2021.742065

Nam JH, Park ES, Won S-Y, Lee YA, Kim KI, Jeong JY, Baek JY, Cho EJ, Jin M, Chung YC, Lee BD, Kim SH, Kim E-G, Byun K, Lee B, Woo DH, Lee CJ, Kim SR, Bok E, Kim YS, Ahn TB, Ko HW, Brahmachari S, Pletinkova O, Troconso JC, Dawson VL, Dawson TM, Jin BK (2015) TRPV1 on astrocytes rescues nigral dopamine neurons in Parkinson's disease via CNTF. Brain 138:3610–3622. https://doi.org/10.1093/brain/awv297

Pajares MI, Rojo A, Manda G, Boscá L, Cuadrado A (2020) Inflammation in Parkinson's disease: mechanisms and therapeutic implications. Cells 9:1687. https://doi.org/10.3390/cells9071687

Pasierski M, Szulczyk B (2022) Beneficial effects of capsaicin in disorders of the central nervous system. Molecules 27:2484. https://doi.org/10.3390/molecules27082484

Pegorini S, Braida D, Verzoni C, Guerini-Rocco C, Consalez GG, Croci L, Sala M (2005) Capsaicin exhibits neuroprotective effects in a model of transient global cerebral ischemia in Mongolian gerbils. Br J Pharmacol 144:727–735. https://doi.org/10.1038/sj.bjp.0706115

Perumal P, Hua Chong DY (2021) Pharmacological evaluation of green chili in Alzheimer's disease. Pharm Res 14:61–70. https://doi.org/10.5530/pres.14.1.10

Peterson LJ, Flood PM (2012) Oxidative stress and microglial cells in Parkinson's disease. Mediat Inflamm 2012:1–12. https://doi.org/10.1155/2012/401264

Pluta R (2022) Alzheimer's disease connected genes in the post-ischemic hippocampus and temporal cortex. Genes 13:1059. https://doi.org/10.3390/genes13061059

Qu W, Yuan B, Liu J, Liu Q, Zhang X, Cui R, Yang W, Li B (2020) Emerging role of AMPA receptor subunit GluA1 in synaptic plasticity: implications for Alzheimer's disease. Cell Prolif 54:e12959. (1–13). https://doi.org/10.1111/cpr.12959

Rempe RG, Hartz AMS, Bauer B (2016) Matrix metalloproteinases in the brain and blood-brain barrier: versatile breakers and makers. J Cereb Blood Flow Metab 36:1481–1507. https://doi.org/10.1177/0271678X16655551

Saha S, Buttari B, Panieri E, Profumo E, Saso L (2020) An overview of Nrf2 signaling pathway and its role in inflammation. Molecules 25:5474. https://doi.org/10.3390/molecules25225474

Sanabria-Castro A, Alvarado-Echeverría I, Monge-Bonilla C (2017) Molecular pathogenesis of Alzheimer's disease: an update. Ann Neurosci 24:46–54. https://doi.org/10.1159/000464422

Siddique YH, Naz F, Jyoti S (2018) Effect of capsaicin on the oxidative stress and dopamine content in the transgenic drosophila model of Parkinson's disease. Acta Biol Hung 69:115–124. https://doi.org/10.1556/018.69.2018.2.1

Sirin S, Nigdelioglu Dolanbay S, Aslim B (2023) Role of plant derived alkaloids as antioxidant agents for neurodegenerative diseases. Health Sci Rev 6:1–11. https://doi.org/10.1016/j.hsr.2022.100071

Sivandzade F, Prasad S, Bhalerao A, Cucullo L (2019) Nrf2 and NF-κB interplay in cerebrovascular and neurodegenerative disorders: molecular mechanisms and possible therapeutic approaches. Redox Biol 21:101059. https://doi.org/10.1016/j.redox.2018.11.017

Španić E, Langer Horvat L, Hof PR, Šimić G (2019) Role of microglial cells in Alzheimer's disease tau propagation. Front Aging Neurosci 11:271. https://doi.org/10.3389/fnagi.2019.00271

Spillantini MG, Schmidt ML, Lee VM-Y et al (1997) α-synuclein in Lewy bodies. Nature 388:839–840. https://doi.org/10.1038/42166

Stefanis L (2012) α Synuclein in Parkinson's disease. Cold Spring Harb Perspect Med 2:a009399 https://doi.org/10.1101/cshperspect.a009399

Stilling RM, Dinan TG, Cryan JF (2014) Microbial genes, brain & behaviour—epigenetic regulation of the gut-brain axis. Genes Brain Behav 13:69–86. https://doi.org/10.1111/gbb.12109

Tabaton M, Piccini A (2005) Role of water-soluble amyloid-β in the pathogenesis of Alzheimer's disease. Int J Exp Pathol 86:139–145. https://doi.org/10.1111/j.0959-9673.2005.00428.x

Tamagno E, Guglielmotto M, Vasciaveo V, Tabaton M (2021) Oxidative stress and beta amyloid in Alzheimer's disease. Which comes first: the chicken or the egg? Antioxidants 10:1479. https://doi.org/10.3390/antiox10091479

Teunissen CE, Chiu M-J, Yang C-C et al (2018) Plasma amyloid-β (Aβ42) correlates with cerebrospinal fluid Aβ42 in Alzheimer's disease. J Alzheimers Dis 62:1857–1863. https://doi.org/10.3233/jad-170784

Trojanowski JQ, Goedert M, Iwatsubo T, Lee VM-Y (1998) Fatal attractions: abnormal protein aggregation and neuron death in Parkinson's disease and Lewy body dementia. Cell Death Differ 5:832–837. https://doi.org/10.1038/sj.cdd.4400432

Váradi C (2020) Clinical features of Parkinson's disease: the evolution of critical symptoms. Biology 9:103. https://doi.org/10.3390/biology9050103

Wang H-Y, Li W, Benedetti NJ, Lee DHS (2003) A7 nicotinic acetylcholine receptors mediate β-amyloid peptide-induced tau protein phosphorylation. J Biol Chem 278:31547–31553. https://doi.org/10.1074/jbc.m212532200

Wang X, Kim J-R, Lee S-B et al (2014) Effects of curcuminoids identified in rhizomes of Curcuma longa on BACE-1 inhibitory and behavioral activity and lifespan of Alzheimer's disease drosophila models. BMC Complement Altern Med 14:1–14. https://doi.org/10.1186/1472-6882-14-88

Wang Q, Qian L, Chen S-H et al (2015a) Post-treatment with an ultra-low dose of NADPH oxidase inhibitor diphenyleneiodonium attenuates disease progression in multiple Parkinson's disease models. Brain 138:1247–1262. https://doi.org/10.1093/brain/awv034

Wang W-Y, Tan M-S, Yu J-T, Tan L (2015b) Role of pro-inflammatory cytokines released from microglia in Alzheimer's disease. Ann Transl Med 3:136. https://doi.org/10.3978/j.issn.2305-5839.2015.03.49

Wang J, Sun B-L, Xiang Y et al (2020) Capsaicin consumption reduces brain amyloid-beta generation and attenuates Alzheimer's disease-type pathology and cognitive deficits in app/PS1 mice. Transl Psychiatry 10:230. https://doi.org/10.1038/s41398-020-00918-y

Wi R, Chung YC, Jin BK, Duan L (2020) Functional crosstalk between CB and TRPV1 receptors protects nigrostriatal dopaminergic neurons in the MPTP model of Parkinson's disease. J Immunol Res 2020:1–11. https://doi.org/10.1155/2020/5093493

Zhang C, Hou L, Yang J et al (2018a) 2,5-hexanedione induces dopaminergic neurodegeneration through integrin AMB2/NADPH oxidase axis-mediated microglial activation. Cell Death Dis 9: 60. https://doi.org/10.1038/s41419-017-0091-7

Zhang Y, Guo O, Huo Y et al (2018b) Amyloid-β induces AMPA receptor ubiquitination and degradation in primary neurons and human brains of Alzheimer's disease. J Alzheimers Dis 62: 1789–1801. https://doi.org/10.3233/jad-170879

Zhao ZX, Wang JF, Wang LL et al (2017) Capsaicin protects against oxidative insults and alleviates behavioral deficits in rats with 6-OHDA-induced Parkinson's disease via activation of TRPV1. Neurochem Res 42:3431–3438. https://doi.org/10.1007/s11064-017-2388-4

Antimicrobial Activity of Capsaicin and Its Derivatives

24

Anjani Saxena, Nitin Puranik, Roshani Kumari, and Sandeep Kumar Verma

Abstract

Chilies contain a special group of alkaloids called capsaicinoids, which are synthesized as secondary metabolites and are responsible for the pungency. Capsaicin and dihydrocapsaicin are the major components of capsaicinoids and generally account for 90% of the pungency of the chilies. Capsaicin acts as a bioactive component of the chili pepper and has the antimicrobial and several other pharmacological properties, such as gastrointestinal effects, pain relief, and beneficial cardiovascular, and cancer prevention. In this chapter, we present antimicrobial activities of different varieties of chili pepper, such as carolina reaper, brain strain yellow, bhut jolokia red, orange blaze, red majesty sweet, and red ruffle pimento. Previously published studies suggest that capsaicin and its related derivatives possesses potent antibacterial, against a variety of Gram-negative and Gram-positive bacterial pathogens, such as *Salmonella paratyphi*, *Escherichia coli*, *Staphylococcus aureus*, and *Streptococcus pyogenes*. Capsaicin also possesses antifungal activity against clinically important fungal pathogens like *Aspergillus parasiticus*, *Candida albicans*, *C. krusei*, *C. parapsilosis*, and *C. tropicalis*. Besides antimicrobial potential, reports also indicate the antihelminthic activity of capsaicin against *Tubifex tubifex*, Indian earthworm, and intestinal roundworm and nematicidal activity against *Meloidogyne javanica* and *Nacobbus* spp.

A. Saxena · S. K. Verma (✉)
Institute of Biological Science, SAGE University, Indore, Madhya Pradesh, India
e-mail: hod.bios@sageuniversity.in

N. Puranik · R. Kumari
ICMR-National Institute for Research in Environmental Health, Bhopal, Madhya Pradesh, India

M. K. Swamy, A. Kumar (eds.), *Capsaicinoids*,
https://doi.org/10.1007/978-981-99-7779-6_24

Keywords

Antimicrobial activity · Capsaicin · *Capsicum annuum* L. · Capsaicinoids · Derivatives · Secondary metabolites

24.1 Introduction

Chilies (*Capsicum annuum* L.), belonging to the Solanaceae family, are commonly used as a vegetable throughout the world. More than 400 varieties of chilies are grown in the entire world. India is the largest production of chilies among other countries, such as China, Thailand, Ethiopia, and Indonesia. Chili is a distinct and widely used spice globally, and it is well recognized for its fiery, pungent flavor (Pawar et al. 2011; Ananthan et al. 2014; Katz 2020; Karim et al. 2021). Although many important phytochemical compounds are found in chili including flavonoids (luteolin, quercetin, myricetin, catechin, kaempferol, apigenin, flavones), phenolic compounds (chlorogenicacis and gallic acid), vitamins (A, C, K, B1, and B6), minerals (phosphorous, zinc, calcium, iron, copper, selenium, and manganese), essential amino acids (linolenic acid, methionine, tryptophan, cysteine, lysine, and phenylalanine), carotenoid content (α-carotene, β-carotene, zeaxanthin, lutein, keto-carotenoid, capsorubin, capsanthin, cryptocapsin, violaxanthin), capsaicin is one of the main components, which is found in the highest quantity in comparison to other compounds (Khan et al. 2014; Sricharoen et al. 2017; Olatunji and Afolayan 2018; Saleh et al. 2018; Salehi et al. 2018; Sachan et al. 2018; Scaldaferro and Moscone 2019; Vázquez-Fresno et al. 2019; Cortés-Ferré et al. 2021). Capsaicin and dihydrocapsaicin, a compound from the chemical group capsaicinoids, are responsible for the pungent nature of chili as well as its many pharmacological and antimicrobial activities (Aza-González et al. 2011; Naves et al. 2019; Tewksbury and Nabhan 2001). It is also a potential agent in pain relief since it binds with the pain receptors and produces a strong burning sensation (O'Neill et al. 2012; Nagy et al. 2014; Srinivasan 2016; Moran and Szallasi 2018; Hall et al. 2020). Chili in addition contains capsanthin, a pigment that is known to pose strong antioxidant potential and gives strong red color to the ripened fruits of chili (del Rocío Gómez-García and Ochoa-Alejo 2013; Hernández-Pérez et al. 2020; Mohd Hassan et al. 2019). Studies revealed that capsaicin and dihydrocapsaicin exert potent antihelminthic, antibacterial, and antifungal properties against several clinically important microbes, such as *Salmonella* spp., *Staphylococcus* spp., *Candida* spp., *Aspergillus* spp., *Tubifex tubifex*, *Ascaris lumbricoides*, *Meloidogyne javanica*, and *Nacobbus* spp. (Lim 2013; Pandey et al. 2012; Mbaebie Oyedemi 2015; Parvez 2017). The present chapter provides an insight of capsaicin from natural and synthetic origin, their structural features, and their various antimicrobial roles.

24.2 Capsaicinoids: A Brief Account

Capsaicinoids are the active metabolites occurring in the members of capsicum family. Some of the major capsaicinoids including capsaicin, nordihydrocapsaicin, dihydrocapsaicin, homocapsaicin, and homodihydrocapsaicin are mostly found in chili peppers (Sanatombi and Sharma 2008; González-Zamora et al. 2013; Guillen et al. 2018; Santos et al. 2020; Jaganathan et al. 2020). Capsaicinoids have a molecular weight of 305.4 Da and a melting temperature of 60–65 °C, and it is a pungent, highly volatile, hydrophobic, colorless, and odorless white crystalline powder. An acid amide of vanillyl amine and branched-chain fatty acids with lengths ranging from 7 to 13 carbons make up the fundamental chemical structure of these molecules. The length of the acyl chain and the benzene ring both play a role in how pungent they are (Tanaka et al. 2009; Gonzalez-Mondragon and Vazquez-Tzompantzi 2011; Reyes-Escogido et al. 2011). Despite the fact that capsaicin and dihydeocapsaicin comprise about 90% of capsaicinoids, the remaining amounts are made up of lower amounts of nordihydeocapsaicin, homocapsaicin, and homodihydrocapsaicin (Rossa et al. 2002; Hamed et al. 2019). A related class of substances called capsinoids, which are non-pungency compunds, also have the beneficial characteristics of capsaicinoids. They trigger receptors in the intestine but not the mouth because of their slightly different structural makeup (Luo et al. 2011; Whiting et al. 2012; Yoneshiro et al. 2012) given that many people find it difficult to swallow capsaicinoids because of their "spiciness"; this quality is especially important. Capsinoids, as opposed to capsaicinoids, are vanillyl alcohol esters containing fatty acids that resemble those of capsaicinoids. This structural difference may possibly be the cause of capsinoids' reduced stability and 1000-fold lower pungency than capsaicinoids (Kobata et al. 2002; Sutoh et al. 2006; Gupta et al. 2022).

Tewksbury et al. (2006, 2008) demonstrated that capsaicinoids are organic substances that are resembled to plant defense systems against fungus and mammalian herbivores. Studies of Stewart Jr et al. (2007) revealed that genetics play a significant role in determining the amount of capsaicin present in different capsicum fruit varieties. Total capsaicin concentrations in cayenne (*Capsicum annuum*) cultivars range from 720 to 4360 ppm, Jalapeno (*C. annuum jalapeno*) cultivars from 1110 to 7260 ppm, and habanero (*C. chinense)* cultivars from 5200. The application of bioregulators, fertilization techniques, environmental growing circumstances, fruit maturity, and genetics are other factors that affect capsaicinoid content (Blum et al. 2003; Stellari et al. 2010; Hill et al. 2013).

The pharmacological and physiological effects of capsaicinoids were previously reviewed by Donnerer et al. (1990). Surh et al. (1995) revealed their analgesic, anti-obesity, anticancer, anti-inflammatory, and antioxidant potential. The prevention of cancer, the gastrointestinal and cardiovascular systems, pain reduction, and weight loss may all benefit from their powerful medicinal potential (Maru et al. 2016; Antonious 2018).

Since ancient times, humans have consumed capsaicinoid, a natural and processed vegetable chemical, without any known negative health effects (Downham and Collins 2000; Meghvansi et al. 2010; Koleva et al. 2012;

Campos-Vega and Oomah 2013). However, a number of toxicological investigations on capsaicin and combinations containing capsaicin have been conducted recently, and contradictory results have been reported (Mózsik et al. 2009; Bley et al. 2012; Díaz-Laviada and Rodríguez-Henche 2014).

The indigenous Mayan people of Mesoamerica have utilized capsicum tissues in preparations of their botanical pharmacopoeia to treat a number of illnesses caused by microbial infections (Cowan 1999; Delelegn 2011; Halberstein 2005; Hayman and Kam 2008). The inhibitory effects of the plain and hot extracts of capsicum against *Bacillus subtilis*, *B. cereus*, *Clostridium tetani*, *C. sporogenes*, and *Streptococcus pyogenes* were observed to vary in strength (Cichewicz and Thorpe 1996; Nick 2002; Sánchez et al. 2013; Doğan et al. 2018; Ekom et al. 2021). Sánchez et al. (2013) demonstrated that according to bioautographic experiments, the primary antibacterial component that is thought to damage the membrane of microorganisms is capsaicin. Studies revealed that *B. subtilis* growth was severely suppressed by high amounts of capsaicin; however *Escherichia coli* and *Pseudomonas solanacearum* growth was found to be slow (Rahman et al. 2021; Bhatti et al. 2022).

24.3 Antimicrobial Properties of Capsaicin

Capsaicin, additionally to its abundant therapeutical and physiological benefits (alleviating pain, preventing cancer, cardioprotection, etc.), has drawn more attention in recent times, because of its antibacterial, antifungal, and antiviral actions (Marini et al. 2015; Menezes et al. 2022). Table 24.1 provides a brief information on different types of antimicrobial properties of capsaicin. According to the experimental studies of Erfanimanesh et al. (2019), capsaicin showed an antiviral activity against *Vibrio cholera*. Similar studies were also conducted to demonstrate the inhibitory activity of capsaicin against *Staphylococcus aureus* and *Porphyromonas gingivalis* and a bactericidal impact against food-borne pathogens, *Helicobacter pylori* and *P. aeruginosa* (Kar et al. 2016; Goci et al. 2021; Menezes et al. 2022). Agarwal et al. (2017) demonstrated antimicrobial efficacy of capsaicin, which was found to have strong inhibitory activity against *S. paratyphi A*, *S. paratyphi B*, and *M. luteus*, with *M. luteus* displaying the greatest susceptibility; similar studies were conducted by Adaszek et al. (2019) in context of capsaicin applications in human and veterinary medicines; Xing et al. (2006) also demonstrated microbes inhibitory properties of capsaicin microcapsules. It is crucial to be aware of secondary antibacterial substances in addition to the known medicines in the current period of antidrug resistance and the advent of pathogenic microbial mutants (Ventola 2015; Chokshi et al. 2019; Singer et al. 2019). In a separate way, the presence of a shared receptor gene enables the identification of a genetic link between eukaryotes and prokaryotes that respond to capsaicin (Rosca et al. 2020; Füchtbauer et al. 2021). While evaluating antibacterial potentials, the inhibition levels against *B. cereus*, *B. subtilis*, *Clostridium sporogenes*, *C. tetani*, and *S. pyogenes* were observed to change between the pure and hot extract of capsicum species. The samples prevented

Table 24.1 Antimicrobial, antifungal, and antiviral activities of Capsaicin

Name of test organisms	Description of activity	Reference
Salmonella paratyphi A, *S. paratyphi B*, *Proteus mirabilis*, and *Micrococcus luteus* nonpathogenic bacterial strains	Capsaicin had antibacterial efficacy against all test microorganisms, with the highest activity against *Micrococcus luteus*.	Agarwal et al. (2017)
Pasteurella multocida, *Klebsiella rhinoscleromatis*, *Staphylococcus pyogenes*, and *Vibrio vulnificus*	The honey/tripolyphosphate/ chitosan mats were enhanced with capsaicin and silver nanoparticles shown antibacterial activities in in vitro and in vivo wound-healing studies.	Al-Musawi et al. (2020)
Salmonella typhi	In vitro, capsaicin extract and pure capsaicin were reported to effectively inhibit *S. typhi* invasion of the Vero cell.	Ayariga et al. (2022)
Candida albicans	Capsaicin inhibits ergosterol production in the cell wall, affecting the structure and integrity of the cells. Most notably, capsaicin increased fluconazole's antifungal activity, and this synergistic action may be advantageous in the prevention of fluconazole resistance and reduced drug dosing.	Behbehani et al. (2023)
Aspergillus parasiticus	According to antifungal and anti-aflatoxigenic experiments, *C. chinense*, *Piper nigrum*, capsaicin, capsaicin, and piperine inhibited *A. parasiticus* growth and aflatoxins synthesis in a dose-dependent manner	Buitimea-Cantúa et al. (2020)
Escherichia coli (ATCC 10536), *B. cereus* (Peru MycA 4), *S. typhi* (Peru Myc 7), *S. aureus* (ATCC 6538), and four yeasts, namely, *C. tropicalis* (YEPGA 6184), *C. albicans* (YEPGA 6379), *C. parapsilosis* (YEPG 6551), and *C. albicans* (YEPG 6138). Candida parapsilosis (ATCC 22019) and *C. krusei* (ATCC 6258).	A carbopol-based formulation containing capsaicin as an active ingredient found effective against all the test bacterial and fungal pathogens.	Goci et al. (2021)
P. aeruginosa	A range of pH-triggered polyethylene glycol/ capsaicin@chitosan, polyvinyl alcohol/capsaicin@chitosan, and alginate/capsaicin@chitosan multilayer films based on capsaicin shown long-term	Hao et al. (2020)

(continued)

Table 24.1 (continued)

Name of test organisms	Description of activity	Reference
	antibacterial activity against *Pseudomonas aeruginosa*.	
Saccharomyces cerevisiae, strain S288C (mat alpha SUC2 mal mel gal2 CUP1	Yeast DNA microarray technologies were to examine the mechanisms of capsaicin's inhibitory effects. The capsaicin treatment elevated the expression of 39 genes out of around 6000 genes considerably. These stimulated genes were divided into four categories: multidrug resistance transporter genes, membrane biosynthesis genes, stress protein genes, and unknown genes. The propensity of the strains that carry triggered genes to evolve indicates that capsaicin gets out of the yeast cells via the PDR5 transporter.	Kurita et al. (2002)
E. coli ATCC 11229	A sol-gel method was used to cover the surface of wool garments with capsaicin, a natural antibacterial agent. After 24 h of incubation at 37 °C, the capsaicin-coated fabric reduced bacterial growth significantly more than the control group (sol-gel-coated fabric without capsaicin). The antibacterial efficiency following laundry was also evaluated. The sol-gel coating approach demonstrated good resistance to washing of capsaicin on fabric.	Liu et al. (2012)
E. coli, *S. aureus*	An in vitro antimicrobial study demonstrated that the 0.10 M formulation of silver capsaicin nanoparticles has significant antibacterial activity, making it a potential antibacterial agent. It also demonstrated that the zone of inhibition in Gram-negative bacteria (*Escherichia coli*) was significantly larger than the zone of inhibition in Gram-positive bacteria (*S. aureus*).	Mahmood et al. (2021)
Streptococcus pyogenes	Capsaicin exhibited bactericidal action against group A streptococci, with the most common MIC being 128 g/mL.	Marini et al. (2015)

(continued)

Table 24.1 (continued)

Name of test organisms	Description of activity	Reference
Trametes versicolor, Gloeophyllum trabeum	A modified cellulose containing 2% capsaicin inhibited test fungus growth very effectively.	Martini et al. (2014)
Escherichia coli, P. solanacearum Pseudomonas solanacearum, Bacillus subtilis, Saccharomyces cerevisiae	High concentration of capsaicin retarded growth of *Escherichia coli, P. solanacearum*, and *B. subtilis.*	Molina-Torres et al. (1999)
Bacillus subtilis	Antibacterial characteristics of capsaicin were found against *B. subtilis.*	Stephen and Kumar (2014)
Faecalibacterium, Akkermansia, Roseburia, Helicobacter, and *Bacteroides* species	*Faecalibacterium* was the most variable of the five bacterial taxa in the capsaicin-treated mouse model.	Wang et al. (2020a)
Escherichia coli	Ferric (Fe 3+) enriched cylindrical organic metal frameworks manufactured as capsaicin nanocarriers and integrated into a gelatin/chitosan matrix to create an antibacterial packaging film demonstrated potent inhibitory activity against *Escherichia coli.*	Zhao et al. (2020)

the growth of biofilms and more completely removed the biofilm created by *C. tropicalis* (Shayan and Saeidi 2013).

Fieira et al. (2013) confirmed the antifungal effects of capsaicin on apple blue mold causing fungus *P. expansum* at both in vitro and in vivo level, where capsaicin considerably inhibits development of *P. expansum* at in vitro setup.

Jones et al. (1997) investigated the growth and development of *Helicobacter pylori* by adding different concentration of capsaicin in the media. Capsaicin was found to reduce the growth of *H. pylori* significantly, and it was dosage-dependent, and the highest inhibition was noticed at 10 g/mL dose. However, this dose failed to inhibit the growth of a human fecal commensal, *E. coli* strain. Bactericidal activity was detected after 4 h. Capsaicin retained bactericidal efficacy when incubated at pH levels as low as 5.4. Consuming chili may thus have a preventive effect against *H. pylori*-associated gastroduodenal illness. Similarly, Tayseer et al. (2020) showed that capsaicin exerts a good antibacterial activity on *H. pylori* using standard antimicrobial sensitivity evaluation approaches and titration checkerboard assay with minimum inhibitory concentration (MIC) of 0.0625 mg/mL, whereas piperine showed MIC as 0.125 mg/mL.

Capsaicin activates autophagy via the transcription factor TFEB, which is a master regulator of autophagosome biogenesis. Capsaicin induced TFEB nuclear localization and the activation of TFEB promotes the transcription of autophagosomal genes. According to Basak et al. (2002), inhibiting autophagy by silencing Atg5 and TFEB encourages the development of microorganisms. Thus, one of the main mechanisms in bacterial clearance is capsaicin-spiker 132-induced autophagy. Moreover, capsaicin prevented the *S. flexneri strain* from proliferating inside cells. Capsaicin's ability to inhibit *S. flexneri* growth was confirmed in an animal model. For the first time, this study demonstrated that *S. flexneri* infection can be prevented by triggering autophagy. Overall, the findings indicate that capsaicin stimulates TFEB in order to promote autophagy and thereby encounter *S. flexneri* infections.

Romero-Luna et al. (2022) reviewed the mechanism of antimicrobial (antibacterial and antifungal) activity of capsaicin. According to the extensive review, bioactive substances from *Capsicum* spp. fruits have antibacterial actions. Antimicrobial peptides (AMP) interact with the components of the cell wall, enhancing membrane permeability and the creation of holes. As a result, AMP enter the bacterial cell and interfere with RNA transcription, impacting protein synthesis. On the other hand, in yeast, AMP cause cell permeabilization (increased Ca^{2+} inflow and K^{+} efflux), pore development, and ergosterol structural changes, resulting in cellular capsule collapse and, eventually, cellular inhibition. AMP can enter the cytoplasm and cause the creation of reactive oxygen species (ROS), thus altering the cellular redox equilibrium.

The antibiofilm action of two phytocompounds, capsaicin and berberine, against *Staphylococcus aureus* isolates from bovine mastitis cases was examined by Mohan et al. (2022). Results revealed that berberine tested positive for antibiotic sensitivity. When coupled with quinolones, capsaicin enhanced the diameter of the zone of inhibition in a dose-dependent way. Berberine and capsaicin were found to have antibiofilm activity when combined with quinolone antibiotics in an in vitro biofilm assay utilizing Congo red agar plate technique. This investigation concluded that antibiotics along with phytoconstituents may lessen microbial resistance mechanisms.

Menezes et al. (2022) investigated the antifungal and antiparasitic activities of the pepper *C. chinense Jacq.*, seeds and peel crude extracts, and its isolated compound, capsaicin. They aimed to determine the capability of crude extract and capsaicin in inhibiting biofilm development, eradicating biofilm, and reducing hemolysin secretion by *Candida* species. The investigation showed how the samples reduced biofilm development and efficiently removed the biofilm generated by *C. tropicalis.* Sub-inhibitory doses of the samples greatly inhibited the hemolytic activity of *C. glabrata* and *C. tropicalis*. When evaluated at higher doses, the materials simply changed host cell viability; however, at nontoxic quantities, they inhibited *Toxoplasma gondii* proliferation.

24.4 Derivatives of Capsaicin: Origin and Synthetic Chemistry

Naturally, capsaicin is obtained from capsicum species; however it involves more time and expense; consequently, it is very vital to develop an easy way to synthesize capsaicin derivatives with equivalent properties and environmental friendliness (Chuichulcherm et al. 2013; Castro-Muñoz et al. 2022). Capsaicin analogues are molecules that have been shown to be TRPV1 receptor agonists by retaining all or part of the aromatic catechol ring in their chemical structure (Basith et al. 2016). Nonetheless, they may possibly or cannot have comparable effects owing to distinct functional groups and a varying number of carbon-carbon double bonds positioned at various points along the chain, which might alter its biological activity and pungency level (Morales-Martínez et al. 2016). Capsaicin derivatives have double bonds in their phenyl, phenolic hydroxyl, amide, and alkyl groups, according to structural studies. Any modification in the functional groups will impact the characteristics of the derivative of capsaicin, and the capsaicin derivative with fewer carbon atoms in the amide side chain has increased fouling resistance and decreased hydrophobicity, rendering it sensitive to deterioration (Wang et al. 2019).

With the time and progressive research, many other derivatives and analogues of capsaicin have been synthesized such as N-(2-methyl-4 hydroxy-5-methylthiobenzene) acetamide, N-(4-hydroxy-3-acetophenonebenzyl) acrylamide, 4-acetamideme thylbenzene-2-benzylphenol, N-(2,5-dihydroxybenzene)acetamide 2-hydroxy-3-(octyloxy)phenyl-5 acrylamidemethylbenzene phenyl methanone, and N-(5-acetamidemethyl benzene-2,4 dihydroxybenzene)acetamide and were synthesized via the Friedel-Crafts (F-C) alkylation reaction. This work on derivatives provides a good foundation for future applications in natural medicine and food additives. The above discussed derivatives were characterized by various methods, including HRMS, IR, and 1HNMR. The authors concluded that capsaicin derivatives have good antibacterial and antioxidant activities (Wang et al. 2019).

24.5 Antimicrobial Properties of Capsaicin Derivatives

With the rise of harmful, antibiotic-resistant bacterial strains of pathogenic microorganisms such as *methicillin-resistant Staphylococcus aureus*, *Pseudomonas aerugenosa*, *Salmonella and Clostridia*, the growing trend of antibiotic resistance is becoming increasingly difficult to combat (Aslam et al. 2018). So, the need of a safe, eco-friendly, and natural antimicrobial agent is very necessary (Arshad and Batool 2017) and, hence due to the typical spicy nature of capsaicin, which is the key element in chilies, has received interest in the field of antimicrobial investigations, particularly in the last two decades (Molina-Torres et al. 1999; Stephen and Kumar 2014). Capsaicin and dihydrocapsaicin (DHC) extracted from *Capsicum annum* fruit powder were found to have varied but high antibacterial activity against 16 different strains of clinically important MDR Gram-negative and Gram-positive bacteria (Oyedemi et al. 2019). In another in vitro investigation, six capsaicin derivatives with phenolic hydroxyl, a benzene ring, and amide structures were synthesized and evaluated for antibacterial activity against *E. coli* and *S. aureus* (Wang et al. 2019).

Wang et al. (2020c) synthesized six derivatives of capsaicin and demonstrated their toxic effects against growth of fouling organisms *Chaetoceros curvisetus*, *Skeletonema costatum*, and *Phaeodactylum tricornutum*. Findings indicated that capsaicin derivatives are fouling resistant, with only a minor amount of algae and muck adhered to the tested panel after 90 days with an eco-friendly algaecide potential.

A variety of new capsaicin derivatives incorporating a sulfonic acid ester moiety were produced in attempt to develop an efficient and broad-spectrum bactericide as per the study of Xie et al. (2022) that some target compounds demonstrated significant antibacterial efficacy against *Xanthomonas oryzae pv. oryzae* (Xoo), *Pseudomonas syringae pv. actinidiae* (Psa), and *Xanthomonas axonopodis pv. citri* (Xac).

Condensation of vanillylamine and acyl chlorides in a biphase $H_2O/CHCl_3$ system yielded seven capsaicin analogues characterized by using spectroscopic approaches such as NMR and IR, as well as CHNS analysis and comparison with literature data (Wang et al. 2009) where all these compounds showed potent antimicrobial activity.

The presence of significant elements, such as valeric acid, lactic acid, butanedioic acid, 5 methoxy, phenylalanine, ethyl ester, hexadecanoic acid, pentanoicacid, 6-methoxy-hexane-2-ol, pentanoicacid, butane, 2,3diol, pentanoic acid, pentanobenzeneacetic acid, 4-hydroxyl, 1,2 benzenedicarboxylic acid ester, 7-hydroxy-7-methyloctanoic acid, and 2,5-furandicarboxylic acid, was confirmed after gas chromatography mass spectrometer analysis of *C. annuum* (concentrated extracts) supported by the potent antibacterial activity against Gram-negative and Gram-positive microorganisms (Oulaï et al. 2018).

The synthesis of two novel boron-capsaicin derivatives with either a short- or long-chain aliphatic tail group was reported in a preliminary investigation employing an iridium catalyzed hydroboration process with pinacolborane by Ramsaywack et al. (2018), which revealed boronate ester groups are located at the tail group's terminal position and are required for the bioactivity of these molecules as well as both compounds were effective against *vancomycin-resistant Enterococcus* (Gram-positive bacteria).

Condensation of 2-aminobenzothiazole with 4-acetamidobenzenesulfonyl chloride derivatives in pyridine-acetic anhydride mixture yielded a series of sulfonamide derivatives with benzothiazole nucleus (Bhusari et al. 2008). All of the test compounds were tested in vitro for antibacterial activity against *E. coli* and *B. subtilis*, antimycobacterial activity against *Mycobacterium tuberculosis* H37Rv strain, and antifungal activity against *C. albicans*. The antibacterial, antifungal, and antimycobacterial activity of the test compounds against the microbiological strains tested was substantial.

Jinbo et al. (2020) synthesized 204 analogues by keeping capsaicin as a lead molecule and reported the most strongest toxicities of benzyl amine derivatives specially 2-fluorobenzyleamine derivative against *Mythemna separata* that showed the insecticidal activity of capsaicin derivatives.

24.6 Current and Future Research on Capsaicin and Its Derivatives

Capsaicin has piqued the interest of many people for millennia because of its numerous culinary and medical applications. Despite its negative consequences, capsaicin is nevertheless utilized as an active principle in a variety of pharmaceutical formulations to treat a variety of human illnesses (Fattori et al. 2016). Furthermore, new research indicates that capsaicin is involved in a greater spectrum of processes than previously thought. Although it is best known for its involvement in nociception and pain, various experimental and clinical studies have shown that it also has a role in other major pathological states such as obesity, cancer, cardiovascular disease, skin disorders, and so on (Chang et al. 2023).

Capsaicin and food-containing capsaicin have been around for thousands of years, but our understanding of how capsaicin affects our bodies has only recently advanced. Capsaicin has helped us understand pathological and physiological processes, as well as the importance of TRPV1 channels (Fischer et al. 2020). Capsaicin remains a hot topic in the scientific community and has a wide range of potential medicinal applications (Lv et al. 2015; Singh et al. 2022). Novel pharmaceutical formulations, the discovery of new analogues, and the targeting of the capsaicin-activated receptor TRPV1 are thus interesting fields needing further exploration (Holzer 2008; Cáceres et al. 2015).

Food firms are looking for specific biomolecules that can add value to novel food formulations in order to meet the current demand for healthier products (Cáceres et al. 2015; Lutz et al. 2019; Zarbà et al. 2020). Capsaicin, the spicy component of chili peppers, has been one of the target biomolecules studied since the 1950s (Castro-Muñoz et al. 2022). It is reported that capsaicin has vital biological properties in human health, such as inhibiting acid secretion and stimulating alkali and mucus secretion and especially gastric mucosal blood flow, which helps to prevent and heal gastric ulcers, and thermoregulation, among many other reported bioactivities (Srinivasan 2016).

In terms of ligand selectivity and sensitivity, the capsaicin receptor TRPV1 stands out among ligand-gated ion channels. However, the molecular interactions that keep the ligand-binding pocket in its permissive conformation are unknown, as is the number of permissive conformations that the ligand-binding pocket can adopt (Vu et al. 2020). Computational modeling and functional studies indicated that these ligands alternate between significantly diverse binding poses yet stabilize the ligand-binding pocket in almost identical permissive conformations; functionally, despite varied potencies, they all provided a stable open state (Meng et al. 2011).

24.7 Conclusions

Spices have been an important part of global civilizations, both for culinary and therapeutic purposes. Among these, chili peppers have been a popular spice for centuries, with numerous species cultivated worldwide. Capsaicin, a phenolic compound found in varying amounts in chili peppers, is responsible for the characteristic pungency and taste associated with them. Capsaicin can also have positive effects on human health, such as pain relief and antibacterial, antifungal, and antihelminthic activities. Capsanthin, a pigment found in chili peppers, also possesses strong antioxidant potential and contributes to their vivid red color. While chili consumption can evoke both pleasant and unpleasant sensations depending on the individual, its therapeutic potential cannot be ignored.

After conducting an extensive review of existing literature, it is apparent that capsaicin and its derivatives have the potential to serve as alternative treatments for bacterial infections caused by multiple-drug-resistant strains, without the need for antibiotics. In vitro studies have shown that capsaicin has synergistic effects when combined with established antibiotics against various pathogens. These findings are encouraging and should prompt further research to investigate the core molecular mechanisms accountable for these antimicrobial effects.

It is also crucial to validate the results of in vitro studies, in vivo models, and clinical trials to determine the effectiveness of capsaicin as a treatment option for bacterial infections. This requires a considerable venture in time and sources; nevertheless the possible benefits of capsaicin-based treatments are too significant to ignore. By exploring alternative treatments such as capsaicin, we can reduce our reliance on antibiotics and decrease the risk of antibiotic resistance development.

References

Adaszek Ł, Gadomska D, Mazurek Ł, Łyp P, Madany J, Winiarczyk S (2019) Properties of capsaicin and its utility in veterinary and human medicine. Res Vet Sci 123:14–19

Agarwal P, Das C, Dias O, Shanbhag T (2017) Antimicrobial property of capsaicin. Int Res J Biol Sci 6(7):7–11

Al-Musawi S, Albukhaty S, Al-Karagoly H, Sulaiman GM, Alwahibi MS, Dewir YH, Soliman DA, Rizwana H (2020) Antibacterial activity of honey/chitosan nanofibers loaded with capsaicin and gold nanoparticles for wound dressing. Molecules 25(20):4770

Ananthan R, Subash K, Longvah T (2014) Assessment of nutrient composition and capsaicinoid content of some red chilies. Int Food Nutr Technol 72:1–4

Antonious GF (2018) Capsaicinoids and vitamins in hot pepper and their role in disease therapy. In: Mozsik G (ed) Capsaicin and its human therapeutic development. IntechOpen, London

Arshad MS, Batool SA (2017) Natural antimicrobials, their sources and food safety. Food Addit 87(1):87

Aslam B, Wang W, Arshad MI, Khurshid M, Muzammil S, Rasool MH, Nisar MA, Alvi RF, Aslam MA, Qamar MU (2018) Antibiotic resistance: a rundown of a global crisis. Infect Drug Resist 11:1645

Ayariga JA, Abugri DA, Amrutha B, Villafane R (2022) Capsaicin potently blocks Salmonella typhimurium invasion of Vero cells. Antibiotics 11(5):666

Aza-González C, Núñez-Palenius HG, Ochoa-Alejo N (2011) Molecular biology of capsaicinoid biosynthesis in chili pepper (Capsicum spp.). Plant Cell Rep 30:695–706

Basak P, Maitra P, Khan U, Saha K, Bhattacharya SS, Dutta M, Bhattacharya S (2022) Capsaicin inhibits Shigella flexneri intracellular growth by inducing autophagy. Front Pharmacol 2022:13

Basith S, Cui M, Hong S, Choi S (2016) Harnessing the therapeutic potential of capsaicin and its analogues in pain and other diseases. Molecules 21(8):966

Behbehani JM, Irshad M, Shreaz S, Karched M (2023) Anticandidal activity of capsaicin and its effect on ergosterol biosynthesis and membrane integrity of Candida albicans. Int J Mol Sci 24(2):1046

Bhatti SA, Hussain MH, Mohsin MZ, Mohsin A, Zaman WQ, Guo M, Iqbal MW, Siddiqui SA, Ibrahim SA, Ur-Rehman S (2022) Evaluation of the antimicrobial effects of Capsicum, Nigella sativa, Musa paradisiaca L., and citrus limetta: a review. Front Sust Food Syst 6:1043823

Bhusari K, Amnerkar N, Khedekar P, Kale M, Bhole R (2008) Synthesis and in vitro antimicrobial activity of some new 4-amino-N-(1, 3-benzothiazol-2-yl) benzenesulphonamide derivatives. Asian J Res Chem 1(2):53–57

Bley K, Boorman G, Mohammad B, McKenzie D, Babbar S (2012) A comprehensive review of the carcinogenic and anticarcinogenic potential of capsaicin. Toxicol Pathol 40(6):847–873

Blum E, Mazourek M, O'connell M, Curry J, Thorup T, Liu K, Jahn M, Paran I (2003) Molecular mapping of capsaicinoid biosynthesis genes and quantitative trait loci analysis for capsaicinoid content in Capsicum. Theor Appl Genet 108:79–86

Buitimea-Cantúa GV, Buitimea-Cantúa NE, Rocha-Pizaña MR, Hernández-Morales A, Magaña-Barajas E, Molina-Torres J (2020) Inhibitory effect of Capsicum chinense and Piper nigrum fruits, capsaicin and piperine on aflatoxins production in Aspergillus parasiticus by downregulating the expression of afl D, afl M, afl R, and afl S genes of aflatoxins biosynthetic pathway. J Environ Sci Health B 55(9):835–843

Cáceres J, Sepúlveda R, Navas C, Latorre R, González-Nilo F (2015) Rational discovery of new capsaicin analogues as TRPV1 activators. BMC Bioinform 8:A8

Campos-Vega R, Oomah BD (2013) Chemistry and classification of phytochemicals. In: Tiwari BK, Brunton NP, Brennan CS (eds) Handbook of plant food phytochemicals. Wiley, Hoboken, pp 5–48

Castro-Muñoz R, Gontarek-Castro E, Jafari SM (2022) Up-to-date strategies and future trends towards the extraction and purification of capsaicin: a comprehensive review. Trends Food Sci Technol 123:161–171

Chang A, Rosani A, Quick J (2023) Capsaicin. [Updated 2023 May 23]. In: StatPearls [Internet]. StatPearls Publishing, Treasure Island, FL. Available from: https://www.ncbi.nlm.nih.gov/books/NBK459

Chokshi A, Sifri Z, Cennimo D, Horng H (2019) Global contributors to antibiotic resistance. J Global Infect Dis 11(1):36

Chuichulcherm S, Prommakort S, Srinophakun P, Thanapimmetha A (2013) Optimization of capsaicin purification from Capsicum frutescens Linn. with column chromatography using Taguchi design. Ind Crop Prod 44:473–479

Cichewicz RH, Thorpe PA (1996) The antimicrobial properties of Chile peppers (Capsicum species) and their uses in Mayan medicine. J Ethnopharmacol 52(2):61–70

Cortés-Ferré HE, Guajardo-Flores D, Romero-De La Vega G, Gutierrez-Uribe JA (2021) Recovery of capsaicinoids and other phytochemicals involved with TRPV-1 receptor to re-valorize chili pepper waste and produce nutraceuticals. Front Sust Food Syst 4:588534

Cowan MM (1999) Plant products as antimicrobial agents. Clin Microbiol Rev 12(4):564–582

del Rocío Gómez-García M, Ochoa-Alejo N (2013) Biochemistry and molecular biology of carotenoid biosynthesis in chili peppers (Capsicum spp.). Int J Mol Sci 14(9):19025–19053

Delelegn S (2011) Evaluation of elite hot pepper varieties (Capsicum species) for growth, dry pod yield and quality under Jimma condition, South West Ethiopia. Jimma University, Jimma

Díaz-Laviada I, Rodríguez-Henche N (2014) The potential antitumor effects of capsaicin. Prog Drug Res 68:181–208

Doğan ANC, Çelik E, Kılıçle PA, Atalay E, Sağlam AG, Doğan A, Otlu S (2018) Antibacterial effect of hot peppers (Capsicum annuum, Capsicum annuum var globriusculum, Capsicum frutescens) on some arcobacter, campylobacter and helicobacter species. Pak Vet J 38(3): 266–270

Donnerer J, Amann R, Schuligoi R, Lembeck F (1990) Absorption and metabolism of capsaicinoids following intragastric administration in rats. Naunyn Schmiedeberg's Arch Pharmacol 342:357–361

Downham A, Collins P (2000) Colouring our foods in the last and next millennium. Int J Food Sci Technol 35(1):5–22

Ekom SE, Tamokou J-D-D, Kuete V (2021) Antibacterial and therapeutic potentials of the Capsicum annuum extract against infected wound in a rat model with its mechanisms of antibacterial action. Biomed Res Int 2021:1–17

Erfanimanesh S, Eslami G, Taherpour A, Hashemi A (2019) Capsaicin inhibitory effects on Vibrio cholerae toxin genes expression. Avicenna J Phytomed 9(3):187

Fattori V, Hohmann MS, Rossaneis AC, Pinho-Ribeiro FA, Verri WA Jr (2016) Capsaicin: current understanding of its mechanisms and therapy of pain and other pre-clinical and clinical uses. Molecules 21(7):844

Fieira C, Oliveira F, Calegari RP, Machado A, Coelho AR (2013) In vitro and in vivo antifungal activity of natural inhibitors against *Penicillium expansum*. Food Sci Technol 33:40–46

Fischer MJ, Ciotu CI, Szallasi A (2020) The mysteries of capsaicin-sensitive afferents. Front Physiol 11:554195

Füchtbauer S, Mousavi S, Bereswill S, Heimesaat MM (2021) Antibacterial properties of capsaicin and its derivatives and their potential to fight antibiotic resistance–a literature survey. Eur J Microbiol Immunol 11(1):10–17

Goci E, Haloci E, Di Stefano A, Chiavaroli A, Angelini P, Miha A, Cacciatore I, Marinelli L (2021) Evaluation of in vitro capsaicin release and antimicrobial properties of topical pharmaceutical formulation. Biomol Ther 11(3):432

Gonzalez-Mondragon E, Vazquez-Tzompantzi E (2011) Chemical and pharmacological aspects of capsaicin. Molecules (Basel, Switzerland) 16(2):1253–1270

González-Zamora A, Sierra-Campos E, Luna-Ortega JG, Pérez-Morales R, Ortiz JCR, García-Hernández JL (2013) Characterization of different capsicum varieties by evaluation of their capsaicinoids content by high performance liquid chromatography, determination of pungency and effect of high temperature. Molecules 18(11):13471–13486

Guillen NG, Tito R, Mendoza NG (2018) Capsaicinoids and pungency in Capsicum chinense and Capsicum baccatum fruits. Pesqui Agropecu Trop 48:237–244

Gupta R, Kapoor B, Gulati M, Kumar B, Gupta M, Singh SK, Awasthi A (2022) Sweet pepper and its principle constituent capsiate: functional properties and health benefits. Crit Rev Food Sci Nutr 62(26):7370–7394

Halberstein RA (2005) Medicinal plants: historical and cross-cultural usage patterns. Ann Epidemiol 15(9):686–699

Hall OM, Broussard A, Range T, Carroll Turpin MA, Ellis S, Lim VM, Cornett EM, Kaye AD (2020) Novel agents in neuropathic pain, the role of capsaicin: pharmacology, efficacy, side effects, different preparations. Curr Pain Headache Rep 24:1–12

Hamed M, Kalita D, Bartolo ME, Jayanty SS (2019) Capsaicinoids, polyphenols and antioxidant activities of Capsicum annuum: comparative study of the effect of ripening stage and cooking methods. Antioxidants 8(9):364

Hao X, Chen S, Qin D, Zhang M, Li W, Fan J, Wang C, Dong M, Zhang J, Cheng F (2020) Antifouling and antibacterial behaviors of capsaicin-based pH responsive smart coatings in marine environments. Mater Sci Eng C 108:110361

Hayman M, Kam PC (2008) Capsaicin: a review of its pharmacology and clinical applications. Curr Anaesth Crit Care 19(5–6):338–343

Hernández-Pérez T, Gómez-García MDR, Valverde ME, Paredes-López O (2020) Capsicum annuum (hot pepper): an ancient Latin-American crop with outstanding bioactive compounds and nutraceutical potential. A review. Compr Rev Food Sci Food Saf 19(6):2972–2993

Hill TA, Ashrafi H, Reyes-Chin-Wo S, Yao J, Stoffel K, Truco M-J, Kozik A, Michelmore RW, Van Deynze A (2013) Characterization of Capsicum annuum genetic diversity and population structure based on parallel polymorphism discovery with a 30K unigene Pepper GeneChip. PLoS One 8(2):e56200

Holzer P (2008) The pharmacological challenge to tame the transient receptor potential vanilloid-1 (TRPV1) nocisensor. Br J Pharmacol 155(8):1145–1162

Jaganathan R, Purushotham B, Radhakrishnan N, Swamy MK (2020) Capsaicin and its potential anticancer mechanisms of action. In: Swamy MK (ed) Plant-derived bioactives: production, properties and therapeutic applications. Springer, Singapore, pp 301–321

Jinbo W, Zhenmeng JI, Zhiqin JI, Shaopeng WEI (2020) Synthesis and insecticidal activity of capsaicin analogues[J]. Chinese J Pest Sci 22(5):892–896. https://doi.org/10.16801/j.issn.1008-7303.2020.0039

Jones NL, Shabib S, Sherman PM (1997) Capsaicin as an inhibitor of the growth of the gastric pathogen Helicobacter pylori. FEMS Microbiol Lett 146(2):223–227. https://doi.org/10.1111/j.1574-6968.1997.tb10197.x

Kar D, Bandyopadhyay S, Dimri U, Mondal DB, Nanda PK, Das AK, Batabyal S, Dandapat P, Bandyopadhyay S (2016) Antibacterial effect of silver nanoparticles and capsaicin against MDR-ESBL producing Escherichia coli: an in vitro study. Asian Pac J Trop Dis 6(10):807–810

Karim K, Rafii MY, Misran AB, Ismail MFB, Harun AR, Khan MMH, Chowdhury M, Nazneen F (2021) Current and prospective strategies in the varietal improvement of chilli (Capsicum annuum L.) specially heterosis breeding. Agronomy 11(11):2217

Katz E (2020) Flavors and colors: the chili pepper in Europe. In: Kaller M, Jacob F (eds) Transatlantic trade and global cultural transfers since 1492. Routledge, New York, pp 30–53

Khan FA, Mahmood T, Ali M, Saeed A, Maalik A (2014) Pharmacological importance of an ethnobotanical plant: Capsicum annuum L. Nat Prod Res 28(16):1267–1274

Kobata K, Kawaguchi M, Watanabe T (2002) Enzymatic synthesis of a capsinoid by the acylation of vanillyl alcohol with fatty acid derivatives catalyzed by lipases. Biosci Biotechnol Biochem 66(2):319–327

Koleva II, van Beek TA, Soffers AE, Dusemund B, Rietjens IM (2012) Alkaloids in the human food chain–natural occurrence and possible adverse effects. Mol Nutr Food Res 56(1):30–52

Kurita S, Kitagawa E, Kim C-H, Momose Y, Iwahashi H (2002) Studies on the antimicrobial mechanisms of capsaicin using yeast DNA microarray. Biosci Biotechnol Biochem 66(3): 532–536

Lim T (2013) Capsicum frutescens. In: Lim T (ed) Edible medicinal and non-medicinal plants, Fruits, vol 6. Springer, Dordrecht, pp 213–239

Liu X, Lin T, Peng B, Wang X (2012) Antibacterial activity of capsaicin-coated wool fabric. Text Res J 82(6):584–590

Luo X-J, Peng J, Li Y-J (2011) Recent advances in the study on capsaicinoids and capsinoids. Eur J Pharmacol 650(1):1–7

Lutz M, Petzold G, Albala C (2019) Considerations for the development of innovative foods to improve nutrition in older adults. Nutrients 11(6):1275

Lv J, Qi L, Yu C, Yang L, Guo Y, Chen Y, Bian Z, Sun D, Du J, Ge P (2015) Consumption of spicy foods and total and cause specific mortality: population based cohort study. BMJ 351:h3942

Mahmood S, Mei TS, Yee WX, Hilles AR, Alelwani W, Bannunah AM (2021) Synthesis of capsaicin loaded silver nanoparticles using green approach and its anti-bacterial activity against human pathogens. J Biomed Nanotechnol 17(8):1612–1626

Marini E, Magi G, Mingoia M, Pugnaloni A, Facinelli B (2015) Antimicrobial and anti-virulence activity of capsaicin against erythromycin-resistant, cell-invasive group a streptococci. Front Microbiol 6:1281

Martini R, Serrano L, Barbosa S, Labidi J (2014) Antifungal cellulose by capsaicin grafting. Cellulose 21:1909–1919

Maru GB, Hudlikar RR, Kumar G, Gandhi K, Mahimkar MB (2016) Understanding the molecular mechanisms of cancer prevention by dietary phytochemicals: from experimental models to clinical trials. World J Biol Chem 7(1):88

Mbaebie Oyedemi B (2015) Antiplasmid and antimicrobial activities of synthetic and natural products from selected medicinal plants. University College London, London

Meghvansi M, Siddiqui S, Khan MH, Gupta V, Vairale M, Gogoi H, Singh L (2010) Naga chilli: a potential source of capsaicinoids with broad-spectrum ethnopharmacological applications. J Ethnopharmacol 132(1):1–14

Menezes RDP, Bessa MADS, Siqueira CDP, Teixeira SC, Ferro EAV, Martins MM, Cunha LCS, Martins CHG (2022) Antimicrobial, antivirulence, and antiparasitic potential of capsicum chinense Jacq. extracts and their isolated compound capsaicin. Antibiotics 11(9):1154

Meng X-Y, Zhang H-X, Mezei M, Cui M (2011) Molecular docking: a powerful approach for structure-based drug discovery. Curr Comput Aided Drug Des 7(2):146–157

Mohan A, Nisha A, Sujith S, Suja Rani S, Thomas N (2022) Antibiofilm activity of berberine and capsaicin in combination with quinolones against staphylococcus aureus from bovine mastitis. J Vet Anim Sci 53(2):253–261

Mohd Hassan N, Yusof NA, Yahaya AF, Mohd Rozali NN, Othman R (2019) Carotenoids of capsicum fruits: pigment profile and health-promoting functional attributes. Antioxidants 8(10): 469

Molina-Torres J, García-Chávez A, Ramírez-Chávez E (1999) Antimicrobial properties of alkamides present in flavouring plants traditionally used in Mesoamerica: affinin and capsaicin. J Ethnopharmacol 64(3):241–248

Morales-Martínez C, Marquez-Aguirre A, Díaz-Martínez E, Rodríguez-González J, Mateos-Díaz J, Esquivel-Solís H, Canales-Aguirre A (2016) The prospective antiobesity effect of capsaicin synthetic analogs: a matter of weight. Med Chem 6:365–371

Moran MM, Szallasi A (2018) Targeting nociceptive transient receptor potential channels to treat chronic pain: current state of the field. Br J Pharmacol 175(12):2185–2203

Mózsik G, Past T, Abdel Salam OM, Kuzma M, Perjési P (2009) Interdisciplinary review for correlation between the plant origin capsaicinoids, non-steroidal antiinflammatory drugs, gastrointestinal mucosal damage and prevention in animals and human beings. Inflammopharmacology 17:113–150

Nagy I, Friston D, Valente JS, Perez JVT, Andreou AP (2014) Pharmacology of the capsaicin receptor, transient receptor potential vanilloid type-1 ion channel. Prog Drug Res 68:39–76

Naves ER, de Ávila SL, Sulpice R, Araújo WL, Nunes-Nesi A, Peres LE, Zsögön A (2019) Capsaicinoids: pungency beyond Capsicum. Trends Plant Sci 24(2):109–120

Nick GL (2002) Medicinal properties in whole foods: the Capsicum fruit. Townsend Lett Dr Patients 227:128–134

O'Neill J, Brock C, Olesen AE, Andresen T, Nilsson M, Dickenson AH (2012) Unravelling the mystery of capsaicin: a tool to understand and treat pain. Pharmacol Rev 64(4):939–971

Olatunji TL, Afolayan AJ (2018) The suitability of chili pepper (Capsicum annuum L.) for alleviating human micronutrient dietary deficiencies: a review. Food Sci Nutr 6(8):2239–2251

Oulaï AC, Djè KM, Eba KP, Adima AA, Kouadio EJP (2018) Chemical composition, antioxidant and antimicrobial activities of Capsicum annuum var. annuum concentrated extract obtained by reverse osmosis. GSC Biol Pharm Sci 05(02):116–125

Oyedemi BO, Kotsia E, Stapleton PD, Gibbons S (2019) Capsaicin and gingerol analogues inhibit the growth of efflux-multidrug resistant bacteria and R-plasmids conjugal transfer. J Ethnopharmacol 245:111871

Pandey SK, Yadav SK, Singh VK (2012) An overview on Capsicum annuum L. J Pharm Sci and Technol 4(2):821–828

Parvez GM (2017) Current advances in pharmacological activity and toxic effects of various capsicum species. Int J Pharm Sci Res 8:1900–1912. [Google Scholar]

Pawar S, Bharude N, Sonone S, Deshmukh R, Raut A, Umarkar A (2011) Chillies as food, spice and medicine: a perspective. Int J Pharm Biol Sci 1(3):311–318

Rahman MM, Rahaman MS, Islam MR, Hossain ME, Mannan Mithi F, Ahmed M, Saldías M, Akkol EK, Sobarzo-Sánchez E (2021) Multifunctional therapeutic potential of phytocomplexes and natural extracts for antimicrobial properties. Antibiotics 10(9):1076

Ramsaywack S, Bos A, Vogels CM, Gray CA, Westcott SA (2018) Preliminary investigations into the synthesis and antimicrobial activities of boron-containing capsaicinoids. Can J Chem 96(12):1065–1070
Reyes-Escogido MDL, Gonzalez-Mondragon EG, Vazquez-Tzompantzi E (2011) Chemical and pharmacological aspects of capsaicin. Molecules 16(2):1253–1270
Romero-Luna HE, Colina J, Guzmán-Rodríguez L, Sierra-Carmona CG, Farías-Campomanes ÁM, García-Pinilla S, González-Tijera MM, Malagón-Alvira KO, Peredo-Lovillo A (2022) Capsicum fruits as functional ingredients with antimicrobial activity: an emphasis on mechanisms of action. J Food Sci Technol 2022:1–11. https://doi.org/10.1007/s13197-022-05578-y
Rosa A, Deiana M, Casu V, Paccagnini S, Appendino G, Ballero M, Dessí MA (2002) Antioxidant activity of capsinoids. J Agric Food Chem 50(25):7396–7401
Rosca AE, Iesanu MI, Zahiu CDM, Voiculescu SE, Paslaru AC, Zagrean A-M (2020) Capsaicin and gut microbiota in health and disease. Molecules 25(23):5681
Sachan A, Kumar S, Kumari K, Singh D (2018) Medicinal uses of spices used in our traditional culture: worldwide. J Med Plants Stud 6(3):116–122
Saleh BK, Omer A, Teweldemedhin B (2018) Medicinal uses and health benefits of chili pepper (Capsicum spp.): a review. MOJ Food Process Technol 6(4):325–328
Salehi B, Hernández-Álvarez AJ, del Mar CM, Martorell M, Ramírez-Alarcón K, Melgar-Lalanne-G, Matthews KR, Sharifi-Rad M, Setzer WN, Nadeem M (2018) Potential phytopharmacy and food applications of Capsicum spp.: a comprehensive review. Nat Prod Commun 13(11): 1934578X1801301133
Sanatombi K, Sharma G (2008) Capsaicin content and pungency of different Capsicum spp. cultivars. Not Bot Horti Agrobot Cluj Napoca 36(2):89–90
Sánchez E, Heredia N, Camacho-Corona MR, García S (2013) Isolation, characterization and mode of antimicrobial action against Vibrio cholerae of methyl gallate isolated from Acacia farnesiana. J Appl Microbiol 115(6):1307–1316
Santos LS, Fernandes CC, Santos LS, Deus IPBD, Sousa TLD, Miranda MLD (2020) Ethanolic extract from Capsicum chinense Jacq. ripe fruits: phenolic compounds, antioxidant activity and development of biodegradable films. Food Sci Technol 41:497–504
Scaldaferro MA, Moscone EA (2019) Cytology and DNA content variation of capsicum genomes. In: Ramchiary N, Kole C (eds) The capsicum genome. Compendium of plant genomes. Springer, Cham. https://doi.org/10.1007/978-3-319-97217-6_4
Shayan S, Saeidi S (2013) Antibacterial and antibiofilm activities of extract Capsicum annuum L on the growth and biofilm formation of common pathogenic strains. Int J Sci Basic Appl Res 5(4): 513–518
Singer AC, Kirchhelle C, Roberts AP (2019) Reinventing the antimicrobial pipeline in response to the global crisis of antimicrobial-resistant infections. F1000Res 8:238
Singh S, Uddin M, Khan MMA, Chishti A, Bhat U (2022) Therapeutic properties of capsaicin: a medicinally important bio-active constituent of chilli pepper. Asian J Pharm Clin Res 15(7): 47–58
Sricharoen P, Lamaiphan N, Patthawaro P, Limchoowong N, Techawongstien S, Chanthai S (2017) Phytochemicals in Capsicum oleoresin from different varieties of hot chilli peppers with their antidiabetic and antioxidant activities due to some phenolic compounds. Ultrason Sonochem 38: 629–639
Srinivasan K (2016) Biological activities of red pepper (Capsicum annuum) and its pungent principle capsaicin: a review. Crit Rev Food Sci Nutr 56(9):1488–1500
Stellari G, Mazourek M, Jahn M (2010) Contrasting modes for loss of pungency between cultivated and wild species of Capsicum. Heredity 104(5):460–471
Stephen BJ, Kumar V (2014) Evaluation of antimicrobial activity of capsaicin extract against B. subtilis species. International Journal of Pharmaceutics and Drug Analysis 2(2):111–116
Stewart C Jr, Mazourek M, Stellari GM, O'Connell M, Jahn M (2007) Genetic control of pungency in C. chinense via the Pun1 locus. J Exp Bot 58(5):979–991

Surh Y-J, Ahn SH, Kim K-C, Park J-B, Sohn YW, Lee SS (1995) Metabolism of capsaicinoids: evidence for aliphatic hydroxylation and its pharmacological implications. Life Sci 56(16): PL305–PL311

Sutoh K, Kobata K, Yazawa S, Watanabe T (2006) Capsinoid is biosynthesized from phenylalanine and valine in a non-pungent pepper, Capsicum annuum L. cv. CH-19 sweet. Biosci Biotechnol Biochem 70(6):1513–1516

Tanaka Y, Hosokawa M, Otsu K, Watanabe T, Yazawa S (2009) Assessment of capsiconinoid composition, nonpungent capsaicinoid analogues, in capsicum cultivars. J Agric Food Chem 57(12):5407–5412

Tayseer I, Aburjai T, Abu-Qatouseh L, Al-Karabieh N, Ahmed W, Al-Samydai A (2020) In vitro anti-Helicobacter pylori activity of capsaicin. J Pure Appl Microbiol 14(1):279–286

Tewksbury JJ, Nabhan GP (2001) Directed deterrence by capsaicin in chillies. Nature 412(6845): 403–404

Tewksbury JJ, Manchego C, Haak DC, Levey DJ (2006) Where did the chili get its spice? Biogeography of capsaicinoid production in ancestral wild chili species. J Chem Ecol 32: 547–564

Tewksbury JJ, Reagan KM, Machnicki NJ, Carlo TA, Haak DC, Penaloza ALC, Levey DJ (2008) Evolutionary ecology of pungency in wild chilies. Proc Natl Acad Sci 105(33):11808–11811

Vázquez-Fresno R, Rosana ARR, Sajed T, Onookome-Okome T, Wishart NA, Wishart DS (2019) Herbs and spices-biomarkers of intake based on human intervention studies–a systematic review. Genes Nutr 14(1):1–27

Ventola CL (2015) The antibiotic resistance crisis: part 1: causes and threats. P T 40(4):277

Vu S, Singh V, Wulff H, Yarov-Yarovoy V, Zheng J (2020) New capsaicin analogs as molecular rulers to define the permissive conformation of the mouse TRPV1 ligand-binding pocket. elife 9:e62039

Wang B, Yang F, Shan Y-F, Qiu W-W, Tang J (2009) Highly efficient synthesis of capsaicin analogues by condensation of vanillylamine and acyl chlorides in a biphase $H_2O/CHCl_3$ system. Tetrahedron 65(27):5409–5412

Wang X, Yu L, Li F, Zhang G, Zhou W, Jiang X (2019) Synthesis of amide derivatives containing capsaicin and their antioxidant and antibacterial activities. J Food Biochem 43(12):e13061

Wang F, Huang X, Chen Y, Zhang D, Chen D, Chen L, Lin J (2020a) Study on the effect of capsaicin on the intestinal flora through high-throughput sequencing. ACS Omega 5(2): 1246–1253

Wang X, Yu L, Liu Y, Jiang X (2020c) Synthesis and fouling resistance of capsaicin derivatives containing amide groups. Sci Total Environ 710:136361

Whiting S, Derbyshire E, Tiwari B (2012) Capsaicinoids and capsinoids. A potential role for weight management? A systematic review of the evidence. Appetite 59(2):341–348

Xie D, Yang Z, Hu X, Wen Y (2022) Synthesis, antibacterial and insecticidal activities of novel capsaicin derivatives containing a sulfonic acid esters moiety. Front Chem 10:929050

Xing F, Cheng G, Yi K (2006) Study on the antimicrobial activities of the capsaicin microcapsules. J Appl Polym Sci 102(2):1318–1321

Yoneshiro T, Aita S, Kawai Y, Iwanaga T, Saito M (2012) Nonpungent capsaicin analogs (capsinoids) increase energy expenditure through the activation of brown adipose tissue in humans. Am J Clin Nutr 95(4):845–850

Zarbà C, Chinnici G, D'Amico M (2020) Novel food: the impact of innovation on the paths of the traditional food chain. Sustainability 12(2):555

Zhao J, Wei F, Xu W, Han X (2020) Enhanced antibacterial performance of gelatin/chitosan film containing capsaicin loaded MOFs for food packaging. Appl Surf Sci 510:145418

Capsaicinoids: Nanoformulations for Improved Therapeutic Applications

25

Shradha Devi Dwivedi, Manju Rawat Singh, Deependra Singh, Satish Dubey, Ashwini Kumar Dixit, Pooja Gadkari, Sanjay J Daharwal, Kamal Shah, Durgesh Nandini Chauhan, Umesh Kumar Patil, and Nagendra Singh Chauhan

Abstract

Chili peppers owe their pungent and spicy flavor to capsaicinoids, bioactive alkaloids that have been extracted from them using various methods over the past few decades. Capsaicinoids act as the most powerful activators of the human pain receptor TRPV1 (transient receptor potential cation channel subfamily V member 1). These compounds are currently obtained through either chemical synthesis or extraction from the fruit. By transferring the biosynthesis of capsaicinoids to a microbial host, it could lead to more efficient production through fermentation and allow for the creation of new compounds with potentially superior properties using synthetic biology. Through biological studies, capsaicinoids have been found to offer several pharmacological benefits, including pain relief, anti inflammation, anticancer properties, cardioprotection, and even aiding in weight loss. The way capsaicinoids interact with the body is

S. D. Dwivedi · M. R. Singh · D. Singh · S. Dubey · P. Gadkari · S. J. Daharwal
University Institute of Pharmacy, Pandit Ravishankar Shukla University, Raipur, Chhattisgarh, India

A. K. Dixit
Department of Botany, Guru Ghasidas Vishwavidyalaya, Bilaspur, Chhattisgarh, India

K. Shah
Institute of Pharmaceutical Research, GLA University, Mathura, Uttar Pradesh, India

D. N. Chauhan
Columbia College of Pharmacy, Raipur, Chhattisgarh, India

U. K. Patil
Phytomedicine and Natural Product Research Laboratory, Department of Pharmaceutical Sciences, Dr. Harisingh Gour Vishwavidyalaya (A Central University), Sagar, Madhya Pradesh, India

N. S. Chauhan (✉)
Drugs Testing Laboratory, Avam Anusandhan Kendra, Raipur, Chhattisgarh, India

M. K. Swamy, A. Kumar (eds.), *Capsaicinoids*,
https://doi.org/10.1007/978-981-99-7779-6_25

directly influenced by its pharmacokinetic properties. It has an unfavorable pharmacokinetic profile characterized by the presence of toxic metabolites and a short half-life, which can harm healthy cells. Studies have indicated that capsaicin tends to accumulate primarily in the liver and intestines, leading to lower concentrations of the medication in other tissues and potentially reducing its efficacy. When capsaicin is taken orally, it is rapidly metabolized by cytochrome P450 enzymes, resulting in a short half-life of only 25 min and the production of active metabolites that could cause mutagenic effects dependent on the dose. All these demerits are overcome via utilizing nanotechnology. Nanotechnology utilizes the distinctive characteristics of materials at the nanoscale, and has gained widespread acceptance in numerous industries due to its ability to create superior and more intelligent products. The nanocarriers, such as liposomes, nanoemulsions, self-assembled nanoparticles, carbon nanotubes, dendrimers, and micelles are utilized to encapsulate or load capsaicinoids. This has resulted in improved biomedical applications, including drug and gene delivery, as well as detection and quantification purposes. This chapter provides a comprehensive information on the roles of capsaicinoid-loaded nanocarriers in therapeutics, detection, and quantification.

Keywords

Capsaicinoid · Pharmacology · Therapeutics · Nanoparticles · Nanotubes · Dendrimers

25.1 Introduction

Red chili peppers containing capsaicinoids are very popular worldwide as food additives and provide a pungent and hot taste. These capsaicinoids are flavor compounds composed mainly of capsaicin (C), dihydrocapsaicin (DHC), nordihydrocapsaicin (n-DHC), homocapsaicin (h-C), and homodihydrocapsaicin (h-DHC), with C and DHC contributing to around 80–90% of the total pungency in most chili peppers. Over 20 different capsaicinoids have been identified in various pepper species. Capsaicinoids are synthesized in the fruits' placenta by the condensation of vanillylamine and medium-chain length fatty acids (Giacalone et al. 2015; Lu et al. 2017).

Capsaicinoids is a group of phenolic alkaloids found mainly in genus *Capsicum*. Globally, a massive importance is given to hot pepper fruits in light of large-scale consumption of its fruits, as well as capsaicinoids utilized as food essences and in the contemporary medical applications. The primary chemical in chilli peppers that gives people a burning sensation is capsaicin, and it is also the most prevalent irritant in hot peppers. Dihydrocapsaicin makes up 22% of all capsaicinoids and has a pungency that is practically identical to that of capsaicin. Around 80–90% of capsaicinoids that occur naturally in spicy peppers were made up of the capsaicin (N-vanillyl-8-methyl-6-nonenamide) as well as dihydrocapsaicin (Cavett et al.

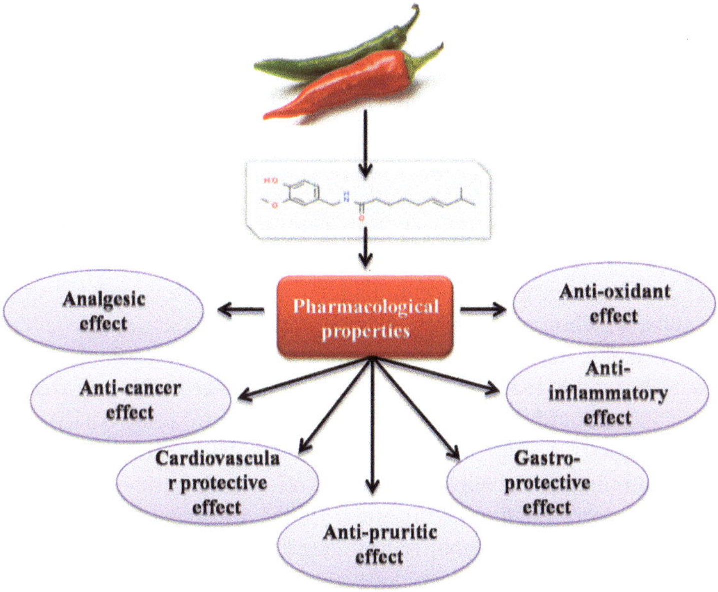

Fig. 25.1 Pharmacological properties of capsaicinoids

2004). Approximately 7% of the combination of capsaicinoids is nordihydrocapsaicin. Homocapsaicin makes up around 1% of all capsaicinoids and has half the pungency of capsaicin. Roughly 1% of all capsaicinoids are homodihydrocapsaicin, and it has about half the pungency of capsaicin. Capsaicin/dihydrocapsaicin ratios range from 1:1 to 2:1 (Hayman and Kam 2008). Each capsaicinoid has a vanilloid component, which is an aromatic ring with a methyl and hydroxyl group, as well as a lengthy chain of hydrocarbon and polar amide group attached. Capsaicinoids exhibit powerful antimutagenic, anticarcinogenic, and antioxidant effects (O'Neill et al. 2012; Jaganathan et al. 2020). From clinical trials, it has been found that the capsaicinoids are not carcinogenic when utilized in combinations of 65% capsaicin and 33% dihydrocapsaicin (Akagi et al. 1998). The capsaicin acts on the transient receptor potential cation channel vanilloid subfamily member 1 (TRPV1) and helps in relieving pain. TRPV1 is found to be the pharmacological target of capsaicin, due to evidences of its analgesic qualities. Capsaicinoids have been demonstrated to have a variety of biological actions like analgesic, anticancer, cardioprotective, anti-obesity, and gastrointestinal illness treatments (Sharma et al. 2013; Jaganathan et al. 2020) (Fig. 25.1). Capsaicinoids have shown to induce apoptosis in cancer cells and inhibit the development of cancer in various parts of the body, including the prostate, skin, breast, colon, lung, and bladder cancers. Oh et al. (2008) discovered that DHC can trigger catalase-mediated autophagy in HCT116 human colon cancer cells, while Ziglioli et al. (2009) found that capsaicin can induce apoptosis in prostate cancer cells via two pathways—a direct pathway (independent of the TRPV1 receptor) and an indirect pathway (dependent on the TRPV1 receptor). Capsaicinoids induced apoptosis in cancer cells through two pathways. In the direct pathway, capsaicinoids acted as a coenzyme Q antagonist to regulate electron transport, causing an excess of reactive oxygen species (ROS) that resulted in cell damage and apoptosis. In the indirect pathway, capsaicinoids interacted with the TRPV1 receptor, leading to an accumulation of Ca^{2+} in cancer cells and ultimately triggering early and late stages of apoptosis. Meanwhile, Choi et al. (2010) conducted a study on the effects of DHC on lung cancer cell lines, treating WI38, H1299, H460, and A549 cells with DHC to investigate its role in inducing autophagy in lung cancer cells. Their results showed that DHC could decrease ROS accumulation and promote autophagy in catalase-sensitive cells, which may help protect against apoptotic and necrotic cell death.

The pharmacokinetic properties of capsaicin directly affect biological reactions to it. Capsaicin has a deprived pharmacokinetic profile that is characterized via toxic metabolites and a brief half-life and is harmful to healthy cells. Analyses of tissue distribution profiles have shown that the liver and intestines are the primary sites of capsaicin accumulation, decreasing the medication concentration in other tissues, which has an impact on the effectiveness of the treatment. When taken orally, capsaicin is more quickly metabolized by cytochrome P450 enzymes, resulting in a small half-life (25 min) and the produce active metabolites that might cause dose-reliant mutagenesis effects (Reilly and Yost 2006; Suresh and Srinivasan 2010; Geng et al. 2016). To overcome these demerits, the capsaicinoids were loaded or encapsulated within the nanocarriers, such as liposomes, nanoparticles (NPS), nanoemulsions, self-assembled nanoparticles (SAN), carbon nanotubes, dendrimers, and micelles, and this has improved biomedical utilization including gene and drug delivery. In addition to the therapeutical application, capsine-loaded nanocarriers are used for the detection and quantification purposes. This chapter provides a comprehensive information on the roles of capsaicinoid-loaded nanocarriers in therapeutics, detection, and quantification.

25.2 Nanotechnological Interventions of Native Capsaicinoid to Overcome Inherent Constraints

Nanotechnology approaches are dominating all over the world due to nanoparticles' smaller sizes (between 1 and 100 nm) and their incredible physical, chemical, and biological properties (Rudramurthy et al. 2016). The word "nanomedicine" is used to describe the use of nanotechnologies in healthcare and medicine (Patra et al. 2018; Rudramurthy and Swamy 2018; Wang et al. 2020). In order to prevent, diagnose, monitor, and cure illnesses, nanomedicine employs technology at the nanoscale and nano-enabled procedures. Nanotechnologies have the potential to significantly advance medical treatments for a number of ailments, including cardiovascular disease, cancer, musculoskeletal conditions, psychiatric and neurodegenerative diseases, bacterial and viral infections, and diabetes. These treatments include tissue-engineered constructs, implants, drug delivery systems, imaging techniques, and diagnostic tools (Sim and Wong 2021).

25.2.1 Nanocarrier Synthesis

The development of different dimensional nanoformulations, including carbon nanotubes (CNPs), ceramic nanoparticles (CNPs), metal nanoparticles (MNPs), quantum dots, and lipidic- and polymeric-based nanoparticles, depends heavily on nanoparticles (NPs) (Rudramurthy and Swamy 2018). These NPs have different morphologies, sizes, shapes, and optical characteristics, but they excel in a variety of applications in many sectors (Bhatia 2016). There are three conventional ways for fabricating NPs with the required form and size based on their assemblies, which are physical methods, chemical-mediated and biological-mediated synthesis of nanoparticle approaches (Swamy et al. 2015; Swamy et al. 2018; Gour and Jain 2019).

25.2.1.1 Physical Methods for Nanoparticle Synthesis

Deposition, ball milling, sputtering, and plasma-dependent processes are used in the physical synthesis of nanomaterials. In most of these approaches, the amount of production of MNPs is quite sluggish. For example, the yield of nanomaterials using ball milling processes is 50% or less. Sputtering produces a large particle size distribution, with only 6–8% of sputtered material being smaller than 100 nm. Laser ablation and plasma methods necessitate significant energy consumption. General size distribution, sluggish manufacturing rate, and excess by-yields, along with significant depletion of energy, are done by utmost of physical processes too costly for realistic commercial uses (Gour and Jain 2019).

25.2.1.2 Chemical Methods of Nanoparticle Synthesis

A number of chemical techniques for nanoparticle (NP) synthesis have been proposed; besides the majority of them are extensively employed to create nanoformulation (e.g., sol-gel method, chemical reduction, pyrolysis, hydrothermal synthesis, microemulsion, polyol synthesis, and chemical vapor deposition). Furthermore, using lethal chemical reagents for synthesis and producing by-products is dangerous for both environment and humans. Therefore, the use of such nanomaterials especially in biological implementation is restricted (Gour and Jain 2019).

25.2.1.3 Biological Methods of Nanoparticle Synthesis

A new discipline called green nanotechnology uses green chemistry to create innovative NPs. A novel option for synthesizing NPs using natural reducing and stabilizing agents is offered by biological techniques. With no use of energy or harmful chemicals, it offers a cost-effective and ecologically beneficial alternative to chemical and physical procedures (Gour and Jain 2019). The bottom-up method of biological production of NPs uses basic single cell organism to complex multicellular living things including bacteria, fungus, actinomycetes, algae, yeast, and plant-based materials (Phanjom and Ahmed 2017; Chengzheng et al. 2018; Swamy et al. 2018; Huang et al. 2019).

Another alternative technique of creating nanoparticles is by microbial mediation. The external and intracellular microbial culture filtrates are employed in this synthesis process as a reducing agent to produce nanoparticles. Some microorganisms have the ability to tolerate metals and flourish in the worst environmental circumstances; these include bacteria, fungus, yeast, and actinomycetes (Swamy et al. 2018). Microbes use these innate characteristics to tolerate, amass, and transform the precursor moieties into the appropriate reducing form (Gour and Jain 2019). For instance, *Bacillus subtilis* was used to create the first bacterial gold nanoparticles (Southam and Beveridge 1994). By using nitrate reductase and anthraquinones, the fungus *Fusarium oxysporum* produced silver nanoparticle (Durán et al. 2005). Silver nanoparticles can be well applied safely to anti-microbial activity especially for microbial keratitis in the eye, when the contact lens gets infected with the bacteria, the effective concentrations of silver nanoparticles on various types of organisms are determined, The bacillus licheniformis is a gram-positive bacteria present as a saprophytic flora over the skin have the capability for adhesion and biofilm formation in contact lens and even on the inner walls of the lens, biofilm is distinct from their planktonic counterparts, since the biofilm prevents resistance to drug therapy, disinfectants and for increasing the pathogenicity. These enzymes have been reported to execute electron shuttle donor operations during the manufacture of silver nanoparticles (AgNPs) from *Bacillus licheniformis* (Kalishwaralal et al. 2008).

Nevertheless, green nanoparticles are another alternative way to traditional nanoparticle synthesis; an easy method should be used for nanoparticle synthesis and manufacture. There were many bioresources (plants, microorganisms) accessible throughout the universe. But using these bioresources is crucial from a synthesis and commercialization viewpoint. In such a scenario, microorganisms may be used successfully; however, handling bacteria, the scale-up procedure, genetic mutation, and obstacles in downstream processing, massive cultivation, and other variables create a block for the synthesis of NPs as well as their use.

From now on, research should proceed logically; using plant parts as a resource for the synthesis of nanoparticles is one such example. Many active substances, including phenols, saponins, alkaloids, tannins, flavonoids, vitamins, terpenoids, polysaccharides, polyphenols, etc., are present in plants. The nanoparticles were stabilized and these components were successfully minimized. Additionally, employing plants as a source for synthesis has benefits such as readily available plant material, low cost, simple scalability for large production, secondary metabolites, and purgative qualities. The synthesis of NPs with proper and effective usage of biological components will result in very stable and well-characterized NPs (Patil and Chandrasekaran 2020).

25.3 Nanotechnology for Capsaicinoids

25.3.1 Nanoparticles

NPs are the most interesting delivery system for medicinal components. NPs have the particle size less than 100 nm that reduced opsonization and the best plasma circulation duration. The smaller size also influences the cellular uptake as particle size shorter than 100 nm offers higher stability in circulatory system, a stronger ability to permeate biological barriers, and a superior capacity to concentrate inside cells, despite the fact that there is no agreement on the ideal size (Vafajoo et al. 2018). As mentioned earlier, the NPs can be synthesized or loaded with the phytomolecules; in both cases, it enhances the therapeutical potential of the loaded moiety and NPs. On the other hand, effective functionalization of the nanoparticle surface enables the monitoring of drug administration by attaching imaging reporters in addition to increasing precise targeting via ligand recognition or enzyme responsiveness. Moreover, by making pharmaceuticals more water-soluble and so reducing the overall amount necessary, vehicles for nanoparticle drug delivery greatly lessen the negative action of medications, notably anticancer treatment (Mohapatra et al. 2011; Vafajoo et al. 2018; Farjadian et al. 2018; Chitkara et al. 2018; Ghasemi et al. 2018).

There are many different cationic polymers that have been used to modify the surfaces of MNPs; nevertheless polyethylenimine (PEI) is unquestionably the most popular one used in gene and drug delivery systems due to its exact positively surface charge that makes it an excellent choice for consistent and predictable electrostatic interactions with negatively charged surfaces and carriers (McBain et al. 2008; Wang et al. 2009; Chertok et al. 2010). There is presently capsaicinoid-associated nanoparticle preparation that demonstrates a good tissue dissemination, beneficial physicochemical contour, higher encapsulation efficiency (EE %), plus minimal toxicity despite the therapeutic potential of capsaicinoids (CAP) and the benefits of its nanoencapsulation. Bovine serum albumin (BSA), a protein polymer that has been broadly employed as a matrix component for the nanoencapsulation of numerous medicines, may be cast off to enhance the CAP pharmacokinetics. The numerous drug-binding spots provided by albumin make it possible to transport both hydrophobic and hydrophilic medications. Additionally, it carries molecules with a variety of structural characteristics, is biodegradable, helps liposoluble medications dissolve, and is robust at the pH range of 4 to 9 and higher temperatures of 60 °C (Takano et al. 2007). Similarly, De Freitas et al. (2018) Similarly, in vitro and in vivo study was demonstrated for anti - inflammatory model the capsaicin were encapsulated within the albumin nanoparticles exhibited antioxidant action for reducing concentrations of TNF-alpha.

Noble nanomaterials, including gold, silver, iron, copper, and zinc nanoparticles, have appealing and distinctive optical features, such as surface plasmon resonance, that can be employed for many therapeutic and diagnostic purposes. Few applications now allow for the usage of metallic nanocarriers. The copper sulfide nanoparticles (CuSNPs) were decorated by antibodies against TRPV1, the resulting complex photothermal switch. The outcomes were determined to be substantial that employed in the forthcoming as a therapeutic tool to slow the progression of atherosclerosis (Gao et al. 2018). Capsaicin was used in another application to bioreduce silver nitrate ($AgNO_3$) to produce AgNPs, and the resulting capsaicin-capped AgNPs (particle mean size 20–30 nm) were discovered to be harmonious with blood types, necessitating certainly not additional research on this complicated nanostructured material (Amruthraj et al. 2015). The combined use of AuNPs chitosan produces unique materials that can be used in a variety of medical applications, including wound healing. Al-Musawi et al. (2020) presented a biocompatible and antimicrobial nano-fibrous wound dressings in skin treatments. They prepared honey and tripolyphosphate (TPP)/chitosan-coated (HTCs) nanofibers, containing capsaicin. AuNPs and capsaicin were loaded in the interior HTCs to form HTCs-AuNP, HTCs-capsaicin, and HTCs-capsaicin and AuNPs nanofibrous materials. The antibiotics difloxacin and chloramphenicol were tested in vitro against *Klebsiella rhinoscleromatis*, *Vibrio vulnificus*, *Staphylococcus pyogenes*, and *Pasteurella multocida*. The cytotoxicity test was used to assess the viability and proliferation of the produced nanofibers. Finally, the wound-closure mechanism was studied in vivo on New Zealand white rabbits. The results show that HTCs-capsaicin and HTCs-AuNPs are more effective at preventing bacterial growth than HTCs and HTCs-AuNP/capsaicin nanofibers and antibiotics. Using magnetic nanoparticles to fine-tune a target site-specific delivery is new and offers various benefits over conventional particles. Researchers were successful in preparing capsaicin magnetic nanoparticles (MNPs) that exhibited a promising and permissible biocompatibility. MNPs showed more toxicity as compared to SK-N-MC cells and the control group. Additionally, research demonstrated the efficient internalization of the delivery system. The in vitro test showed that the produced MNPs are almost completely pH stimuli responsive and release capsaicin under regulated conditions (Rabiee and Rabiee 2018). For site-specific pain treatment a biodegradable poly D, L-lactic-co-glycolic acid (PLGA)-coated capsaicin magnetic NPs (PCMN) provide relief by adjusting the localization of the magnetic particles to the preferred locations using an external magnetic field (Baskaran et al. 2017).

25.3.2 Self-Assembled Nanoparticles

Self-assembled micelles (SAM) are vesicles and are widely used as carriers of medicine and possess a precise active surface. Generally, they are referred to as self-assembled drug delivery systems (SADDS). SADDS provide clear advantages over standard nanoformulations, including excipient-free formulations, increased stability, and reduced burst release phenomena. Some researchers have successfully prepared capsaicin prodrugs that show higher antihyperlipidemic action, better absorption, lesser irritations at gastrointestinal tract (GIT), and enhanced in vitro drug release. Because of the flexibility in structure and acceptable electrostatic potential, the disulfide linkage of prodrugs (capsaicin) generated a reasonably stable SADDS nanoparticles in aqueous medium (Feng et al. 2018). Capsine SADDS NPs had 3.2 times the oral bioavailability of capsaicin alone, with a prolonged T-max between 0.25 and 1 h. In high-fat diet-induced hyperlipidemic rats, Cap-SSVE NPs (capsaicin prodrug nanoparticles) reduced specific lipid profile indicators in both liver tissue and blood samples with raised levels of HDL. The NPs get accumulated mostly in the liver, causing less discomfort in the GIT mucosa. The prepared Cap-SSVE NPs demonstrated improved bioavailability, higher hypolipidemic actions, and decreased mucosal irritations. Till date, very few SAM and SADDS has been formulated for capsaicinoid. Hence, more research has been required in the field of capsaicinoid SADDS.

25.3.3 Nanoemulsions

Nanoemulsions are nano-sized colloidal dispersions generated by dispersing an immiscible oil phase in an aqueous phase. Nanoemulsions are one such system that has been intensively explored and has a highly promising potential in overcoming numerous challenges such as those faced in medicine delivery, pesticide delivery, or the delivery of any other biologically powerful chemical. Because of the increased surface area and energy exchange, any medication turned into a nanoemulsion system has a better effectiveness and serves as a superior drug delivery mechanism. Nanoemulsions offer a wide range of uses in medical, skincare, agriculture, food, and a variety of other sectors because they provide an effective and protective method or vehicle (Naseema et al. 2021). In addition, nanoemulsions have several advantages, including stability against gravity separation, ease of handling, and promising product look and flavor (Pathakoti et al. 2017; Akbas et al. 2018). Nanoemulsions are utilized as carrier systems to increase the durability and bioavailability of hydrophobic compounds such as capsaicin. High energy processes provide several advantages, including better efficiency, the capacity to scale up, and the use of multiple oil and emulsifier types. Akbas et al. (2018) synthesized the capsaicin loaded nanoemulsions made with high-pressure homogenization and ultrasonication methods and assessed their antibacterial effects against *Escherichia coli* and *Staphylococcus aureus* and determined whether they may be employed as potential preservatives. In this work, capsinoid nanoemulsions were made at pH 3.8 and 7.4 and their characteristics were examined in order to simulate acidic and neutral food items. Overall, their results revealed that capsaicin nanoemulsions exhibit worthy antimicrobial activity.

Nanoemulsions offer several advantages as excellent topical drug carriers, including tiny droplet size, ease of manufacture, and elevated drug-loading capacity. These systems have the potential to distribute hydrophobic substances in aqueous medium with enhanced skin permeability and no skin exasperation (Ghiasi et al. 2019). To increase skin permeability and therapeutic benefits, an olive oil nanoemulsion was designed as a carrier for topical administration of capsaicin. The nanoemulsion was created using the self-emulsification process, and it had a tiny droplet size and significant physical stability, making it suitable for topical applications. The results showed that an optimal capsaicin nanoemulsion, particularly in the form of a topical gel, could improve the analgesic and anti-inflammatory effects of capsaicin when compared to commercial formulations, most likely due to the nano-sized particles containing drug agent with improved skin permeation ability. Nigam et al. (2019) formulated a nanoemulsion from capsaicin for its effective analgesic action via topical application. Further, they also demonstrated non-toxicity of capsinoid nanoemulsions confirmed by cytotoxity assay against Neuro-2a neuroblastoma cell line. About 80% of cells were observed to be viable with nanoemulsion prepared from capsaicin.

25.3.4 Carbon Nanotubes

The graphitic carbon tubes known as carbon nanotubes (CNTs) are said to possess exceptional and, in some cases, unmatched qualities. There has been active research finding, like displays of field emission, atomic force, hydrogen storage vehicles, and tips of scanning tunneling microscope, optoelectronics, nanoelectronics, and within composites. These 2D nanostructures are reported to be among the stiffest and strongest fibers known to man, and they also possess remarkable electronic properties along with many other exclusive and advantageous features (Metters et al. 2013).

Depending on the technique of manufacturing, single-walled carbon nanotubes (SWCNT) are essentially one graphene sheet that has been rolled up, with either exposed or bolted ends. They have exceptional strength, have a precise larger surface area per unit mass, and have special optical characteristics. The unit cell, which might have open ends with "armchair," "zig-zag," or "chiral" geometries, determines the structure of each individual graphene tube. Coaxial assemblies of graphene cylinders, multi-walled carbon nanotubes (MWCNT), have lengths of several microns and a diameter that ranges from 2 to 30 nm (Nugent et al. 2001). There have been several sensor-based devices described that take use of the positive properties of CNT integration, including those for point-of-care monitoring for diabetes and asthma as well as ultrasensitive detection of nitrous oxide, carbon monoxide, ammonia, and former gases. Researchers developed the electroanalytical sensors that were tested for the detection of model analytes using innovative screen-printed single-walled carbon nanotube electrodes on flexible polyester substrates. The single-walled carbon nanotube electrodes that are screen printed are also compared to commercially available substitutes and are demonstrated to be very effective for detecting capsaicin at low micromolar levels. Glass carbon electrodes covered with carbon nanotubes were used to create a high sensitivity test that produced good capsaicin detection in a variety of pepper sample types (Metters et al. 2013).

da Silva et al. (2021) found that loading capsine within the multi-walled carbon nanotubes (MWCNT) by redox catalysis helps to quantify the compounds. For the redox catalysis, it is crucial that two reversible redox systems exist following the oxidation of capsine on MWCNT. This may be explained by the development of a revocable surface restricted to quinone system. The obtained analytical curves for modified glassy carbon electrode, to be used in the quantitation of xanthurenic, ascorbic, and uric acids, along with the neurotransmitters like dopamine and epinephrine, displayed linear ranges, correspondingly between 10 and 95, 5–75, 5–115, 50–1150, and 5–70 mol L1. This oxidized capsine-based platform has micromolar sensitivity to the analytes.

25.3.5 Micelles and Dendrimers

Dendrimers and micelles are selected as examples of polymeric carriers. Micelles are nanoscopic core/shell structures made of amphiphilic copolymers, to put it briefly. These have a great chance of delivering chemicals that are bioavailable and hydrophobic. The self-assembly of different polymers into micelles in solutions is accomplished using the solvent-switch approach (direct dissolution method) and the dialysis method (organic solvent-free method). These are mostly employed to integrate medications that are water-soluble. Methods like direct dissolving and dialysis are employed to create polymeric micelles. Capsaicin was produced with a mean size below 50 nm for oral delivery using a sodium cholate/polyvinylpyrrolidone (PVP)/phospholipid mixed micellar system, increasing bioavailability by more than two times and reducing gastrointestinal mucosal irritation (Zhu et al. 2014; Chittepu et al. 2018). Methoxypoly(ethylene glycol)-poly(-ε-caprolactone) (mPEG-PCL) nanoparticles having particle size of 82.54 0.51 nm were used to administer capsaicin orally for a 60-h continuous release. According to pharmacokinetics, there has been a sixfold increase in stomach mucosa irritation (Peng et al. 2015). Dendrimers are composed of branches that resemble trees and have several functional terminals. They are also regarded as monodisperse macromolecules. The growth of these is reliant on cascade areas and is prepared using convergent or divergent techniques. By employing the esterification procedure, triethylamine, PEG 400, and oleoyl chloride were utilized to formulate dendrimers that were then coupled by capsaicin. The final preparation with a particle size of 143.1 nm was shown to be poisonous to zebrafish, but not to MCF-7 or Hep2 cells; however cytotoxicity was observed in these two types of cells (Malar and Bavanilathamuthiah 2015).

25.4 Pharmacological Properties of Capsaicinoid Nanoparticles

Capsaicin has a wide variety of possible applications, ranging from culinary seasonings to medical treatments. In point of fact, capsaicin and a select few of its analogues have been the subject of clinical research that has resulted in the granting of more than a thousand patents. Analgesic, anti-obesity, antipruritic, anti-inflammatory, anti-apoptotic, anticancer, antioxidant, gastroprotective, and neuroprotective are just some of the pleiotropic pharmacological activities that have been ascribed to capsaicin in the past (Wang et al. 2022; Mandal et al. 2023). In addition, new research points to its clinical significance in the treatment of disorders related to the cardiovascular system, such as metabolic syndrome and gastroprotective effects (Xiang et al. 2021; Singh et al. 2022). Because of a shortage of effective medications or some methods may be ineffective or issue with their delivery and absorption in the body for the treatment of these disorders, there is an urgent need for additional research into the pharmacological properties of capsaicin. In this context, researchers have investigated the preclinical impacts of capsaicin on the metabolic syndrome, because capsaicin is an agonist of transient receptor potential vanilloid channel 1 (TRPV1) (Fattori et al. 2016; Panchal et al. 2018). TRPV1 is found in many metabolically active tissues and regulates processes such as adipocyte browning and activation of metabolic modulators, such as AMP-activated protein kinase (AMPK), peroxisome proliferator-activated receptor (PPAR), uncoupling protein 1 (UCP1), and glucagon-like peptide 1 (GLP-1). Capsaicin's modification of these pathways can promote lipid burning, enhance insulin sensitivity, reduce excess weight, and strengthen liver and heart function (Shenoy et al. 2022). However, using capsaicin in metabolic syndrome may be possible in the form of innovative products as chitosan microspheres, liposomes, nanoparticles, or soft gel capsules. These products may be required to carry capsaicin to the intestines and avoid absorption in the stomach (Zhu et al. 2015; Peng et al. 2015). A perfect medication for managing metabolic syndrome might also be created by creating new, orally available TRPV1 agonists that replicate capsaicin's effectiveness but lack its pungency. The CH-19 sweet chilli pepper is one illustration; it contains the three analogues of capsaicin: capsiate, dihydrocapsiate, and nordihydrocapsiate (Sasahara et al. 2010).

Pure capsaicin has poor solubility in water, which contributes to its low bioavailability and permeability. Capsaicin is attached to the surface of metal nanoparticles in a conjugated form to increase bioavailability. Capsaicin's attachment to the surface of silver nanoparticles was reported by using FTIR and TEM analysis. The gold nanoparticles that were created were measured to be 20–30 nm in size. Our research revealed that capsaicin functions as a capping and reducing agent, stabilizing the silver solution for 6 months. In a hemagglutination test, the synthesized SNPs showed excellent ABO blood group compatibility (Fig. 25.2). In vivo studies can be performed on them to test the effectiveness of the capsaicin-capped silver nanoparticles for topical applications (Amruthraj et al. 2015).

In preliminary studies, capsaicin has already been employed as a ligand to activate many ion-channel receptor types (Jung et al. 1999; Zhang et al. 2005; Liu et al. 2020; Domene et al. 2022). Capsaicin-like substances' pharmacological effectiveness is influenced by a variety of variables, including dose, delivery system, and level in target organs. Researchers have developed nanoencapsulation methods to decrease the pungency of capsaicin and facilitate oral delivery. Zhu et al. (2014) examined in vitro and in vivo liposomal encapsulation capabilities as well as oral delivery using film-dispersion technology. Sodium cholate, cholesterol, isopropyl myristate, and phospholipid liposomes were made. Encapsulation inside a liposome increased capsaicin's bioavailability 3.34-fold in vivo. Research conducted on the irritation of the mucosal epithelium indicates that the liposomal system may be safely administered orally. To increase bioavailability of its bioactive components, the capsaicinoids were mixed into regular liposomes as well as penetration enhancer containing vesicles, both of which are forms of phospholipid vesicles (PEVs) (Pappalardo et al. 2021). They were tested on the human monoblastic leukemia U937 cell line and found to do not have genotoxicity. In this study, nitric oxide (NO) and reactive oxygen species (ROS) concentrations inside of cells were examined to determine how well the vesicles controlled inflammatory reactions in vitro.

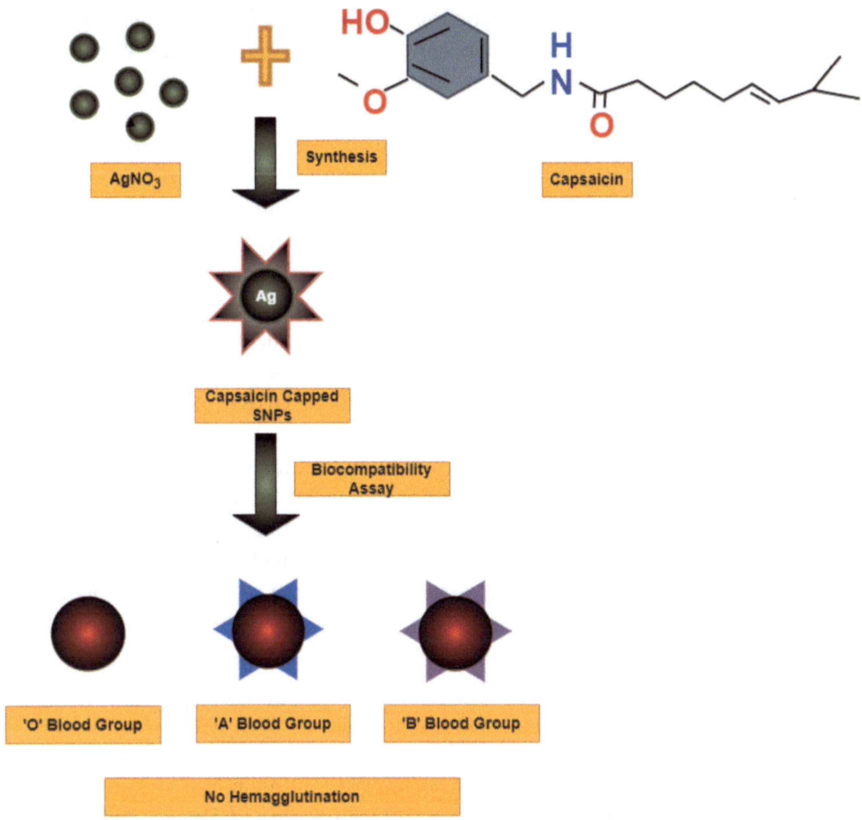

Fig. 25.2 In a hemagglutination test, the synthesized SNPs were compatible with ABO blood groups. (Adapted from Amruthraj et al. 2015)

ROS levels were dramatically lowered by liposomal inclusion in LPS-activated cells treated with the extract.

Kaiser et al. (2015a) constructed core-shell nanocapsules with an oily core and phospholipid stabilization, with or without chitosan covering, to lower the pungency of capsaicin. They looked at the in vitro release of capsaicin-loaded formulations in various physiological mediums. They also investigated the effect of capsaicin encapsulation on the cell viability of buccal cells (TR146). The in vitro release investigation revealed that the systems released capsaicin slowly and monotonically and that the chitosan coating influenced the release profile. The cytotoxic reaction of TR146 cells to capsaicin at a concentration of 500μM, which was visible when the chemical was free, was diminished after it was encapsulated. The sensory study undertaken to determine changes in capsaicin pungency following encapsulation demonstrated that a chitosan covering leads in a reduced threshold of perception of the formulation. The pungency sensation was significantly reduced after capsaicin was nanoencapsulated. When compared to uncoated systems, the inclusion of a chitosan shell around the nanoformulations did not disguise the pungency. Kaiser et al. (2015b) conducted a study on capsaicin to reversibly open cellular tight junctions utilizing a capsaicin-chitosan nanosystem in order to promote the mobility of low-permeability material (FITC-dextran) through a squamous cells monolayer. They compared free capsaicin and capsaicin-loaded chitosan nanocapsules for cytotoxicity toward epithelial MDCK-C7 cells, tight junction integrity, membrane permeability, and cellular uptake. MDCK-C7 cells respond cytotoxically to 500 mM free capsaicin but not when encapsulated. As shown by transepithelial electrical resistance and FITC-dextran-incubated monolayer permeability, nanocapsules altered the way tight junctions interacted with capsaicin. Kaiser et al. (2017) compared the effects of the nanosystem, free capsaicin, and control cells to the capsaicin-free nanoformulation and free capsaicin on the epithelial cell movement and conformation in Madin-Darby canine kidney (MDCK-C7) cultures. Fluorescence microscopy was used to analyze both changes to adhesion molecules protein redistribution and rearrangement of the microtubule. Calcium influx measurements were used to analyze the impact of the nanocomposites and capsaicin on ligand-gated ion activity. It was shown that both completely free capsaicins and chitosan nanocapsules containing and instead of capsaicin inside of them significantly impacted cell membrane movement as well as oriented cell motility. Changes in the roughness of the surface of the cell layer, the integrity of the tight junctions, and cytoskeleton changes are all correlated with an increase in the speed and direction of cell migration.

Sarwa et al. (2015) applied capsaicin topically as an anti-arthritic agent through the transfersomal vesicular system. According to the confocal laser scanning micrographic study, the characterization of the capsaicin-loaded transfersomes reveals their nano-size (94 nm), negative surface charge (14.5 mV), and sufficient structural flexibility, which resulted in 60.34% encapsulation efficiency, penetration across the biomembrane, and 76% of drug release from the vesicular system in 24 h in their intact form. At the same dosage level, authors showed that the preparation has more inhibitory action than the Thermagel commercial formulation. This might be as a result of Thermagel's lower skin barrier permeability compared to their unique transfersomal delivery technology. More information about the usefulness of the liposomal vesicle as a unique carrier system for the topical delivery of such highly irritating chemicals is provided by the enhanced biological system endurance of the developed vesicular formulation.

Ionotropic gelation of cationic trimethyl chitosan was used to make capsaicin-laden trimethyl chitosan nanoparticles (CL TMCS NPs), which were then loaded with capsaicin as a potential medication delivery route for capsaicin. The use of capsaicin-loaded NPs as an adjuvant therapeutic method in the management of HCC may be a novel way to boost the medication's efficiency and resistance to regularly used chemotherapy. This may be the case if the treatment is considered a novel technique to increase resistance. In order to determine the levels of gene expression for Bcl-2, Bax, and MDR-1 mRNA, real-time PCR was utilized. Furthermore, cell viability, immunocytochemistry assays for Bcl-2, Bax, and caspase proteins, as well as TEM and zeta analyzer were utilized in order to characterize the synthesized nanoparticles (Hazem et al. 2021).

Due to various pharmacological activities of capsaicin prodrug, Feng et al. (2018) formulated self-assembled nanoparticles from capsaicin prodrug to enhance their bioavailability and hypolipidemic effects and to reduce mucosal irritations. They recorded a significant result from the self-assembled capsaicin nanoparticles in comparison to free capsaicin. They reported more than three times higher oral bioavailability with a longer Tmax (0.25–1 h). In hyperlipidemia-induced rats, they reported decreased specific lipid profile from the liver tissue and blood samples. Finally, they concluded that the nanoparticles synthesized by capsaicin prodrug are working significantly to reduce mucosal irritations.

Using gel-phase homogenization, fluid-bed granulation, and galactomannan coating, capsaicinoids encapsulated in microbeads (CapF) with diameters ranging from 250 to 400 m were manufactured by Joseph et al. (2021). According to their oral bioavailability, research on Wistar rats demonstrated a higher absorption rate of capsaicinoids encapsulated in microbeads (CapF), which was found to be 19 times greater than that of unformulated capsaicin. It was discovered that CapF does not irritate the stomach mucosa and that it has a substantial anti-obesity impact in rats that are hypercholesteremic and are given a diet that is rich in fat.

In another study, Lorenzoni et al. (2021) used dermal microdialysis to measure capsaicinoid-loaded nanocapsules' in vivo skin penetration (CHNCCaps). Microdialysis showed that CHNCCaps increased capsaicinoids in the dermis by 1.6-fold compared to CHETCaps and commercial creams. CHNCCaps on the skin increased this (after 12 h of its application). The CHNCCaps showed antiallodynic and antihyperalgesic effects from 6 to 96 h within a week of beginning therapy, but the CHETCaps and commercial cream showed effects only at 48 and 96 h. The CHNCCaps showed comparable impacts between 6 and 96 h after administration. After therapy was stopped, CHNCCaps and the commercial cream maintained their antihyperalgesic activity for 6 days.

25.5 Conclusion and Future Perspectives

Nanocarrier-based drug delivery systems have been studied as possible new drug transporters for many decades, with the goal of lowering drug levels in the blood and toxicity for people who take prescription drugs regularly. The distribution of pharmaceuticals has been revolutionized thanks to nanomedicine, despite the fact that studies into its mechanism, safety, and toxicity are still ongoing. Since natural medications are less bioavailable and more water-soluble than synthetic ones, improved methods of drug distribution should investigate the therapeutic properties of natural medicines. The molecular size of a medication and its ability to dissolve in lipids are important factors. This chapter analyzes the literature on nanoformulation-based bioavailability of capsaicinoids. Capsaicinoids have been demonstrated to have a number of beneficial biological effects, including anti-inflammatory, anticancer, antioxidant, pain-relieving, cardioprotective, and anti-obesity properties.

Because of their promising applications in the pharmaceutical sector, natural compounds may be refined into a wide range of nanoforms. Its spiciness prevents it from being used in scientific studies. To counteract this, scientists have created compounds with similar structures but less of a bite. There is a lack of clinical studies, despite the fact that this compound shows great promise for a variety of potential therapeutic uses. If you're trying to find a way to lessen the intense heat of capsaicinoids, this book chapter might be a good place to start. Future research should concentrate heavily on improving capsaicinoids' bioavailability.

References

Akagi A, Sano N, Uehara H, Minami T, Otsuka H, Izumi K (1998) Non-carcinogenicity of Capsaicinoids in B6C3F1 mice. Food Chem Toxicol 36(12):1065–1071. https://doi.org/10.1016/S0278-6915(98)00077-5

Al-Musawi S, Albukhaty S, Al-Karagoly H, Sulaiman GM, Alwahibi MS, Dewir YH, Soliman DA, Rizwana H (2020) Antibacterial activity of honey/chitosan nanofibers loaded with capsaicin and gold nanoparticles for wound dressing. Molecules 25(20):4770

Amruthraj NJ, Preetam Raj JP, Lebel A (2015) Capsaicin-capped silver nanoparticles: its kinetics, characterization and biocompatibility assay. Appl Nanosci 5:403–409. https://doi.org/10.1007/s13204-014-0330-5

Baskaran M, Baskaran P, Arulsamy N, Thyagarajan B (2017) Preparation and evaluation of PLGA-coated capsaicin magnetic nanoparticles. Pharm Res 34:1255–1263

Bhatia S (2016) Nanoparticles types, classification, characterization, fabrication methods and drug delivery applications. In: Natural polymer drug delivery systems. Springer, Cham, pp 33–93. https://doi.org/10.1007/978-3-319-41129-3_2

Cavett V, Waninger EM, Krutak JJ, Eckenrode BA (2004) Visualization and LC/MS analysis of colorless pepper sprays. J Forensic Sci 49(3):469–476

Chengzheng W, Jiazhi W, Shuangjiang C, Swamy MK, Sinniah UR, Akhtar M, Umar A (2018) Biogenic synthesis, characterization and evaluation of silver nanoparticles from Aspergillus Niger JX556221 against human colon cancer cell line HT-29. J Nanosci Nanotechnol 18(5):3673–3681

Chertok B, David AE, Yang VC (2010) Polyethyleneimine-modified iron oxide nanoparticles for brain tumor drug delivery using magnetic targeting and intra-carotid administration. Biomaterials 31(24):6317–6324. https://doi.org/10.1016/j.biomaterials.2010.04.043

Chitkara D, Mittal A, Mahato RI (2018) Molecular medicines for cancer: concepts and applications of nanotechnology. CRC Press, London

Chittepu VC, Kalhotra P, Revilla GI, Velázquez TG (2018) Emerging technologies to improve capsaicin delivery and its therapeutic efficacy. In: Mozsik G (ed) Capsaicin and its human therapeutic development. IntechOpen, London. https://doi.org/10.5772/intechopen.77080

Choi CH, Jung YK, Oh SH (2010) Selective induction of catalase-mediated autophagy by dihydrocapsaicin in lung cell lines. Free Radic Biol Med 49(2):245–257

da Silva LV, dos Santos ND, de Almeida AK, dos Santos DD, Santos AC, França MC, Lima DJ, Lima PR, Goulart MO (2021) A new electrochemical sensor based on oxidized capsaicin/multi-walled carbon nanotubes/glassy carbon electrode for the quantification of dopamine, epinephrine, and xanthurenic, ascorbic and uric acids. J Electroanal Chem 881:114919

Domene C, Darré L, Oakes V, Gonzalez-Resines SA (2022) Potential route of capsaicin to its binding site in the TRPV1 ion channel. J Chem Inf Model 62(10):2481–2489

Durán N, Marcato PD, Alves OL, De Souza GIH, Esposito E (2005) Mechanistic aspects of biosynthesis of silver nanoparticles by several Fusarium oxysporum strains. J Nanobiotechnol 3:8. https://doi.org/10.1186/1477-3155-3

Farjadian F, Moghoofei M, Mirkiani S et al (2018) Bacterial components as naturally inspired nano-carriers for drug/gene delivery and immunization: set the bugs to work? Biotechnol Adv 36(4): 968–985. https://doi.org/10.1016/j.biotechadv.2018.02.016

Fattori V, Hohmann MS, Rossaneis AC, Pinho-Ribeiro FA, Verri WA Jr (2016) Capsaicin: current understanding of its mechanisms and therapy of pain and other pre-clinical and clinical uses. Molecules 21(7):844

Feng Y, Zhu Y, Wan J, Yang X, Firempong CK, Yu J, Xu X (2018) Enhanced oral bioavailability, reduced irritation and increased hypolipidemic activity of self-assembled capsaicin prodrug nanoparticles. J Funct Food 44:137–145. https://doi.org/10.1016/j.jff.2018.03.006

Gao W, Sun Y, Cai M, Zhao Y, Cao W, Liu Z, Cui G, Tang B (2018) Copper sulfide nanoparticle as a photothermal switch for TRPV1 signaling to attenuate atherosclerosis. Nat Commun 9(1): 1–10. https://doi.org/10.1038/s41467-017-02657-z

Geng G, Zheng Y, Meng M, Guo Z, Cao C, Ma X, Du Z, Li J, Duan Y, Du Y (2016) Gingerol reverses the cancer-promoting effect of capsaicin by increased TRPV1 level in a urethane-induced lung carcinogenic model. J Agric Food Chem 64:6203–6211. https://doi.org/10.1021/acs.jafc.6b02480

Ghasemi A, Rabiee N, Ahmadi S et al (2018) Optical assays based on colloidal inorganic nanoparticles. Analyst 143:3249–3283. https://doi.org/10.1039/C8AN00731D

Ghiasi Z, Esmaeli F, Aghajani M, Ghazi-Khansari M, Faramarzi MA, Amani A (2019) Enhancing analgesic and anti-inflammatory effects of capsaicin when loaded into olive oil nanoemulsion: an in vivo study. Int J Pharm 559:341–347. https://doi.org/10.1016/j.ijpharm.2019.01.043

Giacalone M, Forfori F, Giunta F (2015) Chili pepper compounds in the management of neuropathic pain. In: Watson RR, Preedy VR (eds) Bioactive nutraceuticals and dietary supplements in neurological and brain disease prevention and therapy. Elsevier, San Diego, CA, pp 187–195. https://doi.org/10.1016/B978-0-12-411462-3.00020-5

Gour A, Jain NK (2019) Advances in green synthesis of nanoparticles. Artif Cells Nanomed Biotechnol 47(1):844–851

Hayman M, Kam PCA (2008) Capsaicin: a review of its pharmacology and clinical applications. Curr Anaesth Crit Care 19(5–6):338–343. https://doi.org/10.1016/j.cacc.2008.07.003

Hazem NM, ElKashef WF, El-Sherbiny IM, Emam AA, Shaalan D, Sobh M (2021) Anticarcinogenic effects of capsaicin-loaded nanoparticles on in vitro hepatocellular carcinoma. Curr Chem Biol 15(2):188–201. https://doi.org/10.2174/2212796814999201116211648

Huang F, Long Y, Liang Q, Purushotham B, Swamy MK, Duan Y (2019) Safed Musli (Chlorophytum borivilianum L.) callus-mediated biosynthesis of silver nanoparticles and evaluation of their antimicrobial activity and cytotoxicity against human colon cancer cells. J Nanomater 2019:2418785. https://doi.org/10.1155/2019/2418785

Jaganathan R, Purushotham B, Radhakrishnan N, Swamy MK (2020) Capsaicin and its potential anticancer mechanisms of action. In: Swamy MK (ed) Plant-derived bioactives: chemistry and mode of action. Springer, Singapore, pp 301–321

Joseph A, Balakrishnan AM, Mulakal JN, Sivadasan SD, Mohan R, Maliakel B, Madhavamenon KI (2021) A green approach for the sustained-intestinal delivery of red chili (Capsicum annum L) extracted capsaicinoids with enhanced bioavailability. J Funct Foods 85:104658

Jung J, Hwang SW, Kwak J, Lee SY, Kang CJ, Kim WB, Kim D, Oh U (1999) Capsaicin binds to the intracellular domain of the capsaicin-activated ion channel. J Neurosci 19(2):529–538

Kaiser M, Kirsch B, Hauser H, Schneider D, Seuß-Baum I, Goycoolea FM (2015a) In vitro and sensory evaluation of capsaicin-loaded nanoformulations. PLoS One 10(10):e0141017. https://doi.org/10.1371/journal.pone.0141017

Kaiser M, Pereira S, Pohl L, Ketelhut S, Kemper B, Gorzelanny C, Galla HJ, Moerschbacher BM, Goycoolea FM (2015b) Chitosan encapsulation modulates the effect of capsaicin on the tight junctions of MDCK cells. Sci Rep 5(1):1–4

Kaiser M, Pohl L, Ketelhut S, Kastl L, Gorzelanny C, Götte M, Schnekenburger J, Goycoolea FM, Kemper B (2017) Nanoencapsulated capsaicin changes migration behavior and morphology of madin darby canine kidney cell monolayers. In: Nanoencapsulated capsaicin changes migration

behavior and morphology of madindarby canine kidney cell monolayers. PLoS One 12(11): e0187497. https://doi.org/10.1371/journal.pone.0187497
Kalishwaralal K, Deepak V, Ramkumarpandian S, Nellaiah H, Sangiliyandi G (2008) Extracellular biosynthesis of silver nanoparticles by the culture supernatant of Bacillus licheniformis. Mat Lett 62(29):4411–4413. https://doi.org/10.1016/j.matlet.2008.06.051
Liu TY, Chu Y, Mei HR, Chang D, Chuang HH (2020) Two vanilloid ligand bindings per channel are required to transduce capsaicin-activating stimuli. Front Mol Neurosci 12:302
Lorenzoni R, Contri RV, de Lima CK, Barreto F, de Araujo BV, Pohlmann AR, de Miranda AL, Dalla Costa T, Guterres SS (2021) Dermatopharmacokinetic and pharmacodynamic evaluation of a novel nanostructured formulation containing capsaicinoids for treating neuropathic pain. Int J Pharm 596:120294
Lu M, Ho CT, Huang Q (2017) Extraction, bioavailability, and bioefficacy of capsaicinoids. J Food Drug Anal 25(1):27–36
Malar CG, Bavanilathamuthiah (2015) Evaluation of biocompatibility of capsaicin-loaded dendrimersonzebra fish embryos. Int J Drug Deliv Technol 5(2):54–58
Mandal SK, Rath SK, Logesh R, Mishra SK, Devkota HP, Das N (2023) Capsicum annuum L. and its bioactive constituents: a critical review of a traditional culinary spice in terms of its modern pharmacological potentials with toxicological issues. Phytother Res 37(3):965–1002
McBain SC, Yiu HH, Dobson J (2008) Magnetic nanoparticles for gene and drug delivery. Int J Nanomedicine 3(2):169–180. https://doi.org/10.2147/ijn.s1608
Metters JP, Gomez-Mingot M, Iniesta J, Kadara RO, Banks CE (2013) The fabrication of novel screen printed single-walled carbon nanotube electrodes: electroanalytical applications. Sensors Actuators B Chem 177:1043–1052. https://doi.org/10.1016/j.snb.2012.11.078
Mohapatra S, Rout SR, Maiti S, Maiti TK, Panda AB (2011) Monodisperse mesoporous cobalt ferrite nanoparticles: synthesis and application in targeted delivery of antitumor drugs. J Mater Chem 21(25):9185–9193. https://doi.org/10.1039/C1JM10732A
Naseema A, Kovooru L, Behera AK, Kumar KP, Srivastava P (2021) A critical review of synthesis procedures, applications and future potential of nanoemulsions. Adv Colloid Interf Sci 287: 102318. https://doi.org/10.1016/j.cis.2020.102318
Nigam K, Gabrani R, Dang S (2019) Nano-emulsion from capsaicin: formulation and characterization. Mater Today 18:869–878
Nugent JM, Santhanam KSV, Rubio A, Ajayan PM (2001) Fast electron transfer kinetics on multiwalled carbon nanotube microbundle electrodes. Nano Lett 1:87. https://doi.org/10.1021/nl005521z
O'Neill J, Brock C, Olesen AE, Andresen T, Nilsson M, Dickenson AH (2012) Unravelling the mystery of capsaicin: a tool to understand and treat pain. Pharmacol Rev 64(4):939–971. https://doi.org/10.1124/pr.112.006163
Oh SH, Kim YS, Lim SC, Hou YF, Chang IY, You HJ (2008) Dihydrocapsaicin (DHC), a saturated structural analog of capsaicin, induces autophagy in human cancer cells in a catalase-regulated manner. Autophagy 4(8):1009–1019
Panchal SK, Bliss E, Brown L (2018) Capsaicin in metabolic syndrome. Nutrients 10(5):630
Pappalardo I, Santarsiero A, Luca MD, Acquavia MA, Todisco S, Caddeo C, Bianco G, Infantino V, Martelli G, Vassallo A (2021) Exploiting the anti-inflammatory potential of white capsicum extract by the nanoformulation in phospholipid vesicles. Antioxidants (Basel) 10:1683. https://doi.org/10.3390/antiox10111683
Pathakoti K, Manubolu M, Hwang HM (2017) Nanostructures: current uses and future applications in food science. J Food Drug Anal 25(2):245–253
Patil S, Chandrasekaran R (2020) Biogenic nanoparticles: a comprehensive perspective in synthesis, characterization, application and its challenges. J Genet Eng Biotechnol 18(1):1–23. https://doi.org/10.1186/s43141-020-00081-3
Patra JK, Das G, Fraceto LF, Campos EV, Rodriguez-Torres MD, Acosta-Torres LS, Diaz-Torres LA, Grillo R, Swamy MK, Sharma S, Habtemariam S (2018) Nano based drug delivery systems: recent developments and future prospects. J Nanobiotechnol 16(1):1–33

Peng W, Jiang XY, Zhu Y, Omari-Siaw E, Deng WW, Yu JN, Xu XM, Zhang WM (2015) Oral delivery of capsaicin using MPEG-PCL nanoparticles. Acta Pharmacol Sin 36(1):139–148. https://doi.org/10.1038/aps.2014.113

Phanjom P, Ahmed G (2017) Effect of different physicochemical condition on the synthesis of silver nanoparticles using fungal cell filtrate of Aspergillus oryzae (MTCC No. 1846) and their antibacterial effect. Adv Nat Sci Nanosci Nanotechnol 8:045016. https://doi.org/10.1088/2043-6254/aa92bc

Rabiee N, Rabiee M (2018) Magnetic stimuli-responsive cobalt ferrite nanoparticle as theranostic agents for targeted delivery. Curr Nanomater 3(3):160–167. https://doi.org/10.2174/2405461504666190131141236

Reilly CA, Yost GS (2006) Metabolism of capsaicinoids by P450 enzymes: a review of recent findings on reaction mechanisms, bio-activation, and detoxification processes, drug. Metab Rev 38:685–706. https://doi.org/10.1080/03602530600959557

Rudramurthy GR, Swamy MK (2018) Potential applications of engineered nanoparticles in medicine and biology: an update. J Biol Inorgan Chem 23:1185–1204

Rudramurthy GR, Swamy MK, Sinniah UR, Ghasemzadeh A (2016) Nanoparticles: alternatives against drug-resistant pathogenic microbes. Molecules 21(7):836

Sarwa KK, Mazumder B, Rudrapal M, Verma VK (2015) Potential of capsaicin-loaded transfersomes in arthritic rats. Drug Deliv 22:638–646. https://doi.org/10.3109/10717544.2013.871601

Sasahara I, Furuhata Y, Iwasaki Y, Inoue N, Sato H, Watanabe T, Takahashi M (2010) Assessment of the biological similarity of three capsaicin analogs (Capsinoids) found in non-pungent chili pepper (CH-19 Sweet) fruits. Biosci Biotechnol Biochem 74(2):274–278

Sharma SK, Vij AS, Sharma M (2013) Mechanisms and clinical uses of capsaicin. Eur J Pharmacol 720(1–3):55–62. https://doi.org/10.1016/j.ejphar.2013.10.053

Shenoy A, Buttar HS, Dicholkar P, Kaur G, Chintamaneni M (2022) Role of nutraceuticals, functional foods, and spices in the management of metabolic syndrome and related disorders. In: Singh RB, Watanabe S, Isaza AA (eds) Functional foods and nutraceuticals in metabolic and non-communicable diseases. Academic Press, London, pp 583–601

Sim S, Wong NK (2021) Nanotechnology and its use in imaging and drug delivery. Biomed Rep 14(5):1–9

Singh SA, Uddin MO, Khan MM, Chishti AS, Bhat UH (2022) Therapeutic properties of capsaicin: a medicinally important bio-active constituent of chilli pepper. Asian J Pharm Clin Res 15(7): 47–58

Southam G, Beveridge TJ (1994) The in vitro formation of placer gold by bacteria. Geochim Cosmochim Acta 58(20):4527–4530. https://doi.org/10.1016/0016-7037(94)90355-7

Suresh D, Srinivasan K (2010) Tissue distribution & elimination of capsaicin, piperine & curcumin following oral intake in rats. Indian J Med Res 131:682–691

Swamy MK, Akhtar MS, Mohanty SK, Sinniah UR (2015) Synthesis and characterization of silver nanoparticles using fruit extract of Momordica cymbalaria and assessment of their in vitro antimicrobial, antioxidant and cytotoxicity activities. Spectrochim Acta A Mol Biomol Spectrosc 151:939–944

Swamy MK, Rudramurthy GR, Patra JK, Sinniah UR (2018) Microbe-based metallic nanoparticles synthesis and biomedical applications: an update. In: Patra J, Das G, Shin HS (eds) Microbial biotechnology, Application in food and pharmacology, vol 2. Springer Nature, Singapore, pp 395–434

Takano F, Yamaguchi M, Takada S, Shoda S, Yahagi N, Takahashi T, Ohta T (2007) Capsicum ethanol extracts and capsaicin enhance interleukin-2 and interferon gamma production in cultured murine Peyer's patch cells ex vivo. Life Sci 80:1553–1563. https://doi.org/10.1016/j.lfs.2007.01.031

Vafajoo A, Rostami A, Parsa SF et al (2018) Early diagnosis of disease using microbead array technology: a review. Anal Chim Acta 1032:1–17. https://doi.org/10.1016/j.aca.2018.05.011

Wang X, Zhou L, Ma Y, Li X, Gu H (2009) Control of aggregate size of polyethyleneimine-coated magnetic nanoparticles for magnetofection. Nano Res 2(5):365–372. https://doi.org/10.1007/s12274-009-9035-6

Wang Y, Zhang X, Bai Y, Li W, Li X, Xing X, Wang C, Gao L, Yogi M, Swamy MK, Dupadahalli K (2020) Anticancer and antibacterial activities of silver nanoparticles (AgNPs) synthesized from Cucumis melo L. J Nanosci Nanotechnol 20(7):4143–4151

Wang F, Xue Y, Fu L, Wang Y, He M, Zhao L, Liao X (2022) Extraction, purification, bioactivity and pharmacological effects of capsaicin: a review. Crit Rev Food Sci Nutr 62(19):5322–5348

Xiang Q, Guo W, Tang X, Cui S, Zhang F, Liu X, Zhao J, Zhang H, Mao B, Chen W (2021) Capsaicin—the spicy ingredient of chili peppers: a review of the gastrointestinal effects and mechanisms. Trends Food Sci Technol 116:755–765

Zhang N, Inan S, Cowan A, Sun R, Wang JM, Rogers TJ, Caterina M, Oppenheim JJ (2005) A proinflammatory chemokine, CCL3, sensitizes the heat-and capsaicin-gated ion channel TRPV1. Proc Natl Acad Sci 102(12):4536–4541

Zhu Y, Peng W, Zhang J, Wang M, Firempong CK, Feng C, Liu H, Xu X, Yu J (2014) Enhanced oral bioavailability of capsaicin in mixed polymeric micelles: preparation, in vitro and in vivo evaluation. J Funct Foods 8(1):358–366. https://doi.org/10.1016/j.jff.2014.04.001

Zhu Y, Wang M, Zhang J, Peng W, Firempong CK, Deng W, Wang Q, Wang S, Shi F, Yu J, Xu X (2015) Improved oral bioavailability of capsaicin via liposomal nanoformulation: preparation, in vitro drug release and pharmacokinetics in rats. Arch Pharm Res 38:512–521. https://doi.org/10.1007/s12272-014-0481-7

Ziglioli F, Frattini A, Maestroni U, Dinale F, Ciufifeda M, Cortellini P (2009) Vanilloid-mediated apoptosis in prostate cancer cells through a TRPV-1 dependent and a TRPV-1-independent mechanism. Acta Biomed 80(1):13–20

De Freitas GBL, De Almeida DJ, Carraro E, et al.(2018) Formulation characterization and in vitro/in vivo studies of capsaicin-loaded albumin nanoparticles Materials Science and Engineering: C 93:70–79 https://doi.org/10.1016/j.msec.2018.07.064

Akbas E, Soyler B, Oztop MH (2018) Formation of capsaicin loaded nanoemulsions with high pressure homogenization and ultrasonication. LWT 96:266–273. https://doi.org/10.1016/j.lwt.2018.05.043

Pharmaceutical Modifications of Capsaicin to Enhance its Bioavailability

26

Mansi Prashar, Neelima Dhingra, and Satyam Raj

Abstract

Capsaicin or N-vanillyl-8-methyl-6-(E)-none amide, a unique key constituent from the *Capsicum* genus, is a pharmaceutical agent with a wide range of therapeutic applications like diabetes, obesity, cancer, and other inflammatory conditions. Clinically approved capsaicin pharmaceutical formulations are known to associate with certain concentration-dependent side effects, including skin irritation and burning sensations. Further, the major caveats of capsaicin are its short half-life, and it belongs to Biopharmaceutical Classification System (BCS) II on account of its low bioavailability. About 60% of drug candidates fail to reach the market and do not attain their full potential because of low solubility and poor bioavailability behaviour, and thus present the most challenging aspects for new chemical entities in the drug development process. Nano-sizing, micronization, solid dispersions, cyclodextrins, crystal engineering, solid lipid nanoparticles, and additional colloidal drug delivery systems like microemulsions, self-emulsifying and self-microemulsifying systems, and liposomes are some of the newly developed methods to promote the saturation solubility, dissolution rate, and subsequently the bioavailability of BCS Class II. The present chapter aims to summarize the so far adopted technologies to upsurge the BCS parameters of capsaicin and alleviate its associated side effects. The development of new capsaicin formulations with enhanced bioavailability and delivery systems will raise hopes for innovative applications of capsaicin for health complications.

Keywords

Capsaicin · Solubility · Bioavailability · Nanocrystal · Microemulsions

M. Prashar · N. Dhingra (✉) · S. Raj
University Institute of Pharmaceutical Sciences, Panjab University, Chandigarh, India

M. K. Swamy, A. Kumar (eds.), *Capsaicinoids*,
https://doi.org/10.1007/978-981-99-7779-6_26

26.1 Introduction

Capsaicin, an active constituent and secondary metabolite of chili pepper belongs to the economically important genus *Capsicum* of the Solanaceae family. Peppers (*Capsicum* spp.), small spreading shrub of both the pungent (hot or chilli pepper) and non-pungent (sweet pepper) types, are important spice and vegetable crop. Considering the uncertain taxonomy and lack of universal consensus on nomenclature, 200 varieties have been grouped into more than 30 species, and 5 of them *C. annuum*, *C. frutescens*, *C. pubescens*, *C. chinense*, and *C. baccatum* are majority domesticated as shown in Fig. 26.1 (Hernandez-Perez et al. 2020; Suresh and Srinivasan 2010; Al Othman et al. 2011).

Capsaicin was used to spice up cuisines, particularly in South America and Mexico, and slowly it became an essential element of recipes in India, Korea, Ethiopia, Sri Lanka, Malaysia, China, and Thailand. Presently it is being eaten by a quarter of the earth's population every day all over the globe (Clark and Lee 2016). The pungent active ingredient of chilli peppers responsible for pain and burning sensation was first isolated in 1816 by Bucholz in a partially purified crystalline form, followed by isolation of pure crystalline form by Thresh in 1876, who later named it capsaicin. Capsaicin's chemical structure was elucidated by Nelson in 1919 (Nelson 1919), and Späth and Darling (Bode and Dong 2011) successfully synthesized in their laboratories in 1930. Chemically Capsaicin is, N-vanillyl-8-methyl-6-(E)-noneamide (**MF:** $C_{18}H_{27}NO$, **MP:** 62–65 °C, Fig. 26.2); (Barbero et al. 2010) and a colourless, odourless lipophilic solid with a dispersibility in ethanol, acetone, and fatty oils. It is a member of the vanilloid family with compounds such as eugenol (extracted from bay leaves and cloves), vanillin (derived from vanilla), and zingerone (encountered in ginger) and possesses structural

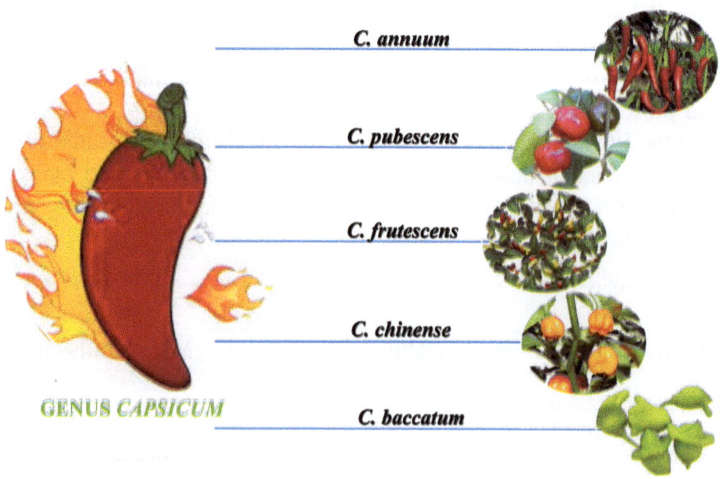

Fig. 26.1 Major domesticated species of *Capsicum*

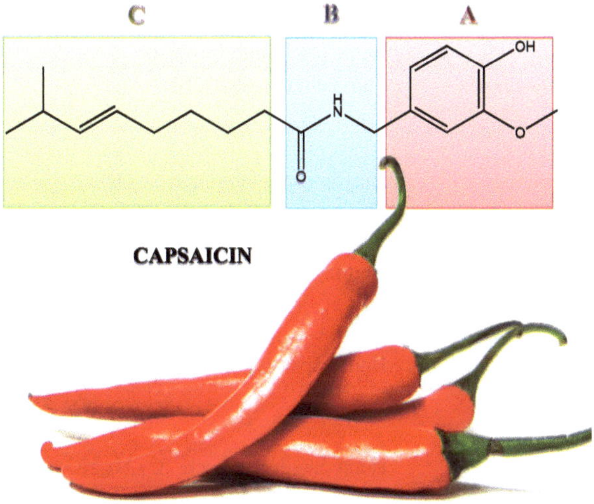

Fig. 26.2 Structural components of the capsaicin. (**a**) Aromatic ring; (**b**) amide bond; and (**c**) hydrophobic side chain

similarities by possessing a polar amide group, an aromatic ring, and a long hydrophobic chain (Hayman and Kam 2008).

A dominant *Capsicum* gene determines the quantity of capsaicin in chili, but the activity of polygenes functioning cumulatively controls the varying degrees of pungency. Scoville heat is a scale that measures the "hotness" of chilli peppers and indicates the number of dilutions with water necessary for the pepper to lose its temperature. Capsaicin's "heat feeling" is caused by its interaction with transient receptor potential vanilloid subfamily member 1 (TRPV1) ion-channel receptors (Reyes-Escogido Mde et al. 2011) previously known as the vanilloid receptor (VR1). These interactions with specific nerve membrane TRPV1 receptor also known to play a vital role in the transmembrane signalling process engaged in inflammatory hyperalgesia, thermal nociception, and a variety of physiological conditions ranging from pain, obesity, diabetes, and inflammatory diseases to infection disorders (Brito et al. 2014; Sharma et al. 2013).

Phenylpropanoid, and fatty acid metabolism are the two pathways that contribute towards the capsaicin biosynthesis and its concentration found to increase gradually with fruit development, which reaches to maximum levels in 40–50 days. *Capsicum* has many more compounds with comparable structure and biological functions to that of capsaicin. They are categorized as the primary capsaicinoids, dihydrocapsaicin, nordihydrocapsaicin, homodihydrocapsaicin, and homocapsaicin (Xiang et al. 2022). Capsaicinoids and dihydrocapsaicin differ by means of presence of a carbon-carbon double bond or saturation of the acyl group and are the two most potent capsaicinoids. Spicy flavor (pungency) of chili pepper fruit is because of capsaicinoids, the active pungent ingredient with wide range of pharmaceutical properties, and 69% of total capsaicinoids is present as capsaicin. Recent studies have further revealed capsiate, nordihydrocapsiate and dihydrocapsiate, and dihydrocapsiconiate and capsiconiate are the non-pungent analogues of capsaicin in the *Capsicum* in the form of capsinoids and capsiconinoids, respectively

Table 26.1 Chemical structure of capsaicinoids, capsinoids, and capsiconinoids

	Name (molecular formula)	Structure
Capsaicinoids	Capsaicin ($C_{18}H_{27}NO_3$)	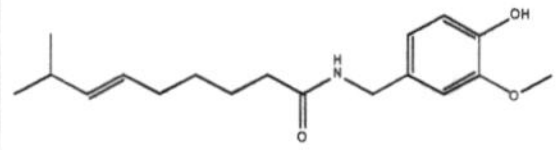
	Dihydrocapsaicin ($C_{18}H_{29}NO_3$)	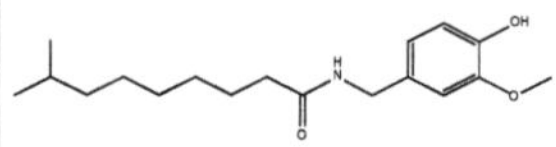
	Nordihydrocapsaicin ($C_{17}H_{27}NO_3$)	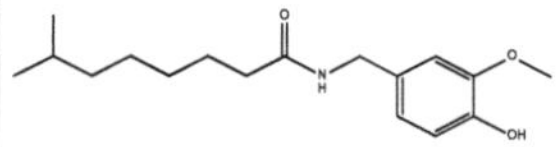
	Homocapsaicin ($C_{19}H_{29}NO_3$)	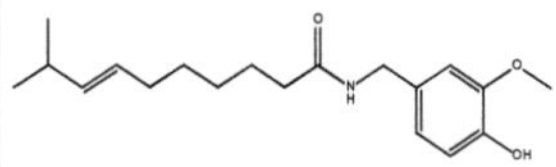
	Homodihydrocapsaicin ($C_{19}H_{31}NO_3$)	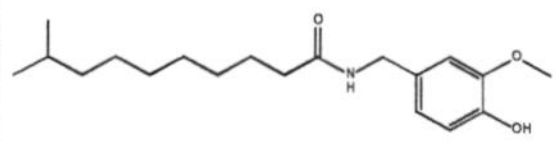
Capsinoids	Capsiate ($C_{18}H_{26}O_4$)	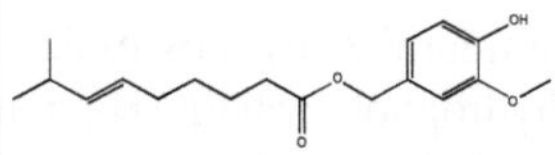
	Dihydrocapsiate ($C_{18}H_{28}O_4$)	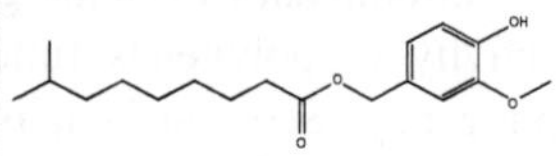
	Nordihydrocapsiate ($C_{18}H_{26}O_4$)	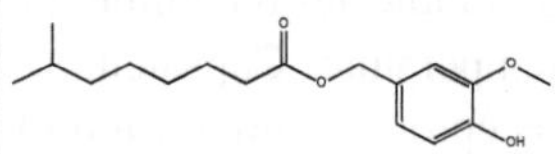
Capsiconinoids	Capsiconiate ($C_{20}H_{28}O_4$)	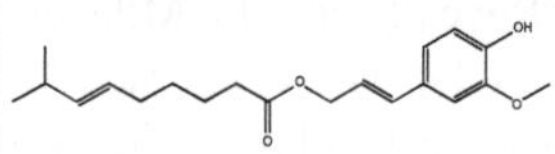
	Dihydrocapsiconiate ($C_{20}H_{30}O_4$)	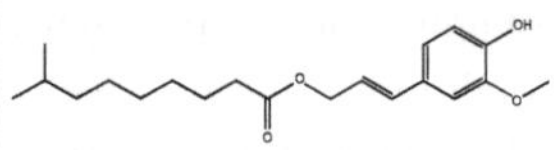

(Hernandez-Perez et al. 2020). Table 26.1 (Suresh and Srinivasan 2010) represents the chemical structure and biosynthetic precursors that differ from capsaicin.

Capsaicin pharmacological effects have been recognized since ancient times, but extensive research in the past one decade helped in determining its more applications in cardiovascular, obesity (Baboota et al. 2014; Baskaran et al. 2016), gastrointestinal tract disorders, and cancer. It has been proposed that capsaicin exerts its antitumorigenic effect by inhibiting oncogenic signalling pathways and tumour promoters or by stimulating tumour-suppressive signalling pathway and associated transcription factors. Furthermore, its synergistic effect with other anticancer agents, has opened the doors to explore the possibility of using it in combination with other chemotherapeutic agents to treat cancer. Different studies conducted over the last few decades (Fig. 26.3) demonstrated its efficiency in weight loss (Yu et al. 2012)

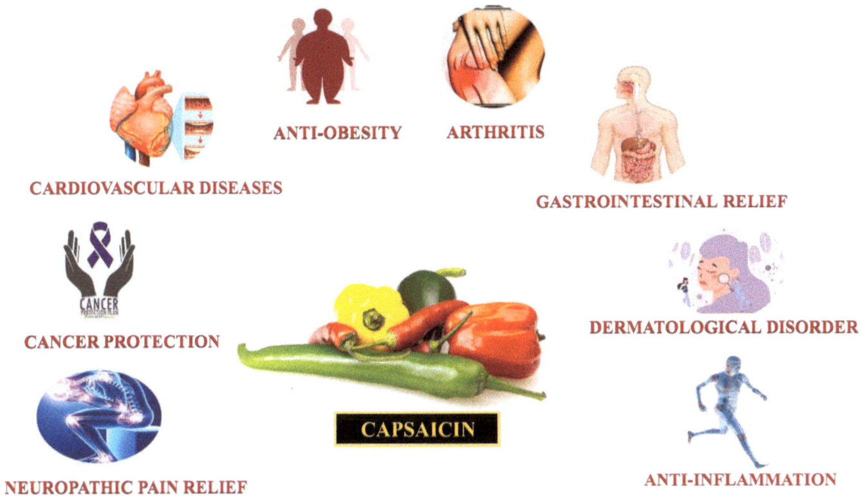

Fig. 26.3 Different therapeutic efficacies of capsaicin

and treating obesity (Ludy et al. 2012). It has been suggested that it decreases adipose tissue in rodents by growing energy and lipid metabolism, along with enhanced secretion of catecholamine from the adrenal medulla in response to the activation of sympathetic nervous system. Additionally, capsaicin has been found to mediate its anti-ulcerous effect by vasodilation and increased gastric mucous blood flow, which is further facilitated by release of nitric oxide and calcium gene-related peptide (CGRP) by TRPV1-bearing cells. But the controversial result to protect or damage gastric mucosa at low or high doses presented the need for its detailed clinical trials in order to describe effective safe dosages (In Ohk Lee et al. 2007).

26.2 Existing Clinically Approved Capsaicin Formulations

The capsaicinoids possess wide-spectrum activity like analgesic (O'Neill et al. 2012), antioxidant, anti-inflammatory (Desai et al. 2013) along with anti-carcinogenic effect that inhibit androgen-dependent (Lee et al. 2011) development of gastric adenocarcinoma and breast, prostate, and colon cancer. But the most frequent clinical application of capsaicin for therapeutic purpose is pain regulation and innumerable studies on extraction and isolation of capsaicin have provided us with number of formulations (Zhao et al. 2016). Since the early 1980s, numbers of the formulations have been designed with low concentration of capsaicin (0.025–0.1%w/w) suitable for frequent topical drug delivery system (Anand and Bley 2011) in the form of lotions, creams, and patches, and some of the existing formulations (Fig. 26.4) include capzasin-HP, Qutenza patches, and leader capsaicin.

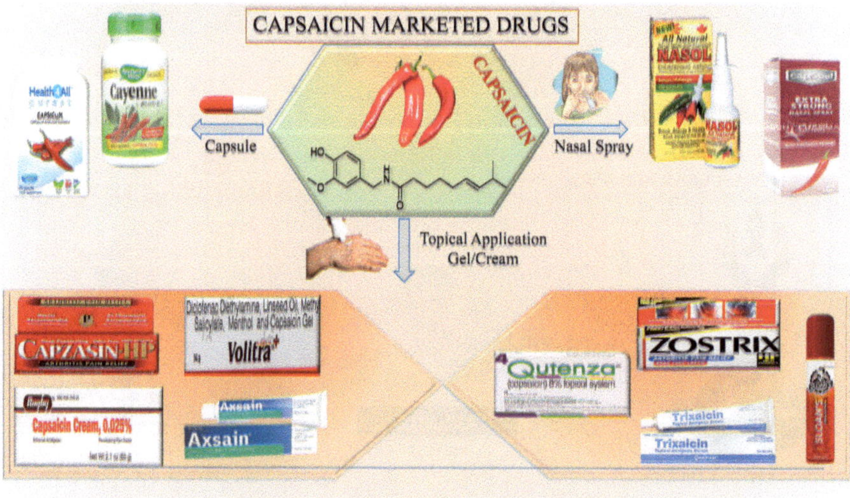

Fig. 26.4 Clinically available capsaicin pharmaceutical formulations

In clinical studies, repetitive applications of these low-concentration patches for 3–5 times/day for a period of 2–6 weeks have shown moderate favourable effects against different pain syndromes, including diabetic neuropathy, chronic musculo-skeletal pain, and post-herpetic neuralgia. Further, high concentration patches with 8% capsaicin made their way to treat HIV neuropathy, post-herpetic neuralgia, and other related disorders with neuropathic pain indications. These duly approved patches in the USA for treatment of neuropathic pain found to deliver capsaicin rapidly into the skin with minimal side systemic effects (Hempenstall et al. 2005; Fattori et al. 2016).

Other *Capsicum annum*-based pharmaceutical formulations include homeopathic preparation eucalyptol and capsicum nasal sprays (Bernstein et al. 2011) useful in the management of non-allergic rhinitis and associated symptoms. Though the intravenous administration (i/v) of capsaicin leads to its rapid entry in the central nervous system, its metabolism has been observed to be low in comparison to the kidney and liver.

For oral consumption of capsaicin (Janssens et al. 2014), capsules made of chili peppers are so far available pharmaceutical formulations. The therapeutic dose for its oral delivery is yet to be identified, but the recommended daily dosage of capsicum with 0.25% capsaicin is listed on labels of commercially available capsules. Capsules have demonstrated improvement in the fat oxidation, appetite suppression (Ludy et al. 2012; Fattori et al. 2016), energy expenditure, and thermogenesis in people.

26.3 Current Status of Capsaicin in the Biopharmaceutical Classification System

In August 2000, the FDA issued guidance with indications of application of the Biopharmaceutics Classification System (BCS) to waiver off in vivo bioequivalence studies. And this guidance sparked the global interest among biopharmaceutical scientists from industry, regulatory agencies, and academia. The BCS guidance was built on a salient scientific background involving the vital properties of the drug substances, viz., solubility, permeability, and dissolution (Gordon L. Amidon et al. 1994). The idea was to introduce the concept of waiving in vivo bioequivalence (BE) studies with specific and relative in vitro experiments to attain bioequivalency of oral fast-release products. And since then, BCS has gained international acceptance in the industry, academic institutions, government agencies, and public authorities. According to the current guidelines provided by USFDA, EMEA, and WHO, drugs can be categorized into one of four BCS classifications within the environment of human pharmaceuticals (Fig. 26.5). Furthermore, it shows that a significant portion of the marketed products (approximately 30%) and around 60–70% of drug molecules in the development stage falls in Class II (Smutzer and Devassy 2016). As a result of these efforts, more than 160 applications with 41.8% (Class I), 20.9% (Class II), and 37.3% (Class III) across the several therapeutic

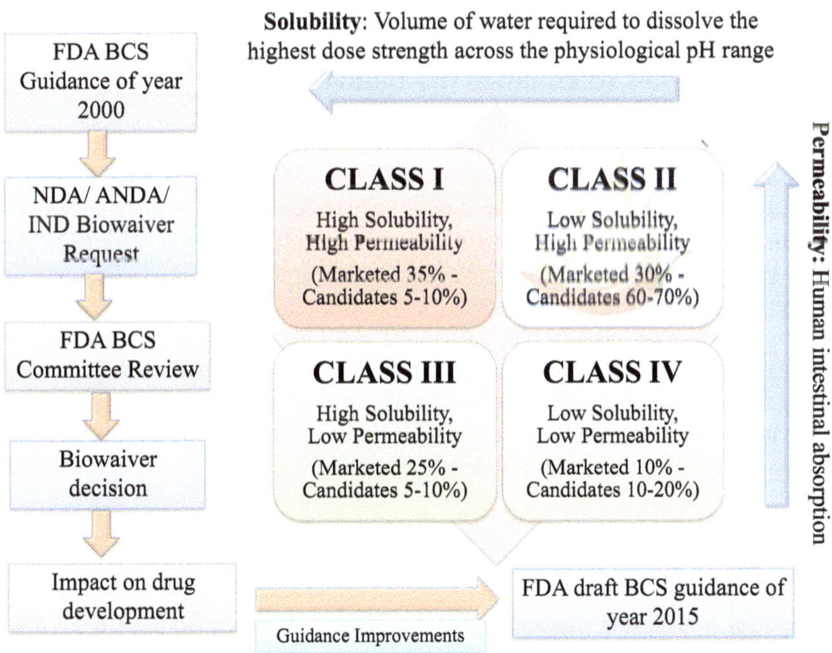

Fig. 26.5 BCS classification system include percentage of marketed drug molecules and impact of USFDA "Biopharmaceutical Classification System" Guidance on Global Drug Development

categories have been accepted or tentatively approved (Nair et al. 2012) in the last dozen years. The benefits of the BCS include reduced drug exposure to a big group of human subjects and, in some scenarios, shortened drug development time with significant cost savings (Nikolakakis and Partheniadis 2017).

Healthcare products appear in different dosage forms like oral, injections, syrups, implants, and sprays in addition to skin products, but usage of oral dosage forms has always been found to be predominant and unparalleled. Majorly existing capsaicin preparation includes cream, lotions, and patch ointments along with conventional oral. And the long half-life of CAP in the lungs and skin than that of liver implies its effective absorption (Mou et al. 2013; Bode and Dong 2011) in these organs and makes it more suitable candidate or efficacious for skin and lung disorders. This topical (dermal and transdermal) application of different formulations aims to have efficient and painless delivery of capsaicin into or through the skin with minimum side effects. However, capsaicin does not possess certain optimal physicochemical properties such as solubility and dissolution rate and achieve sub-therapeutic response with low plasma concentration because of low solubility and slow dissolution or release from the dosage. Further, half-life of the capsaicin is very high because of high first-pass effect after its oral or intravenous administration as compared to topical application avoiding first-pass metabolism. Thus, capsaicin has been categorized as BCS class II candidates as its dissolution rate has become the rate-limiting factor for its absorption along with its short half-life (Reyes-Escogido Mde et al. 2011).

26.4 Strategies to Enhance the Bioavailability

Any drug to be absorbed must be present in the form of an aqueous solution at the site of absorption. And active molecules with low solubility carry a higher risk of failure by not entering to the later stage of pharmaceutical development process. Formulation and development of an effective drug delivery approach for BCS class II and IV have always been a difficult task for any formulator. And discovery and engineering of novel drug entities for formulating capsaicin a BCS Class II candidate presents a significant goal of pharmaceutical researchers to maximize its bioavailability with reduced side effects (Chittepu et al. 2018). Figure 26.6 shows various approaches that have been extensively investigated so far to increase clinical efficacy of capsaicin by improving its delivery, including intraperitoneal, subcutaneous, topical, oral, ocular, and gastrointestinal.

26.4.1 Nano-co-crystal Technology

Principle of transferring the materials into the nano-dimensions have been found to alter their physicochemical characteristics and used in pharmaceutical science to create a new inventive formulation for poorly water-soluble drugs, i.e. drug nanocrystals (NC) (Joshi et al. 2019). NC are composed of 100% drug nanocrystals,

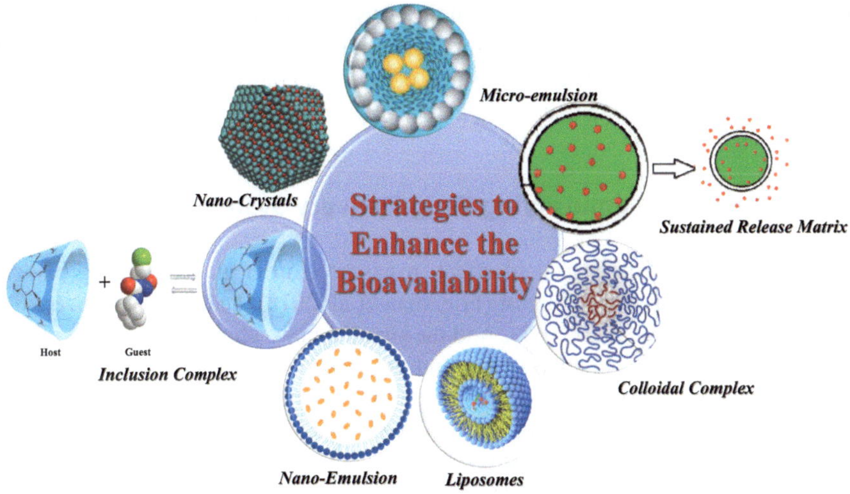

Fig. 26.6 Various technologies used to improve biopharmaceutical properties (Liu et al. 2010)

without any carrier material and dispersed in liquid media to give "nanosuspensions." Surfactants or polymeric stabilizers are dispersed particles of active ingredient in the dispersion media such as water, soluble compounds, or non-aqueous media (e.g. liquid polythene glycol [PEG], oils).

Capsaicin indicates its analgesic, antioxidant, and anti-inflammatory effect (Bode and Dong 2011) by trapping free radicals like peroxyl radicals, oxygen (O_2) molecules, and reactive nitrogen (N_2) species, along with ability to inhibit substance P by agonistic effect on TRPV1 receptors (Smutzer and Devassy 2016). The greater the inhibition of substance P, the less pain perceived, and vice versa; however, sub-therapeutic response has been notified on account of capsaicin low plasma concentration. To address the above said issues and promote its analgesic, antioxidant, and anti-inflammatory properties, Khan et al. group endeavoured to develop and characterize capsaicin nanocrystal formulations.

Capsaicin-loaded nanocrystals were effectively prepared to promote its water solubility by top-down approach utilizing high-speed homogenization and its adhesiveness to skin epidermis layer. Stability studies, drug content, surface charge, thermal analysis, surface morphology, droplet size, solubility profile, polydispensability index, entrapment efficiency, crystalline studies, and release (in vitro) and permeation (ex vivo) studies were performed on newly developed NC formulations. Capsaicin nanocrystals showed improved solubility 12.0%, 18.4%, and 12.0% in different solvents like water, methanol acetate buffer pH 5.5 respectively. Cumulative percent findings showed the 89.94% of the drug release within 24 h of analysis using Tuffryn membrane (Khan et al. 2021) and supported by the increased in vivo anti-inflammatory response in animals as compared to the marketed product (Dicloran®). Improved physicochemical parameters demonstrated the successful design of capsaicin nanocrystals that served as a potential vehicle system for the effective achievement of drugs hydrophobic in nature. This increased solubility of capsaicin may facilitate in developing the cost-effective pharmaceutical formulations that will generate the same or better therapeutic effects but at lower doses and frequencies than currently available products.

26.4.2 Self-Nano-emulsifying Drug Delivery System

Literature indicates the capsaicin effect in gastric (Lin et al. 2018), lung bladder (Anandakumar et al. 2015), breast (Friedman et al. 2018), and prostate cancer cells (Malagarie-Cazenave et al. 2011). And the simultaneous administration of synthetic drug candidates such as daunorubicin and docetaxel with capsaicin (Sanchez et al. 2019; Cheng et al. 2018) is suggested to get the maximum advantage of its anticancer potential. However it could not achieve its full potential because of limited gastric absorption and bioavailability after oral administration (Akiladevi et al. 2020; Zhao et al. 2016; Zhang et al. 2016).

Bhagwat et al. fabricated the isopropyl myristate (IPM) and labrafil-based capsaicin-loaded self-nano-emulsifying drug delivery system (SNEDDS) and assessed for in vitro and in vivo parameters. Oral delivery of BCS class II drugs has been extensively researched using lipid-based isotropic systems. Dispersion of ultrafine (10–200 nm) droplet into gastrointestinal fluids formed the basis of oil/water SNEDDS nano-emulsions (Buya et al. 2020; Abdelmonem et al. 2020; Ding et al. 2019). Fine globules were found to consume less energy and their small size provided increased surface area, which not only maintained the drug to stay in solution but also eliminated the requirement of dissolution, thus causing increase in the bioavailability. Various possible reasons ranging from GI membrane permeability modulation, decline in gastrointestinal decomposition, inhibition of glycoprotein (P-gp) efflux, lymphatic uptake, luminal solubility augmentation were proposed for its improved bioavailability by SNEDDS. SNEDDS found to inhibit the spread of colorectal (HT-29) cancer cells at faster rate than pure capsaicin, which was attributed to 3.6-fold increases in bioavailability with 85.19% release after oral administration of CAP-SNEDDS, in comparison to its pure capsaicin with 6.61% drug release in initial 15 min. Capsaicin was effectively wettable, soluble, and bioavailable against colorectal cancer cells after being entrapped in SNEDDS, with reduced gastric irritability (Bhagwat et al. 2021).

26.4.3 Microemulsion Technology

For a short period, conditions like rheumatoid arthritis, diabetic neuropathy, osteoarthritis, and post-herpetic neuralgia were cured with capsaicin through its intravenous administration (Backonja et al. 2010). Considering its high first-pass effect and short half-life, long-term treatment was recommended with CAP, but it resulted in some adverse events like burning, stinging, and erythema. However later, topical application was suggested as an alternative to retain its therapeutic potential by surpassing the first pass metabolism effect complications.

Reported studies highlight the significance of optically isotropic and thermodynamically stable water-based mixed microemulsion (MC) with oil and surfactants as potential topical therapeutical systems for hydrophobic and hydrophilic drugs including capsaicin (Kogan and Garti 2006). However, the high ratio of surfactants affected the skin's ability to act as a barrier, with increased risk of toxicity and irritation with continued use. It encouraged Zhang et al.'s group to investigate the potential of a novel MC-based formulations without surfactants like benzyl alcohol and short-chain alcohol for the topical drug delivery of capsaicin. First surfactant-free microemulsion system was developed and evaluated for its stability and in vitro permeation studies in the skin using Franz diffusion cells. Surfactant-free MCs were found to be small drop size in comparison to systems based on Tween 80 and afforded higher skin permeation rates in vitro permeation studies in comparison to those of Tween 80 (Pengwei Zhang et al. 2008), hydrogel, cream, and single solvent-based system utilized in these trials.

Another research group Huang et al. further investigated MC methods to improve capsaicinoid nonivamide transdermal delivery. Mixed surfactant-based MC systems were formulated using Tween 80 along with Span 20 and alcohol as a co-surfactant. Isopropyl myristate and water were used as the oil and water phases to establish mixed surfactant-based MCs. These mixed surfactant-based MCs showed 3.7–7.1-fold increase in the transdermal delivery of nonivamide by relative to controls (Huang et al. 2008).

26.4.4 Liposomes Technology

Liposomes have also been explored as a transdermal drug delivery to improve capsaicin bioavailability and alleviate irritation. Liposomes offer distinct advantages that they help in synthesis of biodegradable fine particles, hold high entrapment efficiency, available as drug released carriers, and provide protection to active ingredients from immediate dilution/degradation by enclosing them in liposomes (Slingerland et al. 2012; Tester et al. 2011).

Utilizing sodium cholate, cholesterol, and isopropyl myristate as liposomal agents, capsaicin-loaded liposomes were investigated by Zhu et al. as a nano-formulation with fewer irritants for oral administration of capsaicin with high stability. After structural morphology, an in vitro-in vivo correlation (IVIVC) was established by subjecting capsaicin-loaded liposomes in vitro release, stability studies, encapsulation efficacy, gastric mucosal irritation, and in vivo pharmacokinetics studies. In vivo, liposomal encapsulation made 3.34-fold increase in the relative bioavailability within 72 h in comparison to free capsaicin on account of their enhanced solubility by 69.4%, 80.0%, and 74.3% in phosphate buffer, HCl, and double distilled water, respectively (Zhu et al. 2015). Further, gastric mucosa irritation studies indicated it as a safer oral delivery system. Capsaicin with an average size of 100 nm was also attempted to be encapsulated as solid lipid nanoparticles, with 90% encapsulation recorded efficiency that resulted into long-lasting release of the drug for 14 h. And collective studies by Zhu et al. supported and provided the encapsulated capsaicin liposomes with improved oral bioavailability as a useful drug to cure neuropathic pain.

26.4.5 Inclusion Complex

Growing demand of nutritional and pharmaceutical additive capsaicin necessitated the development of an easy rapid and precise method for its determinations. Existing analytical methods including gas chromatography (Nelson 1919; Suk-Hyun Choi et al. 2006), high-pressure liquid chromatography, capillary electrophoresis, and spectroscopy though are capable of determining capsaicin with improving accuracy and reliability. However, their limited use in an organic medium in a stringent reaction condition employing expensive instrument worked a driving force in the development of more precise simple and cost-effective method, i.e. inclusion complexes for capsaicin determination (Liu et al. 2010).

Cyclodextrins (CDs) the cyclic oligomers comprise of a number of D-glucopyranose units and the β-cyclodextrin (β-CD) the most commonly used agent to formulate inclusion complexes in the pharmaceutical industry is with seven D-glucopyranose units. CDs with hydrophilic outside and hydrophobic inside have hollow truncated cone structure and got the property to change physical, chemical, as well as biological parameters showed by the guest molecules by forming the host-guest complexes. Huang et al. fabricated the complex HPβCD/capsaicin with guest-host ratio of 2:1, wherein 2-hydroxypropyl-β-cyclodextrin (HPβCD) is the hydroxyalkyl derivative of the β-CD. The molecular structure (Fig. 26.7) of HPβCD clearly indicated that hydroxypropyl groups by breaking the

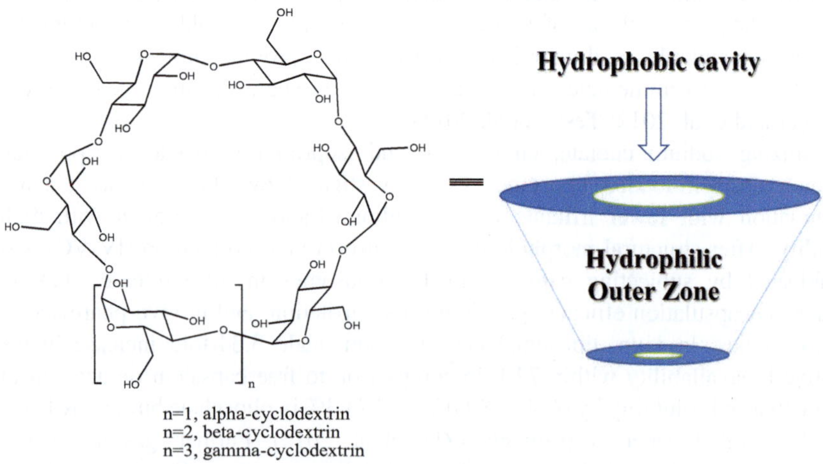

Fig. 26.7 Molecular structure of 2-hydroxypropyl-β-cyclodextrin (HPβCD)

hydrogen bonds (intramolecular) allowed HPβCD to acquire a larger cavity with high water solubility than natural βCD. Improved solubilization capability of the complex and thermal stability by complexation with HPβCD was further apparent in phase solubility diagram (Huang et al. 2019).

Thus significantly enhanced fluorescence intensity by complexation with HPβCD has provided a novel method with a shorter detection limit, superior accuracy and a great practicability for capsaicin determination in aqueous solution.

26.4.6 Colloidal Complex

Continual investigation efforts to improve the oral bioavailability of capsaicin also investigated the nanoencapsulation strategy driven by the extended retention time of drug allowed by its nanocapsules. In the past few years, different formulations of capsaicin nanocapsules ranging from polymeric micelles (Zhu et al. 2014) to polymeric nanoparticles (Peng et al. 2015) and liposomes (Zhu et al. 2015) have been developed, wherein nanoencapsulation especially showed the improvement in in vivo bioavailability in comparison to its native form. The extended retention time has been attributed to the sustained release functionality, nanoscale size, and stabilization supported by the capsule matrix against metabolism (Rollyson et al. 2014). However, these nanocapsules exhibited a major disadvantage in their low payload (<20%), which in turn demanded the large doses to produce the desired therapeutic effects. And its continual use was put on hold because of the rising risk of negative side effects on unnecessary consumption of carrier materials. Further to enhance the bioavailability of capsaicin, a novel strategy, i.e. high-payload submicron capsaicin-chitosan colloidal particle complexes between anionic capsaicin and oppositely charged chitosan, was developed by Tran et al. using electrostatically driven self-assembly complexation technique. Readilly available natural resources and quickly ionization tendancy were some of the salient features to select CHI as a polyelectrolyte, to augment the intestinal absorption of several bioactive chemical entities.

The high payload (75%), and high colloidal stability (85%), with improved solubility observed due to its partially amorphous form was maintained at >5 times saturation solubility for 6 h (Tran and Hadinoto 2017). Capsaicin concentration, ratio of chitosan/capsaicin, and pH were some of the contributing factors for this structural integrity, physical characteristic, and efficiency. Further, absence of any negative effect on the capsaicin's antimicrobial activity indicated the perseverance of its bioactivity its in the novel designed submicroplex with chitosan. The prolonged apparent solubility of the CAP-CHI submicroplex, combined with its appealing mechanical properties and ease of preparation, established nanocapsulation as a feasible oral bioavailability enhancement approach for CAP.

26.4.7 Sustained Release Technology

Therapeutic effect of capsaicin as anticancer (Malagarie-Cazenave et al. 2011), antioxidant (Lee et al. 2011), and anti-inflammatory agent (Desai et al. 2013) can be best achieved in in vivo and oral sustained-release solid (SR) preparations by maintaining the appropriate blood levels of capsaicin, lesser irritation, and solubilizing.

Zhang et al. (2016) designed pellets, where in core was designed with solid-dispersed capsaicin mixed with hydroxypropyl, microcrystalline, and ethyl cellulose using technology of spheronization/extrusion. Physicochemical properties of formulation were evaluated through differential scanning calorimetry, X-ray diffractometry, and scanning electron microscopy followed by relative bioavailability in rabbits, after its oral administration in comparison to solid dispersion and free capsaicin as references. Significant improvement in the bioavailability (oral) by 1.98-fold for matrix and 5.34-fold for sustained-release matrix pellets was observed as compared to the pure capsaicin. A significant correlation between in vitro dissolution and in vivo absorption (IVIVC) of SR matrix pellets strengthened the studies by supporting noticeably increased oral bioavailability of capsaicin as a result of the efficient creation of its sustained-release matrix pellets. Successful preparation of SR matrix pellets as a novel drug delivery system offered therapeutic benefits with less gastrointestinal tract infection side effects after getting freely dispersed and maximum absorbed in the gastrointestinal tract without lowering drug bioavailability (Zhang et al. 2016).

26.5 Conclusions and Future Prospective

Capsaicin and capsaicin containing foods are available for humans over the number of years, and recent understanding of different capsaicin drug targets has provided the molecular basis for its pharmacological action in different conditions. Its significance has been further documented by existing wide range of topical formulations (capsaicin 8% patch) with therapeutic applications to treat neuropathic pain. Unfortunately, the low bioavailability and poor delivery are the main obstacles for its effectiveness and precluded its use in the medicine. The tools explained in this chapter have been used to make capsaicin reach effectively to target sites through different routes of administration. Continual progress in the several strategies could provide capsaicin-based different formulations (other than topical), with significant biopharmaceutical parameter profile in the near future that can be established as capsaicin-based agents as preventive and disease-modifying therapeutics.

References

Abdelmonem R, Azer MS, Makky A, Zaghloul A, El-Nabarawi M, Nada A (2020) Development, characterization, and in-vivo pharmacokinetic study of lamotrigine solid self-nanoemulsifying drug delivery system. Drug Des Devel Ther 14:4343–4362. https://doi.org/10.2147/DDDT.S263898

Akiladevi D, Prakash H, Biju G, Madumitha N (2020) Nano-novel approach: Self Nano Emulsifying Drug Delivery System (SNEDDS)-review article. Res J Pharm Technol 13(2): 983–990. https://doi.org/10.5958/0974-360x.2020.00183.3

Al Othman ZA, Ahmed YB, Habila MA, Ghafar AA (2011) Determination of capsaicin and dihydrocapsaicin in Capsicum fruit samples using high performance liquid chromatography. Molecules 16(10):8919–8929. https://doi.org/10.3390/molecules16108919

Amidon GL, Lennernas H, Shah VP, Crison JR (1994) A theoretical basis for a biopharmaceutic drug classification the correlation of in vitro drug product dissolution and in vivo bioavailability. Pharm Res 12:413–420

Anand P, Bley K (2011) Topical capsaicin for pain management: therapeutic potential and mechanisms of action of the new high-concentration capsaicin 8% patch. Br J Anaesth 107(4):490–502. https://doi.org/10.1093/bja/aer260

Anandakumar P, Kamaraj S, Jagan S, Ramakrishnan G, Asokkumar S, Naveenkumar C, Raghunandhakumar S, Vanitha MK, Devaki T (2015) The anticancer role of capsaicin in experimentally induced lung carcinogenesis. J Pharmacopuncture 18(2):19–25. https://doi.org/10.3831/KPI.2015.18.011

Baboota RK, Singh DP, Sarma SM, Kaur J, Sandhir R, Boparai RK, Kondepudi KK, Bishnoi M (2014) Capsaicin induces "brite" phenotype in differentiating 3T3-L1 preadipocytes. PLoS One 9(7):1–12. https://doi.org/10.1371/journal.pone.0103093

Backonja MM, Malan TP, Vanhove GF, Tobias JK, C102/106 Study Group (2010) NGX-4010, a high-concentration capsaicin patch, for the treatment of postherpetic neuralgia: a randomized, double-blind, controlled study with an open-label extension. Pain Med 11:600–608

Barbero GF, Molinillo JM, Varela RM, Palma M, Macias FA, Barroso CG (2010) Application of Hansch's model to capsaicinoids and capsinoids: a study using the quantitative structure-activity relationship. A novel method for the synthesis of capsinoids. J Agric Food Chem 58(6): 3342–3349. https://doi.org/10.1021/jf9035029

Baskaran P, Krishnan V, Ren J, Thyagarajan B (2016) Capsaicin induces browning of white adipose tissue and counters obesity by activating TRPV1 channel-dependent mechanisms. Br J Pharmacol 173(15):2369–2389. https://doi.org/10.1111/bph.13514

Bernstein JA, Davis BP, Picard JK, Cooper JP, Zheng S, Levin LS (2011) A randomized, double-blind, parallel trial comparing capsaicin nasal spray with placebo in subjects with a significant component of nonallergic rhinitis. Ann Allergy Asthma Immunol 107(2):171–178. https://doi.org/10.1016/j.anai.2011.05.016

Bhagwat DA, Swami PA, Nadaf SJ, Choudhari PB, Kumbar VM, More HN, Killedar SG, Kawtikwar PS (2021) Capsaicin loaded solid SNEDDS for enhanced bioavailability and anticancer activity: in-vitro, in-silico, and in-vivo characterization. J Pharm Sci 110(1): 280–291. https://doi.org/10.1016/j.xphs.2020.10.020

Bode AM, Dong Z (2011) The two faces of capsaicin. Cancer Res 71(8):2809–2814. https://doi.org/10.1158/0008-5472.CAN-10-3756

Brito R, Sheth S, Mukherjea D, Rybak LP, Ramkumar V (2014) TRPV1: a potential drug target for treating various diseases. Cell 3(2):517–545. https://doi.org/10.3390/cells3020517

Buya AB, Beloqui A, Memvanga PB, Preat V (2020) Self-nano-emulsifying drug-delivery systems: from the development to the current applications and challenges in oral drug delivery. Pharmaceutics 12(12):1–55. https://doi.org/10.3390/pharmaceutics12121194

Cheng YY, Hsieh CH, Tsai TH (2018) Concurrent administration of anticancer chemotherapy drug and herbal medicine on the perspective of pharmacokinetics. J Food Drug Anal 26(2S):1–8. https://doi.org/10.1016/j.jfda.2018.01.003

Chittepu VCSR, Kalhotra P, Revilla GIO, Velázquez TG (2018) Emerging technologies to improve capsaicin delivery and its therapeutic efficacy. In: Mozsik G (ed) Capsaicin and its human therapeutic development, vol 3. IntechOpen, London, pp 43–55. https://doi.org/10.5772/intechopen.77080

Choi S-H, Suh B-S, Kozukue E, Kozukue N, Levin CE, Friedman M (2006) Analysis of the contents of pungent compounds in fresh Korean red peppers and in pepper-containing foods. J Agric Food Chem 54:9024–9031

Clark R, Lee SH (2016) Anticancer properties of capsaicin against human cancer. Anticancer Res 36(3):837–843

Desai PR, Marepally S, Patel AR, Voshavar C, Chaudhuri A, Singh M (2013) Topical delivery of anti-TNFalpha siRNA and capsaicin via novel lipid-polymer hybrid nanoparticles efficiently inhibits skin inflammation in vivo. J Control Release 170(1):51–63. https://doi.org/10.1016/j.jconrel.2013.04.021

Ding D, Sun B, Cui W, Chen Q, Zhang X, Zhang H, He Z, Sun J, Luo C (2019) Integration of phospholipid-drug complex into self-nanoemulsifying drug delivery system to facilitate oral delivery of paclitaxel. Asian J Pharm Sci 14(5):552–558. https://doi.org/10.1016/j.ajps.2018.10.003

Fattori V, Hohmann MS, Rossaneis AC, Pinho-Ribeiro FA, Verri WA (2016) Capsaicin: current understanding of its mechanisms and therapy of pain and other pre-clinical and clinical uses. Molecules 21(7):3–33. https://doi.org/10.3390/molecules21070844

Friedman JR, Nolan NA, Brown KC, Miles SL, Akers AT, Colclough KW, Seidler JM, Rimoldi JM, Valentovic MA, Dasgupta P (2018) Anticancer activity of natural and synthetic capsaicin analogs. J Pharmacol Exp Ther 364(3):462–473. https://doi.org/10.1124/jpet.117.243691

Hayman M, Kam PCA (2008) Capsaicin: a review of its pharmacology and clinical applications. Curr Anaesth Crit Care 19(5–6):338–343. https://doi.org/10.1016/j.cacc.2008.07.003

Hempenstall K, Nurmikko TJ, Johnson RW, A'Hern RP, Rice AS (2005) Analgesic therapy in postherpetic neuralgia: a quantitative systematic review. PLoS Med 2(7):628–644. https://doi.org/10.1371/journal.pmed.0020164

Hernandez-Perez T, Gomez-Garcia MDR, Valverde ME, Paredes-Lopez O (2020) Capsicum annuum (hot pepper): an ancient Latin-American crop with outstanding bioactive compounds and nutraceutical potential. A review. Compr Rev Food Sci Food Saf 19(6):2972–2993. https://doi.org/10.1111/1541-4337.12634

Huang YB, Lin YH, Lu TM, Wang RJ, Tsai YH, Wu PC (2008) Transdermal delivery of capsaicin derivative-sodium nonivamide acetate using microemulsions as vehicles. Int J Pharm 349(1–2):206–211. https://doi.org/10.1016/j.ijpharm.2007.07.022

Huang Z, Xu R, Ge X, Cheng J (2019) Complexation of capsaicin with hydroxypropyl-beta-cyclodextrin and its analytical application. Spectrochim Acta A Mol Biomol Spectrosc 223:1–7. https://doi.org/10.1016/j.saa.2019.117278

Janssens PL, Hursel R, Westerterp-Plantenga MS (2014) Capsaicin increases sensation of fullness in energy balance, and decreases desire to eat after dinner in negative energy balance. Appetite 77:44–49. https://doi.org/10.1016/j.appet.2014.02.018

Joshi K, Chandra A, Jain K, Talegaonkar S (2019) Nanocrystalization: an emerging technology to enhance the bioavailability of poorly soluble drugs. Pharm Nanotechnol 7(4):259–278. https://doi.org/10.2174/2211738507666190405182524

Khan BA, Rashid F, Khan MK, Alqahtani SS, Sultan MH, Almoshari Y (2021) Fabrication of capsaicin loaded nanocrystals: physical characterizations and in vivo evaluation. Pharmaceutics 13(6):1–14. https://doi.org/10.3390/pharmaceutics13060841

Kogan A, Garti N (2006) Microemulsions as transdermal drug delivery vehicles. Adv Colloid Interf Sci 123–126:369–385. https://doi.org/10.1016/j.cis.2006.05.014

Lee IO, Lee KH, Pyo JH, Kim JH, Choi YJ, Lee AYC (2007) Anti-inflammatory effect of capsaicin in helicobacter pylori-infected gastric epithelial cells. Helicobacter 12:510–517

Lee TH, Lee JG, Yon JM, Oh KW, Baek IJ, Nahm SS, Lee BJ, Yun YW, Nam SY (2011) Capsaicin prevents kainic acid-induced epileptogenesis in mice. Neurochem Int 58(6):634–640. https://doi.org/10.1016/j.neuint.2011.01.027

Lin RJ, Wu IJ, Hong JY, Liu BH, Liang RY, Yuan TM, Chuang SM (2018) Capsaicin-induced TRIB3 upregulation promotes apoptosis in cancer cells. Cancer Manag Res 10:4237–4248. https://doi.org/10.2147/CMAR.S162383

Liu L, Chen X, Liu J, Deng X, Duan W, Tan S (2010) Determination of capsaicin and dihydrocapsaicin in Capsicum annuum and related products by capillary electrophoresis with a mixed surfactant system. Food Chem 119(3):1228–1232. https://doi.org/10.1016/j.foodchem.2009.08.045

Ludy MJ, Moore GE, Mattes RD (2012) The effects of capsaicin and capsiate on energy balance: critical review and meta-analyses of studies in humans. Chem Senses 37(2):103–121. https://doi.org/10.1093/chemse/bjr100

Malagarie-Cazenave S, Olea-Herrero N, Vara D, Morell C, Diaz-Laviada I (2011) The vanilloid capsaicin induces IL-6 secretion in prostate PC-3 cancer cells. Cytokine 54(3):330–337. https://doi.org/10.1016/j.cyto.2011.03.010

Mou J, Paillard F, Turnbull B, Trudeau J, Stoker M, Katz NP (2013) Efficacy of Qutenza (R) (capsaicin) 8% patch for neuropathic pain: a meta-analysis of the Qutenza Clinical Trials Database. Pain 154(9):1632–1639. https://doi.org/10.1016/j.pain.2013.04.044

Nair AK, Anand O, Chun N, Conner DP, Mehta MU, Nhu DT, Polli JE, Yu LX, Davit BM (2012) Statistics on BCS classification of generic drug products approved between 2000 and 2011 in the USA. AAPS J 14(4):664–666. https://doi.org/10.1208/s12248-012-9384-z

Nelson EK (1919) The constitution of capsaicin, the pungent principle of Capsicum. J Am Chem Soc 41:1115–1121

Nikolakakis I, Partheniadis I (2017) Self-emulsifying granules and pellets: composition and formation mechanisms for instant or controlled release. Pharmaceutics 9(4):1–27. https://doi.org/10.3390/pharmaceutics9040050

O'Neill J, Brock C, Olesen AE, Andresen T, Nilsson M, Dickenson AH (2012) Unravelling the mystery of capsaicin: a tool to understand and treat pain. Pharmacol Rev 64(4):939–971. https://doi.org/10.1124/pr.112.006163

Peng W, Jiang XY, Zhu Y, Omari-Siaw E, Deng WW, Yu JN, Xu XM, Zhang WM (2015) Oral delivery of capsaicin using MPEG-PCL nanoparticles. Acta Pharmacol Sin 36(1):139–148. https://doi.org/10.1038/aps.2014.113

Reyes-Escogido Mde L, Gonzalez-Mondragon EG, Vazquez-Tzompantzi E (2011) Chemical and pharmacological aspects of capsaicin. Molecules 16(2):1253–1270. https://doi.org/10.3390/molecules16021253

Rollyson WD, Stover CA, Brown KC, Perry HE, Stevenson CD, McNees CA, Ball JG, Valentovic MA, Dasgupta P (2014) Bioavailability of capsaicin and its implications for drug delivery. J Control Release 196:96–105. https://doi.org/10.1016/j.jconrel.2014.09.027

Sanchez BG, Bort A, Mateos-Gomez PA, Rodriguez-Henche N, Diaz-Laviada I (2019) Combination of the natural product capsaicin and docetaxel synergistically kills human prostate cancer cells through the metabolic regulator AMP-activated kinase. Cancer Cell Int 19.1–14. https://doi.org/10.1186/s12935-019-0769-2

Sharma SK, Vij AS, Sharma M (2013) Mechanisms and clinical uses of capsaicin. Eur J Pharmacol 720(1–3):55–62. https://doi.org/10.1016/j.ejphar.2013.10.053

Slingerland M, Guchelaar HJ, Gelderblom H (2012) Liposomal drug formulations in cancer therapy: 15 years along the road. Drug Discov Today 17(3–4):160–166. https://doi.org/10.1016/j.drudis.2011.09.015

Smutzer G, Devassy RK (2016) Integrating TRPV1 receptor function with capsaicin psychophysics. Adv Pharmacol Sci 2016:1–17. https://doi.org/10.1155/2016/1512457

Suresh D, Srinivasan K (2010) Tissue distribution & elimination of capsaicin, piperine & curcumin following oral intake in rats. Indian J Med Res 131:682–691

Tester CC, Brock RE, Wu C-H, Krejci MR, Weigand S, Joester D (2011) In vitro synthesis and stabilization of amorphous calcium carbonate (ACC) nanoparticles within liposomes. CrystEngComm 13(12):3975–3978. https://doi.org/10.1039/c1ce05153a

Tran T-T, Hadinoto K (2017) A new solubility enhancement strategy of capsaicin in the form of high-payload submicron capsaicin-chitosan colloidal complex. Physicochem Eng 520:62–71. https://doi.org/10.1016/j.colsurfa.2017.01.069

Xiang Q, Tang X, Cui S, Zhang Q, Liu X, Zhao J, Zhang H, Mao B, Chen W (2022) Capsaicin, the spicy ingredient of chili peppers: effects on gastrointestinal tract and composition of gut microbiota at various dosages. Foods 11(5):1–14. https://doi.org/10.3390/foods11050686

Yu Q, Wang Y, Yu Y, Li Y, Zhao S, Chen Y, Waqar AB, Fan J, Liu E (2012) Expression of TRPV1 in rabbits and consuming hot pepper affects its body weight. Mol Biol Rep 39(7):7583–7589. https://doi.org/10.1007/s11033-012-1592-1

Zhang P, Gao W, Zhang L, Chen L, Shen Q, Wang X, Cuib Y (2008) In vitro evaluation of topical microemulsion of capsaicin free of surfactant. Biol Pharm Bull 12:2316–2320

Zhang Y, Huang Z, Omari-Siaw E, Lu S, Zhu Y, Jiang D, Wang M, Yu J, Xu X, Zhang W (2016) Preparation and in vitro-in vivo evaluation of sustained-release matrix pellets of capsaicin to enhance the oral bioavailability. AAPS PharmSciTech 17(2):339–349. https://doi.org/10.1208/s12249-015-0352-7

Zhao Y, Sun C, Shi F, Firempong CK, Yu J, Xu X, Zhang W (2016) Preparation, characterization, and pharmacokinetics study of capsaicin via hydroxypropyl-beta-cyclodextrin encapsulation. Pharm Biol 54(1):130–138. https://doi.org/10.3109/13880209.2015.1021816

Zhu Y, Peng W, Zhang J, Wang M, Firempong CK, Feng C, Liu H, Xu X, Yu J (2014) Enhanced oral bioavailability of capsaicin in mixed polymeric micelles: preparation, in vitro and in vivo evaluation. J Funct Foods 8:358–366. https://doi.org/10.1016/j.jff.2014.04.001

Zhu Y, Wang M, Zhang J, Peng W, Firempong CK, Deng W, Wang Q, Wang S, Shi F, Yu J, Xu X, Zhang W (2015) Improved oral bioavailability of capsaicin via liposomal nanoformulation: preparation, in vitro drug release and pharmacokinetics in rats. Arch Pharm Res 38(4):512–521. https://doi.org/10.1007/s12272-014-0481-7

MIX
Papier aus verantwortungsvollen Quellen
Paper from responsible sources
FSC® C105338

If you have any concerns about our products,
you can contact us on
ProductSafety@springernature.com

In case Publisher is established outside the EU,
the EU authorized representative is:
Springer Nature Customer Service Center GmbH
Europaplatz 3, 69115 Heidelberg, Germany

Printed by Libri Plureos GmbH
in Hamburg, Germany